Properties of Silicon and Gallium Arsenide

Property	Silicon	Gallium Arsenide
Crystal structure	Diamond	Zincblende
Lattice constant (Å)	5.431	5.646
Distance between neighboring atoms (Å)	2.36	2.44
Atoms or molecules/cm^3	5×10^{22}	2.21×10^{22}
Density (g/cm^3)	2.328	5.32
Melting point (°C)	1412	1238
Thermal expansion coefficient (300 °K)	2.6×10^{-6}	6.86×10^{-6}
Thermal conductivity (W/cm-°C)	1.5	0.46
Energy gap (eV)	1.11	1.435
Relative permittivity	11.9	13.1
Breakdown field strength (V/μm)	≃30	≃40
Intrinsic carrier concentration, cm^{-3} (300 °K)	1.45×10^{10}	8×10^{6}
Effective density of states		
Valence band (cm^{-3})	1.04×10^{19}	7×10^{18}
Conduction band (cm^{-3})	2.8×10^{19}	4.7×10^{17}

VLSI Fabrication Principles

VLSI Fabrication Principles

Silicon and Gallium Arsenide

Sorab K. Ghandhi
Rensselaer Polytechnic Institute

A Wiley-Interscience Publication
John Wiley & Sons
New York Chichester Brisbane Toronto Singapore

Library of Congress Cataloging in Publication Data:

Ghandhi, Sorab Khushro, 1928–
VLSI fabrication principles.

"A Wiley-Interscience publication."
Includes index.
1. Integrated circuits—Very large scale integration.
2. Silicon. 3. Gallium arsenide. I. Title. II. Title:
V.L.S.I. fabrication principles.

TK7874.G473 1982 621.381′71 82-10842
ISBN 0-471-86833-7

Printed in the United States of America

10 9 8 7 6 5 4 3 2 1

To

My Research Students

Preface

This book discusses the basic principles underlying the fabrication of semiconductor devices and integrated circuits. Its emphasis is on processes that are especially useful for Very Large Scale Integration (VLSI) schemes; however, many processes used for discrete devices and medium-density integrated circuits have also been included, since they are the precursors of VSLI technology.

The aim of this book is to provide graduate students and practicing engineers with a body of knowledge of VLSI fabrication principles that will not only *bring* them up-to-date, but also allow them to *stay* up-to-date in this field. This is difficult to do in as fast-moving an area as VLSI. I believe that an effective approach is to concentrate on a broad background in the area, and to emphasize the basic principles governing the direction of developments in this field. As a practicing engineer for 13 years, and a professor and consultant to industry for an additional 19, I believe that this approach will best serve the long-term needs of workers in the area.

My experience as an author has convinced me that the above approach is the correct one for this book. In 1968 I wrote a book on microelectronics* using this approach. It has been a source of deep satisfaction to me that, even after 14 years, it is still considered to be a well-regarded "honest" book, notwithstanding the fact that many new technologies have become firmly established since. The reason for this, I believe, was the book's emphasis on basic principles rather than on the latest technological details.

VLSI Fabrication Principles was written with this idea in mind. Unlike my last volume, however, this book emphasizes fabrication principles. I have avoided discussions of the physics of device operation as much as possible, since this topic is already covered in excellent books† in the area. Even so, this new volume is about one and one-half times the length of the previous one, because of the many advances in the field.

* *The Theory and Practice of Microelectronics*. Published by John Wiley and Sons, New York.
† By way of example, Sze's *Physics of Semiconductor Devices*, 2nd edition. Published by John Wiley and Sons, New York (1981).

This book is about both elemental and compound semiconductors. Silicon is clearly the most widely used material today, and will remain the mainstay of the industry. However, many new materials (gallium arsenide is only one) are under investigation because of their unique capabilities. Gallium arsenide has already advanced technologically to the point where it is being used to make many unique devices and integrated circuits. In terms of both research and advanced development, its exploitation represents the most rapidly growing segment of solid state technology today.

From an academic viewpoint, I believe that a student's valuable time in a university can be put to the best use by acquiring the broadest base of knowledge in the field, rather than by premature specialization in any one specific area of technology. This is particularly true because many of the new problems being uncovered by workers in the latest silicon technology bear a strong resemblance to problems encountered many years ago by workers in gallium arsenide. For example, the anomalous behavior of arsenic diffusion in silicon, which was first explained in 1975, is governed by the same body of mathematics that describes the diffusion of zinc in gallium arsenide (1963). It is my hope, then, that students of this volume will acquire a broad perspective on the subject, and be flexible enough to work in *any* area of semiconductor fabrication technology, as the need arises.

This book emphasizes the basic processes that are involved in integrated circuit fabrication. Each chapter discusses principles common to all semiconductors, with separate sections where necessary to discuss problems and characteristics that are unique to silicon or gallium arsenide. The emphasis is on VLSI techniques; however, in order not to be unduly restrictive, I have also considered techniques that apply to the fabrication of medium and large-scale integrated circuits, as well as of discrete devices.

Chapter 1 describes basic Material Properties, especially those that are important in device processing. This is followed by Chapter 2 on Phase Diagrams, which outlines a valuable tool for investigating the behavior of combinations of two or more materials in intimate contact, when subjected to heat treatments. Chapter 3 describes basic aspects of the Growth of starting materials, and limitations imposed by the different growth technologies.

Chapters 4–6 describe the "anchor" technologies of Diffusion, Epitaxy, and Ion Implantation. Emphasis has been placed on these topics, because they are the key fabrication processes used today. Again, the stress is on the basic principles underlying these processes, rather than on the very latest development in each area. Of necessity, some very new techniques have been omitted, since their eventual role in device fabrication is not yet clear.

Chapter 7 discusses Native Oxide Films, which are grown out of the semiconductor. These play a key role in control of the device surface and thus in its long-term stability. Here, the emphasis is on accepted technologies, and on their limitations.

Deposited Films are discussed in Chapter 8. The large variety of available choices prevents a detailed discussion of every combination. Consequently, these films have been grouped on a functional basis—films for protection and masking, films for interconnections, films for ohmic contacts, and films for Schottky devices. I hope that this approach will provide some cohesiveness to this topic.

Chapter 9 outlines Etching and Cleaning processes. Both wet and dry processes are considered, with special emphasis on the latter, since they represent a clear direction for VLSI fabrication.

Chapter 10 outlines the basic principles of Lithographic Processes. The emphasis here is on photolithography and electron beam lithography. However, promising approaches such as X-ray lithography are considered as well. A discussion of the principles underlying the various resist systems is also included.

Chapter 11, on Device and Circuit Fabrication, is a synthesis of all of the preceding chapters, leading to the fabrication of complete microcircuits. No attempt is made to describe the many combinations of process steps that can be used for this purpose. Instead, basic techniques common to all VLSI schemes are first considered. These include isolation, self-alignment, local oxidation, and planarization. This is followed by a detailed discussion of microcircuits based on the metal-oxide-semiconductor, metal-Schottky gate-semiconductor, and bipolar junction transistor devices. An extensive reference list is provided to guide the reader to many of the important variations that are being considered at the present time, but have not been fully evaluated in terms of performance and cost. Finally, a short Appendix provides the necessary mathematical background for the chapter on diffusion.

Wherever possible, problems have been provided, many of which deal with practical situations and are intended to bring out points not covered in detail in the text. In addition, there are extensive references at the end of each chapter. No attempt has been made to use them to give credit to persons who did the original work; rather, their choice has been based on the need to give the reader means for further study.

A book such as this takes many years to write, and springs from many sources of inspiration and encouragement. It would indeed be difficult to single out all of these for acknowledgment. I know, however, that my primry thanks must go to my many graduate students who provide me with much intellectual stimulation and challenge. They have been exposed to this material, in one form or another, over the past five years. Their many penetrating questions have often led to rethinking and reworking the text over these years.

Next, I must acknowledge the encouragement and understanding I have received from Dr. Donald Feucht and John Benner of the Solar Energy Research Institute. Their generous support of funded research in this area has provided a continuing forum for the development of new ideas and new understanding, which have been incorporated into this book. Discussions with many friends in industry, especially those at the Radio Corporation of

America and the General Electric Company, have added much to the relevance of this book, and are gratefully acknowledged.

A number of my friends and colleagues have been kind enough to read and comment upon chapters of this book in their areas of specialization. These include Drs. B. Jayant Baliga, Ronald J. Gutmann, Shinji Okazaki, Kenneth Rose, Shambhu K. Shastry, and David S. Yaney.

My wife and family have been very supportive during what must have been a trying period for them. Indeed, their pride in this endeavor has done much to make it all very worthwhile.

Finally, much credit and thanks must go to R. Carla Reep for typing the manuscript and also for editing and checking it from its typed version to the page-proof stage. Her participation has greatly reduced my work, and allowed many revisions to be made in order to bring this manuscript into its final form.

S. K. GHANDHI

Niskayuna, New York
November 1982

Contents

Chapter 1	Material Properties	1
Chapter 2	Phase Diagrams and Solid Solubility	49
Chapter 3	Crystal Growth and Doping	81
Chapter 4	Diffusion	111
Chapter 5	Epitaxy	213
Chapter 6	Ion Implantation	299
Chapter 7	Native Oxide Films	371
Chapter 8	Deposited Films	419
Chapter 9	Etching and Cleaning	475
Chapter 10	Lithographic Processes	533
Chapter 11	Device and Circuit Fabrication	567
Appendix	The Mathematics of Diffusion	639
Index		657

VLSI Fabrication Principles

CHAPTER

1

Material Properties

CONTENTS

1.1 CRYSTAL STRUCTURE 3
1.2 CRYSTAL AXES AND PLANES 8
1.3 ORIENTATION EFFECTS 10

1.3.1 Silicon, 10
1.3.2 Gallium Arsenide, 12

1.4 CRYSTAL DEFECTS 14

1.4.1 Point Defects, 14

1.4.1.1 Thermal Fluctuation Effects, 17
1.4.1.2 Vapor Pressure Effects, 20
1.4.1.3 Impurities in Silicon, 23
1.4.1.4 Impurities in Gallium Arsenide, 27

1.4.2 Dislocations, 35

1.4.2.1 Screw Dislocations, 35
1.4.2.2 Edge Dislocations, 37
1.4.2.3 Movement of Dislocations, 38
1.4.2.4 Multiplication of Dislocations, 40
1.4.2.5 Twinning, 41

1.5 ELECTRONIC PROPERTIES OF DEFECTS 42

1.5.1 Point Defects, 42
1.5.2 Dislocations, 44

REFERENCES 46
PROBLEMS 47

Silicon is used almost exclusively in the fabrication of semiconductor devices and modern microcircuits, even though many elements and intermetallic compounds exhibit semiconducting properties. There are many reasons for this choice. Of these, the most important are the following:

1. Silicon is an elemental semiconductor. Together with germanium, it can be subjected to a large variety of processing steps without the problems of decomposition that are ever present with compound semiconductors. For much the same reason its properties can be studied with considerably greater ease than those of compound semiconductors. As a consequence, perhaps more is known today about the preparation and properties of extremely pure single-crystal germanium and silicon than about that of any other element in the periodic table.
2. Silicon has a wider energy gap than germanium. Consequently it can be fabricated into microcircuits capable of operation at higher temperatures than their germanium counterparts. At the present time the upper operating ambient temperature for silicon microcircuits is between 125 and 175°C, and this is entirely acceptable for a large number of military applications.
3. Silicon readily lends itself to surface passivation treatments. This takes the form of a layer of thermally grown silicon dioxide which provides a high degree of protection to the underlying device. Although the fabrication of devices such as metal-oxide-semiconductor (MOS) transistors has emphasized that thermally grown silicon dioxide falls short of providing perfect control of surface phenomena, it is safe to say that the development of this technique [1] resulted in a decisive advantage for silicon over germanium as the starting material in microcircuits. As a result, a significant technological base has been established to take advantage of its characteristics. This includes the development of a number of advanced processes for deposition and doping of silicon layers, as well as sophisticated equipment for forming and defining intricate patterns for very-large-scale integration.

Silicon is not an optimum choice in every respect. For example, the low-field mobility of gallium arsenide is higher than that of silicon, resulting in devices with reduced parasitics and improved frequency response. A more serious disadvantage lies in the fact that silicon is an indirect gap semiconductor. As a consequence, many important electrooptical applications are not possible with silicon devices or microcircuits.

The greatest impact of compound semiconductors has, until now, been in areas where their unique properties allow functions that *cannot* be performed by silicon. These include transferred electron devices, light-emitting diodes, lasers, and infrared photodetectors. Of the many compound semiconductors currently under investigation, gallium arsenide is the most technologically advanced. In recent years it has been shown [2] to have an advantage over silicon in conventional high-speed devices, notably Schottky gate field effect transistors, by a factor* of 3–4. Only recently [3] this speed advantage has been exploited in digital logic. Thus the device fabrication technology for this material is also being rapidly developed, so that both silicon and gallium arsenide form the basis of modern semiconductor devices and integrated circuits.

In this chapter some of the material properties of silicon and gallium arsenide are considered, since these have a bearing both on the fabrication processes that follow as well as on the electrical properties of devices made by these fabrication technologies.

1.1 CRYSTAL STRUCTURE

Both silicon and gallium arsenide belong to the cubic class of crystals. Crystal types belonging to this class exhibit the following structures:

1. *Simple cubic (s.c.) crystals.* This is illustrated in Fig. 1.1*a*. Very few crystals exhibit as simple a structure as this one; an example is polonium, which exhibits this structure over a narrow range of temperatures.
2. *Body-centered cubic (b.c.c.) crystals.* This is illustrated in Fig. 1.1*b*. Molybdenum, tantalum, and tungsten exhibit this crystal structure.
3. *Face-centered cubic (f.c.c.) crystals.* This is illustrated in Fig. 1.1*c*. The structure is exhibited by a large number of elements, such as copper, gold, nickel, platinum, and silver. (The face-centered atoms are shown different from the corner atoms for illustrative purposes only.)
4. *The zincblende structure.* This structure consists of two interpenetrating f.c.c. sublattices, with one atom of the second sublattice located at one fourth of the distance along a major diagonal of the first sublattice. This configuration is illustrated in Fig. 1.2*a* and *b*, where solid dots belong to the first sublattice and open dots to the second. In gallium arsenide, each sublattice contains atoms of only one type (either gallium or arsenic). The *diamond* lattice is a degenerate form of the zincblende structure, with identical atoms in each sublattice. Silicon belongs to this class.

The position of the various atoms in the zincblende lattice can be calculated in multiples of the cube edge a. Thus for the f.c.c. structure the various coordinates for the corner lattice sites are 0, 0, 0; 0, 1, 0; 0, 0, 1; 0, 1, 1; 1, 0, 0; 1, 1, 0; 1, 0, 1; and 1, 1, 1. Coordinates for the face-centered sites are $\frac{1}{2}$, $\frac{1}{2}$, 0; $\frac{1}{2}$, $\frac{1}{2}$, 1; 0, $\frac{1}{2}$, $\frac{1}{2}$; 1, $\frac{1}{2}$, $\frac{1}{2}$; $\frac{1}{2}$, 0, $\frac{1}{2}$; and $\frac{1}{2}$, 1, $\frac{1}{2}$, respectively. For the zincblende lattice it is necessary to include the coordinates of the second

* Ternary compounds such as GaInAs are potentially faster by an additional factor of 3. Their technology is in its infancy, however, and is not considered here.

sublattice, spaced at $\frac{1}{4}, \frac{1}{4}, \frac{1}{4}$ from those of the first. Within the unit cell these coordinates are $\frac{1}{4}, \frac{1}{4}, \frac{1}{4}$; $\frac{3}{4}, \frac{3}{4}, \frac{1}{4}$; $\frac{1}{4}, \frac{3}{4}, \frac{3}{4}$; and $\frac{3}{4}, \frac{1}{4}, \frac{3}{4}$, respectively. In Fig. 1.2*b* these lattice sites are shown different from those of the original sublattice. With reference to this figure, the following comments may be made:

1. The *coordination number* for this lattice is 4, i.e., each atom has four nearest neighbors which belong to a different sublattice. In silicon each

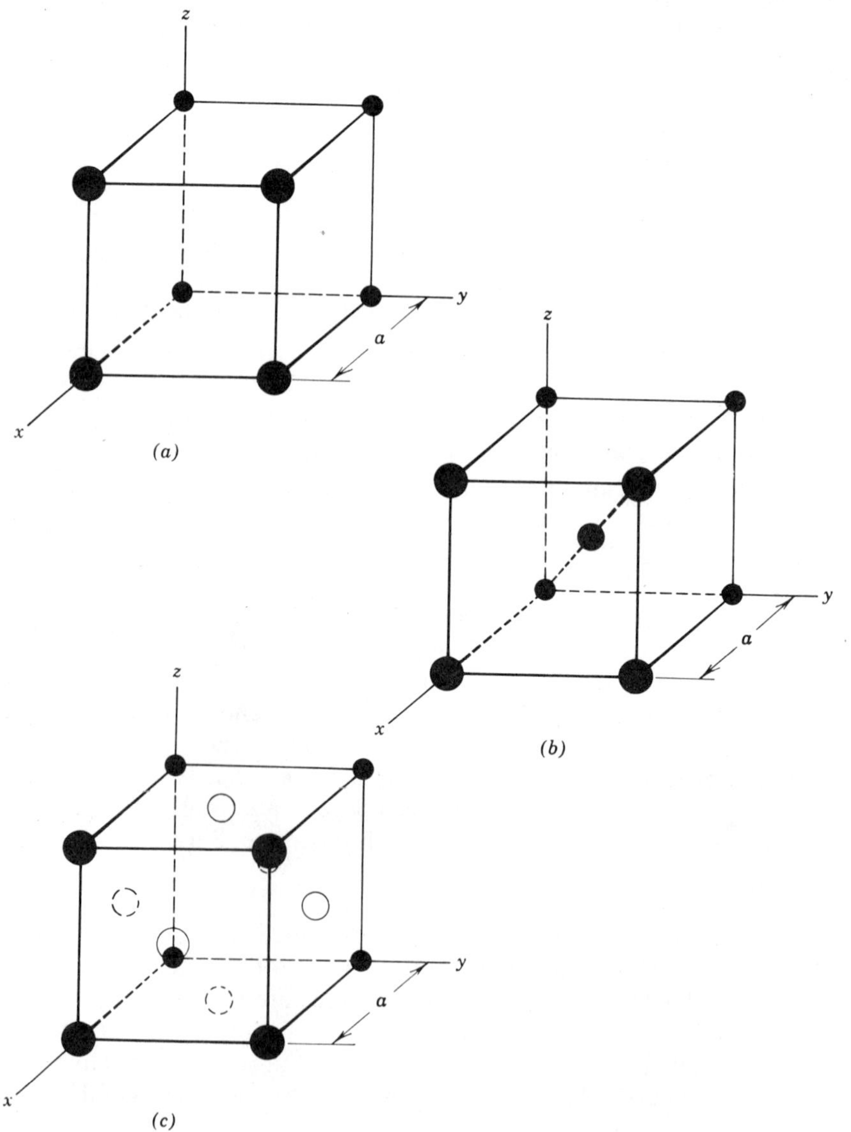

Fig. 1.1 Cubic crystal lattices.

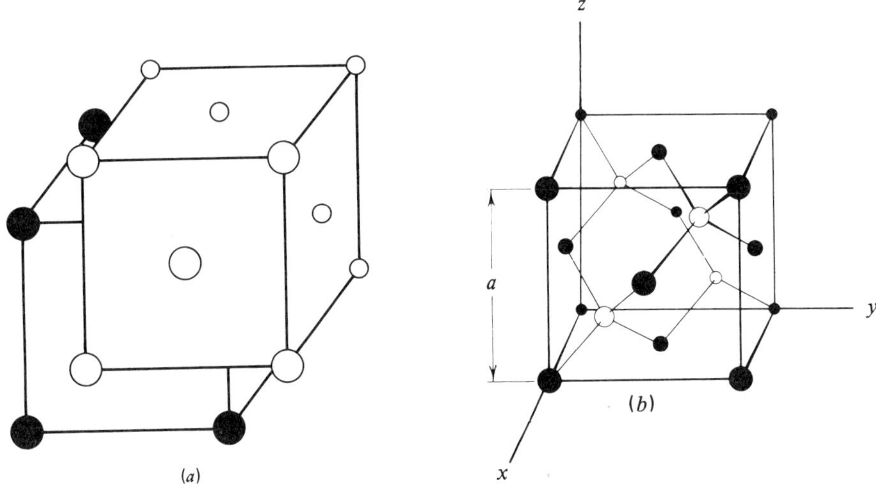

Fig. 1.2 The zincblende lattice.

atom has four valence electrons which provide covalent bonding with these nearest neighbors. In gallium arsenide, however, each arsenic atom (with five valence electrons) has four neighboring gallium atoms, each of which has three valence electrons. In like manner, each gallium atom (with three valence electrons) has four neighboring arsenic atoms, each having five valence electrons. Together, these gallium–arsenic pairs enter into bonding, which is primarily covalent in nature, but also partly ionic. For all practical purposes, gallium arsenide can be considered as a covalent-bonded semiconductor. Figure 1.3 shows an enlarged picture of a subcell with side $a/2$ in order to delineate a *tetrahedral* covalent bond of the type described here.

2. The distance between two neighboring atoms is $(\sqrt{3}/4)a$, where a is the cube edge. For silicon* a = 5.43 Å so that this distance is 2.35 Å. The radius of the silicon atom is thus 1.18 Å, if we assume a "hard sphere" model for atoms. Since each atom is situated within a tetrahedron comprising its four neighbors, this is referred to as the *tetrahedral radius*.

   For gallium arsenide* a = 5.65 Å so that the distance between neighbors is 2.44 Å. The tetrahedral radii of gallium and arsenic are 1.26 Å and 1.18 Å, respectively; together they add up to 2.44 Å.

3. Using the hard sphere model, 34% of the silicon lattice and 33.8% of the gallium arsenide lattice is occupied by atoms. Thus these are relatively loosely packed structures. (By way of comparison the packing density of a f.c.c. crystal is approximately 74%.)

* Room temperature values.

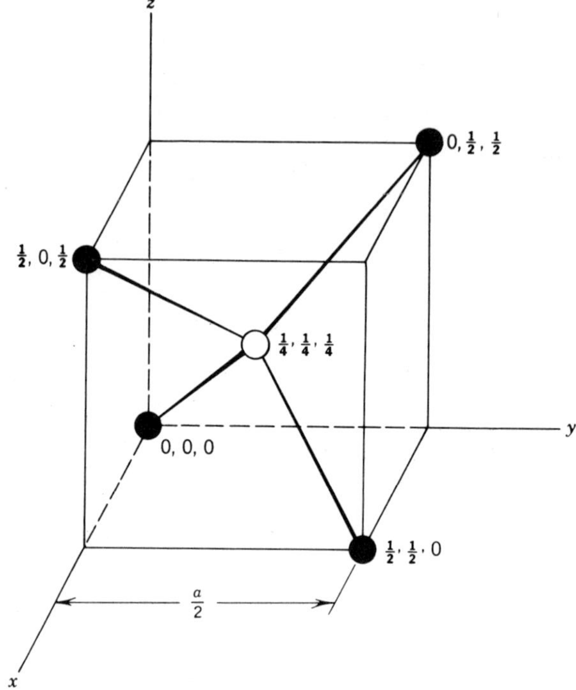

Fig. 1.3 The zincblende subcell.

Table 1.1 lists the *tetrahedral radii** of the various impurities that are commonly introduced into the silicon lattice to control its electronic behavior. If r_0 is the tetrahedral radius of the silicon atom, the radius of the impurity atom may be written as $r_0(1 \pm \epsilon)$. The quantity ϵ is defined as the *misfit factor,* and is indicative of the degree of strain present in the lattice

Table 1.1 Tetrahedral Radii and Misfit Factors for Various Dopants Used with Silicon

Dopant	P	As	Sb	B	Al	Ga	In	Au	Ag
Tetrahedral radius (Å)	1.10	1.18	1.36	0.88	1.26	1.26	1.44	1.5	1.52
Misfit factor	0.068	0	0.153	0.254	0.068	0.068	0.22	0.272	0.29
Type of dopant	*n*-type			*p*-type				Deep lying	

* It should be noted that the radius of an impurity atom in a zincblende lattice is independent of the chemical components of this lattice. The concept of a constant tetrahedral radius is an empirical but very useful one. This radius is, however, not the same as the radius of the atom in its own lattice (i.e., its *ionic radius*), since the internal field conditions are quite different for these cases.

Table 1.2 Tetrahedral Radii and Misfit Factors for Various Dopants Used with Gallium Arsenide

Dopant	S	Se	Te	Sn	C	Ge	Si	Be	Cd	Mg	Zn
Tetrahedral radius (Å)	1.04	1.14	1.32	1.40	0.77	1.22	1.18	1.06	1.48	1.40	1.31
Misfit factor (atom on arsenic site)	0.119	0.034	0.119	0.186	0.347	0.034	0.0	—	—	—	—
Misfit factor (atom on gallium site)	—	—	—	0.111	0.389	0.032	0.063	0.159	0.175	0.111	0.040
Type of dopant	*n*-type			*n*/*p*-type				*p*-type			

as a result of introducing this impurity. It is also an indication of the amount of dopant which can be incorporated into electronically active sites in the lattice.

Impurities used for doping gallium arsenide are listed in Table 1.2. Some of these dopants are incorporated on only one sublattice, corresponding to a single misfit factor. Yet others partially substitute on both lattices, so that two misfit factors describe their effect more appropriately.

1.2 CRYSTAL AXES AND PLANES

Directions in crystals of the cubic class are very conveniently described [4] in terms of Miller notation. Consider, for example, any plane in space, which satisfies the equation

$$\frac{x}{a} + \frac{y}{b} + \frac{z}{c} = 1 \tag{1.1}$$

Here a, b, and c are the intercepts made by the plane at the x, y, and z axes, respectively. Writing h, k, and l as the reciprocals of these intercepts, the plane may be described by

$$hx + ky + lz = 1 \tag{1.2}$$

The Miller indices for such a plane* are written as (hkl). Integral values are usually chosen in multiples of the edge of the unit cell.

Figure 1.4*a* shows a cubic crystal with some of its important planes indicated. Here the plane *ABCD* is designated (110), while the plane *EDC* is designated (111). The (100) and (010) planes are also shown in this figure. Figure 1.4*b* shows an example of how planes with negative indices may be described. Thus the plane *PQRS* is defined by $(0-10)$ and is commonly written as $(0\bar{1}0)$. The plane *RSTU* is written in like manner as $(1\bar{1}0)$.

The atom configurations in many of the Miller planes in a cubic crystal are identical. Thus the planes (001), (010), (100), $(00\bar{1})$, $(0\bar{1}0)$, and $(\bar{1}00)$ are essentially similar in nature. For convenience they are written as the {001} planes.

Figure 1.4*c* shows examples of planes with higher indices. Thus the plane *GHKJ* is denoted by $(1\tfrac{1}{2}0)$ or preferably by (210). Similarly, the plane *HKL* is written in Miller notation as (212).

Planes with higher Miller indices may be sketched by extending these principles. They are not, however, often encountered in discussions of the material properties of semiconductors.

Indices of lattice plane direction (i.e., of the line normal to the lattice plane) are simply the vector components of the direction resolved along the

* It should be noted that (hkl) refers to any one of a series of parallel planes in a cubic crystal. This may be seen by a simple shifting of the origin for the reference axes.

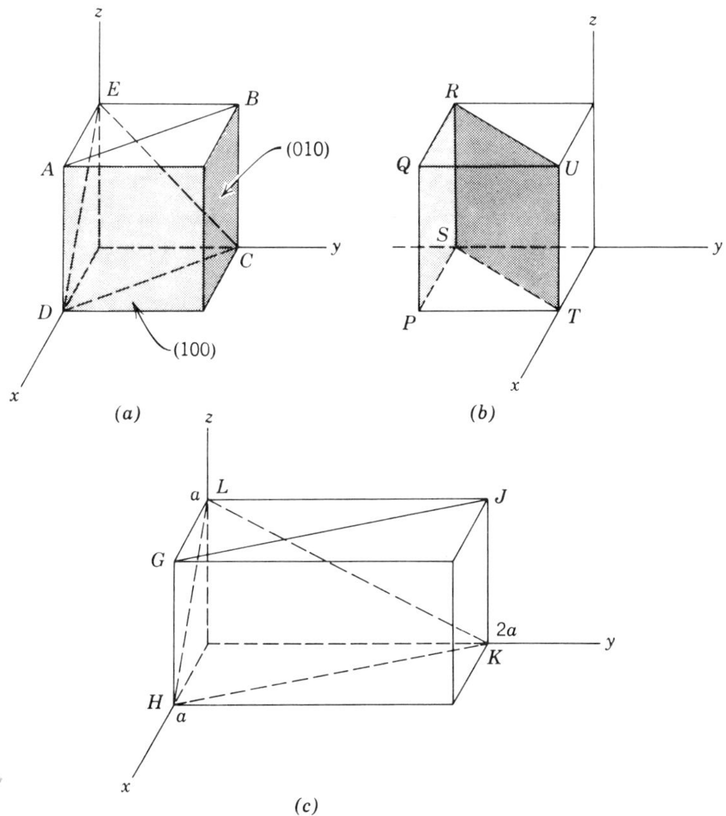

Fig. 1.4 Miller planes.

coordinate axes. Thus the (111) plane has a direction written as [111], and so on. This is an extremely convenient feature of the Miller index system for cubic crystals. For this notation the set of direction axes [001], [010], [100], [00$\bar{1}$], [0$\bar{1}$0], [$\bar{1}$00] is written as ⟨001⟩.

The angle θ included between two planes $(u_1v_1w_1)$ and $(u_2v_2w_2)$ is given by

$$\cos\theta = \frac{u_1u_2 + v_1v_2 + w_1w_2}{\sqrt{(u_1^2 + v_1^2 + w_1^2)(u_2^2 + v_2^2 + w_2^2)}} \tag{1.3}$$

The line describing the intersection of these planes is $[uvw]$, where

$$u = v_1w_2 - v_2w_1 \tag{1.4a}$$

$$v = w_1u_2 - w_2u_1 \tag{1.4b}$$

$$w = u_1v_2 - u_2v_1 \tag{1.4c}$$

The separation between two adjacent parallel planes (hkl) is given by

$$d = \frac{a}{\sqrt{h^2 + k^2 + l^2}} \tag{1.5}$$

This separation is equal to a for the $\langle 100 \rangle$ planes, to $0.707a$ for the $\langle 110 \rangle$ planes, and to $0.577a$ for the $\langle 111 \rangle$ planes. Thus the $\{111\}$ planes are the closest spaced among the low-index planes.

1.3 ORIENTATION EFFECTS

Many fabrication processes are orientation sensitive, that is, they depend on the direction in which the crystal slice is cut. This is to be expected, since many mechanical and electronic properties of the crystal and its surface are orientation dependent. Some of the consequences of crystal orientation are now described.

1.3.1 Silicon

For silicon the $\{111\}$ planes exhibit the smallest separation (3.135 Å). Therefore growth of the crystal along a $\langle 111 \rangle$ silicon direction is most easily accomplished, since it results in the setting down of one atomic layer upon another in closest packed form. From an economic point of view, $\langle 111 \rangle$ silicon is thus the least expensive, and is used in the majority of bipolar* devices and microcircuits today. By the same token, crystal dissolution by alloying is slowest in this direction. Thus parallel-plane alloyed junctions can only be made on $\langle 111 \rangle$ silicon.

Crystal dissolution by chemical etching is also slowest in the $\langle 111 \rangle$ directions. Consequently selective etches will preferentially etch silicon by exposing $\{111\}$ planes. The use of this technique for cutting V-grooves in a (100) oriented silicon slice, as well as for cutting apart a slice into individual dice, is outlined in Chapter 9, together with a number of suitable etch formulations. This forms the basis for many important microelectronic processes (V-MOS, Beam Lead, V-FET, etc.) which are used today.

A common technique for dicing a silicon slice into its individual microcircuits is to scribe its surface with a diamond tool into a rectangular chip pattern.† Once the slice has been weakened in this manner, it is deformed until it breaks apart into individual chips. This occurs, as far as possible, by the propagation of the scribe cracks through the bulk of the silicon, along directions parallel to the natural cleavage planes.

* MOS devices are considered later in this section.

† Increasingly, sawing is used for cutting apart chips, especially for the larger diameter wafers (≥ 100 mm).

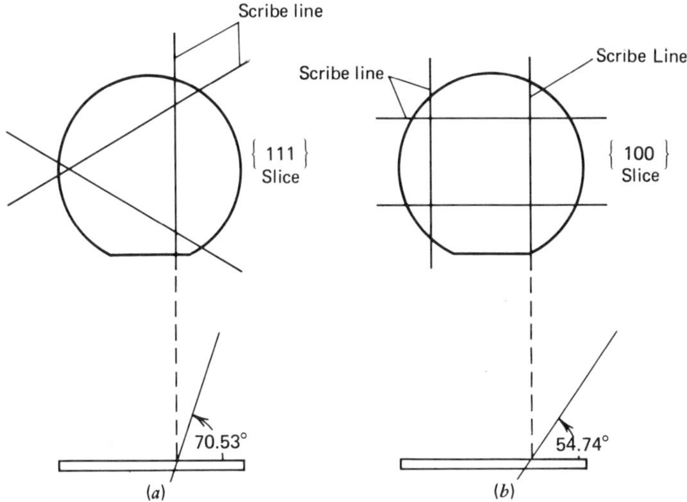

Fig. 1.5 Scribe lines and cleavage planes for {111} and {100} silicon.

The ultimate tensile strength of silicon (0.35×10^{10} dyn/cm^2) is a maximum in the ⟨111⟩ directions. In addition, the modulus of elasticity in the ⟨111⟩ directions is higher than that in the ⟨110⟩ or ⟨100⟩ directions (1.9×10^{12}, 1.7×10^{12}, and 1.3×10^{12} dyn/cm^2, respectively). As a result, silicon tends to cleave on the {111} planes [5].

The {111} planes within a slice meet the (111) plane of the surface at an angle of 70.53°, and along the ⟨110⟩ directions. Thus it is desirable to make scribe lines along the ⟨110⟩ directions, for easy cleaving. In practice, each slice is supplied by the manufacturer with a reference flat ground into it so as to allow the *first* scribe line to be made along an easy cleavage plane. Unfortunately all ⟨110⟩ directions on the {111} surface are mutually at 60° to each other. As a result, the *second* scribe line, which is made at right angles to the first (for rectangular chips), will not be in a ⟨110⟩ direction. Cleaving along this line takes on a jagged, zigzag nature, with each individual jag along one of the ⟨110⟩ directions.

Scribing of (100) silicon does not present this problem. Here the {111} planes intersect the surface at 54.74°, and on ⟨110⟩ directions which are mutually at right angles, so that *both* sets of scribe lines can be made along easy cleavage planes.

Figure 1.5 illustrates this situation for {111} and {100} silicon slices. It is interesting to note that for both cases, the plane of the chip edge is not at right angles to the silicon surface.

It is difficult to grow silicon in the (110) direction, so that this material is not used for conventional applications. However, plates of (110) material*

* See Problem 2 at the end of this chapter.

can be readily cut out of suitably oriented (111) silicon. An unusual property of (110) silicon is that four of the {111} planes intersect its surface at 90°. As a result, this material finds use in some special situations where it is required to etch deep vertical, parallel-sided grooves in silicon [6].

Other processes which are orientation dependent are diffusion and oxidation. These are taken up in detail in Chapters 5 and 7, respectively.

The electronic properties of the silicon surface are related to the density of dangling bonds on the surface, as well as to the bond strength. Typically it has been observed that the surface state density for ⟨100⟩ silicon is lower than that for ⟨111⟩ silicon by a factor of about 3. As a result, ⟨100⟩ silicon is often used for bipolar applications where low $1/f$ (or flicker) noise is required. Furthermore, almost all MOS circuits are built on ⟨100⟩ silicon today. This represents a very large application area, so that the gap between slice costs for these two orientations has been sharply reduced.

1.3.2 Gallium Arsenide

As noted earlier, gallium arsenide comprises two interpenetrating f.c.c. sublattices. One of these is displaced one quarter of the way along the main diagonal of the other, resulting in asymmetry [7]. This is seen in Fig. 1.6, which provides a view of the gallium arsenide lattice in the [110] direction, with the [111] axis (body diagonal) in the plane of the paper. We note that

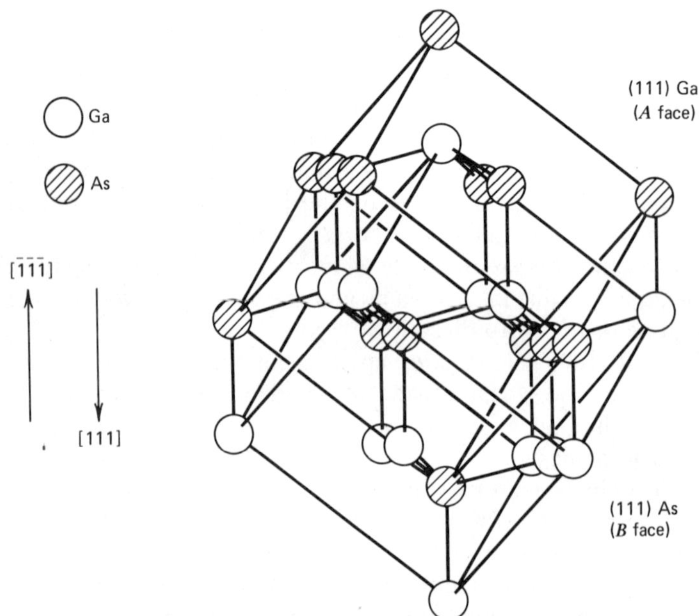

Fig. 1.6 The zincblende lattice observed at right angles to a [111] axis, and along a [110] axis.

the crystal structure consists of hexagonal rings, stacked in the [110] direction, but with different spacings. Assigning the layers to gallium and arsenic, their succession in the [111] direction is Ga-As—Ga-As—Ga-As—, whereas in the [$\bar{1}\bar{1}\bar{1}$] direction it is As-Ga—As-Ga—As-Ga—. In silicon these two directions are identical; with gallium arsenide they can be distinguished from each other, and the [111] axis is a polar axis for this structure. The (111) plane in this figure is referred to as the (111)Ga face, whereas the ($\bar{1}\bar{1}\bar{1}$) plane is called the (111)As face.*

Looking down on the crystal from the top, at the (111)Ga face, we have two possibilities: we can get a gallium atom connected by three covalent bonds to arsenic atoms in the next (lower) layer, with one dangling bond. Alternately, it is possible to have an arsenic atom connected by one covalent bond to a gallium atom in the next (lower) layer, with three dangling bonds. Energetically, the first of these situations is more favored, and the second does not appear to even exist. In fact, crystal growth and etching always seem to occur by the growth or dissolution of such double layers. Looking at the (111)As face from the bottom, the situation is quite different, with arsenic atoms connected by three bonds to a lower layer of gallium, and so on.

In consequence, then, the (111)Ga face has gallium atoms with no free electrons, since all three of their valence electrons are attached. The (111)As face, on the other hand, consists of arsenic atoms, each with two free electrons, since only three of their five valence electrons are attached. The (111)As face is thus more electronically active than the (111)Ga face. Etching of the (111)As face occurs very rapidly with a resultant smooth polish. The (111)Ga face, on the other hand, etches very slowly so that all imperfections become delineated, resulting in a rough surface. This is also true for mechanical processes, where it has been observed that the (111)As face is more readily lapped than the (111)Ga face.

At temperatures below 770°C, surface evaporation occurs more rapidly from the (111)As face, suggesting that the surface energy of atoms on this face is much higher than for the (111)Ga face. Above 800°C, however, the evaporation rates of both surfaces becomes equal. This is indicative of molecular dissociation, and occurs with an activation energy of 112 kcal/mole (4.86 eV/molecule).

Crystal oxidation occurs more readily on the (111)As face than on the (111)Ga face. Again, this is due to the higher electronic activity associated with the (111)As face.

The {111} planes of gallium arsenide consist of alternate layers of group III and group V atoms. These are differently charged, so there is a strong electrostatic attraction between them. As a result it is difficult to cleave gallium arsenide along {111} planes. The situation for the other principal orientations is much simpler. Thus the {110} faces are shared by an equal

* These faces are sometimes called the *A* and *B* faces, respectively.

density of gallium and arsenic atoms. Each atom is attached by one bond to an atom in the lower layer; two bonds connect in the surface plane to two nearest neighbors, leaving a fourth dangling bond. Interatomic forces are thus strong within the {110} planes, but weak between adjacent {110} planes. As a result, this is the preferred cleavage plane for gallium arsenide.

The {100} faces consist of either all gallium or all arsenic atoms. In either case, each atom is attached by two bonds to atoms in the lower layer, leaving two free dangling bonds. The properties of this face do not depend on whether the face is made up of gallium or of arsenic atoms. Some of the {110} cleavage planes intersect the (100) plane at right angles, and along mutually orthogonal directions. Thus perfectly rectangular chips can be cleaved from this material. This property is exploited in the fabrication of laser diodes, where parallel faces are essential.

1.4 CRYSTAL DEFECTS

Any interruption in a perfectly periodic lattice may be called a *defect,* which may take various forms, such as the following:

1. *Point defects.* These include vacancies, interstitials, impurity atoms deliberately introduced for the purpose of controlling the electronic properties of the semiconductor, and impurity atoms which are inadvertently incorporated as contamination during material growth or processing.
2. *Dislocations or line defects.* These are one-dimensional defects, and are made up of a large continuous array of point defects.
3. *Gross defects, such as slip and twin planes.* Here the defect occurs along one or more planes in the crystal.

A study of defects is important since most mechanisms for diffusion and crystal growth are defect induced. Many defects are introduced during the very act of device fabrication. Finally, all types of defects (chemical or otherwise) alter the electrical properties of the semiconductor in which they are present.

1.4.1 Point Defects

Let us consider, at first, the silicon lattice. Here the most elementary point defect is the *vacancy.* This is present when, as a result of thermal fluctuations, an atom is removed from its lattice site and moved to the surface of the crystal. Defects of this type are known as *Schottky defects,* and are associated in silicon with an energy of formation of about 2.3 eV and an energy of migration of about 0.18 eV. Figure 1.7*a* shows the manner in which such a defect occurs in an otherwise regular silicon lattice.

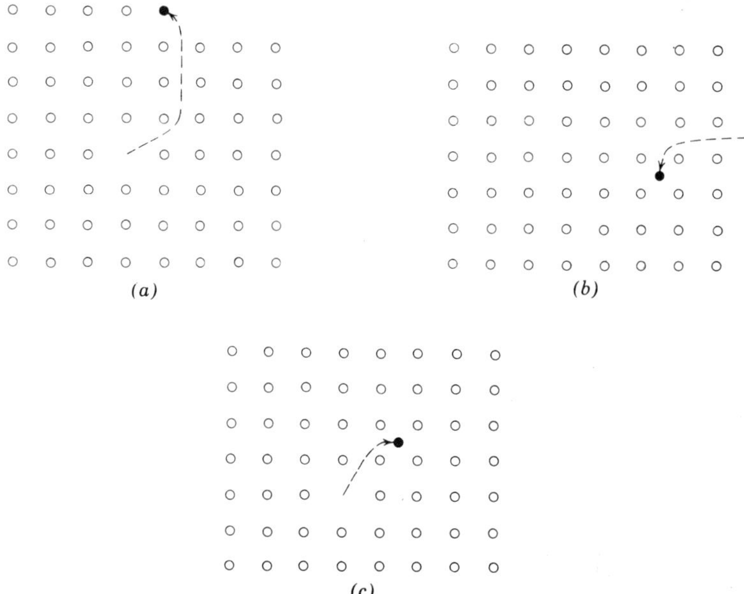

Fig. 1.7 Point defects.

A second elementary point defect that may be present in a crystal lattice is the *interstitial*. Such a defect occurs when an atom becomes located in one of the many interstitial voids within the crystal structure. Figure 1.7*b* shows schematically how this may occur in an otherwise regular crystal lattice. The energy of formation of an interstitial defect is relatively large in close-packed crystal structures. However, in the more loosely packed diamond lattice this is not the case; a value of 1.1 eV is often associated with this defect.

The vacancy–interstitial pair, or *Frenkel defect,* occurs when an atom leaves its regular site in a crystal and takes up an interstitial position, as shown in Fig. 1.7*c*. This interstitial is usually in the vicinity of the newly formed vacancy, so that the energy of formation of Frenkel defects is comparable to that of interstitial defects.

Figure 1.8 shows the unit cell for the diamond lattice. Within this unit cell are the centers of five interstitial voids [8], at $\frac{1}{2}$, $\frac{1}{2}$, $\frac{1}{2}$; $\frac{1}{4}$, $\frac{1}{4}$, $\frac{1}{4}$; $\frac{3}{4}$, $\frac{3}{4}$, $\frac{1}{4}$; $\frac{3}{4}$, $\frac{1}{4}$, $\frac{3}{4}$; and $\frac{1}{4}$, $\frac{3}{4}$, $\frac{3}{4}$. Another three voids, with their centers located at the midpoint of each of the twelve cube edges (each is shared by four unit cells), add up to a total of eight voids per unit cell. Each of these is large enough to contain an atom (again assuming hard spheres), even though there is a constriction in passing from one void to another. Consequently from a purely geometric viewpoint, the interstitial defect can be expected to be quite common in silicon.

Various combinations of these defects can also occur. Thus a single va-

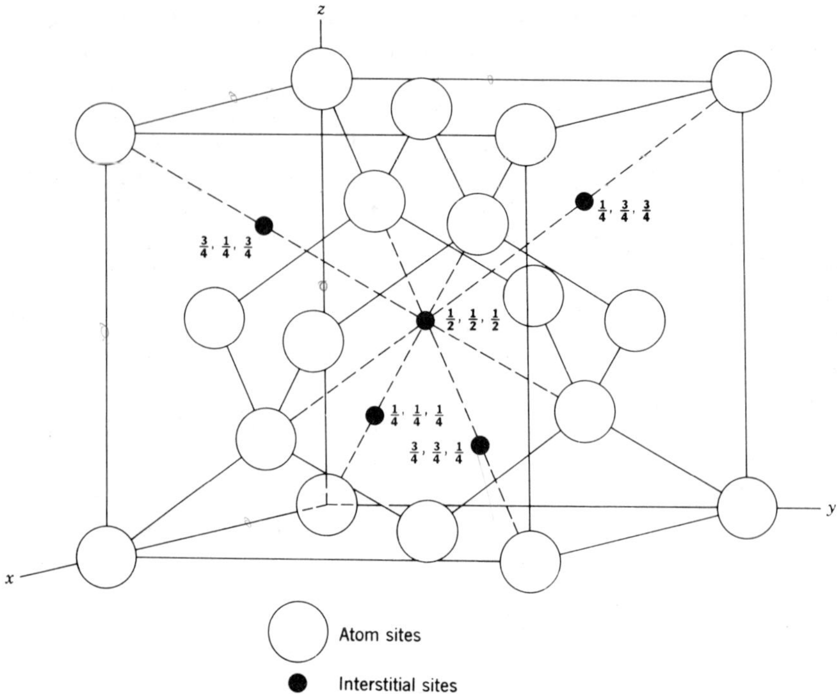

Fig. 1.8 Interstitial sites in the zincblende lattice.

cancy leads to the breaking of four covalent bonds, whereas two vacancies side by side require the breaking of only six bonds. Consequently the energy of formation of a divacancy of this type is less than that required to form two separate vacancies. The divacancy is thus commonly encountered. On the other hand, the di-interstitial is much more difficult, if not impossible, to form.

The classification of point defects is considerably more complex for the gallium arsenide lattice. For example:

1. Schottky defects may exist in the form of either gallium or arsenic vacancies.
2. Voids in gallium arsenide are surrounded by either gallium or arsenic atoms. Thus either a gallium or an arsenic atom can be interstitially located in a gallium or arsenic void. As a result, there are four possible types of interstitials.
3. It is possible for a gallium atom to be located on an arsenic site, or vice versa. These are known as *antistructure* defects.
4. There are four kinds of Frenkel pair possibilities, depending on the type of atom and the type of void into which it is displaced.

By the same token, the number of combinations of defects which can occur is much larger than for silicon. A detailed study of these possibilities is beyond the scope of this book.

An important type of point defect is created by chemical impurities which are intentionally (or unintentionally) introduced into the lattice. Impurity atoms that take up their locations at sites ordinarily occupied by lattice atoms are referred to as *substitutional* impurities. Alternately, *interstitial* impurities are located in the many interstitial voids that are present in the lattice.

Substitutional impurities are usually electronically active, and determine the conductivity type. Interstitial impurities, on the other hand, are usually inactive. This is not always true, however, a notable exception being lithium in silicon, which is interstitial but behaves like a donor.

Many impurities are entirely substitutional or interstitial in nature; yet others exhibit more complex behavior. Thus for gold in the silicon lattice, evidence indicates that about 90% is in active substitutional sites, while the rest is neutral and interstitial. With platinum, about 90% is in active, substitutional sites. The remaining 10% is also active, but its exact lattice location is as yet unidentified. One possibility is that it forms active platinum–silicon complexes in silicon. With nickel in silicon, as much as 99.9% is inactive and in interstitial sites. Finally, we note that the fraction of an impurity which is active is also a function of the doping concentration. This is because the introduction of a dopant at high concentrations is usually accompanied by the generation of strain in the lattice, caused by misfit.

Impurities are sometimes incorporated in the lattice in the form of electronically active complexes. This is particularly so with gallium arsenide, where many impurities exhibit energy levels which have been identified as being associated with impurity–gallium and impurity–arsenic pairs.

The detailed behavior of individual impurities in silicon and gallium arsenide is described in Sections 1.4.1.3 and 1.4.1.4.

1.4.1.1 Thermal Fluctuation Effects [9]

The defect concentration (vacancies, interstitials, Frenkel pairs, etc.) in a semiconductor is caused by thermal fluctuations in the material, and by the vapor pressure of the species surrounding it. With silicon processing, vapor pressure effects are negligible. By way of example, the vapor pressure of silicon is only 6.2×10^{-10} atm at a typical processing temperature of 1100°C.*

The presence of defects in the material changes both the internal energy of the crystal as well as its entropy. Consequently, their equilibrium concentration is a function of the energy of formation and of the equilibrium

* It is for this reason that silicon processing can be carried out readily in an open-tube environment.

temperature. On the other hand, the concentration of a chemical defect is primarily a function of the amount available for introduction into the crystal, and of its solid solubility.

The equilibrium concentration of Schottky defects in silicon may be determined on the assumption that only these effects occur. Let:

N = total number of atoms in a crystal of unit volume (5×10^{22} cm^{-3} for silicon)

n_s = number of Schottky defects per unit volume

E_s = energy of formation of a Schottky defect, i.e., the energy required to move an atom from its lattice site within the crystal to a lattice site on its surface ($\simeq$ 2.3 eV)

The number of ways in which a Schottky defect can occur is given by

$$\mathrm{C}_{n_s}^{N} = \frac{N!}{(N - n_s)!n_s!} \tag{1.6}$$

The entropy associated with this process is

$$\begin{aligned} S &= k \ln (\text{number of ways}) \\ &= k \ln \mathrm{C}_{n_s}^{N} \end{aligned} \tag{1.7}$$

where k is Boltzmann's constant (8.62×10^{-5} eV/°K). The internal energy E is given by n_sE_s. The change in free energy (neglecting volume changes) is

$$F = E - TS \tag{1.8}$$

$$= n_sE_s - kT\,[\ln N! - \ln n_s! - \ln(N - n_s)!] \tag{1.9}$$

The most probable equilibrium condition is the one where the free energy is a minimum with respect to changes in n_s, i.e., the case for

$$\left(\frac{\partial F}{\partial n_s}\right)_{T=\text{const}} = 0 \tag{1.10}$$

Differentiating Eq. (1.9) and setting to zero,

$$E_s = kT\frac{\partial}{\partial n_s}[\ln N! - \ln n_s! - \ln(N - n_s)!] \tag{1.11}$$

The factorial terms may be simplified by using Stirling's formula for the factorial of a large number. Thus

$$\ln x! \simeq x \ln x - x \tag{1.12}$$

so that Eq. (1.11) reduces to

$$E_s = kT \ln \left(\frac{N - n_s}{n_s} \right) \tag{1.13}$$

or

$$n_s = \frac{N}{1 + e^{E_s/kT}} \tag{1.14}$$

$$\simeq Ne^{-E_s/kT} \tag{1.15}$$

The equilibrium concentration of Frenkel defects may be found by an analogous approach. Again it is assumed that only these defects are present in silicon. Let:

N = number of atoms in a crystal of unit volume
N' = number of available interstitial sites per unit volume ($= N$)
n_f = number of Frenkel defects (i.e., vacancy–interstitial pairs) per unit volume
E_f = energy of formation of a Frenkel defect ($\simeq$ 1.1 eV)

A vacancy can occur in $C^N_{n_f}$ ways, and an interstitial in $C^{N'}_{n_f}$ ways. Consequently, a Frenkel defect can occur in $C^N_{n_f} C^{N'}_{n_f}$ ways if the events are assumed to be statistically independent.

The entropy associated with this situation is

$$S = k \ln C^N_{n_f} C^{N'}_{n_f} \tag{1.16}$$

The internal energy is given by $E = n_f E_f$. The change in free energy is thus

$$F = n_f E_f - kT \ln C^N_{n_f} C^{N'}_{n_f} \tag{1.17}$$

As before,

$$\left(\frac{\partial F}{\partial n_f} \right)_{T=\text{const}} = 0 \tag{1.18}$$

in thermal equilibrium. Differentiating Eq. (1.17) and using Stirling's formula gives

$$E_f = kT \ln \left[\frac{(N - n_f)(N' - n_f)}{n_f^2} \right] \tag{1.19}$$

or

$$n_f \simeq \sqrt{NN'}e^{-E_f/2kT} = Ne^{-E_f/2kT} \tag{1.20}$$

Concentrations of point defects in excess of the equilibrium value may be obtained by subjecting the semiconductor to nonequilibrium processes. Thus excessively fast cooling (quenching) can result in a supersaturated concentration of these defects. Nuclear radiation damage also results in increasing the defect concentration over its equilibrium value.

1.4.1.2 Vapor Pressure Effects

The situation with gallium arsenide is quite different to that with silicon. This material melts at 1238°C. Long before this point is reached, however, the surface layers decompose into gallium and arsenic. The vapor pressures of these individual components are quite different, so there is a preferential loss of the more volatile species (arsenic). If processing is carried out in an evacuated ampul, this arsenic goes into its volume until it establishes a sufficient partial pressure to prevent further decomposition of the gallium arsenide. Although gallium and arsenic vacancies are generated by thermal fluctuations as well, the vacancy concentration in this material will be dominated by vapor pressure effects so that thermal fluctuation effects can be neglected.

In its vapor phase, arsenic consists of As, As_2, and As_4. All of these species are present, and in equilibrium over the gallium arsenide. Furthermore, some gallium in the form of vapor is also present. The partial pressure of these species is shown in Fig. 1.9 as a function of temperature [10]. Note that all of these curves are double valued. The upper branch of the arsenic curves and the lower branch of the gallium curves are for conditions over gallium arsenide which is preferentially rich in arsenic. In like manner, the lower branch of the arsenic curves and the upper branch of the gallium curves are for conditions over gallium-rich gallium arsenide.

Under processing conditions, the gallium arsenide will usually be gallium-rich, with predominantly As_2 and As_4 in the vapor phase. Note, however, that at temperatures below 637°C, the partial pressures of arsenic and gallium over gallium arsenide are approximately equal. As a consequence, gallium arsenide evaporates congruently below this point with approximately zero gallium or arsenic vacancy generation. Very often, thermal cleaning of the gallium arsenide is performed under vacuum at about 600°C for a short period of time, to take advantage of this property.

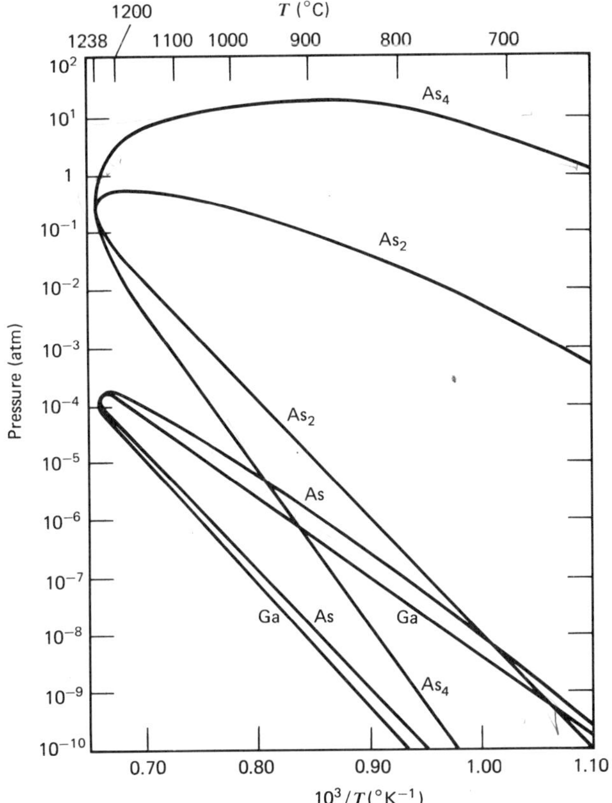

Fig. 1.9 Partial pressures of gallium and arsenic over gallium arsenide, as a function of temperature. From Authur [10]. Reprinted with permission from *Journal of Physics and Chemistry.*

It has been shown [11] that chemical interactions in a solid can be formally treated in the same way as interactions in solutions. Mass-action relationships can thus be applied to solids in order to determine the role of vapor pressure in controlling the vacancy concentration. Thus consider the decomposition reaction and the vacancy formation reactions of gallium arsenide. The decomposition reaction is

$$\mathrm{GaAs}(s) \rightleftharpoons \mathrm{Ga}(g) + \tfrac{1}{2}\mathrm{As_2}(g) \tag{1.21a}$$

assuming that As_2 is the dominant arsenic species. For this reaction the mass-action law relationship gives

$$k_1 = p_{\mathrm{Ga}} p_{\mathrm{As_2}}{}^{1/2} \tag{1.21b}$$

The vacancy formation reactions are

$$\mathrm{Ga}(s) \rightleftharpoons V_{\mathrm{Ga}}{}^{0} + \mathrm{Ga}(g) \tag{1.22a}$$

so that

$$k_2 = [V_{\mathrm{Ga}}{}^{0}]p_{\mathrm{Ga}} \tag{1.22b}$$

and similarly,

$$\mathrm{As}(s) \rightleftharpoons V_{\mathrm{As}}{}^{0} + \tfrac{1}{2}\mathrm{As}_2(g) \tag{1.23a}$$

so that

$$k_3 = [V_{\mathrm{As}}{}^{0}]p_{\mathrm{As}_2}{}^{1/2} \tag{1.23b}$$

Here $[V_{\mathrm{Ga}}{}^{0}]$ and $[V_{\mathrm{As}}{}^{0}]$ are the concentrations of (assumed neutral) gallium and arsenic vacancies, respectively; p_{Ga} and p_{As_2} are the partial pressures of Ga As_2, respectively; and k_1, k_2, k_3 are equilibrium constants. Furthermore, it is assumed that the concentration of interstitials of both gallium and arsenic is negligible.

Combining Eqs. (1.21b) and (1.22b) gives

$$[V_{\mathrm{Ga}}{}^{0}] = \frac{k_2}{k_1} p_{\mathrm{As}_2}{}^{1/2} \tag{1.24}$$

Thus the concentration of gallium vacancies is *directly* proportional to the square root of the partial pressure of As_2. However, the concentration of arsenic vacancies is seen from Eq. (1.23b) to be *inversely* proportional to the square root of the partial pressure of As_2. These relationships can also be written with a fourth-root dependence on the partial pressure of As_4. In both cases, however, the product of the two vacancy concentrations is only a function of temperature, and is independent of the partial pressure of the arsenic.*

Calculations for the arsenic and gallium vacancy concentrations, under equilibrium pressure conditions, have given

$$[V_{\mathrm{Ga}}{}^{0}] = 3.33 \times 10^{18} e^{-0.4/kT} \tag{1.25a}$$

$$[V_{\mathrm{As}}{}^{0}] = 2.22 \times 10^{20} e^{-0.7/kT} \tag{1.25b}$$

These values must be considered as very approximate. Experimental data

* This interesting relationship is similar in nature to the *pn* product in a material, which is also a function of temperature and not of doping.

[12] at 1100°C indicate an uncertainty in $[V_{As}{}^0]$ of less than a factor of 10. No experimental data are available for the gallium vacancy concentration.

1.4.1.3 Impurities in Silicon

At moderate concentrations, impurities belonging to groups V and III of the periodic table are substitutional in nature, and are electronically active. A group V atom has an excess valence electron which does not enter into covalent bonding with neighboring silicon atoms. Consequently, this electron is loosely bound, and is free to participate in the conduction process. Group V atoms are thus referred to as n-type impurities.

An estimate of the binding energy of this electron to its atom may be made by noting that the group V impurity can be represented by a nucleus with a single orbiting electron, i.e., it is hydrogen-like in character. Elementary atomic theory shows that the energy levels of the hydrogen atom are given by

$$E = -\frac{m_0 q^4}{8a^2h^2\epsilon_0{}^2} \tag{1.26a}$$

where ϵ_0 is the permittivity of free space and a takes on the values 1, 2, ... An equivalent situation can be considered for a crystal by replacing the mass of the electron with its effective mass and the permittivity of free space by that of the crystal. Thus

$$E = -\frac{m_n^* q^4}{8a^2h^2(\epsilon\epsilon_0)^2} \tag{1.26b}$$

where ϵ is the relative permittivity of the crystal.

The energy required to remove an electron from the ground state of the hydrogen atom is 13.6 eV. This is known as the first ionization potential. The comparable value in a crystal would thus be

$$E_{\text{ionization}} = \frac{13.6}{\epsilon^2}\frac{m_n^*}{m_0} \tag{1.27}$$

Assuming an effective mass ratio* of about 0.6 and a permittivity of 11.8 for silicon gives an ionization energy of about 0.06 eV. A group III impurity has a hole (the absence of an electron) which is loosely bound to it. Using analogous reasoning, we arrive at a very similar figure for its ionization energy.

* This is roughly the average of the conductivity effective mass ratio and the density-of-states effective mass ratio.

Table 1.3 lists common donors and acceptors for silicon together with their ionization energies. Impurities of this type are referred to as *shallow*, and are almost fully ionized at room temperature. It is worth noting that indium is not generally considered a shallow impurity in silicon.

In principle a shallow impurity may exhibit more than one energy level for each of its charge states. Only one impurity level, however, is normally observed within the forbidden gap for shallow donors and acceptors in silicon. Additional levels, corresponding to the second and higher ionization potentials, have also been observed at low temperatures [13].

Values given in Table 1.3 hold for moderate doping levels. With heavier doping, these levels broaden into bands. In addition, the impurity atoms come closer together (their average distance varies inversely as the cube root of the doping concentration), with a resulting decrease in their potential energy. As a result, the activation energy of the impurity, measured as the minimum energy gap between the impurity level and the appropriate band gap edge, falls. Experimentally the energy gap can be fitted to

$$E_0(N) = E_0(0) - \alpha N^{1/3} \tag{1.28}$$

where $E_0(N)$ is the activation energy at a doping level of N impurity atoms/cm^3, and $E_0(0)$ is the energy gap at low doping levels. For boron and phosphorus, $E_0(0)$ values are 0.08 eV and 0.054 eV, respectively, and α takes on the value of 4.3×10^{-8} eV cm^3 [14]. Thus boron-doped silicon and phosphorus-doped silicon are degenerate [$E_0(N) = 0$ eV] at doping levels in excess of 6.44×10^{18} and 1.98×10^{18} cm^{-3}, respectively.

The hydrogen model is an extremely elementary one. Indeed it is remarkable that it can predict the energy levels of the shallow impurities with any degree of accuracy. Needless to say, the model cannot be extended too far. Thus it cannot explain the ionization energy of indium or of impurities from other groups in the periodic table.

The doping of silicon with elements other than those of groups III and V of the periodic table often gives rise to a somewhat complex energy-level

Table 1.3 Ionization Energies for Group III and V Impurities in Silicon

Dopant	P	As	Sb	B	Al	Ga	In
Acceptor level, distance from valence band (eV)	—	—	—	0.045	0.057	0.065	0.16
Donor level, distance from conduction band (eV)	0.044	0.049	0.039	—	—	—	—
Type of dopant		*n*-type				*p*-type	

Table 1.4 Ionization Energies of Deep Lying Impurities in Silicon

Dopant	Ag	Au	Cu	Mo	Ni	O	Pt	Tl	Zn
Acceptor levels, distance from valence band (eV)	0.89	0.57	0.24 0.37 0.52	0.3	0.21 0.76	—	0.42 0.92	0.26	0.31 0.56
Donor levels, distance from conduction band (eV)	0.79	0.76	—	—	—	0.16	0.85	—	—

structure. In general these impurities exhibit more than one energy level, often of more than one type (i.e., both donor and acceptor levels). Furthermore, these levels are usually found quite deep in the forbidden gap. The deliberate introduction of deep levels is often done in order to reduce minority carrier lifetime in high-speed microcircuits.

The most complex energy-level structure arises [15] for a monovalent impurity atom. Such an atom, in the neutral state, has only one attached electron which provides covalent bonding with its neighboring silicon lattice atoms. When additional electrons are attached to it, it is successively transferred to a more and more negatively charged state. Each additional electron gives rise to a possible new energy level. Since these electrons are attached successively to more negatively charged atoms, it is probable that the value of the associated energy level will continually increase in sequence until it goes beyond the conduction-band edge. At this point the atom will lose this electron to the conduction band, resulting in no further identifiable energy levels.

In addition, a monovalent impurity atom may lose an electron and be promoted to a positive (donor) charge state. Thus it is reasonable to assume that the energy-level structure for a monovalent impurity atom may consist of as many as one donor level and three acceptor levels, progressively spaced in the order of increasing negative charge. In most instances only a few of these various levels are identifiable within the energy gap. A notable exception is gold in germanium, which exhibits all four energy levels.

The ionization energies of some deep lying impurities [16] in silicon are listed in Table 1.4. Many of these are present unintentionally in the starting material. The properties of a few of these, deserving special mention, are now described.

Oxygen. Silicon is commonly grown in silica crucibles, so that a large amount of this impurity is usually present. Depending on the crystal growth technique, this ranges from 10^{16} to 10^{18} cm^{-3}. Most of this oxygen is combined with silicon in the form of large clumps of SiO_2, typically 1 μm in diameter and 1 μm apart [17]. Although inert, it results in a curvature of potential lines in the depletion layer of p–n junctions made in this material. This results in premature breakdown and "soft spots."

Some of the dissolved oxygen is in interstitial sites, in the form of Si_2O complexes. These can be identified by their characteristic absorption peak at 1107 cm^{-1} ($\simeq 9\ \mu m$). Upon heat treatment they convert to another complex, which is active [18], and exhibits a donor level at $E_c - 0.16$ eV. This complex has been identified as SiO_4, and appears during processing at low temperatures (400–500°C), but disappears at higher temperatures. A number of other deep levels have been reported in the literature, but there is little agreement concerning their positions and concentrations. Many have been attributed to oxygen complexes with other impurities such as copper, cobalt, and nickel.

Carbon. Carbon has a high solid solubility ($4 \times 10^{18}\ cm^{-3}$) and is often incorporated in silicon during its chemical purification process. It is electrically inactive, and forms silicon–carbon complexes in the form of microprecipitates. Again, its presence in high concentrations leads to premature breakdown of p–n junctions. A number of metal–carbon complexes have been reported in the literature.

Gold. Gold is of special importance in the technology of silicon, and is commonly used for lifetime reduction in high-speed digital circuits [19]. It has a solid solubility of 2×10^{17} atoms/cm^3, and can be incorporated in large concentrations into silicon without the formation of any complexes. It exhibits a donor level at $E_c - 0.76$ eV and an acceptor level at $E_v + 0.57$ eV, depending on the particular charge state in which it is incorporated. It is commonly referred to as an amphoteric impurity because of this property.

Platinum. Recently platinum has been investigated as a substitute for gold in silicon [20]. It has a solid solubility of about $10^{17}\ cm^{-3}$. About 90% of it is incorporated into substitutional sites, exhibiting an acceptor energy level at $E_v + 0.92$ eV and a donor level at $E_c - 0.85$ eV. The remaining platinum is believed to be in the form of an electronically active complex, and behaves as an acceptor at $E_v + 0.42$ eV. Its capture cross section is extremely large, and accounts for most of the lifetime reduction properties of this dopant.

The energy levels of platinum in silicon are highly asymmetric with respect to their location in the energy gap, and result in very different lifetime characteristics from those obtained with gold in silicon. In particular, the space charge generation lifetime in platinum-doped devices at room temperature is many hundred times longer than the low-level lifetime. Thus platinum doping can reduce low-level lifetime in the neutral regions of a diode without a comparable increase in its leakage current. In contrast, a direct consequence of the symmetrical location of the gold acceptor level is that leakage current in p–n diodes is inversely related to the lifetime in the neutral regions.

Finally, it should be noted that silicon, as obtained commercially, is nearly uncompensated. Thus *n*-type material has a negligibly small amount of shallow or deep acceptors unintentionally incorporated in it. Similarly, the concentration of donors which are unintentionally incorporated in *p*-type silicon is negligibly small. As a result, the mobility of starting silicon, as obtained from various sources, is essentially that of uncompensated material (see Fig. 1.10). A plot of resistivity as a function of carrier concentration is shown in Fig. 1.11.

Although relatively free of compensation, a slice of silicon typically has in it a large amount of inert impurities (such as carbon, oxygen, fluorine, sodium, and calcium). By way of example, Table 1.5 lists the impurity content of a typical silicon crystal, grown by different manufacturing techniques [21].

1.4.1.4 Impurities in Gallium Arsenide [16, 22]

A number of patterns emerge from a study of dopants in gallium arsenide. Thus impurities belonging to group VI will usually be incorporated substitutionally, and on the arsenic sublattice. Consequently, each impurity atom of this type contributes one loosely bound electron for conduction purposes, i.e., it is *n*-type. In like manner, impurities from group II are also substitutional, but are incorporated on the gallium sublattice where they result in *p*-type conduction.

As with silicon, these impurities are shallow, and their ionization energies can be estimated by using the hydrogen model. The effective mass of holes

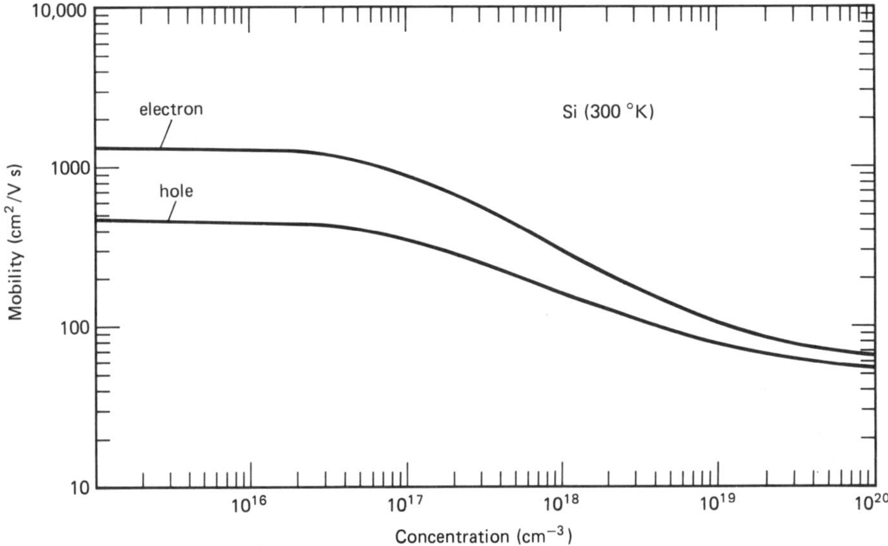

Fig. 1.10 Electron and hole drift mobilities in silicon.

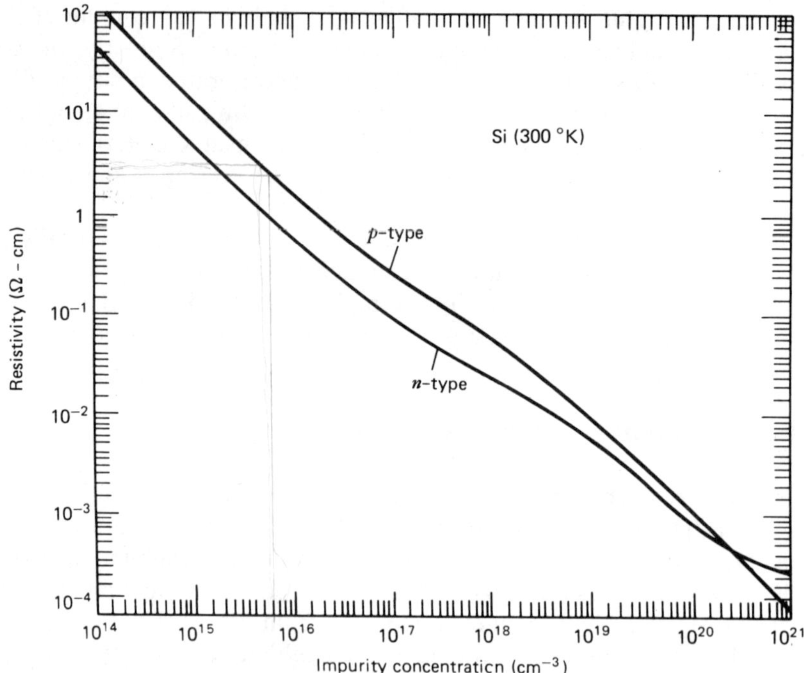

Fig. 1.11 Resistivity versus carrier concentration in silicon.

in gallium arsenide is $m_p \simeq 0.5m_0$, so that the ionization energy for p-type impurities is approximately 0.045 eV. On the other hand, the effective electron mass* is much smaller ($m_n \simeq 0.067m_0$) so that the ionization energy for electrons is only 0.005 eV. As a result, n-type gallium arsenide is degenerate at most useful concentration levels. Typically these n-type impurities are fully ionized, even at liquid nitrogen temperatures (77°K).

Impurities from group IV (carbon, germanium, silicon, tin, lead) are usually incorporated substitutionally into gallium arsenide, partly on each sublattice, depending on the relative vacancy concentrations. Group IV impurities will be n-type on the gallium sublattice, but p-type on the arsenic sublattice. The net free carrier concentration is thus less than the impurity concentration, and is either n- or p-type, depending upon the conditions under which these impurities are incorporated. In some situations this can be used advantageously to obtain successive n-layers and p-layers with the same dopant.

Many impurities, both shallow and deep, are present in the form of complexes with gallium or arsenic. Both active and inactive complexes have been identified in gallium arsenide, and little is known of the manner in which they are incorporated into the lattice. Table 1.6 lists shallow impurities

* For low-field conditions.

Table 1.5 Impurity Content of Typical Silicon Crystals[a]

Impurity	Float Zone	Czochralski
C	2.2×10^{16}	1.5×10^{17}
O	3×10^{16}	4×10^{17}
F	1×10^{16}	2×10^{16}
Na	1.5×10^{15}	10^{15}
Ca	2.71×10^{15}	2×10^{16}
Cl	10^{15}	10^{15}
Fe	$< 10^{15}$	10^{16}
Cu	0.85×10^{16}	1.9×10^{16}
K, Co, Mn, Ni	$< 10^{15}$	$< 10^{15}$

[a] See reference 21.

in gallium arsenide together with their ionization energies and conductivity types. The deep impurities are listed in Table 1.7.

A number of specific impurities are now considered briefly, together with some of their characteristics:

Zinc. This is a very common *p*-type dopant, and is the best studied impurity in gallium arsenide. It is incorporated substitutionally in gallium sites, and can be introduced in concentrations up to 10^{21} cm^{-3}.

Sulfur. This is the most common dopant for *n*-type material, and is incorporated on arsenic sites. Its solid solubility is about 10^{18} cm^{-3} at 900°C.

Table 1.6 Ionization Energies of Shallow Impurities in Gallium Arsenide

		Ionization Energy (eV)	
Impurity	Type	From Conduction Band	From Valence Band
S	*n*	0.0061	
Se	*n*	0.0059	
Te	*n*	0.0058	
Sn	*n*	0.0060	
C	*n*/*p*	0.0060	$\simeq 0.026$
Ge	*n*/*p*	0.0061	0.040
Si	*n*/*p*	0.0058	$\simeq 0.035$
Cd	*p*		0.035
Zn	*p*		0.031
Be	*p*		0.028
Mg	*p*		0.028
Li	*p*		0.023, 0.05

Table 1.7 Ionization Energies of Deep Impurities in Gallium Arsenide[a]

Impurity	Type	Ionization Energy (eV) From Conduction Band	From Valence Band
O	*n*	0.4, 0.75	
Unknown	*n*	0.17	
Co	*p*		0.16, 0.56
Cu	*p*		0.14, 0.24, 0.44
Cr	*p*		0.79
Mn	*p*		0.90
Fe	*p*		0.38, 0.52
Ca	*p*		0.16
Ni	*p*		0.35, 0.42
Au	*p*		0.09
Ag	*p*		0.11

[a] See references 16 and 22.

Selenium and Tellurium. Both are *n*-type and located on the arsenic sublattice. At high concentrations they tend to form compounds with gallium (Ga_2Se_3 and Ga_2Te_3) which are inactive. Evidence of complexes with gallium vacancies have also been observed.

Tin. Although belonging to group IV, tin is almost always *n*-type in gallium arsenide which is grown by vapor transport. However, the electron concentration varies sublinearly with the tin concentration. Thus the tin is located on both types of lattice sites, with an increasing fraction incorporated on arsenic sites at high concentration. Tin-doped gallium arsenide, grown by liquid-phase epitaxy, exhibits a shallow donor level.

Silicon. This is also a group IV element, and is incorporated on both sublattices. It is of special importance because its relative incorporation can be readily controlled in liquid-phase epitaxy to give either *p*-type gallium arsenide by low-temperature processing, or *n*-type gallium arsenide by processing at high temperatures (see Section 2.2.5).

Silicon is present in all gallium arsenide as a contaminant. Often it is in the original materials from which it was made. Processing in silica vessels is also an important contributory factor. The background concentration of undoped gallium arsenide is thus critically dependent on how this contaminant is incorporated into the lattice. Both growth temperature and arsenic overpressure play an important role here, since they determine the relative vacancy concentrations, and hence the incorporation of this impurity. In addition, silicon–oxygen complexes, with the silicon on gallium sites and the oxygen as an interstitial, have been identified as the cause for a number of doping anomalies in gallium arsenide [23].

Carbon. This is also present as a contaminant in gallium arsenide. Although belonging to group IV, it has been observed as a shallow acceptor and as a deep donor, at concentration levels as high as 8×10^{16} cm^{-3}.

Beryllium. Beryllium is a shallow acceptor in gallium arsenide, with a solid solubility of more than 10^{19} cm^{-3}. Recent interest in this dopant has come about because of its use in the ion implantation of gallium arsenide.

Copper. Copper is a deep, triple acceptor in gallium arsenide. It has a high solid solubility (6×10^{18} cm^{-3}) and moves extremely rapidly, even at relatively low processing temperatures (300–400°C). It is often present as a contaminant, and is very effective in reducing the diffusion length of n-type gallium arsenide.

Chromium and Iron. Chromium behaves as a single acceptor, with an impurity level that is extremely close to the center of the energy gap (E_v + 0.79 eV). It has a solid solubility of 1.6×10^{17} cm^{-3}, and can be used to intentionally dope gallium arsenide to make it semi-insulating, with a resistivity of as high as 10^9 Ω cm. This makes it of great importance technologically, since it permits the possibility of using gallium arsenide as an insulating substrate on which active layers of doped gallium arsenide can be grown.

Chromium-doped, semi-insulating gallium arsenide has been extensively studied in recent years. It has been found that only some slices of this material retain their semi-insulating properties upon heat treatment, while others do not [24]. The reasons for this variable behavior are not fully understood. However, it has been shown that they are related to the impurities which are initially present in the material, in addition to the chromium.

Iron can also be used for making semi-insulating gallium arsenide. However, its levels are at E_v + 0.39 eV and E_v + 0.52 eV, so that the highest resistivity obtained by iron doping is 4×10^4 Ω cm. Its solid solubility is in excess of 10^{17} cm^{-3}.

Oxygen. Oxygen can best be described as a problem contaminant. It exhibits a donor level at E_c − 0.75 eV, and has a solid solubility in excess of 10^{17} cm^{-3}. Its presence in p-type gallium arsenide results in raising the resistivity of the material, until semi-insulating behavior is achieved, with a resistivity of 10^8 Ω cm.

The behavior of oxygen in n-type gallium arsenide is somewhat surprising, since the Fermi level in this material is well above the oxygen donor level. Here again, the resistivity increases with increasing oxygen concentration, until the material becomes semi-insulating ($\rho \simeq 10^8\ \Omega$ cm). A number of theories have been advanced to explain this behavior. One theory is based on the fact that *n*-type gallium arsenide is compensated, with silicon as the primary donor or contaminant. Incorporated oxygen combines with this silicon to form inactive silicon–oxygen complexes, thus tying up this donor so as to shift the material toward *p*-type. This in turn allows the remaining oxygen to become ionized and moves the Fermi level toward the center of the gap. Oxygen-doped, semi-insulating gallium arsenide has a relatively high mobility ($\simeq$ 4000 cm^2/V s), which is not what would be expected in a highly compensated material. This would lend strength to the above arguments.

The use of oxygen as a dopant for making semi-insulating gallium arsenide is unfortunately impractical since this impurity is highly mobile at processing temperatures. Its real importance lies in the restrictions it places on the thermal processing of gallium arsenide in an open-air ambient. Such processing poses the ever-present risk of making the material inadvertently semi-insulating. This is perhaps the primary reason why many of the simple device processes available for silicon are not possible with gallium arsenide.

The purification of gallium arsenide is not as advanced as that of silicon, so that many shallow and deep impurities are unintentionally present. Some of these form complexes with gallium or arsenic, some are inactive, while yet others are active and of both *n*- and *p*-type. In addition, they can be incorporated in both gallium and arsenice lattice sites, whose concentrations are a function of the temperature and the ambient partial pressures of the gallium and arsenic vapor species. As a result, all gallium arsenide is compensated, i.e., it has in it both ionized acceptors N_A^- and ionized donors N_D^+. The degree of compensation is very often specified by the ratio $(N_D^+ + N_A^-)/n$, where n is the free electron concentration ($\simeq N_D^+ - N_A^-$). An alternate approach is to specify the *compensation ratio* as N_A^-/N_D^+.

In compensated material the total number of ionized impurities is larger than the electron concentration, so that the mobility is lower than what would be expected for uncompensated material. The total ionized impurity concentration can be determined by analysis of the temperature dependence of the Hall mobility, and by fitting these data to the relevant scatter processes [25]. However, a semi-empirical approach is more commonly used, based on the approximation that the primary mechanism which defines the mobility at 77°K is ionized impurity scatter. This is reasonably true for *n*-type gallium arsenide having an electron concentration of 10^{15} cm^{-3} or higher. A plot of $\mu_{77°K}$, the Hall mobility associated with this scatter mechanism, is shown in Fig. 1.12 as a function of $N_D^+ + N_A^-$ and $n_{77°K}$. This curve can be used to estimate the compensation ratio, since $n_{77°K}$ is also obtained from the Hall measurements at this temperature.

An alternate approach, based on the detailed band structure formulation of gallium arsenide, has also been used to calculate the mobility [26]. Curves

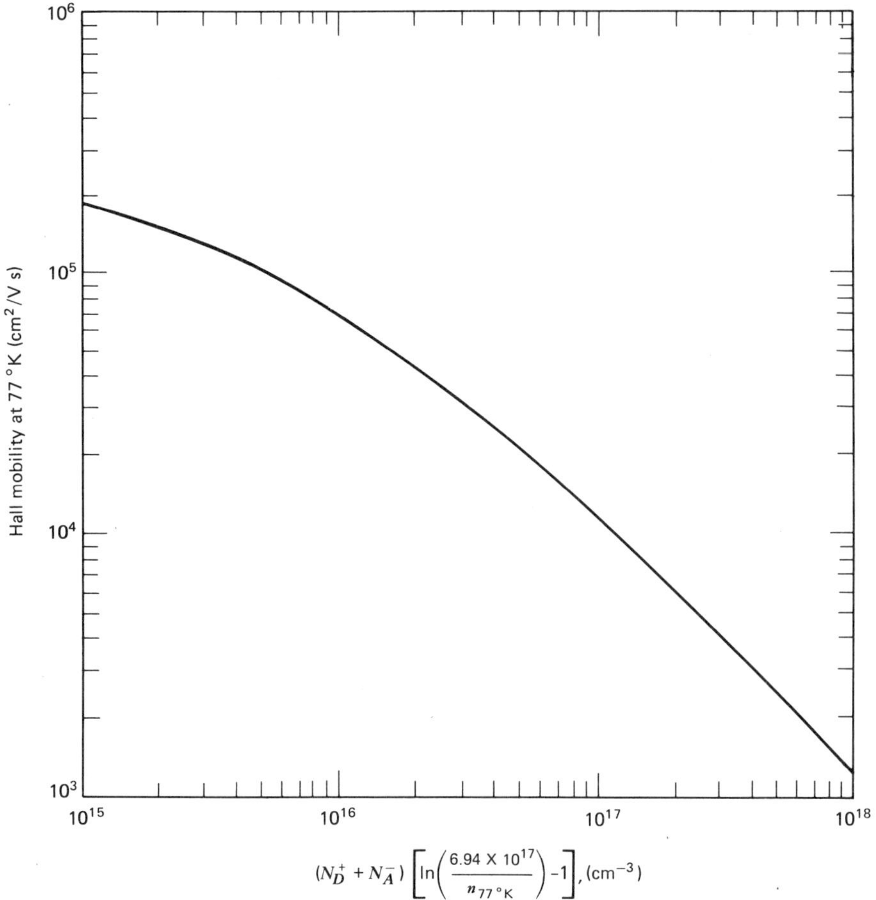

Fig. 1.12 $\mu_{77°K}$, **as a function of** $N_D^+ + N_A^-$ **and** n. **Adapted from Stillman and Wolfe [25].**

for drift mobility* versus free electron concentration, based on this approach, are shown in Fig. 1.13*a* and *b* for 300°K and 77°K, respectively, and for different values of $(N_D^+ + N_A^-)/n$.

A plot of hole mobility in *p*-type gallium arsenide is shown in Fig. 1.14. Studies of this material have not been extensive, since its primary role has been in injecting p^+ contacts to *n*-type gallium arsenide. Recently, however, *p*-type material has become more important because of its use in photovoltaic devices, since its minority carrier diffusion length is larger than that for *n*-type material.

* The Hall mobility is approximately equal to 1.93 times the electron drift mobility for gallium arsenide, where the dominant scatter mechanism is due to ionized impurities. The Hall and drift mobilities of holes are approximately equal.

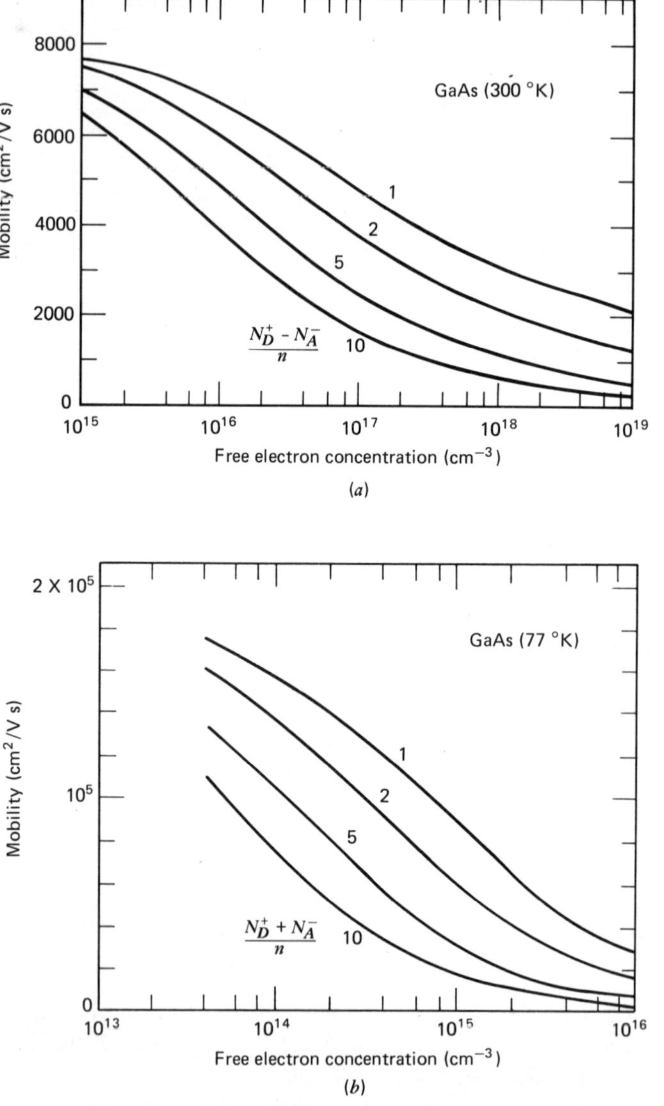

Fig. 1.13 Electron mobility at 77 and 300°K, as a function of ($N_D^+ + N_A^-$)/n. From Rode and Knight [26]. Reprinted with permission from *Physical Review*.

Finally, it is worth noting that gallium arsenide can have in it many inactive impurities, which have only a slight effect on the mobility or the free carrier concentration. Thus high-purity gallium arsenide, with a free electron concentration of 10^{13}–10^{14} cm^{-3}, can contain as much as 10^{16}–10^{17} atoms/cm^3 of carbon and oxygen, in addition to such impurities as aluminum, calcium, potassium, nitrogen, strontium, and tantalum in the $\leq 10^{16}$ cm^{-3}

range. These impurities are generally interstitial, or form complexes which are electronically inactive.

1.4.2 Dislocations

A dislocation is a one-dimensional line defect in an otherwise perfect crystal, and results in a geometric fault in the lattice. It occurs when the crystal is subjected to stresses in excess of the elastic limit, e.g., during its growth from a melt. Although the nature of dislocations is quite complex, they are usually composed of combinations of two basic types—the screw dislocation and the edge dislocation. A simple cubic lattice is considered in the following sections. The diamond lattice is considerably more complex; the general properties of dislocation types, however, are very similar to those of the cubic lattice.

1.4.2.1 Screw Dislocations

Figure 1.15 shows the manner in which a regular crystal lattice may be subjected to shear stresses in order to establish a screw dislocation. Imagine that the crystal is cut along the plane *ABCD*, which is one of its regular lattice planes, and let the two halves of the crystal on either side of this plane be subjected to shearing forces that are sufficiently large to cause them to be separated by one atomic spacing. The line of the screw dislocation so formed is *AD*, since this marks the boundary in the plane *ECBF* which divides the perfect crystal from the imperfect.

The strain energy associated with a screw dislocation may be calculated by considering a cylinder of material, of length l, with axis *AD* and inner and outer radii R_i and R_o, respectively, as shown in Fig. 1.16. It is assumed

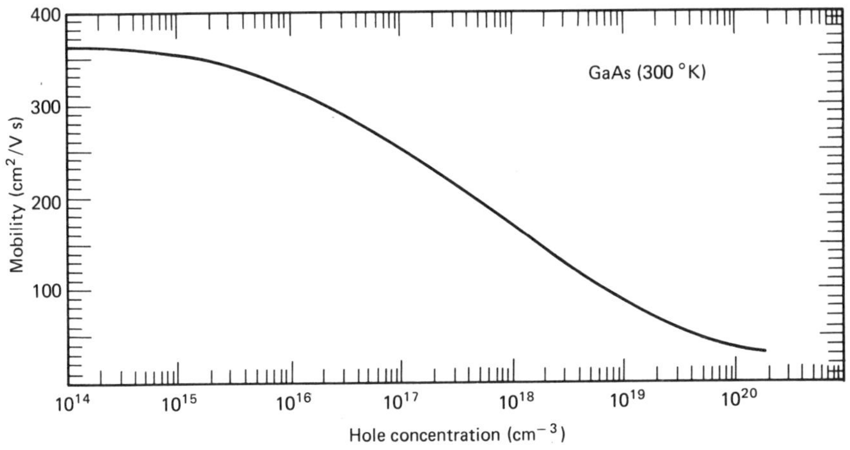

Fig. 1.14 Hole mobility in gallium arsenide.

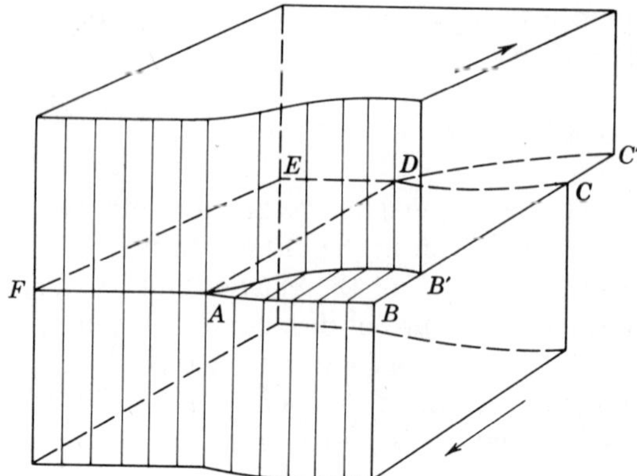

Fig. 1.15 Screw dislocation.

that the crystal behaves as an elastic solid within the cylinder defined by these radii. Let:

b = amount of shear present in a shell of radius r and thickness dr
μ = shear modulus ($\simeq 7.9 \times 10^{11}$ dyn/cm^2 for silicon)

Then the elastic shear strain is given by $b/2\pi r$. The elastic energy of the

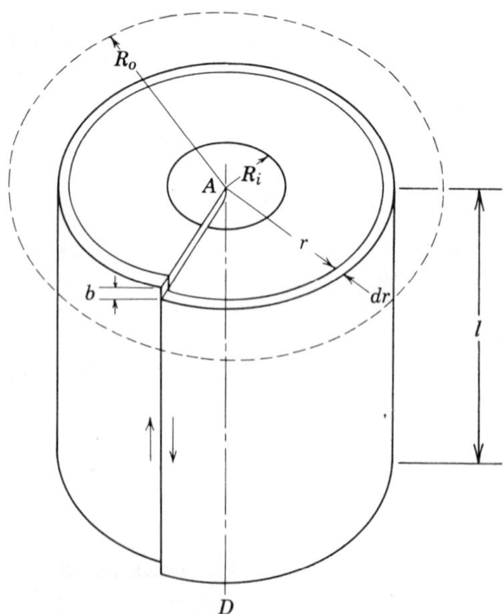

Fig. 1.16 Details of a screw dislocation.

shell dE_{μ} is given by

$$dE_{\mu} = \tfrac{1}{2}\mu(\text{shear strain})^2\, dV \tag{1.29}$$

where dV is the volume of the shell. Hence

$$dE_{\mu} = \left(\frac{\mu b^2 l}{4\pi}\right)\left(\frac{dr}{r}\right) \tag{1.30}$$

and

$$E_{\mu} = \left(\frac{\mu b^2 l}{4\pi}\right)\ln\left(\frac{R_o}{R_i}\right) \tag{1.31}$$

If this is the only dislocation in an infinite volume of material, $R_o = \infty$ and the energy associated with it is infinite. In practice, however, crystals would contain many dislocations randomly distributed. As a result, their strain fields are also randomly distributed and cancel each other at distances approximately equal to the mean distance between them. In typical crystals, R_o is about 10^5 atom spacings. The inner radius limit R_i is set by the fact that a region of atomic dimensions can no longer be considered as an elastic continuum, and the theory of elasticity ceases to hold. As a consequence, it is reasonable to eliminate the inner 4–5 atoms from consideration.

Practical values of the ratio R_o/R_i are usually taken around 10^4. With the use of this value the strain energy for a screw dislocation in silicon may be calculated as about 10–19 eV/atom length. (By way of comparison, values for aluminum and diamond are 3.1 and 29 eV, respectively.)

1.4.2.2 Edge Dislocations

An edge dislocation is shown in Fig. 1.17. Here an extra half-plane of atoms, *ABCD*, is present in the otherwise regular lattice, with most of the distortion concentrated around the line *AD*. An edge dislocation of this type is created by applying a shearing force along the face of the crystal, parallel to a major crystallographic plane. When this force exceeds that required for elastic deformation, the upper half of the crystal moves by a slip mechanism. The plane along which slip occurs is commonly referred to as a *slip plane*.

The strain energy associated with an edge dislocation can be shown [27] to be given by

$$E_{\mu} = \left[\frac{\mu b^2 l}{4\pi(1-\nu)}\right]\ln\left(\frac{R_o}{R_i}\right) \tag{1.32}$$

where ν is Poisson's ratio ($\simeq 0.3$ for both silicon and gallium arsenide). Its magnitude is thus approximately 50% larger than that for a screw dislocation.

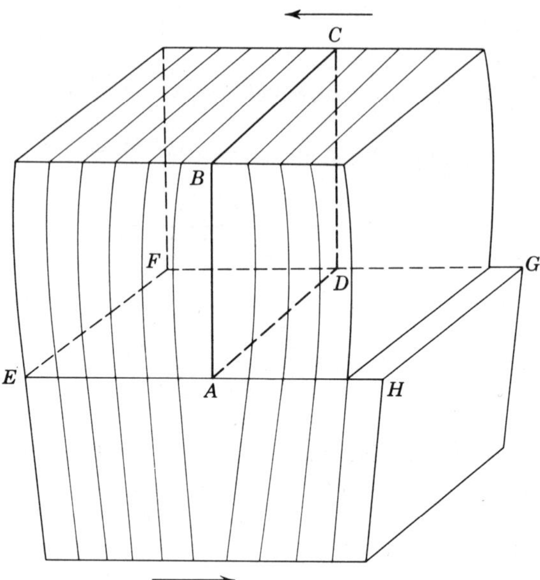

Fig. 1.17 Edge dislocation.

In view of the large energies of formation for both basic dislocation types, it must be concluded that their equilibrium concentration is negligible, i.e., their presence is due to nonequilibrium processes such as those associated with the growth, freezing, or quenching of the crystal.

1.4.2.3 Movement of Dislocations

Figure 1.18 indicates the manner in which an edge dislocation may move completely through a crystal. The mechanism for such a movement is called *slip*. It is characteristic of the slip mechanism that it results in movement along planes of high atomic density where opposing forces are at a minimum.

The displacement of a screw dislocation also takes place along a slip plane. In Fig. 1.15 this slip plane is given by *ABCD*. The end result of such a displacement is identical to the movement of an edge dislocation, even though the strain pattern is different.

In addition to slip, *climb* is an alternate method by which a dislocation can move in a crystal. For an edge dislocation, such as that shown in Fig. 1.17, climb of the plane *ABCD* takes place at right angles to the slip plane *EFGH*. Figure 1.19 shows that this may occur as the result of the movement of atoms* out of the plane *ABCD*. Both substitutional or interstitial atoms may be involved in this process. Intuitively, it is seen that the energy of formation associated with such a process is on the same order of magnitude

* Alternately, climb may also occur by atoms moving into the plane.

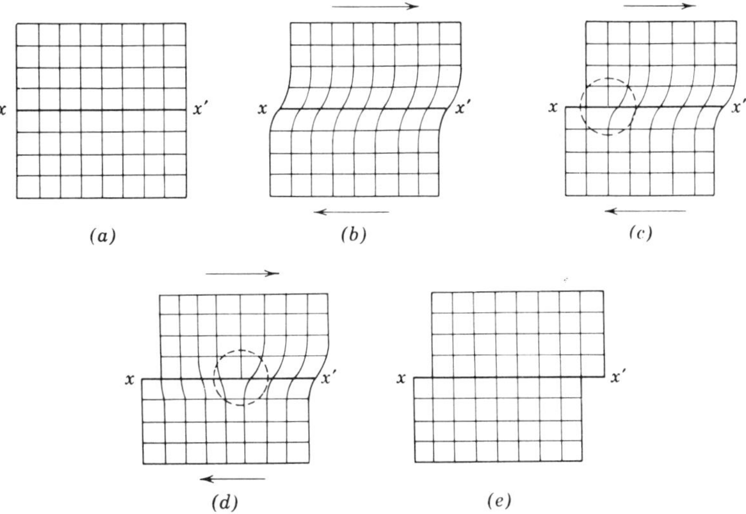

Fig. 1.18 Crystal movement along a slip plane.

as that for the energy of migration ($\simeq$ 0.18 eV) of a point defect. In fact, it is somewhat less, since the migration of these atoms is aided by the stress field surrounding the dislocation.

Climb in a screw dislocation occurs by a complex motion. Here the screw dislocation line twists itself into a helix, which can then climb. The actual movement of dislocations in a crystal is made up of combinations of these and other types of movements.

The energy of movement of a dislocation has been shown to be 0.15 eV/atom spacing for silicon. This is the energy barrier that must be overcome in order for a dislocation to move in a crystal. A comparison with the energy of formation of a dislocation (about 10–19 eV) shows that it is extremely easy to induce dislocation motion in a crystal by thermal means, even though it is almost impossible to create a dislocation in this manner. Thus one of the more important problems of crystal growth and device processing is to avoid (or minimize) the formation of dislocations in the first place. Alter-

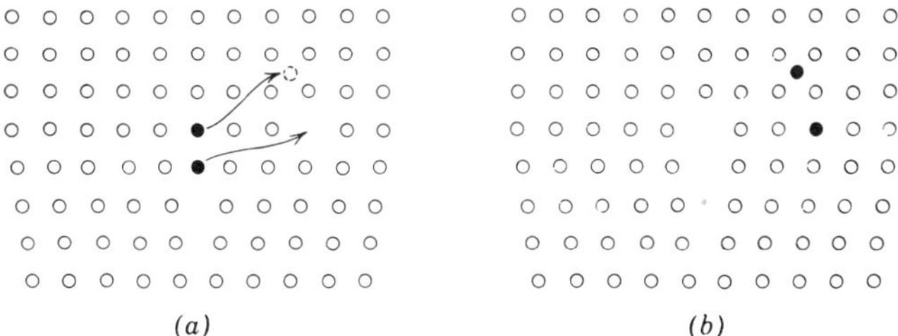

Fig. 1.19 Climb of an edge dislocation.

nately, if such dislocations are unavoidable, they can sometimes be relieved by annealing. During crystal growth, techniques are available for inducing these dislocations to grow out of the crystal, leaving behind a relatively dislocation-free lattice.

1.4.2.4 Multiplication of Dislocations

There is considerable evidence indicating that dislocation multiplication occurs in a crystal in addition to dislocation movement. Examination of deformed crystals has shown that this is indeed the case, and various mechanisms have been suggested for this multiplication. Figure 1.20 shows a model by which this can occur [28]. Consider a dislocation, as shown in Fig. 1.20*a*. Under the application of a stress F, the dislocation tends to expand along its length by climb. If, however, it is restricted at xx', possibly by the presence of some obstruction such as an oxygen or metallic cluster, it will tend to bow out of its slip plane, as shown in Fig. 1.20*b*. In doing so it becomes longer and requires a greater stress to maintain its new radius. A critical condition is reached at which the dislocation line becomes semicircular. For a stress in excess of that required for this condition, the dislocation becomes unstable and progresses as shown in Fig. 1.20*c* and *d*. Eventually it returns to its original form by the collapse of the cusp, as shown in Fig. 1.20*e*, leaving an expanding loop in addition to it. The process now repeats itself, resulting in a number of dislocations from a single dislocation source of this type.

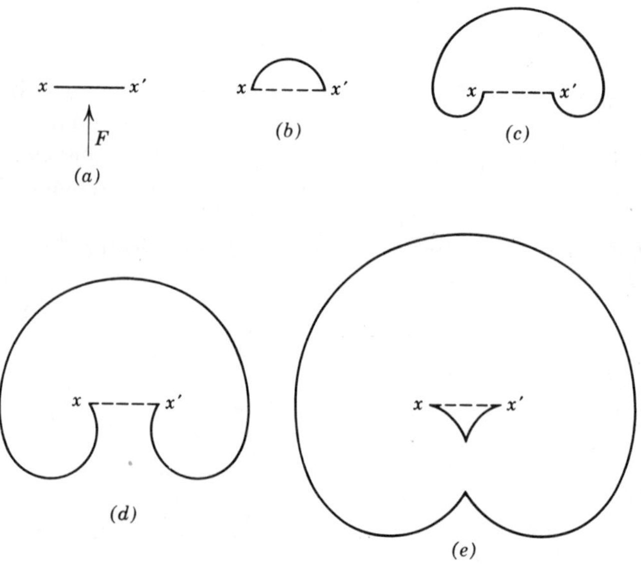

Fig. 1.20 Mechanism for dislocation multiplication.

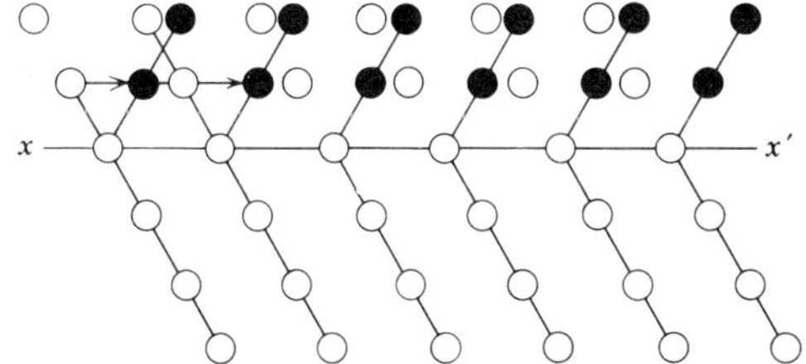

Fig. 1.21 A twinned structure.

1.4.2.5 *Twinning*

Twinning is one form of gross defect that may occur in a crystal. Its presence is usually indicative of material that has a high dislocation content and is not suited for the fabrication of devices and microcircuits. Consequently the subject will be treated very briefly in this text.

Twinning occurs when one portion of a crystal lattice takes up an orientation with respect to another, the two parts being in intimate contact over their bounding surfaces. This bounding surface is called the *twinning plane*. Figure 1.21 shows a two-dimensional representation of twinned and untwinned parts of a crystal. For this case, atoms along xx' are common to both twinned and untwinned sections, and the twinning plane is sometimes referred to as the *composition plane*.

Experimental evidence shows that excessive twinning is encountered if the material is restricted during its growth from a melt. Thus crucible grown materials are highly prone to this defect.

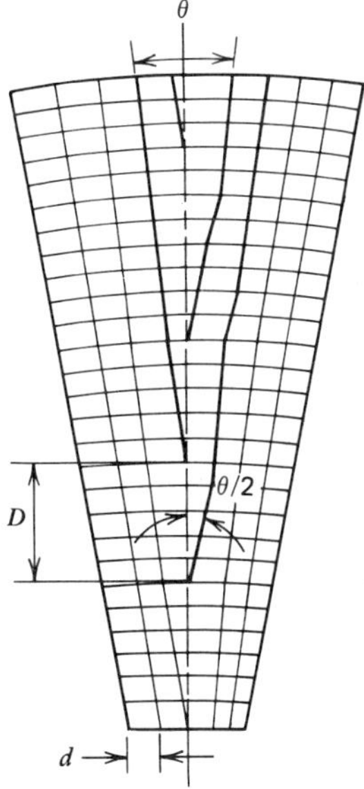

Fig. 1.22 Model for a low-angle grain boundary.

The twinning plane in a crystal is a region with a large concentration of broken, or unsatisfied, bonds. This region is known as a *grain boundary,* and the actual number of such broken bonds is related to the angle of the grain boundary. This is shown in Fig. 1.22 for a symmetric grain boundary with an angle θ. If d is the atomic spacing, the distance between broken bonds is given by

$$D = \frac{d}{2\sin(\theta/2)}. \tag{1.33}$$

If $\theta \leq 5°$, this distance is on the order of $\geq$11 lattice spacings. This is comparable to the lattice spacing for dopant atoms with a concentration of 3.3×10^{19} cm^{-3}. As a result, a crystal having a *low-angle grain boundary* of this type can still be considered as a coherent structure. The term *lineage* is used for grain boundaries where $\theta \leq 1°$. Here any loss in coherency due to this type of grain boundary can be ignored.

1.5 ELECTRONIC PROPERTIES OF DEFECTS

The deliberate insertion of chemical defects into the semiconductor lattice is the basis for the fabrication and control of electrical properties of semiconductor devices and microcircuits. This subject has been discussed in Sections 1.4.1.3 and 1.4.1.4. However, attention must also be paid to the electronic behavior of defects that are unintentionally present in the crystal.

1.5.1 Point Defects

In silicon the presence of a vacancy in a crystal results in four unsatisfied bonds which would ordinarily be used to bind the atom to its tetrahedral neighbors. Thus a vacancy tends to be acceptor-like in behavior. The addition of each electron to this vacancy results in successively higher values of energy level because of the large mutual electrostatic repulsion present between them. It is highly improbable, however, that such a vacancy will exhibit as many as four energy levels within the band gap. An acceptor level at 0.4 eV from the conduction-band edge (E_v + 0.71 eV), and a second acceptor level at 0.11 eV from the conduction-band edge (E_v + 1.0 eV), have been positively identified in silicon by a number of workers. A donor level at 0.35 eV from the valence-band edge (E_c − 0.76 eV) has also been identified, and is considered to be due to a distorted bond configuration.

In like manner, an interstitial has four valence electrons that are not involved in covalent binding with other lattice atoms and which may be lost to the conduction band. As a result it should exhibit donor-like behavior, and one or more levels within the energy gap. A singly ionized donor level, at 0.2 eV above the valence-band edge (E_c − 0.91 eV) has been identified here.

Many complex vacancy–interstitial combinations are also electronically active in silicon. Thus electron irradiation in the 1–2-MeV range gives rise [29] to four energy levels at E_v + 0.27 eV, E_c − 0.17 eV, E_c − 0.23 eV, and E_c − 0.41 eV. Annealing for 36 h at 300°C alters the defect structure, with the last two levels converting to one level at E_c − 0.36 eV.

The energy levels associated with vacancies and interstitials in silicon are deep. They serve as localized centers for minority carrier recombination, and result in a fall in the lifetime. In general there is an inverse relationship between the concentration of deep levels and the minority carrier lifetime. This has been verified by numerous experiments on material with "as grown" as well as with induced defects (by electron and nuclear radiation, and by plastic deformation techniques). In fact, the deliberate introduction of deep levels into silicon by electron irradiation provides a new and highly promising technique for controlling lifetime in high-power semiconductor devices.

Both arsenic and gallium vacancies are present in gallium arsenide, their concentration being determined by the overpressure of arsenic during processing. Unfortunately, the identification of specific defects has proved very difficult because of the abundance of residual impurities, and the ease of contamination during these studies. Table 1.8 lists some of these defects, as obtained [30] from annealing experiments in the 600–1100°C temperature range.

There is much uncertainty concerning the electronic nature of these vacancies. A number of workers [31] have concluded that, of themselves, they are inactive. Recent data [32] would indicate, however, that arsenic vacan-

Table 1.8 Point Defects in Gallium Arsenide[a]

Location from Valence-Band Edge at 5°K (eV)	Type of Vacancy, and Possible Impurity Associated with it
1.49	V_{As} (Si)
1.40	V_{As}
1.37	V_{Ga} (Cu)
1.35	V_{Ga}
1.20	O_i
1.02	V_{Ga} V_{As} (Si)
0.81	V_{Ga}
0.70	V_{As} (O)
0.58	V_{Ga}

[a] See reference 30.

cies behave as deep donors, whereas gallium vacancies exhibit deep acceptor-like behavior.

A considerable body of experimental evidence points to the presence of acceptor-like complexes which are formed when donors combine with gallium vacancies. This model has been used to explain the compensation behavior of gallium arsenide when doped with both group IV and group VI impurities [33]. In addition defects or complexes associated with the arsenic vacancies have also been postulated to explain the low-temperature annealing behavior of heat-treated n-gallium arsenide [34].

Vacancies in gallium arsenide often serve as sites for other impurities or contaminants which then exhibit electronic behavior. Silicon, by way of example, is a common contaminant in gallium arsenide which is grown in quartz boats. Belonging to group IV, this impurity behaves p-type when incorporated into arsenic vacancies, but n-type when incorporated into gallium vacancies.

1.5.2 Dislocations

Many of the electronic properties of dislocations are similar to those of an ensemble of point defects. Thus there is considerable evidence to show that dislocations in silicon are acceptor-like, because of the presence of dangling (or unfilled) bonds at the edge of the half-plane comprising the dislocation. However, it is possible to obtain [35] a diffusional rearrangement of the atoms which reduces the formation of dangling bonds of this type. This would explain the absence of strong acceptor-like properties for these dislocations.

The behavior of dislocations in gallium arsenide is not firmly established. Some workers have noted that they serve as recombination centers, and greatly lower the internal quantum efficiency in light-emitting diodes; others have found that dislocations cause filamentary emission, and result in accelerated degradation of laser devices [36].

Dislocations take the form of a line of defects, with an associated distortion of the energy-band structure in their vicinity. Thus they behave as anisotropic scattering centers for carriers because of the extended space charge region surrounding them. Furthermore, this distortion of the energy-band structure leads to the formation of trapping sites for free holes and electrons, and a concurrent reduction in minority carrier lifetime, in proportion to their concentration.

The most important property of a dislocation is that it interacts with chemical and other point defects in its neighborhood. This interaction exists between the localized disturbance, due to impurity atoms, and the strain field in the vicinity of the dislocation. Its extent is directly proportional to the misfit between the impurity and the lattice atom. Thus the presence of a dislocation is usually associated with an enhanced rate of impurity diffusion

Table 1.9 Properties of Silicon and Gallium Arsenide at 300°K

Property	Silicon	Gallium Arsenide
Crystal structure	Diamond	Zincblende
Lattice constant (Å)	5.43	5.64
Tetrahedral distance between neighboring atoms (Å)	2.36	2.44 (1.26 for Ga 1.18 for As)
Atoms or molecules (cm^{-3})	5×10^{22}	2.21×10^{22}
Atomic or molecular weight	28.09	144.63
Density (g/cm^3)	2.328	5.32
Dielectric constant	11.9	13.1
Melting point (°C)	1412	1238
Thermal expansion coefficient	2.6×10^{-6}	6.86×10^{-6}
Thermal conductivity (W/cm °C)	1.5	0.46
Specific heat (J/g °C)	0.7	0.35
Thermal diffusivity $\left(\frac{\text{thermal conductivity}}{\text{density} \times \text{specific heat}}\right)$ (cm^2/s)	0.9	0.44
Energy gap (eV)	1.11 $\left(1.17 - \frac{4.73 \times 10^{-4}\, T^2}{T + 636}\right)$	1.435 $\left(1.522 - \frac{5.88 \times 10^{-4}\, T^2}{T + 300}\right)$
Intrinsic carrier concentration (cm^{-3})	1.45×10^{10}	8×10^6
Effective density of states		
Valence band (cm^{-3})	1.04×10^{19}	7×10^{18}
Conduction band (cm^{-3})	2.8×10^{19}	4.7×10^{17}
Conductivity effective mass		
Holes	$0.38\ m_0$	$0.5\ m_0$
Electrons	$0.26\ m_0$	$0.067\ m_0$ (light) $0.35\ m_0$ (heavy)

in its neighborhood, leading to the formation of diffusion pipes. Often the presence of the dislocation results in the segregation of metallic impurities in its vicinity. Taken together, they lead to such problems as excessive leakage and premature breakdown in semiconductor junctions made on this material.

Many of the important properties of silicon and gallium arsenide are assembled in Table 1.9.

REFERENCES

1. M. M. Atalla, E. Tannenbaum, and E. J. Scheibner, Stabilization of Silicon Surfaces by Thermally Grown Oxides, *Bell Sys. Tech. J.* **38,** 749 (1959).
2. J. V. Dilorenzo, GaAs FET Development–Low Noise and High Power, *Microwave J.* **39** (Feb. 1978).
3. B. M. Welch, Advances in GaAs LSI/VLSI Processing Technology, *Solid State Tech.* **95** (Feb. 1980).
4. C. Kittel, *Introduction to Solid State Physics,* Wiley, New York, 1966.
5. K. E. Bean and P. S. Gleim, The Influence of Crystal Orientation on Silicon Semiconductor Processing, *Proc. IEEE* **57,** 1469 (1969).
6. K. E. Petersen, Fabrication of an Integrated Planar Silicon Ink-Jet Structure, *IEEE Trans. Electron Dev.* **ED-26,** 1918 (1979).
7. O. Madelung, *Physics of III–V Compounds,* Wiley, New York, 1964.
8. R. G. Rhodes, *Imperfections and Active Centers in Semiconductors,* Pergamon, Macmillan, New York, 1964.
9. J. H. Brophy, R. M. Rose, and J. Wulff, *The Structure and Properties of Materials,* vol. II, *Thermodynamics of Structure,* Wiley, New York, 1964.
10. J. R. Arthur, Vapor Pressures and Phase Equilibria in the Ga-As System, *J. Phys. Chem. Solids* **28,** 2257 (1967).
11. H. Reiss, C. S. Fuller, and F. J. Morin, Chemical Interactions Among Defects in Germanium and Silicon, *Bell Sys. Tech. J.* **35,** 535 (1956).
12. H. R. Potts and G. L. Pearson, Annealing and Arsenic Overpressure Experiments on Defects in Gallium Arsenide, *J. Appl. Phys.* **37,** 2098 (1966).
13. W. Kohn, Shallow Impurity States in Silicon and Germanium, *Solid State Phys.* **5,** 257 (1957).
14. G. L. Pearson and J. Bardeen, Electrical Properties of Pure Silicon and Silicon Alloys Containing Boron and Phosphorus, *Phys. Rev.* **75,** 865 (1949).
15. W. Shockley and J. R. Last, Statistics of the Charge Distribution for a Localized Flaw in a Semiconductor, *Phys. Rev.* **107**(2), 392 (July 15, 1957).
16. A. G. Milnes, *Deep Impurities in Semiconductors,* Wiley, New York, 1973.
17. K. V. Ravi, The Heterogeneous Precipitation of Silicon Oxides in Silicon, *J. Electrochem. Soc.* **121,** 1090 (1974).
18. A. Kanamori, Annealing Behavior of the Oxygen Donor in Silicon, *Appl. Phys. Lett.* **34,** 287 (Feb. 15, 1979).
19. W. W. Bullis, Properties of Gold in Silicon, *Solid State Electron.* **9,** 143 (1966).
20. K. P. Lisiak and A. G. Milnes, Platinum as a Lifetime Control Deep Impurity in Silicon, *J. Appl. Phys.* **46,** 5229 (1975).
21. A. Mayer, The Quality of Starting Silicon, *Solid State Tech.* **38** (Apr. 1972).
22. D. L. Partin, A. G. Milnes, and L. F. Vassamillet, Hole Diffusion Lengths in VPE GaAs and $GaAs_{0.6}P_{0.4}$ Treated with Transition Metals, *J. Electrochem. Soc.* **126,** 1584 (1979).
23. M. E. Weiner and A. S. Jordan, Analysis of Doping Anomalies in GaAs by Means of a Silicon-Oxygen Complex Model, *J. Appl. Phys.* **43,** 1767 (1972).
24. D. C. Look, The Electrical Characterization of Semi-Insulating GaAs, *J. Appl. Phys.* **48,** 5141 (1977).
25. G. E. Stillman and C. M. Wolfe, Electrical Characterization of Epitaxial Layers, *Thin Solid Films* **31,** 69 (1976).
26. D. L. Rode and S. Knight, Electron Transport in GaAs, *Phys. Rev. B,* 3 (Apr. 15, 1971).

27. H. F. Matare, *Defect Electronics in Semiconductors,* Wiley-Interscience, New York, 1971.
28. F. C. Frank and W. T. Read, Jr., Multiplication Processes for Slow Moving Dislocations, *Phys. Rev.* **79,** 722 (1950).
29. A. O. Evwaraye and B. J. Baliga, The Dominant Recombination Centers in Electron-Irradiated Semiconductor Devices, *J. Electrochem. Soc.* **124,** 913 (1977).
30. L. L. Chang, L. Esaki, and R. Tsu, Vacancy Association of Defects in Annealed GaAs, *Appl. Phys. Lett.* **19,** 143 (1971).
31. H. C. Casey, Jr., Diffusion in the III-V Compound Semiconductors, Chap. 6 in *Atomic Diffusion in Semiconductors,* D. Shaw, Ed., Plenum, New York, 1973.
32. S. Y. Chiang and G. L. Pearson, Properties of Vacancy Defects in GaAs Single Crystals, *J. Appl. Phys.* **46,** 2986 (1975).
33. D. T. J. Hurle, Revised Calculation of Point Defect Equilibria and Non-Stoichiometry in Gallium Arsenide, *J. Phys. Chem. Solids* **40,** 613 (1979).
34. J. Nishizawa, H. Otsuka, S. Yamakoshi, and K. Ishida, Nonstoichiometry of Te-Doped GaAs, *Jpn. J. Appl. Phys.* **13,** 46 (1974).
35. J. Hornstra, Dislocations in the Diamond Lattice, *J. Phys. Chem. Solids* **5,** 129 (1958).
36. P. M. Petroff, Point Defects and Dislocation Climb in III–V Compound Semiconductors, *J. Phys. Colloque Suppl.* **40,** 201 (1979).

PROBLEMS

1. Compute the number of atoms/cm^2 for the principal planes of gallium arsenide. Identify these atoms and their positions by means of a sketch.
2. We wish to chemically delineate a chip with vertical walls in (110) silicon. This is done by opening a window in an oxide mask on the surface and using an anisotropic etch which exposes {111} planes. Indicate all the directions of the edges of this oxide cut, and also all the planes delineated by this etch.

   A flat is provided on the slice to aid in making the first cut. What is the plane of this flat? What is the direction of the line made by the intersection of this flat with the (110) surface?
3. It is possible for a tetrahedral stacking fault to be initiated during the growth of an epitaxial layer on (111) silicon. This fault usually begins at the substrate–layer interface, and propagates along {111} planes until it reaches the surface.

   Sketch the outline of a stacking fault within a unit cell, and identify its various {111} planes. Determine the thickness of the epitaxial layer in terms of the length of one side of the stacking fault, as it penetrates to the surface.
4. Stacking faults also arise during epitaxial growth on (100) silicon. Again, these propagate along {111} planes. Repeat Problem 3 for this situation.
5. When indium arsenide is epitaxially grown on gallium arsenide, a 2-μm region is formed in which the film accommodates to the substrate lattice. What is the defect concentration in this region, assuming that it is uni-

form, and that growth is on the (100) plane. Note that this will give you a pessimistic number, because effects such as lattice distortion or strain relief have not been considered. Assume that $a = 6.058$ Å for indium arsenide.

6. What is the equilibrium concentration of Schottky defects in silicon at 300 and 1500°K. Repeat for Frenkel defects and compare the numbers. Comment on the results.
7. Assuming that arsenic vacancies in gallium arsenide are donor-like, and that gallium vacancies are acceptor-like, show that the functional dependence of the concentration of these vacancies on the arsenic overpressure is the same as that given in Eqs. (1.23) and (1.24).

CHAPTER

2

Phase Diagrams and Solid Solubility

CONTENTS

2.1 UNITARY DIAGRAMS 50
2.2 BINARY DIAGRAMS 51

2.2.1 The Lever Rule, 51
2.2.2 The Phase Rule, 52
2.2.3 Isomorphous Diagrams, 53
2.2.4 Eutectic Diagrams, 55
2.2.5 Congruent Transformations, 58
2.2.6 Peritectic and Other Reactions, 62
2.2.7 Phase Diagrams for Oxide Systems, 68

2.3 SOLID SOLUBILITY 69
2.4 TERNARY DIAGRAMS 71

2.4.1 Isothermal Sections, 71
2.4.2 Congruently Melting Compounds, 74
2.4.3 Degrees of Freedom, 74
2.4.4 Some Ternary Systems of Interest, 75

REFERENCES 79

A number of different materials are used in the fabrication of semiconductor devices and microcircuits. As a consequence, combinations of two or more of these are often encountered at various points within the structure. By way of example the electronic properties of silicon and gallium arsenide are controlled by the introduction of small amounts of donor and acceptor elements into the lattice, whereas ohmic contacts are made by the alloying of various metals to the semiconductor. Occasionally combinations of materials are inadvertently formed during heat treatment or during the storage of devices at elevated temperature. The most notorious of these, formed at the interface between gold leads and aluminum-bonding pads, results in the so-called *purple plague,** which sets an upper limit to the temperature at which many silicon microcircuits can be stored.

Similar problems occur with gallium arsenide, since nearly all metals react with either gallium or arsenic at relatively low temperatures. Thus, the choice of a metal for a Schottky barrier on this semiconductor is limited by the subsequent processing temperatures to which it is subjected.

In this chapter the behavior of these combinations is described by means of phase diagrams [1–3]. This behavior is important since it often determines the nature and choice of the fabrication process. In addition, it provides a clue as to the problems that may arise when certain combinations of materials are used together.

A *phase* is defined as a state in which a material may exist, which is characterized by a set of uniform properties. If these phases are presented for equilibrium conditions, the resulting diagram is called an *equilibrium diagram.* Since equilibrium conditions are attained at rates that are much slower than the freezing rate, most diagrams involving one or more solid phases are usually called *phase diagrams* and represent quasi-equilibrium conditions.

2.1 UNITARY DIAGRAMS

These diagrams show the phase change in a single element as a function of temperature and pressure. They also apply to compounds† that undergo no chemical change over the range for which the diagram is constructed.

In its simplest form, the unitary diagram consists of three lines that intersect at a common point, thus delineating three areas on a two-dimensional plot. Figure 2.1 shows such a diagram for water. The common point, referred to as a *triple point,* is invariant for the system and defines the temperature and pressure at which solid, liquid, and gaseous phases are all in equilibrium with one another. $0A$, $0B$, and $0C$ are univariant lines. Water can coexist

* This topic is treated in Section 2.2.6.

† The term *component* is often used to denote elements and compounds of this type interchangeably.

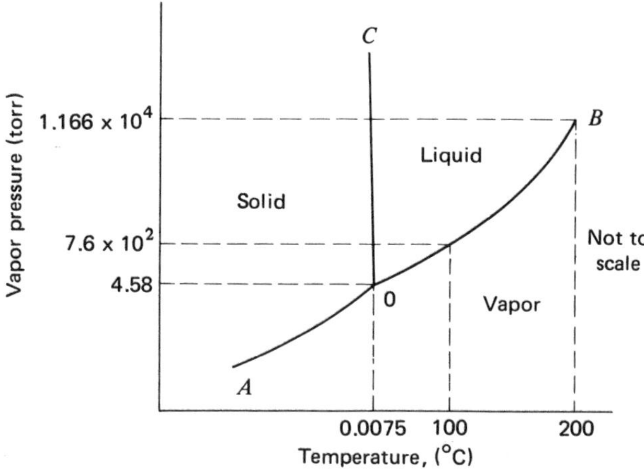

Fig. 2.1 Unitary phase diagram.

in two phases for any pressure–temperature combination represented by these lines.

2.2 BINARY DIAGRAMS

These phase diagrams show the relationship between two components as a function of temperature. The second variable, pressure, is usually set at 1 atm. In this way a relatively complex three-dimensional representation is avoided.

Figure 2.2 shows one of the many different types of binary diagrams that are encountered in practice. Here the abscissa represents various compositions of two components A and B, usually specified in atom percent or weight percent. Each end represents a pure component, which may be an element or a compound.

2.2.1 The Lever Rule

At any temperature, the equilibrium composition of the two single phases that make up a two-phase region may be determined as follows (see Fig. 2.2). Consider a melt of initial composition C_M (the percentage weight of B in the melt). Let this melt be cooled from some temperature T_1 to a temperature T_2, corresponding to a point in the two-phase region. Let:

W_L = weight of liquid at this temperature

W_S = weight of solid (in the β phase, for this example)

C_L, C_S = composition of the liquid and solid, respectively (percentage amount of B by weight)

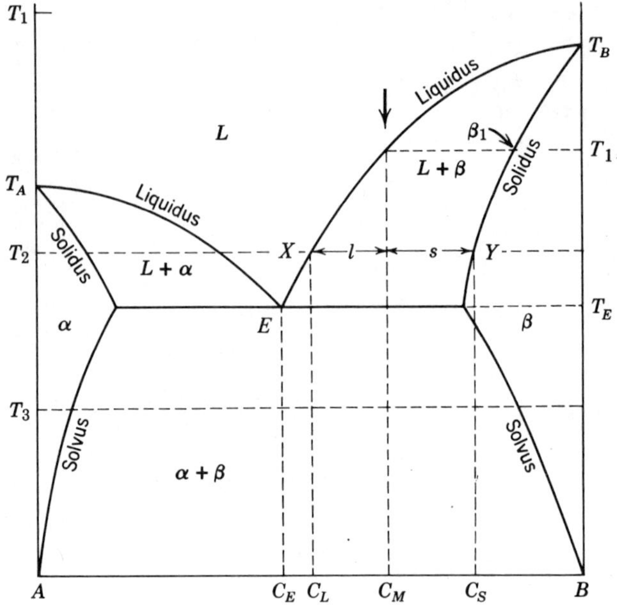

Fig. 2.2 Binary phase diagram.

Then $W_L C_L$ and $W_S C_S$ are the weights of B in the liquid and solid, respectively. But the total weight of B is $(W_L + W_S)C_M$. Hence by the conservation of matter

$$\frac{W_S}{W_L} = \frac{C_M - C_L}{C_S - C_M} = \frac{l}{s} \tag{2.1}$$

where l and s are the length of the two lines measured from the C_M ordinate to the boundaries of the two-phase region. This is known as the *lever rule* and is directly applicable to the analysis of compositional changes during the freezing of a crystal from the melt. The line XY is commonly referred to as a *tie line*.

2.2.2 The Phase Rule

The correct interpretation of phase diagrams is greatly helped by knowledge of the *phase rule*. This rule, which is based on thermodynamic considerations, states that for any system in thermal equilibrium, the sum of the number of phases P and the number of degrees of freedom F is related to the number of components C by

$$P + F = C + 2 \tag{2.2}$$

Here the degrees of freedom are the number of variables that can be independently changed while still preserving a specific phase. For example, $P + F = 3$ for a single-component diagram of the type shown in Fig. 2.1. For water in its liquid phase ($P = 1$) we have 2 degrees of freedom, i.e., *both* pressure *and* temperature can be independently changed, and still maintain water in this liquid phase. Along $0B$, however, $P = 2$ (liquid and vapor) so that we have 1 degree of freedom. Now *either* pressure *or* temperature (but not both) can be independently varied while this two-phase coexistence is still preserved. At 0, $P = 3$, and $F = 0$, i.e., there is a unique temperature–pressure combination where water coexists in all three phases.

For a binary phase diagram, $F + P = C + 2 = 4$. If we confine our discussions to systems at atmospheric pressure (as assumed for Fig. 2.2), then $F = 3 - P$, since 1 degree of freedom is preassigned to pressure. Furthermore, we note that composition is assigned a single degree of freedom, since X_A (the fraction of A) and X_B (the fraction of B) are related by $X_A + X_B = 1$. In Fig. 2.2 regions L, α, and β are one-phase regions each, so that $F = 2$ (*both* temperature *and* composition can be independently varied). In the two-phase regions $L + \alpha$, $L + \beta$, $\alpha + \beta$, we have $F = 1$. These are univariant regions; at any given temperature, for example, the compositions of both the liquid and the solid phases are fixed by the lever rule. Thus *either* temperature *or* composition can be independently varied in this region, but not both. At the point E, $P = 3$ (L, α, β) so that there are 0 degrees of freedom; E is an invariant point whose temperature and composition are *both* fixed.

As a result of the phase rule, we note that a binary diagram can only have regions of one and two phases. (Degenerated three-phase regions, represented by a point such as E, are also permitted.) Furthermore, two-phase regions cannot be next to each other. It follows that, in traversing a binary phase diagram from one side to another at any constant temperature, all single-phase regions are separated by two-phase regions as the composition is varied. By way of example, the phases that exist at T_2 are α, $L + \alpha$, L, $L + \beta$, and β in succession. On the other hand, the phases that are present at T_3 are α, $\alpha + \beta$, and β in succession.

Various types of binary phase diagrams are encountered in practice, depending on the components involved and their degree of miscibility in the solid and liquid states. Some of these are now described.

2.2.3 Isomorphous Diagrams

The isomorphous diagram is characteristic of components that are completely soluble in each other. It has been empirically found that phase diagrams of this type are restricted to binary systems in which the components are within 15% of each other in atomic radius, have the same valence and crystal structure, and have no appreciable difference in electronegativity. As a consequence, not many binary systems belong to this class; some examples are copper–nickel, silver–palladium, and gold–platinum. Limited solubility of components is by far the more common occurrence in binary systems.

Silicon and germanium are very similar in structure and atomic properties. As expected, they are completely miscible in both the liquid and the solid phases, and are characterized by the isomorphous phase diagram of Fig. 2.3. Here α represents the full range of solutions of germanium and silicon; a single-crystal diamond lattice material can be grown for any composition. In this lattice, however, atom sites will be randomly occupied by silicon and germanium, while preserving a specific overall composition ratio. It must be emphasized that this material is quite different from a compound, such as gallium arsenide, where lattice sites are specifically assigned to either gallium or arsenic.

Gallium arsenide and gallium phosphide are also completely miscible, as is seen in the GaAs-GaP phase diagram of Fig. 2.4. Again, a solution comprising any ratio of gallium arsenide to gallium phosphide can be grown in single-crystal form [4]. In this case, however, the column V sites will be occupied randomly by arsenic or phosphorus atoms with the specified overall composition. Solutions of two-compound semiconductors are sometimes referred to as *mixed-compound semiconductors* or *ternary compounds,* and are usually written in the compositional form $GaAs_xP_{1-x}$.

Many of the properties of mixed-compound semiconductors vary monotonically between the two extremes. Thus, the energy gap of $GaAs_xP_{1-x}$ varies monotonically with x from that of gallium arsenide to that of gallium

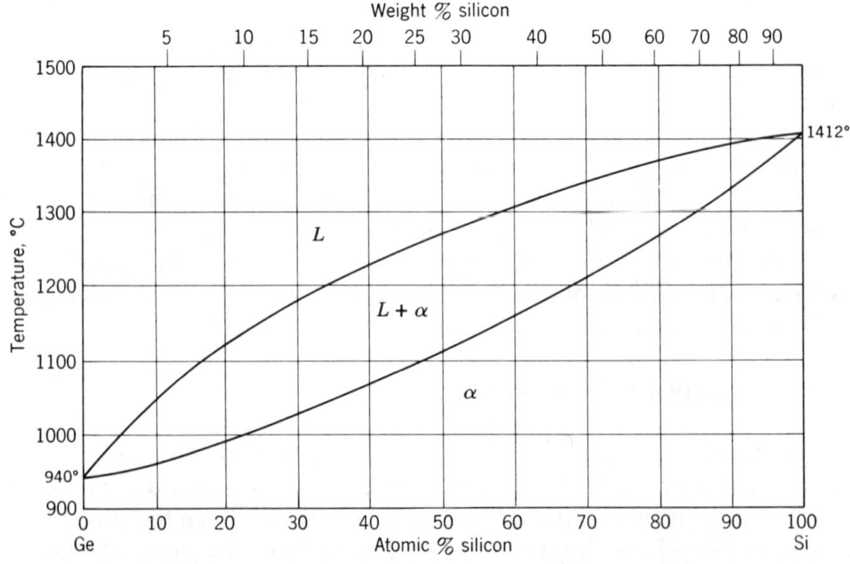

Fig. 2.3 The germanium–silicon system. From M. Hansen and A. Anderko, *Constitution of Binary Alloys* [1], 1958. Used with permission of the McGraw-Hill Book Company.

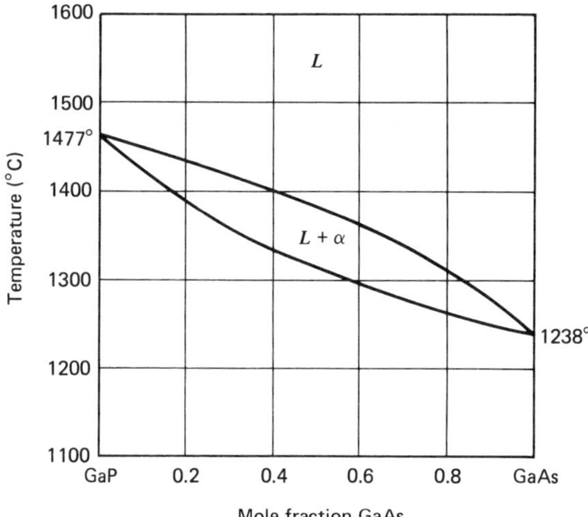

Fig. 2.4 The GaAs–GaP system. From Antypas [4]. Reprinted with permission of the publisher, The Electrochemical Society, Inc.

phosphide. An important technological application of this property is in the fabrication of light-emitting diodes in the visible part of the spectrum.

2.2.4 Eutectic Diagrams

A eutectic diagram results when the addition of either component to a melt lowers its overall freezing point, as shown in Fig. 2.2. Here the freezing point of the molten mixture has a minimum value T_E, below T_A and T_B. This minimum value is known as the *eutectic point,* and the corresponding mixture C_E is called the *eutectic composition*. Eutectic systems usually occur when two components are completely miscible in the liquid phase but are only partly soluble in the solid state. Most semiconductor systems fall into this class.

Referring to Fig. 2.2, consider the cooling of a melt of initial composition C_M. As the temperature is reduced below T_1, a solid of composition β_1 first freezes out. With falling temperature, the liquid composition moves along the liquidus line, becoming richer in A, until the eutectic temperature T_E is reached. At this point the melt is of eutectic composition C_E and freezes isothermally to form the $\alpha + \beta$ phase. Thus the final solid consists of β-phase aggregates in an $(\alpha + \beta)$-phase mixture of eutectic composition.

Figure 2.5 shows the lead–tin system which exhibits this type of characteristic. Here the eutectic point has a 38% lead–62% tin composition by weight and a eutectic temperature of 183°C. On the other hand, the freezing points of pure lead and pure tin are 327 and 232°C, respectively. Lead–tin combinations, of eutectic composition, are used as solders in electronic

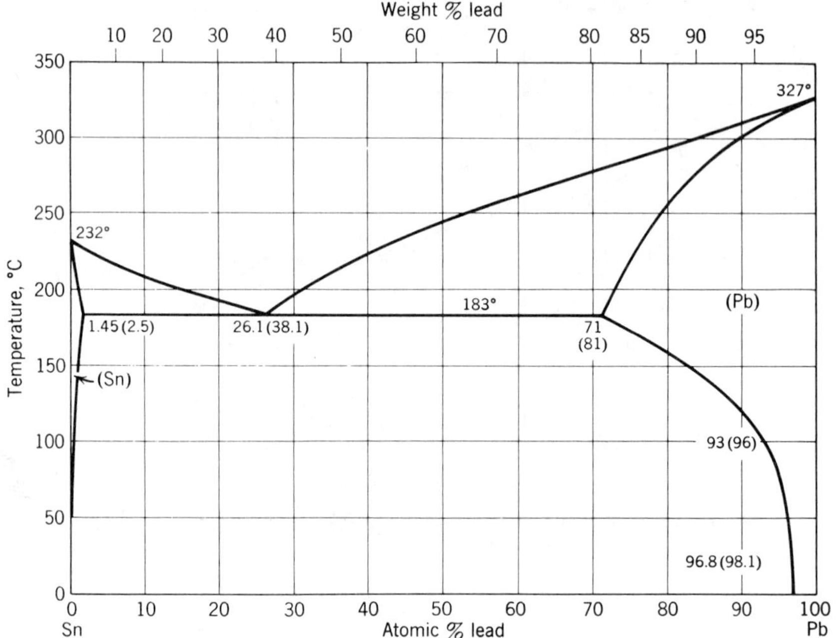

Fig. 2.5 The lead–tin system. From M. Hansen and A. Anderko, *Constitution of Binary Alloys* [1], 1958. Used with permission of the McGraw-Hill Book Company.

circuit wiring. A 95% lead–5% tin solder, with a melting point of 310–314°C, is commonly used in the die mounting of power transistors.

Variations of the eutectic diagram of Fig. 2.2 are often seen in practice, depending on the nature of the terminal solid solutions. Thus in the lead–tin system the terminal solid solubility of tin in lead is significant (19%), as is that of lead in tin (2.5%). On the other hand, the terminal solid solubility of silicon in aluminum is 1.65%, while that of aluminum in silicon is so small (<0.1%) that the β phase cannot be indicated in this diagram. Figure 2.6 shows the phase diagram for this system, which is of importance in the fabrication of ohmic contacts* to *p*-type and degenerate *n*-type silicon. Aluminum is often used for the fabrication of Al-Si Schottky diodes in high-speed digital logic circuits. Here subsequent processing can alter the interface and cause changes in diode characteristics. Thus the phase diagram gives useful information about potential problems which can arise during the fabrication of circuits incorporating these devices.

The gold–silicon system is shown in Fig. 2.7. Here both terminal solid solubilities are too small to be indicated on the diagram. The sharply depressed eutectic temperature of this combination leads to its widespread use in the die bonding of discrete silicon devices and microcircuits to substrates.

* This topic is covered in Chapter 8.

An alternative combination, which provides a better wetting action, is gold and germanium, whose phase diagram is shown in Fig. 2.8. In this case the resulting bond is a ternary Au-Ge-Si alloy. Gold alloys formed during this process are extremely hard and strong, and cannot be used in power devices which are subject to significant thermal cycling. Lead–tin solders are more suitable here because of their plastic behavior.

Occasionally binary systems may exhibit a eutectic composition that is very close to one of the components. Figure 2.9 shows the gallium–germanium

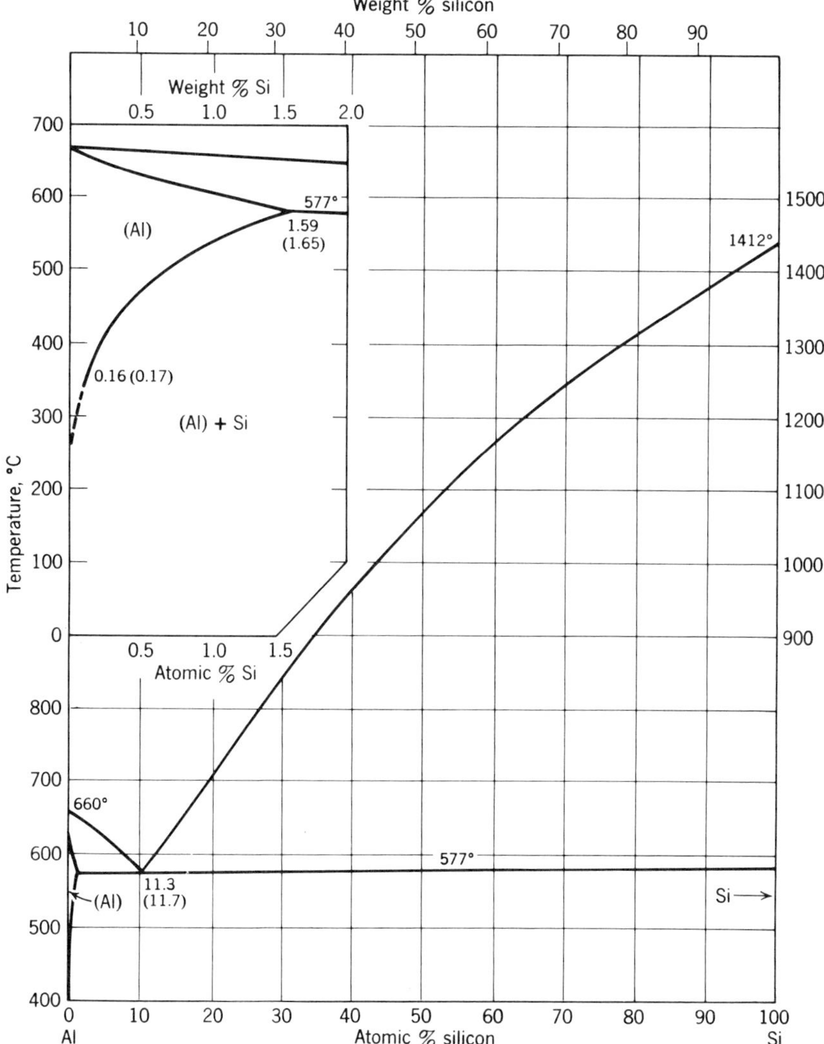

Fig. 2.6 The aluminum–silicon system. From M. Hansen and A. Anderko, *Constitution of Binary Alloys* [1], 1958. Used with permission of the McGraw-Hill Book Company.

system which belongs to this class. Germanium is often used as an ohmic contact material to *n*-type gallium arsenide. The effect of heat treatment during contact formation may be studied with the aid of this diagram. However, the Ga-As-Ge ternary phase diagram (Fig. 2.28) is sometimes used for this purpose.

2.2.5 Congruent Transformations

In many complex systems, one or more intermediate compounds may be formed at specific temperatures and compositions. In contrast to mixtures, which exist at the microscopic level, these intermediate compounds exist on an atomic scale. They usually occur within an extremely narrow compositional range, corresponding to very small departures from stoichiometry. Consequently, they are indicated on the phase diagram in the form of discrete vertical lines, as shown for the gallium–arsenic system of Fig. 2.10. Since these lines appear to represent a change from one phase directly into another, without any apparent alteration in composition, they are called *congruent transformations*. The formation of a congruently melted compound in this manner effectively isolates the system on each side of it. Thus Fig. 2.10 can be considered as a Ga-GaAs system and a separate GaAs-As system, each of which is seen to be eutectic in nature. In this manner it is possible to break up relatively complex phase diagrams into a number of simpler ones.

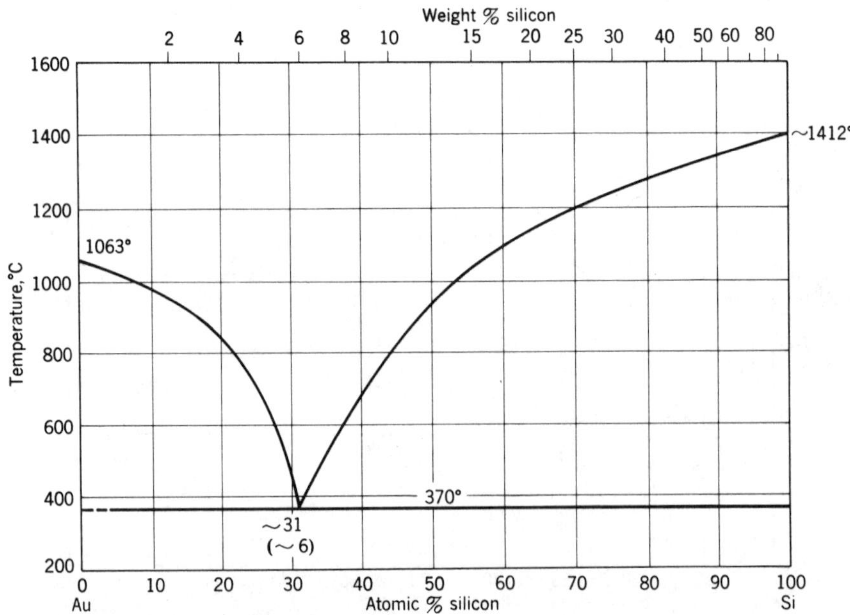

Fig. 2.7 The gold–silicon system. From M. Hansen and A. Anderko, *Constitution of Binary Alloys* [1], 1958. Used with permission of the McGraw-Hill Book Company.

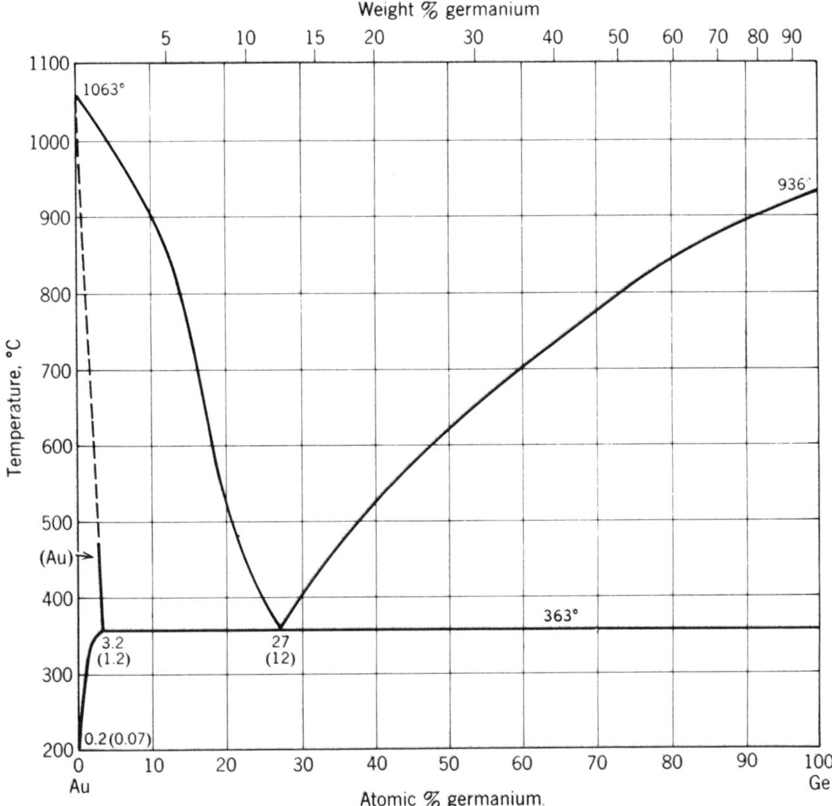

Fig. 2.8 The gold–germanium system. From M. Hansen and A. Anderko, *Constitution of Binary Alloys* [1], 1958. Used with permission of the McGraw-Hill Book Company.

A vertical line, drawn to indicate a congruently melted compound, is somewhat of an oversimplification. In general, congruent melting will appear as a region of finite width if a sufficiently expanded compositional scale is used. In materials such as gallium arsenide, the presence of large concentrations of vacancies and interstitials of both gallium and arsenic (with their associated different free energies) results in melting at a point slightly removed from that of the ideal 1:1 stoichiometric composition. A greatly expanded phase diagram around this region, estimated on the assumption of neutral vacancies [5], is sketched in Fig. 2.11. We note from this diagram that the solidification of gallium arsenide can occur in a region of excess gallium or arsenic,* resulting in the formation of excess arsenic or gallium vacancies, respectively.

As mentioned in Section 1.5.1, there is some doubt as to whether vacan-

* The stoichiometry range for gallium arsenide is between 49.998 and 50.009 at.% of arsenic.

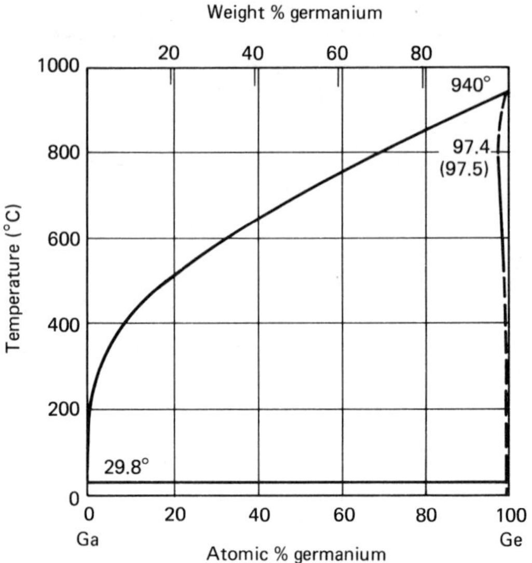

Fig. 2.9 The germanium–gallium system. From M. Hansen and A. Anderko, *Constitution of Binary Alloys* [1], 1958. Used with permission of the McGraw-Hill Book Company.

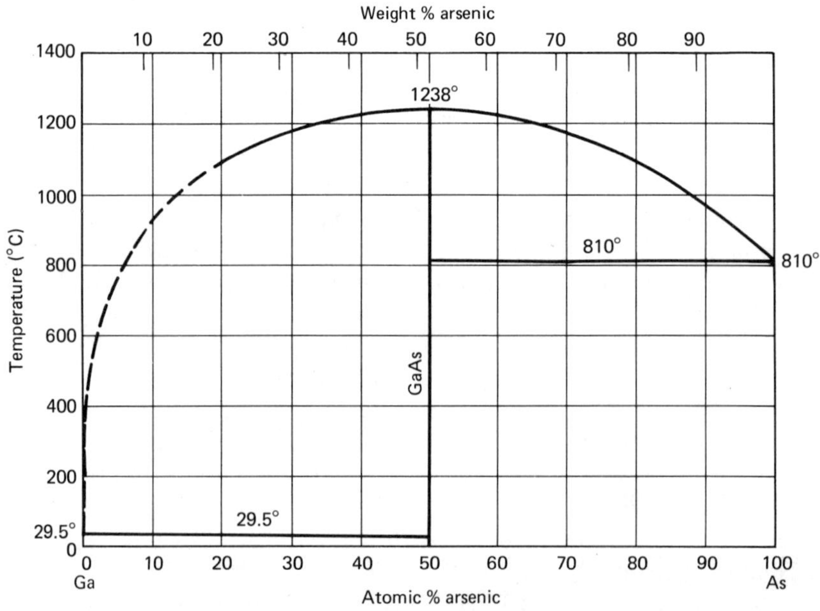

Fig. 2.10 The gallium–arsenic system. From M. Hansen and A. Anderko, *Constitution of Binary Alloys* [1], 1958. Used with permission of the McGraw-Hill Book Company.

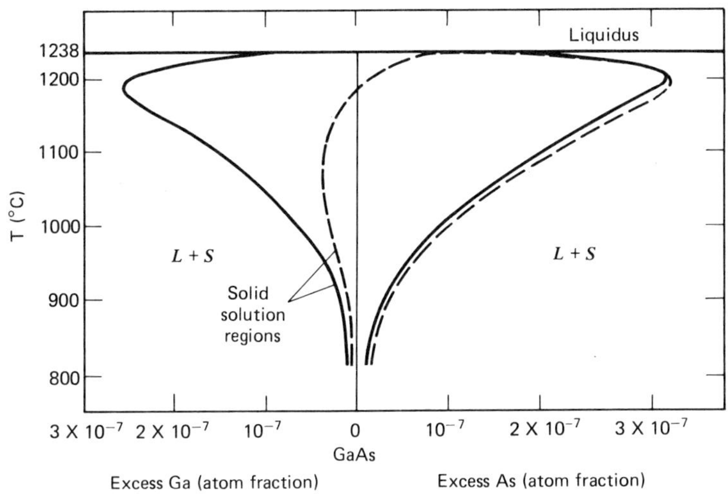

Fig. 2.11 Details of the gallium arsenide phase diagram.

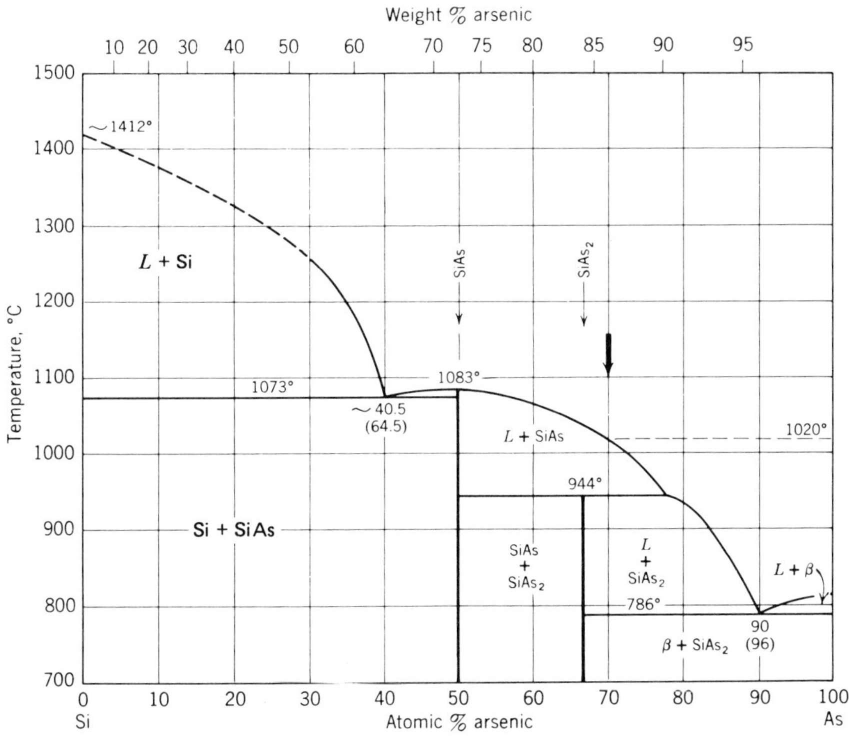

Fig. 2.12 The arsenic–silicon system. From M. Hansen and A. Anderko, *Constitution of Binary Alloys* [1], 1958. Used with permission of the McGraw-Hill Book Company.

cies in gallium arsenide are of themselves ionized. However, many of these excess vacancies can become occupied by impurity atoms which are either intentionally or unintentionally present during crystal growth with a resultant shift in the conductivity. Furthermore, the amount of shift in conductivity, as well as its direction (i.e., toward *p*- or *n*-type) will be determined by the temperature at which solidification occurs from the melt. This has important consequences in crystal growth and epitaxy, as shown in Chapters 3 and 5.

2.2.6 Peritectic and Other Reactions

A peritectic reaction is yet another way in which an intermediate compound can occur in a binary system. By way of example, Fig. 2.12 shows the

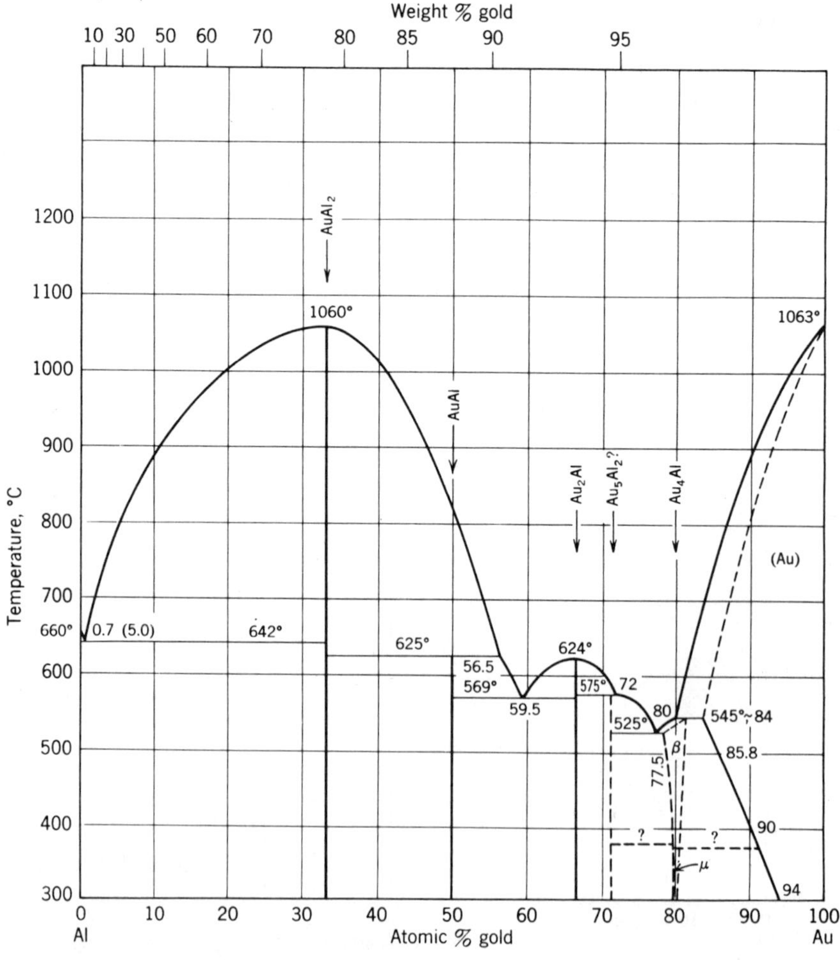

Fig. 2.13 The aluminum–gold system. From M. Hansen and A. Anderko, *Constituion of Binary Alloys* [1], 1958. Used with permission of the McGraw-Hill Book Company.

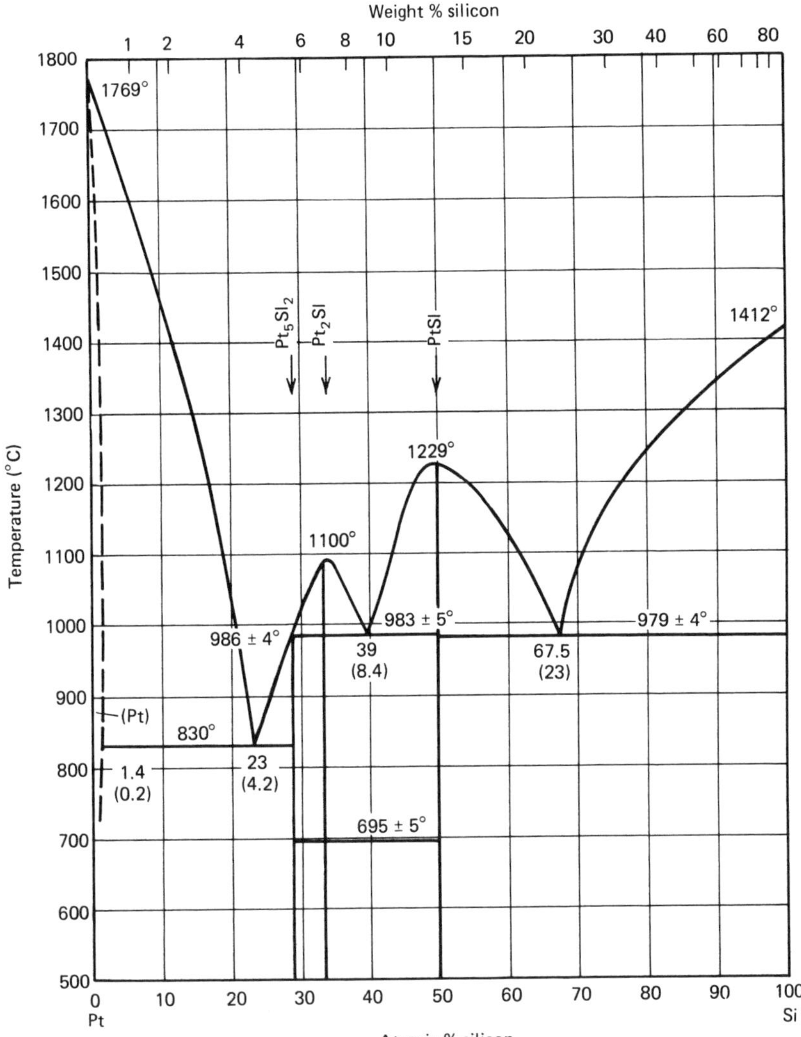

Fig. 2.14 The platinum–silicon system. From M. Hansen and A. Anderko, *Constitution of Binary Alloys* [1], 1958. Used with permission of the McGraw-Hill Book Company.

arsenic–silicon system, in which the compound $SiAs_2$ is formed by a peritectic reaction, between SiAs and a liquid. Note that both these compounds are depicted by vertical lines, since they occur within a narrow compositional range. A eutectic point also exists for ths SiAs-As system, corresponding to a 96% arsenic–4% silicon composition by weight.

Consider what happens if a mixture of 86% arsenic–14% silicon by weight is cooled from some high temperature. At and below 1020°C solid SiAs is precipitated from the melt, which becomes arsenic-rich until the temperature

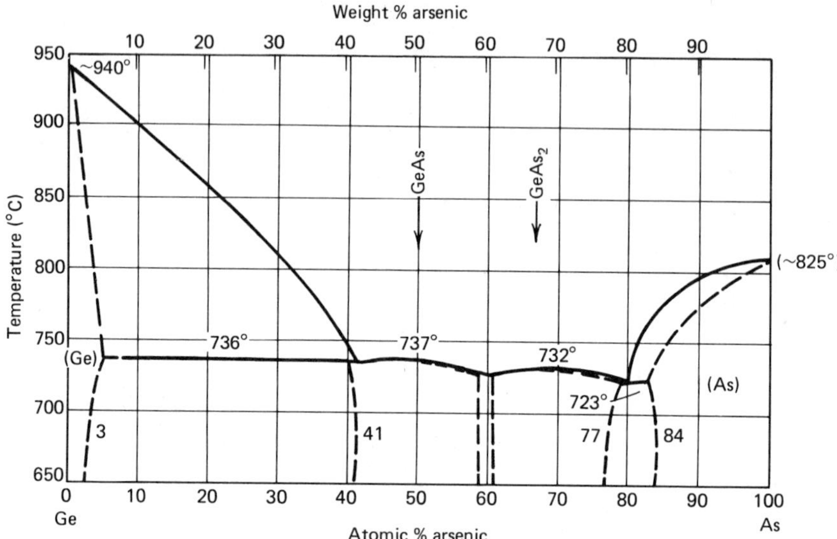

Fig. 2.15 The germanium–arsenic system. From M. Hansen and A. Anderko, *Constitution of Binary Alloys* [1], 1958. Used with permission of the McGraw-Hill Book Company.

reaches 944°C. At this point the liquid composition is 90% arsenic–10% silicon by weight. With a further reduction in temperature, the solid SiAs combines with some of this excess liquid to form a liquid + $SiAs_2$ phase. At 786°C and lower, both the $SiAs_2$ and the β phases precipitate from the liquid to form a solid β + $SiAs_2$ phase, as shown.

Cooling through a peritectic temperature causes the formation of a nonequilibrium structure by the process of "surrounding." Here the $SiAs_2$ is formed by the surrounding of a solid SiAs core by the liquid. Under normal cooling conditions, this $SiAs_2$ layer creates a barrier to the diffusion of liquid to the SiAs, resulting in a reaction that proceeds at an ever-decreasing rate. Thus a peritectic reaction is usually accompanied by the formation of relatively large (micron-size) precipitates.* Peritectic phases usually occur as part of more complex phase diagrams. A few of these are now described.

Figure 2.13 shows the phase diagram of the aluminum–gold system, which is of importance in evaluating the bonding of gold leads to aluminum contact pads on microcircuits. Of the many possible intermetallic compounds that are indicated here, the most significant [6] are $AuAl_2$, a dark purple, strongly bonding, highly conductive compound, and Au_2Al, a tan-colored, brittle, poorly conducting compound. This latter compound† has a melting point of 624°C. Nevertheless significant compound formation occurs at 300°C and

* The anomalous behavior of arsenic doping at high concentrations can be explained by the formation of As-Si clusters by this reaction.

† For many years, the tan Au_2Al was not noticed in the presence of the purple $AuAl_2$. Thus the term *purple plague* was wrongly given to this phenomenon, and still persists in the literature.

higher. This is a serious cause for lead-attachment failure in microcircuits which are stored for periods of time at these temperatures [7].

Experiments have shown that the formation of Au_2Al is enhanced by the presence of silicon. Although the phase diagram of the ternary Al-Au-Si system has not been established, it can be expected that the effect [7] of silicon addition to the aluminum–gold system produces simple eutectic lowering of the phase diagram. Thus the formation of these compounds may be expected to be accelerated in the presence of the silicon. This has been observed experimentally.

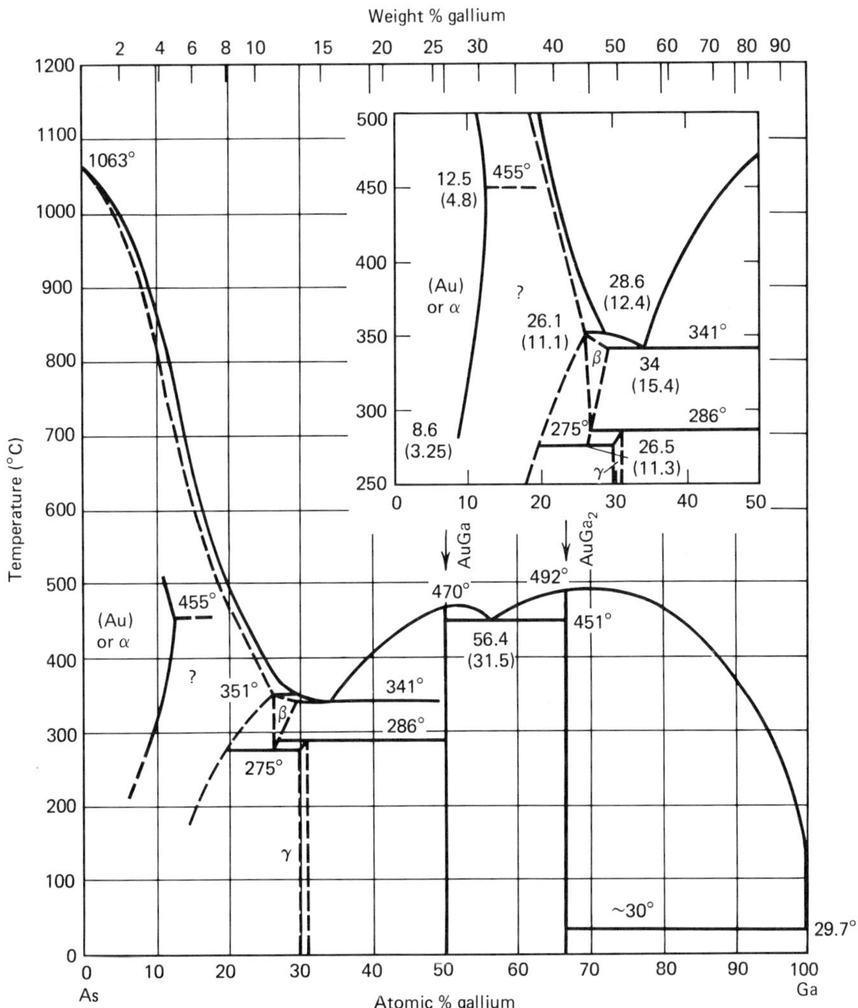

Fig. 2.16 The gold–gallium system. From M. Hansen and A. Anderko, *Constitution of Binary Alloys* [1], 1958. Used with permission of the McGraw-Hill Book Company.

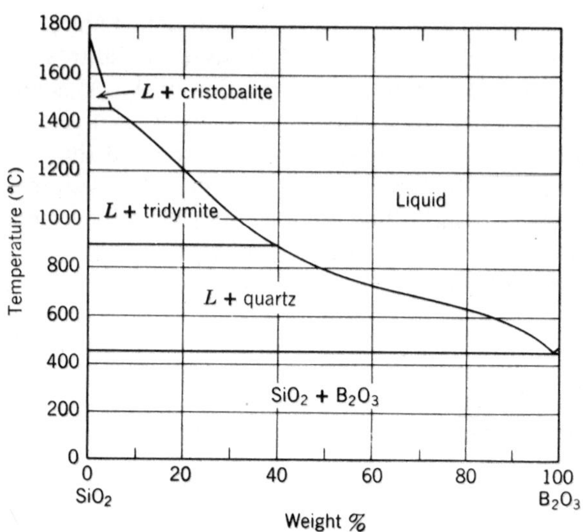

Fig. 2.17 The B_2O_3–SiO_2 system. From Rockett and Foster [9]. Reprinted with permission from the American Ceramic Society, Inc.

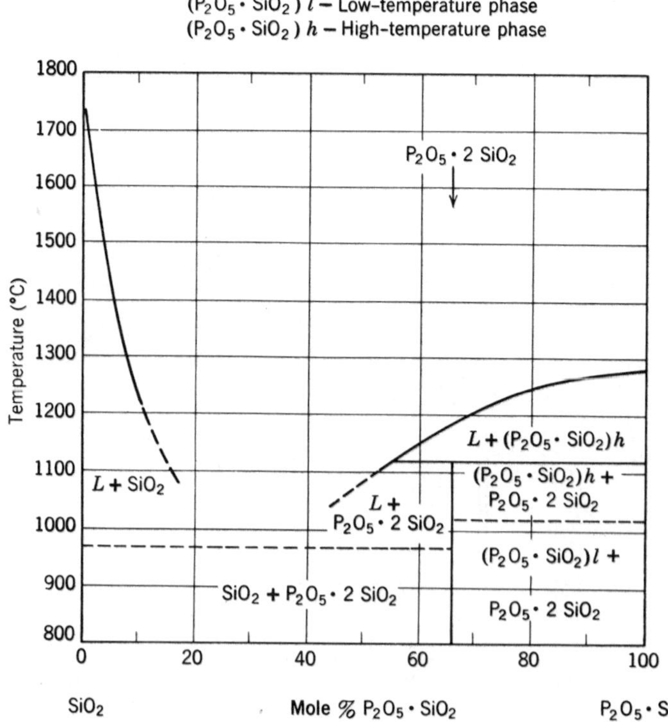

Fig. 2.18 The P_2O_5–SiO_2 system. From Tien and Hummel [10]. Reprinted with permission from the American Ceramic Society, Inc.

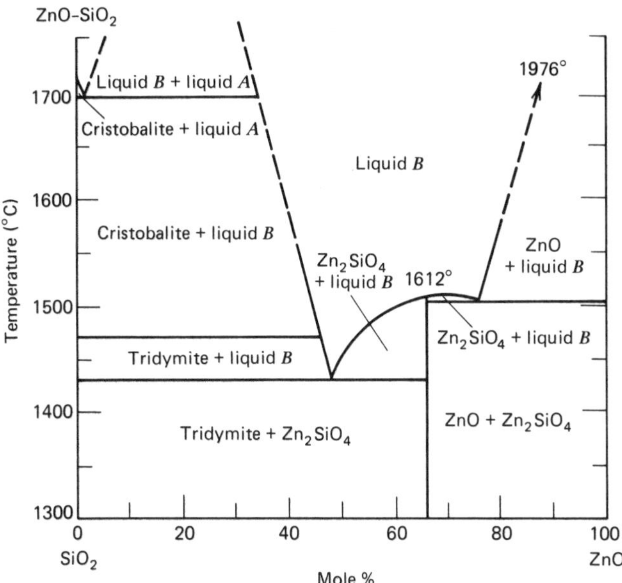

Fig. 2.19 the $ZnO-SiO_2$ system. From Bunting [11]. Reprinted with permission from the American Ceramic Society, Inc.

Figure 2.14 shows the platinum–silicon system, and indicates the phases involved in the formation of a PtSi barrier layer on a silicon surface. Typically, a layer of platinum, 500 Å thick, is deposited on the silicon which is then heated to about 600°C for 15 min. This suffices to complete* the reaction and form PtSi, even though the melting point of this compound is 1229°C. This PtSi layer serves as a base which is followed by successive layers of titanium, platinum, and gold. This complex metallization scheme forms the basis of the beam lead system [8], which is described in Chapter 8.

A common contacting scheme for *n*-type gallium arsenide consists of depositing a layer of germanium, followed by a layer of gold to which leads can be readily bonded. The phase diagram of the germanium–arsenic system, showin Fig. 2.15, together with that for the germanium–gallium system (Fig. 2.9) and the gold–gallium system of Fig. 2.16, is useful in evaluating potential problems here. The phase diagram for the ternary Ga-As-Ge system is shown in Fig. 2.28. The Ga-As-Au system is shown in Fig. 2.27.

In addition to eutectic and peritectic behavior, a number of other less common reactions may occur in binary systems. Thus a single phase can cool to form two phases in one of three possible ways, as follows:

1. Monotectic: $L_1 \xrightarrow{\text{cooling}} \alpha + L_2$.

* It should be emphasized that phase diagrams provide no information about the kinetics of a reaction.

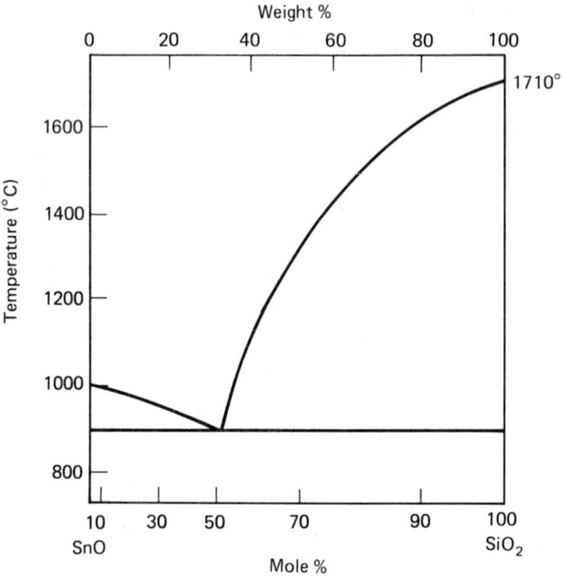

Fig. 2.20 the SnO–SiO_2 system. Adapted from Toropov, et al. [12].

2. Eutectic: $L \xrightarrow{\text{cooling}} \alpha + \beta$
3. Eutectoid: $\gamma \xrightarrow{\text{cooling}} \alpha + \beta$

Here L, L_1, and L_2 represent liquid phases and α, β, and γ represent solid phases, including compounds.

In addition, three possibilities occur when two phases react to form a third, different phase, as follows:

1. Syntectic: $L_1 + L_2 \xrightarrow{\text{cooling}} \beta$
2. Peritectic: $L + \alpha \xrightarrow{\text{cooling}} \beta$
3. Peritectoid: $\alpha + \gamma \xrightarrow{\text{cooling}} \beta$

Examples of some of these reactions are shown in the phase diagrams of Figs. 2.12–2.16.

2.2.7 Phase Diagrams for Oxide Systems

Phase diagrams involving oxide systems are especially important in both silicon and gallium arsenide technology. Thus boron doping of silicon is carried out from a B_2O_3-SiO_2 source which is deposited on the silicon slice prior to diffusion. The phase diagram of Fig. 2.17 is relevant to this process. Although of a relatively straightforward eutectic type, a number of structural phase change regions are seen here as the SiO_2 goes through its cristobalite and tridymite phases [9].

Figure 2.18 shows the phase diagram for the P_2O_5-SiO_2 system. Phosphosilicate glass has considerable importance in device technology [10], since it is used for phosphorus doping of silicon, for gettering to improve lifetime in silicon devices, as a passivation layer over silicon microcircuits, as a diffusion mask for gallium arsenide, and as a cap for open-tube diffusions in gallium arsenide.

Figures 2.19 and 2.20 show phase diagrams [11, 12] for ZnO-SiO_2 and for SnO-SiO_2. Both of these find use during the *p*- and *n*-doping, respectively of gallium arsenide.

2.3 SOLID SOLUBILITY

It has been noted that many binary systems exist in which the terminal solid solubility of one component in the other is extremely small. This is usually the case for donor and acceptor impurities in silicon. As a consequence, it is necessary to expand greatly the scale of the phase diagram in order to show this important region. Figure 2.21 shows this part of the phase diagram, and is typical of most impurities in silicon. The solid solubility of these impurities is seen to increase with temperature, reach a peak value, and fall off rapidly as the melting point of the silicon is approached. This is commonly referred to as a *retrograde* solid-solubility characteristic. Figure 2.22 shows values for various commonly used dopants in silicon [13, 14]. As seen from

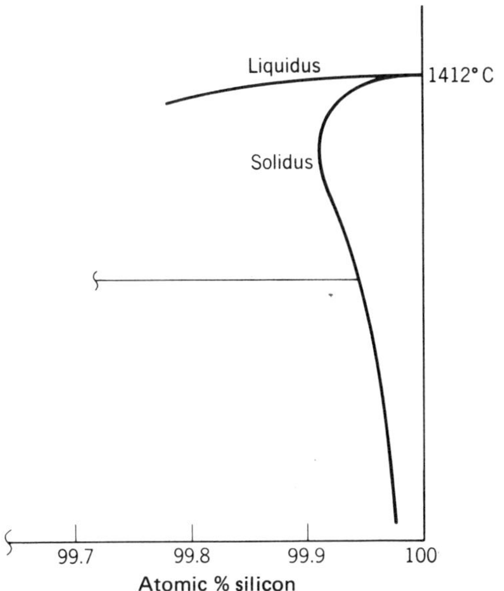

Fig. 2.21 Retrograde solubility characteristic.

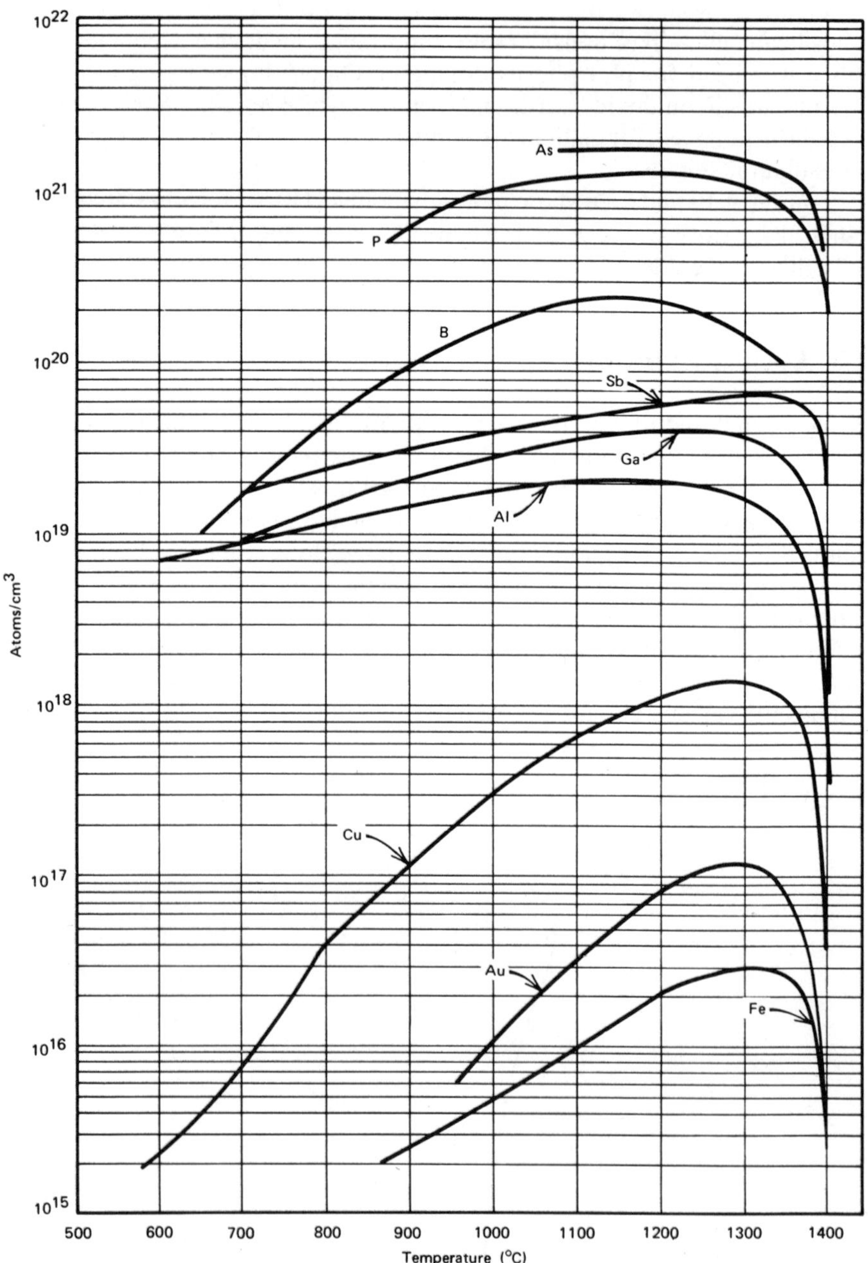

Fig. 2.22 Solid solubility of impurities in silicon.

this figure, nearly all of these elements indicate a retrograde behavior in their solubility characteristics.

2.4 TERNARY DIAGRAMS

These diagrams are needed when three components are involved. It is customary to sketch phase diagrams for such systems as three-dimensional plots, with each dimension corresponding to a binary system of two of these three components. Very few systems of this type have been studied in detail. However, even fragmentary information, which is available for many such systems, is useful in evaluating the potential problems which can arise during device processing.

Ternary phase diagrams are required for systems with three components. They are of special use in the study of the behavior of III–V semiconductors such as gallium arsenide, in the presence of an additional component. In these systems, $F + P = 3 + 2 = 5$. Since 1 degree of freedom is usually assigned to pressure, $F + P = 4$ under ordinary circumstances. In addition, we note that two compositions must be assigned in order to determine the third, since $X_A + X_B + X_C = 1$.

A ternary system, where all alloys solidify eutectically, is shown in Fig. 2.23*a* and illustrates many of the characteristics of these diagrams [15]. We note that this is a three-dimensional plot, and is rather complex to construct, even for such a simple situation.

The base of this diagram is an equilateral triangle, as shown in Fig. 2.24, with each side corresponding to a binary composition, usually on an atom fraction basis. Any line, parallel to a side of this triangle, represents compositions in which the fraction of the component in the opposite vertex is held constant. (For example, all compositions with 25% *A* are represented by the line *XY* in Fig. 2.24.) Any line through a vertex represents all compositions with a fixed ratio of the components in the other two vertices. Thus compositions where $B\!:\!C = 3\!:\!1$ are represented by the line *AZ*. The composition at 0 is thus 25% *A*, 56.25% *B*, and 18.75% *C*.

Another interesting property of the equilateral triangle is seen by dropping perpendiculars to the three sides from the point 0. Let the lengths of these perpendiculars be *a, b, c,* drawn as shown to the sides opposite the vertices *A, B, C,* respectively. Then the composition represented by the point 0 is given by $A\!:\!B\!:\!C = a\!:\!b\!:\!c$. Furthermore, $a + b + c = \sin 60°$. For this example, $a = 0.25 \sin 60°$, $b = 0.5625 \sin 60°$, and $c = 0.1875 \sin 60°$, so that the composition defined by 0 is 25% *A*, 56.25% *B*, and 18.75% *C*, as before.

2.4.1 Isothermal Sections [16]

Ternary phase diagrams of the type shown in Fig. 2.18 are usually drawn in simplified form as a series of isothermal sections taken at different temperatures of interest. Sections of this type are shown in Fig. 2.23*b*, and correspond to the temperatures indicated in Fig. 2.23*a*. Very few such systems have been completely determined. Nevertheless, many sections can be drawn qualitatively from a knowledge of the behavior of the individual binary systems *A-B*, *B-C*, and *C-A*. A study of these sections shows the following:

1. All two-phase regions are enclosed by four boundaries. Two opposite boundaries are adjacent to three-phase regions, or may be boundaries of the phase diagram. The other two opposite boundaries separate this region from one-phase regions. In addition, contact is made with two-phase regions at each vertex.
2. All three-phase regions are triangular, bounded by two-phase regions on three sides. In addition, they are in contact with one-phase regions at each vertex.

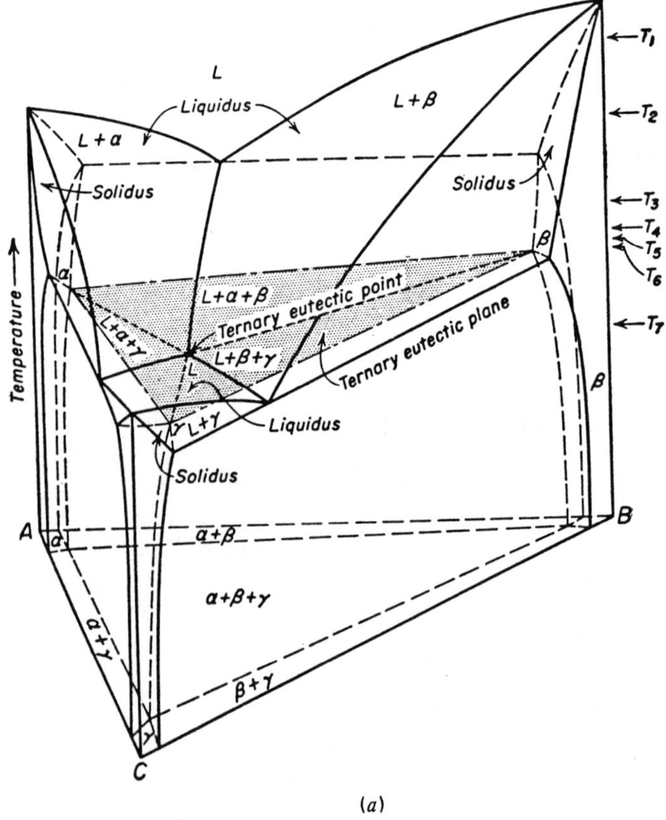

(*a*)

Fig. 2.23 Ternary phase diagram. (*a*) Isometric construction. (*b*) Isothermal sections. From F. N. Rhines, *Phase Diagrams for Metallurgy* [15], 1956. Used with permission of the McGraw-Hill Book Company.

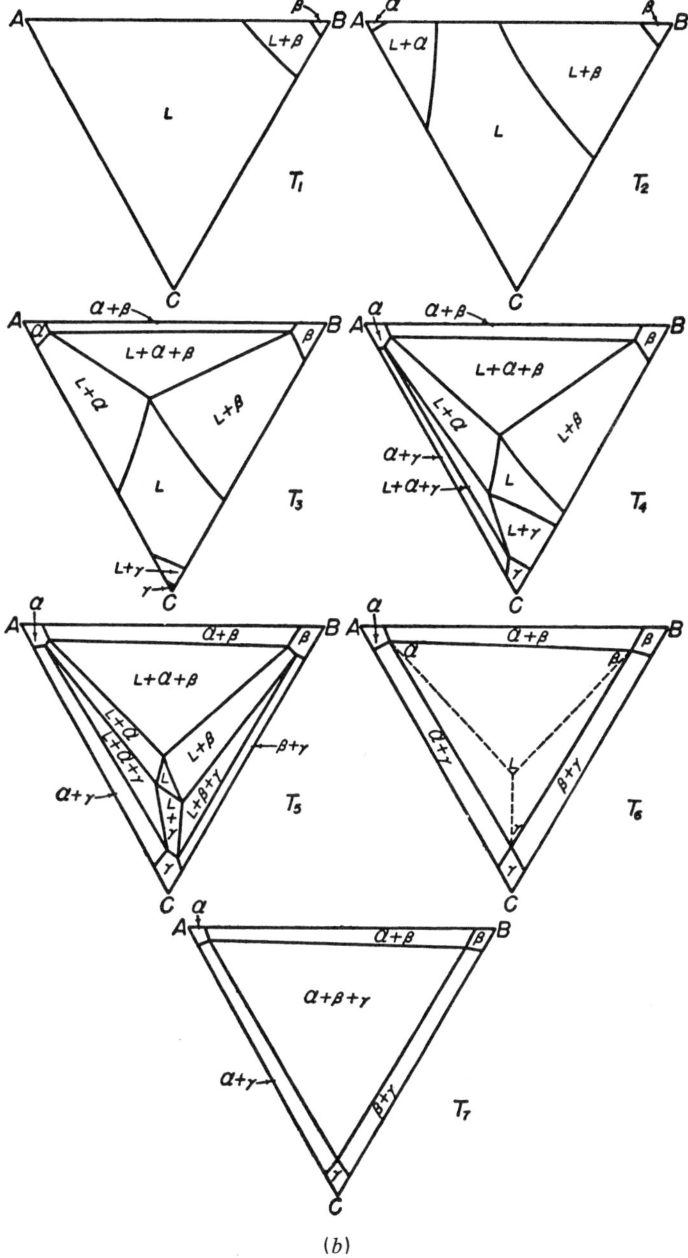

(b)

Fig. 2.23 (Continued)

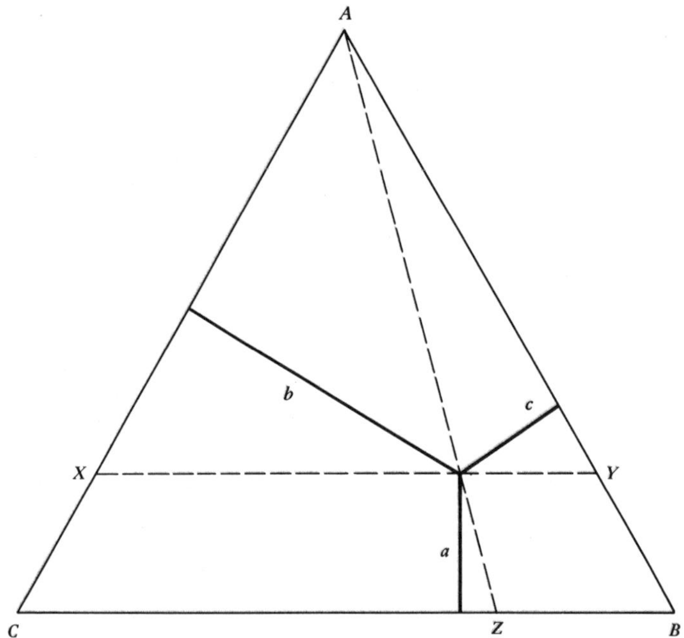

Fig. 2.24 Compositions in the ternary phase diagram.

2.4.2 Congruently Melting Compounds

These are always present in the gallium-arsenic-impurity system. Examples are binary compounds such as GaAs or Zn_3As_2 (in the Ga-As-Zn system). The presence of congruently melting compounds of this type serves to divide the ternary system into subsidiary eutectic systems.

2.4.3 Degrees of Freedom

For any isothermal section, $F = 3 - P$ since *both* pressure *and* temperature are preassigned. In a one-phase region, as shown in Fig. 2.25 (such as L, α, β, γ), there are 2 degrees of freedom. Both X_A *and* X_B (and hence X_C), can be varied independently over this region. For a two-phase region ($L + \alpha, L + \beta, L + \gamma, \alpha + \beta$, etc.) only 1 degree of freedom is possible. Thus only one of the components can be varied. This is illustrated in Fig. 2.25. Here a given value of X_A results in a liquid of composition X' along the boundary *ab*. This fixes, via a tie line, the composition of the solid solution β at X'' so that the system is fully defined.

In ternary diagrams these tie lines may be straight or curved, and must be determined experimentally. In general, however, their positions are usually estimated by noting that they vary gradually from one boundary to the other, without crossing each other. Furthermore, they must run between two one-phase regions. A series of such tie lines is shown in Fig. 2.25.

Figure 2.25 also shows a three-phase region, $L + \beta + \gamma$. This is an invariant region, with no degrees of freedom. For this temperature the composition of the liquid and solid phases (L, β, and γ) is fixed at the values given by the vertices of the triangle bounding this region. Thus the partial pressure of these three components is fixed by temperature alone. Compositions of this type are particularly attractive as diffusion sources for this reason. The application of ternary considerations to diffusion is considered in Chapter 4.

2.4.4 Some Ternary Systems of Interest

It is customary to show these systems in the form of a contour map, looking down on the composition triangle, with isotherms as contour lines. Also indicated on these diagrams are unique features (eutectics, peritectics, compounds, etc.) which aid in constructing specific isothermal projections as required. The valleys of the liquidus are also projected onto the composition triangle, with arrows indicating decreasing temperature.

The diagrams for gallium arsenide, with silver, gold, germanium, sulfur, tin, tellurium, or zinc as the third component, are shown in Figs. 2.26–2.31 [16–18]. Most of these diagrams are incomplete, with many lines estimated from the appropriate binary diagrams. Nevertheless, these diagrams are useful for assessing problems which might arise during the growth [19] or alloying of gallium arsenide, and also for evaluating conditions during diffusions [20] in gallium arsenide. This latter topic is taken up in Chapter 4,

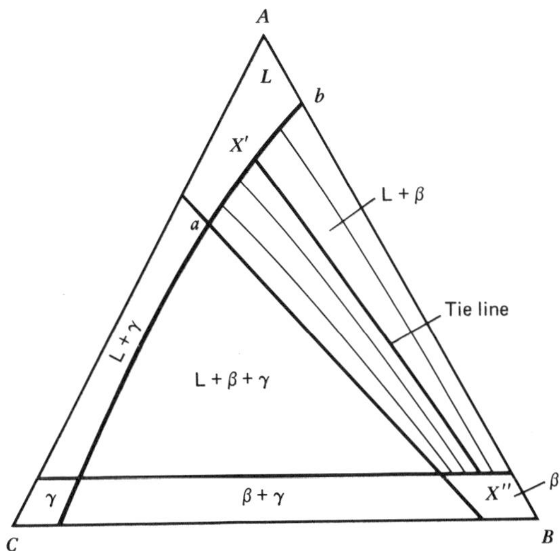

Fig. 2.25 Diagram illustrating regions of one, two, and three phases.

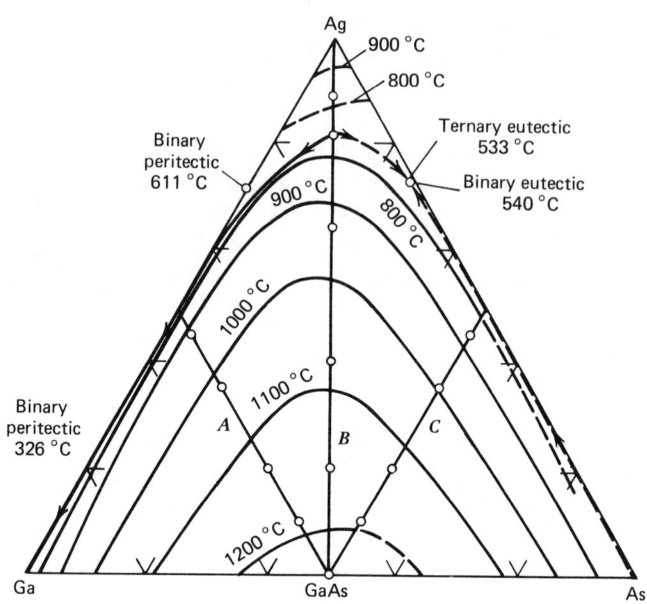

Fig. 2.26 The ternary system Ga–As–Ag. From Panish [18]. Reprinted with permission of the publisher, The Electrochemical Society, Inc.

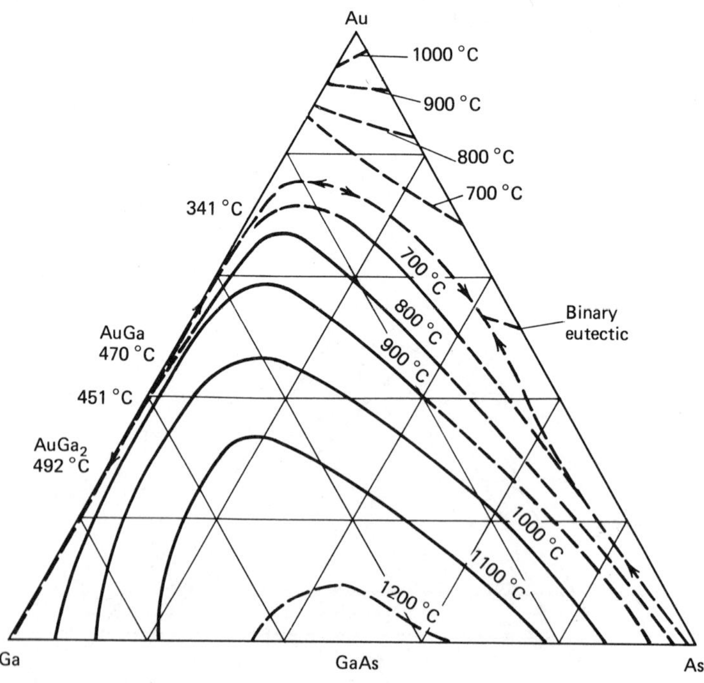

Fig. 2.27 The ternary system Ga–As–Au. From Panish [18]. Reprinted with permission of the publisher, The Electrochemical Society, Inc.

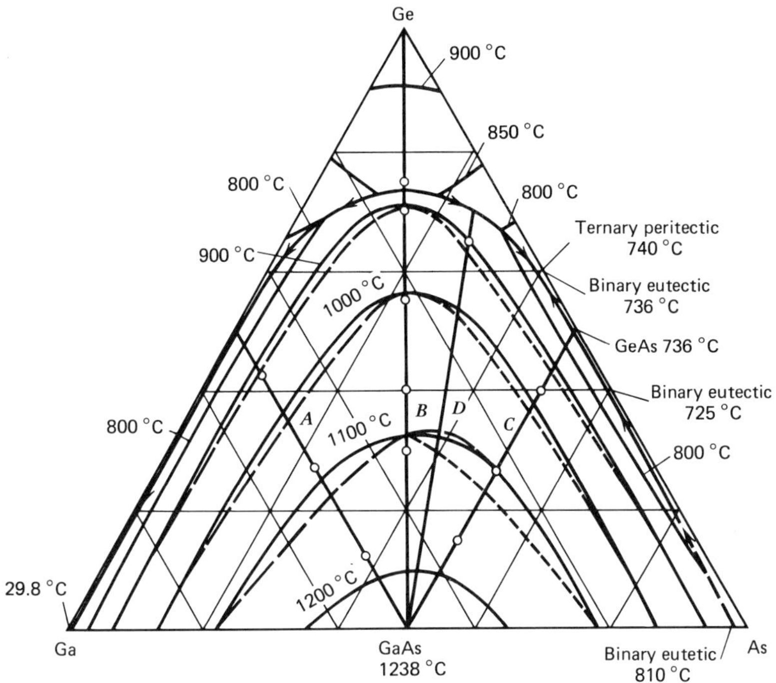

Fig. 2.28 The ternary system Ga–As–Ge. From A. M. Alper, Ed., *Phase Diagrams,* vol. III [16], 1970. Used with permission of the Academic Press, Inc.

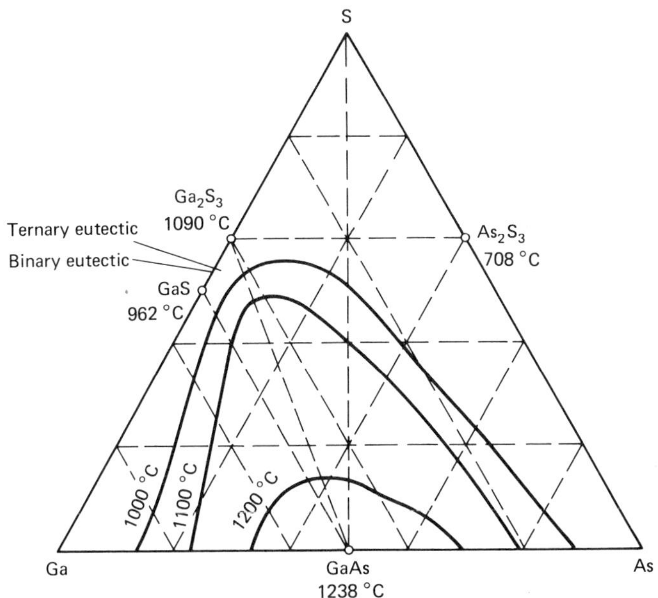

Fig. 2.29 The ternary system Ga–As–S. From Matino [17]. Reprinted with permission of the publisher, Pergamon Press, Ltd.

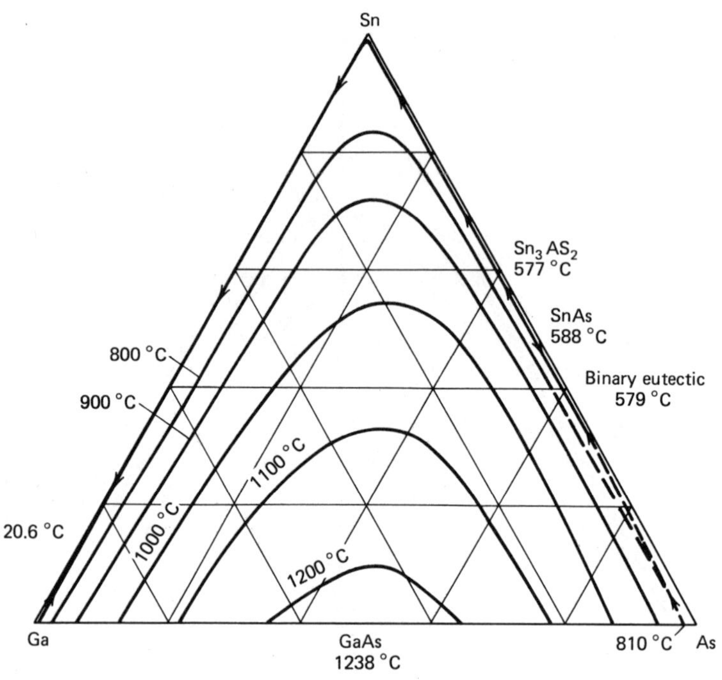

Fig. 2.30 The ternary system Ga–As–Sn. From A. M. Alper, Ed., *Phase Diagrams,* vol. III [16], 1970. Used with permission of the Academic Press, Inc.

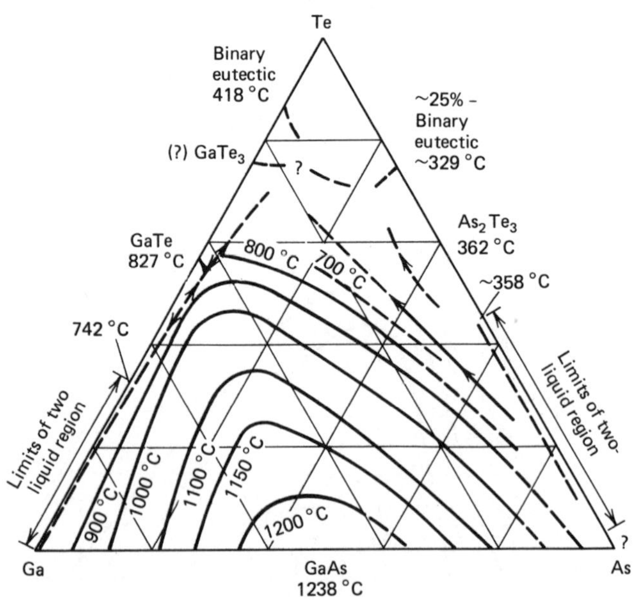

Fig. 2.31 The ternary system Ga–As–Te. From A. M. Alper, Ed., *Phase Diagrams,* vol. III [16], 1970. Used with permission of the Academic Press, Inc.

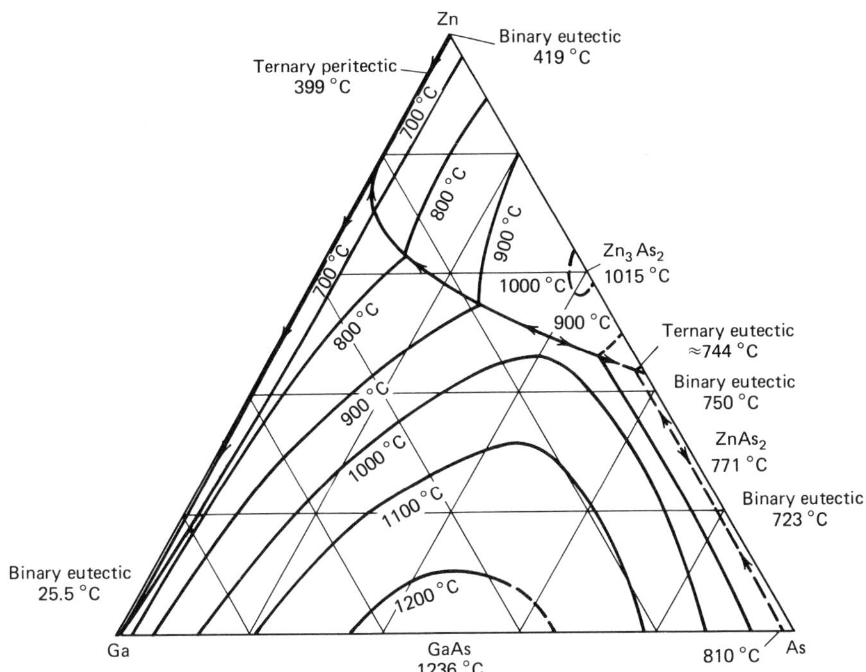

Fig. 2.32 The ternary system Ga–As–Zn. From A. M. Alper, Ed., *Phase Diagrams,* vol. III [16], 1970. Used with permission of the Academic Press, Inc.

where a series of isothermal sections are shown for the Ga-As-Zn and for the Ga-As-S systems.

REFERENCES

A large collection of phase diagrams for binary alloys may be found in [1] and in its more up-to-date supplement [2]. A compendium of phase diagrams for refractory oxides may be found in [3]. Phase diagrams for ternary systems of interest to gallium arsenide processing technology are scattered throughout the technical literature.

1. M. Hansen and K. Anderko, *Constitution of Binary Alloys,* McGraw-Hill, New York, 1958.
2. R. P. Elliot, *Constitution of Binary Alloys* (A Supplement to [1]), McGraw-Hill, New York, 1965.
3. E. M. Levin Ed., Phase Diagrams for Ceramists, Am. Ceramics Soc. (1956).
4. G. A. Antypas, The GaP-GaAs Ternary Phase Diagram, *J. Electrochem. Soc.* **117,** 700 (1970).
5. H. C. Yeh, Interpretation of Phase Diagrams, in *Phase Diagrams,* Vol. 1, A. M. Alper, Ed., Academic, New York, 1970.
6. R. Schmidt, Mechanism of Lead Failures in Thermocompression Bonds, *IRE Trans. Electron Dev.* **ED-9,** 506 (1962).

7. B. Selikson and T. Longo, A Study of Purple Plague and Its Role in Integrated Circuits, *Proc. IEEE* **52,** 1638 (1964).
8. J. H. Forster and J. B. Singleton, Beam-Lead Sealed Junction in Integrated Circuits, *Bell Lab. Rec.* **44,** 312 (Oct.–Nov. 1966).
9. T. J. Rockett, and W. R. Foster, Phase Relations in the System Boron Oxide-Silica, *J. Am. Ceram. Soc.* **48,** 75 (1965).
10. T. Y. Tien, and R. A. Hummel, The System SiO_2-P_2O_5, *J. Am. Ceram. Soc.* **45,** 422 (1962).
11. E. N. Bunting, Phase Equilibria in the System SiO_2-ZnO_2, *J. Am. Ceram. Soc.* **13,** 5 (1930).
12. N. A. Toropov, V. P. Barzakovskii, V. V. Lapin, and N. N. Kurtseva, *Handbook of Phase Diagrams of Silicate Systems,* Vol. I, *Binary Systems* (Translated from Russian), Israel Program for Scientific Translations, Jerusalem, 1972.
13. F. A. Trumbore, Solid Solubility of Impurity Elements in Germanium and Silicon, *Bell Sys. Tech. J.* **39,** 205 (1960).
14. G. L. Vick and K. M. Whittle, Solid Solubility and Diffusion Coefficients of Boron in Silicon, *J. Electrochem. Soc.* **116,** 1142 (1969).
15. F. N. Rhines, *Phase Diagrams in Metallurgy,* McGraw-Hill, New York, 1956.
16. M. B. Panish, The Use of Phase Diagrams for Compound Semiconductors, in *Phase Diagrams,* Vol III, A. M. Alper, Ed., Academic, New York, 1970.
17. H. Matino, Reproducible Sulfur Diffusions in GaAs, *Solid State Electron.* **17,** 35 (1974).
18. M. B. Panish, Ternary Condensed Phase Systems of Gallium and Arsenic with Group 1B Elements, *J. Electrochem. Soc.* **114,** 516 (1967).
19. K. Pande, D. Reep, A. Srivastava, S. Tiwari, J. M. Borrego, and S. K. Ghandhi, Device Quality Polycrystalline Gallium Arsenide on Germanium/Molybdenum Substrates, *J. Electrochem. Soc.* **126,** 300 (1979).
20. H. C. Casey, Jr., and M. B. Panish, Reproducible Diffusion of Zinc into GaAs: Application of the Ternary Phase Diagram and the Diffusion and Solubility Analyses, *Trans. Met. Soc. AIME,* **242,** 406 (1968).

CHAPTER

3

Crystal Growth and Doping

CONTENTS

3.1 STARTING MATERIALS 82
3.2 GROWTH FROM THE MELT 84

3.2.1 The Bridgman Technique, 85
3.2.2 The Czochralski Technique, 86

3.2.2.1 Liquid Encapsulation, 87
3.2.2.2 Direct Compounding, 88
3.2.2.3 Requirements for Proper Crystal Growth, 88

3.3 DOPING IN THE MELT 90

3.3.1 Rapid-Stirring Conditions, 92
3.3.2 Partial-Stirring Conditions, 95
3.3.3 Constitutional Supercooling, 98

3.4 PROPERTIES OF MELT-GROWN CRYSTALS 98
3.5 SOLUTION GROWTH 100
3.6 ZONE PROCESSES 101

3.6.1 Zone Refining, 102
3.6.2 Doping by Zone Leveling, 104
3.6.3 Neutron Doping of Zone-Refined Crystals, 105

3.7 PROPERTIES OF ZONE-PROCESSED MATERIAL 107
REFERENCES 108
PROBLEMS 108

This chapter describes common techniques for growing single-crystalline silicon and gallium arsenide of a quality suitable for device and microcircuit fabrication. Doping of the starting materials in the melt is also considered, followed by a discussion of the properties of material grown by these methods.

3.1 STARTING MATERIALS

Polycrystalline silicon is used as the starting material from which device quality single crystals are grown. This element is commercially synthesized by heating silica and carbon in an electric furnace. The resulting material is only 95–97% pure. Further purification is usually carried out by converting to trichlorosilane and chemically processing this compound until it is ultrapure. Semiconductor grade silicon is made by its thermal decomposition in a hydrogen atmosphere.

Elemental, chemically pure gallium and arsenic are used as the starting materials for the synthesis of polycrystalline gallium arsenide. The formation of this compound, with near perfect stoichiometry, is complicated by the highly different vapor pressures of its individual components, and is now described.

The melting point of gallium arsenide is 1238°C. At this temperature the direct reaction of gallium and arsenic is exothermic, and proceeds with considerable violence. Consequently, this reaction is usually carried out in a slow, controlled manner.* In addition, gallium arsenide must be synthesized in an overpressure of arsenic in order to avoid simultaneous decomposition into its separate constituents. The appropriate arsenic pressure over molten gallium arsenide is seen from Fig. 1.9 to be about 1.0 atm. This pressure can be established by conducting the reaction in a sealed tube maintained slightly above 1238°C, with a precisely measured quantity of arsenic. A better technique, however, is to provide what is essentially an infinite supply of arsenic which is maintained at a lower temperature. For this situation, the pressure in the tube is the equilibrium vapor pressure of the volatile component (arsenic) at the lower temperature. Values for these vapor pressures can be displayed [1] in a pressure–temperature (P–T) diagram, as shown in Fig. 3.1. Here the solid curve shows the pressure of gaseous arsenic over gallium arsenide, which is necessary to prevent decomposition of the compound, as a function of the temperature at which the compound is maintained. The dashed curve in this figure shows the pressure of gaseous arsenic over solid arsenic for this same temperature range as obtained from standard vapor pressure tables. From these curves it is seen that the required arsenic overpressure to avoid decomposition of the gallium

* The direct compounding of gallium arsenide from its elements is quite recent. This technique is described in Section 3.2.2.2.

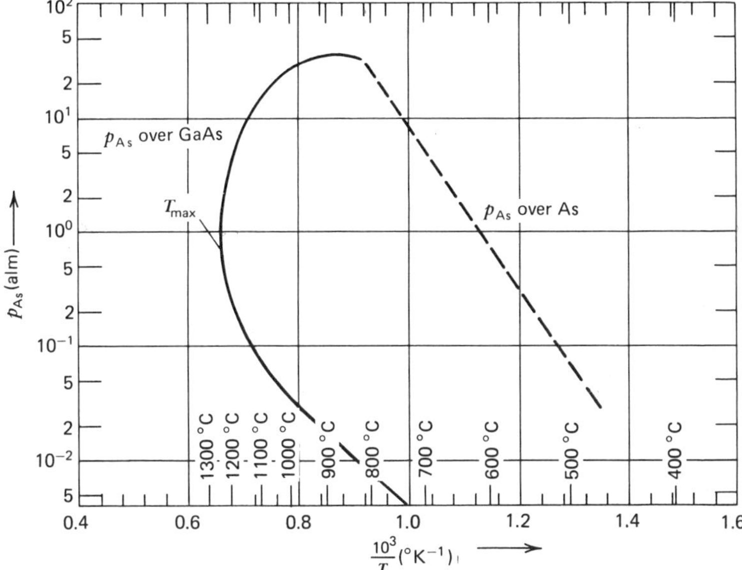

Fig. 3.1 ***P–T*** **diagram for gallium arsenide. Adapted from Boomgaard and Scholl [1].**

arsenide can be achieved by keeping the elemental arsenic at about 600–620°C while the gallium arsenide is maintained at about 1240–1260°C.

Figure 3.2 shows an evacuated sealed-tube configuration in a two-temperature furnace that can be used for this purpose. The sealed tube is made of quartz; the boats are made of quartz, but preferably of graphite which has a closer thermal expansion match to gallium arsenide. Graphite boats are usually coated with silicon carbide, boron nitride, or pyrolytic graphite in order to minimize contamination of the gallium arsenide.

In operation, one boat is loaded with a charge of pure gallium, with the arsenic kept in a separate boat. Next the tube is evacuated, sealed, and brought up to system temperature, as shown in Fig. 3.2. This results in the transport of arsenic vapor to the gallium, with its conversion into gallium arsenide in a slow, controlled manner. Such a process typically takes many hours to accomplish, for a starting charge of 500 g of gallium.

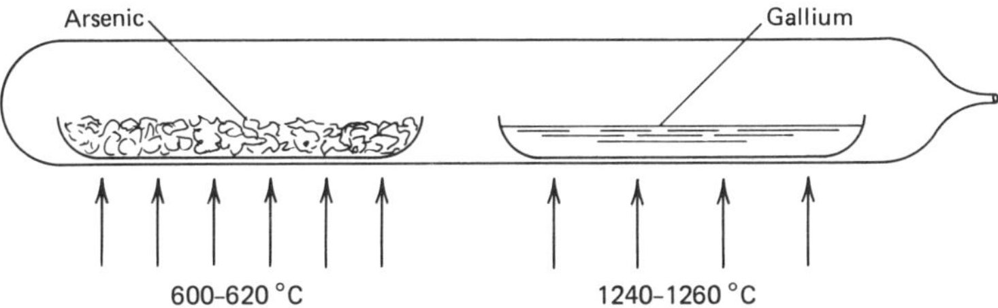

Fig. 3.2 Sealed-tube system for gallium arsenide synthesis.

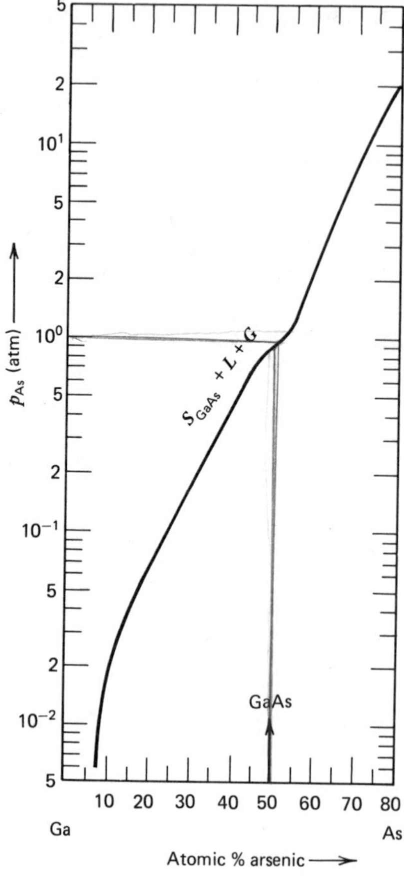

Fig. 3.3 ***P–x*** **diagram for gallium arsenide. Adapated from Boomgaard and Scholl [1].**

A pressure–composition (*P–x*) diagram can also be drawn by combining the *P–T* diagram with the phase diagram (*T–x*) for the gallium–arsenic system (Figs. 2.10 and 2.11). This diagram is shown in Fig. 3.3 and emphasizes the necessity for maintaining extremely close control of the arsenic overpressure in order to avoid a departure from stoichiometry. This is usually done by selecting the precise arsenic temperature by a trial-and-error process.

Gallium arsenide, formed in this manner, is generally polycrystalline in nature. Processes for single-crystal growth use this as the starting material. Elemental silicon is used as the starting material for the growth of single-crystal silicon.

3.2 GROWTH FROM THE MELT

There are two basic techniques for crystal growth from the melt: the Bridgman technique and the Czochralski technique. Silicon is almost universally grown by the latter technique, which is becoming increasingly popular for gallium arsenide. Much gallium arsenide is still grown by the Bridgman method, however. Both of these methods are now described.

3.2.1 The Bridgman Technique [2]

This consists of melting the polycrystalline semiconductor in a long boat. The melt is now cooled from one end, which is usually necked down in order to restrict nucleating during freezing to a (hopefully) single event. This allows a single grain to propagate at the liquid–solid interface. Eventually the melt is frozen to take up this crystalline structure. A crucible that is suitable for crystal growth by this technique often has a seed crystal placed in the narrow end in order to establish a specific crystallographic orientation.

Figure 3.4 shows a horizontal Bridgman system with a traveling heater which allows the growth of gallium arsenide in the requisite overpressure of arsenic. This is sometimes called the gradient-freeze technique [3]. Typical values for temperature gradient are 10°C/cm, with a heater travel of 5 μm/s. The choice of boat material presents somewhat of a problem since no wetting must occur at its walls. Both vitreous carbon and carbon-coated quartz boats are used to minimize this wetting effect. Carbon and graphite boats are distinctly superior to quartz, because of their relatively close thermal expansion match to gallium arsenide.

An alternate approach, which is a modification of the methods shown in Figs. 3.2 and 3.4, combines the synthesis of the compound and the growth of the single crystal in a single operation. This approach is more economical than the separate processes outlined here, and is often used in commercial practice.

Bridgman methods for silicon can be carried out in a furnace with only one temperature zone. They are rarely used, however, because of the great ease by which silicon can be grown by the Czochralski technique.

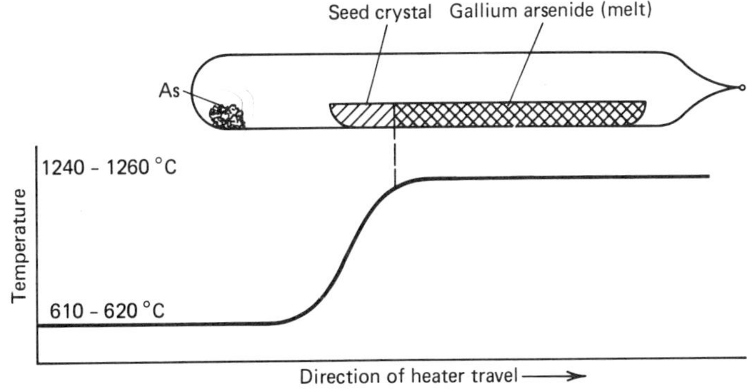

Fig. 3.4 Gradient-freeze system.

3.2.2 The Czochralski Technique

Both silicon and gallium arsenide expand upon freezing, so that boat grown material tends to be highly defected in character. This problem can be almost completely eliminated by use of the Czochralski technique [4] for crystal growth. Here, as shown in Fig. 3.5, the melt is contained in a quartz (for silicon) or graphite (for gallium arsenide) crucible and kept in a molten condition by inductive or resistive heating. A seed crystal, suitably oriented, is suspended over the crucible in a chuck. For growth the seed is inserted into the melt until its end is molten. It is now slowly withdrawn, resulting in a single crystal which grows by progressive freezing at the liquid–solid interface. A pull rate of about 10 μm/s is typical for both silicon and gallium arsenide. Provisions are also made to rotate the crystal, and sometimes the crucible as well, during the pulling operation.

For silicon the entire assembly is enclosed within a fused quartz envelope, which is water cooled and flushed with an inert gas such as helium or argon.

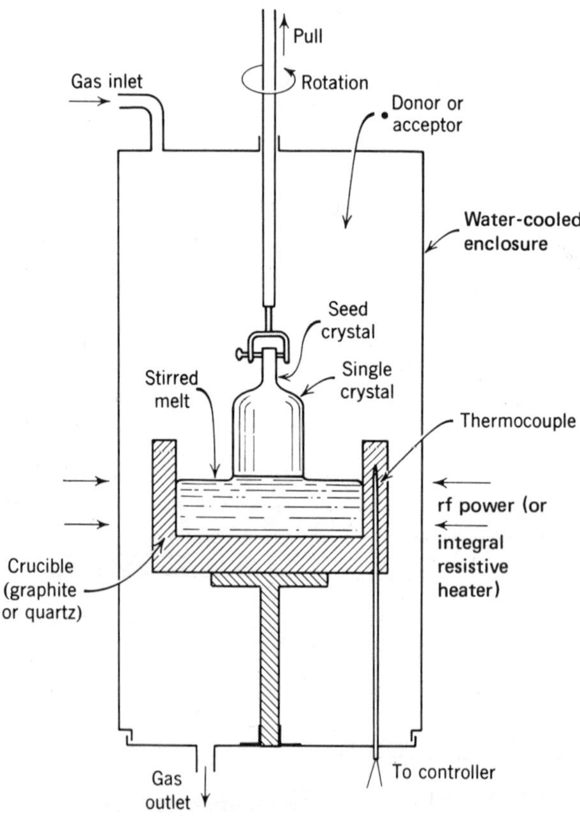

Fig. 3.5 Czochralski crystal growing apparatus.

With gallium arsenide, on the other hand, it is important to prevent decomposition of the melt during crystal growth by maintaining an overpressure of ≃1.0 atm of arsenic. This can be done by making the melt slightly arsenic-rich, and by maintaining the entire chamber at high temperature, to prevent arsenic condensation on its walls. This is extremely difficult to accomplish in a practical system.

Extreme care must be taken to prevent the leakage of oxygen into the system during gallium arsenide growth, if semi-insulating material is to be avoided. With silicon, however, a large amount of oxygen is incorporated by dissolution of the quartz vessel in which this material is usually grown. This oxygen is in the form of clumps which are electronically inactive (see Section 1.4.1.3), and also in interstitial sites which are active.

Thermal conditions are continually changing during crystal growth, so a feedback control system is necessary to maintain the temperature of the melt to within $\pm\frac{1}{2}$°C. If H_i is the rate of heat input to the system and H_o is the rate of heat loss, then the difference $H_i - H_o$ is largely accounted for by the latent heat of crystallization L. Writing ρ as the density and A as the cross section of the grown crystal, the heat-balance equation is

$$H_i - H_o = AL\rho \frac{dx}{dt} \tag{3.1}$$

where dx/dt is the pull rate. From this equation it is seen that the crystal diameter can be controlled by adjusting the thermal input rate and the pull rate.

3.2.2.1 *Liquid Encapsulation*

Czochralski growth of gallium arsenide is seriously hampered by the need for maintaining an arsenic overpressure of about 1.0 atm during growth. This can be done by keeping the entire chamber hot, at about 600–620°C. Unfortunately, arsenic is highly reactive at these temperatures, thus greatly restricting the choice of materials for this purpose. The *liquid encapsulation* technique has been developed [5] to avoid these problems, and consists of using a cap layer of an inert liquid to cover the melt. This cap prevents decomposition of the gallium arsenide as long as the pressure on its surface is higher than 1.0 atm. This pressure can be readily provided by means of an inert gas such as helium or argon.

Requirements on this cap are that it float on the gallium arsenide surface without mixing, and be chemically stable. In addition, it should be optically transparent so that the crystal can be viewed during growth, and impervious to the diffusion of arsenic (which is the volatile component). Boron trioxide is most commonly used in the form of a layer about 0.5 cm thick, although materials such as $BaCl_2$ and $CaCl_2$ have also been used successfully. An additional advantage of using a B_2O_3 cap is that a thin film of this material remains on the gallium arsenide as it comes out of the melt. This prevents loss of arsenic from the solidified, but still hot, crystal.

The liquid encapsulation technique is found to result in almost no boron or oxygen contamination of the gallium arsenide. Typically, the incorporated boron content is in the 0.01–0.1 ppm range. Furthermore it has also been noted that impurities initially present in the B_2O_3 layer remain preferentially in this layer, rather than becoming incorporated into the gallium arsenide. However, quartz crucibles cannot be used in this method, because the B_2O_3 cap dissolves SiO_2, which is then incorporated as silicon into the melt. As much as 0.1% silicon can be dissolved in this manner. Both graphite and alumina, however, do not suffer from this problem and are sometimes used for crucible materials.

3.2.2.2 Direct Compounding

Many of the contaminants in gallium arsenide are incorporated during the compound formation operation described in Section 3.1. Thus the avoidance of this step, by in-situ compounding in the Czochralski system, is highly desirable.

The vapor pressure of arsenic (over arsenic) at the melting point of gallium arsenide is about 60 atm. Until recently this has prevented direct compounding. However, the availability of crystal pullers, which can operate at as high as 150 atm, in combination with the liquid encapsulation technique, makes this process feasible.

Here [6], stoichiometric quantities of high-purity gallium and arsenic (99.9999% purity) are placed in a pyrolytic boron nitride crucible, which serves as a liner for a heated graphite container. Vacuum baked B_2O_3 is placed on its surface and the system pumped down to remove all residual water vapor and volatile oxides. The system is now pressurized to 3 atm of nitrogen and heated to 450°C, at which point the B_2O_3 melts. Next the pressure is raised to about 60 atm and the crucible heated slowly to the melting point of gallium arsenide. Rapid reaction of the gallium and arsenic takes place around 700°C during this process.

Eventually the pressure is reduced to 1 atm. A single crystal is grown by inserting a seed crystal in the melt, and pulling in the manner described earlier. High-purity material, of 10^8–10^9 Ω cm resistivity, can be grown by this approach [7].

3.2.2.3 Requirements for Proper Crystal Growth

An important advantage of the Czochralski process over the Bridgman technique is that the crystal is not in contact with the crucible at its liquid–solid interface. It is therefore possible to grow crystals with a high degree of perfection because of the absence of restraining forces during freezing. Nevertheless attention must be paid [8] to a number of factors during the crystal growth. Thus crystal growth proceeds by the successive addition of layers of atomic planes at the liquid–solid interface. Since the {111} planes are the most closely packed, growth is easiest in the ⟨111⟩ directions for both silicon and gallium arsenide. It is for this reason that {111} material is commonly used in bipolar silicon device and integrated circuit technology.

Crystal growth in the ⟨100⟩ direction is highly desirable for gallium arsenide, which is difficult to cleave along {111} planes (see Section 1.3.2). In contrast, {100} gallium arsenide cleaves readily along the {110} planes. Cleavage along some of these planes results in rectangular dice with parallel faces.

The {100} planes are also used for silicon-based low-threshold MOS microcircuits and in low-noise operational amplifiers. Considerable care must be taken during growth in this orientation, in order to avoid any accidental thermal or mechanical shock, which can result in the crystal changing its orientation by twinning, with subsequent growth in a different direction [9].

Based on geometric and energy considerations, twinning will usually occur with a common {111} plane. If crystal growth is in the ⟨111⟩ directions, the twin plane will be at an angle of 70.53° to the growth plane, so it will work its way out of the ingot as growth proceeds. This new growth can be shown to be along one of the ⟨511⟩ directions. In practice, a (111) crystal which has twinned will almost invariably twin again and proceed as a $(\bar{1}\bar{1}\bar{1})$. Thus growth in the ⟨111⟩ direction tends to be self-correcting.

A crystal growing in the ⟨100⟩ direction can also twin along {111} planes, which are at 54.74° to the growth plane, with the twinned section growth in one of the ⟨221⟩ directions. Although the twin plane will again work itself out of the crystal, further growth will usually continue along the ⟨221⟩ directions, provided no further twinning occurs. With gallium arsenide such ingots can eventually be cut in the ⟨100⟩ direction, resulting in noncircular slices. The state of the silicon technology is sufficiently advanced so that this material is rejected.

Crystals growing in the ⟨110⟩ directions can also twin along {111} planes, with further growth in some of the ⟨411⟩ or ⟨110⟩ directions. Some of these {111} twin planes are at right angles to the growth plane, and so are propagated throughout the entire length of the crystal. Consequently, it is extremely difficult to grow {110} material that is free from twinning.

The liquid–solid interface must be maintained as flat as possible, and at right angles to the direction of crystal growth. The importance of accurate orientation of the seed crystal must be emphasized at this point. In general, the interface may be flat, or take the form of a meniscus which may either be concave up or concave down. Appropriate selection of the pull and spin rates results in a nearly flat interface. This is highly desirable in order to minimize radial forces during freezing, since they result in a radial distribution of defects in the final slice. The flatness of the interface is enhanced by increasing the spin rate. Unfortunately, this also increases the corrosive effect of the melt on the crucible walls.

Silicon crystals are often grown with a meniscus that is slightly Ω-shaped. This has been experimentally found to result in a greatly increased allowable pull rate while still giving crystals with acceptable quality. With gallium arsenide, however, a curved interface results in large radial temperature differences because of its relatively low thermal conductivity. This can set up a condition where stresses in the crystal can exceed the elastic limit, resulting in a large concentration of dislocations and other crystal imperfections.

It has been noted in Chapter 1 that the energy of formation of a dislocation is on the order of 10–20 eV, while its energy of propagation is lower by a factor of 100. Consequently, every effort is made to begin with a very small, accurately aligned, dislocation-free crystal as the seed. In addition, it has been found that, with a rapid growth rate, it is possible to make dislocations propagate preferentially out of the crystal. Hence it is common practice to start a crystal growth with a rapid pull rate and later to slow it down to the actual rate at which the majority of the crystal is grown.

3.3 DOPING IN THE MELT

This is usually done by adding a known mass of dopant to the melt in order to obtain the desired composition. Raw dopants are not used since the amounts to be added are unmanageably small, except for heavily doped material. Moreover, their physical characteristics are often quite different from those of the melt (e.g., elemental phosphorus cannot be added to molten silicon without disastrous results). It is common practice to add the dopant in the form of a highly doped powder of about 0.01 Ω cm resistivity. In this manner both problems are avoided.

Boron and phosphorus are the most common dopants for *p*- and *n*-type silicon, respectively. The variety of selections for gallium arsenide is much larger. Thus zinc and cadmium are commonly used for *p*-type gallium arsenide; tellurium, selenium, and silicon are used for *n*-type material; chromium is the invariable choice for semi-insulating material.

The addition of impurities to the melt, accompanied by stirring, results in a doped liquid from which the crystal is grown. In general, the concentration of the solute will be quite different in the solid and liquid phases of a crystal because of energy considerations. Consider a crystal at any given point during its growth. Writing C_S and C_L as the concentration of solute (by weight) in the solid and the liquid phases, respectively, in the immediate vicinity of the interface, a distribution coefficient k may be defined, where

$$k = \frac{C_S}{C_L} \tag{3.2}$$

Table 3.1 gives values [10] of k for the commonly used dopants for silicon. To an approximation, these values are independent of concentration. It is

Table 3.1 Equilibrium Distribution Coefficients for Dopants in Silicon

Dopant	P	As	Sb	B	Al	Ga	In	Au
Distribution coefficient k	0.32	0.27	0.02	0.72	1.8×10^{-3}	7.2×10^{-3}	3.6×10^{-4}	2.25×10^{-5}
Atomic weight	30.97	74.92	121.75	10.81	26.98	69.72	114.82	196.97
Type of Dopant	n-type			p-type				Deep lying

seen that k is normally less than unity and ranges from as high as 0.72 for boron to as low as 2.25×10^{-5} for gold. Values of distribution coefficients for gallium arsenide [11] are given in Table 3.2. Again, most of these are seen to be less than unity.

Since k is less than unity, excess solute is thrown off at the interface between the melt and the crystal. Consequently, the melt becomes increasingly solute-rich as crystal growth progresses, resulting in a crystal of varying composition. The precise nature of this composition may be determined for different growing conditions.

3.3.1 Rapid-Stirring Conditions [12]

Assume that rapid stirring is involved during very slow growth of the crystal. Then the solute in the immediate vicinity of the freezing interface is immediately removed into the melt. In addition, it is assumed that there is no diffusion* of the solute within the crystal during the growth process. Let:

W_M = initial weight of melt
C_M = initial concentration of solute in melt (by weight)

At a specified point in the growth process, when a crystal of weight W has been grown, let:

C_L = concentration of solute in liquid (by weight)
C_S = concentration of solute in crystal (by weight)
S = weight of solute in melt

Consider an element of the crystal of weight dW. During its freezing, the weight of the solute lost from the melt is $C_S dW$. Thus

$$-dS = C_S dW \tag{3.3}$$

At this point the weight of the melt is $W_M - W$, and

$$C_L = \frac{S}{W_M - W} \tag{3.4}$$

Combining these equations and substituting $C_S/C_L = k$,

$$\frac{dS}{S} = -K \frac{dW}{W_M - W} \tag{3.5}$$

* This is reasonable, since the diffusion coefficient of impurities is about six orders of magnitude larger in the melt than in the solid crystal.

Table 3.2 Equilibrium Distribution Coefficients for Dopants in Gallium Arsenide

Dopant	S	Se	Te	Sn	C	Ge	Si	Be	Mg	Zn	Cr	Fe
Distribution coefficient k	0.3–0.5	0.1–0.3	0.059	0.08	0.2–0.8	0.01	0.14–2.0	3	0.1	0.4–1.9	5.7×10^{-4}	1×10^{-3}
Atomic weight	32.06	78.96	127.60	118.69	12.01	72.59	28.09	9.01	24.31	65.37	52	55.85
Type of dopant	*n*-type				*n*/*p*-type			*p*-type			Semi-insulating	

The initial weight of the solute is $C_M W_M$. Consequently Eq. (3.5) may be integrated as

$$\int_{C_M W_M}^{S} \frac{dS}{S} = k \int_0^W \frac{dW}{W_M - W} \tag{3.6}$$

Solving and combining with Eq. (3.4), gives

$$C_S = kC_M \left(1 - \frac{W}{W_M}\right)^{k-1} \tag{3.7}$$

Figure 3.6 illustrates the crystal composition, described by this expression, for a range of values of k. It is seen that, as crystal growth progresses, the composition continually increases from an initial value of kC_M. In addition, high values of k are seen to result in considerably more uniform crystal composition during growth than are low values.

The process of freezing from a doped melt may also be described in terms of a phase diagram. By way of example, Fig. 3.7 shows a greatly expanded version of the solid-solution region for a semiconductor B which is doped with A. Starting with a melt of composition C_M (percentage weight of the solute in the melt), the crystal begins to freeze out initially at a composition kC_M, where k is the ratio of the slopes of the liquidus and solidus curves (assumed straight lines). With further cooling, the crystal composition moves along the solidus line until the composition of the solid solution is C_M. During this freezing, the melt composition is initially C_M and moves along the liquidus curve until its composition becomes C_M/k. It should be noted

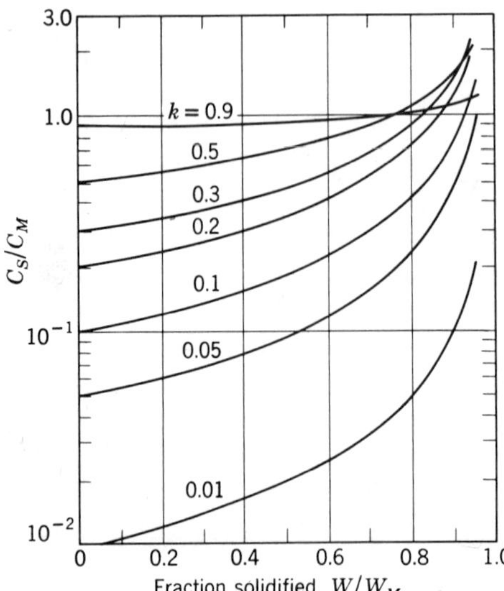

Fig. 3.6 Crystal composition during growth.

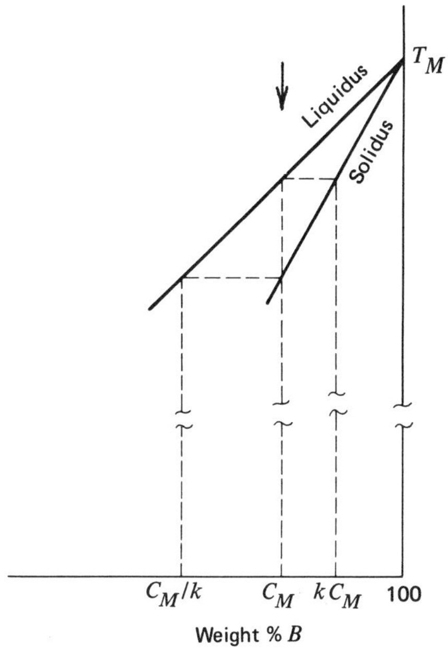

Fig. 3.7 Phase diagram for crystal doping.

that k is the distribution coefficient described in Section 3.3 and is thus given by the ratio of the slopes of the liquidus and solidus curves in the vicinity of the melt composition.

The validity of Eq. (3.7) breaks down as W/W_M approaches unity. In this region, the melt becomes excessively rich in the solute, and the crystal composition is determined by its solid-solubility characteristic. Eventually the solid growth becomes polycrystalline.

3.3.2 Partial-Stirring Conditions

For realistic growth and stirring conditions, the rejection rate of the solute atoms at the interface is higher than the rate at which they can be transported into the melt. Consequently, the solute concentration at the interface builds up in excess of the concentration in the melt. This results in a crystal with a doping concentration in excess of that obtained for the case of full stirring.

The crystal composition may be determined by postulating the presence of a thin stagnant layer of liquid immediately adjacent to the liquid–solid interface, through which solute atoms flow by diffusion alone. Equilibrium conditions prevail beyond this layer. Figure 3.8 shows the liquid–solid interface and the solute concentration beyond this interface. The stagnant layer thickness is denoted by δ, and the distribution coefficient k is given by C_S/C_L'. One consequence of this stagnant layer is that the concentration of the melt at the interface exceeds the equilibrium concentration.

An effective distribution coefficient may now be defined for partial-stir-

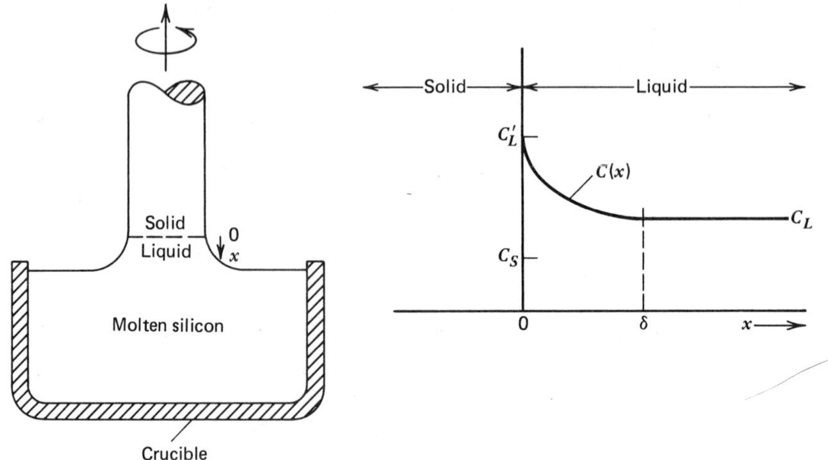

Fig. 3.8 Partial-stirring conditions.

ring conditions, such that $k_e = C_S/C_L$. This coefficient is larger than k (for $k < 1$) and depends on the growth parameters of the crystal. Let:

R = growth rate for crystal, i.e., rate of movement of liquid–solid interface

D = diffusion constant for solute atoms in liquid ($\simeq 5 \times 10^{-5}$ cm²/s for most situations)

The equation governing the diffusion of solute atoms in the layer may now be written. Noting that the amount of solute rejected from the solid is equal to that gained by the liquid, the stationary distribution is given by

$$D \frac{d^2C}{dx^2} + R \frac{dC}{dx} = 0 \tag{3.8}$$

so that

$$C = Ae^{-Rx/D} + B \tag{3.9}$$

and

$$\frac{dC}{dX} = -\frac{AR}{D} e^{-Rx/D} \tag{3.10}$$

where A and B are constants of integration.

The first boundary condition is that $C = C_L'$, at $x = 0$. The second boundary condition is obtained by noting that the sum of the impurity fluxes at a boundary must be zero. Again assuming that diffusion of solute atoms in the solid is negligible compared with diffusion in the liquid, this condition may be written as

$$D\left(\frac{dC}{dx}\right)_{x=0} + (C_L' - C_S)R = 0 \tag{3.11}$$

so that

$$\left(\frac{dC}{dx}\right)_{x=0} = -\frac{R}{D}(C_L' - C_S) \tag{3.12}$$

Substituting these boundary values into Eq. (3.9) and noting that $C = C_L$ at $x = \delta$, gives

$$\frac{C_L - C_S}{C_L' - C_S} = e^{-R\delta/D} \tag{3.13}$$

Substituting for k and k_e and rearranging terms, gives

$$k_e = \frac{k}{k + (1 - k)e^{-R\delta/D}} \tag{3.14}$$

Finally, the crystal composition for partial stirring may be derived from the results for complete stirring by substituting k_e for k in Eq. (3.7), so that

$$C_S = k_e C_M \left(1 - \frac{W}{W_M}\right)^{k_e - 1} \tag{3.15}$$

Values of k_e are higher than k and approach unity for large values of the normalized growth parameter $R\delta/D$. Uniform crystal composition is thus approximated with a high pull rate and a low spin rate (since δ is inversely related to spin rate). In practice, these growth parameters are set by considerations outlined in Section 3.2.2.3. Thus the pull rate is optimized to grow crystals of the desired diameter with low dislocation concentration. The spin rate is normally kept low to prevent corrosion of the crucible walls by the melt, with its subsequent contamination.

It is common commercial practice to program* the growth parameters so as to obtain uniform composition over a large fraction of the crystal growth. This is particularly true for silicon which is used in MOS technology, because of the relatively tight requirements that are placed on its starting resistivity.

* See Problem 4 at the end of this chapter.

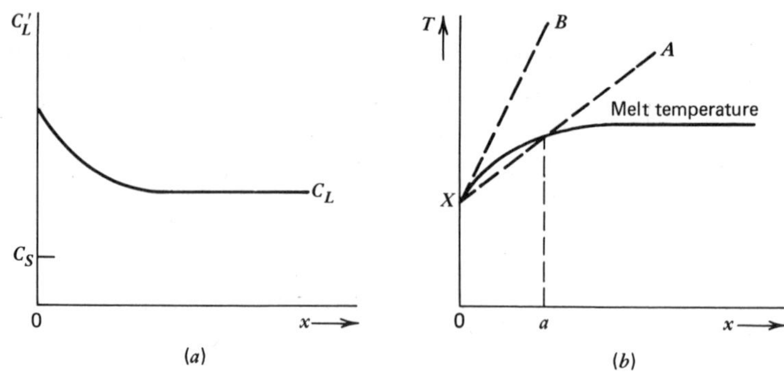

Fig. 3.9 Constitutional supercooling.

3.3.3 Constitutional Supercooling

The presence of dopants in a melt greatly reduces the rate at which a crystal can be grown without the loss of crystallinity. Consider, again, the process of crystal growth in a partial-stirring situation where $k_e < 1$. Here the growing crystal results in a pileup of impurity at the liquid–solid interface. The amount of this pileup is a function of the melt volume and the degree of stirring. In addition, it increases with higher growth rate. Too great a pileup can lead to supersaturation in this region, which can cause spurious nucleation and polycrystalline growth. Such a condition is known as *constitutional supercooling*.

The growing conditions under which constitutional supercooling can occur, or can be avoided, are illustrated in Fig. 3.9. Here Fig. 3.9*b* is a plot of the melting temperatures of the compositions shown in Fig. 3.9*a*, on the assumption that an increase in doping concentration results in a linear fall in the melting point (see Fig. 3.7). Thus the curved regions of both figures are exponential in character, for the steady-state growth condition shown.

The actual temperature in the liquid is usually a linear function of distance* and is shown for one specific growth rate by the dashed line *XA*. Here supersaturation of the melt occurs for a distance *a* beyond the interface, with the resulting polycrystalline growth. The line *XB* shows the temperature in the liquid for a second, slower growth rate. Here no region of supercooling is present, and single-crystalline growth is the result.

From the above it is seen that the growth rate of a doped crystal is considerably less than if the crystal were undoped. Moreover, this difference is even larger for dopants (and stirring conditions) with lower values of k_e.

3.4 PROPERTIES OF MELT-GROWN CRYSTALS

All single-crystal silicon material that is used in the fabrication of discrete devices and microcircuits is grown by the Czochralski technique. Gallium

* This assumes that heat transfer from the melt to the (cooler) crystal is by conduction.

arsenide, on the other hand, has conventionally been grown by the Bridgman method. Consequently, much of this material is available. However, Czochralski systems, in combination with liquid encapsulation techniques, are rapidly growing in popularity in this area as well. For both approaches, growth is a trade-off between a number of interrelated (and sometimes conflicting) considerations. It is not surprising, therefore, that extreme variations in gallium arsenide crystal quality are the rule rather than the exception.

The state of silicon technology is highly advanced while that of gallium arsenide is primitive in comparison. At the present time silicon ingots of 20 kg weight and 100–125 mm diameter are routinely grown, whereas the comparable numbers for gallium arsenide are 500 g and 25–75 mm diameter. In addition, most achievable specifications for silicon are considerably better (and tighter) than those for gallium arsenide.

Slices cut from single-crystal material are usually evaluated on the basis of the following properties:

Dislocation Content. It is possible to obtain Czochralski-grown silicon that is entirely free from dislocations. In practice the dislocation content of acceptable material ranges from zero to as high as 10^3 dislocations/cm^2. In addition, crystals which freeze along a curved interface show corresponding radial patterns of dislocation concentration. The dislocation content of Czochralski-grown gallium arsenide crystals is generally 10^4–10^5 cm^{-2}. Values for Bridgman material are a factor of 2–10 lower.

Resistivity. Values as high as 100 Ω cm can be achieved with silicon. Here the upper limit is set by residual contamination from the high-purity quartz crucible in which the material is grown. Typically, starting silicon for bipolar microcircuits is specified at 3–10 Ω cm *p*-type (with many manufacturers setting no upper limit), whereas 5–10 Ω cm *n*- or *p*-type is used in MOS circuits. The starting material for discrete devices is typically 0.01–0.001 Ω cm of both *p*- and *n*-types.

Commercially available melt-grown gallium arsenide is heavily contaminated by the crucible since carbon and silicon are active impurities. Fortunately, most requirements call for material that is heavily doped. Tellurium, selenium, and silicon are common *n*-type dopants, in the 10^{17}–10^{18} cm^3 range, whereas cadmium and zinc are common *p*-type dopants in the 10^{16}–10^{18} cm^{-3} range.

Chromium-doped gallium arsenide is unique in that it is semi-insulating and forms the basis for most gallium arsenide integrated circuits, as well as for a number of discrete devices such as field effect transistors. Typically doped to 10^{17} chromium atoms/cm^3, this material has a starting resistivity of 10^8–19^9 Ω cm. Frequently, this resistivity drops to as low as 10^3–10^4 Ω cm during processing [13]. This behavior has been associated with the presence of contaminants in the starting material, and is not fully understood at the present time.

Melt-grown gallium arsenide, made by in-situ compounding, can achieve a starting resistivity of 10^8 Ω cm when undoped, and 10^9 Ω cm when chromium doped. In addition, these materials have been shown to be thermally stable. Consequently, it is highly probable that all gallium arsenide (undoped semi-insulating, chromium-doped semi-insulating, and also *p*- and *n*-doped material) will eventually be made by this technique.

Radial Resistivity. Radial variations in resistivity may run as high as ± 30–50% on high-resistivity silicon slices. For MOS and bipolar silicon microcircuits, a radial resistivity variation of ± 5–10% is the maximum that can be tolerated. Comparable values for gallium arsenide have not been established. However, this parameter is generally not a problem, since the starting material is only used as a substrate for the active layer which is epitaxially grown or ion-implanted.

Oxygen Content. The presence of oxygen in silicon is due to the corrosive action of the melt on the walls of the silicon crucible. Oxygen concentrations of 10^{16}–10^{18} atoms/cm^3 are obtained, the higher values corresponding to the higher spin rates. Most of the oxygen is present in the form of clumps, while some is in electronically active, interstitial sites (see Section 1.4.1.3). Oxygen tends to pin the extremities of dislocations, which can now readily multiply by the mechanism described in Section 1.4.2.4. Thus a high oxygen content is usually accompanied by a large dislocation concentration.

The behavior of oxygen in gallium arsenide has been described in Section 1.4.1.4. Its presence is undesireable; fortunately it is avoidable since gallium arsenide reacts only slightly with the walls of the silica container. In addition, the use of suitably coated graphite crucibles can almost completely eliminate this problem.

3.5 SOLUTION GROWTH

The basic techniques of crystal growth by freezing from a melt have the disadvantage that this melt must be cooled from a relatively high temperature (1412°C for silicon and 1238°C for gallium arsenide). An alternate approach, which is an extension of this melt-growth technique, is to grow the crystal from a solution of the desired material in a solvent. The underlying principle can be described by means of the simple phase diagram of Fig. 3.10 for semiconductor B (silicon or gallium arsenide) and its solvent A. Let T_A and T_B be the melting points of A and B, respectively. Then the semiconductor B can be grown (in β form) by freezing a solution of composition C_M, which is liquid at a temperature that is much lower than T_B. Tin is a convenient solvent for silicon since it is inert in this semiconductor. Gallium is suitable for gallium arsenide since it is one of its constituent elements.

Single crystals of B can be grown in this manner so that it is, in theory, a low-temperature alternative to melt growth. However, growth rates for

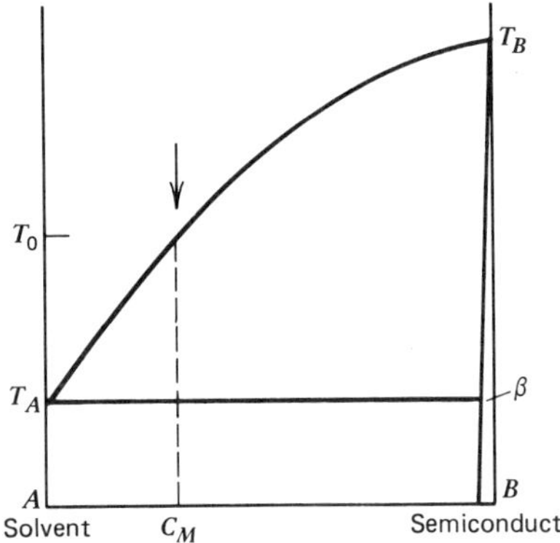

Fig. 3.10 Phase diagram for solution growth.

this technique must be kept extremely small in order to avoid constitutional supercooling, since the heat of solution is usually much larger than the heat of fusion. Consequently, it is restricted to the growth of thin epitaxial films where the low growth rate becomes an advantage, since it allows control of the layer thickness. This technique, referred to as *liquid-phase epitaxy,* is considered in some detail in Chapter 5.

3.6 ZONE PROCESSES [14]

Zone processes can be used to grow material with higher purity than is normally obtained by Czochralski or Bridgman techniques. They require a rod of cast starting material and hence are more expensive than melt-growth techniques. Their use is thus restricted to situations where this increased cost is justified.

In silicon technology, a rod of this semiconductor is held in a vertical position, and rotated during the operation. A small zone (typically 1.5 cm long) of the crystal is kept molten by means of a radio-frequency heater which is moved so that this floating zone traverses the length of the bar. A seed crystal is provided at the starting point where the molten zone is initiated, and arranged so that its end is just molten. As the zone traverses the bar, single-crystal silicon freezes at its retreating end and grows as an extension of the seed crystal. Figure 3.11 shows a schematic of the necessary equipment. As with the Czochralski process, the bar is enclosed in a cooled silica envelope in which an inert atmosphere is maintained.

Zone processes have also been used for gallium arsenide growth. This technique, combined with liquid encapsulation, is potentially capable of producing large diameter ingots, but has met with only partial success [15] at the present time.

The primary advantage of zone-processed silicon over Czochralski material is that its oxygen content is less by a factor of 10–100, and it is thus

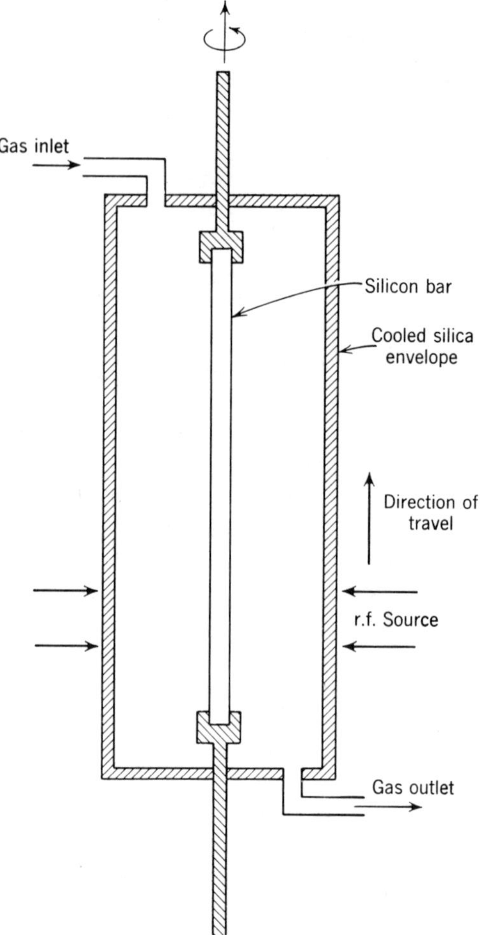

Fig. 3.11 Float zone apparatus.

relatively free of oxygen clumps. In addition, the absence of a crucible allows the growth of material with less dissolved impurities in it. It is used for semiconductor devices which must support reverse voltages in excess of 750–1000 V [16] for these reasons. Recently it has received increased attention for use in very-large-scale integration (VLSI) schemes where the requirement for clump-free material is considerably more stringent than can be met by the Czochralski method.

Zone processes are primarily used for refining silicon, which is subsequently doped. These processes are now described.

3.6.1 Zone Refining

In this process the charge is usually in the form of a cast silicon rod, of doping concentration C_M (by weight), with a seed crystal juxtaposed at one end. The molten zone is initiated at this end, and passes along the bar. Referring to Fig. 3.12, let:

L = length of molten zone, at a distance x along bar
S = weight of solute present in molten zone at any given time
A = cross section area of bar
ρ = specific gravity

As the zone advances by dx, the amount of solute added to it at its advancing end is $C_M A\rho\, dx$. The amount of solute removed from it at the retreating end is $(k_e S/L)\, dx$, where k_e is the effective distribution coefficient. Thus

$$dS = C_M A\rho\, dx - \frac{k_e S}{L}\, dx \tag{3.16}$$

This equation is subject to the boundary value

$$S = C_M A\rho L \qquad \text{at} \qquad x = 0 \tag{3.17}$$

Solving, gives

$$S = \frac{C_M A\rho L}{k_e}[1 - (1 - k_e)e^{-k_e x/L}] \tag{3.18}$$

But C_S, the concentration of the solute in the crystal at the retreating end, is given by

$$C_S = \frac{k_e S}{A\rho L} \tag{3.19}$$

so that

$$C_S = C_M[1 - (1 - k_e)e^{-k_e x/L}] \tag{3.20}$$

Figure 3.13 shows the doping concentration as a function of distance, for

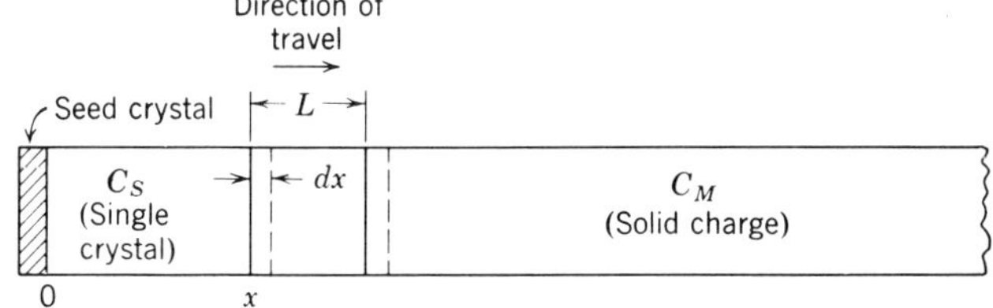

Fig. 3.12 Model for zone refining.

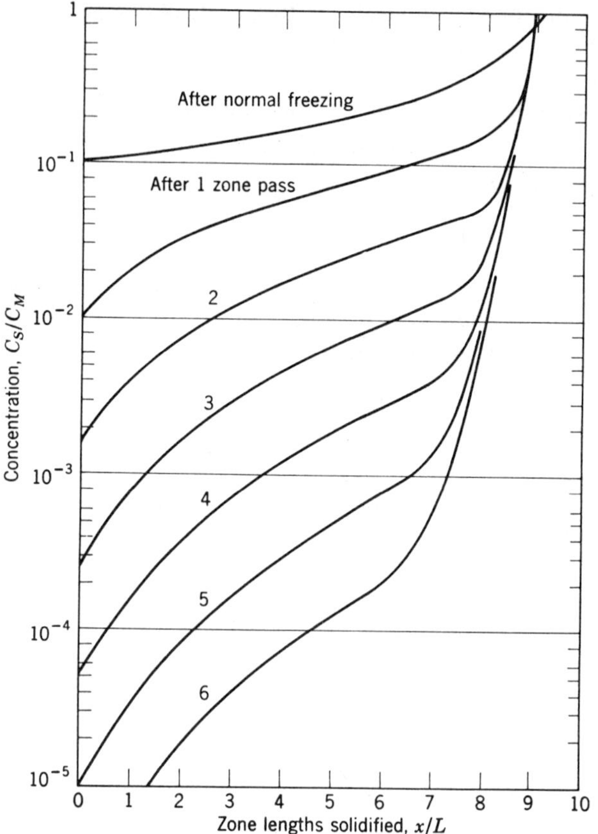

Fig. 3.13 Doping profiles with multiple passes and k_e = 0.1.

a relatively high value of $k_e(= 0.1)$, and for a single pass of the zone along the length of the bar. Note that this process is ideally suited to crystal refinement. This is seen by the relative compositions obtained in a series of successive passes of the zone, also shown in this figure. Often zone refining is speeded up by the use of more than one molten zone.

3.6.2 Doping by Zone Leveling

Zone leveling is a technique by which a zone-refined bar can be doped. Consider a bar of undoped material with a molten zone of length L. Initially, let this zone contain a charge of doping concentration C_I. A seed crystal is also juxtaposed, as illustrated in Fig. 3.14. Using the same notation as in Section 3.6.1, we may rewrite Eq. (3.16) as

$$dS = \frac{k_e S}{L}\,dx \tag{3.21}$$

since $C_M = 0$. At $x = 0$,

$$S = C_I A \rho L \tag{3.22}$$

Solving Eq. (3.21) and inserting this boundary value, gives

$$S = C_I A L \rho e^{-k_e x/L} \tag{3.23}$$

But

$$C_S = \frac{k_e S}{A \rho L} \tag{3.24}$$

Hence

$$C_S = k_e C_I e^{-k_e x/L} \tag{3.25}$$

From this equation it is seen that, for low values of k_e, very uniformly doped crystals may be obtained in a single pass. In addition, even better uniformity can be achieved if the zone length is programmed to be directly proportional to x.

The technique of zone reversal may also be used and provides a significant improvement in the uniformity of the composition. In this method the direction of the zone (or zones) is reversed after it has traversed the bar. Thus deviations in composition resulting from the first pass are compensated by the reverse pass. This results in very uniform compositions, except for the two end zones where all the remaining impurities eventually freeze out.

3.6.3 Neutron Doping of Zone-Refined Crystals

A major problem with float zone material is the presence of radial microresistivity variations. These take the form of large, abrupt resistivity changes, which occur over intervals 50–500 μm wide and can be detected only by high-resolution, spreading resistance probe techniques.

Microresistivity variations are generally considered to be caused by a misalignment of the rotational axis of the crystal and the thermal axis of symmetry of the crystal growth apparatus. They can be reduced, but not eliminated, by careful mechanical design of the system.

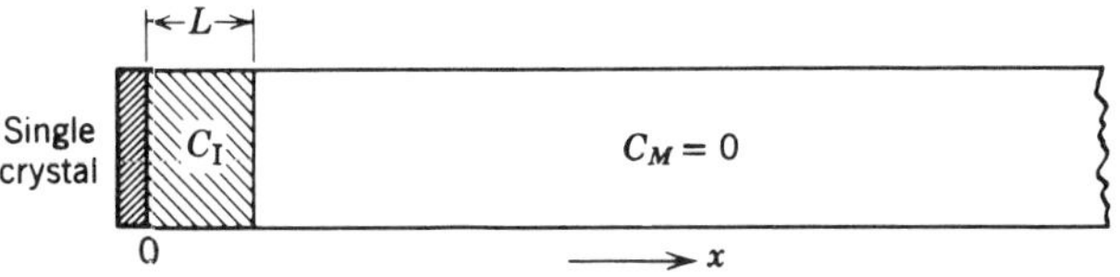

Fig. 3.14 Model for zone doping.

Microresistivity variations can be virtually eliminated by a crystal preparation technique known as *neutron transmutation doping* [17]. This is based on the fact that bombardment of silicon with thermal neutrons gives rise to the formation of finite amounts of phosphorus, which is an n-type dopant. Specifically, the ^{30}Si isotope form occurs as a component of native silicon, with a concentration of about 3%. Neutron bombardment causes this ^{30}Si to change to ^{31}Si, liberating gamma rays. Simultaneously the ^{31}Si undergoes a transmutation to phosphorus (^{31}P) with the liberation of β rays, having a half-life of 2.62 h.

A secondary process results in the conversion of ^{31}P to ^{32}P, which transmutes to ^{32}S. This reaction has a half-life of 14.3 days and is present in significant quantities only for heavy doping. Typically, neutron doping down to 5–10 Ω cm can be accomplished with a cool-down period of 3–4 days, and be completely safe by international standards.

Since the penetration range of neutrons in silicon is roughly 90–100 cm, doping is very uniform throughout the slice. Thus starting with slices of float zone silicon having an average resistivity well in excess of what is required, it is possible to obtain uniformly doped slices, free from both macro- and microresistivity variations. Figure 3.15 shows spreading resistance R_S data

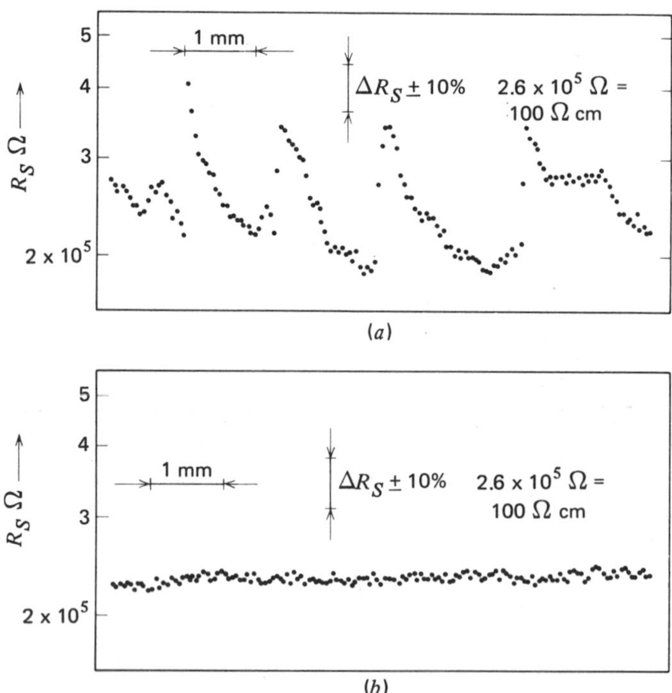

Fig. 3.15 Microresistivity variations (*a*) Conventional growth. (*b*) Neutron doped. From Janus and Malmros [18]. © 1976 by the Institute of Electrical and Electronic Engineers, Inc.

for a typical* 100 Ω cm (nominal) slice of melt-doped silicon, as well as for a slice prepared by the neutron doping technique. Typically, radial resistivity variations can be reduced to below ±1% by this technique.

Annealing for as little as 2 min at 750°C has been found [18] to be sufficient to stabilize the neutron-doped silicon against resistivity changes. Radiation damage of crystals does not present a problem, since normal crystal processing during device fabrication suffices to anneal out all defects created in this manner.

Neutron transmutation doping is used extensively in high-voltage semiconductor devices at the present time. However, this technique has also been used to control the resistivity of epitaxial layers that are required for bipolar VLSI circuits.

3.7 PROPERTIES OF ZONE-PROCESSED MATERIAL

Because of its two-step nature (i.e., from melt to cast rod to single crystal), the zone process is more expensive than the Czochralski process. Only silicon is available in zone-processed form at the present time.

Dislocation Content. Typically, values range from 10^3 to 10^4 dislocations/cm^2. This is because the region near the retreating solid–liquid interface (where the crystal is grown) is highly stressed owing to the weight of the molten zone. New techniques have been developed permitting a reduction in the dislocation content by a factor of 10–100.

Resistivity. The composition of zone-refined silicon can be better controlled than that of the Czochralski type, since no outside matter is added during growth and the chamber can be made quite free from contamination. In fact, it is possible to conduct the process in a vacuum and purify the crystal of already present volatile impurities such as zinc. Finally, repeated processing is possible by successive passes of the molten zone. As a consequence, silicon with resistivities in excess of 4000 Ω cm for n-type material (and 12,000 Ω cm for p-type material) can be obtained by this process.

Radial Resistivity. Variations of this parameter are comparable to those obtained with Czochralski crystals in the same resistivity range. More serious, however, is the presence of microresistivity variations. The use of neutron doping can reduce variations in radial resistivity to under ±1% for 5–10 Ω cm n-type silicon, which is useful as a starting material for MOS microcircuits.

* The data given here are for material that was intended for high-voltage devices. Further neutron doping, to the 5–10 Ω cm level, would further reduce the radial resistivity variations across the slice.

Oxygen Content. Float zone silicon can be made with one to two decades less oxygen than Czochralski material. Although primarily used for high-voltage devices, this material, in combination with neutron doping, is an increasingly attractive candidate for MOS-based VLSI circuits.

REFERENCES

1. J. van den Boomgaard and K. Scholl, The P-T-x Phase Diagrams of the Systems In-As, Ga-As, and In-P, *Philips Res. Rep.* **12,** 127–140 (1957).
2. C. H. L. Goodman, Ed., *Crystal Growth: Theory and Techniques,* Vol. 1, Plenum, New York, 1974.
3. N. B. Hannay, Ed., *Semiconductors,* Reinhold, New York, 1959.
4. G. K. Teal and J. B. Little, Growth of Germanium Single Crystals, *Phys. Rev.* **78,** 647 (1950).
5. J. B. Mullin, B. W. Straughan, and W. S. Brickell, Liquid Encapsulation Techniques: The Use of an Inert Liquid in Suppressing Dissociation During the Melt Growth of InAs and GaAs Crystals, *J. Phys. Chem. Solids* **26,** 782–784 (1965).
6. T. R. AuCoin, R. L. Ross, M. J. Wade, and R. O. Savage, Liquid Encapsulated Compounding and Czochralski Growth of Semi-Insulating Gallium Arsenide, *Solid State Tech.,* p. 59 (Jan. 1979).
7. H. M. Hobgood, G. E. Eldridge, D. L. Barrett, and R. N. Thomas, High Purity Semi-Insulating GaAs for Monolithic Microwave Integrated Circuits, *IEEE Trans. Electron Dev.* **ED-28,** 140 (1981).
8. R. G. Rhodes, *Imperfections and Active Centres in Semiconductors,* Pergamon, Macmillan, New York 1964.
9. W. R. Runyan, *Silicon Semiconductor Technology,* McGraw-Hill, New York, 1965.
10. F. A. Trumbore, Solid Solubilities of Impurity Elements in Germanium and Silicon, *Bell Sys. Tech. J.* **39,** 205 (1960).
11. A. G. Milnes, *Deep Impurities in Semiconductors,* Wiley, New York, 1973.
12. H. E. Bridgers et al., *Transistor Technology,* Vol. 1, Van Nostrand, New York, 1958.
13. P. F. Lindquist, A Model Relating Electrical Properties and Impurity Concentrations in Semi-Insulating GaAs, *J. Appl. Phys.* **48,** 1262 (1977).
14. W. G. Pfann, *Zone Melting,* Wiley, New York, 1958.
15. E. S. Johnson, Liquid Encapsulated Floating Zone Melting of GaAs, *J. Crystal Growth* **30,** 249 (1975).
16. S. K. Ghandhi, *Semiconductor Power Devices,* Wiley, New York, 1977.
17. H. A. Herrmann and H. Herzer, Doping of Silicon by Neutron Radiation, *J. Electrochem. Soc.* **122,** 1568 (1975).
18. H. M. Janus and O. Malmros, Application of Thermal Neutron Irradiation for Large Scale Production of Homogeneous Phosphorus Doping of Float Zone Silicon, *IEEE Trans. Electron Dev.* **ED-23,** 797 (1976).

PROBLEMS

1. Diffusions into gallium arsenide are often carried out in a sealed ampul with an arsenic overpressure to prevent decomposition. Junction depths

for diffusions in gallium arsenide are a function of partial pressures, set up in this ampul.

Zinc diffusions are to be carried out in an ampul having a volume of 100 ml at 800°C. The arsenic overpressure is to be 0.2 atm (arsenic exists almost entirely as As_2 at 800°C). Calculate the amount of arsenic that is to be placed in the ampul. What would happen if you omitted the arsenic?

2. It is desired to grow gold-doped and boron-doped crystals by the Czochralski method, such that the concentration of dopant in each case is 10^{17} atoms/cm^3 when one-half of the crystal is grown. Assuming that growth is undertaken under rapid-stirring conditions, compare the relative amounts of gold and boron that should be added to a pure silicon melt on an atom ratio basis. Comment on the feasibility of this approach to doping.
3. An antimony-doped crystal is required to have a resistivity of 1.0 Ω cm when one half of the crystal is grown. Assuming that a 100-g pure silicon charge is used, what is the amount of 0.01 Ω cm antimony-doped silicon that must be added. (Assume that the electron mobility is 1500 cm^2/V).
4. A small silicon bar is doped with both boron and antimony. One end of the bar is inserted into a furnace and melted. The bar is now withdrawn. Sketch the doping profile that results from the regrowth process and calculate the base width of the resulting transistor. Assume equilibrium values for k, and that the original antimony to boron atom ratio was 33.
5. Neglecting the effects of impurity distribution in the final zone, show that the limiting value of the solute concentration in a zone-refined crystal is given by

$$C_S = Ae^{Bx}$$

where

$$A = \frac{C_M Bl}{e^{Bl} - 1}$$

$$K = \frac{BL}{e^{Bl} - 1}$$

Here L is the length of the molten zone and l is the total length of the bar.

CHAPTER

4

Diffusion

CONTENTS

4.1 THE NATURE OF DIFFUSION 113

4.1.1 Interstitial Movement, 115
4.1.2 Substitutional Movement, 116
4.1.3 Interstitial–Substitutional Movement, 117

4.2 DIFFUSION IN A CONCENTRATION GRADIENT 118

4.2.1 The Diffusion Coefficient, 120
4.2.2 Field-Aided Motion, 121
4.2.3 Interaction with Charged Defects, 123
4.2.4 The Dissociative Process, 126

4.3 IMPURITY BEHAVIOR: SILICON 128

4.3.1 Substitutional Diffusers, 128
4.3.2 Interstitial–Substitutional Diffusers, 132

4.4 IMPURITY BEHAVIOR: GALLIUM ARSENIDE 139
4.5 THE DIFFUSION EQUATION 140

4.5.1 The D = Constant Case, 141

4.5.1.1 Diffusion from a Constant Source, 141
4.5.1.2 Diffusion from a Limited Source, 143
4.5.1.3 The Two-Step Diffusion, 146
4.5.1.4 Successive Diffusions, 148

4.5.2 The $D = f(N)$ Case, 149

4.5.2.1 Interstitial–Substitutional Diffusers, 151
4.5.2.2 Substitutional Diffusers, 152

4.5.3 Lateral Diffusion Effects, 155

4.6 DIFFUSION SYSTEMS 157

4.6.1 Choice of Dopant Source, 160
4.6.2 Contamination Control, 163

4.7 DIFFUSION SYSTEMS FOR SILICON 164

4.7.1 Boron, 165
4.7.2 Phosphorus, 168
4.7.3 Arsenic, 170
4.7.4 Antimony, 171
4.7.5 Gold and Platinum, 171

4.8 SPECIAL PROBLEMS IN SILICON DIFFUSION 172

4.8.1 Redistribution during Oxide Growth, 172
4.8.2 Boron in Silicon, 174
4.8.3 Arsenic in Silicon, 176
4.8.4 Phosphorus in Silicon, 177
4.8.5 The Emitter-Push Effect, 180
4.8.6 Preferential Doping with Gold and Platinum, 182

4.9 DIFFUSION SYSTEMS FOR GALLIUM ARSENIDE 184

4.9.1 Ternary Considerations, 184
4.9.2 Sealed Tube Diffusions, 185
4.9.3 Open Tube Diffusions, 193
4.9.4 Masked Diffusions, 195

4.10 EVALUATION TECHNIQUES FOR DIFFUSED LAYERS 196

4.10.1 Junction Depth, 196
4.10.2 Sheet Resistance, 198
4.10.3 Surface Concentration, 200
4.10.4 Diffused Layers in Silicon, 200

4.10.4.1 High-Concentration Diffused Layers, 200

4.10.5 Diffused Layers in Gallium Arsenide, 208

REFERENCES 209
PROBLEMS 212

Diffusion provides an important means of introducing controlled amounts of chemical impurities into the silicon lattice [1, 2]. In addition, diffusion techniques are ideally suited to batch processing, where many slices are handled in a single operation.

The laws governing the diffusion of impurities [3] were established many years before the development of semiconductors. It is only since reasonably defect-free materials of high purity have become available, however, that the method has become generally accepted for semiconductor device fabrication. At the present time, diffusion is a basic process step in the fabrication of both discrete devices and microcircuits.

A number of simplifying assumptions can be made in the development of diffusion theory as it applies to semiconductor device fabrication. Thus the single-crystal nature of the solid in which diffusion takes place often allows the effects of grain boundaries to be ignored. Because a very small number of impurities are involved, dimensional changes during the diffusion process are not considered. Finally, the almost exclusive use of parallel-plane device and circuit structures results in considerable simplification in the mathematics. As a consequence, diffusion theory is a reasonably good approximation to diffusion practice.

4.1 THE NATURE OF DIFFUSION

Diffusion describes the process by which atoms move in a crystal lattice. Although this includes self-diffusion phenomena, our interest is in the diffusion of impurity atoms that are introduced into the lattice for the purpose of altering its electronic properties. In addition to concentration gradient and temperature, geometrical features of the crystal lattice (such as crystal structure and defect concentration) play an important part in this process.

The wandering of an impurity in a lattice takes place in a series of random jumps [4]. These jumps occur in all three dimensions, and a flux of diffusing species results if there is a concentration gradient. The mechanisms by which jumps can take place are described as follows.

Interstitial Diffusion. This is illustrated in Fig. 4.1*a*. Here impurity

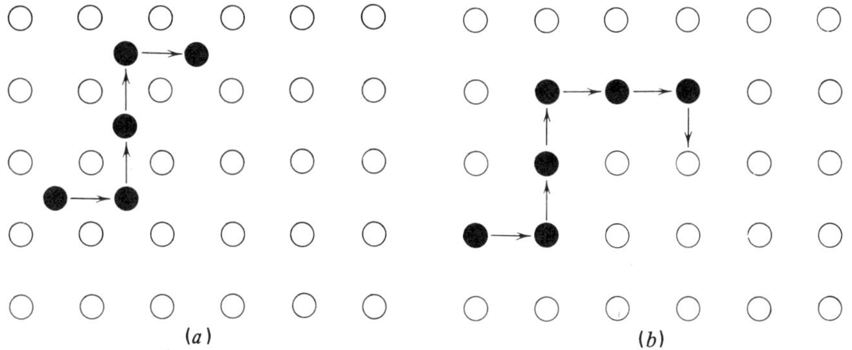

Fig. 4.1 Diffusion by jumping processes.

atoms move through the crystal lattice by jumping from one interstitial site to the next. They may start at either lattice or interstitial sites and may finally end up in either type of site. However, interstitial diffusion requires that their jump motion occur from one interstitial site to another adjacent interstitial site.

Interstitial diffusion can also occur by a *dissociative mechanism*. In this case, impurity atoms occupy both substitutional as well as interstitial sites. However, they only move at a significant rate when in interstitial sites (by interstitial diffusion). The dissociative mechanism, by which a substitutional atom can become an interstitial, is the controlling factor for this process. Copper, nickel, and gold move in silicon by this mechanism, as do zinc, cadmium, and copper in gallium arsenide.

Substitutional Diffusion. Here (see Fig. 4.1*b*) impurity atoms wander through the crystal by jumping from one lattice site to the next, thus substituting for the original host atom. However, it is necessary that this adjacent site be vacant, that is, vacancies must be present to allow substitutional diffusion to occur. These are provided by the mechanisms discussed in Section 1.4.1. Since the equilibrium concentration of vacancies is quite low, it is reasonable to expect substitutional diffusion to occur at a much slower rate than interstitial diffusion. This is indeed the case.

A modified version of substitutional diffusion is the *interstitialcy mechanism*. Here the diffusing atom moves by pushing one of its nearest substitutional neighbors into an adjacent interstitial site, and taking up the substitutional site which is made vacant in this manner. Again, diffusion is dependent on the availability of vacant substitutional sites, and occurs at a slower rate than interstitial diffusion.

Interchange Diffusion. This occurs when two or more atoms diffuse by an interchange process. Such a process is known as a *direct interchange* process when it involves two atoms and as a *cooperative interchange* when a larger number are involved. The probability of occurrence of interchange diffusion effects is relatively low.

Combination Effects. Combinations of these mechanisms may occur within a crystal. For example, a certain fraction of impurity atoms may diffuse substitutionally and the rest interstitially, resulting in a two-stream process. In addition, some of the diffusion atoms may finally end in substitutional sites, with others in interstitial sites.

Diffusion also occurs by movement along dislocations and grain boundaries. Theories for this type of behavior are at a rudimentary stage. However, it has been established that these processes are anisotropic, being much faster in directions parallel to the dislocation core or to the grain boundary edge, than at right angles to it. Furthermore, atom movement is three or more orders of magnitude faster than that obtained by the lattice diffusion processes outlined above.

4.1.1 Interstitial Movement

Interstitial voids are arranged tetrahedrally in the zincblende lattice (see Fig. 1.8). Although some of these are occupied by point defects, their equilibrium concentration is low, even at normal diffusion temperatures (700–1100°C). Consequently, nearly all interstitial sites are available for receiving impurity atoms as they wander through the lattice.

Each tetrahedral void can readily accommodate an interstitial atom. However, there is a slight constriction between these voids, so that interstitially located impurity atoms can be considered to jump from one void to the next by squeezing through the intervening constrictions in the lattice. For this situation the energy barrier for interstitial diffusers is periodic in nature, as shown in Fig. 4.2. Let:

E_m = interaction energy involved, in eV

T = temperature of lattice, in °K

ν_0 = frequency of lattice vibrations, about 10^{13}–10^{14} s^{-1}; this is the frequency with which atoms strike the potential barrier depicted in Fig. 4.2

ν = frequency with which thermal energy fluctuations occur with sufficiently large magnitude to overcome the potential barrier

Assuming a Boltzmann energy distribution, the probability that an atom has an energy in excess of E_m is given by $e^{-E_m/kT}$. Atoms in the zincblende lattice can jump from one interstitial site to the next in four different ways. Assuming that these jumps are uncorrelated, and that all sites are vacant, the frequency of jumping is given by

$$\nu = 4\nu_0 e^{-E_m/kT} \tag{4.1}$$

Typical values for E_m are about 0.6–1.2 eV. Thus an interstitial impurity

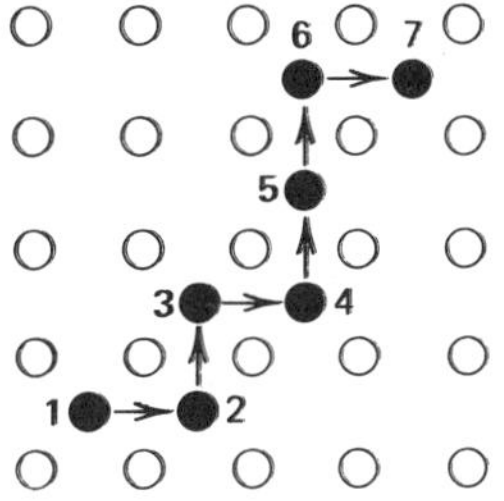

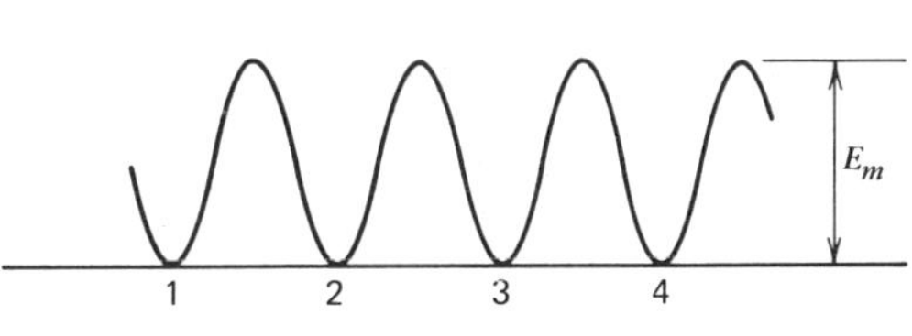

Fig. 4.2 Interstitial diffusion by jumping.

atom will jump from one void to another at a rate of about once every minute at room temperature. The rate is considerably higher at typical diffusion temperatures (700–1100°C).

4.1.2 Substitutional Movement

The jumping of substantial diffusers requires first that there be available vacant sites into which they can move. These vacant sites are Schottky defects, with an energy of formation given by E_s. Thus the atom fraction of such defects in a crystal is $e^{-E_s/kT}$.

Moreover, the jump process requires the breaking of certain covalent bonds, and the making of yet others, as shown in Fig. 4.3. If the potential barrier associated with this process is E_n, then the probability of atoms having a thermal energy in excess of this value is $e^{-E_n/kT}$. Finally, each lattice site has four tetrahedrally situated neighbors, so that each jump can be made in one of four different ways. Again assuming that these jumps are uncorrelated,*

$$\nu = 4\nu_0 e^{-E_s/kT} e^{-E_n/kT} \tag{4.2}$$

$$= 4\nu_0 e^{-(E_n+E_s)/kT} \tag{4.3}$$

Values of $E_n + E_s$ predicted from this simple theory are found to be slightly larger than those actually observed. This is because the binding energy between an impurity atom and its neighboring atom is less than that between two adjacent atoms in the lattice. Consequently, the energy of formation of a vacancy next to an impurity atom is less than that of forming a vacancy at any other site. This is borne out by the experimental fact that values of $E_n + E_s$ range from 3 to 4 eV for substitutional impurities in silicon, while the activation energy of self-diffusion in silicon is 5.13 eV. From Eq. (4.3) it is seen that the jump rate of a substitutional impurity at room temperature is about once every 10^{45} years (as compared to the jump rate of about once every minute for interstitial diffusers).

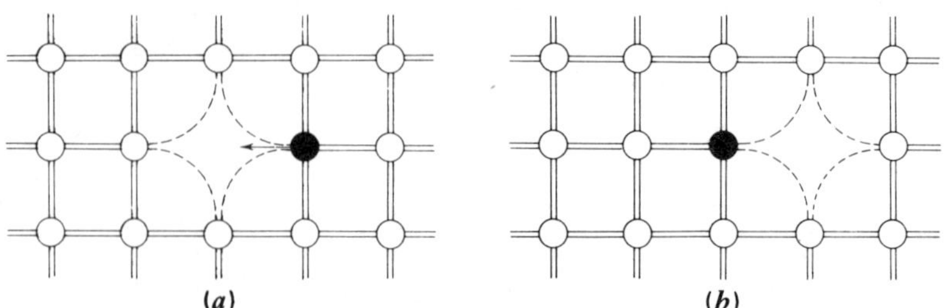

Fig. 4.3 Substitutional jumps.

* A correlation factor of 0.5 is sometimes used in these computations.

4.1.3 Interstitial–Substitutional Movement

Many impurities dissolve both substitutionally and interstitially in the host lattice, and diffuse by the combined motion of these two species. For such situations, let N_s and N_i be the substitutional and interstitial solubilities, respectively, at the diffusion temperature. Usually the solubility of the substitutional component is much higher than that of the interstitial component. On the other hand, the rates of movement of these components through the lattice are very different, with the impurity comparatively immobile when in substitutional sites, but free to move rapidly when it is interstitial. Assuming *independent* motion of these two types of diffusers, the impurity will spend $N_i/(N_i + N_s)$ of its time in an interstitial state, and $N_s/(N_i + N_s)$ in a substitutional state. It follows that the effective jump frequency ν_{eff} is given by

$$\nu_{\text{eff}} = \frac{\nu_s N_s}{N_i + N_s} + \frac{\nu_i N_i}{N_i + N_s} \tag{4.4}$$

where ν_s and ν_i are the substitutional and interstitial jump frequencies, respectively. Typically, the second term is much larger than the first.

Atom movement by two *independent* streams rarely occurs in silicon or gallium arsenide. Here the two streams are *interdependent,* because of the possibility of dissociation of a substitutional atom into an interstitial atom and a vacancy. This reaction can be written as

$$S \rightleftharpoons I + V \tag{4.5}$$

Again, the interstitial species is a rapid diffuser, whereas the substitutional is essentially immobile. Now, however, the diffusion is mainly controlled by the rate of production of the interstitial species by this dissociative mechanism.

If the crystal contains many defects, the vacancy concentration readily attains its equilibrium value, and the impurity spends $N_i/(N_i + N_s)$ of its time in an interstitial state, so that

$$\nu_{\text{eff}} \simeq \frac{\nu_i N_i}{N_i + N_s} \tag{4.6}$$

As seen from this equation, the jump frequency is close to that for the interstitial.

For low-dislocation material, however, the formation of a substitutional atom requires the generation of new vacancies which must come from the surface of the slice, i.e., by a relatively slow process. If n_v is the vacancy concentration, then the number of vacancies which are unoccupied is $n_v/(n_v + N_s)$, so that the effective jump frequency becomes

$$\nu_{\text{eff}} = \frac{\nu_s n_v}{n_v + N_s} \tag{4.7a}$$

If the diffusion of vacancies is the rate limiter, then a more appropriate value of the effective jump frequency is

$$\nu_{\text{eff}} = \frac{\nu_v n_v}{n_v + N_s} \tag{4.7b}$$

Here ν_v is the jump frequency associated with vacancies, which is usually somewhat smaller than ν_s.

In summary, the characteristics of interstitial–substitutional diffusion via a dissociative mechanism are:

1. A jump rate that rapidly increases with defect concentration.
2. A jump rate that is a function of the vacancy concentration.
3. A jump rate that is highly dependent on the impurity concentration.

Impurity movement by an interstitial–substitutional process is thus seen to have many of the characteristics of interstitial diffusion. Impurities which utilize this process are often included in the general class of interstitial diffusers.

4.2 DIFFUSION IN A CONCENTRATION GRADIENT

So far we have only considered the statistics of random jump motion in the zincblende lattice. In the presence of a concentration gradient, this motion results in a net transport of impurities. The law governing this motion is now considered for impurities which diffuse substitutionally. However, the arguments presented here can be extended to interstitial diffusers as well.

Substitutional diffusion takes place by jump motion between tetrahedral sites of spacing d. A single jump has projections of length $d/\sqrt{3}$ along each of the crystal axes. Consider a crystal of cross section A, divided up by a series of parallel planes at right angles to the [100] axis. (Impurity motion in other directions follows along similar lines of reasoning.) Let the spacing between planes be $d/\sqrt{3}$. Figure 4.4 depicts this arrangement, with the crystal divided into layers 1, 2, etc. Then each layer contains atoms whose tetrahedral neighbors are in adjacent layers on either side.

Let n_1 and n_2 be the number of atoms in layers 1 and 2, respectively, and

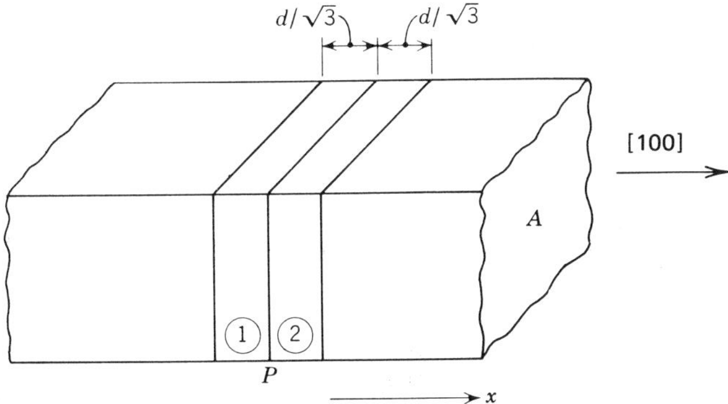

Fig. 4.4 Diffusion due to a concentration gradient.

N_1 and N_2 the volume concentrations. Then

$$N_1 = \frac{\sqrt{3}n_1}{Ad} \tag{4.8a}$$

$$N_2 = \frac{\sqrt{3}n_2}{Ad} \tag{4.8b}$$

For diffusion in this lattice, atoms in any one layer must jump into neighboring layers. Each atom has four such neighbors, two in the layer to the left and two in the layer to the right. Thus in a single jump period of time $1/\nu$, half of the moving atoms jump right while the other half jump left, on the average.

The net flow of atoms across the plane P in a direction x is given by

$$\begin{aligned} \frac{\Delta n}{\Delta t} &= \frac{(n_1 - n_2)/2}{1/\nu} \\ &= \frac{\nu}{2}\frac{Ad}{\sqrt{3}}(N_1 - N_2) \end{aligned} \tag{4.9}$$

But

$$\frac{\Delta N}{\Delta x} = \frac{N_2 - N_1}{d/\sqrt{3}} \tag{4.10}$$

so that

$$\frac{\Delta n}{\Delta t} = -A\frac{\nu d^2}{6}\frac{\Delta N}{\Delta x} \tag{4.11}$$

Writing the flux density j as the time rate of change of the number of impurities per unit area, Eq. (4.11) reduces to

$$j = -\frac{\nu d^2}{6}\frac{\Delta N}{\Delta x} \tag{4.12}$$

Defining a diffusion constant D such that

$$D = \frac{\nu d^2}{6} \tag{4.13}$$

Eq. (4.12) may be rewritten in partial derivative form as

$$j = -D\frac{\partial N}{\partial x} \tag{4.14}$$

where j is the flux density, in atoms/cm^2 s, D the diffusion coefficient, in cm^2/s, N the volume concentration, in atoms/cm^3, and x the distance, in cm. This is Fick's first law. It states that, under diffusion conditions, the flux density is directly proportional to the concentration gradient.

Similar arguments apply to interstitial diffusers, except that jumps between interstitial voids must be considered. Since each void also has four nearest neighbors, the same line of reasoning applies.

4.2.1 The Diffusion Coefficient

The behavior of the diffusion coefficient may now be determined. Substituting Eq. (4.13) into Eq. (4.1), the diffusion coefficient for an interstitial impurity is given by

$$\begin{aligned} D &= \frac{\nu d^2}{6} = \frac{4\nu_0 d^2}{6} e^{-E_m/kT} \\ &= D_0 e^{-E_m/kT} \end{aligned} \tag{4.15}$$

In like manner the diffusion coefficient for a substitutional impurity is obtained by combining Eqs. (4.13) and (4.3). Thus

$$\begin{aligned} D &= \frac{4\nu_0 d^2}{6} e^{-(E_n+E_s)/kT} \\ &= D_0 e^{-(E_n+E_s)/kT} \end{aligned} \tag{4.16}$$

Both D_0 and the exponential term are functions of temperature. However, the variation of D with temperature is largely controlled by changes in the exponential factor. As a result, D_0 is usually considered to be constant over a range of a few hundred degrees of diffusion temperature.

4.2.2 Field-Aided Motion

In addition to diffusion, ionized impurities can move in the presence of a drift field. This motion results in a drift component of velocity in the direction of the electric field. If the random scattering of ionized impurities in the lattice is represented by a restraining force that is directly proportional to the drift velocity, the equation of motion for an impurity atom of valence Z may be written as

$$F = Zq\mathscr{E} = m^* \frac{dv}{dt} + \alpha v \tag{4.17}$$

where F is the force on the impurity ion, Zq its charge, $\mathscr{E}$ the electric field, v the average velocity, m^* the effective mass, and α a factor of proportionality.

Note that the sign of the charge on the impurity ion determines the direction of this force. Thus a positive force is exerted on a positively charged impurity in a positive $\mathscr{E}$ field.

In steady state a drift velocity v_d is attained, such that

$$v_d = \frac{Zq}{\alpha} = \mu\mathscr{E} \tag{4.18}$$

where μ is defined as the mobility of the impurity ion in $cm^2/V\ s$.

The movement of impurities due to drift and diffusion may be assumed to be independent events. Consequently, in the presence of a field, the net flow of atoms across the plane P in the direction x (see Fig. 4.4) is given by modifying Eq. (4.9) so that

$$\frac{\Delta n}{\Delta t} = \frac{\nu}{2} \frac{Ad}{\sqrt{3}} (N_1 - N_2) + \mu N\mathscr{E}A \tag{4.19}$$

Thus the flux density is given by

$$\begin{aligned} j &= -\frac{\nu d^2}{6} \frac{\Delta N}{\Delta x} + \mu N\mathscr{E} \\ &= -D \frac{\partial N}{\partial x} + \mu N\mathscr{E} \end{aligned} \tag{4.20}$$

in partial derivative form.

This electric field can be internally generated during the diffusion of substitutional impurities at high doping levels, if the semiconductor remains

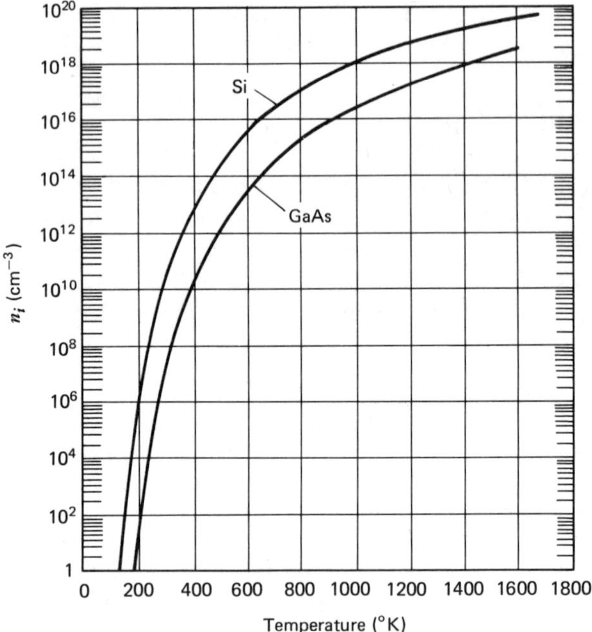

Fig. 4.5 Intrinsic carrier concentration.

extrinsic at diffusion temperatures. That this often occurs is seen from Fig. 4.5, which displays the intrinsic carrier concentration for silicon and gallium arsenide as a function of temperature.

The effect of this field is now described for the diffusion of an n-type impurity. Here the impurity is ionized, with an electron concentration n. These electrons have a much higher diffusion rate than the impurity atoms and tend to outrun them, creating a space charge which causes the field to build up. The direction of the $\mathscr{E}$ field due to the space charge associated with the positively charged impurity ions and the mobile electrons aids the drift of these ions in the direction of the diffusing flux, resulting in an increased rate of impurity movement through the lattice. A similar conclusion can be reached for p-type impurities.

The magnitude of this field-aided motion may be calculated [5] by noting that the impurity flux for this situation is described by

$$j = -D\frac{\partial N}{\partial x} + \mu n \mathscr{E} \tag{4.21}$$

Substituting

$$D = \frac{kT}{q}\mu \tag{4.22a}$$

and

$$\mathscr{E} = -\frac{kT}{q}\frac{1}{n}\frac{dn}{dx} \tag{4.22b}$$

gives the flux density as

$$j = -D\left(1 + \frac{dn}{dN}\right)\frac{\partial N}{\partial x} \tag{4.23}$$

so that the impurity moves with an effective diffusion constant D_{eff}, where

$$D_{\text{eff}} = D\left(1 + \frac{dn}{dN}\right) \tag{4.24}$$

For an n-type impurity

$$\frac{n}{n_i} = \frac{N}{2n_i} + \left[\left(\frac{N}{2n_i}\right)^2 + 1\right]^{1/2} \tag{4.25}$$

so that

$$\frac{dn}{dN} = \frac{1}{2}\left\{1 + \left[1 + \left(\frac{2n_i}{N}\right)^2\right]^{-1/2}\right\} \tag{4.26}$$

Thus the effect of this *field enhancement factor* is to (at most) double the magnitude of the diffusion constant under field-free conditions. Consequently, this mechanism cannot account for the large increase in diffusivity with doping concentration that has been experimentally observed for substitutional diffusers. Here interaction with charged defects can play a dominant role, as seen in the next section.

4.2.3 Interaction with Charged Defects

The theory of substitutional diffusion is based on the interaction of the impurity ion with vacancies in the zincblende lattice. The charge states of these vacancies can strongly influence the nature of this interaction, and hence the effective diffusion coefficient.

It can be shown [6] that any "flaw" in a semiconductor lattice can usually be represented by a series of energy levels, in the sequence E^+, E^-, E^{2-}, E^{3-}, . . . , in addition to being neutral. In general, all deep levels, as well as vacancies, exhibit one or more of these charge states, and in this sequence. Let us assume that vacancies can be represented in this way, corresponding to V^+, V^-, V^{2-}, V^{3-}, in addition to V^0, for the sake of specificity.

These vacancies will interact with the diffusing impurity ions in different ways. Consequently, the jump statistics, and hence the diffusion constant and activation energy associated with each impurity–vacancy combination, will be different. However, if it is assumed that no correlation exists between the separate jump processes, the effective diffusion constant will be given by the sum of these terms taken over their respective contributions.

Consider the dilute case, where an extremely small amount of impurity diffuses in this lattice. Diffusion under this condition can be characterized by intrinsic diffusion coefficients D_i^0, D_i^+, D_i^-, D_i^{2-}, and D_i^{3-}, where these terms are associated with I-V^0, I-V^+, I-V^-, I-V^{2-}, and I-V^{3-} pair interactions, respectively (and I represents the impurity). It follows that the overall intrinsic diffusivity is given* by

$$D_i = D_i^0 + D_i^+ + D_i^- + D_i^{2-} + D_i^{3-} \tag{4.27}$$

Diffusion under extrinsic conditions results in a displacement of the Fermi level, and a change in the concentration of the various defect species. Consequently, the diffusivity for this condition is given [7] by

$$D = D_i^0 \frac{[V^0]}{[V^0]_i} + D_i^+ \frac{[V^+]}{[V^+]_i} + D_i^- \frac{[V^-]}{[V^-]_i} + D_i^{2-} \frac{[V^{2-}]}{[V^{2-}]_i} + D_i^{3-} \frac{[V^{3-}]}{[V^{3-}]_i} \tag{4.28}$$

where $[V^0]$, $[V^+]$, $[V^-]$, $[V^{2-}]$, and $[V^{3-}]$ refer to the atom fractions of these species under extrinsic conditions, and $[V^0]_i$, $[V^+]_i$, $[V^{2-}]_i$, $[V^{3-}]_i$ pertain to intrinsic conditions.

The occupation statistics for vacancies must be determined in order to apply this equation. To do so, it must be recognized that the neutral vacancy concentration will remain the same throughout the semiconductor, since it is unaffected by the electric field due to the impurity concentration gradient. It follows that the extrinsic concentration of neutral vacancies will be equal to the concentration in intrinsic material, which is only a function of temperature. On the other hand, the concentration of the various ionized states is still determined by the position of the Fermi level, so that the total defect concentration changes with shifts in its position.

The ionized vacancy ratios can be determined using the law of mass action. Degeneracy factors are ignored in this analysis. For the V^+ case we have

$$V^0 + h^+ \rightleftharpoons V^+ \tag{4.29}$$

* The right-hand side of this equation is sometimes multiplied by 0.5, which is the correlation factor for tracer self-diffusion in the zincblende lattice.

so that

$$K_1 = \frac{[V^+]}{p[V^0]} \tag{4.30}$$

Under intrinsic conditions, $p = p_i = n_i$, so that

$$K_1 = \frac{[V^+]_i}{n_i[V^0]_i} \tag{4.31}$$

Combining these equations, and noting that $[V^0] = [V^0]_i$, gives

$$\frac{[V^+]}{[V^+]_i} = \frac{p}{n_i} \tag{4.32}$$

In like manner, the reaction for a vacancy with the $-r$ charge state is given by

$$V^0 + re^- \rightleftharpoons V^{-r} \tag{4.33}$$

so that

$$K_2 = \frac{[V^{-r}]}{n^r[V^0]} = \frac{[V^{-r}]}{n_i^r[V^0]_i} \tag{4.34}$$

and

$$\frac{[V^{-r}]}{[V^{-r}]_i} = \left(\frac{n}{n_i}\right)^r \tag{4.35}$$

Thus extrinsic diffusion is characterized by

$$D = D_i^0 + D_i^+\left(\frac{p}{n_i}\right) + D_i^-\left(\frac{n}{n_i}\right) + D_i^{2-}\left(\frac{n}{n_i}\right)^2 + D_i^{3-}\left(\frac{n}{n_i}\right)^3 \tag{4.36}$$

This equation can be further modified by a field enhancement factor (Section 4.2.2) which takes into consideration the internally generated electric field term.* Thus the effective diffusivity is given by

$$D_{\text{eff}} \simeq h\left[D_i^0 + D_i^+\left(\frac{p}{n_i}\right) + D_i^-\left(\frac{n}{n_i}\right) + D_i^{2-}\left(\frac{n}{n_i}\right)^2 + D_i^{3-}\left(\frac{n}{n_i}\right)^3\right] \tag{4.37}$$

* This term is often ignored in diffusion calculations.

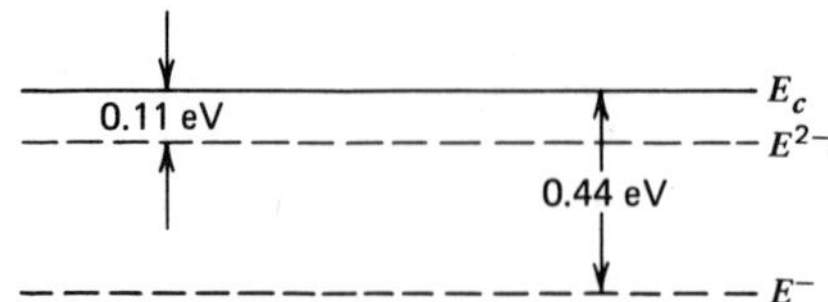

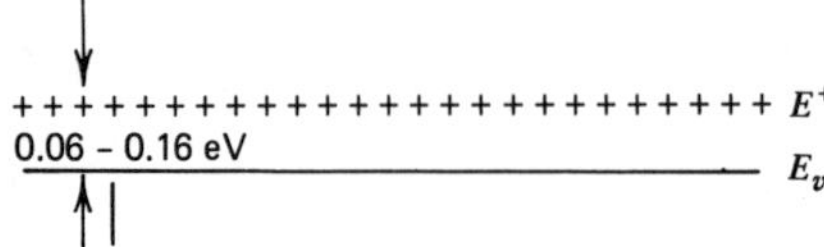

Fig. 4.6 Energy-band diagram for vacancies in silicon.

where, from Eq. (4.26),

$$h = 1 + \left[1 + \left(\frac{2n_i}{N}\right)^2\right]^{-1/2} \tag{4.38}$$

Not all of these terms apply to every situation, so that only relevant terms must be considered. The situation with gallium arsenide is relatively simple, since both arsenic and gallium vacancies are usually considered to be neutral. For this semiconductor,

$$D_{\text{eff}} = hD_i^0 \tag{4.39}$$

Silicon, on the other hand, exhibits the charge states V^+, V^-, and V^{2-}, in addition to being neutral. Consequently,

$$D_{\text{eff}} = h\left[D_i^0 + D_i^+\left(\frac{p}{n_i}\right) + D_i^-\left(\frac{n}{n_i}\right) + D_i^{2-}\left(\frac{n}{n_i}\right)^2\right] \tag{4.40}$$

Figure 4.6 shows [8] the energy-band diagram for vacancies in silicon. From this figure it is seen that only I-V^+ interactions are present during the diffusion of p-type impurities, since the Fermi level is close to the valance-band edge. By the same reasoning, the diffusion of n-type impurities is dominated by interactions with the negatively charged species. Thus Eq. (4.40) is greatly simplified when applied to specific diffusion situations.

4.2.4 The Dissociative Process

We have noted earlier that many impurities can exist in both interstitial as well as substitutional sites in the zincblende lattice. The substitutional solubility of these impurities is usually greater than the interstitial.* On the

* The behavior of copper in silicon is an exception to this rule.

other hand, the interstitial jump rate is many decades larger than that of the substitutional component. Consequently, diffusion in a concentration gradient is dominated by the movement of the interstitial, and by its concentration.

Consider what happens in a region of high concentration. Here, since the interstitial atom moves much faster than the substitutional one, it follows that the N_i/N_s ratio will fall below the equilibrium value. The dissociative process, defined by

$$S \rightleftharpoons I + V \tag{4.41}$$

will now come into play in order to reestablish equilibrium. Thus Eq. (4.41) is driven from left to right in regions of high concentration, and from right to left in regions of low concentration. In view of the fact that the interstitial species is a fast diffuser whereas the substitutional one is (by comparison) immobile, the rate of diffusion will be enhanced in regions of high concentration, and retarded in regions of low concentration.

The diffusivity of an impurity of this type can be calculated for two limiting situations. The first of these, where the vacancy concentration is unchanged from its equilibrium value, occurs if the semiconductor is highly dislocated. Here it follows from Eq. (4.6) that

$$D_{\text{eff}} \simeq \frac{D_i' N_i}{N_i + N_s} \tag{4.42}$$

where D_i' is the interstitial diffusivity.

The second is the limiting case of low dislocation material, where the rate limiter is the delivery of vacancies for the dissociative mechanism. Here it can be assumed that N_i does not change from its equilibrium value. For this case it follows from Eq. (4.7) that

$$D_{\text{eff}} = \frac{D_v n_v}{n_v + N_s} \tag{4.43a}$$

$$\simeq \frac{D_v n_v}{N_s} \tag{4.43b}$$

where n_v and D_v are the vacancy concentration and diffusivity, respectively.

In silicon the diffusion of vacancies is synonymous with the substitutional movement of silicon atoms. Thus D_v is the coefficient of self-diffusion, which is given by

$$D_v = 9 \times 10^3 \, e^{-5.13/kT} \tag{4.44}$$

The physical interpretation of D_v in gallium arsenide is less clear, since each gallium atom has arsenic as its nearest neighbors, and vice versa. Thus substitutional motion for these atoms would necessitate a large number of antistructure defects which are not observed in practice. Other mechanisms, such as interstitial–substitutional diffusion, have been proposed, but the situation has not been resolved at the present time. From annealing studies of gallium arsenide [9], the activation energy of diffusion of gallium and arsenic vacancies is given by

$$D(V_{\mathrm{Ga}}) = 2.1 \times 10^{-3}\, e^{-2.1/kT} \tag{4.45a}$$

$$D(V_{\mathrm{As}}) = 7.9 \times 10^{3}\, e^{-0.4/kT} \tag{4.45b}$$

4.3 IMPURITY BEHAVIOR: SILICON

Impurities belonging to groups III and V of the periodic table are substitutional diffusers in silicon. Their motion is thus strongly influenced by the number and the charge state of lattice point defects. They include aluminum, boron,* gallium, and indium (p-type), as well as antimony, arsenic, and phosphorus (n-type). With the exception of indium, all are shallow impurities in silicon. They are technologically important since they are used in junction formation.

Impurities which move interstitially in silicon belong to group I and group VIII of the periodic table. These include alkali metals such as lithium, potassium, and sodium, and gases such as argon, helium, and hydrogen. These take up interstitial sites in silicon and are usually electronically inactive. A notable exception is lithium, which is n-type. These impurities are of little interest in microcircuit fabrication technology, and are not considered further.

Most transition elements (cobalt, copper, gold, iron, nickel, platinum, and silver, for example) diffuse by an interstitial–substitutional mechanism, and end up in both types of sites, the difference in solubility between these sites being substantial. The dissociation of the substitutional into an interstitial and a vacancy is the primary mechanism involved during this diffusion process. All of these impurities reduce the minority carrier lifetime in silicon, and are thus important contaminants. A few are used intentionally for this purpose, in high-speed saturated logic microcircuits.

4.3.1 Substitutional Diffusers

Figure 4.7 shows the diffusion coefficient D as a function of temperature, for substitutional impurities in silicon. Data of this type are of an average nature, independent of concentration, and are obtained by diffusing the impurities into silicon wafers of known (but opposite type) background concentration and by making accurate measurements of the p–n structures thus

* It has been proposed that boron diffuses by the interstitialcy mechanism, which is similar in character to substitutional diffusion.

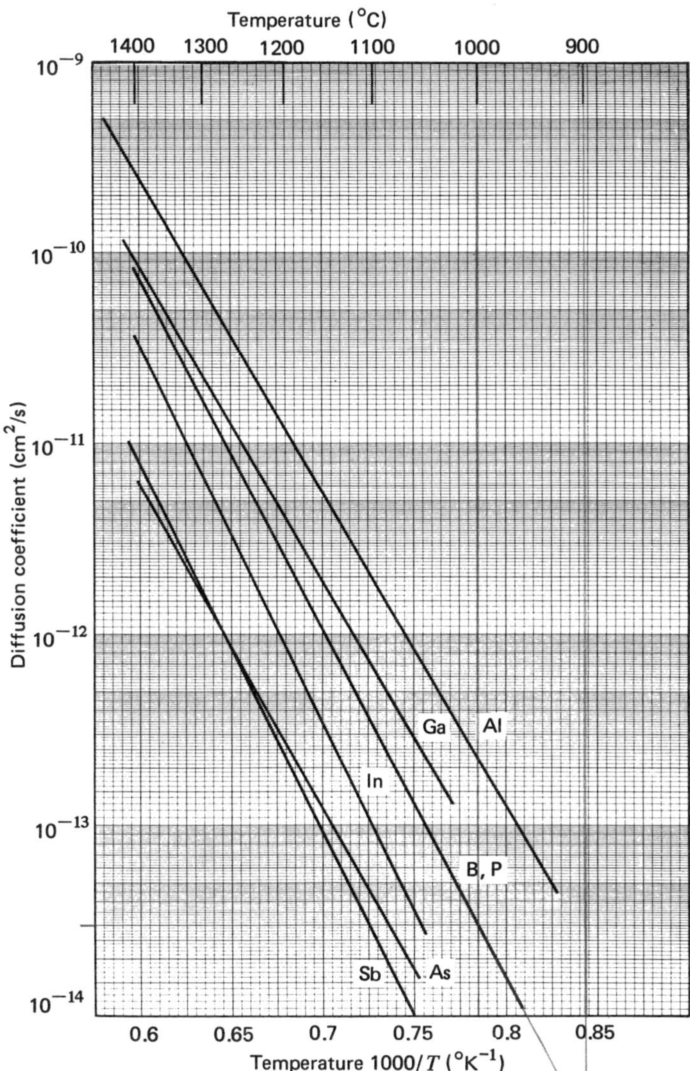

Fig. 4.7 Average diffusion coefficients for substitutional diffusers in silicon.

formed. This experimental technique essentially duplicates the fabrication process for p–n junctions. Consequently, the results are reasonably accurate for many junction-formation situations. However, they become increasingly poor when used to determine doping profiles for shallow junctions with high surface concentration. Table 4.1, which lists values of the intrinsic diffusivity and activation energy for commonly used impurities, with their separate (I-V) contributions, is more useful in these situations.

Figure 4.8 [10] shows the intrinsic diffusion coefficients for boron in

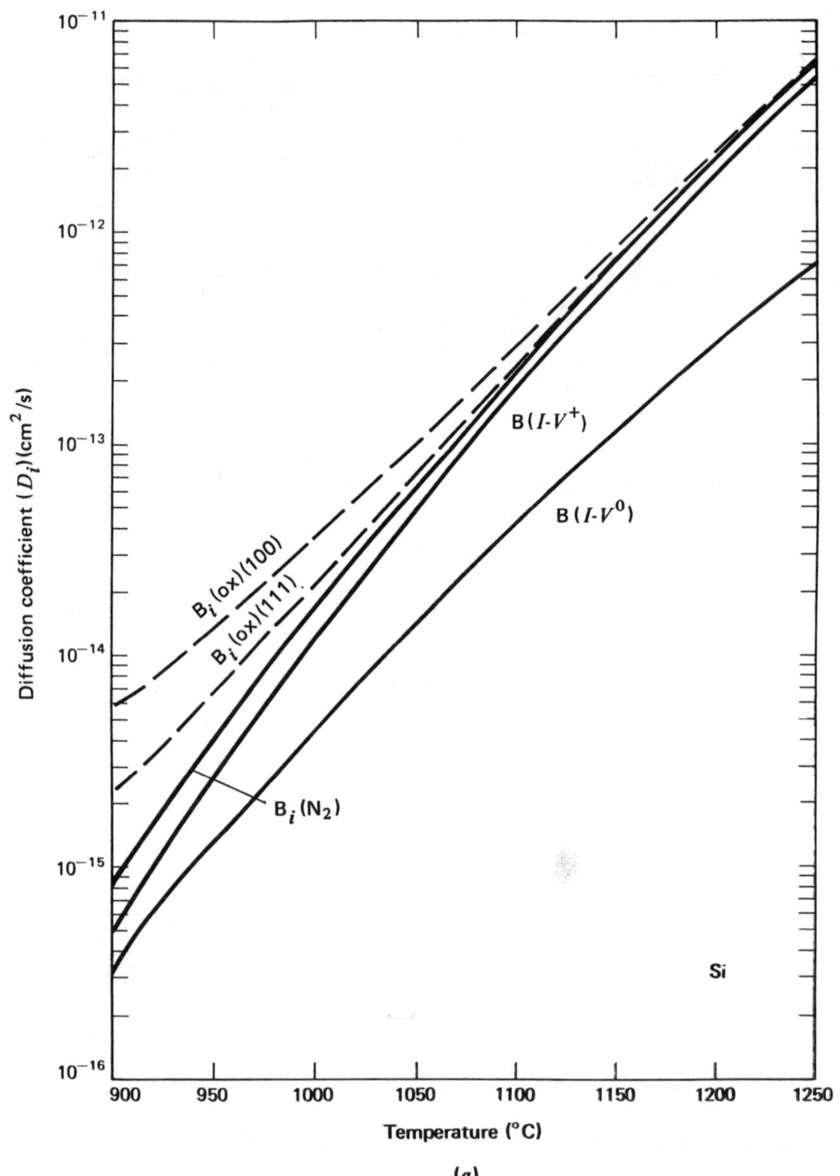

(a)

Fig. 4.8 Diffusivities of boron in silicon. From R. Colclaser, *Microelectronics: Processing and Device Design* [10], 1980. Reprinted with permission of the publisher, John Wiley & Sons.

silicon. Comparable data for arsenic and phosphorus are shown in Fig. 4.9. The application of these data to junction formation is outlined in Section 4.5.2.

The diffusion of boron in silicon is isotropic when carried out in a nitrogen ambient. Its diffusivity increases significantly for an oxidizing ambient, and becomes a function of crystal orientation [11, 12]. It is highest for boron diffusions performed in (100) silicon, and somewhat lower for (111) silicon. Curves for $D_i = D_i^0 + D_i^+$ under these conditions are also shown in Fig. 4.9, together with the curve for D_i in a nonoxidizing ambient (nitrogen).

The anomalous diffusion of boron in oxidizing ambients is of technological

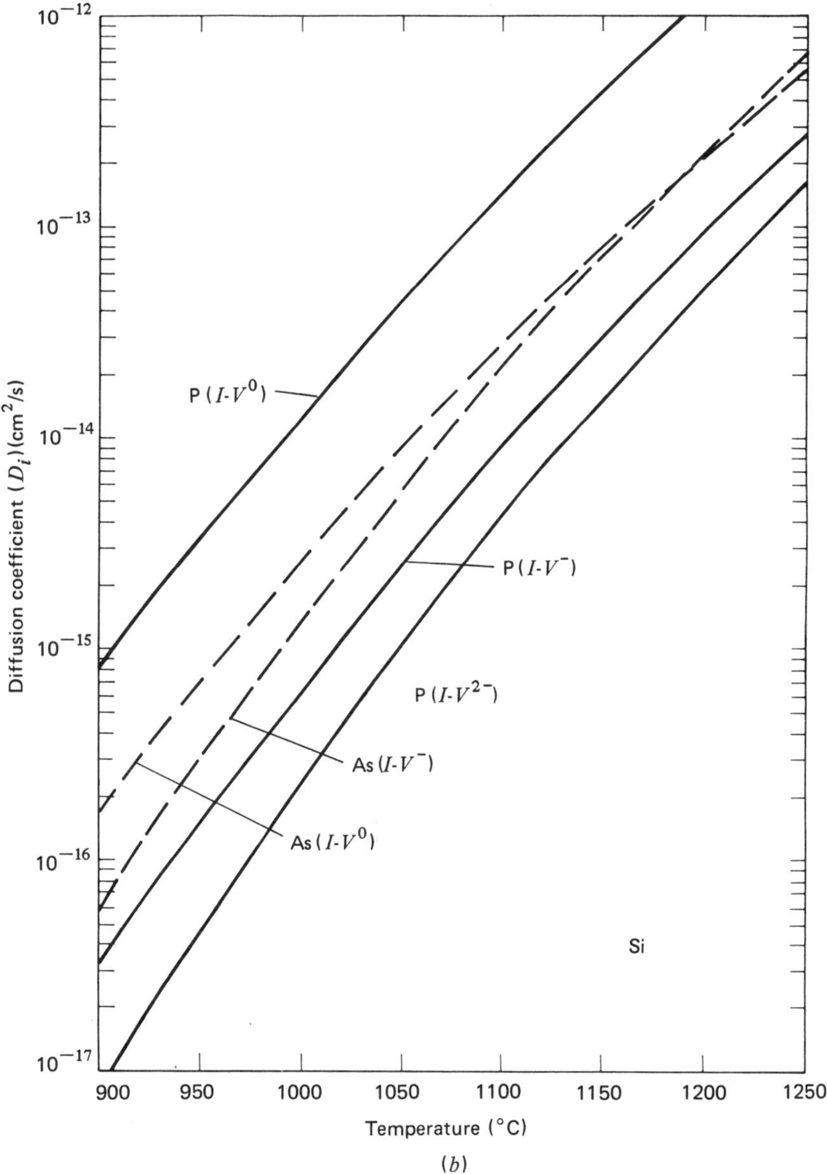

(b)

Fig. 4.9 Diffusivities of arsenic and phosphorus in silicon. From R. Colclaser, *Microelectronics: Processing and Device Design* [10], 1980. Reprinted with permission of the publisher, John Wiley & Sons.

Table 4.1 Intrinsic Diffusivities and Activation Energies of Substitutional Diffusers in Silicon[a]

		P	As	Sb	B	Al	Ga
D_i^0	D_0	3.85	0.066	0.214	0.037	1.385	0.374
	E_0	3.66	3.44	3.65	3.46	3.41	3.39
D_i^+	D_0	—	—	—	0.76	2480	28.5
	E_0	—	—	—	3.46	4.20	3.92
D_i^-	D_0	4.44	22.9	13	—	—	—
	E_0	4.0	4.1	4.0	—	—	—
D_i^2	D_0	44.2	—	—	—	—	—
	E_0	4.37	—	—	—	—	—

[a] D_0 in cm^2/s; E_0 in eV. See reference 8.

importance in the fabrication of low-threshold MOS circuits, which are often made on (100) silicon. It can be understood by noting that the oxidation rate for silicon is orientation dependent. Incomplete oxidation of silicon during diffusion in an oxidizing ambient generates excess silicon interstitials over their equilibrium value. This, in turn, enhances the diffusivity of boron which moves by the interstitialcy mechanism, and also makes it orientation dependent. (The presence of an oxide that is already on the surface of the silicon does not affect the diffusivity of boron. This tends to reinforce the above arguments, since oxygen incorporation into silicon from such layers is relatively insignificant).

Arsenic and phosphorus are commonly used in situations where heavy doping is required. Here band gap narrowing effects are important, and result in altering the intrinsic carrier concentration at these doping levels. Values of n_{ie}, the intrinsic electron concentration for solid-solubility-limited diffusions with these impurities, are shown in Fig. 4.10. The solid-solubility limit, as well as the electrically active component of these impurities, are shown in Fig. 4.11 as a function of diffusion temperature.

4.3.2 Interstitial–Substitutional Diffusers

Table 4.2 shows the diffusion constant and the activation energy of diffusion for a number of interstitial–substitutional diffusers in silicon. It must be emphasized that the data here are by no means definitive, many of them being in the nature of average values and estimates. Reliable values for their diffusivities are difficult to obtain for the following reasons:

1. These impurities usually diffuse about five to six orders of magnitude faster than substitutional diffusers, so that considerable error can occur from out-diffusion effects when the specimens are cooled to room temperature. Attempts at rapid quenching usually result in the generation

of a large number of crystal defects and obscure the interpretation of data.

2. Their movement is described by a diffusion coefficient which is a function of both concentration and temperature. Thus the data in Table 4.2 are an attempt to cast their behavior in a simplified form, which only holds for substitutional diffusers.

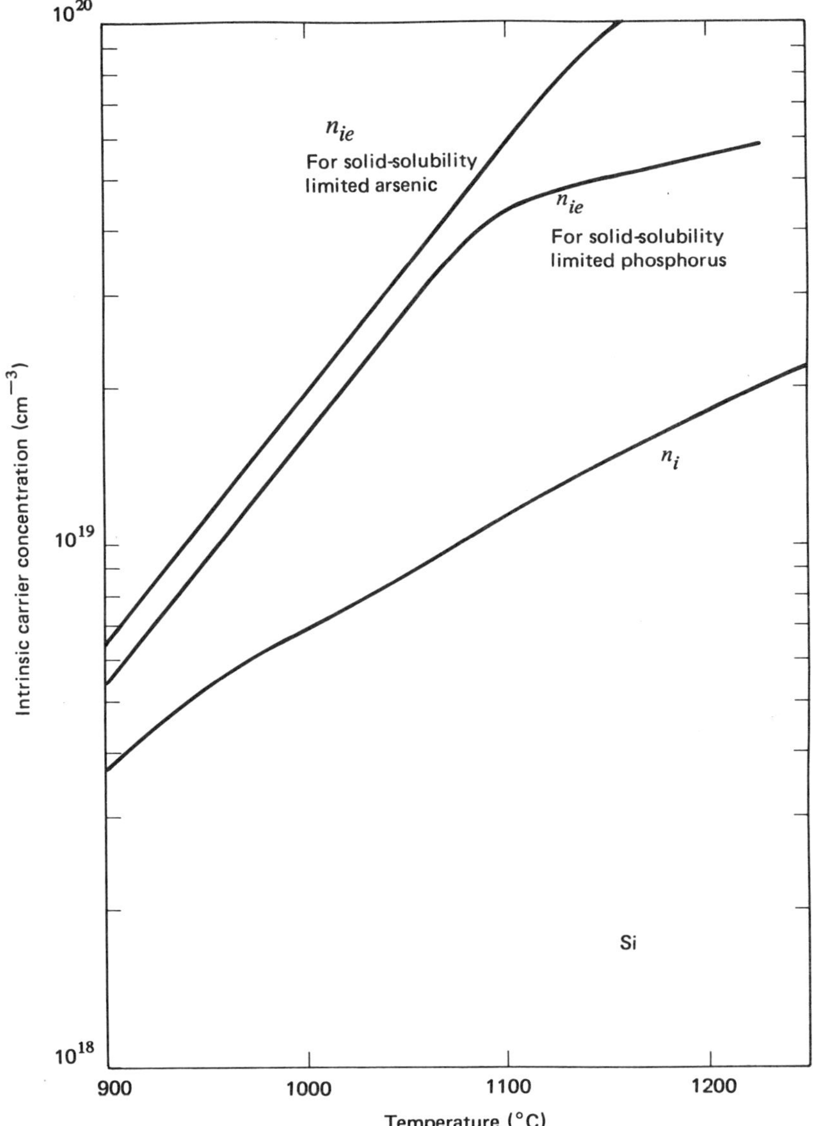

Fig. 4.10 Intrinsic carrier concentration for heavy doping conditions. From R. Colclaser, *Microelectronics: Processing and Device Design* [10], 1980. Reprinted with permission of the publisher, John Wiley & Sons.

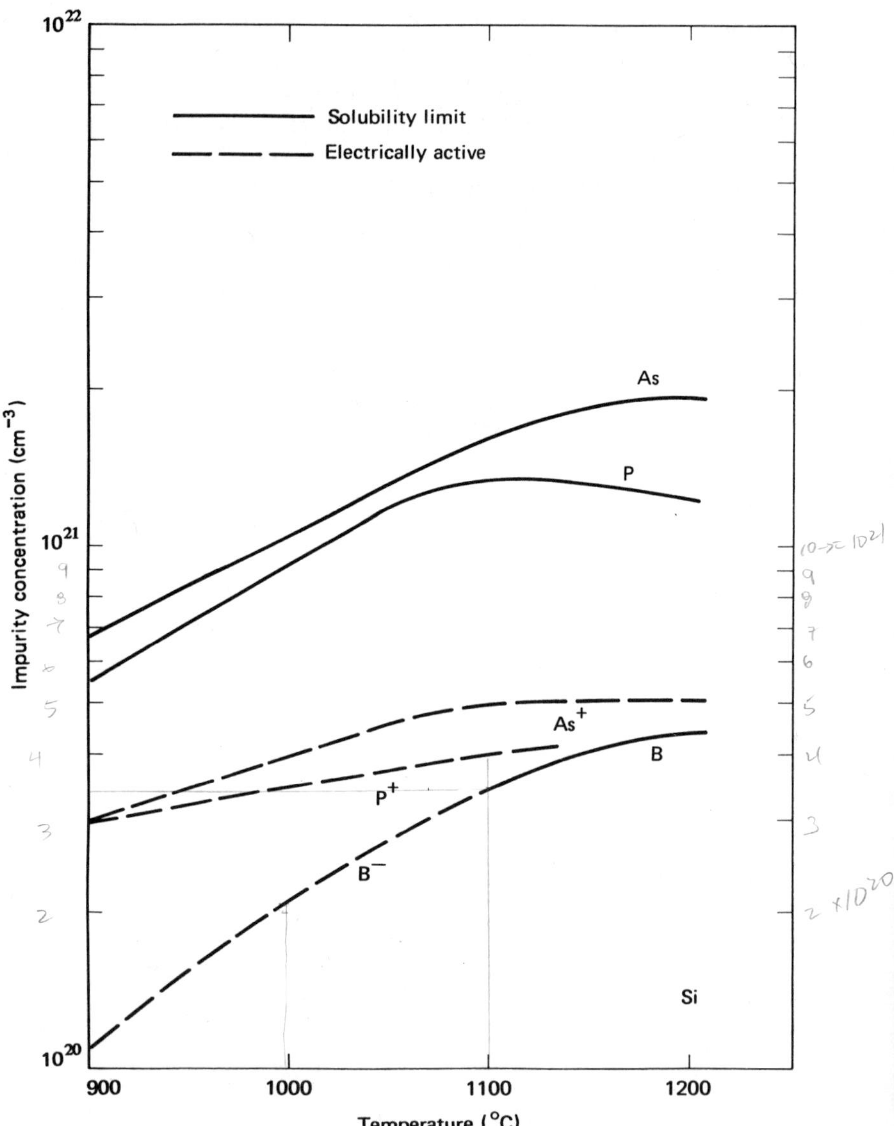

Fig. 4.11 Solubility limits for boron, arsenic, and phosphorus. From R. Colclaser, *Microelectronics: Processing and Device Design* [10], 1980. Reprinted with permission of the publisher, John Wiley & Sons.

3. On freezing, most interstitial–substitutional diffusers end up in both electronically active and electronically inactive sites. The fractions of each type differ widely from element to element. Thus about 90% of gold terminates in active sites, while the corresponding figure for nickel is only 0.1%. The analysis of the results is complex, since analytical

Table 4.2 Diffusion Constant and Activation Energy of Diffusion for Interstitial–Substitutional Diffusers in Silicon[a]

Impurity	D_0 (cm^2/s)	E_0 (eV)
Ag	2×10^{-3}	1.6
Au	1.1×10^{-3}	1.12
Co	0.16	1.12
Cu	4×10^{-2}	1.0
Fe	6.2×10^{-3}	0.87
Ni	1.3×10^{-2}	1.4
O	0.21	2.44
S	0.92	2.2
Zn	0.1	1.4

[a] See reference 2.

techniques such as resistivity measurements only provide information on the part that is electronically active. On the other hand, secondary ion mass spectrometry techniques result in information on the entire impurity content.

4. The electronically active part of all these dopants exhibits one or more deep levels. Thus their average diffusion parameters cannot be measured by *p–n* junction techniques, as for substitutional dopants.
5. The interaction energy between an interstitial–substitutional diffuser and the strain field associated with a dislocation tends to favor clustering in the neighborhood of this dislocation. Thus the diffusion process is dominated by the defect nature of the crystal, and the experimental data are difficult to interpret meaningfully. In addition, many of these impurities form compounds with silicon over certain ranges of diffusion temperature. These tend to segregate in clusters in the silicon lattice and are often electronically inactive. Material doped with these impurities is usually sensitive to heat treatments.

As a result of the above, the data of Table 4.2 must be considered as highly tentative.

Most interstitial–substitutional diffusers exhibit one or more deep lying levels in silicon. Their intentional introduction in high-speed digital circuits reduces the minority carrier lifetime and constitutes an important fabrication step. Gold is today the most commonly used impurity for this purpose. Its energy levels are shown in Fig. 4.12*a*; Table 4.3 lists its capture properties [13, 14].

The use of gold is based on the fact that it has a relatively high solid solubility in silicon (see Fig. 2.22) so that it can be controlled over a wide

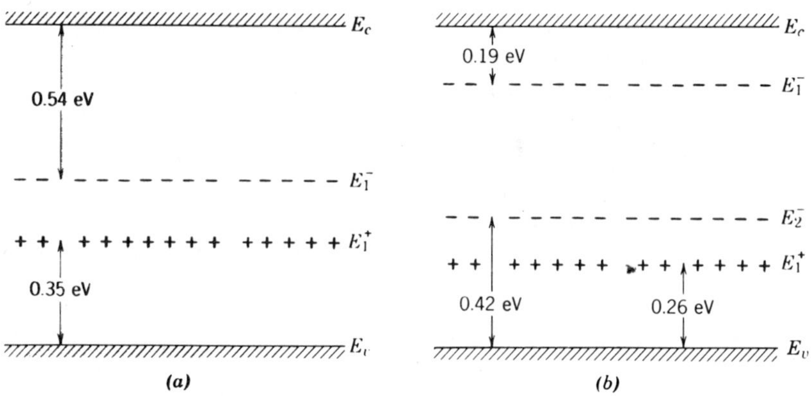

Fig. 4.12 Energy levels of (*a*) gold and (*b*) platinum in silicon.

range of values. Moreover, it does not form any compounds with silicon (as seen from the phase diagram of the gold–silicon system in Fig. 2.7). Thus its behavior is free from anomalous effects that may be caused by the formation or decomposition of these compounds. Such effects are commonly encountered with many interstitially diffusing impurities.

The solid solubilities for gold in interstitial and substitutional sites are shown in Fig. 4.13 [15]. Diffusivities associated with these components are shown in Fig. 4.14. Its behavior in silicon follows that of interstitial–substitutional diffusers. Consequently, its diffusivity is a strong function of the defect concentration for the particular slice, in addition to the doping level and the temperature.

Gold diffusion is generally the last step in the fabrication of high-speed digital microcircuits. Thus diffusion takes place into a wafer in which various doped regions are already present, each of different defect concentration, depending on the type and amount of impurity in it. Moreover, the solid solubility of gold in n^+-silicon is enhanced because of ion-pairing effects.

Table 4.3 Capture Rate Constants for Gold and Platinum in Silicon[a]

	Capture Rate Constants (cm³/s)	
	Au [13]	Pt [16, 17]
Electron capture at $E_1{}^+$	6.3×10^{-8}	2.2×10^{-8}
Hole capture at $E_1{}^+$	2.4×10^{-8}	1.2×10^{-9}
Electron capture at $E_1{}^-$	1.65×10^{-9}	2.4×10^{-9}
Hole capture at $E_1{}^-$	1.15×10^{-7}	1.5×10^{-7}
Electron capture at $E_2{}^-$	—	3.2×10^{-7}
Hole capture at $E_2{}^-$	—	2.7×10^{-5}

[a] See references 14 and 16.

As a result, the doping profile in the wafer cannot be calculated with any degree of accuracy.

In actual practice no attempt is made to establish a specific doping profile, and the wafer is doped by maintaining it at an elevated temperature for what is, for all practical purposes, an infinitely long time (15–30 min). Thus varying amounts of gold are present in the entire microcircuit. The diffusion temperature is altered to adjust the gold concentration in the silicon. However, the actual concentration is also a function of such variables as the manner in which the wafer is cooled to room temperature. The choice of the actual diffusion temperature is determined by trial and error, with a desired end effect in mind.

Platinum has recently been investigated as a substitute for gold in silicon [16]. Like gold, it has a relatively high solid solubility, and diffuses predom-

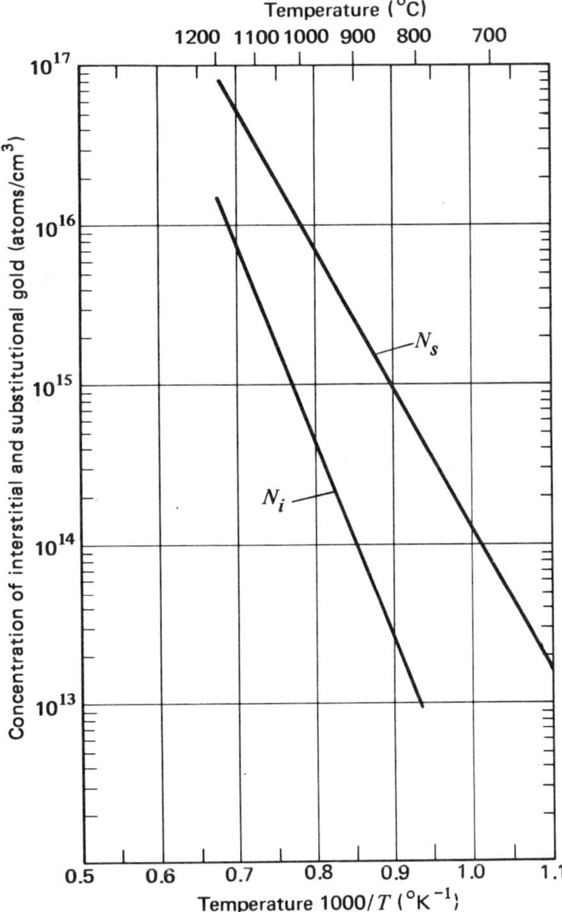

Fig. 4.13 Interstitial and substitutional solubilities of gold in silicon. Adapted from Wilcox and LaChapelle [15].

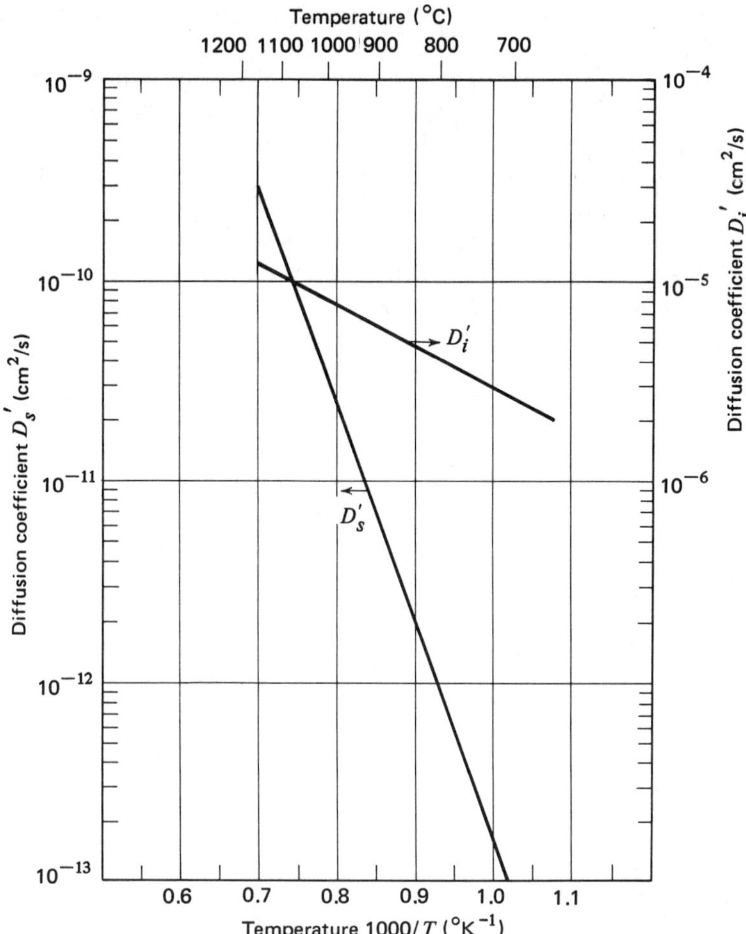

Fig. 4.14 Interstital and substitutional diffusion coefficients of gold in silicon. Adapated from Wilcox and LaChapelle [15].

inantly by an interstitial–substitutional mechanism. Its diffusion constants are as yet undetermined.

About 90% of the diffused platinum ends up in substitutional sites and is electronically active, with a donor level E_1^- and an acceptor level E_1^- as shown in Fig. 4.12*b*. Unlike gold, however, the remainder is also active, and exhibits an acceptor level E_2^-. This is the dominant level for lifetime control. The capture properties of platinum are listed in Table 4.3.

The impurity levels of platinum in silicon are highly asymmetric, whereas one gold level (the acceptor) is almost in the center of the gap. As a result, platinum doping can be used to control the lifetime in silicon [17] without a comparable increase in the leakage current, as is the case with gold. Diffusions are usually conducted from platinum compounds which are made up in the form of spin-on dopants. Techniques for platinum diffusion closely parallel those for gold, and diffusion times are very similar.

4.4 IMPURITY BEHAVIOR: GALLIUM ARSENIDE

The fundamental mechanisms for the diffusion of impurities in gallium arsenide have not been firmly established at the present time. Nevertheless, there is a wealth of experimental data (some conflicting in nature) from which the basic diffusion characteristics can be enunciated. Only one system, zinc in gallium arsenide, has been studied in detail, and many of the results have been extrapolated to other impurities.

Radioactive tracer analysis of the diffusion of gallium and arsenic in gallium arsenide gives the following values for the diffusion coefficients:

$$D_{\mathrm{Ga}} = 0.1e^{-3.2/kT} \tag{4.46a}$$

$$D_{\mathrm{As}} = 0.7e^{-5.6/kT} \tag{4.46b}$$

These very different rates indicate that movement occurs along separate sublattices, with each atom moving on its own sublattice. This argument has been extended to impurities in gallium arsenide. In general these impurities are also considered to diffuse by movement on their sublattices. Typically, impurities from group II (beryllium, cadmium, magnesium, mercury, and zinc) are shallow, p-type, and move along the gallium sublattice. This cannot be accomplished by substitutional diffusion alone, since each gallium site has four arsenic sites as its nearest neighbors. It is generally accepted that these impurities move by an interstitial–substitutional mechanism, with rapid movement in high-concentration regions and slow movement in regions of low concentration.

Impurities from group VI (selenium, sulfur, and tellurium) are shallow, n-type, and move along the arsenic sublattice. Here a divacancy mechanism has been proposed for this type of movement. This approximates substitutional behavior, so that these impurities are extremely slow diffusers. Impurities from group IV (carbon, germanium, silicon, and tin) can be either n- or p-type, depending on the sublattice on which they are preferentially located. It is assumed that they move on both sublattices, but no attempts have been made to propose a detailed diffusion model for them. These have also been found to be extremely slow diffusers. Finally, most other impurities are deep in nature, often exhibiting multiple levels. No information is available on their mechanism of diffusion. In spite of these many uncertainties, it is common practice to cast the diffusion data in the Arrhenius form, $D = D_0e^{-E_0/kT}$. Values for D_0 and E_0 are given in Table 4.4 for various impurities.

Vacancies play a dominant role in diffusion processes in gallium arsenide,

Table 4.4 Diffusion Constant and Activation Energy of Diffusion for Impurities in Gallium Arsenide[a]

Impurity	D_0 (cm²/s)	E_0 (eV)
Au	2.9×10^{1}	2.64
Be	7.3×10^{-6}	1.2
Cr	4.3×10^{3}	3.4
Cu	3×10^{-2}	0.53
Li	5.3×10^{-1}	1.0
Mg	2.6×10^{-2}	2.7
Mn	6.5×10^{-1}	2.49
O	2×10^{-3}	1.1
S	1.85×10^{-2}	2.6
Se	3.0×10^{3}	4.16
Sn	3.8×10^{-2}	2.7
Hg	$D = 5 \times 10^{-14}$	@ 1000°C
Te	$D = 10^{-13}$	@ 1000°C
	$= 2 \times 10^{-12}$	@ 1100°C

[a] See reference 2.

since the p- and n-type impurities ultimately reside in lattice sites. It is interesting to note, however, that these vacancies are generally considered to be neutral or un-ionized, although this point has been questioned by some workers. No information is available on the role of impurity–point defect interactions in determining the diffusion constant.

4.5 THE DIFFUSION EQUATION

Fick's second law may be derived by applying considerations of continuity to Eq. (4.14). Consider the flow of particles in a crystal of cross section A, between planes P_1 and P_2, separated by dx, as shown in Fig. 4.15. The rate of accumulation of particles in the region between planes is $A(\partial N/\partial t)\,dx$. This can also be written as the difference between the fluxes flowing into and out of the region. The flux entering the region at P_1 is Aj and the flux leaving the region at P_2 is $A(j + dj)$. The net flux entering the region is thus $-Adj$. Hence

$$A \frac{\partial N}{\partial t}\, dx = -A\, dj \tag{4.47}$$

But $dj = (\partial j/\partial x)\, dx$. Hence, from Eq. (4.14),

$$\frac{\partial N}{\partial t} = \frac{\partial}{\partial x}\left(D \frac{\partial N}{\partial x}\right) \tag{4.48a}$$

This expression is in a more convenient form than Eq. (4.14) since it involves only volume concentrations. It can be used to obtain the impurity distribution for a variety of diffusion conditions.

4.5.1 The D = Constant Case

The diffusion of shallow impurities in silicon, at relatively low concentrations (n or $p < n_i$ at the diffusion temperature), can be approximated by assuming that the diffusion coefficient is concentration independent. Making this simplication, Fick's second law reduces to

$$\frac{\partial N}{\partial t} = D \frac{\partial^2 N}{\partial x^2} \tag{4.48b}$$

Where N is the volume concentration, in atoms/cm^3, D the diffusion coefficient, in cm^2/s, x the distance, in cm, and t the diffusion time, in s. In the presence of an electric field, this equation is modified to

$$\frac{\partial N}{\partial t} = D \frac{\partial^2 N}{\partial x^2} - \mu \mathscr{E} \frac{\partial N}{\partial x} \tag{4.48c}$$

Equation (4.48b) may be solved for a number of situations which arise in practice. Computations have also been made by several authors for a variety of other diffusion conditions. Some of these have been listed in references at the end of this chapter [18–22].

4.5.1.1 *Diffusion from a Constant Source*

This situation occurs when a wafer is exposed to an impurity source of constant concentration during the diffusion period. The concentration of the impurity is such as to result in a surface concentration of N_0 in the silicon. Here the impurity concentration at any given distance and time is $N(x, t)$,

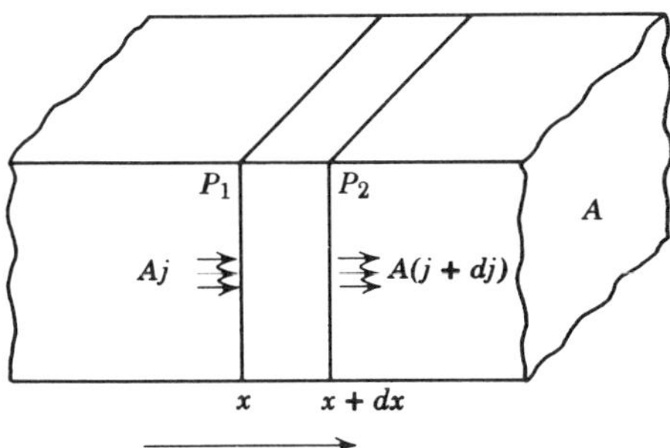

Fig. 4.15 Diffusing flux.

such that

$$N(x, t) = N_0 \text{erfc} \frac{x}{2\sqrt{Dt}} \tag{4.49}$$

where N_0 is the impurity concentration at the silicon surface, in atoms/cm^3, D the value of diffusion constant for the specific diffusion temperature, in cm^2/s, x the penetration depth, in cm, and t the diffusion time, in s.

Figure 4.16 shows a sketch of this concentration for various diffusion times. The most significant characteristic of this type of diffusion is that the surface concentration is constant whereas the diffusion depth increases with time. A normalized plot of Eq. (4.49) is shown in Fig. 4.17. A diffusion of this type is referred to as a *complementary error function* (erfc) diffusion.

The erfc diffusion is performed by exposing the slice to a constant concentration of the impurity during the entire process. At low levels the surface concentration is given by the value that is in equilibrium with the surrounding gas. With increasing source concentration, the surface concentration rises until it is ultimately set by the solid solubility of the impurity for that specific diffusion temperature. The emitter diffusion of a bipolar silicon transistor is usually of this type. Here a high surface concentration is desired, necessitating the use of high diffusion temperatures (1100–1200°C).

Since the diffusion equation is linear, superposition may be used to calculate the effect of a finite background. For example, a p–n junction is fabricated by diffusing a p-type impurity into an n-type wafer of background concentration N_C, as shown in Fig. 4.16. The effective doping concentration at any point is given by the difference of acceptor and donor concentrations, whichever is larger. If the p-type impurity profile is given by

$$N(x, t) = N_0 \text{ erfc} \frac{x}{2\sqrt{Dt}} \tag{4.49}$$

the junction is located at x_j, such that

$$N_0 \text{ erfc} \frac{x_j}{2\sqrt{Dt}} = N_C \tag{4.50}$$

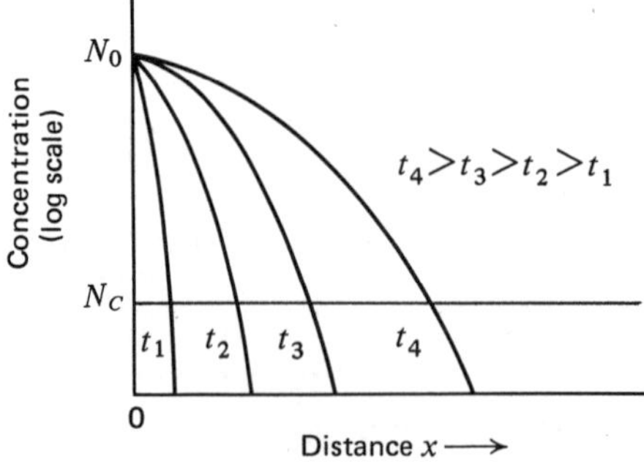

Fig. 4.16 Constant-source diffusion profiles.

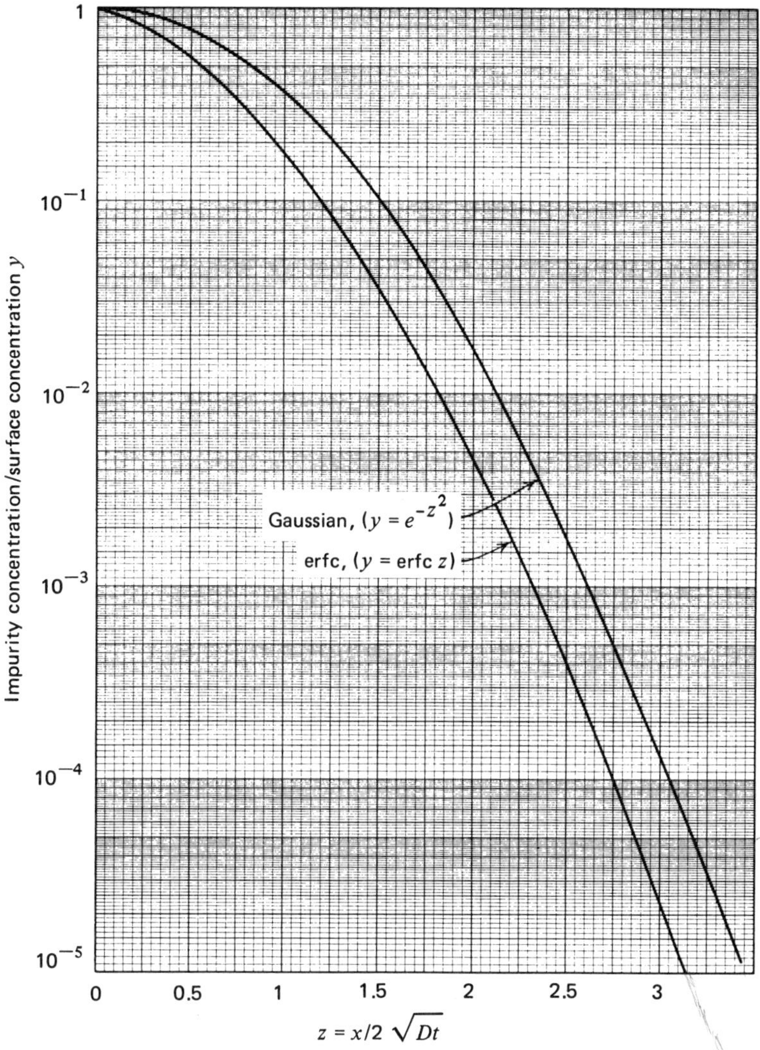

Fig. 4.17 The erfc and gaussian profiles.

4.5.1.2 Diffusion from a Limited Source

Here a finite quantity of the diffusing matter is first placed on the wafer. Diffusion proceeds from this limited source, and it is assumed that all of this matter is consumed. The impurity concentration resulting from this type of diffusion is given by

$$N(x, t) = \frac{Q_0}{\sqrt{\pi D t}} e^{-(x/2\sqrt{Dt})^2} \tag{4.51}$$

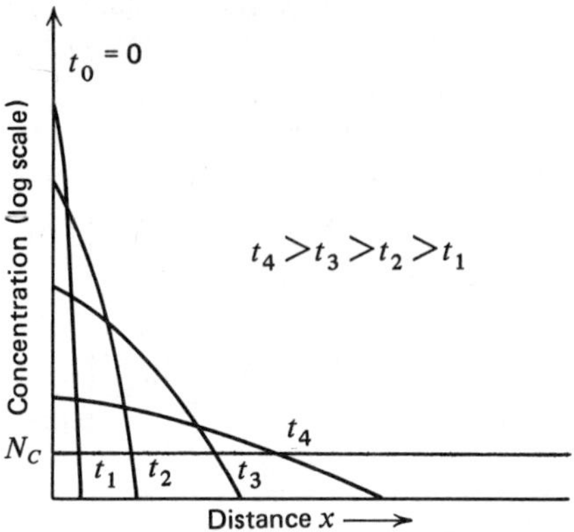

Fig. 4.18 Limited-source diffusion profiles.

where Q_0 is the amount of matter placed on the surface prior to diffusion, in atoms/cm^2, D the diffusion coefficient, in cm^2/s, x the diffusion distance, in cm, and t the diffusion time, in s. A diffusion of this type is known as a *gaussian diffusion*.

Figure 4.18 shows a sketch of the diffusion profile for various diffusion times. The effect of a background concentration N_C in establishing the position of the junction is also shown here. A normalized plot of Eq. (4.51) is shown in Fig. 4.17.

A comparison of Figs. 4.16 and 4.18 shows that the significant difference between these two types of diffusion lies in the fact that the surface concentration of the gaussian diffusion changes for varying diffusion times, whereas that of the erfc diffusion does not. Otherwise they are essentially similar and can be approximated by exponential profiles at concentration levels that are two or more orders of magnitude below the surface concentration.

Limited source diffusion is ideally suited for those cases where a relatively low value of surface concentration is required in conjunction with a high diffusion depth.* The base diffusion of a bipolar silicon transistor presents this type of situation. Surface concentrations encountered here are below those needed for the emitter diffusion, whereas the diffusion depth is higher.

Figure 4.19 shows a bipolar silicon transistor, formed by successive gaussian and erfc diffusions into a wafer with a collector background concentration given by N_C. Typical values for a modern high-speed digital transistor are also indicated.

The quantity of matter required for a gaussian diffusion is considerably less than a monolayer. Thus special techniques must be used for its accurate placement on the slice. A direct approach is to use ion implantation methods,

* Although the low surface concentration can also be obtained by diffusion from a constant source at low diffusion temperature, the amount of time required for the desired penetration depth would be prohibitive.

which are described in detail in Chapter 6. An indirect approach, which can also be used, is to conduct a constant-source diffusion at a low temperature (≃900°C) for a short time. By this process, known as *predeposition*, a small quantity of the impurity is transported into an extremely thin layer *in* the silicon slice. Since the penetration depth is negligibly small, this provides a suitable approximation to placing the matter *on* the surface. The actual amount of matter transported during the predeposition phase is given by (see Appendix)

$$Q_0 = 2N_{01}\left(\frac{D_1 t_1}{\pi}\right)^{1/2} \tag{4.52}$$

where Q_0 is the amount of matter entering the silicon during predeposition, in atoms/cm^2, N_{01} the surface concentration at the predeposition temperature, in atoms/cm^3, D_1 the diffusion coefficient at the predeposition temperature, cm^2/sec, and t_1 the predeposition time, in s.

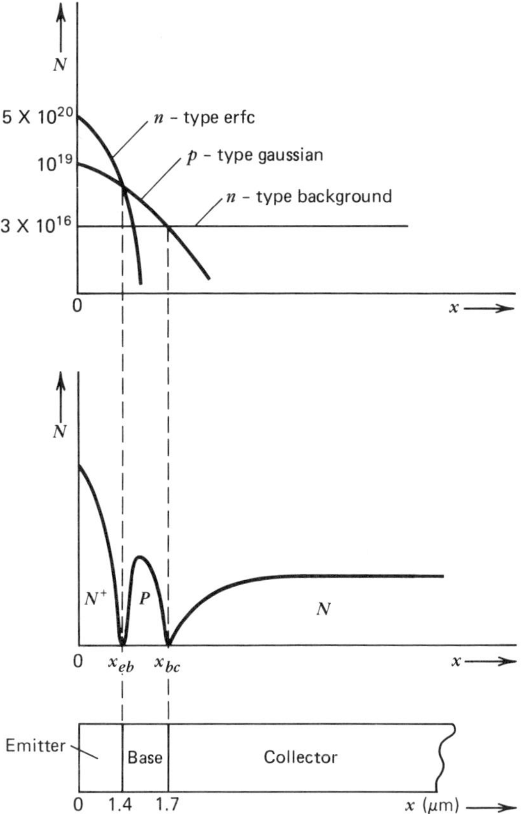

Fig. 4.19 Bipolar transistor formation by successive diffusions.

The external source of impurity is now removed by dissolving all impurity-bearing compounds on the surface in hydrofluoric acid, and the slice is subjected to a high-temperature drive-in phase for an appropriate time and temperature. The final impurity concentration is

$$N(x, t_1, t_2) = \frac{2N_{01}}{\pi}\left(\frac{D_1 t_1}{D_2 t_2}\right)^{1/2} e^{-(x/2\sqrt{D_2 t_2})^2} \tag{4.53}$$

where the subscript 2 refers to drive-in parameters.

Equation (4.53) assumes an extremely short predeposition phase compared with the drive-in phase. Thus it holds as long as

$$\sqrt{D_1 t_1} \ll \sqrt{D_2 t_2} \tag{4.54}$$

4.5.1.3 The Two-Step Diffusion

This diffusion technique is an outgrowth of the last process. It presents so many advantages over the previous methods, however, that it is the most common diffusion process today for silicon devices and microcircuits.

The two-step diffusion is initiated by first conducting a constant-source diffusion at a low temperature for a short time. The impurity supply is shut off and the drive-in phase initiated in an oxidizing ambient. This results in the formation of a surface oxide on the silicon, preventing the further in-diffusion of impurities which are present *on* the slice and inhibiting the out-diffusion of impurities which were already predeposited *in* the slice. The final impurity concentration is a function of the predeposition and drive-in parameters. If the subscripts 1 and 2 are used to denote predeposition and drive-in, respectively, an erfc diffusion results when $D_1 t_1 \gg D_2 t_2$. Conversely, a gaussian profile results if $D_1 t_1 \ll D_2 t_2$.

In a practical situation neither of these inequalities holds. For this case the resulting distribution has been obtained [23] for the analogous heat-flow problem. These results may be modified to give the final impurity distribution as

$$N(x, t_1, t_2) = \frac{2N_{01}}{\pi}\int_0^U \frac{e^{-\beta(1+U^2)}}{1+U^2}\,dU \tag{4.55}$$

where

$$U = \left(\frac{D_1 t_1}{D_2 t_2}\right)^{1/2} \tag{4.56}$$

and

$$\beta = \left(\frac{x}{2\sqrt{D_1 t_1 + D_2 t_2}}\right)^2 \tag{4.57}$$

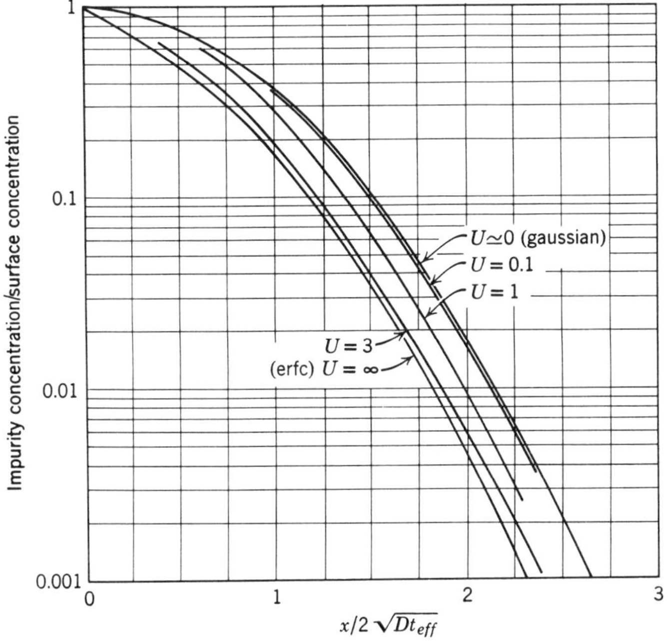

Fig. 4.20 Profiles for the two-step diffusion process.

In addition, it can be shown that the final surface concentration is given by

$$N_{02} = \frac{2N_{01}}{\pi} \tan^{-1} U \tag{4.58}$$

Hence

$$\frac{N(x, t_1, t_2)}{N_{02}} = \frac{1}{\tan^{-1} U} \int_0^U \frac{e^{-\beta(1+U^2)}}{1 + U^2}\, dU \tag{4.59}$$

Table 4.5 gives values of the integral of Eq. (4.59). In addition, Fig. 4.20 shows normalized concentration profiles for two-step diffusions with various values of U as well as the limiting cases of erfc and gaussian diffusions.

The two-step process may be used to approximate an erfc diffusion profile without the accumulation of large amounts of impurity on the surface of the silicon. Thus it avoids the possibility of surface damage. A slow oxidation process,* using dry oxygen, is used for this purpose. In addition, this process can be used to approximate a gaussian diffusion while avoiding the necessity of exposing the surface of the slice at any time. A fast oxidation process, using steam, is more desirable here. Junctions are always formed beneath

* Oxidation processes are covered in detail in Chapter 7.

Table 4.5 Values of the Smith Integral[a]

U \ β	0·1	0·2	0·3	0·4	0·5	0·6	0·7	0·8	0·9	1·0	1·1	1·2
0.1	0.09015	0.08155	0.07376	0.06672	0.06035	0.05459	0.04938	0.04467	0.04040	0.03655	0.03306	0.02990
0.2	0.17838	0.16119	0.14566	0.13162	0.11894	0.10748	0.09713	0.08777	0.07931	0.07167	0.06477	0.05853
0.3	0.26295	0.23723	0.21403	0.19310	0.17422	0.15719	0.14182	0.12795	0.11545	0.10416	0.09398	0.08479
0.4	0.34254	0.30837	0.27761	0.24993	0.22501	0.20259	0.18240	0.16422	0.14786	0.13314	0.11988	0·10794
0.5	0.41626	0.37374	0.33557	0.30132	0.27058	0.24299	0.21822	0.19599	0.17603	0.15812	0.14203	0.12759
0.6	0.48366	0.43290	0.38751	0.34692	0.31062	0.27814	0.24908	0.22308	0.19982	0.17900	0.16036	0.14368
0.7	0.54464	0.48580	0.43340	0.38673	0.34515	0.30809	0.27505	0.24562	0.21937	0.19596	0.17508	0.15645
0.8	0.59940	0.53264	0.47347	0.42100	0.37447	0.33317	0.29652	0.26398	0.23508	0.20940	0.18657	0.16628
0.9	0.64829	0.57380	0.50812	0.45017	0.39903	0.35385	0.31393	0.27864	0.24742	0.21979	0.19532	0.17365
1.0	0.69176	0.60975	0.53784	0.47475	0.41935	0.37066	0.32783	0.29013	0.25693	0.22765	0.20183	0.17903
1.1	0.73033	0.64100	0.56318	0.49529	0.43600	0.38415	0.33877	0.29900	0.26411	0.23348	0.20655	0.18286
1.2	0.76448	0.66808	0.58465	0.51232	0.44950	0.39486	0.34726	0.30574	0.26946	0.23772	0.20991	0.18553
1.3	0.79470	0.69148	0.60276	0.52634	0.46035	0.40327	0.35377	0.31078	0.27336	0.24074	0.21225	0.18734
1.4	0.82144	0.71164	0.61797	0.53781	0.46901	0.40979	0.35870	0.31449	0.27616	0.24286	0.21385	0.18855
1.5	0.84509	0.72899	0.63069	0.54714	0.47586	0.41482	0.36238	0.31720	0.27815	0.24431	0.21492	0.18933
1.6	0.86601	0·74388	0.64130	0.55469	0.48123	0.41865	0.36511	0.31914	0.27953	0.24530	0.21562	0.18983
1.7	0.88454	0.75666	0.65010	0.56076	0.48542	0.42153	0.36710	0.32051	0.28048	0.24595	0.21607	0.19014
1.8	0.90095	0.76759	0.65739	0.56562	0.48865	0.42369	0.36854	0.32147	0.28112	0.24638	0.21636	0.19033
1.9	0.91549	0.77693	0.66340	0.56948	0.49114	0.42529	0.36956	0.32213	0.28154	0.24665	0.21653	0.19045
2.0	0.92838	0.78491	0.66833	0.57254	0.49303	0.42646	0.37029	0.32258	0.28182	0.24682	0.21664	0.19051
2.5	0.97404	0.81009	0.68228	0.58029	0.49735	0.42887	0.37165	0.32335	0.28225	0.24707	0.21678	0.19059
3.0	0.99920	0.82094	0.68698	0.58234	0.49825	0.42928	0.37183	0.32343	0.28229	0.24708	0.21679	0.19059
∞	1.02834	0.82795	0.68892	0.58291	0.49843	0.42933	0.37184	0.32343	0.28229	0.24709	0.21679	0.19059

U \ β	1.3	1.4	1.5	1.6	1.7	1.8	1.9	2.0	2.5	3.0	4.0	5.0
0.1	0.02705	0.02446	0.02213	0.02002	0.01811	0.01638	0.01481	0.01340	0.00811	0.00491	0.00180	0.00066
0.2	0.05289	0.04779	0.04319	0.03903	0.03527	0.03187	0.02880	0.02603	0.01568	0.00945	0.00343	0.00125
0.3	0.07651	0.06903	0.06228	0.05620	0.05071	0.04575	0.04128	0.03725	0.02228	0.01333	0.00477	0.00171
0.4	0.09720	0.08752	0.07881	0.07097	0.06391	0.05756	0.05183	0.04668	0.02766	0.01640	0.00577	0.00204
0.5	0.11462	0.10297	0.09251	0.08312	0.07468	0.06711	0.06030	0.05419	0.03178	0.01866	0.00645	0.00224
0.6	0.12875	0.11538	0.10340	0.09268	0.08308	0.07448	0.06678	0.05988	0.03475	0.02021	0.00688	0.00236
0.7	0.13982	0.12499	0.11174	0.09992	0.08936	0.07993	0.07150	0.06398	0.03677	0.02120	0.00712	0.00242
0.8	0.14824	0.13219	0.11790	0.10519	0.09387	0.08379	0.07481	0.06680	0.03806	0.02180	0.00724	0.00244
0.9	0.15444	0.13741	0.12230	0.10889	0.09699	0.08642	0.07702	0.06867	0.03885	0.02213	0.00730	0.00245
1.0	0.15889	0.14109	0.12535	0.11141	0.09907	0.08814	0.07844	0.06985	0.03931	0.02231	0.00733	0.00246
1.1	0.16200	0.14361	0.12739	0.11307	0.10041	0.08923	0.07933	0.07056	0.03956	0.02240	0.00734	0.00246
1.2	0.16411	0.14529	0.12872	0.11412	0.10125	0.08989	0.07985	0.07098	0.03969	0.02244	0.00735	0.00246
1.3	0.16552	0.14638	0.12956	0.11478	0.10176	0.09028	0.08016	0.07122	0.03976	0.02246	0.00735	0.00246
1.4	0.16643	0.14706	0.13008	0.11517	0.10205	0.09051	0.08033	0.07134	0.03979	0.02247	0.00735	0.00246
1.5	0.16700	0.14749	0.13039	0.11540	0.10222	0.09063	0.08042	0.07141	0.03980	0.02247	0.00735	0.00246
1.6	0.16736	0.14774	0.13057	0.11552	0.10231	0.09070	0.08046	0.07144	0.03981	0.02247	0.00735	0.00246
1.7	0.16757	0.14789	0.13067	0.11559	0.10236	0.09073	0.08049	0.07146	0.03981	0.02247	0.00735	0.00246
1.8	0.16770	0.14797	0.13073	0.11563	0.10239	0.09075	0.08050	0.07147	0.03982	0.02247	0.00735	0.00246
1.9	0.16777	0.14802	0.13076	0.11565	0.10240	0.09075	0.08050	0.07147	0.03982	0.02247	0.00735	0.00246
2.0	0.16781	0.14804	0.13078	0.11566	0.10240	0.09076	0.08051	0.07147	0.03982	0.02247	0.00735	0.00246
2.5	0.16786	0.14807	0.13079	0.11567	0.10241	0.09076	0.08051	0.07147	—	—	—	—
3.0	0.16786	0.14807	0.13079	0.11567	0.10241	0.09076	0.08051	0.07147	—	—	—	—
∞	0.16786	0.14807	0.13079	0.11567	0.10241	0.09076	0.08051	0.07147	0.03982	0.02247	0.00735	0.00246

[a] From reference 23.

the protective layers and show excellent breakdown characteristics. Finally, the glass may be used as a mask for the next diffusion step.

4.5.1.4 Successive Diffusions

It is often required to calculate the total effect of diffusion during a series of temperature cycles. By way of example the emitter-diffusion step takes place after the base drive-in. Thus the impurities in the base are subjected to one set of time and temperature values during the base drive-in phase and to a second set during the emitter-diffusion phase. If a gold-diffusion step follows, the base is subjected to yet another time and temperature cycle. To compute the total effect of these cycles, it is necessary to obtain an effective Dt product for this region. The effective Dt product is given by

$$(Dt)_{\text{eff}} = \sum D_1 t_1 + D_2 t_2 + D_3 t_3 + \cdots \tag{4.60}$$

where t_1, t_2, t_3, . . . are the different diffusion times, and D_1, D_2, D_3, . . . are the appropriate diffusion constants in effect during these times.

Cooperative diffusion effects can also occur, especially in situations where a high-concentration diffusion is performed. Often this results in enhancing the atom movement of earlier diffusions. This process is discussed more fully in Section 4.8. [5.]

4.5.2 The $D = f(N)$ Case

The greater majority of diffusion situations are described by a concentration dependent diffusivity. Calculations of the doping profile for these cases are not available in closed form. However, computer-derived solutions have been obtained for many cases of interest.

Most of these computations have been made for a constant-source diffusion, so that the surface concentration and diffusivity are constant for any specific diffusion temperature. All of these result in profiles that are more abrupt than those given by the erfc function, for which D is independent of N.

Consider the situation where the diffusivity can be written directly as some power of the concentration, such as

$$D = K_1 D_0 N \tag{4.61a}$$

$$D = K_2 D_0 N^2 \tag{4.61b}$$

$$D = K_3 D_0 N^3 \tag{4.61c}$$

These equations can be rewritten in normalized form as

$$D = D_{\text{sur}} \left(\frac{N}{N_{\text{sur}}} \right) \tag{4.62a}$$

$$D = D_{\text{sur}} \left(\frac{N}{N_{\text{sur}}} \right)^2 \tag{4.62b}$$

$$D = D_{\text{sur}} \left(\frac{N}{N_{\text{sur}}} \right)^3 \tag{4.62c}$$

Here N_{sur} is the surface concentration and D_{sur} is the value of the diffusion constant at this concentration. For these cases it can be shown (see Section A.3 in the Appendix) that Fick's second law can be rewritten as an ordinary differential equation, and solved by numerical computation techniques. These solutions, for a constant-source diffusion, are shown in Fig. 4.21 [24] together with the solution for D = constant. Tabulated values for these distributions are given in Table A.2.

From this figure it is seen that doping profiles are extremely steep at low concentrations ($\ll N_{sur}$). Consequently, highly abrupt junctions are formed when diffusions are made into a background of opposite impurity type. Furthermore, the steepness of the doping profile results in a junction depth which is given by the value for which $N \simeq 0$, i.e., it is almost independent of the background concentration.

From Table A.2 it is seen that the junction depth for these types of diffusions is given by

$$x_j = 1.616(D_{sur}t)^{1/2}, \qquad D \propto N \tag{4.63a}$$

$$x_j = 1.092(D_{sur}t)^{1/2}, \qquad D \propto N^2 \tag{4.63b}$$

$$x_j = 0.872(D_{sur}t)^{1/2}, \qquad D \propto N^3 \tag{4.63c}$$

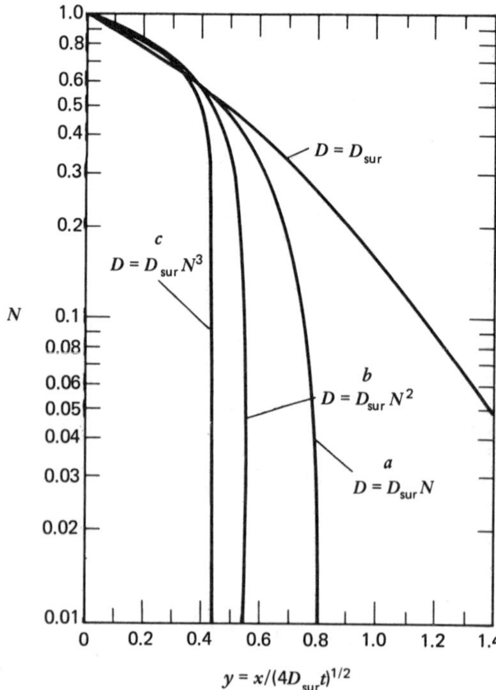

Fig. 4.21 Normalized diffusion profiles for concentration dependent diffusions. From Weisberg and Blanc [24]. Reprinted with permission of the American Physical Society.

In some situations, dopant behavior can be characterized by a combination of the forms described by Eqs. (4.62*a*)–(4.62*c*). For these cases the junction depth occurs at some intermediate point. Specific situations which arise in practice are now described.

4.5.2.1 Interstitial–Substitutional Diffusers

Consider a dopant whose movement can be described by an interstitial–substitutional mechanism. Here these different types of jumps may be considered to be uncorrelated. Moreover, since the diffusion coefficient has been shown to be linearly related to the jump frequency, it follows from Eq. (4.48*a*) that

$$\frac{\partial}{\partial t}(N_s + N_i) = \frac{\partial}{\partial x}\left(D_s' \frac{\partial N_s}{\partial x} + D_i' \frac{\partial N_i}{\partial x}\right) \qquad (4.64a)$$

where N_s and N_i are the substitutional and interstitial concentrations, and D_s' and D_i' are the substitutional and interstitial diffusivities, respectively. Since $N_s > N_i$, this equation reduces to

$$\frac{\partial N_s}{\partial t} \simeq \frac{\partial}{\partial x}\left(D_s' + D_i' \frac{\partial N_i}{\partial N_s}\right) \frac{\partial N_s}{\partial x}\,] \qquad (4.64b)$$

Comparison with Eq. (4.48*a*) gives the effective diffusivity as

$$D \simeq D_s' + D_i' \frac{\partial N_i}{\partial N_s} \simeq D_i' \frac{\partial N_i}{\partial N_s} \qquad (4.65)$$

since $D_i' \gg D_s'$. The relationship $\partial N_i/\partial N_s$ is dependent on the nature of the charge species involved in the dissociation reaction between the interstitial and the substitutional.

The behavior of zinc in gallium arsenide can be described in this way. Here it is generally accepted that substitutional zinc is a singly ionized acceptor, whereas interstitial zinc is a singly ionized donor. Moreover, the gallium vacancy is neutral. For this situation, the dissociation reaction can br written as

$$M_s^- \rightleftharpoons V_{Ga}{}^0 + M_i^+ + 2e^- \qquad (4.66)$$

where M_s^- and M_i^+ represent the substitutional and interstitial impurities, and $V_{Ga}{}^0$ is the vacancy. Applying the mass-action principle, gives

$$K_1 = \frac{n^2[Zn_i^+][V_{Ga}{}^0]}{[Zn_s^-]} \qquad (4.67)$$

Since the hole concentration is due to the substitutional zinc, and $np = n_i^2$,

$$N_i = \frac{K_2 N_s^3}{[V_{Ga}^0]} \tag{4.68}$$

Differentiating this equation, and combining with Eq. (4.65), gives

$$D \simeq \frac{D_i' \, K_3 N_s^2}{[V_{Ga}^0]} \tag{4.69}$$

For any specific diffusion temperature, the surface concentration and the surface diffusivity are fixed, as is the gallium vacancy concentration. Consequently, Eq. (4.69) can be rewritten in the normalized form

$$D \simeq D_{sur} \left(\frac{N}{N_{sur}} \right)^2 \tag{4.70}$$

the same as Eq. (4.62*b*). Diffusion profiles of zinc in gallium arsenide closely follow curve *b* in Fig. 4.21, which represents this case, and result in a junction depth of 1.092 $(D_{sur}t)^{1/2}$.

Note that the N^2 dependence arises from the fact that the dissociation reaction involves a charge change of +2. It can be shown that the assumption of a doubly ionized donor state for Zn_i would result in an N^3 dependence of the diffusivity. It has been suggested that a small fraction of the zinc is a doubly ionized donor, and diffuses according to this law [25]. As a consequence, the junction depth is slightly less than that given by Eq. (4.63*b*).

4.5.2.2 Substitutional Diffusers

Substitutional diffusers are used for junction formation in silicon technology. Here we have pointed out that deep junctions, with relatively low surface concentration, can be characterized by a constant diffusivity. However, VLSI applications require the use of shallow, high-concentration junctions. These are more correctly described by the concentration dependent diffusivity given by Eq. (4.40). Solution of the resulting diffusion equation in closed form is difficult, if not impossible, for many of these cases. These are now considered, together with some approximate solutions.

Arsenic in Silicon. The concentration dependent diffusivity of arsenic in silicon is described by Eq. (4.40), in which the interaction of As(I-V^0) and As(I-V^-) must be considered, so that

$$D = h \left[D_i^0 + D_i^- \left(\frac{n}{n_{ie}} \right) \right] \tag{4.71}$$

where n_{ie} is the intrinsic electron concentration for solid-solubility-limited diffusions at high concentration. The field enhancement factor can be approximated by 2 for these diffusions. Assuming that the electron concentration is given by the doping concentration N, and noting that $N \gg n_{ie}$, this equation can be rewritten as

$$D \simeq 2D_i^- \left(\frac{N}{n_{ie}}\right) \tag{4.72}$$

This is of the same form as Eq. (4.61a), and results in the abrupt doping profile shown by curve a in Fig. 4.21. The junction depth is thus independent of the background concentration, and is given by

$$x_j = 1.616\,(D_{\text{sur}}t)^{1/2} \tag{4.73}$$

From Eq. (4.72),

$$D_{\text{sur}} = 2D_i^- \left(\frac{N_{\text{sur}}}{n_{ie}}\right) \tag{4.74}$$

so that

$$x_j = 2.29 \left(\frac{N_{\text{sur}}}{n_{ie}}\right)^{1/2} (D_i^-\,t)^{1/2} \tag{4.75}$$

where (see Table 4.1)

$$D_i^- = 22.9e^{-4.1/kT} \tag{4.76}$$

A closed-form solution to Eq. (4.72) can also be written in polynomial form as

$$N = N_{\text{sur}}(1 - 0.87Y - 0.45Y^2) \tag{4.77}$$

where

$$Y = x(4D_{\text{sur}}t)^{-1/2} \tag{4.78}$$

and results in the same junction depth. This equation can be readily manipulated [26] to determine additional properties of arsenic-diffused layers in silicon (see Section 4.10.4).

Boron in Silicon. Boron diffusions are often made with a low surface concentration, so that a constant diffusivity can be used to characterize them. In some situations, however, high concentrations are required. Here the interaction of $B(I\text{-}V^0)$ and $B(I\text{-}V^+)$ must be considered in Eq. (4.40), so that

$$D = h\left[D_i^0 + D_i^+\left(\frac{p}{n_i}\right)\right] \tag{4.79}$$

It is not possible to make simplifying approximations in this case, although the field enhancement factor can usually be neglected. Qualitatively, however, the boron profile can be expected to be more abrupt than the erfc (i.e., the D = constant) case, but somewhat less abrupt than the arsenic profile (i.e., the case where $D \propto N$).

Empirical data on a large number of high-concentration ($N_{sur} \geq 2 \times 10^{19}$ cm^{-3}) boron diffusions made into background material that is doped to less than 10^{18} cm^{-3} have shown [27] that the junction depth can be approximated by

$$x_j \simeq 2.45\left(\frac{N_{sur}}{n_i}\right)^{1/2}(Dt)^{1/2} \tag{4.80}$$

where

$$D = 3.17e^{-3.59/kT} \tag{4.81}$$

It has also been shown that, for these conditions, the doping profile can be closely approximated by

$$N \simeq N_{sur}(1 - Y^{2/3}) \tag{4.82}$$

where

$$Y = \left(\frac{x^2}{6D_{sur}t}\right)^{3/2} \tag{4.83}$$

This polynomial approximation can be used in the evaluation of diffused layers (see Section 4.10.4).

A plot of this profile is shown in Fig. 4.22. Also shown for comparison are the profiles for arsenic in silicon, and the erfc function which holds for the constant diffusivity case. Note that the boron profile is intermediate to these two.

Phosphorus in Silicon. The effective diffusion constant for phosphorus in silicon involves interactions of $P(I\text{-}V^0)$, $P(I\text{-}V^-)$, and $P(I\text{-}V^{2-})$. As a result, the doping profile is dominated by dissociation effects which come into play as the Fermi level sweeps past the V^{2-} level [28]. Straightforward solution

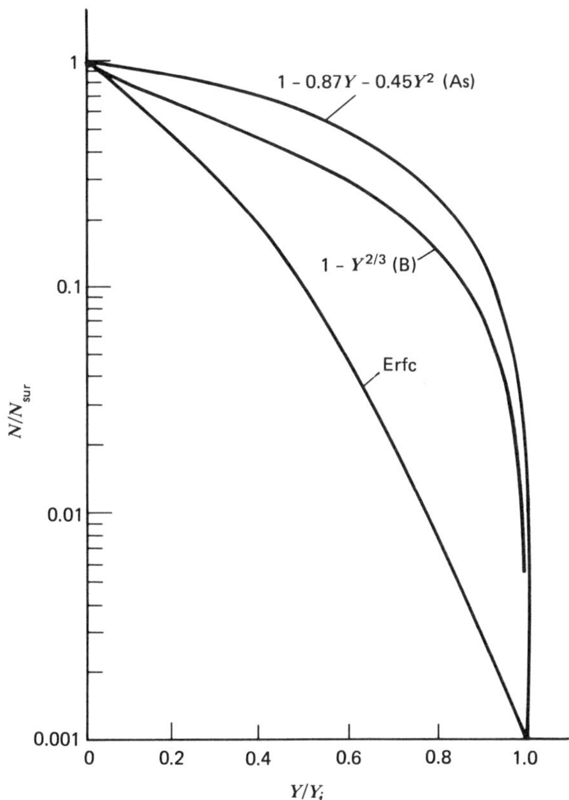

Fig. 4.22 Normalized diffusion profiles for boron and arsenic in silicon.

of the diffusion equation cannot be used to describe this situation. The detailed nature of the doping profiles for this purity is described in Section 4.8.4.

4.5.3 Lateral Diffusion Effects

In practice, diffusion in microcircuits is always carried out through windows cut in the mask that is placed on the slice. The one-dimensional diffusion equation represents a satisfactory means of describing this process, except at the edge of the mask window. Here the dopant source provides impurities which diffuse at right angles to the semiconductor surface as well as parallel to it (i.e., laterally).

Contours of constant doping concentration [29] resulting from this situation are shown in Fig. 4.23*a* for constant-source diffusing conditions, on the assumption of a concentration independent diffusion constant. These contours are, in effect, a map of the location of the junctions created by diffusing into various background concentrations. It is seen from this figure

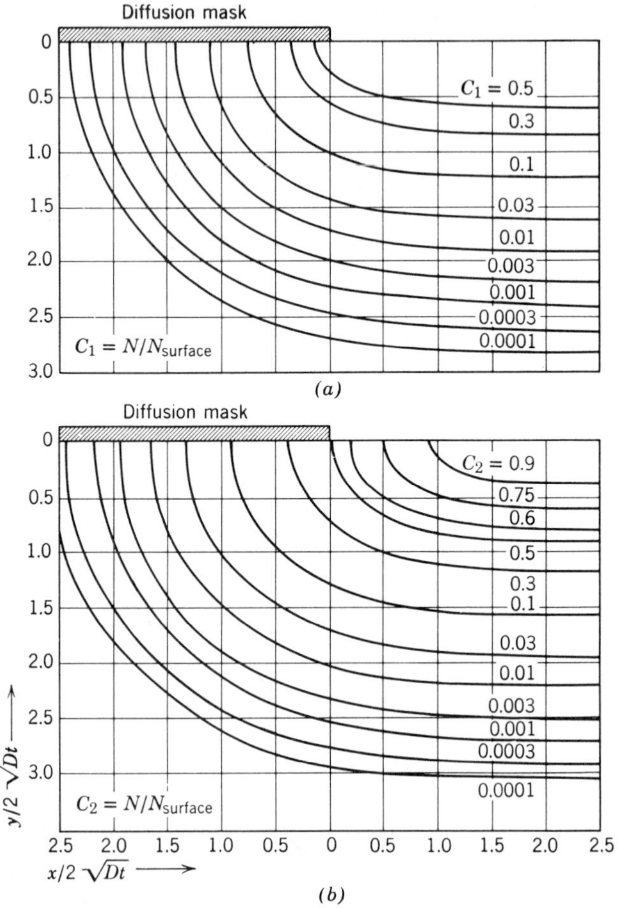

Fig. 4.23 Diffusion contours at the edge of an oxide window.

that the lateral penetration is about 75–85% of the penetration in the vertical direction for concentrations that are two or more orders of magnitude below the surface concentration.

Figure 4.23*b* shows the contours of constant doping concentration for limited-source diffusions, again assuming a concentration independent diffusion constant. Here too, the penetration ratio (lateral to vertical) is about 75–85% for background concentrations that are two or more orders of magnitude below the surface concentration. Note that the depletion of this source results (for some cases) in actually terminating the contours within the window. This is of no consequence in practice, since a junction is usually formed by doping into a background that is considerably below the surface concentration.

More recently detailed solutions have been made [30] for the case where the diffusion constant is concentration dependent. These solutions are of special interest for shallow diffusions with high surface concentration, of the type encountered in many VLSI applications. Here it has been shown that, in addition to the more abrupt doping profile in both directions, the lateral junction penetration is greatly reduced. Typically the ratio of lateral to vertical penetration for these diffusion conditions ranges from 65 to 70%.

In some VLSI situations the width of a diffusion window may become so narrow so that it must be treated as a line source. A constant-source diffusion through such a slot is approximately cylindrical in nature for this limiting case, and is represented by a gaussian profile in the radial direction. Here

$$N(r, t) = \frac{\Delta}{2\pi Dt} e^{-r^2/4Dt} \tag{4.84}$$

where r is the radius and Δ is the number of impurities per unit length of the slot. For a narrow slot of width W,

$$\Delta \simeq Q_0 W \tag{4.85}$$

where Q_0 is the dopant density on the silicon surface, in atoms/cm^2. Analysis of this equation shows that junctions formed in this manner will have a smaller depth than those made through a wide slot, which approximates a parallel-plane source.

4.6 DIFFUSION SYSTEMS

There are many common aspects of diffusion technology for both silicon and gallium arsenide. This section deals with these aspects. The special problems of these semiconductors are considered at a later point.

The basic requirements of any diffusion system are that a means be provided for bringing the diffusing impurity in contact with a suitably prepared slice, and that this process be maintained for a specific time and at a specific temperature. Within the broad framework of these requirements, the following additional features are also desired:

1. The surface concentration should be capable of being controlled over a wide range of values, up to the solid-solubility limit of the impurity.
2. The diffusion process should not result in damage to the surface of the slice. This is an extremely important requirement, because the entire microcircuit is fabricated within the first few micrometers below this surface.
3. After diffusion, material residing on the surface of the slice should be capable of easy removal, if so desired.
4. The system should give reproducible results from one diffusion run to the next and from slice to slice within a single run.
5. The system should be capable of processing a number of slices at a time.

Diffusion systems may be of either sealed- or open-tube type. In the sealed system the slices and dopant are enclosed in a clean, evacuated quartz tube prior to heat treatment. After diffusion the slices are removed by breaking the tube. Systems of this type can be easily maintained free from contamination. Until recently, however, their use for silicon diffusions has been restricted to the laboratory because of the inconvenience of sealing and unsealing these tubes. Arsenic diffusions are a notable exception (see Section 4.7.3).

Sealed-tube systems operate by thermal evaporation of the dopant source, transport in the gas phase, adsorption on the surface of the semiconductor and the tube walls, and eventual diffusion of the dopant into the slices [31]. As a result, surface conditions can influence the outcome of the diffusion by altering the sticking probability, which determines the amount of material exchange. The residual gas pressure in the ampul can also affect the diffusion process. Pressures in excess of 10 torr cause restriction of the dopant transport, presumably by covering the surface with foreign atoms. On the other hand, pressures below 0.01 torr can result in a mean free path for dopant atoms which exceeds the capsule dimensions, and lead to shadowing effects. Typically, a pressure of around 1 torr is used for these processes.

Surface concentrations obtained in sealed-tube diffusion can be made to approach the solid-solubility limit for the impurity at the diffusion temperature, provided that the dopant surface area is large compared to the rest of the system. This is achieved by using a granular dopant source. In addition, the diffusion time must be sufficient so that both slice and wall coverage attain steady-state conditions. Thus long, deep diffusions are favored for this process.

The open-tube method is favored for practical diffusion systems. Here slices are placed in a clean diffusion tube made of high-purity fused quartz. A separate diffusion furnace, diffusion tube, and slice carrier are reserved specifically for each impurity because they are contaminated with it during the process. Insertion is done (either by hand or mechanically) from one end of the tube, whereas the other end is used for the flow of gases or impurities in vapor form. This method is capable of handling a number of slices at one time and is considerably more convenient than the sealed-tube system.

The process of inserting slices into a diffusion tube, as well as their removal, results in radial thermal gradients. These can cause crystal damage in the form of dislocations and slip, and a resulting fall in the minority carrier lifetime. In extreme situations, loss of planarity results, causing the slice to take on a "potato-chip" shape.* These problems have become increasingly severe with the trend toward the use of large diameter (>75 mm) slices. One approach to alleviating them is to insert and remove the slice carrier in a

* On the other hand, slow cooling results in an improved lifetime in heavily doped regions, since it prevents precipitation in the diffused layer.

slow, controlled manner, using a mechanical puller. A pull rate on the order of 10 cm/min is typical.

The use of a mechanical puller has the disadvantage that wafers at the front end of the slice carrier are diffused for a longer time than those at the rear. A better approach is to insert the slice carrier into a relatively cool (≈600°C) furnace, and bring it up to the diffusion temperature at a linear rate. Modern diffusion furnaces use a programmed controller to perform this ramp-up function, as well as a ramp-down control at the end of the diffusion. Typical ramp rates are 3–10°C/min for high-temperature diffusions.

Consider a furnace whose temperature is ramped down so that

$$T = T_0 - Ct \tag{4.86}$$

where T_0 is the initial temperature and C is the ramp rate. Then it can be shown (see Section A.2.1) that the ramp-down process is equivalent to an extra diffusion time of kT_0^2/CE_0 at the initial diffusion temperature, where E_0 is the activation energy of diffusion and k is Boltzmann's constant. This is usually a small fraction of the total diffusion time.* The problem of ramp-up can be handled in an analogous manner.

Diffusion furnaces are usually operated at temperatures ranging from 600 to 1200°C, and are equipped with feedback controls to maintain a central flat zone with a $\pm\frac{1}{2}$°C tolerance. The length of this zone varies from 10 cm for laboratory systems to 75 cm for industrial units.

A number of techniques are used to bring the impurity atoms in contact with the slices. By far the most common consists of transporting an impurity-bearing compound (usually its oxide) to the surface of the wafers in an inert carrier gas. At operating temperature a reversible reaction occurs between the compound and the semiconductor, resulting in an equilibrium concentration of impurity atoms from which diffusion takes place. At typical diffusion temperatures, this produces an excess supply of the dopant. Thus the surface concentration is set by its solid-solubility limit at the diffusion temperature, so that this temperature must be tightly controlled. Diffusions made in this manner have excellent reproducibility from run to run. Furthermore, they proceed out of a liquid layer that provides a considerable degree of protection to the semiconductor surface.

Diffusion can also proceed from an elemental impurity† which is deposited on the slice. Often this impurity–semiconductor system exhibits a eutectic point below the diffusion temperature, so that an alloy interface results, with diffusion taking place from a dopant source of essentially infinite concentration. Thus the surface concentration is again set at the solid-solubility limit of the impurity in the semiconductor at this diffusion temperature. In this case, however, massive damage of the surface occurs, so that this

* See Problem 3 at the end of this chapter.

† A case in point is the diffusion of gold in silicon.

approach is restricted to diffusions which can be made from the back side of the wafer.

4.6.1 Choice of Dopant Source

The choice of dopant source depends on the way the dopant is presented to the semiconductor slice. The simplest technique consists of painting a slurry of the dopant source and an inert oxide (such as BaO, CaO, MgO, or SiO_2) in polyvinyl alcohol on the semiconductor surface. This *paint-on* method is too crude for modern device and microcircuit fabrication, and gives extremely poor control of the surface concentration.

The *spin-on* technique is an improvement over this method. Here the slice is held in a vacuum chuck and rotated at high speed (2500–5000 rpm). A drop of the dopant mixture is next applied to form a thin layer (≈5000 Å) across the slice by means of centrifugal force. With appropriate attention to proper viscosity control, relatively uniform layers of dopant can be obtained in this manner. Recently a number of acyloxysilanes and alkylsiloxanes have been used instead of the inert oxide and binder combination [32, 33]. These provide better viscosity control, and thermally decompose into SiO_2 during bake out at 200°C.

Spin-on techniques are versatile, and almost any dopant for silicon or gallium arsenide can be applied in this manner. Their degree of control is poor by modern microcircuit standards, so that they are only used in applications where dopant control is not critical.

Solid-source techniques can also be used for transporting the dopant to the slices. Figure 4.24 shows a sketch of the diffusion system used for this method. Here a platinum boat is used to hold a source of the dopant oxide upstream from the carrier with the semiconductor wafers. In operation, the carrier gas transports vapors from this source and deposits them on the semiconductor slices. In these systems, source shutoff is usually accomplished by moving the dopant source to a colder region of the furnace.

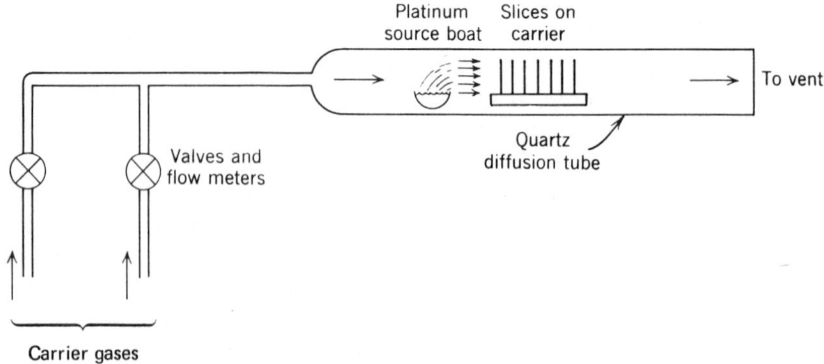

Fig. 4.24 Solid-source diffusion system.

The success of this technique depends critically on the vapor pressure of the source. In some cases this necessitates a two-temperature furnace, with the source maintained at a lower temperature than is used for the diffusions. Often, however, the source boat and the slices may be conveniently maintained at the same temperature, avoiding the need for a two-zone furnace.

Solid sources can also be directly placed on the semiconductor slice by chemical vapor deposition techniques, which are described in Section 8.2. These have the disadvantage of requiring an extra processing step; however, this approach avoids the necessity for uniform transport of dopant in vapor form to the slice. As a result, these sources are used in some VLSI applications, where tight diffusion control over large diameter slices is important. Furthermore, slices using this dopant source can be stacked closer, resulting in a higher throughput. An additional flexibility is that they can be patterned to provide selective diffusions on the semiconductor slice.

Deposited solid sources allow the use of very dilute dopant concentrations by adjusting the dopant-to-binder ratio. Diffusion from these sources results in a surface concentration which is controlled by the concentration of the dopant in the oxide and not by its solid-solubility limit in the semiconductor. This technique can be used to provide a low surface concentration by a one-step process, and thus avoids the necessity of the predeposit-and-diffuse sequence described in Section 4.5.1.3. An alternate technique, that of ion implantation, is increasingly used for this purpose.

Liquid sources can also be used. Here a carrier gas is bubbled through the liquid which is transported in vapor form to the surface of the slices. It is common practice to saturate this gas with the vapor so that the concentration is relatively independent of gas flow. Surface concentration is thus entirely set by the temperature of the bubbler and of the diffusion system (see Fig. 4.25). Additional lines are also provided for other gases in which the diffusion is performed.

Liquid-source systems are extremely convenient since the doping process can be readily initiated (or terminated) by control of the gas through the bubbler. In addition, the amount of dopant transported to the slices is relatively easy to control in these systems, by adjustment of the bubbler temperature. Finally, a number of halogenic dopant compounds are available as liquids. Use of these dopant sources greatly reduces heavy metal contamination in diffusion systems.

Gaseous sources are even more convenient than the liquid ones. Although it is possible to control the surface concentration by adjusting the gas flow, it is common practice to use an excess dopant gas concentration. Thus these systems are relatively insensitive to the gas-flow rate. As with the liquid system, an initial reaction results in the formation of the dopant oxide which is transported onto the semiconductor slice, where diffusion takes place. Figure 4.26 shows the schematic for a typical diffusion system using a gaseous dopant source. Here provision is made for an ambient carrier gas in

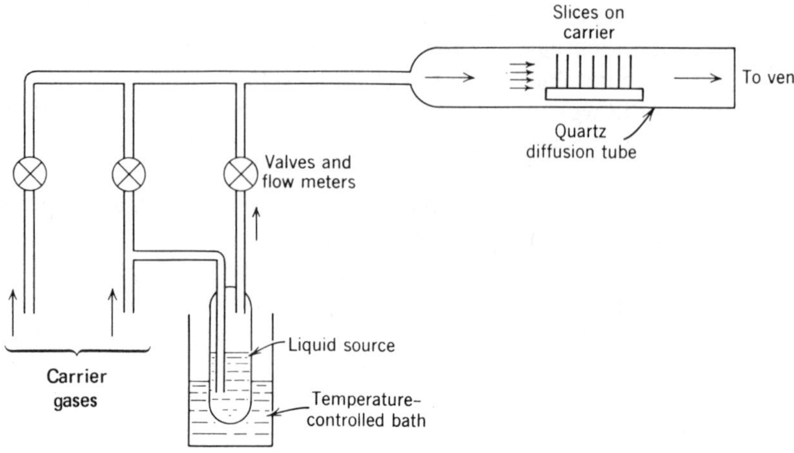

Fig. 4.25 Liquid-source diffusion system.

which the diffusion takes place. In addition, a chemical trap is often incorporated to dispose of unreacted dopant gases, which are often highly dangerous.

Impurity doping can also be done by *ion implantation*. This is a technique by which impurity atoms, traveling at high energy, are made to impinge on the semiconductor and are thus deposited in it. Details of this important process are treated in Chapter 6. This approach to semiconductor placement of the dopant on (or rather, in) the semiconductor has a number of significant characteristics. First, extremely high-purity sources are available, since in-situ mass spectrometric techniques are used to purify them. Next, impurity deposition is done by scanning the ion beam over the slice, so that it is uniform over large diameter slices. Finally, in-situ dopant monitoring

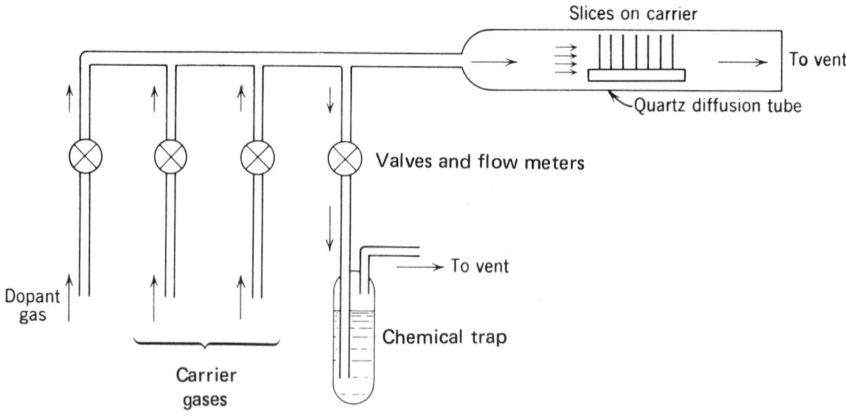

Fig. 4.26 Gaseous-source diffusion system.

techniques are available, so that a dopant accuracy of 1% is readily achievable (as compared to 5–10% which is the best that can be achieved by the approaches mentioned earlier). In addition, this control is available over many decades of dopant concentration, so that ion implantation can replace the predeposition step (see Section 4.5.1.3) that is required for diffusions with low surface concentration.

A disadvantage of this process is its high initial cost. In addition, the technique of halogenic doping cannot be readily exploited in this approach. Finally, ion implantation techniques result in considerable damage to the semiconductor. This damage is relatively easy to remove for light ion doses, followed by a high-temperature diffusion step. Conversely, heavily doped shallow regions, which require high ion doses and relatively low subsequent diffusion temperatures, are difficult to achieve by this means.

4.6.2 Contamination Control

The prevention of unwanted impurities, as well as their removal, plays an important part in any semiconductor process. This is particularly true for open-tube diffusions, where the slices are subjected to high-temperature processing. It goes without saying that all gases and dopant sources should be of semiconductor grade. In addition, the chemical cleaning of the semiconductor is of special importance, and is discussed in detail in Chapter 9.

The diffusion furnace itself is a major source of impurity contamination. Commercial diffusion tubes, made of fused quartz, are typically 96–97% pure so that they must be cleaned upon installation, and also regularly during service. Typically this cleaning is accomplished by flowing anhydrous hydrogen chloride gas through them for 15–30 min at diffusion temperatures. This tends to leach out impurities which are removed by conversion to their more volatile halides.

The firebricks in a diffusion furnace are a major source of alkali ion contamination. In addition, since they are hotter than the semiconductor slices, there is a thermal gradient which assists the transport of these ions into the diffusion tube, by rapid movement through the quartz walls. The use of a high-purity, high-density ceramic liner, made of mixtures of alumina and zirconia, is common practice since this serves as a barrier to the transport of these ions. Liners of this type are especially important in MOS-based fabrication processes, where sodium contamination critically affects the threshold stability of active devices.

A variety of techniques can be used [34] for the control of heavy metallic impurities (such as iron, copper, gold, etc.) The unwanted incorporation of these impurities results in a fall in minority carrier lifetime, and in an increase in the leakage current in p–n junctions. Thus their removal is important for both bipolar as well as field effect devices.

One approach here is to use dopant sources which are halogenic compounds. With these the halogen is liberated during diffusion, and reacts with any metallic impurities in the gas stream, as well as with impurities within the semiconductor that reach its surface during their rapid movement at high temperatures. This reaction converts them to their more volatile halides, which leave the system by incorporation into the gas stream. The use of a halogenic dopant source thus effectively getters the semiconductor. These sources must be used with care, however, since they can result in local dissolution of the semiconductor, and cause its pitting.

A second approach to the control of metallic impurities is to use backside gettering processes. These consist of damaging this surface so that it acts as a sink for metallic impurities during the subsequent diffusion step. This can be done by using a deposited layer of silicon nitride directly on the semiconductor surface, which creates a dislocation network as a result of high interfacial stress. Ion implantation with argon, or laser-induced damage serves the same purpose, and is also used.

Films of phosphosilicate glass (PSG) have been found to be extremely effective for the control of metallic impurities in silicon fabrication technology [35] and are in routine use at the present time. Here heat treatment of the semiconductor on which these layers are placed results in the creation of a diffusion-induced stress well ahead of the diffusing front. This region serves as a sink for metallic impurities. In addition the glass serves to permanently trap a significant amount of these contaminants so that they cannot reenter into solution upon cooling.

4.7 DIFFUSION SYSTEMS FOR SILICON

Diffusion systems are of the open-tube type, with quartz being the most common tube material. Increasingly, tubes of semiconductor grade silicon are being used for this purpose. Although considerably higher in initial cost, they are extremely clean, and are impervious to the movement of contaminants such as sodium through their walls. In addition, their useful life is very long compared to quartz tubes since they do not devitrify during use and are impervious to thermal shock. Thus they are cost effective in the long run.

The *p*-type impurities are aluminum, boron, gallium, and indium. Figure 2.22 indicates the maximum solid solubilities that are attainable using these impurities. Of the different choices, the only practical one is boron, since indium is actually a moderately deep lying impurity in silicon, with an acceptor level at 0.16 eV above the valence band. Furthermore, neither aluminum or gallium can be masked by silicon dioxide. Although other forms of masking are possible, they have not received much commercial use in microcircuit technology.

The *n*-type impurities are antimony, arsenic, and phosphorus. None of these exhibits any undesirable characteristic that precludes their use in microcircuit fabrication. As seen from Fig. 2.22, they are all highly soluble in silicon; consequently, the choice of impurity depends on the specific application for which each is uniquely suited. Thus,

1. The diffusion constant of phosphorus is almost the same as that of boron, and is about ten times larger than that of arsenic or antimony. It is used in the majority of diffusion applications, because it is both uneconomical and undesirable to operate diffusion systems for longer periods of time than necessary.
2. The low diffusion constant of arsenic and antimony make these impurities ideal for use in early phases of device fabrication, since they are relatively less affected by subsequent fabrication steps than are the faster substitutional diffusers such as phosphorus.
3. Arsenic systems are capable of considerably higher useful surface concentration than their antimony counterparts. This is because the arsenic atom provides an excellent fit to the silicon lattice. Most arsenic systems utilize highly poisonous, volatile sources and require elaborate precautions for safe handling. In addition, surface depletion of arsenic by evaporation is often a serious problem during diffusion. Antimony is somewhat easier to use than arsenic, because of its lower vapor pressure. It is occasionally chosen on the mistaken assumption that it is less dangerous.

Specific aspects of both *p*- and *n*-type diffusion processes are now considered.

4.7.1 Boron

Boron has a diffusion coefficient of about 10^{-12} cm^2/s at 1150°C. It has a high solid solubility and can be diffused with an active surface concentration as large as 4×10^{20} atoms/cm^3. Thus it is suited for a wide range of diffusion requirements.

The tetrahedral radius of boron in silicon is 0.88 Å, corresponding to a misfit factor $\epsilon = 0.254$. As a result, the presence of large amounts of boron in the silicon lattice is accompanied by strain-induced defects which lead to considerable crystal damage. This sets an upper limit of about 5×10^{19} atoms/cm^3 to the impurity concentration that can actually be achieved in practical structures, with the rest being electronically inactive.

Elemental boron is inert at temperatures up to and exceeding the melting point of silicon. Consequently, diffusion is brought about by means of a surface reaction between boron trioxide (B_2O_3) and the silicon, as given by

$$2B_2O_3 + 3Si \rightleftharpoons 4B + 3SiO_2 \tag{4.87}$$

Excessive amounts of B_2O_3 lead to the formation of silicides and other compounds of boron on the surface of the slices. This *boron skin* causes a dark brown stain, is electrically insulating in character, and often results in device failure due to open contacts.

Once formed, boron skin is difficult to remove in any acid. However, wet oxidation of the silicon can be used to convert this layer into a borosilicate glass which can then be dissolved in hydrofluoric acid. At any rate, it is far easier to avoid this skin by carrying out diffusions in a weakly oxidizing atmosphere (3–10% oxygen by volume). This promotes the formation of some SiO_2 and thus reduces this tendency to skin formation.

Paint-on methods for boron diffusion involve the use of mixtures of B_2O_3 and SiO_2 in polyvinyl alcohol. Mixtures of carborane and an alklysiloxane, which provides superior viscosity control, have also been used in spin-on sources. Here an initial bake out before diffusion results in their decomposition into B_2O_3 and SiO_2, respectively.

Disks of boron nitride are commonly used [36, 37] as a solid source for boron diffusion. These are of the same (or slightly larger) diameter as the silicon slices. Typically, they must be preoxidized at 750–1100°C for about 30 min in order to form a thin skin of B_2O_3 on their surface, which serves as the diffusion source. Diffusion is carried out by placing these disks and the silicon wafers in a slice carrier, with a uniform spacing between them, so that each disk is placed between the working surfaces of two silicon slices. Although no carrier gas is necessary, a flow of about 1 liter/min of dry nitrogen is commonly used to prevent the back diffusion of airborne contaminants into the diffusion tube. Since there is so little flow of gas between the slices during this process, diffusions are reproducible from run to run, and are extremely uniform across the diameter of the slices.

Spacing between the wafers and the silicon surface is usually kept at about 2–3 mm, and must be held constant for run-to-run consistency. In addition, preoxidation of wafers must be done on a regular schedule, depending on the frequency of usage, if consistent results are to be maintained.

There are a number of limitations to the use of solid sources of this type. First, the B_2O_3 on the silicon slice is not diluted with SiO_2, so that boron skin often occurs with this type of source. One approach to preventing this is to introduce a small amount of oxygen into the carrier gas. This leads to nonuniformity across the slices, since the oxidation rate is governed by radial gas flow between them.

A second problem is the sticking of the boron nitride slices to the silica carrier. The use of mixtures of boron nitride in a SiO_2 binder minimizes this problem and also reduces the possibility of boron skin formation. An alternate approach is to use slices consisting of mixtures of B_2O_3 with materials such as BaO, CaO, MgO, or Al_2O_3. All of these have very high negative free energies of formation as compared to SiO_2, so that they are not reduced to elemental barium, calcium, magnesium, or aluminum, respectively, during diffusion.

Diffusions from boron nitride sources generally result in high surface concentrations. Their doping profiles follow closely the shape described in Section 4.5.2.2, rather than the erfc function [38].

Various liquid sources may be used in the system shown in Fig. 4.25. In each case a preliminary reaction results in the deposition of B_2O_3 on the silicon slices. Some sources in use are the following.

Trimethylborate (TMB). The preliminary oxidizing reaction is

$$2(CH_3O)_3B + 9O_2 \xrightarrow{900°C} B_2O_3 + 6CO_2 + 9H_2O \tag{4.88}$$

The vapor pressure of TMB is usually controlled by refrigeration since it is extremely volatile at room temperatures.

Boron Tribromide. Here the reaction is

$$4BBr_3 + 3O_2 \rightarrow 2B_2O_3 + 6Br_2 \tag{4.89}$$

Boron tribromide is a halogenic source, so that it can be used for the gettering of metallic impurities during diffusion. Since bromine is a reaction product, provision must be made for venting the system. In addition, there is a tendency for pitting to occur if excessive vapor concentrations are used or if the carrier gas is not sufficiently oxidizing in nature.

Gaseous sources are used in the configuration of Fig. 4.26. Again, B_2O_3 is formed as a primary reaction product. Common boron sources are the following.

Diborane. This is a highly poisonous, explosive gas. It is used in a 99.9% argon dilution by volume, which is reasonably convenient to handle. The preliminary oxidizing reaction for the system is

$$B_2H_6 + 3O_2 \xrightarrow{300°C} B_2O_3 + 3H_2O \tag{4.90}$$

Since only water is released from this reaction, the method is not prone to the pitting effects experienced with boron tribromide. In addition, venting is not required. It is necessary, however, to install an input trap with a weak hydrochloric acid solution to capture any unused diborane in the gas lines.

Higher surface concentrations can be obtained, especially for low-temperature (800–900°C) predeposition, by using carbon dioxide gas [39] instead of oxygen. Here

$$B_2H_6 + 6CO_2 \rightarrow B_2O_3 + 6CO + 3H_2O \tag{4.91}$$

This is because less silica is formed in this process, since carbon dioxide is a weaker oxidizing agent than oxygen. Note, however, that carbon monoxide is liberated as a reaction product, so it is necessary to vent this system.

Boron Trichloride. Here the preliminary reaction is

$$4BCl_3 + 3O_2 \rightarrow 2B_2O_3 + 6Cl_2 \qquad (4.92)$$

Again, system venting is necessary, as well as care to avoid halogen pitting effects at high concentrations.

It is considerably more difficult to obtain uniform diffusions from one end of the slice carrier to the other with BCl_3 than with BBr_3. This is because the oxidizing reaction of BCl_3 takes a long time (as much as 100 s in some situations), whereas that for BBr_3 is relatively short (on the order of 5 s for similar operating parameters). Consequently, all the BBr_3 is rapidly converted to B_2O_3, and is available for use as a diffusion source, before delivery over the entire length of the slice carrier. With BCl_3, on the other hand, much of the B_2O_3 is not available to slices at the leading end of the carrier because of its slow oxidation rate [40].

The BCl_3 reaction is greatly accelerated in the presence of water vapor. It is quite customary to provide this by introducing a small amount of hydrogen into the furnace during the predeposition phase, together with the oxygen flow.

4.7.2 Phosphorus

Phosphorus has a diffusion coefficient which is comparable to that of boron. Its misfit factor is considerably lower (0.068 as compared to 0.254), so that an active carrier concentration of 10^{21} atoms/cm^3 can be achieved in practical structures.

The most commonly used compound that participates in phosphorus diffusion is phosphorus pentoxide. At diffusion temperatures, the surface reaction from which diffusion proceeds is given by

$$2P_2O_5 + 5Si \rightleftharpoons 4P + 5SiO_2 \qquad (4.93)$$

Since phosphorus pentoxide is hygroscopic, other phosphorus-bearing compounds are commonly used. Nevertheless, they ultimately result in the transport of P_2O_5 to the surface of the silicon slice. This readily combines with SiO_2 to form a phosphosilicate glass that is liquid at diffusion temperatures. The phase diagram for the SiO_2-P_2O_5 system is presented in Fig. 2.18 and shows those regions that are relevant to the diffusion process.

Spin-on techniques have been used successfully for phosphorus diffusions. Here organic compounds of phosphorus, such as triphenylphosphate [$(C_6H_5O)_3PO$] are used in vehicles of the type described earlier for boron diffusions. Again, a bake-out phase is required to convert the spin-on dopant into a phosphosilicate glass, prior to insertion of slices into the diffusion furnace.

Phosphorus diffusion can be conducted in a number of systems similar to those used for boron. A brief description of these follows.

Solid-Source Systems. It is quite difficult to obtain close control of surface concentration with the use of a phosphorus pentoxide source, because of its hygroscopic nature. Other sources, such as ammonium monophosphate ($NH_4H_2PO_4$) and ammonium diphosphate [$(NH_4)_2H_2PO_4$], do not suffer from this problem and are used at source temperatures from 450 to 900°C. Again the preliminary reaction appears to result in the delivery of P_2O_5 to the silicon slices.

Solid sources of the disk type can also be used, and have many of the advantages (and disadvantages) outlined earlier for boron. These usually consist of hot-pressed $NH_4H_2PO_4$ or $(NH_4)_2H_2PO_4$ in inert ceramic binders. One problem with these sources is that they release pure P_2O_5 during diffusion. An alternate source, which avoids this problem, consists of 25% of silicon pyrophosphate in an inert binder [41]. This source directly converts to a P_2O_5-SiO_2 glass at diffusion temperatures, and is released as a vapor from the surface of the disk.

Liquid-Source Systems. The most popular of these systems uses phosphorus oxychloride ($POCl_3$) in the temperature range of 0–40°C. An oxidizing gas mixture is used during the predeposition phase, resulting in the formation of P_2O_5 at some point in the system before the diffusion zone. The preliminary reaction is

$$4POCl_3 + 3O_2 \rightarrow 2P_2O_5 + 6Cl_2 \tag{4.94}$$

The presence of oxygen during the predeposition phase aids in preventing halogen pitting effects, which are noticeable only at high surface concentrations. The system is free from moisture problems and allows a high degree of control of surface concentration by adjusting the bubbler temperature [42].

An alternate liquid source is phosphorus tribromide. This dopant source has somewhat superior gettering properties to $POCl_3$ and is being increasingly used for this reason.

The last diffusion in the fabrication of bipolar transistor microcircuits consists of an *n*-type emitter diffusion. Use of either of these halogenic dopant sources at this point is highly advantageous since it getters contaminants from all previous high-temperature steps.

Gaseous Systems. The most commonly used gas is phosphine, which is both highly toxic and explosive. It is relatively convenient to handle in dilute form (with 99.9% N_2 or Ar). A slightly oxidizing carrier gas is used, the preliminary reaction being

$$2PH_3 + 4O_2 \rightarrow P_2O_5 + 3H_2O \tag{4.95}$$

As with other reactions, P_2O_5 is delivered to the silicon slices. Although it is not necessary to provide system venting to the output, an acidic CuCl trap must be used in the inlet gas lines to remove any unused gas. Characteristics of the system are similar to those of the diborane system.

4.7.3 Arsenic

Arsenic has a diffusion coefficient which is about one-tenth that of boron or phosphorus. Consequently, its use is dictated for those cases in which it is important that the dopant be relatively immobile with subsequent processing. It is also desirable for shallow diffusions where its abrupt doping profile makes control of the junction depth more precise.

The tetrahedral radius of arsenic is identical to that of silicon, so that it can be introduced in large concentration without causing lattice strain. An active electron concentration of 2×10^{21} cm^{-3} is achievable with this dopant.

Arsenic diffusion is usually performed [43] from the reaction of arsenic trioxide (As_2O_3) and silicon. A source temperature of 150–250°C is commonly used for this solid source. At diffusion temperature,

$$2As_2O_3 + 3Si \rightleftharpoons 3\,SiO_2 + 4As \tag{4.96}$$

In actual operation, vapors of the oxide are transported into the diffusion zone by means of a carrier gas of almost pure nitrogen, since arsenic is readily masked by the presence of oxide on the silicon slices. However, 0.5% of oxygen is often added to prevent staining caused by the formation of an arsenic layer at the oxide–silicon interface. Surface concentrations obtained by this technique generally do not exceed $(2–3) \times 10^{19}$ cm^{-3}, because of rapid depletion by evaporation from the silicon surface.

Spin-on dopants have been used with arsenic and consist of the arsenosiloxanes in an appropriate organic binder. Upon bake out at 250 °C, these react to form an arsenosilicate glass which acts as the source of this dopant. Higher doping concentration can be achieved by this method, since the glass serves as an evaporation barrier to the arsenic during diffusion.

Gaseous systems, using arsine (AsH_3), have been successfully operated. They are more convenient than systems using As_2O_3, but do not result in surface concentrations above 3×10^{19} cm^{-3}.

Sealed-tube technology has also been used successfully for arsenic [44]. Here silicon slices are stacked within a quartz tube together with the arsenic source, which consists of doped silicon powder having an arsenic concentration of about 3%. A short, high-temperature bake is used to remove traces of water vapor from the tube, at which point it is evacuated and sealed. After diffusion the tube is cut open.

The sealed-tube approach has many advantages over the open-tube system for this dopant. Surface concentrations of as high as 10^{21} cm^{-3} are achievable because of the complete absence of source evaporation. In addition, uniformity from slice to slice is better than 2% because of the closed nature of the system. Finally, a large number of wafers can be diffused at a single time, since there are no problems of dopant depletion from one end of the carrier to the other.

Ion implantation is also used for the predeposition of arsenic before drive-in. A high-energy implant serves to deposit this dopant deep into the semiconductor surface, and thus minimizes its depletion during the drive-in step.

4.7.4 Antimony

Antimony is often used as an alternative to arsenic because of its comparable diffusion coefficient. Its misfit factor in silicon is 0.153, and its electronically active surface concentration is limited to (2–5) $\times$ 10^{19} atoms/cm^3. However, antimony has a lower vapor pressure than arsenic, so that surface concentrations close to this value can be achieved in open-tube systems.

The diffusion of antimony into silicon is usually performed from the trioxide Sb_2O_3, which is a solid source [43]. A two-zone diffusion furnace is required, with the Sb_2O_3 maintained at 600–650°C. Some systems have also been operated with the tetroxide Sb_2O_4 at a source temperature of about 900°C. Care must be taken to operate these systems in an almost oxygen-free ambient, because antimony is relatively easily masked by small traces of SiO_2 on the silicon surface.

Liquid-source systems have also been used with antimony pentachloride (Sb_3Cl_5) in a bubbler arrangement, and a carrier gas of oxygen. This results in the delivery of antimony trioxide to the silicon surface, from which the diffusion takes place. Stibine (SbH_3) cannot be used in gaseous systems since it is unstable.

4.7.5 Gold and Platinum

The reasons for the choice of gold over other impurities exhibiting deep lying levels have already been outlined in Section 4.3.2. Since gold is an extremely rapid diffuser (with an effective diffusion coefficient about five orders of magnitude larger than that of boron or phosphorus), it is usually the last impurity to be introduced into the microcircuit. Sometimes the gold diffusion and emitter drive-in are carried out simultaneously to obtain a saving in process time.

Gold diffusion is usually carried out from the element, which is vacuum evaporated on the silicon in a layer about 100 Å thick. Diffusion proceeds from a liquid gold–silicon alloy, and results in silicon damage to a depth of many microns. For this reason, gold diffusion is always performed from the back of the wafer, remote from the actual microcircuit.

Diffusion is normally carried out in the temperature range of 800–1050°C. The concentration is controlled by means of this temperature while the time (about 10–15 min) is more than enough to cause the gold to diffuse through the entire microcircuit. The precise diffusion temperature is selected on a cut-and-try basis to obtain the desired end result. In-process monitoring of the collector–emitter voltage of a test transistor, driven into saturation, is often used to determine the correct process time [45].

Gold-doped silicon slices must be removed rapidly from the diffusion furnace and brought to room temperature in a manner that is repeatable from run to run. The repeatability of this quenching cycle is important in ensuring consistent results because out-diffusion effects are significant with fast diffusers. As a result, gold doping is perhaps the most difficult process to control in semiconductor manufacture.

Platinum diffusion is presently in a developmental stage, with primary interest in its use in semiconductor power devices. The behavior of this dopant in silicon is poorly understood. As a result, the usual approach for diffusion is to use techniques that have worked with gold [17]. Unlike gold, however, results with platinum diffusion from the element are highly variable, so that spin-on dopants are more commonly used. Typically, these consist of platinum chloride in an alkyloxysilane.

4.8 SPECIAL PROBLEMS IN SILICON DIFFUSION

This section discusses a number of problems and anomalies encountered during silicon diffusion. Many of these have become increasingly important with the trend toward large-scale integration, and the resultant use of shallow diffusions with high surface concentration. A few of these problems have been analyzed by closed-form solutions, while computer-aided models have been used extensively [46] for the others.

4.8.1 Redistribution During Oxide Growth

In microcircuit fabrication, diffusions are made through windows cut in a mask on the surface of the silicon. The most convenient mask is one of silicon dioxide, grown out of the slice itself by subjecting it to an oxidizing gas at elevated temperatures. This can be done in a separate step after a diffusion. Alternately, the two-step diffusion process may be used to grow it simultaneously with the drive-in phase. This latter technique is preferred, because it not only provides immediate protection for the silicon surface, but also saves an additional masking step. In addition, it minimizes the time the slice is maintained at elevated temperatures and so reduces the undesired movement of already diffused impurities.

Common to both of these processes is the fact that some of the impurity-doped silicon is consumed to form an oxide. This effect has its parallel in the freezing of a doped crystal from a melt, as described in Chapter 3, and results in a redistribution of impurities. The extent of this redistribution is a function of the rate at which the silicon is consumed and of the relative diffusion coefficients and solid solubilities of the impurities in silicon and silicon dioxide.

The ratio of the equilibrium concentration of the impurity in silicon to its equilibrium concentration in the oxide is the distribution coefficient for the impurity in the Si-SiO_2 system. Experimentally derived values [47] for this distribution coefficient m are about 20 for gallium and 10 for phosphorus, antimony, and arsenic. Values of m for boron are a function of crystal orientation, and are given [48] by

$$m_{(100)} \simeq 33\, e^{-0.52/kT} \tag{4.97a}$$

$$m_{(111)} \simeq 20\, e^{-0.52/kT} \tag{4.97b}$$

At typical diffusion temperatures, m ranges from 0.15 to 0.3 for {100} silicon.

For the case in which $m < 1$, the growing oxide takes up the impurity from the silicon. Thus for boron diffusion, impurity depletion occurs on the silicon side of the interface. The extent of this depletion also depends on the rate at which the boron diffuses through the SiO_2 layer to its surface and escapes into the gaseous ambient. The diffusion rate in SiO_2 is orders of magnitude below that in silicon; hence boron depletion is dominated by distribution effects.

For the case in which $m > 1$, the growing oxide rejects the impurity. If the diffusion through the oxide is slow, as with phosphorus, this results in a pileup of the impurity at the silicon surface. If, however, the impurity diffuses rapidly through the oxide (as is the case with gallium), the net result can still be an impurity depletion.

Finally, even if $m = 1$, the process of oxidation results in impurity depletion in the silicon. This is because the silicon roughly doubles in volume on forming silicon dioxide. Thus some diffusion will occur from the highly doped silicon into the relatively lightly doped oxide.

Figure 4.27 shows the impurity equilibrium concentrations that may occur in uniformly doped silicon in the presence of a growing oxide surface. For practical diffusions the problem is more complicated since the concentration is not uniform. In addition, the impurity is diffusing both into the silicon and into the growing oxide at the same time.

Figure 4.28 shows the results of calculations [49] for boron drive-in with simultaneous oxidation, and $m = 0.1$. These data are for typical drive-in situations which follow a predeposition with a Dt product of 3.6×10^{-12} cm^2. In addition, it is assumed that the diffusion coefficients of boron in silicon and silicon dixoide are 0.6×10^{-12} and 1.02×10^{-13} cm^2/s, re-

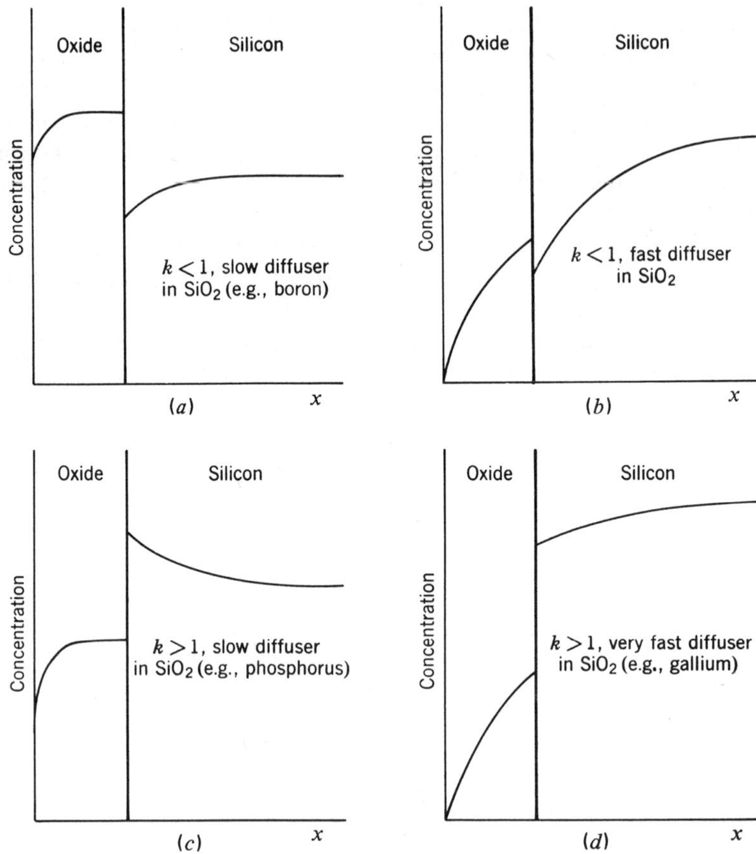

Fig. 4.27 Redistribution effects.

spectively, at the drive-in temperature. It is seen from this figure that redistribution effects cannot be ignored for boron. In a typical practical situation involving a 45 min drive-in at 1200°C (15 min in dry oxygen followed by 30 min in wet oxygen), the surface concentration is approximately 50% of its value if redistribution effects were absent. Such departures from theoretical predictions are usually handled with the aid of correction curves developed for the specific conditions under which the diffusion run is made, or by computer-aided process control.

In practice the pileup effect in phosphorus diffusion is relatively small and may be ignored. This is because of the fortuitous cancellation of the redistribution and diffusion effects under typical processing conditions.

4.8.2 Boron in Silicon

Boron diffusion is isotropic when carried out in a nitrogen ambient. Its diffusivity increases significantly in an oxidizing ambient, however, and becomes a function of crystal orientation [11, 12]. It is highest for boron diffusions performed in (100) silicon and somewhat lower for diffusions in (111) silicon. Curves of $D_i = D_i^0 + D_i^+$ are shown in Fig. 4.29, together with the curve for D_i in a nonoxidizing ambient, and illustrate this effect.

This anomalous behavior can be explained along the following lines. We have shown that boron movement in silicon is dominated by interactions between the impurity atom and silicon vacancies in the charge states V^0 and V^+. In addition, a significant fraction of boron movement is by an interstitialcy mechanism. Here the boron atom moves by pushing one of its nearest silicon neighbors into an adjacent interstitial site, and taking up the substitutional site which has been vacated in this manner. Thus boron–interstitial interaction must also be considered in a study of boron diffusion.

At a given diffusion temperature, there exists an equilibrium concentration of silicon interstitials. Interaction with these is implicitly taken into consideration for diffusion in a nonoxidizing ambient. The situation is quite different when diffusion and oxidation occur simultaneously. It has been proposed [50] that incomplete oxidation at the Si-SiO_2 interface results in the formation of silicon interstitials in excess of the equilibrium value, at concentrations which are dependent on the oxide growth rate and the crystal orientation. Since boron diffusion is due in part to the interstitialcy mech-

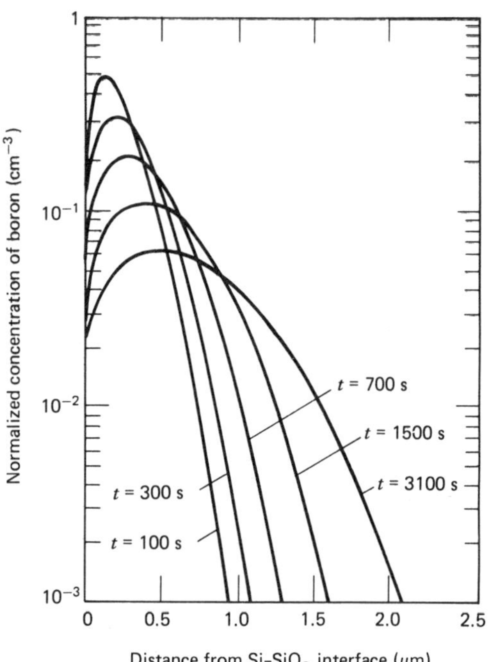

Fig. 4.28 Redistribution effects with boron in an oxidizing ambient. From Kato and Nishi [49]. Reprinted with permission from the *Japanese Journal of Applied Physics.*

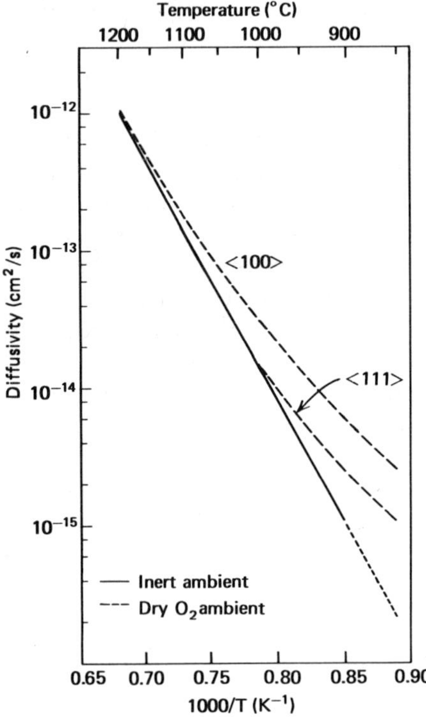

Fig. 4.29 Boron diffusivity in an oxidizing ambient. From Antoniadis, Gonzales, and Dutton [12]. Reprinted with permission of the publisher, The Electrochemical Society, Inc.

anism, it follows that an increased concentration of silicon interstitials results in an enhanced diffusion rate. Moreover, this rate will further increase if diffusion is carried out under faster oxidizing conditions [51].

4.8.3 Arsenic in Silicon

The low diffusion coefficient of arsenic, coupled with its abrupt doping profile, makes it ideally suited for shallow diffused structures. Thus the use of this dopant in VLSI applications is becoming of increasing importance.

Anomalous behavior, noted by a number of workers, is seen in a fall in the conductivity of arsenic diffused (1100–1200°C) layers which are subsequently heat treated in the 500–900°C range. The conductivity of these layers can be fully recovered by reprocessing at high temperatures (1100–1200°C), so that this behavior is not due to loss of arsenic by out-diffusion from the silicon surface. Although not understood at the present time, possible reasons for this effect are that it is due to the formation of As-As clusters or to the formation of $SiAs_2$, both of which are electronically inactive.

A small diffusion enhancement effect has been noted [52] for arsenic in an oxidizing ambient. This has been attributed to interactions of arsenic with excess silicon interstitials produced during the oxidation process. The effect, however, is slight compared to that observed for boron under these conditions.

4.8.4 Phosphorus in Silicon

The anomalous behavior of shallow phosphorus diffusions in silicon was first observed [53] during a study of high-concentration diffusions in the 0.5–1 μm range. Typically, it was found that the doping profile was of the shape depicted in Fig. 4.30, which displays both the total and the electronically active phosphorus concentrations. Essential features of the electronically active phosphorus profile [28] are an initial flat-top region followed by a kink at around 10^{20} cm^{-3}, and then by an extensive tail beyond this point. Furthermore, it is seen that only a small fraction of the phosphorus is electronically active at high concentrations. Empirically, the electron concentration is given by

$$N_{\text{tot}} = n + 2.04 \times 10^{-41} n^3 \tag{4.98}$$

for values of N_{tot} from 10^{19} to 10^{21} cm^{-3}.

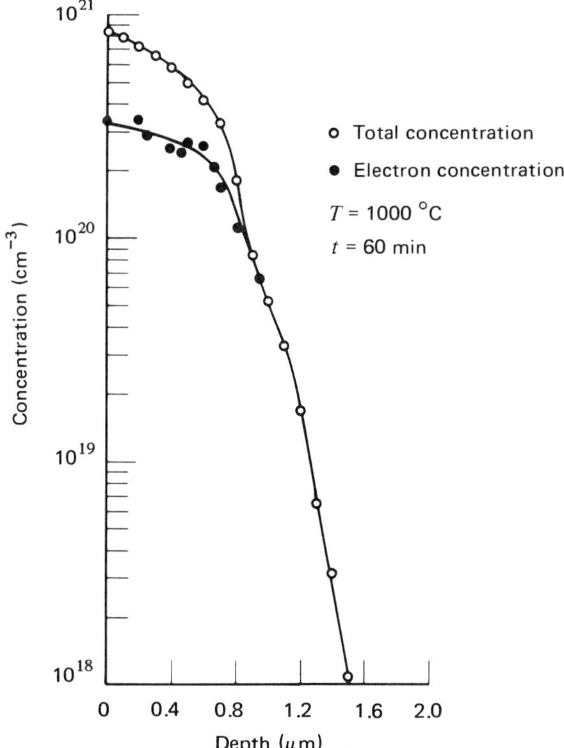

Fig. 4.30 High-concentration phosphorus diffusion profile. From Fair and Tsai [28]. Reprinted with permission of the publisher, The Electrochemical Society, Inc.

An idealized version of the phosphorus doping profile is shown in Fig. 4.31. Its essential features are a surface concentration of n_s, a kink at n_e, and a tail beyond this point. These features can be explained by considering phosphorus interactions with vacancies in the silicon lattice, which bring about enhanced diffusion effects. Terms of interest are P(I-V^0), P(I-V^-), and P(I-V^{2-}), so that the diffusivity is given by

$$D = h\left[D_i^0 + D_i^-\left(\frac{n}{n_i}\right) + D_i^{2-}\left(\frac{n}{n_i}\right)^2\right] \tag{4.99}$$

At high concentrations, the dominant term is associated with the P(I-V^{2-}) interactions. Furthermore, because there is essentially no change in concentration in this region, the field enhancement factor is unity, and the diffusivity is thus given by

$$D_i \simeq D_i^{2-}\left(\frac{n}{n_i}\right)^2 \tag{4.100}$$

where

$$D_i^{2-} = 44.2\, e^{-4.37/kT} \tag{4.101}$$

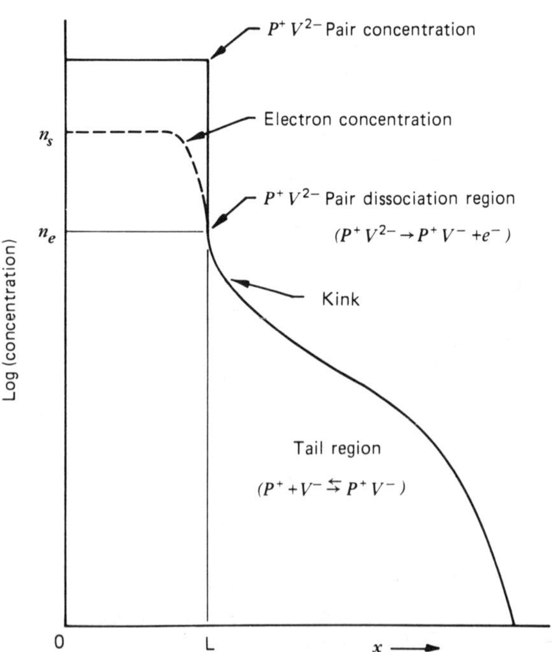

Fig. 4.31 Idealized phosphorus diffusion profile. From Fair and Tsai [28]. Reprinted with permission of the publisher, The Electrochemical Society, Inc.

The doping profile for this concentration dependent diffusivity is extremely abrupt, as shown in curve b in Fig. 4.21. Eventually the concentration falls to n_e, and the Fermi level in the semiconductor sweeps past the V^{2-} level (0.11 eV below the conduction-band edge). The impurity–vacancy pair dissociates, so that

$$\mathrm{P}(I\text{-}V^{2-}) \rightleftharpoons \mathrm{P}(I\text{-}V^{-}) + e^{-} \tag{4.102}$$

The binding energy of the $\mathrm{P}(I\text{-}V^{-})$ pair has been shown to be about 0.3 eV less than for $\mathrm{P}(I\text{-}V^{2-})$, so that further dissociation to P^{+} and V^{-} is favored. Many vacancies are released by this process, and move rapidly into the tail region. The diffusivity in this region is thus given by

$$D_{\text{tail}} = D_i^0 + D_i^- \frac{[V^-]}{[V^-]_i} \tag{4.103}$$

The second term may be evaluated by considering the formation and dissociation processes involved during phosphorus diffusion. Upon entry, the formation reaction of $\mathrm{P}(I\text{-}V^{2-})$ pairs is given by

$$\mathrm{P}^{+} + V^{2-} \rightleftharpoons \mathrm{P}(I\text{-}V^{2-}) \tag{4.104}$$

Applying the mass-action principle at the surface, where $n = n_s$, gives

$$\frac{[\mathrm{P}(I\text{-}V^{2-})]}{[\mathrm{P}(I\text{-}V^{2-})]_i} = \left(\frac{n_s}{n_i}\right)^3 \tag{4.105}$$

At the point where the Fermi level crosses the V^{2-} level, i.e., the point at which $n = n_e$, the dissociation reaction can be written as

$$\mathrm{P}(I\text{-}V^{2-}) \rightleftharpoons \mathrm{P}^{+} + V^{-} + e^{-} \tag{4.106}$$

Applying the mass-action principle at $n = n_e$, gives

$$\frac{[V^-]}{[V^-]_i} = \left(\frac{n_i}{n_e}\right)^2 \frac{[\mathrm{P}(I\text{-}V^{2-})]}{[\mathrm{P}(I\text{-}V^{2-})]_i} \tag{4.107}$$

Combining with Eq. (4.105),

$$\frac{[V^-]}{[V^-]_i} = \frac{n_s^3}{n_e^2\, n_i} \tag{4.108}$$

This term must be modified to take into account the enhanced dissociation that occurs because of the lower binding energy of the $\mathrm{P}(I\text{-}V^{-})$ pair. It has been shown that this results in

$$\frac{[V^-]}{[V^-]_i} = \frac{n_s^2}{n_e^2 n_i}(1 + e^{0.3/kT}) \tag{4.109}$$

The tail region diffusivity is thus given by

$$D_{\text{tail}} = D_i^0 + D_i^- \left[\frac{n_s^2}{n_e^2 n_i}(1 + e^{0.3/kT})\right] \tag{4.110}$$

where

$$D_i^0 = 3.85\, e^{-3.66/kT} \tag{4.111a}$$

and

$$D_i^- = 4.44\, e^{-4.0/kT} \tag{4.111b}$$

Experimental values of D_{tail} are shown in Fig. 4.32 and agree well with the values calculated from Eq. (4.110).

4.8.5 The Emitter-Push Effect

This effect, sometimes referred to as *emitter dip* or *enhanced diffusion under the emitter*, results in a bipolar transistor whose cross section is shown in Fig. 4.33. Experimentally, it has been noted that the parameter δ in this figure varies approximately inversely with the unenhanced base diffusion depth, ranging from 0.2 to 0.4 μm for values of x_{bc} from 2 to 0.5 μm, respectively. Thus it is particularly important in shallow structures of the type used in VLSI applications, and has been the subject of much study.

This enhanced diffusion effect can be attributed to two mechanisms. The first of these can be studied by noting that the flux of base impurities consists of both a concentration dependent term as well as a field dependent term, as given by Eq. (4.21). In the diffusion of a single species, this field dependent term is self-produced, and aids the movement of the impurity. The emitter diffusion, however, sets up its own electric field, which acts against the base concentration gradient. If sufficiently large, diffusion "up" the gradient is possible, and can even result in a dip in the base doping profile. This simple model is illustrated in Fig. 4.34. Here Fig. 4.34*a* shows the initial base doping profile, Fig. 4.34*b* shows the electric field produced by a high-concentration emitter diffusion, and Fig. 4.34*c* shows the base doping profile after emitter diffusion.

This model does not predict the large base diffusion enhancement which has been shown to extend well below the emitter. In addition, the enhance-

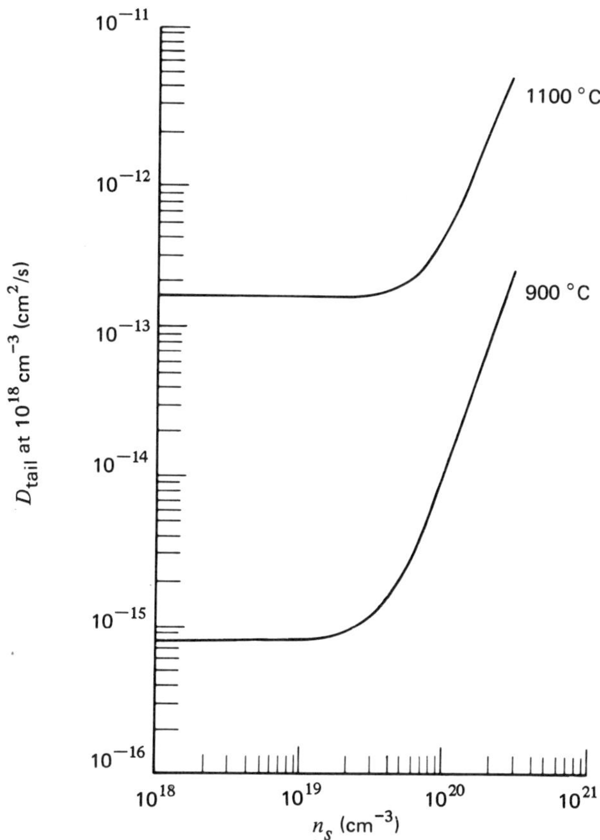

Fig. 4.32 Phosphorus difffusivity in the tail region. From Fair and Tsai [28]. Reprinted with permission of the publisher, The Electrochemical Society, Inc.

ment occurs in all directions. A second mechanism, which can account for these factors, is that enhanced diffusion most probably occurs because of point defects generated in the emitter region, which can migrate a considerable distance. This has been shown in Section 4.8.4 to produce a tail region in high-concentration phosphorus diffusions, where the dominant process

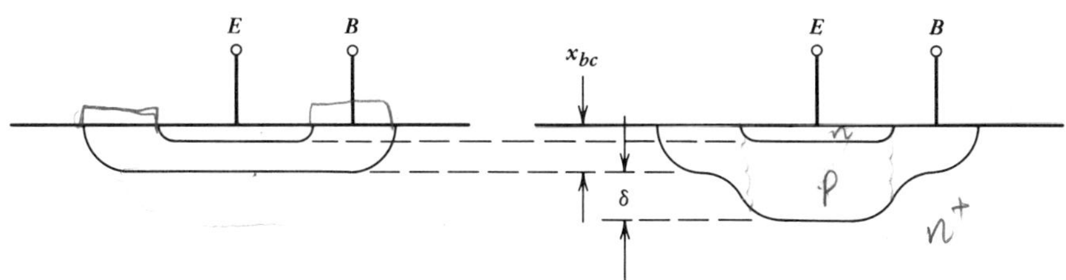

Fig. 4.33 Enhanced diffusion under the emitter.

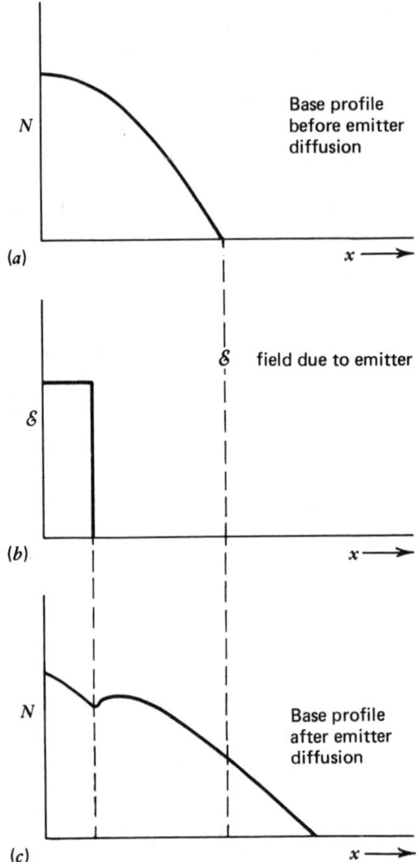

Fig. 4.34 Model for base profile resulting from emitter diffusion.

is dissociation of the P(I-V^-) pair and the generation of vacancies. These effects are shown in Fig. 4.35 for a transistor with a phosphorus-doped emitter and a boron-doped base. Also shown are the results of calculations which take both these mechanisms into account [8].

A dip in the base doping profile is also observed when arsenic is used for the emitter diffusion. However, no extended push effect is seen in this case. This is because the diffusion of arsenic into silicon does not involve the pair dissociation mechanism outlined in Section 4.8.4.

4.8.6 Preferential Doping with Gold and Platinum

As described in Section 4.3.2, the diffusion coefficient of gold is a strong function of the defect content of the silicon into which it is introduced. Gold diffuses preferentially in those regions of the slice that have a high concentration of defects, that is, in the region in which the microcircuit has already been fabricated. Even though gold is introduced from the opposite side of the wafer, its concentration is often found to be higher in these regions of the microcircuit than it is in the intervening regions which are relatively lightly doped.

The solid solubility of gold in silicon increases with both p- and n-type doping, because of interaction with vacancies [6]. This effect is especially significant in n-type silicon. If $[\mathrm{Au}]_i$ is the solid solubility of gold in intrinsic silicon, and [Au] is the solid solubility in silicon with a doping concentration N_D, it can be shown [55] that

$$[\mathrm{Au}] \simeq [\mathrm{Au}]_i \left(\frac{N_D}{n_i}\right) \tag{4.112}$$

for $N_D \gg n_i$. At typical gold diffusion temperatures, $n_i \simeq 10^{19}\ \mathrm{cm}^{-3}$.

The effectiveness of gold doping is thus related to the type of layer through which the gold moves. A factor which further inhibits the movement of gold through a heavily phosphorus-doped layer is the formation of a stable compound, Au_2P_3. Platinum, on the other hand, experiences no observable blockage effects, although stable compounds of platinum and phosphorus (as well as platinum and silicon) can be formed in the process.

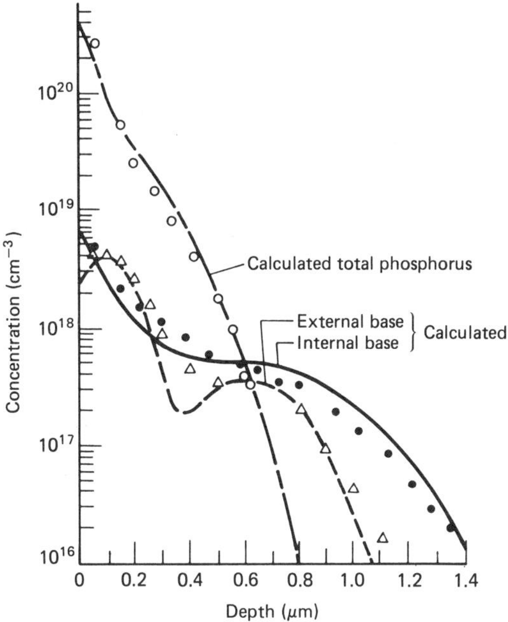

Fig. 4.35 Doping profiles for a transistor with a phosphorus-doped emitter and a boron-doped base. From Fair [8]. Originally presented at the Fall 1977 Meeting of the Electrochemical Society, Inc. held in Atlanta, Georgia. Reprinted with permission of The Electrochemical Society, Inc.

4.9 DIFFUSION SYSTEMS FOR GALLIUM ARSENIDE

Diffusion systems for gallium arsenide are based on the same considerations as those for silicon. Here, too, the diffusion process has as its primary requirements the need to bring an impurity in contact with the slice, and maintain it there for a fixed time and at an elevated temperature. With gallium arsenide, however, this results in the loss of arsenic by decomposition or by evaporation. Consequently, techniques must be used to prevent this occurrence. These include diffusion in sealed ampuls with a sufficient overpressure of arsenic, and diffusion in open-tube furnaces with cap layers of materials such as silica, phosphosilicate glass, or silicon nitride. Much of the work in this area has been done with sealed ampuls, with a known weight of arsenic (or an arsenic-bearing compound) to establish a suitable arsenic overpressure. These systems involve the necessity of sealing and breaking open these ampuls, so that there is much interest in open-tube systems because of their inherent convenience.

A second problem is in the choice and amount of dopant source. It has been found that excessive amounts of elemental sources often result in surface damage to the gallium arsenide because of the formation of reaction products. This problem is worse for some dopants than for others. It has usually been minimized by using dilute concentrations of the dopants, often as alloys or compounds of gallium or arsenic ($ZnAs_2$, GaS, and Zn-Ga alloys, for example). A great variety of early diffusion studies were made in this manner, with a wide scatter in diffusion data, by as much as one to two orders of magnitude [56, 57].

Significant advances in the understanding and control of diffusion in gallium arsenide have been made, once attention was focused on the importance of ternary considerations [58] in these systems. A number of authors have shown that reproducible diffusions can be made in sealed-tube systems, if attention is paid to these considerations [25, 59, 60].

More recently the use of doped oxide sources with cap layers has resulted in diffusion processes which can be conducted in an open-tube environment [61, 62]. In some cases, the degree of reproducibility has exceeded that which is conventionally obtained in silicon technology. Thus the prognosis for this approach is very promising. Even so, the more recent work with gallium arsenide diffusion has been limited to studies of only a few dopants.

4.9.1 Ternary Considerations

During diffusion, the impurity and the gallium arsenide surface with which it is in contact must reach an equilibrium composition, which determines the partial pressure of the three components and hence the vacancy concentrations. These quantities are crucial to establishing the conditions for the diffusion process. In a ternary system, at any given temperature and pressure, it follows from the phase rule that $F = 3 - P$, where P is the number of condensed phases. This rule can be used to predict what happens when gallium arsenide is placed in contact with the vapor during diffusion. The behavior of zinc in gallium arsenide is considered [58] as a case in point, since it has been well studied. Here Fig. 4.36*a* shows an isothermal section at 1000°C for this system (see Fig. 2.31). From this figure we note that diffusion sources represented by compositions in the region A are liquid. Sources with compositions in regions B and C involve a liquid in equilibrium with a solid. Diffusions with either of these types of sources are undesirable since they result in liquid formation on both the source and the gallium arsenide, thus destroying the surface flatness of the semiconductor.

Figure 4.36*b* shows an isothermal cut at 775°C. Here regions such as D_1 and D_2 have three condensed phases and thus no degrees of freedom. Consequently, the arsenic and zinc vapor pressures are fixed by the source composition and temperature, and independent of the weight of the source or the volume of the ampul. Note, however, that one condensed phase is a liquid in each case, so that the use of sources in these regions, and at this diffusion temperature, will probably lead to poor surfaces.

Figure 4.36*c* shows an isothermal cut at 700°C. Compositions in the regions labeled B, D, and G are all invariant. However, the G regions are bounded by only solid phases. Diffusion is most desirable with sources in these regions since they lead to undamaged surfaces, in addition to being independent of the amount of the source or the volume of the ampul. Here it should be noted that compositions in G_1 will result in higher arsenic vapor pressures than compositions in G.

Finally, it is possible to use diffusion sources whose compositions fall on the lines connecting gallium arsenide and Zn_3As_2 or $ZnAs_2$. These have no liquid phase, but are bounded by two solid phases so that $F = 1$. Consequently, control of the source weight and ampul volume is necessary for reproducible diffusions.

4.9.2 Sealed-Tube Diffusions

Both *p*- and *n*-type diffusions have been made into gallium arsenide by sealed-tube techniques. Most of the work with *p*-type diffusions has been confined to the use of zinc; these systems are now considered.

Sealed-tube systems for zinc require both arsenic and zinc sources. Elemental sources can be used; more commonly, however, compounds of zinc and arsenic, or alloys of Zn-Ga-As are employed to obtain control of the arsenic and zinc pressures.

The surface concentration can be determined by an analysis of the incorporation reaction for zinc vapor into gallium arsenide, in the form of substitutional zinc. This reaction is given as

$$Zn(g) + V_{Ga}{}^0 \rightleftharpoons Zn_s^- + h^+ \qquad (4.113)$$

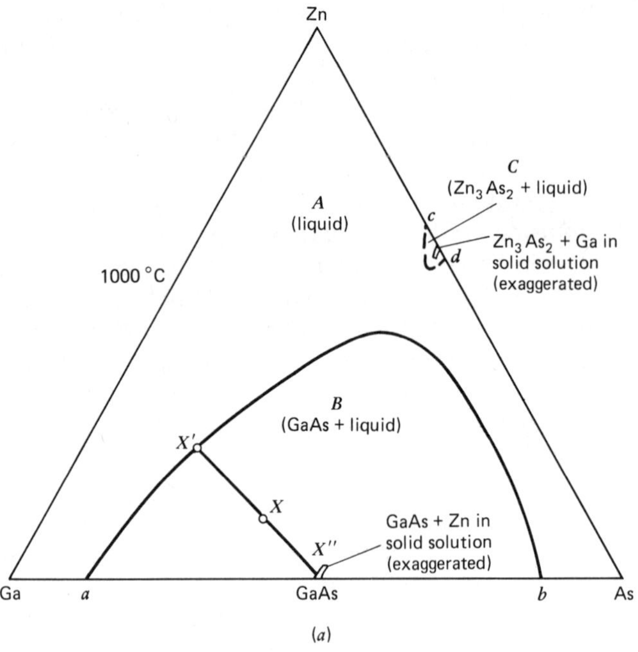

(a)

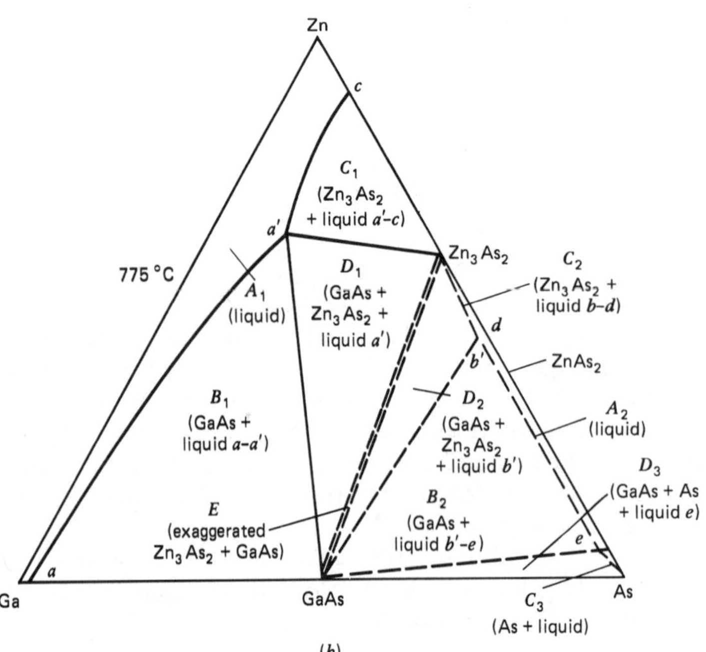

(b)

Fig. 4.36 Isothermal sections of the Ga–As–Zn ternary phase diagram.

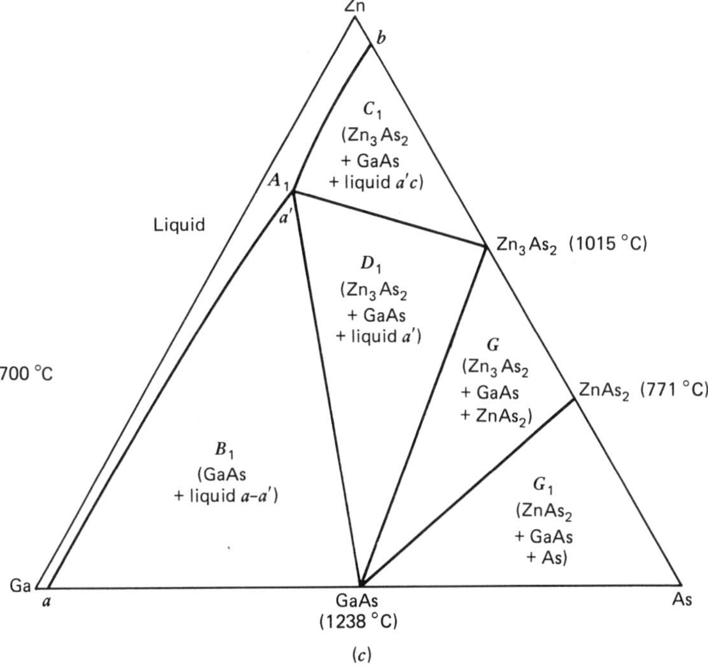

(c)

Fig. 4.36 (Continued)

Applying the mass-action principle,

$$K_1 = \frac{[\mathrm{Zn}_s^-]\,p}{p_{\mathrm{Zn}}\,[V_{\mathrm{Ga}}^{0}]} \tag{4.114}$$

where K_1 is an equilbrium constant. But

$$p = [\mathrm{Zn}_s^-] \tag{4.115}$$

so that

$$[\mathrm{Zn}_s^-]^2 = K_1 p_{\mathrm{Zn}}[V_{\mathrm{Ga}}^{0}] \tag{4.116}$$

Assuming that arsenic is present as As_4 during diffusion, it can be shown from Section 1.4.1.2 that

$$[V_{\mathrm{Ga}}^{0}] = K_2 p_{\mathrm{As}_4}{}^{1/4} \tag{4.117}$$

so that

$$[\mathrm{Zn}_s^-] = (K_1K_2)^{1/2}\,p_{\mathrm{Zn}}{}^{1/2}\,p_{\mathrm{As}_4}{}^{1/8} \tag{4.118}$$

Thus the surface concentration is proportional to the square root of the zinc partial pressure, and is a very weak function (eighth root) of the arsenic pressure.

Figure 4.37 shows measured values [2] of the surface concentration as a function of the atom fraction of zinc in the vapor. The square-root dependence is seen here to hold for doping concentrations below 10^{19} cm^{-3}. Above this point, however, a slight departure from this law is observed. This can be explained by noting that the above analysis holds for low-concentration diffusions, where the mass-action principle applies. However, it has been shown [63] that equations of the same form can be developed by introducing the concept of a hole activity coefficient γ_p linking the Boltzmann and Fermi statistics that apply to the low- and high-concentration cases, respectively. This activity coefficient is unity at low concentrations, but falls rapidly to about 0.5 at hole concentrations above 3×10^{19} cm^{-3} (for diffusions made at 700°C), and remains constant beyond this point.

The effect of zinc and arsenic overpressures on the junction depth can also be determined. We have shown that

$$D = \frac{K_3 D_i' N_s^2}{[V_{Ga}^0]} \tag{4.69}$$

where D_i' is the interstitial diffusion coefficient and N_s is the substitutional

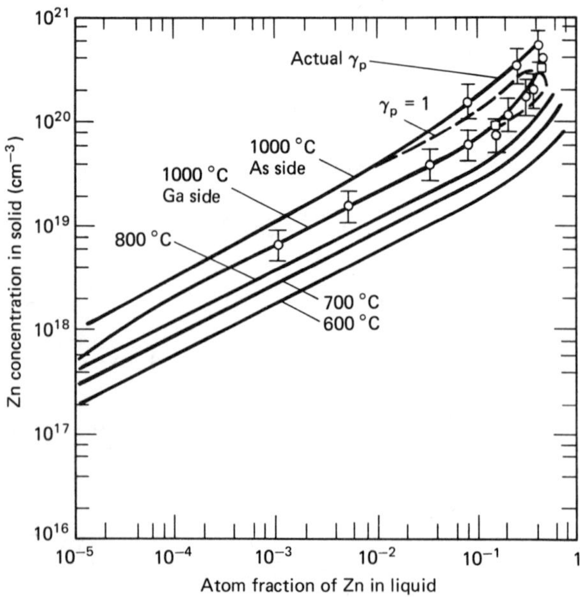

Fig. 4.37 Zinc surface concentration as a function of the atom fraction of zinc in the vapor [2]. Adapted from Shaw [2].

zinc concentration. This equation can be rewritten as

$$D = \frac{K_3 D_i' [\text{Zn}_s^-]^2}{[V_{\text{Ga}}{}^0]} \tag{4.119}$$

Combining with Eq. (4.116), gives

$$D = K_1 K_3 p_{\text{Zn}} D_i' \tag{4.120}$$

Thus the diffusion constant is directly proportional to the zinc pressure, but is independent of the arsenic pressure.

As noted in Section 4.5.2.1, the zinc-doping profile is abrupt, and results in a junction depth that is dependent on the surface concentration, but is relatively independent of the background concentration. This junction depth has been given as

$$x_j = 1.092(D_{\text{sur}} t)^{1/2} \tag{4.63b}$$

where D_{sur} is the diffusion constant at the surface. It follows therefore that

$$x_j = 1.092(K_1 K_3 p_{\text{Zn}} D_i' t)^{1/2} \tag{4.121}$$

Thus the junction depth is independent of the arsenic pressure. This has been observed [58] in a comparison of diffusions made with a Ga-As-Zn alloy source to those made with an elemental zinc source. The zinc source provided an increase in p_{Zn} by a factor of approximately 10, and a decrease in p_{As_4} by a factor of approximately 10^8. However, junction depths were larger by a factor of only 10 with this source, and not reproducible from run to run. In contrast, the Ga-As-Zn source, whose composition (5 : 50 : 45 atom %) placed it in region *G* of Fig. 4.36*c*, gave extremely reproducible junctions with properties that were not sensitive to either the source weight or the ampul volume.

The effect of surface concentration on the junction depth can now be determined. Combining Eqs. (4.118) and (4.121), gives

$$x_j = 1.092 \left(\frac{K_3}{K_2}\right)^{1/2} p_{\text{As}_4}{}^{-1/8} [\text{Zn}_s^-] (D_i' t)^{1/2} \tag{4.122}$$

so that the junction depth is linearly proportional to the surface concentration, and only weakly dependent on the arsenic pressure. A series of sealed-tube junctions, made with an elemental zinc source [64], have shown that this is indeed the case.

Zinc diffusions in gallium arsenide have also been made [59] using Zn_3As_2, ZnAs_2, and ZnAs_2 + As sources. Use of these sources results in interrelated values of arsenic and zinc overpressure, so that the equations derived earlier must be applied with great care. By way of example, these sources provide a decreasing zinc pressure, but an increasing arsenic pressure, in the above order. Although

$$[\mathrm{Zn}_s^-] = (K_1K_2)^{1/2}\, p_{\mathrm{Zn}}{}^{1/2}\, p_{\mathrm{As}_4}{}^{1/8} \tag{4.118}$$

the overall effect of changing both these pressures was a small ($\simeq$ 20%) increase in the zinc incorporation in these experiments [59].

Sealed-tube zinc diffusions have also been made in gallium arsenide whose surface was coated with a layer of sputtered silica [65]. Here a 6500-Å layer of SiO_2 was used to protect the gallium arsenide surface from dissociation at diffusion temperatures, as well as to prevent surface damage due to alloying with the zinc. A zinc–gallium alloy served to establish the partial pressure of zinc; no additional arsenic was included in the ampul.

Results of these experiments showed a reduction in surface concentration to about 25% of what was obtained on uncoated wafers which were diffused at the same time. Thus the SiO_2 is highly transparent to the zinc. In addition, the SiO_2 prevented loss of arsenic from the gallium arsenide during heat treatment. Consequently, the surface finish of the coated wafers was protected by this process.

***n*-Type.** The *n*-type impurities in gallium arsenide are sulfur, tellurium, and selenium. Very little work has been done with these. Even the nature of their diffusion mechanism has not been established at the present time. All of these impurities exhibit extremely low values of diffusivity, which are comparable to those obtained for substitutional movement. Moreover, doping profiles for these impurities are of the erfc type, as is the case for substitutional diffusers. A divacancy model has been proposed to account for this behavior. In contrast, *p*-type impurity motion is more closely approximated by interstitial diffusion, with diffusivity values that are five to six decades larger, and an abrupt doping profile.

Some sealed-tube work has been done using sulfur as the dopant. Here it has been proposed [66] that the dominant vapor species is Ga_2S, and that the sulfur vapor density is directly proportional to the partial pressure of this compound. The incorporation reaction for sulfur from Ga_2S is given by

$$\mathrm{Ga_2S}(g) + V_{\mathrm{As}} \rightleftharpoons S_{\mathrm{As}}{}^{+} + 2\mathrm{Ga}(g) + e^{-} \tag{4.123}$$

Application of the mass-action principle results in a surface concentration which is linearly proportional to $p_{\mathrm{Ga_2S}}$, and thus to the sulfur vapor density. This has been observed to be the case.*

* The molecular species of sulfur are S_2, S_6, and S_8. Direct incorporation of any of these into the gallium arsenide would result in a surface concentration that is proportional to the $\frac{1}{2}$, $\frac{1}{6}$, and $\frac{1}{8}$ power of the sulfur vapor density, respectively.

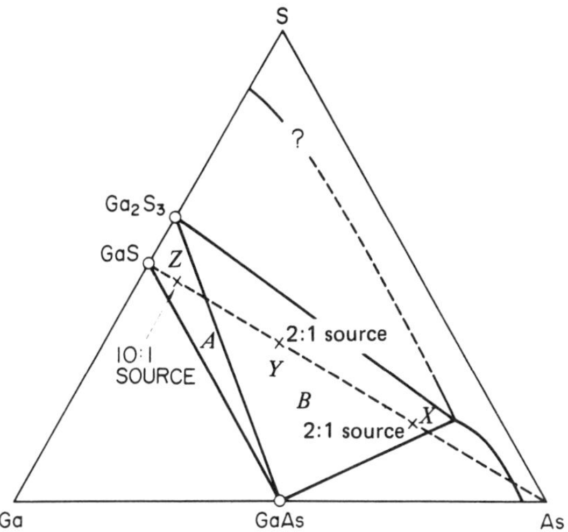

Fig. 4.38 Isothermal section of the Ga–As–S ternary phase diagram at 910°C [60]. Reprinted with permission of the publisher, Pergamon Press, Ltd.

It has been experimentally observed that the diffusivity of sulfur increases as the square root of the arsenic partial pressure. Again, the divacancy mechanism has been invoked as a possible explanation. However, the reasons for this behavior are not firmly established at the present time.

Reproducible sulfur diffusions have been made [60] using mixtures of gallium sulfide and arsenic at temperatures of 860 and 910°C. The appropriate isothermal section of the phase diagram is shown in Fig. 4.38. The phase diagram of the Ga-As-S ternary system is shown in Fig. 2.29.

Compositions were selected to fall into the invariant regions *A* and *B* of Fig. 4.38, so that the diffusion was independent of the weight of the dopant or the volume of the ampul. These sources were made up of mixtures of gallium sulfide and arsenic in 1:2, 2:1, and 10:1 atomic ratios, respectively; they are labeled *X*, *Y*, and *Z* in Fig. 4.38.

Curves of junction depth versus the square root of time are shown in Fig. 4.39 for the three sources, and for diffusions at 860°C. These are seen to be straight lines, so that values of *D* can be extracted from them. Values of *D* at 860°C were 10^{-12} cm^2/s for sources *X* and *Y*, and 2×10^{-13} cm^2/s for source *Z*. This is to be expected, since the arsenic pressure for source *Z* is considerably lower than that for the other sources.

Although tin is a group IV impurity, it behaves as a donor when diffused into gallium arsenide. Sealed-tube diffusions, using elemental tin as a vapor source, have been made in this manner. More recently, however, a doped oxide source, formed by the oxidation of tetramethyltin and silane,* has been used for this purpose. This source prevents loss of arsenic from the

* Details of source preparation are provided in Chapter 8.

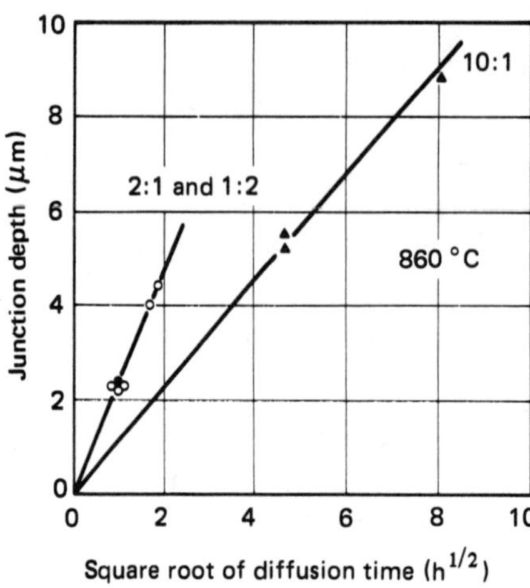

Fig. 4.39 Junction depth versus square root of time for sulfur diffusions. From Matino [60]. Reprinted with permission of the publisher, Pergamon Press, Ltd.

semiconductor during diffusion, and thus protects the surface, even when no arsenic overpressure is established in the ampul.

Figure 4.40 shows [67] the junction depth versus time characteristics for an SnO_2-SiO_2 source of this type, for diffusions carried out at 1010°C. As with other *n*-type impurities, it is seen that high temperatures and long diffusion times are required to achieve a significant penetration depth. A diffusivity of 1.8×10^{-13} cm²/s was obtained for this source at 1010°C.

An interesting feature of Fig. 4.40 is that the junction depth was initially proportional to the square root of time. With longer times, however, an anomalous falloff in junction depth was observed. This has been attributed

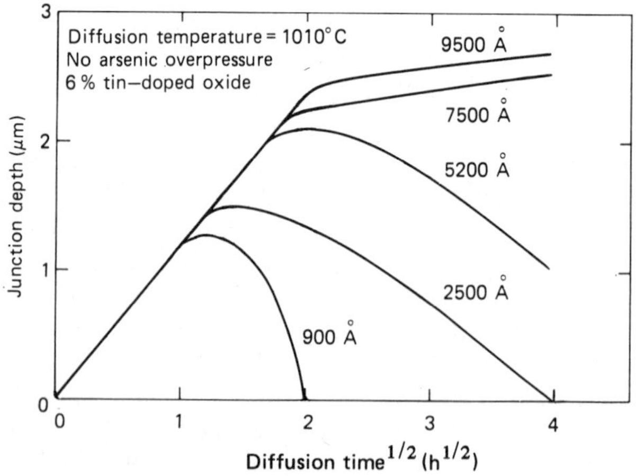

Fig. 4.40 Junction depth versus square root of time for tin diffusions at 1010°C. From Baliga and Ghandhi [67]. Reprinted with permission of the publisher, The Electrochemical Society, Inc.

to depletion effects, as verified from the fact that thicker oxides were linear in their $x_j \propto t^{1/2}$ behavior for longer diffusion times.

4.9.3 Open-Tube Diffusions

Two techniques have been used for making open-tube *p*-type diffusions with zinc. The first of these requires that diffusions be carried out with a gas such as arsine to provide the necessary arsenic overpressure. Diffusions of this type, using diethylzinc as the dopant source, have been made in gallium arsenide [68].

Although this approach avoids the necessity of using sealed ampuls, it has many disadvantages when compared to its silicon counterpart. Thus both arsine and diethylzinc are highly dangerous, and are only slightly consumed during the diffusion process. In addition, decomposition of the arsine gas results in a large amount of arsenic deposits on the walls, as well as arsenic dust in the diffusion tube. Consequently, elaborate venting precautions must be taken with systems of this type. An alternate approach, which consists of using doped oxides with cap layers, is considerably more convenient. Diffusions using this approach can be carried out in an inert gas ambient, and are as convenient as diffusions in silicon.

In *p*-type diffusions of this sort, a ZnO-SiO_2 source was grown on the gallium arsenide slice by the simultaneous oxidation of diethylzinc and silane [62]. A composition with 20–40% zinc oxide by weight was found to be suitable for this purpose. This represents an excess of zinc, so that neither the layer composition nor its thickness was important. Typical growth temperatures for these layers ranged from 300 to 400°C.

A cover layer of phosphosilicate glass, formed by the oxidation of phosphine and silane gases, was next deposited at 350°C. This served to prevent loss of zinc from the source during diffusion,* and also to prevent the gallium arsenide from decomposition.

Diffusions were made in an open-tube furnace with a small flow of nitrogen gas. The ambient gas had no effect on the diffusion process, as seen from the fact that diffusions made in a 50% N_2, 50% H_2 ambient gave the same results. Undamaged surfaces were obtained after both cap and doped oxide layers were removed in buffered hydrofluoric acid (see Chapter 9).

Figure 4.41 [69] shows that the junction depth varied linearly with the square root of time over a wide range of diffusion temperatures. It is also seen that the diffusion rate was extremely rapid, as compared to that for *n*-type impurities. By way of example it is seen from this figure that $D_{sur} = 1.6 \times 10^{-12}$ cm^2/s at 600°C. It is interesting to note that approximately the same value of diffusivity was obtained with sulfur diffusions at 860°C, and with tin diffusions at 1010°C.

Open-tube zinc diffusions have also been made using layers of pure zinc

* A back-side cap can also be used if protection of this gallium arsenide surface is desired.

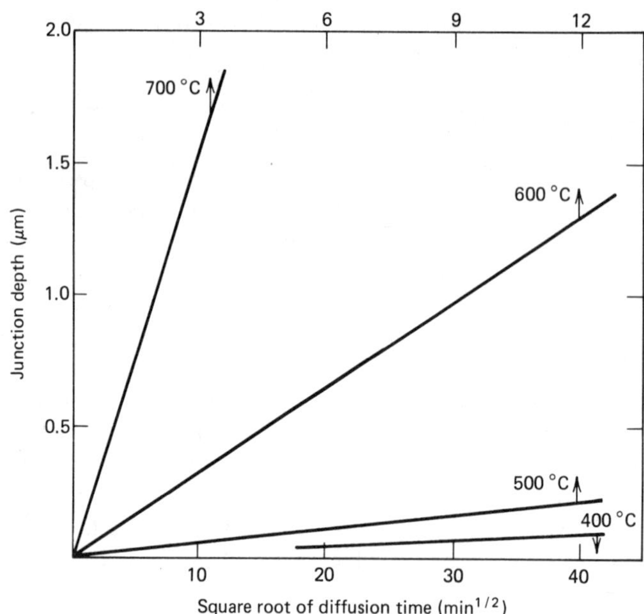

Fig. 4.41 Junction depth versus square root of time for open-tube zinc diffusions. From Field and Ghandhi [69]. Reprinted with permission of the publisher, The Electrochemical Society, Inc.

oxide, covered with phosphosilicate glass [70]. Here, however, it was necessary to provide a barrier layer of undoped silica between the zinc oxide and the gallium arsenide, to prevent damage to the semiconductor surface during diffusion. A major disadvantage of this approach was that the junction depth was now a strong function of the barrier layer thickness. Thus it was difficult to control on a run-to-run basis.

Open-tube tin diffusions have been made using a spin-on SnO_2-SiO_2 source, covered with a chemical vapor deposited layer of silica [61]. Results with this technique were highly variable, most probably because of the poor compositional control of the spin-on source. In addition, it is probable that the undoped SiO_2 cap induced considerable stress in the gallium arsenide at the diffusion temperature and so affected the diffusion process.

Both of these problems have been avoided [71] by growing a chemical vapor deposited layer of SnO_2-SiO_2 by the simultaneous oxidation of tetramethyltin and silane gas. In addition, the use of a phosphosilicate cap greatly reduced the interface stress during diffusion, because of its better thermal expansion match to the gallium arsenide.

Diffusions were made at 800°C for 30 min and resulted in a 10^{17} cm^{-3} surface concentration with a junction depth of 0.1 μm. The average mobility of these layers was 3000–4000 cm^2/V s, so that they were useful as starting layers for the fabrication of microwave field effect transistors.

It should be noted that careful control of the doped oxide composition was necessary for this application, since the surface concentration was well below the equilibrium solubility limit for this diffusion temperature. High surface concentration (10^{18}–10^{19} cm^{-3}) diffusions, which are solid-solubility limited, can be readily achieved without careful attention to this factor.

4.9.4 Masked Diffusions

Masked diffusions are required in integrated circuit fabrication technology. These can be made in gallium arsenide, using deposited films* of either silica, phosphosilicate glass, or silicon nitride as mask materials. An anodically grown oxide of gallium arsenide has also been used for some low-temperature diffusions.

All of these materials have both advantages and disadvantages so that the ideal mask is not available. Deposited silica films are certainly the worst. They do not inhibit the decomposition of gallium arsenide at elevated temperatures, since they permit the rapid diffusion of gallium through them. They are relatively transparent to both zinc and tin, so thick layers ($\geq$ 0.5 μm) must be used for masking purposes. Finally, they are very poorly matched to gallium arsenide in their thermal expansion characteristics, so that thick films often crack during thermal processing, or damage the semiconductor.

Thermal expansion problems can be greatly reduced by the use of PSG [72]. Moreover, this glass effectively blocks the transport of both zinc and tin, so that thinner films (0.1 μm) can be used for masking. Its blocking characteristics for gallium are not established at the present time. However, all indications are that it is considerably better than silica in this respect.

Silicon nitride is an almost impermeable barrier to zinc, tin, and gallium. However, it has a bad thermal expansion match to gallium arsenide, and must be used in extremely thin layers (300–500 Å) to avoid problems during processing. Its deposition technology is considerably more difficult than that of either silica or PSG.

A number of workers have noted extremely large lateral diffusion effects (lateral to vertical ratios of as much as 50:1), and developed techniques for reducing them to manageable proportions (2:1). Yet others have reported that these effects are not present with some dopants, but are with others. No definitive work has been done in this area, but some general observations can be made at the present time.

1. These effects are confined to the first 50–100 Å of the gallium arsenide surface.
2. They are sensitive to interfacial stress between the mask and the semiconductor. Close matching of the respective thermal expansion coefficients greatly reduces the lateral diffusion [73].

* Techniques for the growth of these films are described in Chapter 8.

3. Lateral diffusion effects are also sensitive to interface reactions between the mask and the gallium arsenide surface. Such reactions are most severe with silica and least with silicon nitride. PSG is somewhat intermediate; lateral diffusion effects are reduced as the phosphorus content of the glass is increased.
4. Lateral diffusion effects are worst when there is a large available reservoir of the dopant, since this provides the driving force for the diffusion. Thus both ion implantation and doped oxide techniques are effective in minimizing this problem.

In summary, then, there are no magic dopants that work and others that do not. Rather, the technique by which the masked diffusion is made appears to be the dominant factor in the extent of this lateral diffusion.

4.10 EVALUATION TECHNIQUES FOR DIFFUSED LAYERS

Many techniques are available for evaluating the properties of a diffused layer. The most straightforward method is to profile the layer by secondary ion mass spectrometry [74]. An alternate method is to diffuse the sample with a radioactive isotope and evaluate the radiation intensity upon removal of successive layers. This allows the detailed nature of the diffusion profile to be obtained for the entire impurity content, under the assumptions that the isotope atoms disperse uniformly among the normal atoms of the diffusing impurity and that the diffusion coefficient of the isotope is the same as that of the impurity [1].

Details of these techniques are beyond the scope of this book. However, knowledge of the impurity profile permits the use of Boltzmann–Matano techniques (see Section A.4.1 in the Appendix) to extract the value of the diffusion coefficient on a point-by-point basis, as a function of the concentration. Their primary disadvantage is that they are not suited for the rapid monitoring of diffusion processes during microcircuit fabrication. In addition, they provide information on the entire impurity content and not on the electronically active part with which we are concerned.

The properties of the diffused layer are readily determined by measurements of masked patterns on p–n junctions formed by diffusion into a slice of known background concentration, but opposite impurity type. This method gives results which bear directly on the end goal of the diffusion process, and it is commonly used. Details of this evaluation technique now follow.

4.10.1 Junction Depth

To determine the junction depth, a small chip of the diffused slice is lapped on an angle so as to expose the actual junction. This angle is on the order of $\frac{1}{2}°$, so that the junction region is visually magnified. Next the junction is delineated by means of a selective etch. A number of such etches are in use, a common one consisting of a solution of copper sulfate in dilute hydrofluoric acid. In operation the acid serves to dissolve surface oxides; the copper sulfate selectively plates the region so as to delineate the junction.

An alternate technique consists of grinding a cylindrical groove into the wafer, as shown in Fig. 4.42. Once again, this visually magnifies the junction region, which is delineated by means of a selective etch. If R is the radius of the tool used for forming the groove, then the junction depth x_j is given by

$$x_j = (R^2 - b^2)^{1/2} - (R^2 - a^2)^{1/2} \tag{4.124a}$$

where a and b are indicated in the figure. In addition, if R is much larger than a and b, then

$$x_j \simeq \frac{a^2 - b^2}{2R} \tag{4.124b}$$

The grooving technique is more commonly used because of its greater convenience and ease of execution.

Junction-depth measurements based on geometric considerations can lead to considerable error if the lapping angle is not precisely known. Consequently, an interferometric method is preferred, with the upper surface of the chip serving as a reference plane. An optical flat is placed on this chip, as shown in Fig. 4.43, and is vertically illuminated by collimated monochromatic light. The resulting fringe pattern gives a direct measure of the vertical depth in wavelengths of the illuminating source. For a sodium vapor lamp whose spectral radiation is concentrated at 5895.93 and 5889.96 Å (the

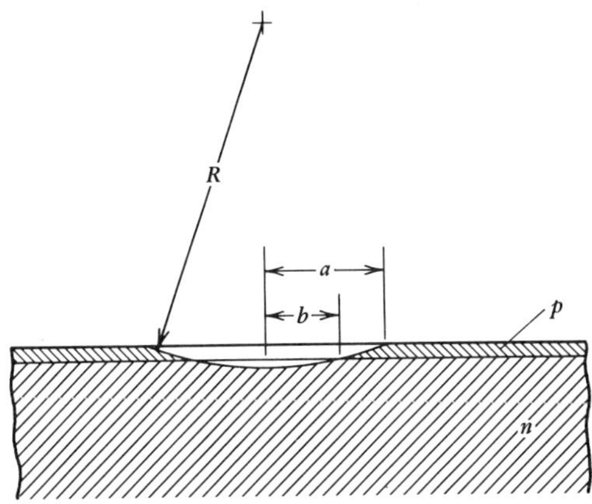

Fig. 4.42 Junction Beveling Technique.

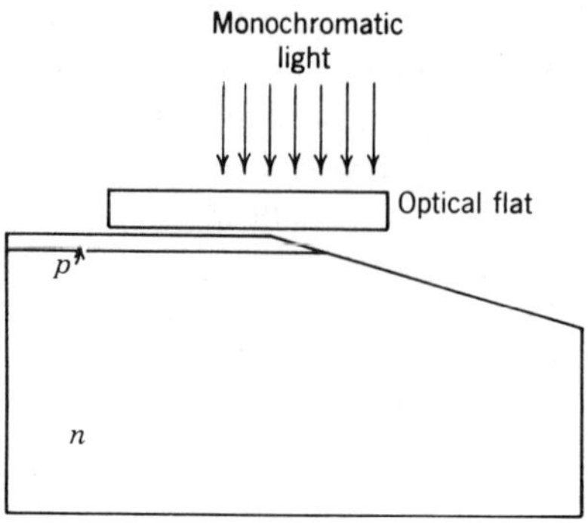

Fig. 4.43 Junction-depth measurement.

D_1 and D_2 lines, respectively), the distance between fringes is approximately 0.29 μm.

Multiple-beam interferometers are also commonly employed. Their use results in sharply delineating the fringes, so that they are more accurate. Electron-beam-induced current (EBIC) methods are also used in conjunction with a scanning electron microscope [74] for junction-depth measurements.

4.10.2 Sheet Resistance

It is not possible to specify the specific resistivity of a diffused layer, because it is inhomogeneous. For a layer of this type, a sheet resistance is more appropriate. Consider a rectangular layer of diffused material of length l, width w, and thickness t. If its resistance is measured across the faces of width w and thickness t, then

$$R = \frac{\rho(t)}{t}\frac{l}{w} \tag{4.125}$$

where $\rho(t)$ is the specific resistivity of the material in Ω cm and varies with depth. This equation may be more conveniently rewritten in the form

$$R = R_S \frac{l}{w} \tag{4.126}$$

where R_S is defined as the sheet resistance of the layer, in Ω. If the layer takes the form of a square sheet, its resistance is given by R_S, regardless of its actual dimensions. Hence the sheet resistance is often specified in "ohms per square."

In microcircuit fabrication, the sheet resistance of a diffused layer can be directly measured if this layer is made in the patterned shape shown in Fig. 4.44. Here, as shown for a bipolar transistor, a p-type base diffusion is made into an n-type collector region. After this diffusion is conducted,

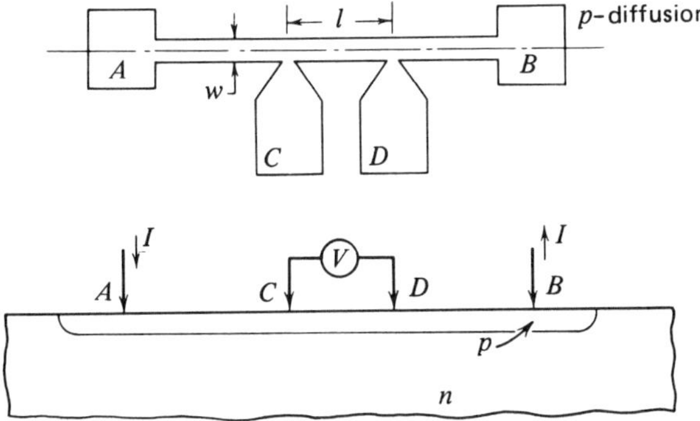

Fig. 4.44 Base sheet resistance test pattern.

a constant current is applied across the points AB, and the voltage developed across CD is read with the aid of a high-impedance voltmeter. With reference to Fig. 4.44,

$$R_S = \frac{V}{I}\frac{w}{l} \tag{4.127}$$

Since w/l is known for a specific test pattern, the sheet resistance is directly found.

Figure 4.45 shows the manner in which the sheet resistance of the emitter diffusion (n-type) is determined. Here an initial p-type region is formed during the base-diffusion step, and the n-type emitter diffusion is made within it. The sheet resistance of this layer is measured as described for the base diffusion. Patterns of this type are usually made as part of a test chip; one or more of these placed on each slice, and used for diagnostic purposes. Often these test patterns are located in the dead space between microcircuit chips.

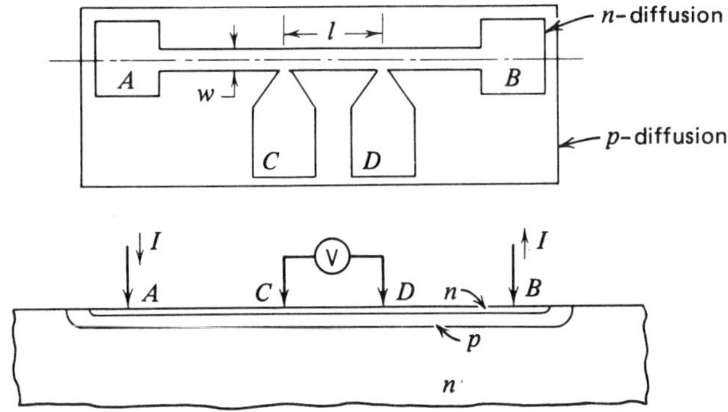

Fig. 4.45 Emitter sheet resistance test pattern.

4.10.3 Surface Concentration

The surface concentration can be determined from the sheet resistance and the junction depth. For a diffusion made into a background concentration N_B, the sheet resistance of the diffused layer is given by R_S, where

$$R_S = \left[\int_0^{x_j} q\, \mu(x)\, [N(x) - N_B]\, dx \right]^{-1} \tag{4.128}$$

Here $\mu(x)$ is the mobility for the majority carrier in the diffused layer and is a function of the concentration at any given depth. This equation can be solved to determine the surface concentration, provided that the doping profile is known, as well as the functional dependence of majority carrier mobility on the doping. A number of special cases are now described.

4.10.4 Diffused Layers in Silicon

Both high- and low-concentration diffusions are used in silicon microcircuit fabrication technology. Layers with low surface concentration can usually be fitted to the erfc or the gaussian diffusion profiles shown in Fig. 4.17. High-concentration diffusions can also be approximated by the erfc profile provided that they are deep.

Computer-derived solutions have been obtained [75] for both erfc and gaussian layers in silicon, for different values of background concentration. These are displayed in Figs. 4.46 and 4.47, and interrelate the surface concentration, the sheet resistance, and the junction depth for a wide variety of background concentrations.

The resistivity of base layers under an emitter can also be determined by integrating the $q\mu N$ product over the base layer width. Computer-derived curves of this type are also available in the literature [75].

4.10.4.1 High-Concentration Diffused Layers

Many diffusions that are used in VLSI fabrication technology are made with high surface concentration. The doping profiles for these diffusions are described in Section 4.5.2. Most profiles are abrupt, so that the junction depth is relatively independent of the background concentration. Profile parameters for these diffusions are now described.

Boron. High-concentration boron diffusions in silicon have been described in Section 4.5.2.2. Here an average hole mobility (55 $cm^2/V\ s$) can be assigned to layers with surface concentration in excess of $5 \times 10^{19}\ cm^{-3}$, resulting in straightforward relationships between R_S, x_j, and N_{sur}. These are obtained [27] if the background concentration is assumed to be less than 1% of the surface value. These relationships are compiled in Table 4.6.

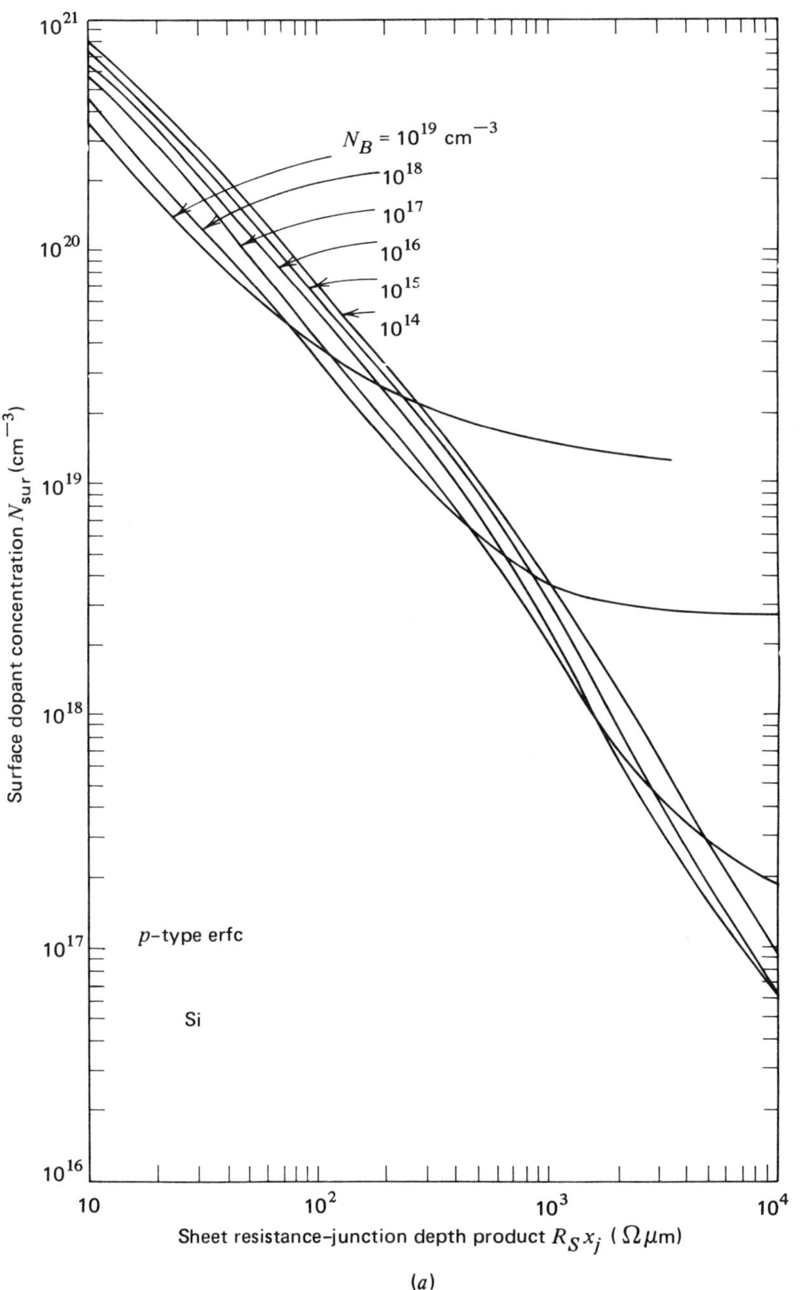

Fig. 4.46 Erfc-type diffusions into silicon. Adapted from Irvin [75].

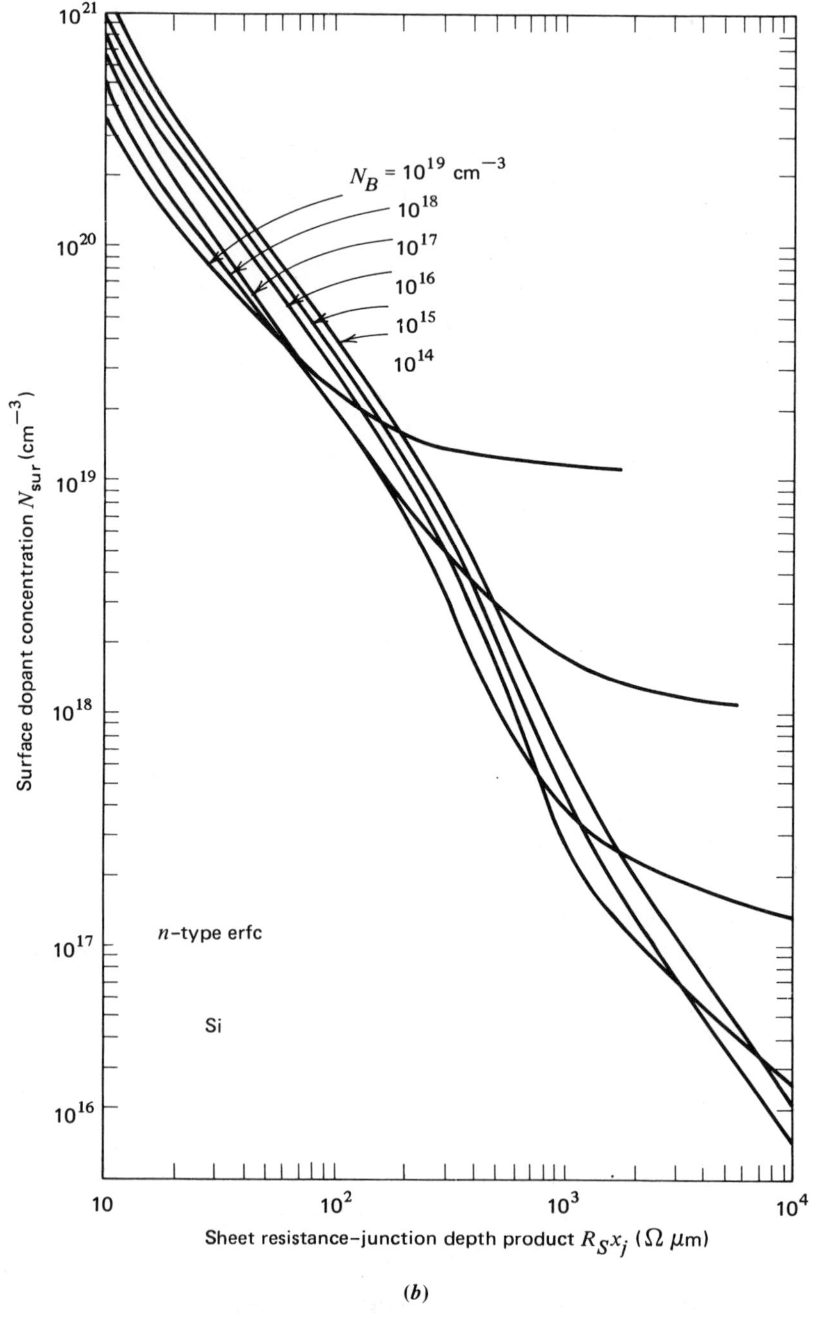

$N_B = 10^{19}$ cm^{-3}
10^{18}
10^{17}
10^{16}
10^{15}
10^{14}
n-type erfc
Si
Surface dopant concentration N_{sur} (cm^{-3})
Sheet resistance–junction depth product $R_S x_j$ (Ω μm)
10^{21}
10^{20}
10^{19}
10^{18}
10^{17}
10^{16}
10
10^2
10^3
10^4

(*b*)

Fig. 4.46 (Continued)

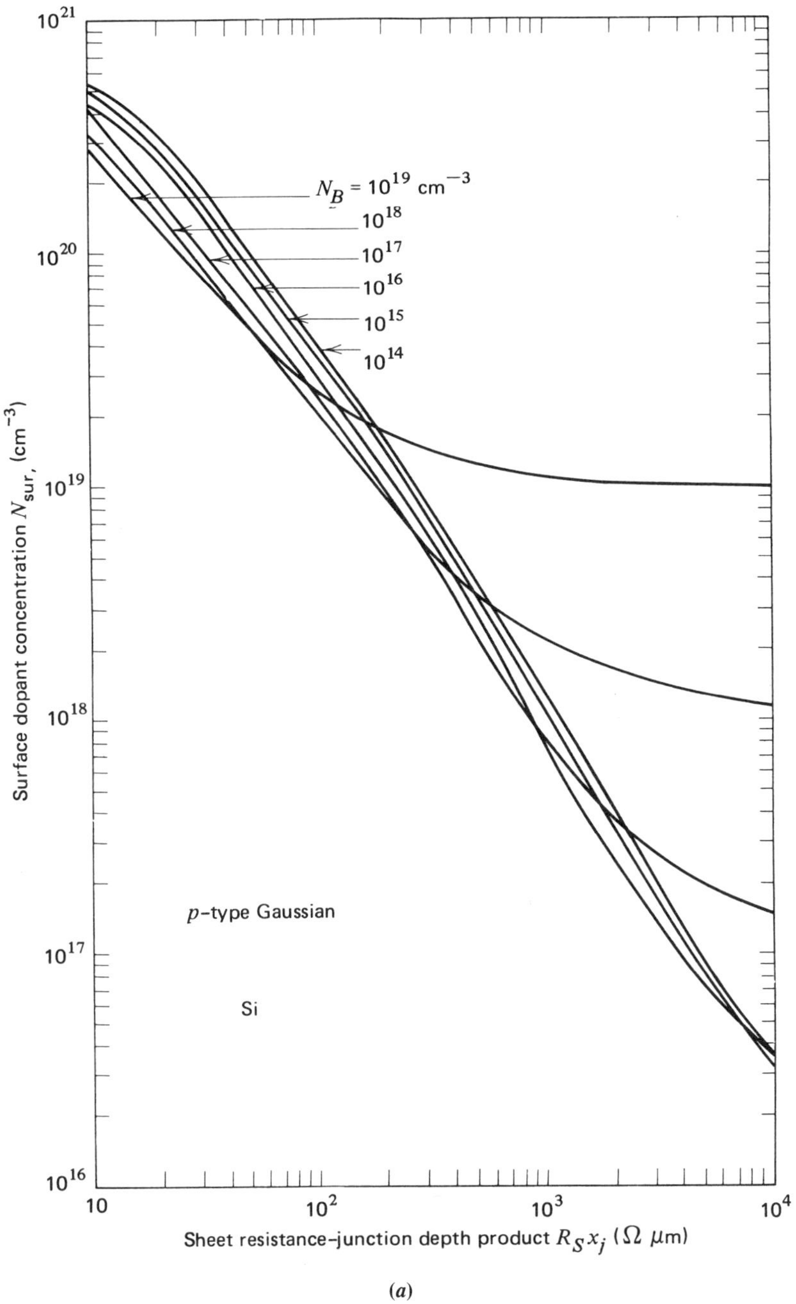

(*a*)

Fig. 4.47 Gaussian diffusions into silicon. Adapted from Irvin [75].

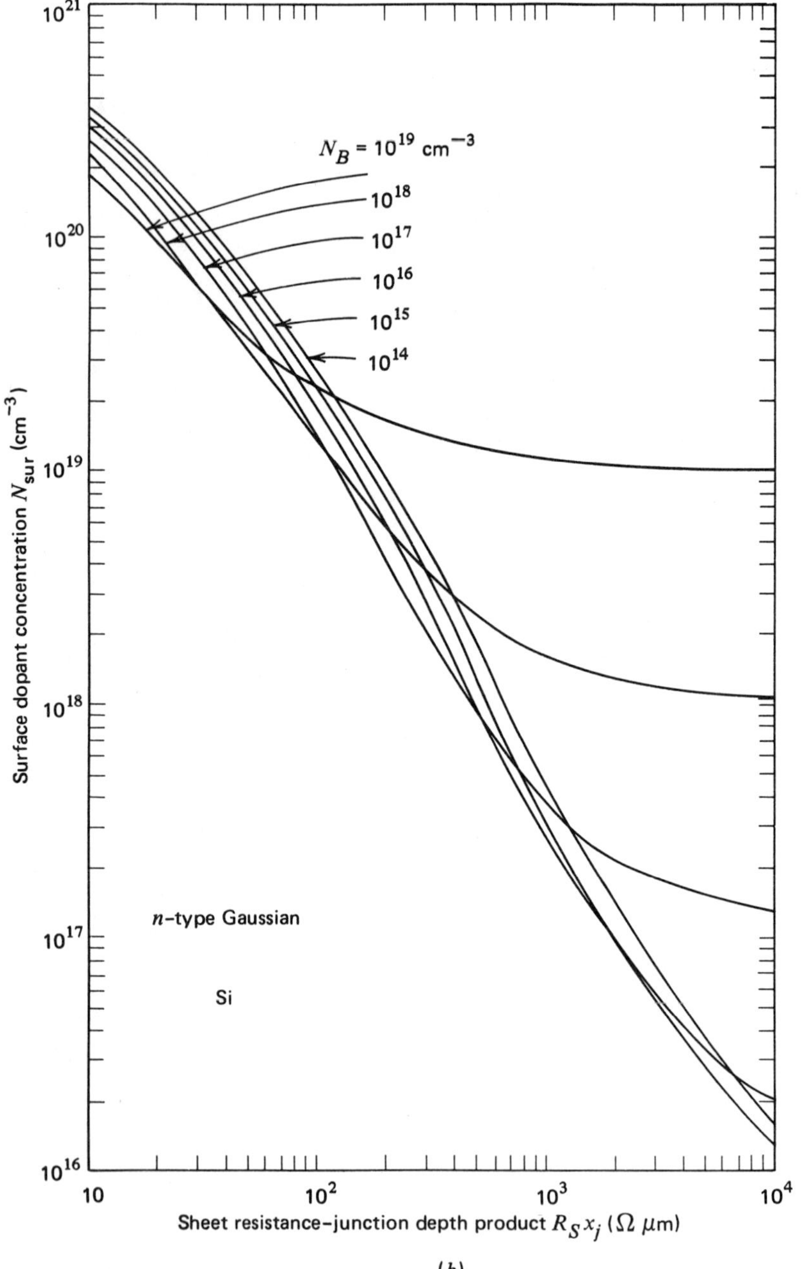

$N_B = 10^{19}$ cm^{-3}
10^{18}
10^{17}
10^{16}
10^{15}
10^{14}
n-type Gaussian
Si
Surface dopant concentration N_{sur} (cm^{-3})
10^{21}
10^{20}
10^{19}
10^{18}
10^{17}
10^{16}
10
10^2
10^3
10^4
Sheet resistance–junction depth product $R_S x_j$ (Ω μm)

(b)

Fig. 4.47 (Continued)

Table 4.6 Properties of High-Concentration Boron Diffusions in Silicon[a]

	Diffused from a Constant-Source Concentration	Implanted with a Dose of Q_0 ions/cm^2 and Diffused for t s
x_j (cm)	$2.45\left(N_{sur}\dfrac{D_i t}{n_i}\right)^{1/2}$	$Q_0\,(0.4N_{sur})^{-1}$
N_{sur} (cm^{-3})	$2.78 \times 10^{17}(R_S x_j)^{-1}$	$0.53\left(Q_O{}^2\dfrac{n_i}{D_i t}\right)^{1/3}$
D_i (cm^2/s)	$3.17e^{-3.59/kT}$	

[a] See reference 27.

Arsenic. This impurity results in extremely steep doping profiles of the type shown in curve *a* of Fig. 4.21. Their abrupt character allows an average value of mobility to be assigned to this layer, and interrelations of x_j, R_S, and N_{sur} have been developed [76, 77] based on this assumption. The polynomial approximation to the doping profile (see Section 4.5.2.2), as well as the assumption that the junction depth is independent of background concentration, were used to derive the formulas listed in Table 4.7.

Phosphorus. The anomalous doping profile for high-concentration phosphorus diffusions has been described in Section 4.8.4. A systematic analysis of its detailed features has permitted the necessary interrelations between R_S, x_j, and N to be extracted [78]. These results are now presented without proof.

Table 4.7 Properties of High-Concentration Arsenic Diffusions in Silicon [76, 77]

	Diffused from a Constant-Source Concentration	Implanted with a Dose of Q_0 ions/cm^2 and Diffused for t s
x_j (cm)	$2.29\left(N_{sur}\dfrac{D_i^- t}{n_{ie}}\right)^{1/2}$	$2\left(Q_0\dfrac{D_i^- t}{n_{ie}}\right)^{1/3}$
R_S (Ω)	$1.56 \times 10^{17}\,(N_{sur}\,x_j)^{-1}$	$\dfrac{1.7 \times 10^{10}}{Q^{7/9}}\left(\dfrac{n_{ie}}{D_i^- t}\right)^{1/9}$
N_{sur} (cm^{-3})	N_{sur}	$Q_0\,(0.45x_j)^{-1}$ $6.26 \times 10^{15}\,(R_S x_j)^{-3/2}$
D_i^- (cm^2/s)	$22.9e^{-4.1/kT}$	

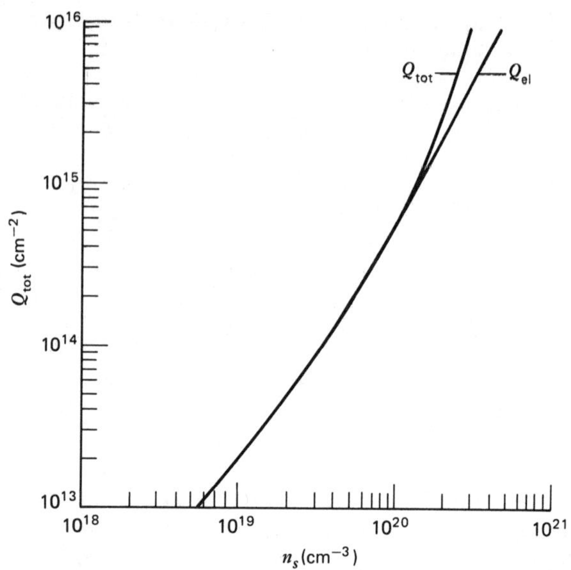

Fig. 4.48 n_s **versus** Q_{el} **and** Q_{tot} **for phosphorus diffusion in silicon. From Fair [78]. Reprinted with permission of the publisher, The Electrochemical Society, Inc.**

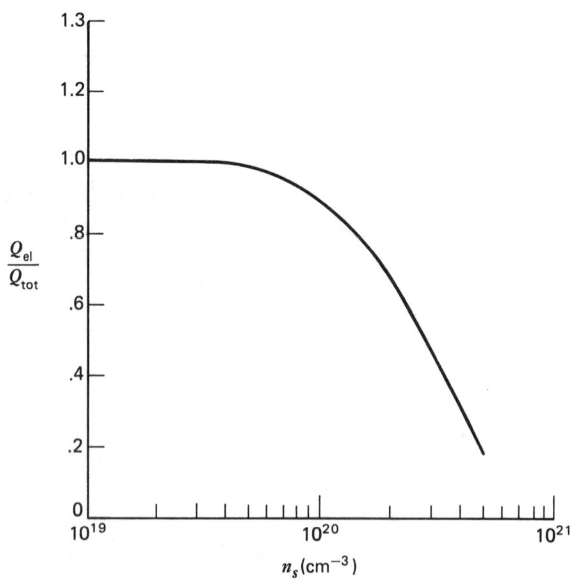

Fig. 4.49 n_s **versus** Q_{el}/Q_{tot} **for phosphorus diffusion in silicon. From Fair [78]. Reprinted with permission of the publisher, The Electrochemical Society, Inc.**

1. The surface electron concentration n_s and the surface impurity concentration N_{tot} are related by

$$N_{tot} = n_s + 2.04 \times 10^{-41} n_s^3 \qquad (4.129)$$

2. The relationship of n_s and the total number of diffused impurities per unit area Q_{tot} is shown in Fig. 4.48. Also shown is the graph for Q_{el}, the electronically active component of Q_{tot}. Figure 4.49 shows the fraction Q_{el}/Q_{tot} as a function of n_s.
3. The sheet resistance is related to the junction depth and to Q_{el} by

$$R_S = [q(75\ Q_{el} + 1.8 \times 10^{20}\ x_j)]^{-1} \qquad (4.130)$$

Profile parameters can be obtained by use of the above data. However, the approximate nature of these relationships must be stressed.

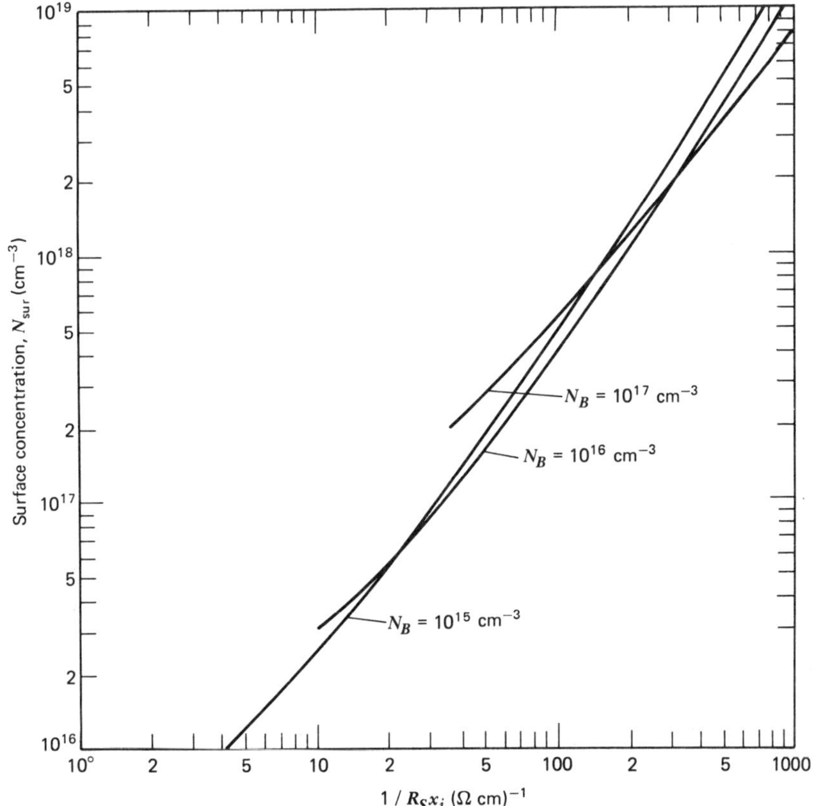

Fig. 4.50 ***n*-Type erfc diffusions into gallium arsenide. From Baliga [79]. Reprinted with permission of the publisher, Pergamon Press, Ltd.**

4.10.5 Diffused Layers in Gallium Arsenide

The behavior of n-type dopants in gallium arsenide is similar to that of substitutional diffusers, and is characterized by an erfc doping profile. This is especially true for low-concentration ($< 10^{19}$ cm^{-3}) diffusions, which are commonly required with these dopants. Figure 4.50 shows computer-derived curves, which relate the profile parameters for these junctions, for substrates with varying background concentrations [79].

p-type dopants, such as zinc, are characterized by a steep doping profile, and relatively high surface concentrations ($\geq 10^{20}$ cm^{-3}). Figure 4.51 shows curves relating N_{sur} and x_j with the average conductivity of zinc-diffused layers in gallium arsenide [80]. These curves are based on sheet resistance measurements of these layers for a wide variety of diffusion conditions, using elemental zinc as well as alloy sources in a sealed-tube system.

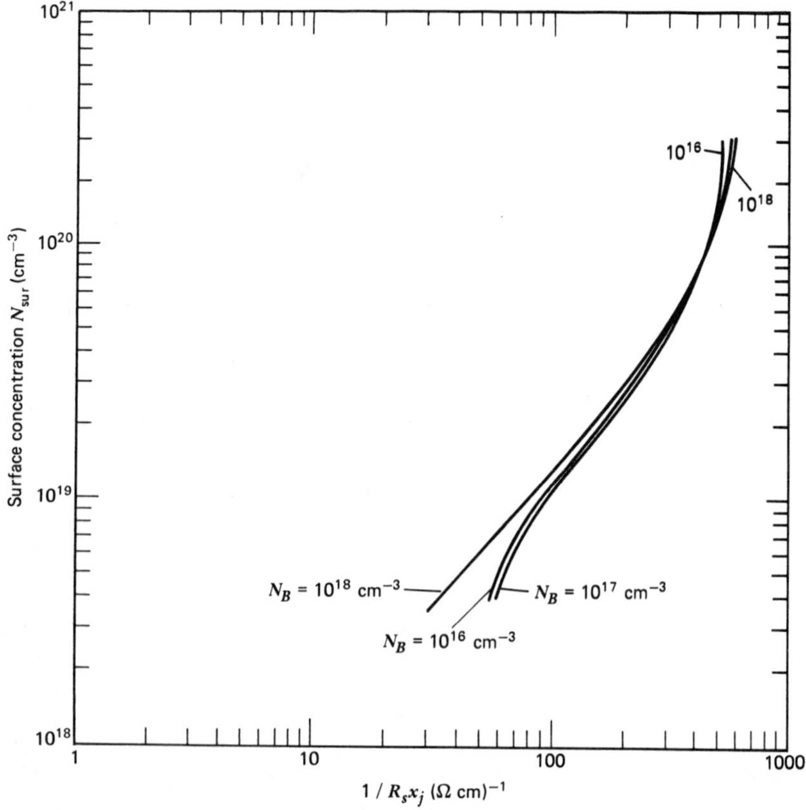

Fig. 4.51 ***p*-Type zinc diffusions into gallium arsenide. Adapted from Kendall [80].**

REFERENCES

1. B. I. Boltaks, *Diffusion in Semiconductors*, Academic, New York, 1963.
2. D. Shaw, Ed., *Atomic Diffusion in Semiconductors*, Plenum, New York, 1973.
3. W. Jost, *Diffusion in Solids, Liquids and Gases*, Academic, New York, 1962.
4. N. B. Hannay, Ed., *Semiconductors*, Reinhold, New York, 1960.
5. S. M. Hu and S. Schmidt, Interactions in Sequential Diffusion Processes in Semiconductors, *J. Appl. Phys.* **39,** 4272 (1968).
6. W. Shockley and J. T. Last, Statistics of the Charge Distribution for a Localized Flaw in a Semiconductor, *Phys. Rev.* **107**(2), 392 (1957).
7. D. Shaw, Self- and Impurity-Diffusion in Ge and Si, *Phys. Stat. Sol.* (*b*) **72,** 11 (1975).
8. R. B. Fair, Recent Advances in Implantation and Diffusion Modelling for the Design and Process Control of Bipolar ICs. *Semiconductor Silicon*, Electrochem. Soc., Princeton, NJ, 968, 1977.
9. S. Y. Chiang and G. L. Pearson, Properties of Vacancy Defects in GaAs Single Crystals, *J. Appl. Phys.* **46,** 2986 (1975).
10. R. A. Colclaser, *Microelectronics: Processing and Device Design*, Wiley, New York, 1980.
11. G. Masetti, S. Solmi, and G. Soncini, Temperature Dependence of Boron Diffusion in (111), (110), and (100) Silicon, *Solid State Electron.* **19,** 545 (1976).
12. D. A. Antoniadis, A. G. Gonzalez, and R. W. Dutton, Boron in Near Intrinsic ⟨100⟩ and ⟨111⟩ Silicon under Inert and Oxidizing Ambients—Diffusion and Segregation, *J. Electrochem. Soc.* **125,** 813 (1978).
13. J. M. Fairfield and B. V. Gokhale, Gold as a Recombination Center in Silicon, *Solid State Electron.* **8,** 685 (1965).
14. W. M. Bullis, Properties of Gold in Silicon, *Solid State Electron.* **9,** 143 (1966).
15. W. R. Wilcox and T. J. LaChapelle, The Mechanism of Gold Diffusion in Silicon, *J. Appl. Phys.* **35,** 240 (1964).
16. K. P. Lisiak and A. G. Milnes, Platinum as a Lifetime Control Deep Impurity in Silicon, *J. Appl. Phys.* **46,** 5229 (1975).
17. M. D. Miller, Differences between Platinum and Gold-Doped Silicon Power Devices, *IEEE Trans. Electron Dev.*, **ED-23,** 1279 (1976).
18. F. M. Smits and R. C. Miller, Rate Limitation at the Surface for Impurity Diffusion in Semiconductors, *Phys. Rev.* **104,** 1242 (1956).
19. W. R. Rice, Diffusion of Impurities During Epitaxy, *Proc. IEEE* **52,** 284 (1964).
20. T. I. Kucher, The Problem of Diffusion in an Evaporating Solid Medium, *Sov. Phys. Solid State* **3,** 401 (1961).
21. R. L. Batdorf and F. M. Smits, Diffusion of Impurities into Evaporating Silicon, *J. Appl. Phys.* **30,** 259 (1959).
22. R. B. Allen, H. Bernstein, and A. D. Kurtz, Effect of Oxide Layers on the Diffusion of Phosphorus into Silicon, *J. Appl. Phys.* **31,** 334 (1960).
23. R. C. T. Smith, Conduction of Heat in the Semi-Infinite Solid with a Short Table of an Important Integral, *Austral. J. Phys.* **6,** 129 (1953).
24. L. R. Weisberg and J. Blanc, Diffusion with Interstitial-Substitutional Equilibrium. Zinc in GaAs, *Phys. Rev.* **131,** 1548 (1963).
25. A. Luque, J. Martin, and G. L. Araujo, Zn Diffusion in GaAs under Constant As Pressure, *J. Electrochem. Soc.* **123,** 249 (1976).

26. R. B. Fair and J. C. C. Tsai, The Diffusion of Ion-Implanted Arsenic in Silicon, *J. Electrochem. Soc.* **122,** 1689 (1975).

27. R. B. Fair, Boron Diffusion in Silicon: Concentration and Orientation Dependence, Background Effects, and Profile Estimation, *J. Electrochem. Soc.* **122,** 800 (1975).

28. R. B. Fair and J. C. C. Tsai, A Quantitative Model of Phosphorus in Silicon and the Emitter Dip Effect, *J. Electrochem. Soc.* **124,** 1107 (1977).

29. D. P. Kennedy and R. R. O'Brien, Analysis of the Impurity Atom Distribution Near the Diffusion Mask for a Planar *p–n* Junction, *IBM J. Res. Dev.* **9,** 179 (May 1965).

30. D. D. Warner and C. L. Wilson, Two-Dimensional Concentration Dependent Diffusion, *Bell Sys. Tech. J.* **59,** 1 (1980).

31. A. Kostka, R. Gereth, and K. Kreuzer, A Physical and Mathematical Approach to Mass Transport in Capsule Diffusion Processes, *J. Electrochem. Soc.* **120,** 971 (1973).

32. K. D. Beyer, A New Paint-On Diffusion Source, *J. Electrochem. Soc.* **123,** 1556 (1976).

33. J. A. Becker, Silicon Wafer Processing by Application of Spun-On Doped and Undoped Silica Layers, *Solid State Electron.* **17,** 87 (1974).

34. S. K. Ghandhi, *Semiconductor Power Devices*, Wiley, New York, 1977.

35. S. P. Murarka, A Study of the Phosphorus Gettering of Gold in Silicon by Use of Neutron Activation Analysis, *J. Electrochem. Soc.* **123,** 765 (1976).

36. D. Rupprecht and J. Stach, Oxidized Boron Nitride Wafers as an In-Situ Boron Dopant for Silicon Diffusions, *J. Electrochem. Soc.* **120,** 9 (1973).

37. Technical Literature, Carborundum Corp., Graphite Products Div., P.O. Box 577, Niagara Falls, NY 14302, 1975.

38. J. Stach and A. Turley, Anomalous Boron Diffusion in Silicon from Planar Boron Nitride Sources, *J. Electrochem. Soc.* **121,** 722 (1974).

39. H. Nakamura, Diborane-Carbon Dioxide System for Boron Diffusion into Silicon at Low Temperature, *J. Appl. Phys. Jpn.* **11,** 761 (1972).

40. V. Geiss and E. Fröschle, Mass Spectrometric Investigation of the Reaction Velocities of BCl_3 and BBr_3 with Oxygen and Water Vapor in a Diffusion Furnace, *J. Electrochem. Soc.* **123,** 133 (1976).

41. N. Jones, D. M. Metz, J. Stach, and R. E. Tressler, A Solid Planar Source for Phosphorus Diffusion, *J. Electrochem. Soc.* **123,** 1565 (1976).

42. R. M. Burger and R. P. Donovan, *Fundamentals of Silicon Integrated Device Technology*, Vol. I, Prentice Hall, Englewood Cliffs, NJ, 1967.

43. W. Runyan, *Silicon Semiconductor Technology*, McGraw Hill, New York, 1965.

44. J. S. Sandhu and J. L. Reuter, Arsenic Source Vapor Pressure Kinetics and Capsule Diffusion *IBM J. Res. Dev.*, 464 (Nov. 1971).

45. C. J. Uhl, A Gold Diffusion Process for Reducing Switching Times in Diffused Silicon Transistors, *Western Elec. Eng.* **7,** 18 (Jan. 1963).

46. D. A. Antoniadis and R. W. Dutton, Models for Computer Simulation of Complete IC Fabrication Processes, *IEEE J. Solid State Circuits* **SC-14,** 412 (1979).

47. A. S. Grove, O. Leistko, and C. T. Sah, Redistribution of Acceptor and Donor Impurities during Thermal Oxidation of Silicon, *J. Appl. Phys.* **35,** 2695 (1964).

48. J. W. Colby and L. E. Katz, Boron Segregation at the Si-SiO_2 Interface as a Function of Temperature and Orientation, *J. Electrochem. Soc.* **123,** 409 (1976).

49. T. Kato and Y. Nishi, Redistribution of Diffused Boron in Silicon by Thermal Oxidation, *J. Appl. Phys. Jpn.* **3,** 377 (1964).

50. S. M. Hu, Defects in Silicon Substrates, *J. Vac. Sci. Tech.* **14,** 17 (1977).

51. A. M-R. Lin, D. A. Antoniadis, and R. W. Dutton, The Oxidation Rate Dependence of

Oxidation-Enhanced Diffusion of Boron and Phosphorus in Silicon, *J. Electrochem. Soc.* **128,** 1131 (1981).

52. D. A. Antoniadis, A. M. Lin, and R. W. Dutton, Oxidation-Enhanced Diffusion of Arsenic and Phosphorus in Near-Intrinsic ⟨100⟩ Silicon, *Appl. Phys. Lett.* **33,** 1030 (1978).
53. E. Tannenbaum, Detailed Analysis of Thin Phosphorus-Diffused Layers in p-Type Silicon, *Solid State Electron.* **3,** 123 (1961).
54. C. L. Jones and A. F. W. Willoughby, Studies of the Push-Out Effect in Silicon, Pt. I, *J. Electrochem. Soc.* **122,** 1531 (1975); Pt. II, *ibid* **123,** 1531 (1976).
55. S. F. Cagnina, Enhanced Gold Solubility Effect in Heavily n-Type Silicon, *J. Electrochem. Soc.* **116,** 498 (1969).
56. L. R. Weisberg, Diffusion in Gallium Arsenide, *Trans. Met. Soc. AIME* **230,** 291 (1964).
57. F. A. Cunnell and C. H. Gooch, Diffusion of Zinc in Gallium Arsenide, *J. Phys. Chem. Solids* **15,** 127 (1960).
58. H. C. Casey, Jr., and M. B. Panish, Reproducible Diffusion of Zinc into GaAs: Application of the Ternary Phase Diagram and the Diffusion and Solubility Analyses, *Trans. Met. Soc. AIME* **242,** 406 (1968).
59. K. K. Shih, High Surface Concentration Zn Diffusion in GaAs, *J. Electrochem. Soc.* **123,** 1737 (1976).
60. H. Matino, Reproducible Sulfur Diffusion into GaAs, *Solid State Electron.* **17,** 35 (1974).
61. Y. I. Nissim, J. F. Gibbons, C. A. Evans, Jr., V. R. Deline, and J. C. Norberg, Thermal Diffusion of Tin in GaAs from a Spin-On SnO_2/SiO_2 Source, *Appl. Phys. Lett.* **37,** 89 (1980).
62. S. K. Ghandhi and R. J. Field, Precisely Controlled Shallow p^+-Diffusions in GaAs, *Appl. Phys. Lett.* **38,** 267 (1981).
63. H. C. Casey, Jr., M. B. Panish, and L. L. Chang, Dependence of the Diffusion Coefficient on the Fermi Level: Zinc in Gallium Arsenide, *Phys. Rev.* **162,** 660 (1967).
64. L. L. Chang, The Junction Depth of Concentration-Dependent Diffusion: Zinc in III–V Compounds, *Solid State Electron.* **7,** 853 (1964).
65. S. R. Shortes, O. A. Kang, and E. C. Wurst, Jr., Zinc Diffusion in GaAs through SiO_2 Film, *Trans. Met. Soc. AIME* **230,** 300 (1964).
66. A. B. Y. Young and G. L. Pearson, Diffusion of Sulfur in Gallium Phosphide and Gallium Arsenide, *J. Phys. Chem. Solids* **31,** 517 (1970).
67. B. J. Baliga and S. K. Ghandhi, Planar Diffusions in Gallium Arsenide from Tin-Doped Oxides, *J. Electrochem. Soc.* **126,** 135 (1979).
68. M. Dohsen, J. Kasahara, Y. Kato, and N. Watanabe, GaAs J-Fet, Formed by Localized Zn Diffusion, *IEEE Trans. Electron Dev. Lett.* **EDL-2,** 157 (1981).
69. R. J. Field and S. K. Ghandhi, An Open Tube Method for Diffusion of Zinc into GaAs, *J. Electrochem. Soc.* **129,** 1567 (1982).
70. J. R. Shealy, B. J. Baliga, and S. K. Ghandhi, Open Tube Diffusion of Zinc in Gallium Arsenide, *IEEE Trans. Electron Dev. Lett.* **EDL-1,** 119 (1980).
71. R. Agarwal, Characterization of Tin Diffusions in Gallium Arsenide, Masters thesis, Rensselaer Polytechnic Institute, Troy, N.Y. (1982).
72. B. J. Baliga and S. K. Ghandhi, PSG Masks for Diffusion in Gallium Arsenide, *IEEE Trans. Electron Dev.* **ED-19,** 761 (1972).
73. B. J. Baliga and S. K. Ghandhi, Lateral Diffusion of Zinc and Tin in Gallium Arsenide, *IEEE Trans. Electron Dev.* **ED-21,** 410 (1974).
74. J. I. Goldstein, H. Yakowitz, Eds., *Practical Scanning Electron Microscopy and Ion Microprobe Analysis*, Plenum, New York, 1975.

75. J. C. Irvin, Resistivity of Bulk Silicon and Diffused Layers in Silicon, *Bell Sys. Tech. J.* **41,** 387 (1962).

76. R. B. Fair, Profile Estimation of High Concentration Diffusions in Silicon, *J. Appl. Phys.* **43,** 1278 (1972).

77. R. B. Fair and J. C. C. Tsai, Profile Parameters of Implanted-Diffused Arsenic Layers in Silicon, *J. Electrochem. Soc.* **123,** 583 (1976).

78. R. B. Fair, Analysis of Phosphorus-Diffused Layers in Silicon, *J. Electrochem. Soc.* **125,** 323 (1978).

79. B. J. Baliga, Conductivity of Complementary Error Function *n*-type Diffused Layers in Gallium Arsenide, *Solid State Electron.* **20,** 321 (1977).

80. D. L. Kendall, Diffusion, in *Semiconductors and Semimetals*, Vol. 4, R. K. Willardson and A. C. Beer, Eds., Academic, New York, 1968.

PROBLEMS

1. A diffusion furnace operating at 1000°C has a ± 1°C tolerance. What is the corresponding tolerance on the diffusion depth, assuming a gaussian diffusion?
2. A *p*-diffusion is made into an *n*-region such that $N_{\text{sur}} = 1000N_B$. Show that the junction depth is proportional to $(Dt)^{1/2}$ and determine the factor of proportionality. A constant-source diffusion may be assumed.
3. A diffusion furnace is ramped up (from 600°C) for 15 min, held at 1100°C for 30 min, and ramped down to 600°C in 15 min. Calculate the effective diffusion time, assuming phosphorus in silicon.
4. The following diffusions are performed into *n*-type silicon with background concentration of 3×10^{16} atoms/cm^3: (a) a 10-min constant-source boron predeposition at 900°C, followed by a 30-min drive-in at 1100°C; (b) a 20-min constant-source phosphorus diffusion at 1100°C. Determine the surface concentrations and junction depths that result from this process. Sketch the doping profiles on semilog paper. Hence determine the net impurity gradients at the emitter and the collector in atoms/cm^4.
5. A base diffusion is made into *n*-type silicon ($N_C = 3 \times 10^{16}$ atoms/cm^3) using the same schedule as in Problem 4. Ignoring redistribution effects, determine the surface concentration and junction depth, and develop a schedule for a constant-source diffusion that will result in the same surface concentration and junction depth. Hence verify the statement that the base diffusion should be carried out from a limited source.
6. Determine the effective diffusion coefficient of gold as a function of temperature. Highly intrinsic material may be assumed.
7. Show that the logarithm of the sheet resistance after a predeposition is (approximately) a linear function of reciprocal temperature. What is the activation energy associated with this Arrhenius plot?

CHAPTER

5

Epitaxy

CONTENTS

5.1 VAPOR-PHASE EPITAXY (VPE) 216

5.1.1 Reactant Transport, 217
5.1.2 Reactant Flux at the Substrate, 219
5.1.3 Reaction at the Substrate, 221
5.1.4 Nucleation and Growth, 222
5.1.5 Steady-State Growth, 225
5.1.6 System Design Aspects, 227

5.2 VPE PROCESSES FOR SILICON 231

5.2.1 Chemistry of Growth, 231
5.2.2 Epitaxial Systems and Processes, 233
5.2.3 In Situ Etching before Growth, 234
5.2.4 Impurity Redistribution during Growth, 235

5.2.4.1 Lateral Autodoping, 242

5.2.5 Selective Epitaxy, 243

5.2.5.1 Pattern Shift, 244

5.2.6 Growth Imperfections, 245

5.3 VPE PROCESSES FOR GALLIUM ARSENIDE 248

5.3.1 General Considerations, 248
5.3.2 Growth Strategy, 251
5.3.3 The Ga-$AsCl_3$-H_2 Process, 251
5.3.4 The Ga-AsH_3-HCl-H_2 Process, 253
5.3.5 The $(CH_3)_3Ga$-AsH_3-H_2 Process, 254
5.3.6 Doping, 257

5.3.7 Orientation Effects, 260

5.3.7.1 Selective Epitaxy, 262

5.3.8 Growth Imperfections, 263

5.4 LIQUID-PHASE EPITAXY (LPE) 264

5.4.1 Choice of Solvent, 265
5.4.2 Nucleation, 267
5.4.3 Growth, 269

5.4.3.1 Constitutional Supercooling, 273

5.4.4 In Situ Etching, 273
5.4.5 Doping, 274

5.5 LPE SYSTEMS 276

5.5.1 Hydrodynamic Considerations, 276
5.5.2 Vertical Systems, 277
5.5.3 Horizontal Systems, 278
5.5.4 Growth Imperfections, 280

5.6 HETEROEPITAXY 281

5.6.1 Silicon on Sapphire, 281
5.6.2 Aluminum Gallium Arsenide on Gallium Arsenide, 283

5.6.2.1 VPE Growth, 283
5.6.2.2 LPE Growth, 284

5.7 EVALUATION OF EPITAXIAL LAYERS 285

5.7.1 Sheet Resistance, 285
5.7.2 Mobility and Carrier Concentration, 288
5.7.3 Impurity Profile, 291

REFERENCES 292
PROBLEMS 296

Once an ingot is grown, it is cut into slices by a diamond saw. These slices, after suitable mechanical polishing with an abrasive grit, are chemi-mechanically polished to remove surface damage. These are the starting materials of which semiconductor devices and microcircuits are made. For many applications, the starting slice serves merely as a mechanical support on which is first grown an upper layer of the appropriate resistivity and conductivity type, in which the device or microcircuit is fabricated. This process of growth, by which an amount of material is set down upon a crystalline substrate while the overall single-crystal structure is still preserved, is known as *epitaxy*. An exception to this approach is the MOS-based microcircuit, which is conventionally made on the bulk slice. Increasingly, however, VLSI applications require that these circuits be fabricated in epitaxial layers as well.

In its most general form, the term epitaxy can apply to any two materials of different crystalline structures and orientations. The growth of crystalline (100) silicon (diamond lattice) on $(01\bar{1}2)$ sapphire (hexagonal) is an example of this type of *heteroepitaxy* [1, 2]. Most other examples consist of growing one zincblende or diamond lattice material on another, and with the same crystal orientation. Examples are germanium (diamond) on gallium arsenide (zincblende), and gallium aluminum arsenide on gallium arsenide (both zincblende).

A number of important problems must be surmounted in heteroepitaxy. First, the substrate must be physically and chemically inert to the growth environment, and capable of being prepared with a damage-free surface. Next, there must be chemical compatibility between the materials, as evidenced by their phase diagrams. Otherwise great care must be taken to avoid compound formation, and/or massive dissolution of one layer by the other [3]. In addition, the layer and the substrate should be closely matched in thermal expansion characteristics, to prevent the formation of excess stress upon cooling to the ambient. This stress can lead to the formation of dislocations at the interface, or even the breaking of the structure. Finally, the layer and the substrate should be closely matched in their lattice parameters. Interestingly, this last requirement is not as serious an impediment to epitaxial growth as originally predicted, especially with thick layers where considerable accommodation can take place in the first few micrometers.

In view of the above difficulties, it is no surprise that the art of heteroepitaxy is still in its infancy. However, many of these problems are absent during epitaxial growth where the layer and substrate are both of the same material. Here the compatibility problems are relatively minor, and epitaxy (or, more correctly, *autoepitaxy*) is routinely used. In fact, it generally results in layers of better quality than the starting melt-grown material, since epitaxy is often conducted at temperatures well below the melting point.

Epitaxial layers in silicon microcircuits are typically about 2–5 μm thick for high-speed digital applications, and 10–20 μm thick for linear circuit applications. A wider range is used with gallium arsenide, from 0.1 μm for field effect transistors to as much as 100 μm for some transferred electron device structures.

There are a number of ways in which epitaxial growth can be achieved, the most direct being the physical transport of material to a heated substrate. This can be accomplished by vacuum evaporation or by molecular beam epitaxy (MBE). These techniques are identical in principle, but critically different in their operational details. Specifically, delivery rates for the material to the substrate are many orders of magnitude slower for MBE, as are the corresponding growth rates. Of necessity, MBE systems must operate at much higher vacuum levels (three to four decades lower pressure) to avoid background contamination. In addition, high-purity materials are possible with MBE by using mass spectrometric means for in-situ purification of the evaporant source material. As a result, MBE is a very promising field and is the subject of much research [4]. This is especially true for the area of III–V compounds and mixed III–V compounds where the precise control of multiple impurities, combined with low-temperature deposition processes, are mandatory. However, its commercial future is by no means clear at the present time. Consequently, the subject is not treated in this text.

At the present time, the two most important techniques for epitaxy are vapor-phase growth and liquid-phase growth. These are discussed in detail in this chapter.

5.1 VAPOR-PHASE EPITAXY (VPE)

There are many different ways in which the growth of a crystalline film can be accomplished by vapor-phase epitaxy (VPE). Of necessity, material transport must be in the form of a vapor, either of the element(s), or of its compound(s). In both cases, these species flow toward the substrate in the form of a vapor. Two possibilities now exist: they may be adsorbed (or chemisorbed) on the substrate surface where they react to form the element(s) of which the layer is made up; alternately, one or more may thermally convert to this form before being adsorbed on the surfaces. The surface temperature is high, and the species are free to move by surface diffusion. Typically, the activation energy of surface diffusion is 25–50% of the bulk diffusion energy. Thus the surface diffusion rate is five to eight decades faster than the bulk diffusion rate at typical growth temperatures.

Crystal growth on foreign substrates begins by the coalescence of deposits to form stable nuclei. These expand by the attachment of *adatoms* (or *admolecules*) to atomic steps or *kinks** in the crystal lattice. Growth on a crystalline surface, as in the case of epitaxy, may also be initiated in this manner. In all probability, however, the direct adsorption of species, and their migration to kink sites, results in growth by advancement of these steps across the crystal surface. This process continues as the epitaxial film builds up.

Growth can also occur by the direct formation of stable nuclei in the gas phase [6]. This process is not common; however, once initiated, these nuclei provide a surface on which rapid coalescence can occur, with the formation of large particles in the gas phase. This is highly undesirable since it leads

* Note that gallium arsenide has both gallium and arsenic kink sites. Silicon, on the other hand, has only one type because of its elemental nature.

to defects in the growing layer, and eventually to polycrystalline growth. An additional problem is the creation of dust in the reactor, which further destroys layer quality.

Desorption of reactants and reaction products also takes place into the gas stream. Most of these are carried off to the exhaust; however, some reenter into the chemical processes outlined here.

5.1.1 Reactant Transport

Reactants that are useful for VPE must be available in high purity, either as gases or as liquids with a high vapor pressure. A bubbler arrangement can be used with the latter to provide the necessary vapor for this reaction. Hydrogen gas is commonly used as a vapor transport medium and also as a diluent gas in VPE systems. Gas flow may be measured directly by a flow gauge consisting of a vertical, tapered tube with a spherical ball inside it. The position of the ball in this tube is related to both dynamic and viscous forces which operate against the force of gravity. Consequently, flow is not known with any degree of precision. Typical values which can be achieved are $\pm 10\%$ for large flow rates to $\pm 30\%$ at low flows. However, flow gauges are simple and trouble-free, and can be set from run to run with a precision of about $\pm 2.5\%$.

A widely accepted technique for gas flow monitoring is the mass flow controller. Here the gas flows through a heated tube, causing a change in temperature across its ends, which is directly related to its thermal capacity. Sophisticated electronic sensing, together with feedback operation of a control valve, make these units insensitive to changes in inlet or exhaust pressure, and provide a measure of the absolute flow rate with a precision of ± 1–2%. In addition, they can be adjusted by electronic means, and so allow the use of computer control of the process.

A bubbler arrangement is used with liquids. This requires control of the bubbler temperature, which sets its vapor pressure, as well as the inlet gas flow and the exhaust pressure.

It is commonly assumed that the carrier gas gets saturated with the reactant vapor in passing through the liquid in a bubbler. This is only true if the residence time of the bubbles in the liquid is long compared to the diffusion time of the vapor into them. Thus attention must be paid to these factors for consistent results from run to run.*

Solid reactants can also be used. These are transported by preheating to the vapor phase, or by in-situ conversion to a volatile species before transport to the substrate.

A number of the chlorides of silicon are reasonably stable and can be readily transported in the vapor phase. Of these, silicon tetrachloride ($SiCl_4$) is by far the most commonly used. Trichlorosilane ($SiHCl_3$) and dichloro-

* See Problem 1 at the end of this chapter.

silane (SiH_2Cl_2) have recently found favor because they allow epitaxial growth at lower temperatures than the $SiCl_4$ reaction. Figure 5.1 shows the vapor pressure of these reactants as a function of temperature. An alternative material, silane gas (SiH_4), can be used in halogen-free systems; it has some unique features because of this fact.

The situation is somewhat more varied for the growth of gallium arsenide. Here arsine (AsH_3) is a commonly used gaseous source, while arsenic trichloride ($AsCl_3$) is a liquid source. Both of these are stable at room temperature. Elemental arsenic can also be used. However, it must be separately heated to provide the necessary vapor (As_2 or As_4) which is required for its transport.

Stable sources for gallium are the metalorganic compounds trimethylgallium [$(CH_3)_3Ga$] and triethylgallium [$(C_2H_5)_3Ga$]. Both of these are liquids at room temperature. Elemental gallium can also be used, but it must be separately reacted with a halogen such as chlorine or hydrogen chloride to convert it into a vapor source. Gallium arsenide itself can also be used as the source material. However, it also requires the use of transport agents (usually halogens, although other agents have been used in some earlier work) to provide the sources that are necessary for VPE.

Vapor pressures of reactants used in gallium arsenide epitaxy are shown in Fig. 5.2. Many combinations of sources are possible for gallium arsenide.

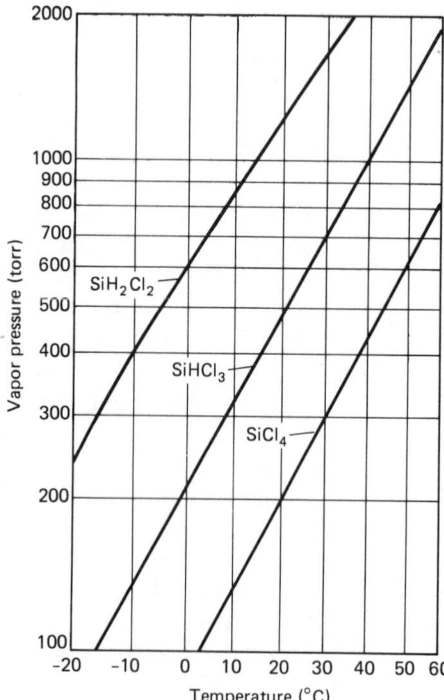

Fig. 5.1 Vapor pressure versus temperature for reactants used in silicon epitaxy.

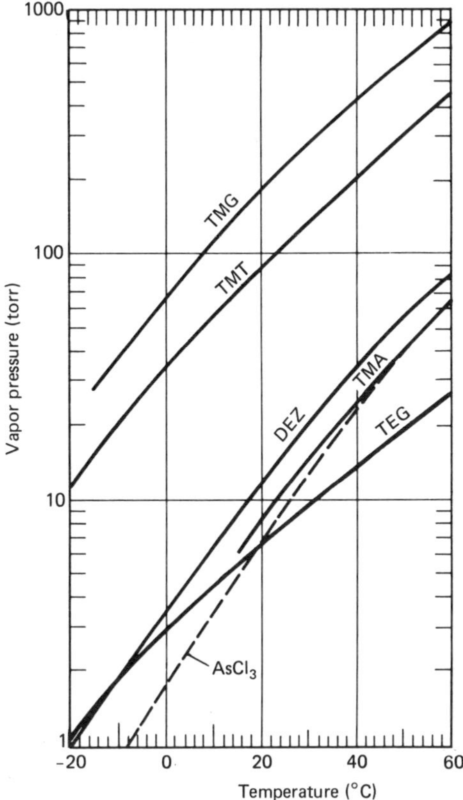

Fig. 5.2 Vapor pressure versus temperature for reactants used in gallium arsenide epitaxy.

Of these, the Ga-$AsCl_3$ combination is unique since both of these starting materials are available in six nines (99.9999%) purity or higher. The $(CH_3)_3Ga$-AsH_3 system is also unique because of its flexibility, and its complete absence of halogens.

5.1.2 Reactant Flux at the Substrate

The reactant is delivered to the substrate by transport in the gas stream, with a finite velocity. At the substrate this velocity must be zero because of friction. This is a reasonable assumption, since otherwise there would be an infinitely large velocity gradient at this boundary. A logical extension of this argument is that there is a region, next to the substrate, over which the gas velocity is extremely low. This region represents a *stagnant* or *boundary layer* through which the reactant species must diffuse in order to reach the semiconductor surface. Figure 5.3 shows the character of this boundary layer [7] for slices placed parallel to the gas stream (horizontal reactor) as

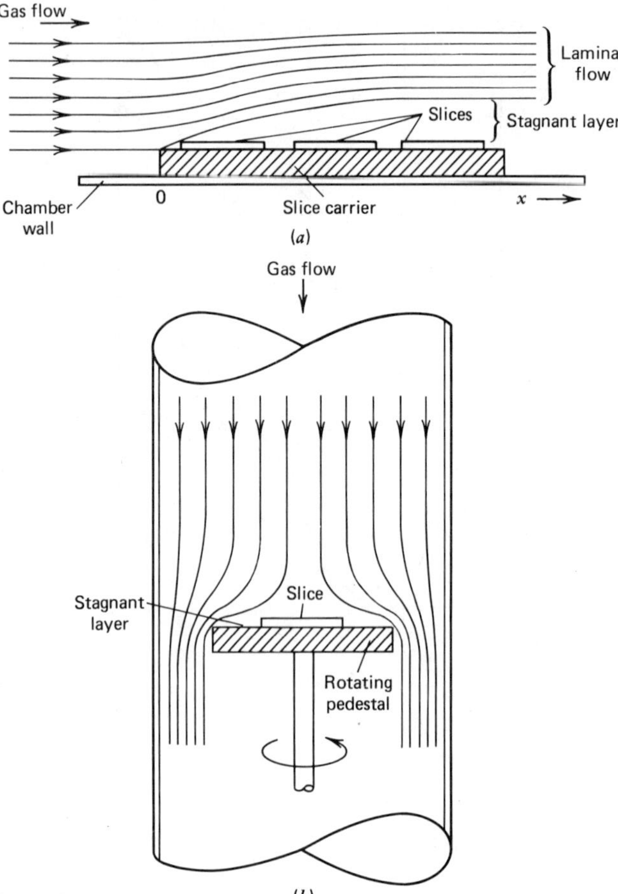

Fig. 5.3 Stagnant layer configurations.

well as for a slice placed normal to the direction of gas flow (vertical reactor).*

The gas flow in a pipe is characterized by a dimensionless, empirical Reynolds number N_R, given by

$$N_R = \frac{dv\rho}{\mu} \tag{5.1}$$

where d is the pipe diameter, v the velocity, μ the absolute viscosity, and ρ the fluid density. The viscosity of hydrogen, which is commonly used as

* Note that, in this context, the terms 'horizontal' and 'vertical' do not refer to the physical position of the semiconductor slices, but to the relative orientation of the axes of the slices with respect to the gas flow.

a carrier gas, is about 200 × 10^{-6} g/cm s at 700°C and increases to about 250 × 10^{-6} g/cm s at 1200°C. Its density is approximately 2.5 × 10^{-5} g/cm^3 at 700°C and 1.65 × 10^{-5} g/cm^3 at 1200°C.

Experimentally, values of N_R between 2000 and 3000 represent the transition betweem laminar and turbulent flow. In a typical reactor many factors, such as entrance conditions and initial disturbances, determine the initial state of fluid flow. Nevertheless, N_R is usually two decades below the critical value, so that flow is generally considered to be laminar, once the gases have traveled some distance in the reactor. Under these conditions, the width of the stagnant film can be shown [8] to be equal to $(\mu x/\rho v)^{1/2}$, where x is the distance marked in Fig. 5.3a.

The actual flux of reactant species arriving at the substrate is a complex function [9] of the temperature and partial pressure at its surface and in the boundary layer. However, a simple expression can be developed in terms of N_g, the concentration of reactant species in the gas stream, which results in a concentration N_0 at the substrate. Assuming diffusion of the reactant through the boundary layer, the flux density of diffusing impurities is given by

$$j = -D\frac{dn}{dx} = -\frac{D}{\delta}(N_g - N_0) \tag{5.2}$$

where j is the number of molecules of reactant per unit area per unit time, δ the effective width of the boundary layer, and D the diffusion constant. Typically, $D \propto T^m$, where m is between 1.75 and 2.

The relationship of Eq. (5.2) is often referred to as Henry's law, which states that the mole fraction of a species dissolved in a gas at any time is proportional to the partial pressure exerted by that species at that temperature. It can be written as

$$j = h(N_g - N_0) \tag{5.3}$$

where h is the gas-phase mass transfer coefficient.

From Eqs. (5.2) and (5.3) it is seen that the reactant flux density is inversely proportional to the boundary layer thickness. Control of this layer, and its uniformity over the slices, is an important consideration in the design of practical chemical vapor deposition reactors.

5.1.3 Reaction at the Substrate

Reaction proceeds once the reactant (or reactants) arrive at the substrate. If this reaction is assumed to be linear, then the flux of reaction products is given by

$$j = kN_0 \tag{5.4}$$

where k is the surface reaction rate constant. This parameter can be described by

$$k = k_0 e^{-\Delta E_a/Rt} \tag{5.5}$$

where ΔE_a is the activation energy of the process. For most surface activated processes, this is on the order of 25–100 kcal/mole.*

5.1.4 Nucleation and Growth

Nucleation is a process during which molecules of the reactants combine to form isolated nuclei which attach to points on the surface. The simplest nucleation process, known as *homogeneous nucleation,* is one that can come about by direct condensation out of the gas phase, in the absence of other objects such as surfaces. Although rarely occurring in practice, this serves as the starting point for further analysis.

If P_∞ is the equilibrium vapor pressure over the solid phase, then the solid will neither condense nor evaporate if exposed to this pressure. If, however, the pressure is higher, i.e., the vapor is supersaturated, conditions for nucleation may prevail. If P is the actual vapor pressure, the free energy difference per atom between the vapor at these pressures is given by

$$G_v = kT \int_P^{P_\infty} \frac{dP}{P} = -kT \ln \frac{P}{P_\infty} \tag{5.6}$$

If ΔG_v is the change in energy per unit volume, then

$$\Delta G_v = -nkT \ln \frac{P}{P_\infty} \tag{5.7}$$

where n is the number of atoms per unit volume of the semiconductor.

During nucleation, a number of small embryos are formed. These coalesce and grow† in size as the conversion proceeds from the vapor phase to the solid phase. There is a release of free energy (ΔG_v per unit volume) during this conversion process. At the same time, however, the formation of a surface between these phases involves an increase of free energy (γ per unit area of surface).

The total free energy change for an embryo, i.e., the activation energy required for its homogeneous nucleation, is thus given by [10]

$$\Delta G_{\text{homo}} = (\tfrac{4}{3}\pi r^3)\Delta G_v + (4\pi r^2)\gamma \tag{5.8}$$

* 1 kcal/mole = 0.0434 eV/molecule.

† The possibility of nucleus shrinkage has been treated elsewhere [10], and is not considered here.

Here spherical embryos, of radius r, have been assumed. Note that γ is always positive, but ΔG_v is negative for any spontaneous reaction. From this equation it is seen that the change in free energy is maximized for embryos of critical radius r_{crit}, where

$$r_{\text{crit}} = -\frac{2\gamma}{\Delta G_v} \tag{5.9}$$

Moreover, the free energy change leading to the formation of these embryos is given by

$$\Delta G_{\text{crit}} = \frac{16\pi\gamma^3}{3\Delta G_v{}^2} \tag{5.10}$$

Embryos below this size will lower their free energy change by becoming smaller, whereas those that are larger than r_{crit} will lower their free energy by growing. Note that large values of P/P_∞ result in an increase in the value of ΔG_v, thus reducing both the critical radius for nucleation as well as the critical change in free energy.

In theory it is not possible to calculate the concentration of critical nuclei by Boltzmann statistics. Nevertheless, this can be done if a quasi-equilibrium condition is assumed. Thus if n_0 is the number of particles of the new solid phase, we may write the number of critical nuclei as

$$n_{\text{crit}} = n_0 e^{-\Delta G_{\text{crit}}/kT} \tag{5.11}$$

Finally, if τ is the average lifetime for critical nuclei, then the nucleation rate R_n is given by

$$R_n = \frac{n_0}{\tau} e^{-\Delta G_{\text{crit}}/kT} \tag{5.12}$$

In most situations, nucleation takes place directly on a substrate, rather than in the gas phase. Such a process is known as *heterogeneous nucleation*. Here the presence of the substrate greatly alters the situation because of the possibility of "wetting" this surface. In addition, it leads to better crystal quality since the layer is constituted directly on the surface of the slice and not in the gas phase.

Let γ_{NS}, γ_{NV}, and γ_{SV} be the surface energies associated with nucleus–surface, nucleus–vapor, and surface–vapor, respectively. Again assuming spherical nuclei, three situations are possible. No wetting will occur if γ_{NS} is larger than $\gamma_{\text{NV}} + \gamma_{\text{SV}}$, since this process would require an increase in the net free energy. On the other hand, complete wetting occurs if γ_{SV} is larger than $\gamma_{\text{NS}} + \gamma_{\text{NV}}$, since the spreading out of the nucleus now lowers

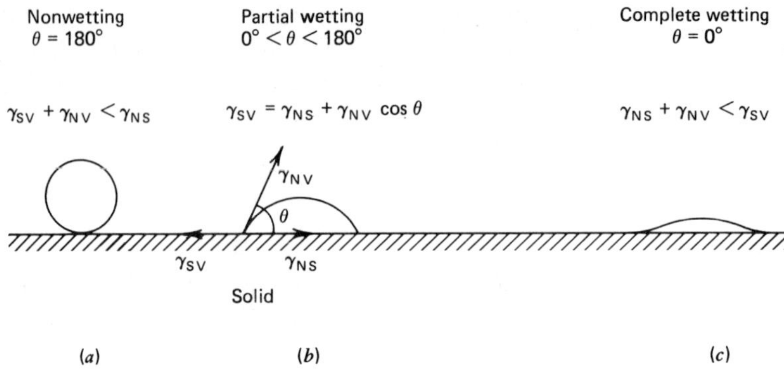

Fig. 5.4 Wetting characteristics of surfaces.

the net free energy. These situations are shown in Fig. 5.4*a* and *b*. The intermediate situation of partial wetting occurs when force equilibrium is established, i.e., when

$$\gamma_{SV} = \gamma_{NS} + \gamma_{NV} \cos \theta \tag{5.13}$$

where θ is the contact angle (Fig. 5.4*c*).

Writing the respective surface areas as S_{NS} and S_{NV}, the surface energy of this nucleus is given by $\gamma_{NS}S_{NS} + \gamma_{NV}S_{NV} - \gamma_{SV}S_{NS}$. Using this expression and Eq. (5.8), it can be shown that the critical energy for heterogeneous nucleation ΔG_{het} is related to the value for homogeneous nucleation by

$$\Delta G_{het} = \Delta G_{homo} \left[\frac{(2 + \cos \theta)(1 - \cos \theta)^2}{4} \right] \tag{5.14}$$

Thus as long as wetting takes place ($\theta \simeq 0°$), the substrate greatly reduced the activation energy for nucleation. As a consequence, the vast majority of reactions involving phase change nucleate heterogeneously.

Equation (5.14) illustrates the importance of surface preparation on epitaxial growth. Thus the presence of any cavities or imperfections increases the surface area on a local basis. This results in a large decrease in the activation energy of nucleation at this point.

The equation describing the rate of heterogeneous nucleation is of the same form as Eq. (5.12), except that a smaller critical energy is involved. For both situations, therefore, the nucleation rate increases very rapidly as the saturation ratio P/P_∞ increases. As a consequence, careful attention must be paid to the partial pressures of the reactants, the temperature, and the gas flow conditions in the reactor, if high-quality epitaxial layers are desired.*

* Note that ΔG_v approaches zero as the growth temperature approaches the melting point. Consequently, the nucleation rate falls rapidly with increasing temperature, leading to improved crystal quality.

The above arguments can be used to explain the initial stages of growth on a foreign substrate, and the start of epitaxial growth on a semiconductor, since it always has some native oxide on its surface. Some workers have proposed that these nuclei spread across the surface until it is fully covered, and that epitaxial growth proceeds on a layer-by-layer basis. This theory cannot account for the large growth rates observed in practice (0.5–2 μm/min).

It is generally accepted that continuous growth on a substrate results from the impingement of reactant species on its surface. The incident flux of these species is

$$j = \beta\,(2\pi mkT)^{-1/2}P_s \tag{5.15}$$

where P_s is the vapor pressure of the reactant species at the substrate, j is the number of atoms/cm^2 s and β is a proportionality factor. These species move by surface diffusion until attachment; alternately, they are lost to the gas stream by desorption. The desorption energy, in turn, is a function of the crystal surface to which they attach; species adsorbed on a low energy face (with a low binding energy) will have a high probability of desorption, and vice versa. In general, the closest packed planes have the lowest surface energies, and hence the lowest growth rates. This is observed in silicon epitaxy, where growth is about 40% slower in the {111} direction than in the {100} direction. With gallium arsenide, however, the noncentrosymmetric nature of the crystal tends to modify this conclusion somewhat (see Section 5.3.7).

Adsorbed species move rapidly by surface diffusion until they finally come to rest, to form a growing layer. Here, attachment is favored at a "kink site," where the species can attach by three bonds, rather than to the substrate face or to a ledge, where attachment is by one (or two) bonds only.

This model leads to more rapid growth than the nucleation model, but also results in a layer-by-layer process, with an unrealistically small growth rate. However, it can be extended to allow continuous advances of the growing layer by assuming that growth occurs in a spiral motion along screw-type dislocations. This model for crystal growth results in realistic values of growth rate, and is commonly accepted at the present time [5]. Microscopic observation of spiral growth for many crystalline materials has given considerable support to this theory.

5.1.5 Steady-State Growth

Steady-state growth requires that the fluxes described by Eqs. (5.3) and (5.4) be equal. Combining these equations and writing n as the number of atoms (or molecules) of grown material in a unit volume ($n \simeq 5 \times 10^{22}$ atoms/cm^3 for silicon, and 2.22×10^{22} molecules/cm^3 for gallium arsenide), the growth rate of the epitaxial layer is given by

$$\frac{dx}{dt} = \frac{j}{n} = \frac{N_g}{n}\left(\frac{hk}{h + k}\right) \tag{5.16}$$

At low temperatures,

$$\frac{dx}{dt} = \frac{kN_g}{n} \tag{5.17}$$

Now the growth is *reaction-rate limited,* and proceeds under kinetic control.

At high temperatures,

$$\frac{dx}{dt} = \frac{hN_g}{n} \tag{5.18}$$

and the reaction is said to be *mass-transfer limited,* at these temperatures.

Open-tube epitaxy is generally not carried out under thermal equilibrium conditions. Nevertheless, thermodynamics provides some indication of the nature of the reactions as a function of temperature. Consider first the case where this reaction is endothermic, i.e., involves a positive heat of reaction, and growth is favored with increasing temperature. Here the growth in the kinetic control region will fall off with decreasing temperature. With increasing temperature, however, mass transfer will limit the growth rate so that it will be essentially constant. Figure 5.5*a* illustrates a growth characteristic of this type, which is typical for all silicon systems described in this chapter. Note that the mass-transfer region exhibits a small falloff with decreasing temperature, on the order of 3–8 kcal/mole. This is because the

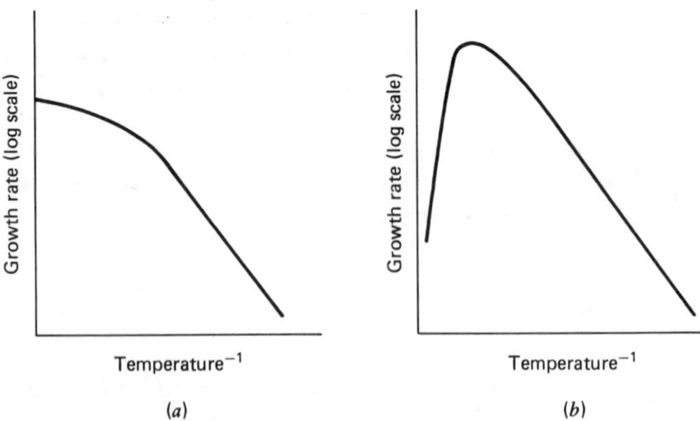

Fig. 5.5 Growth rate versus temperature for a system whose formation reaction is (*a*) endothermic and (*b*) exothermic.

gas diffusion constant itself varies slightly with temperature, as mentioned earlier.*

Figure 5.5*b* illustrates the growth characteristic of systems where the formation reaction is exothermic. Here the reaction rate limits growth at low temperatures, as before. At elevated temperatures, however, the system is usually in the mass-transfer limited region. Here the reaction itself is the rate limiter, and causes a decrease in the growth rate with increasing temperature. Gallium arsenide systems of the type described in Sections 5.3.3 and 5.3.4 (the halide and hydride processes) exhibit this characteristic. The organometallic process, described in Section 5.3.5, is also exothermic in character. However, the formation reaction takes place under conditions that are far removed from thermal equilibrium, so that thermodynamics does not provide any clues to its nature.

Finally, it is worth noting that many situations arise where the overall system behavior is not easily definable into two separate regions of this type. Thus caution must be used in interpreting growth data in terms of the physical initiating process. In practice, most epitaxial reactors are operated in the mass flow control region. A consequence is that the growth rate is relatively unaffected by minor variations on the substrate surface, or by the substrate orientation.

5.1.6 System Design Aspects

Consideration of the principles of VPE lead to a series of guidelines for reactor design and operation. The first condition for obtaining a high-quality layer is that of preserving a high level of cleanliness during the entire process. Thus chemical precleaning operations must be carried out in a dust-free environment immediately before loading the reactor. Reactor chambers must be made of high-purity quartz, with stainless steel and monel hardware used for all gas handling apparatus. High-purity gases must be used for growth and flushing operations. Hydrogen gas, of 99.9999% purity, is commonly used as the carrier gas in both silicon and gallium arsenide systems. Cold wall (radio frequency heated) reactors have the advantage over hot wall (resistance heated) systems in that thermal gradients in them are from the slice to the walls. As a result, the movement of contaminants is away from the slice, rather than toward it.

Provisions must be made for *in situ* cleaning of the substrate just before epitaxial growth. This is an important step if high-quality layers are desired. Typically, some type of halogen reaction with the semiconductor is used for this purpose.

Most vapor-phase epitaxial systems are operated in the mass flow control regime, so that considerable care must be taken to ensure precise control

* See Problem 2 at the end of this chapter.

of the flux of reactant species at the slice. This often involves the use of closed-loop mass flow controllers to precisely control the partial pressure of the reactant vapor. In addition, tight temperature control of the bubbler is necessary for systems in which the reactant species is a liquid. Normally the bubbler is designed so that the carrier gas is fully saturated with the vapor under flow conditions.

Premixing of gases must be done upon (or just before) entry into the reaction chamber, where laminar flow conditions generally prevail. Here careful attention must be paid to fluid flow considerations in order to minimize disturbances to the laminar flow [12]. Often the edge of the substrate is chamfered to present a smooth profile to the incoming gases. Wafers are also recessed in the slice carrier for the same reason.

Provision is often made for vacuum pump-down of the reactor, especially in gallium arsenide systems which are extremely sensitive to atmospheric leaks. This is useful for leak checking, and also greatly aids in flushing and backfilling the reactor prior to start-up. Hydrogen is the most commonly used carrier gas so that considerable care must be taken to ensure hazard-free operation. Finally, provision must be made for adequate venting of the various lines and of the exit port. Gaseous effluents are usually thoroughly scrubbed and burned before being exhausted to the atmosphere.

Three types of reactors are in common use, the simplest having its reaction chamber in the form of a horizontal quartz tube. In this reactor the flow of gases is parallel to the surface of the wafers, as seen in Fig. 5.6*a*. Here the laminar flow of gases is deflected over the slice carrier edge, resulting in a boundary layer whose width increases as the square root of distance, measured from the leading edge. Thus the flux of arriving species, and hence the growth rate, fall off with distance. This falloff is further aggravated by the depletion of the reactant species as they move down the chamber. Both of these effects can be greatly reduced by confining the gas flow to a reaction chamber of rectangular cross section, and by tilting the slice carrier a few degrees (1.5–3° is typical for silicon epitaxy, whereas 5–15° is more commonly used with gallium arsenide) so that the gas velocity is forced to increase as it travels down the chamber [13]. Further improvements in uniformity result from higher flow rates of carrier gas, thus increasing the average velocity. (This is because the thickness of the stagnant layer at any point is inversely proportional to the square root of the gas velocity.)

The vertical reactor, shown in a single-slice version in Fig. 5.6*b*, lessens this problem by making the gas flow at right angles to the surface of the slices. In addition, the slice holder is rotated during growth, resulting in improved uniformity. Disadvantages of this system are its increased complexity, and also its inability to handle many slices at any one time. In addition, convection effects result in increased turbulence, and grossly alter the laminar flow conditions over the substrate.

The barrel reactor, shown in Fig. 5.6*c*, is suitable for high-volume production. Here slices are held (by gravity) in niches along the slightly sloping

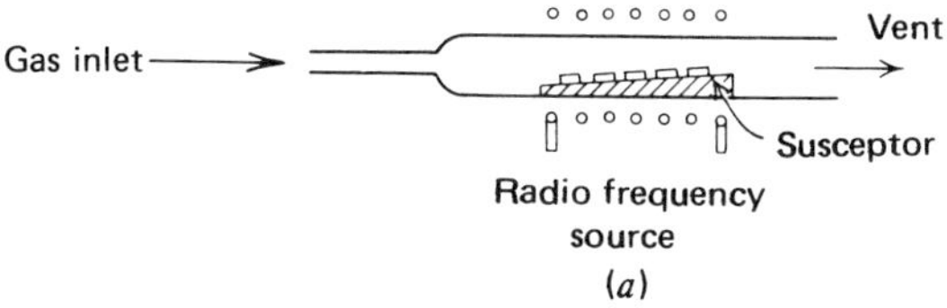

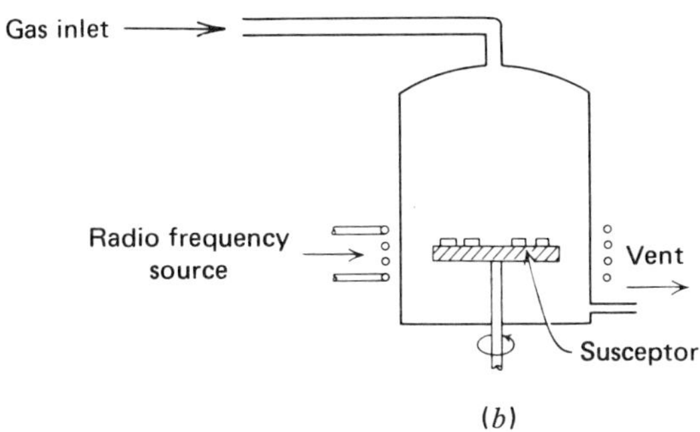

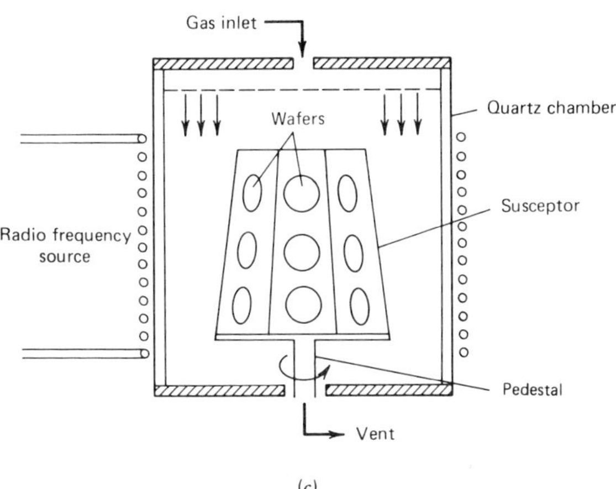

Fig. 5.6 Epitaxial reactor configurations: (*a*) horizontal reactor; (*b*) vertical reactor; (*c*) barrel reactor.

vertical wall of a large cylindrical slice carrier. Thus it is essentially an expanded adaptation of the horizontal reactor of Fig. 5.6*a*, since the gas flow is parallel to the slices. Here, too, the tilt of the slice carriers allows an increase of the gas velocity to compensate for boundary layer and reaction depletion effects [14].

The growth temperature, and its control, are generally of secondary importance in epitaxial reactors, which usually operate in the mass-transfer-limited regime. A $\pm 5°C$ tolerance is normally considered to be adequate, and systems often use open-loop temperature control. Temperature measurement is done by optical or infrared pyrometry; Thermocouples are usually avoided in high-temperature systems ($>1000°C$) in order to prevent contamination of the reactor environment.

The actual operating temperature is a compromise between competing factors. Ideally, higher growth temperatures result in greater surface mobility, and thus lead to improved crystal quality. However, this may be offset by the creation of increased temperature gradients across the slice, leading to increased radial stresses. In some cases this can result in significant dislocation motion, or even in bowing and slip. These problems tend to get worse with increasing slice diameter. An additional problem is that of interdiffusion between the substrate and the epitaxial layer, which becomes progressively more serious with increasing temperature. This is especially severe in situations where thin epitaxial layers are required.

One technique for greatly reducing these problems, and also for improving deposition uniformity [15], is to operate the reactor chamber at low pressures, in the 20–200 torr range.* An immediate benefit of this approach is to increase the flow velocity in the chamber (for the same consumption of carrier gas at atmospheric pressure). A second benefit is that convection currents, which disturb the laminar flow conditions, are greatly reduced by this means.

Often the amount of diluent gas is also reduced in low-pressure chemical vapor deposition (LPCVD) systems of this type. This reduces the hydrogen adsorption on the semiconductor surfaces [16] and allows high-quality layers to be grown at temperatures that are 100–200°C lower than for atmospheric pressure systems. Other advantages of LPCVD systems are that the boundary layer width also increases at reduced pressures (for the same gas velocity). This causes a significant reduction of autodoping effects, to be described in Section 5.2.4.

LPCVD systems are not without their problems, however. Thus the thermal resistance between slices and substrate is increased by low-pressure operation, so that there is often a 100–150°C difference in their corresponding temperatures. In addition, these systems are considerably more complex than their atmospheric pressure counterparts. This is primarily due to the fact that all pumping and gas pressure control equipment must be capable of operating in a continuous flow of highly corrosive chemicals, and maintenance problems are especially acute for these systems. Nevertheless, they are proving to be cost effective, and are becoming increasingly popular.

* The lower limit is set by the need to obtain a reasonable growth rate, and also to avoid a glow discharge in the reaction chamber which is usually heated by a radio frequency source.

5.2 VPE PROCESSES FOR SILICON

The vapor-phase epitaxial growth of silicon is highly advanced in comparison to that of gallium arsenide, with many large automated systems in commercial use. $SiCl_4$ is commonly used in these systems and its chemistry of growth well understood. At the same time, systems using $SiHCl_3$ and SiH_2Cl_2 are gaining in popularity, since they can be operated at somewhat lower temperatures. Silane processes, although unique, are difficult to implement, and only found in special situations.

5.2.1 Chemistry of Growth

The processes involved in the hydrogen reduction of the chlorosilanes $SiCl_4$, SiH_3Cl_3, or SiH_2Cl_2 are qualitatively similar in nature. Mass spectrometric studies [17] have shown that all result in the formation of H_2, HCl, and $SiCl_2$ in great quantity in the vapor phase, in addition to the above reactants themselves. Thus the primary effect of introducing any of these reactants, in addition to hydrogen, into the hot zone is to form a mixture of HCl and $SiCl_2$. Typically this reduction reaction occurs at some point where the gas temperature exceeds 800°C. The actual amounts of these components increases with the chlorine-to-hydrogen ratio and also with the operating temperature. In all cases SiH_2Cl_2, $SiHCl_3$, and $SiCl_4$ are also formed, but their partial pressures fall with increased operating temperatures. Upon arrival at the silicon slice, which is maintained at about 1150–1250°C, the $SiCl_2$ is adsorbed on its surface. The reaction taking place on the surface is given by

$$2SiCl_2 \rightleftharpoons Si + SiCl_4 \tag{5.19}$$

For *this* reaction to take place, it is necessary that two molecules of $SiCl_2$ be involved in the presence of a host body (the substrate). Thus the reaction is surface catalyzed and is a heterogeneous one.

The resulting $SiCl_4$ is desorbed into the gas stream, where it again undergoes reduction. The overall reaction may be written as

$$SiCl_4 + 2H_2 \rightleftharpoons Si + 4HCl \tag{5.20}$$

It is important to note that this is a reversible reaction, the end product of which is either the growth of a silicon layer or the etching of the substrate. The exact nature of this reaction is a function of the mole fraction of $SiCl_4$ in H_2. Experimental data for a typical $SiCl_4$ system are shown in Fig. 5.7 and illustrate this effect. As seen from these data, etching occurs at concentrations in excess of 0.28 mole fraction of $SiCl_4$ for this reactor system.

Although all of these reactions are qualitatively similar, there are a number

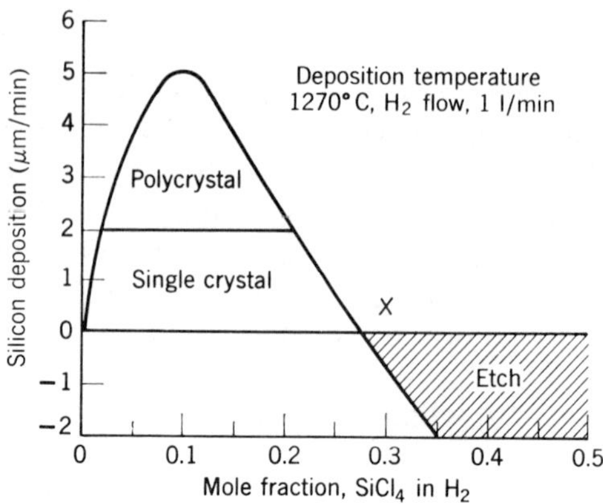

Fig. 5.7 Deposition rate for the $SiCl_4$ process.

of differences which are important in epitaxy. Thus there is almost no formation of $SiCl_2$ below 1000°K when $SiCl_4$ is used, whereas roughly comparable amounts are formed with either $SiHCl_3$ or SiH_2Cl_2. As a result, successful silicon epitaxy can be carried out with these reactants at temperatures 100–200°C lower. However, all three chlorosilanes are equally effective in $SiCl_2$ production by 1300°K.

Another difference is in the efficiency of the reaction, i.e., in the ratio of the amount of deposit to the amount of silicon in the incoming reactant. Here SiH_2Cl_2 has the highest efficiency, and $SiCl_4$ the lowest, with $SiHCl_3$ at some intermediate value. Thus SiH_2Cl_2 is the most attractive for high-volume production of epitaxial silicon. It is often used as the starting material in the production of bulk semiconductor grade silicon for this reason.

$SiCl_4$ has the advantage that it is the most stable chlorosilane and is also quite insensitive to the presence of small amounts of oxidizing species. $SiCl_4$ systems are relatively trouble-free in operation, because of this reason. They also have the longest history of operational use, and are well understood.

The SiH_4 reaction is quite different, and essentially more direct. Here silicon is formed by the pyrolytic decomposition of silane, SiH_4. The reaction is given by

$$SiH_4 \xrightarrow{500°C} Si + H_2 \tag{5.21}$$

Unlike the previous situation, this is not a reversible reaction. Furthermore the complete absence of hydrogen chloride gas as an end product avoids the necessity of elaborate venting. An additional advantage is that the process can be carried out at relatively low temperatures, thus avoiding many of the out-diffusion problems that arise with the chloride processes.

There are two problems inherent to the silane process. The first comes about because some degree of homogeneous nucleation is possible with this reaction. Here it is extremely difficult to avoid some gas-phase nucleation of silicon when this process is used, so that it can result in poor morphology or even in polycrystalline growth.* This problem can be reduced, however, by operating the reactor at significantly higher gas velocities than are common for chloride systems, and by careful control of the deposition temperature. Second, silane is more sensitive to oxidation than any of the chlorosilanes. As a result, silane systems must be extremely free of oxidizing impurities in order to avoid the formation of silica dust, and the consequent deterioration of layer quality. Because of these problems, they are not commonly used at the present time. Increasingly, however, the development of high-frequency structures has emphasized the need for thin silicon epitaxial layers which much be grown on heavily doped substrates and with an abrupt doping profile. The use of the silane process is almost mandated for these situations.

5.2.2 Epitaxial Systems and Processes

Silicon epitaxy is carried out in reaction chambers of the type shown in Fig. 5.6. The horizontal reactor is the most commonly used system, with new and larger systems tending toward the barrel design. The vertical reactor system is primarily used in pilot line and laboratory situations.

The silicon formation reactions are endothermic, so that the extent of deposit increases with temperature [18]. An induction heated system, with essentially cold walls, is used here, with the deposits confined to the slices and the slice carrier. Infrared lamp heating is also used, notably in some barrel reactor systems, where it is considerably simpler and more cost effective than radio frequency heating. A further advantage claimed for this type of heater is that it reduces temperature gradients in the slices, and thus lessens the problems of slip and bowing during epitaxy. Cold wall systems of these types have the added advantage of establishing a thermal gradient from the slices toward the reactor wall. The transport of foreign matter from the walls to the slices is avoided in this manner.

Slices are placed on an electrically conductive *susceptor* which is heated by radio frequency. The most common form of susceptor is a graphite block, which is coated with one or more layers of pyrolytic graphite, silicon carbide, or boron nitride to reduce contamination in the reactor vessel. Slices are usually recessed in these susceptors in order to minimize turbulences in the reactant flow stream. A series of vertically mounted susceptors is used in the barrel reactor system.

Growth temperatures in $SiCl_4$ systems can be as high as 1300°C. Typically, however, growth is carried out in the 1150–1250°C range with this reactant.

* Advantage is taken of this characteristic for the growth of polycrystalline silicon films (see Chapter 8). Here silane is the most commonly used reactant source. Typically, this reaction is conducted at low temperatures ($\simeq$ 600°C) with reasonably fast growth rates (1 μm/min) to promote polycrystalline growth.

Systems using $SiHCl_3$ and SiH_2Cl_2 are commonly operated in the 1100–1200°C range, whereas silane systems are usually operated around 1000°C. The vapor pressure of the various chlorosilanes is shown in Fig. 5.1.

Initially, the growth rate of the epitaxial layer increases linearly* with the mole fraction of the reactant. However, the quality of the layer becomes progressively worse, and eventually polycrystalline. A further increase of the reactant flux eventually leads to the etching condition, as shown in Fig. 5.7 for a $SiCl_4$ system. In practice a growth rate of about 1.0 μm/min results in reasonably defect-free single-crystal layers.

Impurity doping is done at the same time as the epitaxial growth. Although solid and liquid dopants can be used, gaseous diborane (B_2H_6), phosphine (PH_3), and some arsine (AsH_3) are far more convenient. These dissociate to form the elemental dopant impurity in the vicinity of the slices. Gas mixtures are ordinarily used with hydrogen as the diluent to allow reasonable control of flow rates for the desired doping concentration.

The resistivity of the grown layer is varied by controlling the dopant concentration in the gas phase. Accurate control of the gas flow rate is critical to successful doping. Dopant incorporation is essentially linear with the partial pressure of the dopant gas, up to concentrations as high as 10^{18} cm^{-3}. At higher concentrations, however, there is a slight falloff in the dopant incorporation with phosphine. Diborane, on the other hand, has a slightly superlinear incorporation characteristic.

In addition to the above, the growth rate of the layer is slightly altered by dopant incorporation. Enhanced growth occurs with diborane, whereas phosphorus and arsenic tend to retard the growth. This behavior has been ascribed to catalytic effects due to the dopant, and is not well understood at the present time.

5.2.3 In Situ Etching before Growth

Perhaps the single most important condition for good epitaxial growth is the use of a completely damage-free, oxide-free surface. Thus the silicon slices are mechanically lapped and chemically polished to remove, as much as possible, the regions of surface damage caused by slicing of the ingot. After chemical polishing, slices are inserted into the reactor and flushed in dry hydrogen gas at a temperature of about 1200°C. This serves to reduce all traces of native oxide that may have been formed while transferring the slices into the reactor.

Once the hydrogen cleanup is over, anhydrous HCl gas, diluted with H_2, is introduced into the system. Etching proceeds by the conversion of surface silicon to volatile $SiCl_4$ by the reaction of Eq. (5.20). Correct adjustment of the etchant mixture results in extremely high-quality, optically flat finishes.

* This assumes a constant gas velocity in the reaction tube (i.e., a constant boundary layer thickness over the silicon slices).

A 1–5% mole concentration of HCl is typical, and provides an etch rate of about 0.5–2 μm/min. Excessive HCl concentration in the etching gas leads to halogen pitting of the silicon slice, and must be avoided.

The success of etching, and of the subsequent epitaxial growth, is a critical function of the quality of the etchant gas. Traces of water, nitrogen, or hydrocarbons can lead to the formation of oxides, nitrides, and carbides of silicon, respectively; these, in turn, serve as sites for the initiation of defects in the epitaxial layer. The commercial availability of acceptably pure, dry, HCl gas has played an important role in establishing its common usage in this critical step.

One drawback of using hydrogen chloride gas is that polishing etching occurs at temperatures of 1100–1250°C, so that there can be considerable out-diffusion from the substrate during this process. Alternative etchants, such as HI, Cl_2, HBr, H_2O, and H_2S, have been used to reduce this problem. Recently sulfur hexafluoride has been found to be suitable, resulting in a polishing etch at temperatures as low as 1060°C [19], while still providing a significant etch rate (approximately 0.2 μm/min, which is comparable to that obtained with HCl at 1200°C). The etching reaction for this gas is

$$4Si(s) + 6SF_6(g) \rightarrow SiS_2(g) + 3SiF_4(g) \quad (5.22)$$

This reaction is an irreversible one, whose end products are both volatile at etching temperatures.

An additional advantage of SF_6 is that it is nontoxic and noncorrosive. Furthermore, although sulfur exhibits deep levels, its incorporation into the silicon during this process has been found to be less than 10^{13} cm^{-3}. As a result it has little effect on the minority carrier lifetime of the resulting devices. Finally, the commercial availability of this gas, in pure form, is an important factor in its practical use.

5.2.4 Impurity Redistribution during Growth

Redistribution of impurities occurs during epitaxial growth, resulting in departures from the ideal, abruptly discontinuous, doping profile. This impurity redistribution is primarily a function of two sets of factors, those related to the epitaxial growth process of itself, and those related to diffusion processes. Furthermore, diffusion effects can occur in the substrate and in the film, both during epitaxial growth, as well as during all subsequent high-temperature processes. We first concern ourselves with processes which are present during the actual epitaxial growth, and are sometimes referred to as *autodoping*.

The process of autodoping can be described in general terms. During epitaxy, both silicon and dopant are removed from the substrate by thermal evaporation (as is the case for growth from SiH_4), or by evaporation and chemical etching, which occur during growth from the chlorosilanes. Furthermore, this removal may or may not involve preferential leaching of either the silicon or the dopant. Thus thermal evaporation would tend to be preferential, whereas etching would not. Note that the sources of silicon and dopant may be the substrate itself (front and back surface, as well as the sides) or the susceptor which is silicon coated, either before or during the growth process.*

Both the silicon and the dopant diffuse through the boundary layer, and thus modify the composition of the incoming gas stream. This changes the composition of the deposited material from that which would be present if there were no reverse flux of these constituents.

As a result of this redistribution process, the doping concentration of an epitaxial layer varies during its growth until an equilibrium situation is eventually reached for layers of sufficient thickness. With thin epitaxial layers, however, it is possible for the layer to have a continually varying impurity concentration over its entire thickness as a result of this mechanism.

A relatively simple analysis can be made of this process if it is assumed that the only loss of silicon and dopant (by thermal evaporation or chemical reaction) is from the front surface of the silicon substrates. Here we expect the effects of autodoping to be greatest during the initial growth, and diminish as the layer builds up.

Assume a substrate concentration N_S, and that an undoped silicon layer is to be grown. For this situation it can be shown that the impurity distribution in the grown layer is given by [20]

$$N_E(x) = N_S e^{-\phi x} \tag{5.23}$$

where $N_E(x)$ is the impurity concentration, in atoms/cm^3 at x cm from the interface, N_S the impurity concentration of the substrate, in atoms/cm^3, x the distance from interface, in cm, and ϕ the growth constant, experimentally determined, in cm^{-1}. The growth constant ϕ is a function of the dopant, the reaction, the reactor system, and the process. For example, arsenic is more readily evaporable by thermal means than boron or phosphorus; ϕ is smaller for chlorosilane reactions than for the silane process; ϕ is related to the thickness of the boundary layer and is larger for large values of layer thickness.

Finally, it must be recognized that the diffusional transport of evaporants into the gas stream also occurs during the pregrowth period, when the substrates are maintained at growth temperature in a hydrogen carrier gas ambient, but *before* the introduction of reactant species. Thus thermal evaporation *prior* to actual epitaxial growth is an important factor in determining redistribution effects during initial growth of the layer.

Figure 5.8 shows a plot of ϕ as a function of growth temperature, for a

* Note that the reactor walls are not a source of evaporant, since silicon epitaxy is carried out in a cold wall system.

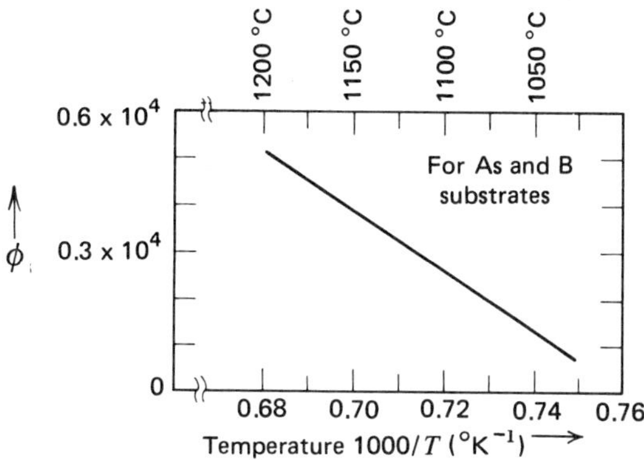

Fig. 5.8 Growth constant for arsenic substrates in the $SiCl_4$ process. Adapted from Kahng, Thomas, and Manz [21].

horizontal reactor system using arsenic substrates and a $SiCl_4$ transport process, and gives some idea of the magnitude of this constant [21]. Solution of Eq. (5.23) shows that doping in the epitaxial layer falls off as the substrate becomes covered with the less doped crystalline silicon, until eventually an undoped layer is achieved. In practice, however, this doping limits to some finite value, which represents the minimum background concentration for the reactor (typically 10^{14}–10^{15} impurity atoms/cm^3). This is because of the effects of continuing evaporation from the other surfaces of the substrates (especially from the back face), as well as from the susceptor. Measures for reducing the background doping level include precoating of the entire susceptor, as well as the back surface of the substrates, with silicon nitride or polycrystalline silicon. An improvement of one to two decades in background concentration can be achieved in this way.

The theory outlined above can be extended to the situation of epitaxial growth from a doped gas stream onto an undoped substrate. Assuming that the dopant concentration in the layer is the same as that in the incoming gas, we obtain

$$N_E(x) = N_{E_\infty}(1 - e^{-\phi x}) \tag{5.24}$$

where N_{E_∞} is the doping concentration if equilibrium conditions are reached during the epitaxial growth (i.e., for an infinitely thick layer). In actual situations, where a doped layer is to be grown on a doped substrate, the final impurity distribution is given by the superposition of these two cases.

Figure 5.9*a* shows the application of this theory to the growth of a lightly doped layer on a heavily doped substrate of the same impurity type, and is representative of the situation which occurs during epitaxial growth over a buried layer in a microcircuit, or during the growth of an epitaxial layer for a discrete device. It is assumed that the substrate concentration is N_S, and the concentration in the epitaxial layer, if made infinitely thick, is N_{E_∞}. As

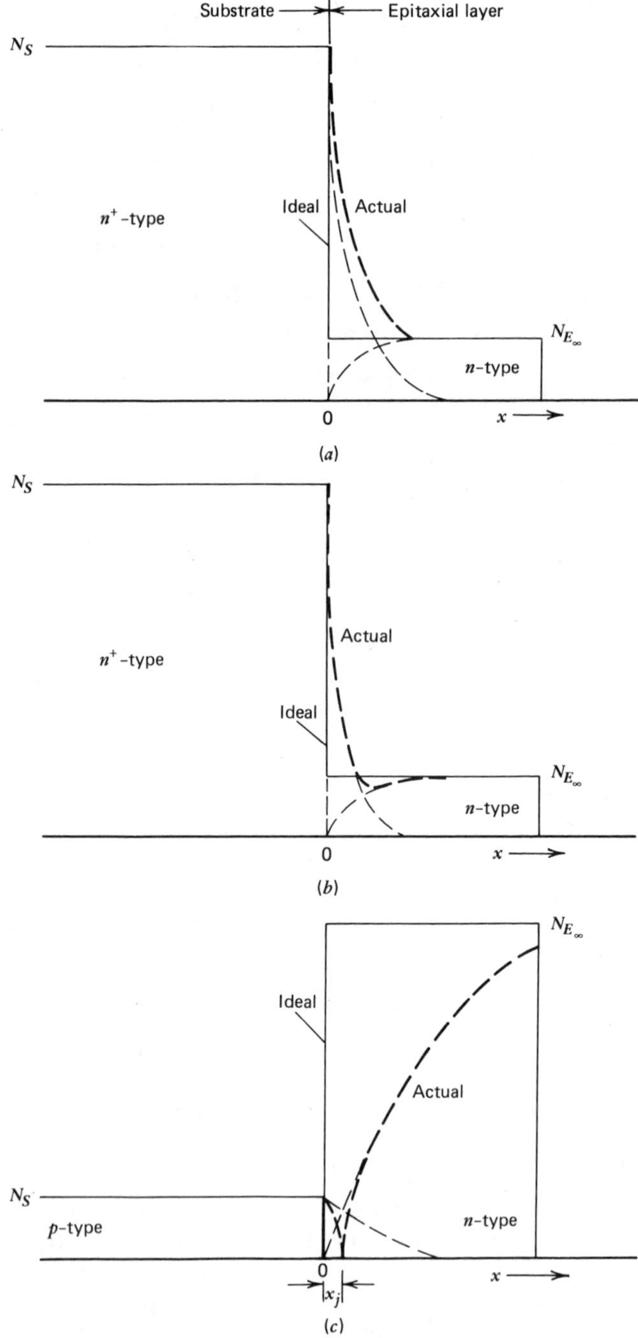

Fig. 5.9 Effect of autodoping on the impurity profile.

seen from this figure, the ideal step of impurity concentration is not realized. In addition, it is possible that the final steady-state value of $N_{E\infty}$ may not be attained for sufficiently thin epitaxial layers. Figure 5.9*b* shows what can happen if the substrate is extremely heavily doped. Here it is possible to obtain a dip in the epitaxial layer doping concentration, close to the substrate–epitaxial layer interface.

Figure 5.9*c* shows the situation when a heavily doped layer is epitaxially grown on a lightly doped substrate of opposite impurity type. This case is commonly encountered in microcircuit fabrication processes. Here departure from the ideal results in a shift in the position of the junction delineated by the two layers. This shift, known as *junction lag,* can be shown to be inversely proportional to the growth parameter ϕ. Thus the junction lag may be reduced by growing the epitaxial layer at elevated temperatures.

An alternate theory can be developed [22] to explain redistribution effects, based on the premise that diffusion processes in the growing film are primarily responsible for variations from the ideal, abrupt doping profile. Here, too, the problem can be broken up into two cases, and the final distribution obtained by superposition. This theory gives reasonable results for the case when the film growth rate V is much higher than the diffusion rate, i.e., $V \gg \sqrt{Dt}$, so that the epitaxial film can always be considered to be infinitely thick compared with the extent of the region affected by solid-state diffusion. Hence diffusion can be considered as proceeding into an infinite solid. With this in mind, consider the case of the growth of an intrinsic layer on a doped substrate. Let:

$N_S(x)$ = impurity concentration in epitaxial layer, in cm^{-3}
N_S = initial impurity concentration in substrate, in cm^{-3}
x = distance along epitaxial layer, in cm, with origin taken at substrate–epitaxial layer interface

Then

$$N_E(x) = \frac{N_S}{2}\left(1 - \operatorname{erf}\frac{x}{2\sqrt{D_S t}}\right) \tag{5.25}$$

where D_S is the diffusion constant for the substrate impurity during epitaxy.

The growth of a doped layer on an intrinsic substrate can be treated in a similar manner. Here let $N_E(x)$ be the impurity concentration and $N_{E\infty}$ the impurity concentration at the surface of the epitaxial layer, both in cm^{-3}. Then

$$N_E(x) = \frac{N_{E\infty}}{2}\left(1 + \operatorname{erf}\frac{x}{2\sqrt{D_E t}}\right) \tag{5.26}$$

where D_E is the diffusion coefficient for the epitaxial layer impurity. At distances removed from the substrate–film interface, such that $x \gg 2\sqrt{D_E t}$, this equation reduces to

$$N_E(x) \simeq N_{E\infty} \tag{5.27}$$

The final impurity distribution is given by the superposition of Eq. (5.25) and either Eq. (5.26) or Eq. (5.27), with due attention paid to the impurity type. Using these relationships, the doping profiles in the epitaxial layer are obtained as shown in Fig. 5.10*a–c*. These profiles are very similar to those obtained on the assumption that only redistribution effects are important in the growing layer.

Both of the above theories have been used to satisfactorily describe dopant behavior in thick (≥5 μm) layers, of the type commonly used in transistor–transistor logic (TTL) and in linear integrated circuits. An initial rapid drop near the epitaxial layer–substrate interface is primarily due to out-diffusion effects. A more gradual fall-off, beyond this point, results from autodoping effects. The combined redistribution effects dominate the first 2 μm of epitaxial layer growth, and allowance is made for this effect in the device design. Increasingly, however, the need for faster devices has necessitated the use of structures with thin (≤2 μm) epitaxial layers. These are usually grown by the silane process because of the lower temperatures involved here, and because of the irreversibility of the process. Arsenic-doped buried layers are commonly used, since this impurity is a slow diffuser in silicon. In addition, the arsenic atom is a perfect fit to the silicon lattice, so that heavily doped layers are relatively damage-free. Unfortunately, arsenic has a high vapor pressure and is readily evaporated from a silicon surface during epitaxy. As a result, redistribution effects are particularly severe for this impurity.

A detailed study of this problem requires that both autodoping and diffusion effects be considered. This includes consideration of the following processes:

1. Diffusion of the impurity from the substrate surface, both into its bulk as well as into the growing epitaxial layer
2. Diffusion of evaporants (silicon and impurity) from the surface of the epitaxial layer into the gas stream
3. Diffusion of evaporants from other surfaces into the gas stream
4. Reverse diffusion of reactants from the gas stream to the silicon surface, where growth proceeds

Computer-aided solutions of the coupled diffusion equations which describe these processes have been made [23, 24]. This technique allows the simulation of changes in the process in order that critical steps can be

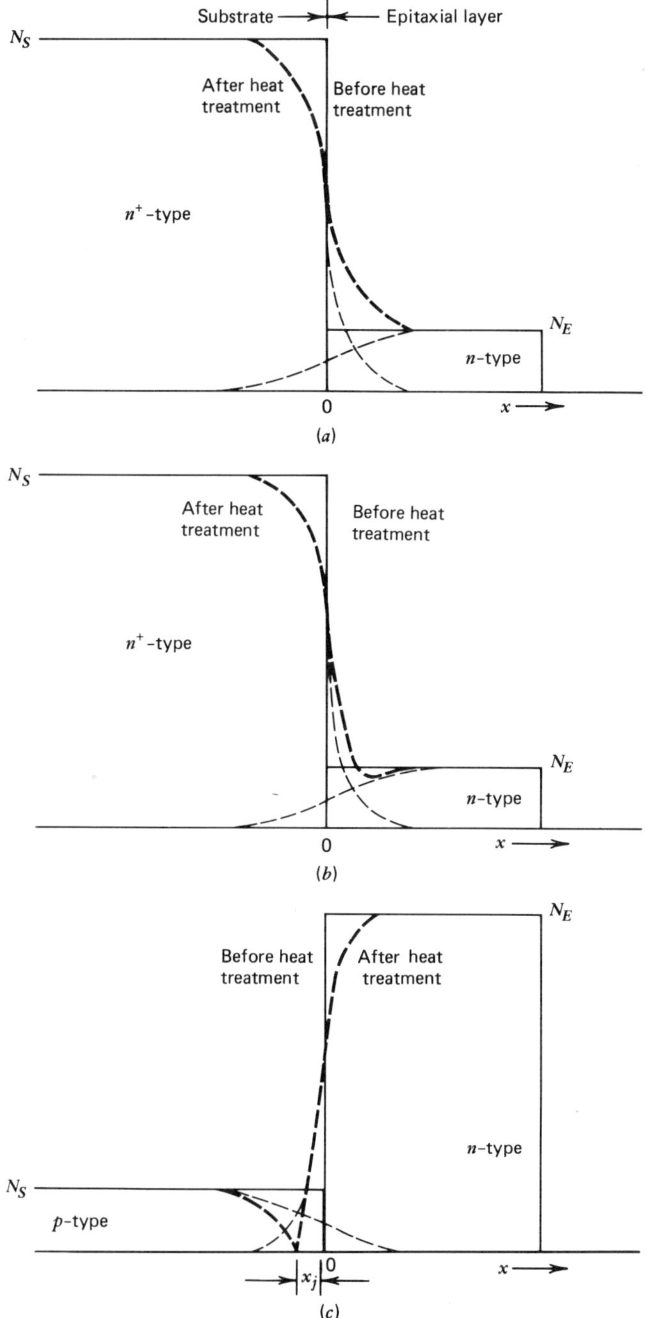

Fig. 5.10 Effect of diffusion on the impurity profile.

exposed, and modified. One such step, not often considered, is the high-temperature rest period after hydrogen chloride etching of the substrate, and before layer growth. Thermal evaporation during this period has been found to have a significant effect on impurity redistribution of the epitaxial layer. Specifically, redistribution effects are reduced as both the time and the temperature of this rest period are increased, so as to allow purging of the boundary layer region prior to growth.

5.2.4.1 Lateral Autodoping

Redistribution effects outlined in the previous section are essentially of a one-dimensional nature, that is to say, the silicon substrate behaves as a large, doped surface area from which evaporants leave as if from an infinite plane. This is a reasonably accurate model, since substrate diameters are typically 75–125 mm, whereas the width of the stagnant layer is about 5 mm. However, an additional problem arises during the fabrication of bipolar microcircuits. Here it is customary to begin by diffusing a pattern of small n^+-buried layers into a slice of lightly doped, p-type material (3–10 Ω cm). This serves as the substrate for the growth of the n-epitaxial layer in which subsequent process steps are carried out. During epitaxy, each n^+-buried layer acts as a two-dimensional source from which evaporant diffuses into the surrounding medium.* This results in a redistribution of impurities over the adjoining lightly doped substrate region (*lateral autodoping*), in addition to conventional autodoping over the buried layer. In extreme situations, a slice can serve as a dopant source for slices that are downstream from it.

The problem of lateral autodoping is especially serious in high-speed microcircuits, which require the growth of very thin lightly doped epitaxial layers on substrates with heavily n^+-buried layers. Its effect is illustrated in Fig. 5.11 for the growth of an epitaxial layer by the $SiCl_4$ process, on slices with a number of arsenic-doped buried layer regions [25]. In Fig. 5.11*a* the doping profile is shown directly over the buried layer region. The profile at a point 0.1 mm away from this region, and over the lightly doped substrate, is shown in Fig. 5.11*b*. As seen here, a significant amount of lateral doping is observed for this experimental situation.

Lateral autodoping occurs primarily by the transport of evaporants *within* the stagnant layer. There is a definite (though small) velocity gradient in this region, especially near the gas-flow-stream–layer interface. As a consequence, regions which are downstream are more heavily doped than those that are upstream. The extent of this effect may be characterized by the maximum value of doping which occurs in a region that is some fixed distance away from the buried layer. Referring to this value as N_{max}, it has been experimentally shown that N_{max} falls with increasing deposition temperature

* The situation is analogous to that of diffusion through a window in an oxide, as described in Section 4.5.3.

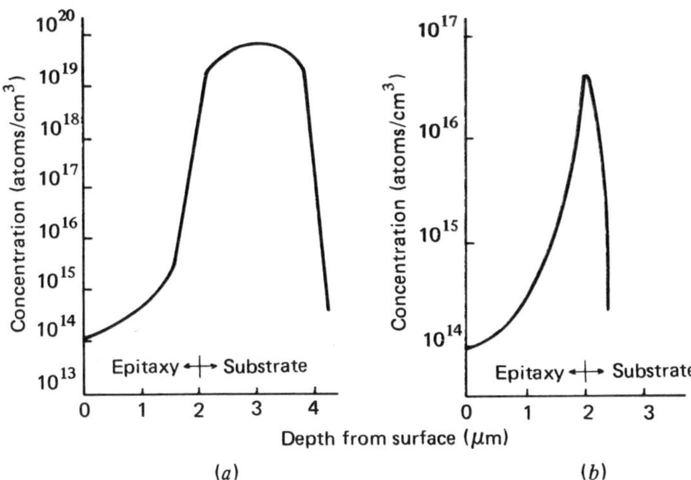

Fig. 5.11 Vertical and lateral autodoping behavior for an arsenic-doped substrate in the $SiCl_4$ process. From Srinivasan [25]. Reprinted with permission of the publisher, The Electrochemical Society, Inc.

or with decreasing growth rate. This is very similar to what is obtained for vertical redistribution effects.

Lateral autodoping is sensitive to the nature of the prebake* period, and falls as either its time or its temperature is increased. This can be explained by noting that diffusion of the dopant to the silicon surface, and its evaporation into the gas stream, are the primary effects that occur during this process. Thus the prebake step can be used to scavenge the evaporants in the stagnant layer, prior to epitaxial growth [25].

A significant reduction of lateral autodoping effects can be accomplished by operating the reactor at reduced pressure [15, 16]. Although this increases the width of the boundary layer, the diffusivity increases more rapidly.† Thus the flux of evaporants which is returned to the gas stream is increased, aiding their probability of removal from the substrate environment. In addition, for the same reason, the prebake period is more effective in purging the boundary layer in a low-pressure system.

5.2.5 Selective Epitaxy

Epitaxial growth approximately follows the surface contour of the substrate, and completely covers it. There are a number of interesting device situations, however, where epitaxy is required only in selected regions. One example

* This is the rest period between in situ etching of the slice and the actual deposition of the epitaxial layer.

† See Problem 4 at the end of this chapter.

consists of photolithographically opening windows in an oxide layer over the slice, and filling these with epitaxial material. Typically, growth of this type is carried out under low supersaturation conditions in order to control the crystal morphology at the edge of the window. Often additional HCl gas is used to reduce the buildup of polycrystalline silicon on the oxide surface.

Selective epitaxy can also be achieved on substrates without masking, if the substrate preparation and reactor conditions are such that the growth is nonuniform in a prescribed manner. One approach is to operate the reactor in the kinetic regime, where the growth rate is sensitive to both temperature as well as crystallographic orientation. A second approach is based on the thermodynamics of reversible processes, such as those involving chloride transport. Here growth is generally carried out at high reactant concentrations, corresponding to very low growth rates (point $\times$ in Fig. 5.7). Alternately, the addition of HCl gas can be used to simulate this reactor condition [26].

One example of this approach is a situation where deep grooves, cut in a silicon surface, can be refilled without masking. Grooves of this sort can be formed by using (110) material, delineating cuts in the mask along the $[\bar{1}10]$ direction, and using a preferential etch to expose the $\{111\}$ planes. These planes are at right angles to the surface so that deep grooves with vertical faces can be etched in this manner.

For this situation it is possible to adjust reactor conditions (by controlling the chlorine-to-silicon source ratio) so that the growth rate on the surface is zero. The temperature within the groove is higher, however, so that conditions for growth exist here. As a result, growth proceeds until the grooves are filled, and then ceases.

5.2.5.1 Pattern Shift

In general, epitaxial growth will not precisely follow the contour of the substrate, since a nonplanar slice exposes a number of different surfaces. This presents a problem in the fabrication of bipolar transistor microcircuits. Here a buried layer diffusion is made into windows in the slice prior to the growth of the active layer. This leaves an optically sharp pattern of indentations in the plane surface, due to the oxidation and subsequent photolithographic process. It is important that this pattern be carried through to the surface of the epitaxial layer, since it is used for the alignment of succeeding masks. Typically, however, it suffers from a variety of spatial distortions and displacements, which are referred to as *pattern distortion* and *pattern shift,* respectively. In extreme cases, complete obliteration of the pattern, known as *washout,* can occur. Although these problems are somewhat lessened with thin epitaxial layers, the requirements on such layers is comparatively more tight, since they are used in high-density applications.

Reactor conditions during growth, as well as crystal orientation, both affect the extent of these problems. Extensive studies have been made with

(111) silicon on which layers were grown by the $SiCl_4$ process [27]. Here it has been shown that the amount of pattern shift is reduced when the growth temperature is increased. In addition, silicon that is misoriented by 2–4° from the ⟨111⟩ directions shows a reduction in pattern shift over precisely oriented material. Finally, these effects are minimized if the misorientation is toward the nearest (110) plane. For (100) silicon, on the other hand, best results are obtained when the slice is precisely oriented [28]. Here, however, a misorientation of as little as 0.5° can lead to a significant displacement of the pattern.

Conservative design practice allows for an angular pattern shift of 45° due to epitaxial growth. Thus a substrate pattern, after epitaxy, can be assumed to shift a maximum of x_0 in all directions, where x_0 is the thickness of the epitaxial layer. Successive masks must be designed to take this uncertainty in the position of the buried layer pattern into account.

Both pattern shift and washout are minimized at high deposition temperatures and reduced growth rates. Furthermore, their effect is less in systems where the chlorine-to-silicon ratio is lower. Thus the use of SiH_2Cl_2 reduces these problems as compared to $SiCl_4$. The silane process, which is chlorine-free, essentially eliminates this problem. However, pattern distortion effects become very large when this reactant is used [29].

5.2.6 Growth Imperfections

A number of different types of imperfections may be present in epitaxially grown layers. Some of these are crystallographic in nature and result in a loss of periodicity in the crystal lattice. Yet others are gross defects, usually arising from improper handling or cleaning procedures.

Stacking faults are perhaps the most important types of crystallographic defects that occur in epitaxial silicon films. They can be made suitable for microscopic examination by etching,* and appear as equilateral triangles and lines on the surface of layers grown on {111} substrates. The orientation of the sides of these stacking faults is in the ⟨110⟩ directions for this type of substrate. In addition, all the lines and the sides of these triangles are usually of equal length and are directly proportional to the thickness of the layer. Stacking faults on {100} substrates are generally square in shape, with their sides oriented along the ⟨110⟩ directions. Here, too, the length of each side is directly proportional to the layer thickness.

The growth of stacking faults can be described by considering epitaxial growth, in its most elementary form, as the ordered deposition of atomic planes, one at a time, with nucleation centers for a fresh plane being formed only after the last plane has been completed. If, however, there is a small growth area which is mismatched with respect to the substrate, the regularity of the layers is disrupted. On disruption, the layers continue to grow in this

* A variety of etches, suitable for this purpose, are described in Chapter 9.

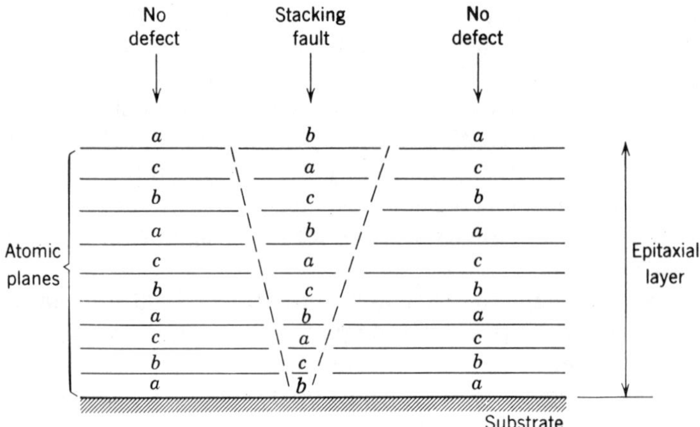

Fig. 5.12 Formation of a stacking fault.

new sequence within the fault. This is shown in Fig. 5.12 for {111} layers in an f.c.c. lattice. For this case the stacking sequence of atomic planes is *abcabc* . . ., whereas that of the diamond lattice is *aa'bb'cc'aa'bb'* . . . Thus the growth mechanisms are essentially similar, although somewhat easier to visualize for the f.c.c. lattice. Here the defect propagates from layer to layer, retaining its shape and increasing in size with each successive atomic plane. The various faces of the defect all fall upon inclined {111} planes, resulting in the tetrahedron of Fig. 5.13. From geometrical considerations, $h = \sqrt{\frac{2}{3}}\, s$, where h is the height of the stacking fault, and s is the length of one side of the triangle on the surface.

Stacking faults in (100) silicon also propagate along {111} planes. They intersect the surface plane in the ⟨110⟩ directions and are mutually perpendicular. Thus the fault is in the form of a pyramid with a square base, and appears on the surface in the form of oriented lines and squares.

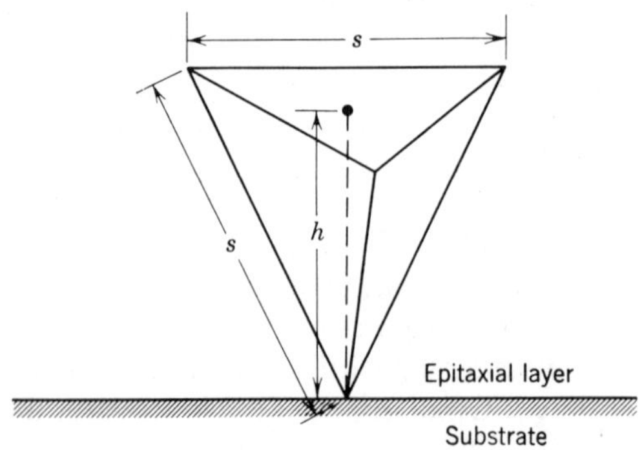

Fig. 5.13 Stacking fault geometry on (111) silicon.

Stacking faults generally originate from the substrate surface, where inhomogeneities such as dislocation loops and precipitated impurities act as sites [30]. Surface measurements of these faults are often used to check the thickness of the epitaxial layer. One of the primary causes of stacking faults are patches of SiO_2 on the surface of the crystal. Since the Si-Si spacing in the lattice is 2.35 Å and the Si-O-Si spacing for silicon polyhedra is 3.05 Å, the presence of traces of oxide results in steps which are not equal to an integral number of atomic plane spacings. Periodicity is maintained in the lattice during epitaxial growth by the initiation of stacking faults that start at the edge of the oxide and grow into the depositing layer. Experimental verification of this theory has been noted in the fact that the concentration of stacking faults is critically dependent on the effectiveness of the hydrogen reduction step. (Note that the subsequent HCl etch does not remove SiO_2.)

Scanning electron microscope studies have been made [31] to investigate the electronic activity of stacking faults, i.e., their ability to act as recombination centers. Results are inconclusive; however, much evidence of activity has been noted at the corners of the triangles (or squares) formed where the defect penetrates through to the surface. This is most probably the result of impurity segregation at these points of stress concentration. Thus it is generally considered that stacking faults, taken by themselves, do not significantly alter the electronic properties of the epitaxial layer. They give rise, however, to inhomogeneities in the diffusion of impurities and act as nucleation centers for metal precipitates. *p–n* junctions built in faulted regions often exhibit soft reverse-breakdown characteristics [32], with breakdown initiated by the formation of microplasmas at the edge of these faults. The elimination of stacking faults is thus especially important as devices become smaller and more densely packed on a microcircuit. Basic techniques for their reduction are those that result in the elimination of surface defects. In addition, misorientation of the crystal surface by 1–3° exposes a large number of layers to which atoms can attach during epitaxy. This allows a faster growth rate for the same layer quality without a corresponding increase in stacking fault density. Thus techniques for the reduction of pattern shift in (111) silicon are also effective in reducing stacking faults for this orientation.

A second type of imperfection is the *tripyramidal defect*. This is the result of crystal growth by twinning, and is initiated by carbon contamination (in the form of β-SiC microprecipitates). In (111) silicon, it appears as a rosette-shaped cluster of three pyramids with hexagonal bases oriented in the ⟨110⟩ directions. Faults of this type are not common; they are rarely observed in (100) silicon, but this is most probably because (111) silicon has been extensively used for bipolar microcircuits.

Many gross defects can also occur during the growth of an epitaxial layer. The majority of these are caused by leaks in the system, the presence of foreign matter during epitaxial growth, and surface scratches caused by pregrowth handling. Their presence can usually be detected by optical means and often leads to rejection of the slice. Defects of this type are described here very briefly.

Substrate scratches act as sites for the initiation of stacking faults and result in lines of such faults on the surface of the layer.

Pits, voids, and *spikes* are caused by small particles of silicon or silicon dioxide in the reactor during the epitaxial growth.

Haze takes the form of a cloudy appearance and results from a leaky system or from improper cleaning and solvent removal procedures prior to epitaxial growth.

Orange-peel appearance is sometimes caused by preferential etching during the in-situ step prior to layer growth.

All of these gross defects are indicative of poor fabrication technique and can be corrected by taking suitable precautions.

5.3 VPE PROCESSES FOR GALLIUM ARSENIDE

Epitaxial growth of gallium arsenide is sometimes carried out in sealed-tube systems, because of the ease of maintaining controlled vapor pressures and equilibrium conditions by this method. In this section, however, we confine ourselves to a study of open-tube processes which are most commonly used because of their convenience. These involve vapor-phase transport of the active species, followed by some form of chemical conversion in the gas stream, and by a growth reaction at the substrate. Reactions are usually under nonequilibrium conditions, so that chemical thermodynamics provides only a general guide to the actual processes that are involved.

5.3.1 General Considerations

Gallium arsenide decomposes into gallium and arsenic upon evaporation. Consequently, its direct transport in the vapor phase is not possible. Transport of its separate components must therefore be considered. Elemental arsenic has a sufficiently high vapor pressure so that it can be transported by direct sublimation. However, transport in the form of its chloride ($AsCl_3$ is a liquid which can be used in a bubbler arrangement) or its hydride (AsH_3 is a gas) is more common because of convenience. Of these compounds, $AsCl_3$ is the most stable and is available in very high purity (99.9999%). It is a poisonous, corrosive liquid with a melting point of $-18°C$ and a boiling point of 130.2°C. Its vapor pressure is shown as a function of temperature in Fig. 5.2.

Arsine is a toxic, colorless gas which burns spontaneously upon exposure to air. It is typically used in a 5–50% dilution in ultrapure hydrogen gas. It is a somewhat less pure source of arsenic (99.999%), with silicon as its major contaminant. This presents a problem with gallium arsenide, since silicon is a shallow impurity. Recently, improved methods have been developed for arsine purification, and special grades of this gas are now available where the silicon contamination level is kept to a minimum.

Gallium is also available in ultrapure form (99.9999%). However, its vapor pressure is vanishingly small at growth temperatures, so that it must be transported in the form of a volatile compound. Gallium chloride (GaCl) is commonly used for this purpose, and must be formed by a secondary reaction at some point upstream from the substrates.* An alternate approach, rapidly gaining popularity, is the use of trimethylgallium $(CH_3)_3Ga$ (TMG). This organometallic compound of gallium is a colorless, pyrophoric liquid which melts at −15.8°C and boils at 55.7°C. It can be transported in vapor form by means of a bubbler held at −10°C, and is available in sufficient purity (99.999%) to be used for gallium arsenide epitaxy. Triethylgallium $(C_2H_5)_3Ga$ (TEG) can also be used as an alternate organometallic source for gallium. This compound is a liquid, with a melting point of −82.3°C and a boiling point of 143°C. The vapor pressures of these compounds are shown in Fig. 5.2.

Hydrogen is the most commonly used carrier gas. It is readily produced in 99.9999% purity by passing the low-cost commercial grade gas through a palladium–silver purifier.

Oxygen is a deep lying impurity in gallium arsenide. As a result, gallium arsenide epitaxial systems must be extremely leak-tight for satisfactory operation. Sometimes the entire gas handling apparatus is placed in a box which is flushed with an inert gas to prevent the accidental introduction of this impurity. In addition, elaborate pump-down and back-fill techniques are often used for leak checking and flushing prior to growth.

Both horizontal and vertical reactors are employed for gallium arsenide. In addition, both radio frequency and resistance heated systems are in use. Systems using an elemental gallium source are generally of the resistance heated, hot wall type, with at least two temperature zones (one for the source and one for the deposition). The formation reaction in these systems is exothermic, with deposition occurring in the cooler regions (i.e., the substrate).

Cold wall, radio frequency heated systems are generally used in reactors with a single temperature zone. They have the advantage of low thermal mass and allow the use of processes which require sequencing of the temperature. They are especially suited for systems in which the formation reaction is kinetically controlled. Here their cold wall minimizes premature reaction of the constituent gases prior to arrival at the hot substrate.

It is somewhat more difficult to obtain a uniform growth rate in gallium arsenide systems than in silicon systems [33]. Although the same requirements of maintaining a uniform boundary layer are present, there is often the additional complication that reactants are unstable at temperatures well below the growth temperature. In some instances competitive reactions can

* Note that GaCl is only stable at elevated temperatures.

also occur. As a result, gallium arsenide systems today usually process a few (one to ten) wafers at a time, and operate under relatively high gas velocities to balance the growth rate between slices. Some of these problems can be alleviated by using low-pressure chemical vapor deposition techniques; these will no doubt become more popular in the near future.

Epitaxial gallium arsenide is usually grown on (100) substrates. This allows the fabrication of rectangular chips whose edges are {110} planes, which are preferred for easy cleavage. This orientation is mandatory for laser fabrication, where the parallelism of opposite faces is a device requirement.

Gallium arsenide epitaxy is sometimes carried out on its {111} faces. Although these are the most closely spaced, growth is extremely difficult because of the difference in growth conditions for successive layers [34]. Thus the simple crystal model of Fig. 5.14*a* shows that growth of the layer *A–A* requires the making of three bonds, while leaving one bond dangling. However, growth of the next layer (*B–B*) requires the making of one bond with three bonds left dangling. In contrast, it is seen from Fig. 5.14*b* that the growth of each layer on a (100) surface is identical, i.e., two bonds are made, leaving two dangling. Thus there is an inherent barrier to nucleation during growth in the ⟨111⟩ directions, which is not present in the ⟨100⟩ directions. This is of special significance when epitaxial growth is carried

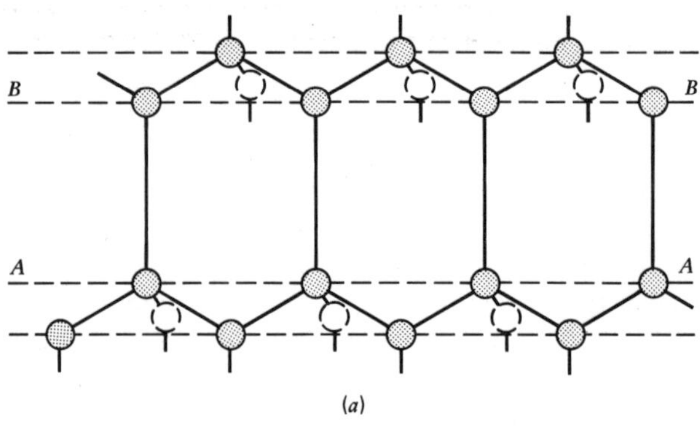

(*a*)

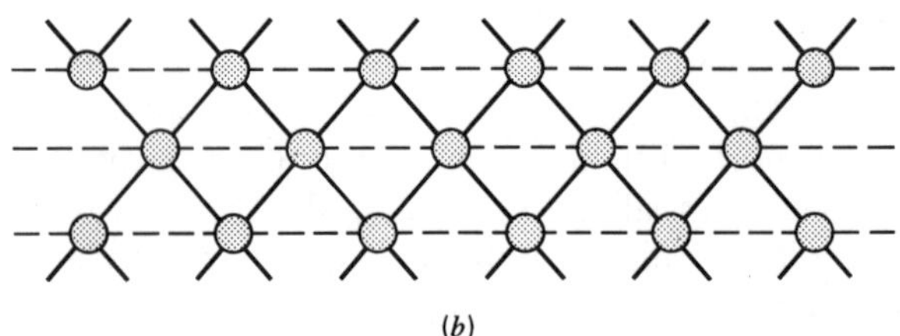

(*b*)

Fig. 5.14 (*a*) Growth on (111) plane. (*b*) Growth on (100) plane.

out at temperatures that are well below the melting point, as is the case with gallium arsenide.*

Gallium arsenide growth in the [111] direction is usually carried out on its arsenic face. This is because the (111)As face is more electronically active than the (111)Ga face, and so can be readily etched to a smooth polish. Differences in growth rates on these faces are discussed in Section 5.3.7.

Both (100) and (111) gallium arsenide are sometimes purchased with a 2–3° misorientation toward the nearest [110] direction, as is done with silicon. The evidence that this misorientation has an advantage in reducing stacking faults during layer growth is inconclusive.

5.3.2 Growth Strategy

A common strategy for good growth is applicable to all gallium arsenide VPE systems. Thus from stoichiometric considerations it is desirable to grow gallium arsenide in the 650–850°C temperature range. Growth at the high end of this temperature range results in excessive gallium vacancy formation. On the other hand, the reduced surface mobility at the low end creates a tendency for films to become polycrystalline. Systems whose thermochemistry allows growth to occur in this range are the most successful.

There must always be sufficient arsenic overpressure to prevent thermal decomposition of the substrate and of the growing film. Moreover, the arsenic-to-gallium mole ratio must be controlled to prevent the formation of unconsumed gallium, which can otherwise lead to the formation of droplets, hillocks, and whiskers on the growing surface [35]. Excess arsenic, on the other hand, remains in vapor form (As_2 and As_4) and so has a zero sticking coefficient.

Substrates must be extremely clean and free from mechanical damage prior to growth. An in-situ etch step is highly desirable for this purpose. Finally, it is desirable that the growth conditions be mass transport limited, since operation in the kinetic control regime is orientation dependent, and also tends to accentuate surface defects, resulting in poor crystal morphology.

5.3.3 The Ga-$AsCl_3$-H_2 Process [36]

This is sometimes referred to as the *halide process* because of the nature of the arsenic source. Its use results in epitaxial layers with the highest available purity to date, since all of the starting materials are readily available in 99.9999% purity or better.

A two-zone, resistance heated, hot wall reactor is used for growth in this system. The gallium is held in a graphite crucible in the source zone, as

* Typically, gallium arsenide is grown at 700–750°C, and melts at 1238°C. In contrast, silicon melts at 1412°C and is usually grown at 1200°C. Thus silicon epitaxy in the ⟨111⟩ directions is relatively easy.

shown in Fig. 5.15. Growth occurs downstream in the deposition zone where the substrates are laid on a quartz or graphite carrier. During growth, part of the hydrogen is sent through an $AsCl_3$ bubbler which is maintained in the 0–20°C range. Both hydrogen and the $AsCl_3$ vapor enter the system where they react upstream from the gallium source. Here

$$AsCl_3 + \tfrac{3}{2}H_2 \xrightleftharpoons{425°C} \tfrac{1}{4}As_4 + 3HCl \tag{5.28}$$

These reaction products flow over a gallium source boat which is typically held at 800–850°C. Gallium arsenide is formed and dissolves in this gallium source:

$$Ga + \tfrac{1}{4}As_4 \rightleftharpoons GaAs \tag{5.29}$$

Simultaneously, the HCl gas resulting from the first reaction serves to transfer this gallium arsenide to the substrate, as follows:

$$GaAs + HCl \rightleftharpoons GaCl + \tfrac{1}{2}H_2 + \tfrac{1}{4}As_4 \tag{5.30}$$

From thermodynamic data [37] it can be shown that this reaction is driven from left to right at temperatures above 800°C, i.e., it is a decomposition reaction for the gallium arsenide. Note that GaCl is unstable at room temperature. Thus this reaction serves to prepare it in situ, so that it can be transported to the substrate. There is some evidence of formation of both $GaCl_2$ and $GaCl_3$ during this process as well. However, their effect on the growth is of second order.

The substrate is held at a lower temperature, typically 750°C. Here the same reaction occurs, except that it is now driven in the opposite direction so that it is a deposition reaction for gallium arsenide. The net effect is to transfer gallium arsenide from the source boat, where it is formed, to the substrate on which it is epitaxially grown.

Thermodynamic data show that the growth of gallium arsenide by this method involves a negative heat of reaction (i.e., it is exothermic). As a result, a hot wall system must be employed with growth occurring on the

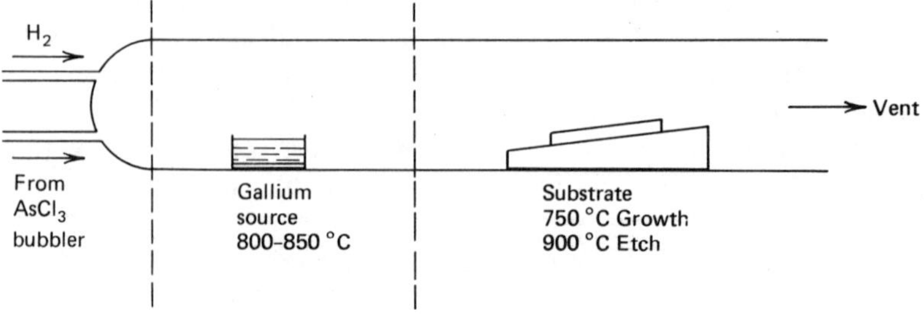

Fig. 5.15 The halide process.

colder susceptor. Furthermore, growth takes place by a surface catalyzed (heterogeneous) reaction so that excellent crystal quality can be obtained. Typical growth rates of 0.2–0.5 μm/min are achieved in systems of this type.

The growth versus temperature characteristic of this system takes the form of Fig. 5.5*b* because of the exothermic nature of the reaction. Here the growth rate falls with decreasing temperature due to reaction rate limitations, i.e., kinetic control. As can be expected, the growth rate is orientation dependent in this temperature region.* The growth rate also falls with *increasing* temperature because of the exothermic reaction, even if the reaction is mass-transfer limited in this region (as is usually the case). However, as is characteristic for mass-transfer control, the growth rate is not orientation dependent in this temperature range.

There are two important considerations in the use of this system. First, the gallium source must be fully saturated with arsenic before consistent operation can be obtained. As a result it is customary to saturate† this source *before* the actual deposition run is made. Typically, this is done by passing the $AsCl_3$ vapor over the source for as much as 24 h, until a crust of gallium arsenide is formed on its surface. Next, the substrate must be in situ etched prior to epitaxial growth. This can be accomplished by raising its temperature above 800°C, so as to drive the reaction of Eq. (5.30) from left to right. An alternate approach is to inject HCl or $AsCl_3$ into the system beyond the source boat, but before the deposition zone. Either technique results in the production of a substrate that is free from mechanical damage, and greatly improves the quality of the subsequent epitaxial layer. The HCl technique is considerably more convenient, since temperature control of a hot wall furnace (with a large thermal mass) is not necessary. However, it is a potential source of contamination and $AsCl_3$ should be used if high-purity layers are desired.

Argon can also be used as a carrier gas in place of hydrogen and results in a somewhat simpler set of reactions, since it serves only as a transport and diluent medium. However, its use is not popular because of the unavailability of low-cost argon at the purity level that is achievable with hydrogen.

5.3.4 The Ga-AsH_3-HCl-H_2 Process [38]

This is sometimes referred to as the *hydride process* because of its use of arsine gas. It differs from the last in that the fluxes of gallium and arsenic species are formed independently. This allows greater control in the vapor phase, and hence a wider control of the deposition parameters. The flux of arsenic is formed by the decomposition of AsH_3,

$$AsH_3 \xrightleftharpoons{400°C} \tfrac{1}{4}As_4 + \tfrac{3}{2}H_2 \tag{5.31}$$

* This topic is covered in Section 5.3.7.

† The solubility of arsenic in gallium is 4 at.% at 850°C.

The flux of gallium species is formed by passing HCl gas over a heated gallium source, resulting in GaCl:

$$Ga + HCl \rightleftharpoons GaCl + \tfrac{1}{2}H_2 \tag{5.32}$$

Finally, both these fluxes arrive at the substrate where epitaxial growth occurs by means of the following reaction:

$$GaCl + \tfrac{1}{4}As_4 + \tfrac{1}{2}H_2 \rightleftharpoons GaAs + HCl \tag{5.33}$$

This is the same reaction as that for the halide process, i.e., Eq. (5.30). Thus both systems have many similarities, and require almost identical reactor conditions. Here, too, the reaction is exothermic, so that a hot wall reactor is used, with two temperature zones. Typically, the gallium source is held at 775–800°C, and the substrate is maintained at 750°C. HCl gas enters the system and passes over the gallium source, which must be presaturated with gallium arsenide. The arsine gas, however, is injected downstream from the gallium source, where it provides the arsenic flux (see Fig. 5.16). In-situ etching is done by increasing the HCl flow beyond the gallium source.

The growth characteristics of this system are very similar to those that are observed with the halide process. Here, too, the reaction is surface catalyzed so that excellent quality layers are obtained, in the growth range of 0.2–0.5 μm/min.

In the halide system, the primary process variable is the mole fraction of $AsCl_3$, which is varied by altering the carrier gas flow through the bubbler. As a result, the arsenic-to-chlorine ratio of this system is fixed at 0.33. In the hydride process, however, this ratio can also be varied over a wide range, giving an additional degree of freedom. Thus the system is considerably more flexible, and is the preferred approach for most situations where carrier concentrations in excess of 10^{14}–10^{15} cm^{-3} are acceptable [39]. Lower carrier concentrations, below 10^{13} cm^{-3} are achievable in the halide system.

5.3.5 The $(CH_3)_3Ga$-AsH_3-H_2 Process [40]

This is one of the many *organometallic* processes which have been the object of much recent work. Its attractiveness is largely due to the fact that trimethylgallium (TMG) is a gallium source which can be readily transported in vapor form by a bubbler arrangement maintained in the −10 to 0°C range. Thus the flux of gallium reactant can be delivered conveniently and independently, without the necessity of introducing a halogen (see Fig. 5.17).

The flux of arsenic is delivered independently from an arsine source, and hydrogen is used as a carrier gas. The overall reaction leading to the growth

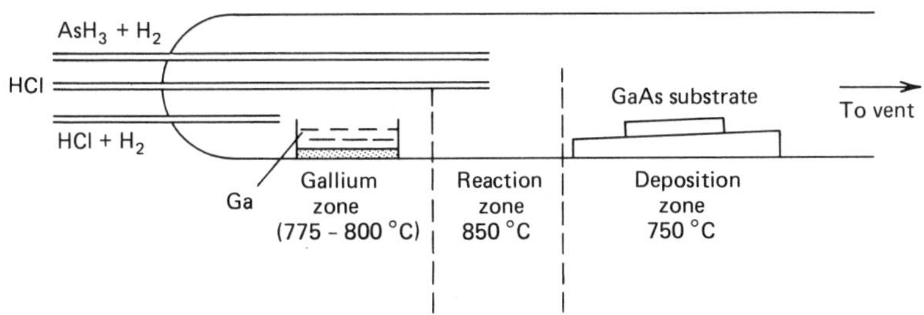

Fig. 5.16 The hydride process.

of gallium arsenide is thus

$$(CH_3)_3Ga + AsH_3 \rightarrow GaAs + 3CH_4 \tag{5.34a}$$

Very little is known about the details of this reaction. However, evidence indicates that TMG decomposes by the successive removal of CH_3 groups upon heating. However, the bond strength of $(CH_3)Ga$ is sufficiently high so that, in all probability, this species arrives at the substrate by diffusion through the boundary layer. In like manner, arsine gas most probably arrives with one hydrogen atom intact, i.e., as AsH.

It has been proposed that gallium arsenide is formed by the chemisorption of these species, and by their surface diffusion to gallium and arsenic kink sites respectively. Here, loss of the methyl radical can occur by combining with hydrogen to form methane or by combining with a second methyl radical to form ethane (C_2H_6). It is also possible that

$$CH_3Ga + AsH \rightleftharpoons CH_3GaAsH \rightarrow GaAs + CH_4 \tag{5.34b}$$

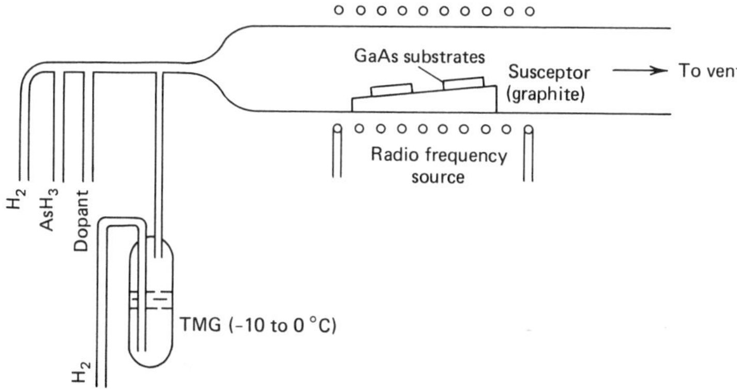

Fig. 5.17 The organometallic process.

and complex species of this type have been identified. In both cases, however, the growth of gallium arsenide is a surface catalyzed process.

The reactions described here are irreversible, and can be used to grow gallium arsenide over a wide range of temperatures (500–1100°C). Higher temperatures favor better crystal quality, but lead to excessive vacancy formation. This results in a high undoped carrier concentration because of contaminant incorporation.* Typically, 650–750°C is considered the optimum range for crystal growth, and undoped carrier concentrations in the $(1–5) \times 10^{14}$ cm^{-3} range can be achieved in this way.

The TMG is the limiting species, so that the growth rate is a linear function of its partial pressure. Values of As/Ga from 8 to 30 are used in operating systems in order to ensure complete reaction with the gallium, and growth rates of 0.1–0.5 μm/min can be obtained by independent control of the partial pressure of TMG. The partial pressure of arsine gas (which is approximately equal to the partial pressure of As_2 at these temperatures [41]), is usually many decades higher than the equilibrium pressure of arsenic over gallium arsenide at growth temperatures.

Methane and possibly ethane gas are a by-product of this reaction. However, they are stable at typical growth temperatures and do not lead to significant carbon doping by dissociation. Moreover, the large excess of hydrogen carrier gas serves to further reduce the possibility of carbon contamination by this mechanism. Nevertheless, carbon is an important contaminant in all organometallic systems.

Like the chloride process, the TMG-AsH_3 reaction is also exothermic. It must be carried out in a cold wall, radio frequency heated system in order to avoid decomposition of the starting compounds before they enter the reaction zone. The growth rate also falls off with decreasing temperature. Some degree of orientation dependence has been noted here [42]. However, this effect is not as pronounced as for the halide and hydride systems, because of the pyrolytic nature of the reaction. A slight falloff in growth rate at temperatures above 800°C has also been observed, and is probably due to increased desorption of the arsenic species at these temperatures. The growth rate at these temperatures is orientation-dependent as well, thus reinforcing this conclusion.

Growth over the 600–800°C temperature range occurs in the mass-transfer-limited region, with a growth rate that is relatively independent of both temperature and substrate orientation effects. In addition, the reaction is surface catalyzed so that excellent crystal quality can be achieved.

The lack of reversibility in this system has the disadvantage that the reaction cannot be controlled to obtain *in situ* etching of the substrate. However, this etching step can be performed in a separate process, by the introduction of either hydrogen chloride gas or $AsCl_3$ vapor into the reaction

* Silicon, which is an impurity in the arsine source, is a possible cause for doping the layers *n*-type at these temperatures.

chamber [43]. In both cases, the etching reaction must be carried out at temperatures around 900°C to obtain mass flow control. Otherwise etching is orientation dependent, and leads to faceting and fish scaling of the surface. Layers grown on in situ etched substrates have been found to have improved mobility over unetched layers, with a very narrow epitaxial layer–substrate transition region [44]. This is especially important for the growth of thin layers of the type required for high-speed field effect transistors.

Triethylgallium (TEG) has also been used for gallium arsenide epitaxy [45]. Claims for the use of this compound are that it results in higher purity films (lower undoped concentration) than are obtained with TMG, because of its lower organic–metal bond strength. This leads to reduced carbon incorporation in the epitaxial layer.

Some work has also been done [46] with "mixed" systems, using dopants such as gallium diethylchloride, $Ga(C_2H_5)_2Cl$. However, the inavailability of this material in sufficient purity has led to inconclusive results, so that this work has not been aggressively pursued.

5.3.6 Doping

From an applications point of view, n-type gallium arsenide is by far the most interesting, and there is a need for growing this material over a wide carrier concentration range. In contrast, p-type material is primarily required in heavily doped regions for p^+–n junction formation, and also for making ohmic contact to lightly doped p-layers.

The doping of n-type gallium arsenide films above 10^{16} cm^{-3} is done by the introduction of impurities in vapor form during epitaxy. Below this concentration, however, doping control is usually effected by means of the inherent contaminants which are present in the starting materials. Of these, silicon is found as an impurity in all arsenic compounds, and is the most important. This group IV impurity is amphoteric in nature; its incorporation into group III lattice sites results in n-type behavior. Conversely, its incorporation into group V sites results in p-type behavior.

An increase in temperature during the growth of gallium arsenide results in an increase in the formation of both gallium and arsenic vacancies, and a consequent increase in contaminant incorporation. It is generally observed that the primary contaminant is silicon; although amphoteric, this dopant is found to selectively dope the layer n-type. This is because vapor-solid equilibrium determines the concentration of silicon incorporated in lattice vacancies. For VPE, silicon incorporation into gallium sites is $\simeq 1.6 \times 10^3$ times that into arsenic sites. Consequently, silicon doping is invariably n-type in VPE gallium arsenide. By the same token, growth at lower temperatures results in layers that are less n-type. The reduction in the downward direction is limited, however, since it becomes harder to achieve good crystal quality at low temperatures.

The effect of arsenic overpressure depends on the detailed nature of the growth reactions. In the organometallic process, the primary effect of increasing the arsenic partial pressure is to increase the gallium vacancy concentration, resulting in greater silicon incorporation into gallium sites, and making the layer more *n*-type. Conversely, reduction of the arsenic overpressure reduces the doping level, and the material eventually becomes *p*-type. A limit on this process is set, however, by the need for having enough arsenic to avoid surface decomposition, as well as to avoid the formation of unreacted gallium. Moreover, the desorption rate of hydrocarbons from the growing surface is reduced, so that the possibility of carbon contamination increases. Thus, films grown below 550°C have been found to be heavily carbon contaminated; interestingly, very little carbon contamination has been observed in films grown at excessively high temperatures, in the 800–1100°C range [42].

Competing reactions occur in systems in which gallium transport is by means of a halogen such as chlorine. Here, increases of the arsenic and chlorine (or HCl) overpressures go hand in hand. The reaction of chlorine with the silicon forms compounds which are relatively stable at growth temperatures, so that this impurity becomes unavailable for doping the layer. As a result, the doping concentration rapidly falls as the partial pressure of the chlorine bearing species is increased [47]. This is true for the halide process in which $AsCl_3$ is used, as well as for the hydride process in which the halogen is HCl gas. In fact, the reduction of free-carrier concentration by means of the halide partial pressure represents an important feature of these systems, which is not present with the organometallic process.

The use of halogens has not proved satisfactory for controlling the doping level in the organometallic process. This is because films grown in the presence of either HCl or $AsCl_3$ become increasingly compensated as their *n*-type doping concentration falls [44]. It has been proposed that the introduction of the halogen assists in cleaving the methyl radical, and thus incorporates excess carbon into the growing layer.

Oxygen (or water vapor) can also be introduced into the reactor during growth, for the purpose of tying up the silicon contaminant in the form of SiO_2. A reduction of one or two decades in carrier concentration can be obtained in this manner. However, there is danger of oxygen incorporation into the films, with the possibility of rendering them semi-insulating. Ammonia has proved to be superior in this regard, and is equally effective in removing the silicon in the form of Si_3N_4. Furthermore, any nitrogen incorporation in the film represents less of a problem, since it is inactive (as opposed to oxygen, which is a deep donor). None of these approaches are in common use at the present time.

The hydrides of sulfur, selenium, and tellurium are used for *n*-type doping at higher concentration levels. Of these, hydrogen sulfide has the highest long-term stability, especially if stored in stainless steel or aluminum tanks. Hydrogen sulfide is a covalent compound with a large negative free energy of formation. It is relatively stable at reactor temperatures. In the presence of chlorine, however, the hydrogen–sulfur bond is easily broken. As a result, sulfur doping is readily accomplished in both the halide and the hydride processes, but is quite difficult with the organometallic technique, which is halogen-free. Typically, for the same partial pressure, sulfur incorporation is some 20–50 times less for this system than for the others.

One technique for improving the sulfur incorporation in organometallic systems is to use sulfur monochloride (S_2Cl_2) as a dopant source. This halide can be transported in a bubbler held at 0–30°C. It decomposes more readily than H_2S at reactor temperatures, and results in sulfur doping that is comparable to what is obtained by the halide and hydride processes [48]. Its use is restricted to heavily doped layers because of the increased carbon incorporation which comes about due to the presence of the chlorine radical.

Recently a number of organometallic compounds have become available for doping purposes. These can be used with either the metalorganic or the halide and hydride processes. They include diethylzinc (DEZ) and diethylcadmium (DECd) (p-type) as well as tetramethyltin (TMT) and diethyltelluride (n-type). Although liquids, these materials are available in high dilutions as gases with hydrogen as the diluent. These are more convenient to use since this avoids the necessity of a bubbler arrangement.

Semi-insulating gallium arsenide can be made by doping with oxygen, iron, or chromium. Of these, oxygen-doped gallium arsenide is very sensitive to thermal processing and is of little technological importance. Iron doping, though relatively stable, results in p-type layers with a maximum resistivity in the 10^4–10^5 Ω cm range. Chromium, on the other hand, can be used to produce melt-grown material in the 10^8–10^9 Ω cm range, and is also relatively stable.

Chromium is a deep acceptor at $E_v + 0.79$ eV. It can compensate n-type gallium arsenide, provided its background concentration is sufficiently low. Typically, material with a donor concentration of up to 5×10^{15} cm^{-3} can be fully compensated by counterdoping with this impurity.

Chromium doping has been done by the addition of chromyl chloride [49] during VPE. Vapors of this material are transported by means of a bubbler arrangement into the reactor where they first react with hydrogen as

$$3CRO_2Cl_2 + 3H_2 \rightarrow CrO_3 \cdot Cr_2O_3 + 6HCl \qquad (5.35)$$

The CrO_3 component is subsequently reduced to form chromium, which dopes the growing gallium arsenide layer. The by-product of this reaction is water vapor, so it is possible that material doped in this way also has some incorporated oxygen. Nevertheless, its properties are relatively unchanged by thermal treatment, so that this oxygen is inactive (possibly in the form of Cr-O complexes), if at all present. A specific resistivity of 10^8 Ω cm has been achieved in this manner, using the chloride process for gallium arsenide growth.

5.3.7 Orientation Effects

We have noted that substrate orientation has a strong effect on the deposition rate, especially at low temperatures where the process is kinetically controlled. An additional complication with gallium arsenide comes about because of its noncentrosymmetric nature. Thus among the low-index planes the (111) Ga face contains only gallium atoms, whereas the (111) As face has only arsenic atoms. These faces have completely different chemical properties as well as growth properties, because of their different adsorption characteristics for species which are involved in the gallium arsenide formation reaction.

Orientation effects have been extensively studied for chemical vapor deposition systems where the gallium is transported by means of its chlorides. In these systems, GaCl, which is stable at the deposition temperature, is formed in a preliminary reaction. Reaction of this compound on the surface produces the epitaxial deposit, so that the surface plays a major role in controlling the growth rate.

Figure 5.18 shows the temperature dependence of the growth rate as a

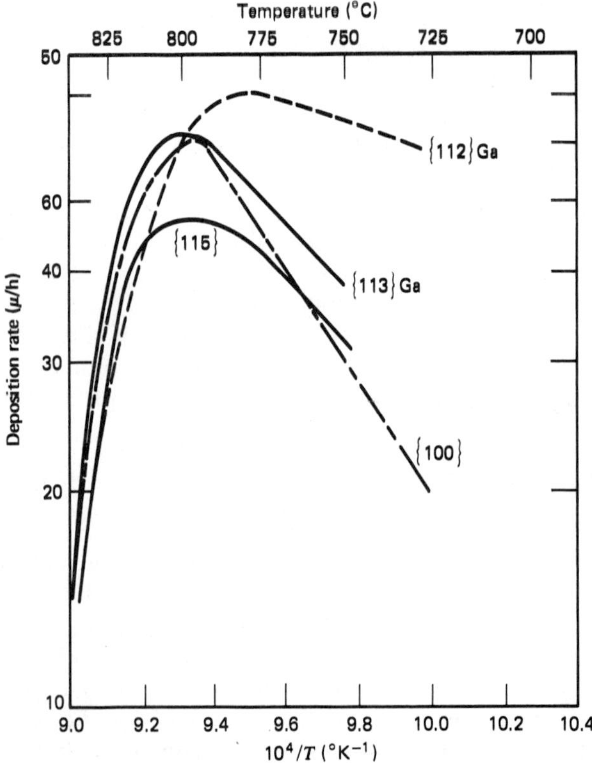

Fig. 5.18 Temperature dependence of growth rate for the halide process. From Shaw [50]. Reprinted with permission of the publisher, The Electrochemical Society, Inc.

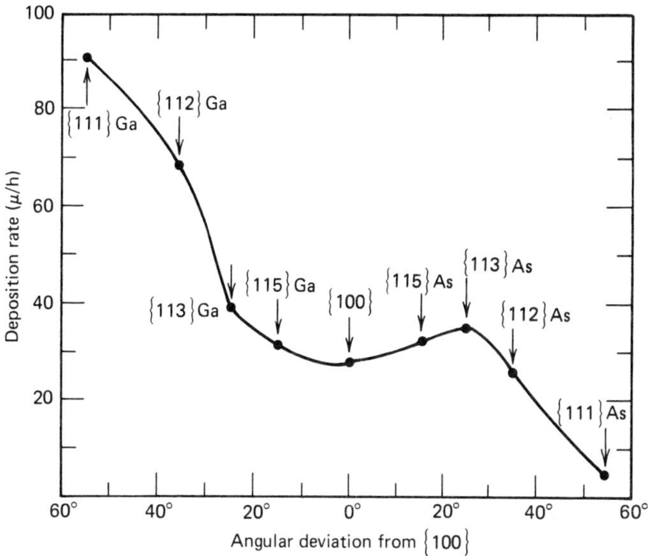

Fig. 5.19 Orientation dependence of growth rate for the halide process. From Shaw [50]. Reprinted with permission of the publisher, The Electrochemical Society, Inc.

function of crystal orientation [50]. Here it is seen that growth in the low-temperature kinetic control region is highly sensitive to crystal orientation. At high temperatures, in the mass-transfer limited region, rates for all orientations become similar. Moreover, the growth rate falls with increasing temperature. This is to be expected, since the exothermic nature of the formation reaction shifts the equilibria toward less deposition (and ultimately etching) at increased temperatures.

Figure 5.19 shows the growth rate at low temperatures for a large number of different crystal orientations. From this figure it is seen that growth on the arsenic-rich surfaces is slower than on gallium-rich surfaces. Thus the addition of gallium atoms appears to be more difficult than arsenic addition to surfaces of the same geometry and packing. This has been explained by postulating [51, 52] that the addition of a gallium atom requires attachment of GaCl to an arsenic site to form an intermediate compound AsGaCl. This is followed by the subsequent detachment of the chlorine, and its desorption from the surface. The addition of arsenic atoms, on the other hand, does not involve this rate-limiting step, since it occurs by direct adsorption on the surface.

Very few studies of orientation dependent growth have been made for the organometallic process. Early work [41] indicated that such effects should be absent since the chemical processes were largely kinetic in nature, with the growth dominated by gas-phase phenomena rather than by the nature of the surface.

Recent studies of the chemistry of organometallic reactions [53] has established the formation of a variety of intermediate compounds such CH_3Ga

and $(CH_3)GaAsH$. It has been proposed that these species are adsorbed on the surface and enter into the growth reaction. Thus the surface does play a role in these adsorption and reaction processes, and will consequently affect the growth rate as well.

Growth at low temperatures, in the reaction-rate-limited region, falls off with temperature, with an activation energy of 16–23 kcal/mole [42]. This is probably due to the heterogenous decomposition of arsine, which is the dominant process at these temperatures. Here, growth in the (111) Ga direction was found to be more rapid than for the (111) As face, as is the case for the halide and hydride processes.

A falloff in growth rate at high temperatures (>800°C) has also been noted, and has been attributed to desorption of the arsenic species from the surface, as well as gas-phase depletion effects. Here, however, the growth rate for the (111) As face has been found to be somewhat higher than for the (111) Ga face.

Impurity incorporation in the halide system is also orientation dependent, as is shown in the experimental data of Table 5.1 [54]. Here too, this behavior has been explained by selective adsorption of dopants on arsenic and gallium-rich surfaces in the presence of the halogen. No comparable data are available for the organometallic process.

5.3.7.1 Selective Epitaxy

Epitaxy in selected regions of a gallium arsenide slice can be performed by depositing through a mask etched in its surface, and permits a number of unique possibilities for device structures. Another approach is to epitaxially grow gallium arsenide in selected regions without the aid of a mask. This has the advantage that the reactor system does not become contaminated by the mask material. Of greater importance, however, is the fact that the epitaxial growth is of better quality, since it avoids the tendency for polycrystalline growth at the mask boundaries.

Table 5.1 Dopant Incorporation in VPE–GaAs

	Dopant Incorporation Ratio		
Dopant	(111) As / 100	(111) As / (111) Ga	(111) As / 110
No intentional doping	11	15	28
Zn	0.43	0.2	0.49
Te	7.4	20	15
Se	6.3	—	4.4
Sn	2.3	—	—
S	1.4	—	—

An example of the first approach is the use of low pressure CVD (LPCVD) using the organometallic process [55]. Here gallium arsenide growth was carried out on windows cut in the surface of an SiO_2 covered substrate. Epitaxy was confined to the exposed surface. However, the quantity of material which grew in these windows was found to be almost equal to the amount that would have grown in the absence of SiO_2. The most probable reason for this is that gallium, formed on the SiO_2 surface, diffuses faster to the gallium arsenide than it reacts with arsenic. Consequently, all growth is confined to the semiconductor surface.

Growth on foreign surfaces is relatively simple with the organometallic process. This would lend further support to the argument that the decomposition reactions associated with this process take place in the vapor phase.

The halide process has also been used for selective epitaxy without masking the gallium arsenide surface. Here V-grooves, etched in (100) substrates, were formed by delineating a pattern in the [110] direction, and etching to expose the low-growth (111) As faces. During epitaxy, the groove filled from the bottom until eventually the material reached the (100) surface by the formation of successively shallow grooves, corresponding to planes of progressively slower growth [56].

Selective epitaxy can also be carried out so as to completely cover over a masked pattern on a gallium arsenide substrate. This allows the formation of a coherent single crystal layer which can be subsequently removed from its parent substrate and used in thin film form [57].

All of the above techniques are applicable to halide and hydride systems, because of their highly orientation-dependent growth characteristics. They are not possible with the organometallic process, which is relatively insensitive in this regard, because of its pyrolytic nature.

5.3.8 Growth Imperfections

Presently available gallium arsenide substrates are poorer in their crystal quality than starting silicon, so that the quality of gallium arsenide epitaxial layers can be expected to be correspondingly worse. Furthermore, gallium arsenide is made up of two components with very different vapor pressures, so that defects are often nonstoichiometric in character. One such defect, the *hillock* [35], is often seen in the form of a well-defined fault surrounded by a streamline pattern of material. Hillocks have been observed in gallium arsenide grown by all the processes described earlier. They appear to originate at the start of layer growth, as evidenced by a uniformity of size and shape, much like stacking faults on silicon substrates. However, their shape is not as distinct, because of partial dissolution due to excess gallium. Typically, hillocks are very nonuniform in composition, with regions having a gallium excess of as much as 35%.

It is generally felt that hillocks are formed when the rate of incident atoms from the gas phase is higher than the rate at which atoms are added to the substrate surface. They are usually controlled by lowering the growth rate; in particular by reducing the flux of the gallium reactant species.

Another common type of defect is the *growth pyramid* [58]. This is due to twinning, and is often initiated at the substrate surface, at the beginning of epitaxial growth. Pyramids also tend to have ill-defined sides, because of partial dissolution due to excess gallium. An in situ etch, prior to growth, greatly reduces their incidence.

Whisker formation is also encountered if an excessive amount of the gallium species is present. These whiskers are usually observed when growth is attempted at very low temperatures.

5.4 LIQUID-PHASE EPITAXY (LPE)

Liquid-phase epitaxy (LPE) involves the growth of epitaxial layers on crystalline substrates by direct precipitation from the liquid phase [59]. Although many applications of this technique have been described in the literature, the widest use of LPE has been in the growth of compound semiconductors such as gallium arsenide from solutions of gallium metal. The reason for its popularity is seen with reference to the phase diagram of Fig. 5.20. Here the conventional approach to growth of a crystal of composition B (or rather β) is direct freezing from a melt of this material, which is held at T_B. An alternate approach, often referred to as *solution growth*, is to begin with a melt of composition C_0, i.e., a solution of B and A, and to cool it from an initial temperature of T_0, which is well below T_B. This considerable reduction in temperature is an important advantage in the growth of gallium arsenide, where it is necessary to maintain an appropriate overpressure of arsenic to prevent its decomposition.

Boules of single-crystal material can be grown in this manner, by the insertion of a seed in the cooling solution. However, growth rates by this method are extremely slow compared to growth from the melt, so that the

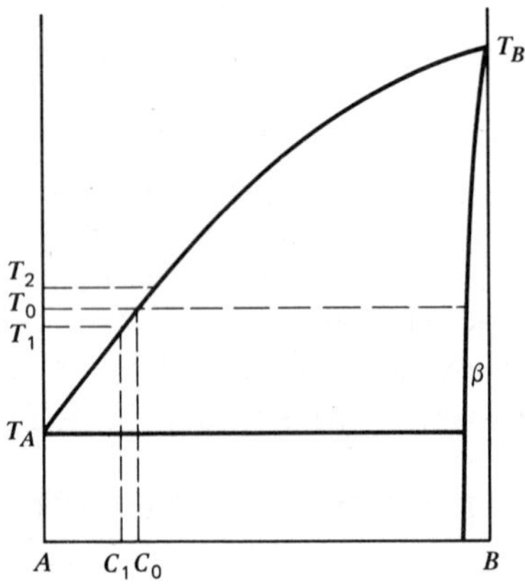

Fig. 5.20 Phase diagram illustrating liquid-phase epitaxy.

latter approach is usually preferred for bulk crystal growth. On the other hand, the method is ideally suited to the growth of thin epitaxial layers on single-crystal substrates, where a slow growth rate can be a distinct advantage.

A second advantage of operation at reduced temperatures is the ability to grow gallium arsenide with a near-stoichiometric composition. This can be achieved by operating in the 600–900°C range. In contrast, gallium arsenide which is grown from the melt is highly arsenic-rich (see Fig. 2.11), and is usually heavily doped because of contaminants which are incorporated into the accompanying excess of gallium vacancies.

Next, growth of the layer can be readily preceded by an in situ dissolution (or etch) step, so that the substrate–layer interface can be kept free of damage. This is an important requirement for the growth of high-quality layers. Finally, the distribution coefficients of most deep lying impurities are low compared to those for shallow impurities, as seen from Table 3.2. Consequently, epitaxial layers grown by LPE have lower deep level incorporation than those grown by VPE. They are thus at the present time the preferred choice for optical devices.

The disadvantages of LPE are primarily related to the difficulty of obtaining layers of uniform thickness, the difficulty of avoiding growth instabilities, and the necessity of wiping off the solvent after the layer is grown. Thus gallium inclusions, ridged and wavy surfaces, and incomplete surface coverage are typical problems encountered with this process. In addition, the technique is not suited for the growth of submicron layers, of the type required for high-speed field effect transistors and integrated circuits, although there has been some limited success along these lines [60]. VPE processes are more readily adapted to high-volume production techniques and to large area coverage at the present time.

LPE is not commonly used in silicon technology, since this semiconductor will tolerate high-temperature processing with relative ease. However, some new device applications [61], which require the low-temperature growth of silicon layers with minimal autodoping, have brought about an interest in this technique.

5.4.1 Choice of Solvent

It is necessary that the material to be grown be capable of dissolving in a solvent, and that this solution melt at a temperature that is well below the melting point of the semiconductor. A number of binary and pseudobinary* systems can be used for this purpose. However, the most suitable ones are those of the eutectic type which are free from compound formation.

Epitaxial layers, formed by freezing from the solution, will be saturated by the solvent. This is indicated as the β solution in Fig. 5.20. Ideally, this

* Systems such as A-GaAs fall into this category.

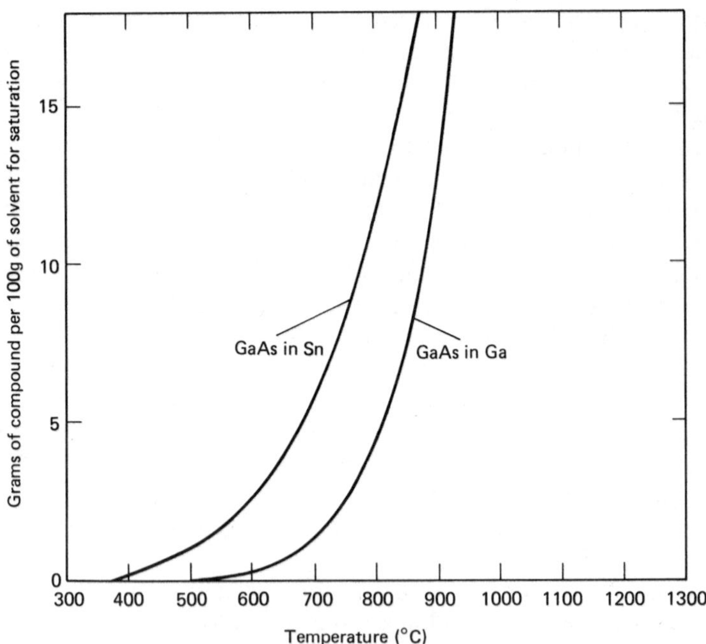

Fig. 5.21 Saturation curves for the Ga–GaAs and Sn–GaAs systems. Adapted from B. R. Pamplin, Ed., *Crystal Growth* [5], 1980.

requires that the solid solubility of the solvent in the grown crystal be very small, or that the solvent be an electrically inactive species. Examples of systems with active solvents are Ga-Si, Al-Si, Ge-GaAs (p-type), and Sn-GaAs (n-type). Inactive systems for silicon* include Sn-Si. Inactive systems for gallium arsenide include Pb-GaAs and Ga-GaAs. Of these, the latter is by far the most common, since the solvent is one of the components of this compound semiconductor.

The actual choice of solvent also depends on such factors as its availability in highly pure form, its vapor pressure at growth temperature, and its reactivity with the crucible in which the LPE growth is performed. Both tin and gallium are near-ideal solvents in these respects. Tin is the preferred inactive solvent for silicon, whereas gallium is almost universally used with gallium arsenide. In addition, gallium has the unique advantage of acting as a getter for many deep lying impurities in the solution.

The Ga-GaAs system can be considered as one in which gallium is the solvent and gallium arsenide is the solute. An alternate viewpoint, that gallium is the solvent and arsenic is the solute, is sometimes used for de-

* Systems such as In-Si, Au-Si, and Ag-Si are often considered inactive, even though the resulting silicon is doped to its solid-solubility limit with these deep impurities. This is because the concentration of these dopants at growth temperatures (600–900°C) is negligible.

scribing the LPE of gallium arsenide. For any given temperature, there is a specific amount of arsenic that can saturate the melt, to result in an equilibrium solution at that temperature. Typically, this is on the order of a few mole %. Figure 5.21 shows [5] the amount of gallium arsenide that must be added to 100 g of gallium for saturation at any given temperature. Also shown in this figure are comparable data for the Sn-GaAs system. A solubility diagram for the Sn-Si system [62] is shown in Fig. 5.22.

5.4.2 Nucleation

A number of different approaches can be used for LPE. In all cases, however, the substrate is held in contact with the solution, and conditions are established so that this solution becomes supersaturated by the solvent. Once a critical supersaturation is achieved, the solute will freely nucleate, resulting in nuclei formation in the solution and on the crucible walls, in addition to the substrate. Thus it is desirable to maintain a thermal gradient in the system, so that the solute can nucleate in the region where its solubility is the lowest, i.e., the coldest region. Ideally, this is the immersed substrate, and growth proceeds by heterogeneous nucleation on it.

The most common technique for establishing these conditions is to place the substrate in equilibrium with a saturated solution, and then to cool the system so as to achieve a supersaturated condition. An alternate approach is to add additional solute to a saturated solution while keeping it at constant temperature. (By way of example, the addition of gallium to a saturated Sn-GaAs solution will decrease the solubility of arsenic, and tin-doped

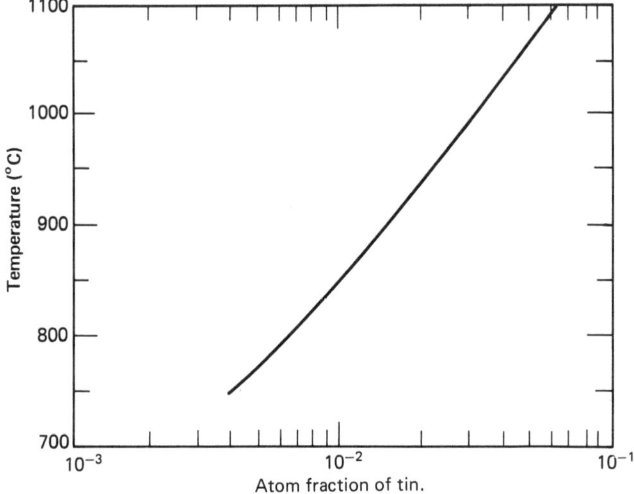

Fig. 5.22 Saturation curves for the Sn–Si system. Adapted from Thurmond and Kowalchik [62].

GaAs has been grown in this manner.) Techniques of this type are more complex than the use of temperature control, and are not commonly employed.

The initial stages of nucleation are quite similar to those described for VPE. Using the same line of argument, embryos which achieve a critical radius will lower their energy by growing larger, until eventually the nuclei coalesce to form an atomic layer. The probability of formation of a critical nucleus is given by Eq. (5.11)) as

$$n_{\text{crit}} = n_0\, e^{-\Delta G_{\text{crit}}/kT} \tag{5.11}$$

where ΔG_{crit} is the free energy change required for heterogeneous nucleation. With VPE, the addition of atoms to the embryo is relatively rapid, since they come directly from the vapor phase. With LPE, on the other hand, the process is limited by diffusion through the solute. For gallium arsenide, grown in an arsenic–gallium solution, the diffusivity of arsenic through gallium is the important controlling factor [63]. For silicon, grown from a tin–silicon solution, we must consider the diffusion of silicon in tin. Thus the rate of addition of atoms into the nucleus is reduced by $e^{-E_d/kT}$, where E_d is the activation energy for diffusion. As a consequence, the nucleation rate is given by

$$R_n = Ke^{-(\Delta G_{\text{crit}} + E_d)/kT} \tag{5.36}$$

where K is a proportionality factor.

This behavior differs from that of VPE in two respects. First, since nucleation occurs close to the melting point, $\Delta G_v \rightarrow 0$ and the nucleation rate is slower. Next, the limiting effects of diffusion through the solvent further reduce this nucleation rate. As a result, it is considerably easier to obtain control of nucleation in LPE than in VPE.

The growth of a layer by LPE is analogous to the process of doping from a melt, described in Section 3.3. Consider here the situation of growth of B (β) from a solution of A and B. The phase diagram of Fig. 5.20 illustrates this situation. A melt with an initial composition C_0 is used. Assume that a substrate is inserted into the melt at temperature T_0 and is brought in equilibrium with it. The temperature of the system is now lowered to T_1 so that a solid (β) freezes out of the solution.* At the same time, the composition of the solution shifts to C_1. With specific reference to gallium arsenide, where Fig. 5.20 represents the Ga-GaAs phase diagram (A = Ga and B = GaAs), the process of cooling the arsenic-saturated solution of gallium causes the arsenic to precipitate in the form of gallium arsenide. At the same time, the solution becomes more gallium-rich, shifting its composition from

* Thus the whole operation takes place in a relatively small region of the phase diagram.

C_0 to C_1. Finally, since gallium arsenide exists as a compound in a very narrow stoichiometric range, the β solution is essentially this semiconductor.

5.4.3 Growth

Growth proceeds by the transport of material from the bulk solution to the edge of the boundary layer, through which it must diffuse before it can be adsorbed on the surface. Here it moves until it ultimately attaches to a kink in a step on the surface. Most often the process of diffusion to the edge of the boundary layer can be ignored since the melt volume is usually many times the volume of the grown layer. Diffusion through the boundary layer, on the other hand, is generally the limiting factor under most conditions of LPE growth.

Consider a gallium arsenide system, in order to illustrate this process. Here a solution of gallium with a few atomic percent arsenic (the solute) is used, so that growth proceeds by the diffusion of arsenic in gallium. In the 700–950°C range, the diffusion constant for this process is given by [64]

$$D = 8.6 \times 10^{-4} e^{-3240/kT} \quad \text{cm}^2/\text{s} \tag{5.37}$$

Let C_s be the concentration of the solute in the bulk solution, and C_e the equilibrium solute concentration at the surface of the substrate. Assuming a linear gradient through the boundary layer, of thickness δ, the flux of the solute is given by

$$j = \frac{D(C_s - C_e)}{\delta} \tag{5.38}$$

If ρ is the density of the grown layer, the growth rate is given by [65]

$$R = \frac{DC_e \sigma}{\rho\delta} \tag{5.39}$$

Here the supersaturation σ is defined as

$$\sigma = \frac{C_s - C_e}{C_e} \tag{5.40}$$

The actual growth strategy will determine the appropriate values of C_e and σ, since these are a function of the temperature of the solution and its supersaturation.

One basic approach for growing a LPE layer of gallium arsenide is to begin with a solution of arsenic in gallium, which is saturated at a temperature T_0. The temperature is now lowered by a few degrees to T_1 so that the melt becomes supersatured. Typically, $T_0 \simeq$ 5–20°C, although growth can be

achieved with much lower values of temperature differential. The sample, which is also at T_1, is now inserted into the melt where it is held indefinitely. Initially, growth occurs because of the supersaturation in the melt. During growth, however, the solution becomes depleted of arsenic. Thus the supersaturation, as well as the growth rate, fall with increasing time.

This technique, known as the *step-cooling method*, is well suited for the growth of layers with controlled thickness, as well as for extremely thin layers where thickness control is important. Growth rates are generally quite low ($\simeq$0.01 μm/min), and layers grown in this manner are extremely smooth.

A second technique is known as *equilibrium cooling*. Here the solution is initially saturated with arsenic at T_0, and has a composition C_0 which is in equilibrium with it. A substrate, also at T_0, is inserted into the solution, and the temperature of the system is lowered slowly. This drives the solution into a supersaturated state, from which it proceeds toward equilibrium by the loss of arsenic, i.e., by the growth of gallium arsenide. Growth will thus continue until the substrate is removed from the solution.

A detailed analysis of these processes must take into consideration [64] the gradients in composition, temperature, and density of the gallium solvent which arise during growth. In addition, the outflow of latent heat from the surface can affect the thermal conditions in the system. These and other factors have been considered elsewhere [65, 66]. Here an approximate analysis is undertaken [67, 68], after the following assumptions have been made:

1. The solution and the substrate are isothermal.
2. The liquid and the substrate are in equilibrium at the surface, so that the solute (arsenic) concentration is given by the liquidus curve.
3. The bulk concentration and the diffusion constant do not change during the growth. These approximations are reasonable, because of the large volume of the chamber, and the small temperature changes that are involved during LPE.

Figure 5.23 shows the concentration of the solute in the bulk C_0 and at the liquid–substrate interface. For the LPE of gallium arsenide, the distribution coefficient of arsenic is less than unity. Consequently, the growing layer rejects this component, so that $C_0 > C_1$, as shown here.

Diffusion of arsenic takes place through the boundary layer and results in growth at the surface. The magnitude of $C(0, t)$ depends on the growth process used.

The one-dimensional diffusion equation for this system is

$$\frac{\partial C}{\partial t} = D\frac{\partial^2 C}{\partial x^2} + R\frac{\partial C}{\partial x} \tag{5.41}$$

where R is the velocity of the moving interface. A reasonable approximation

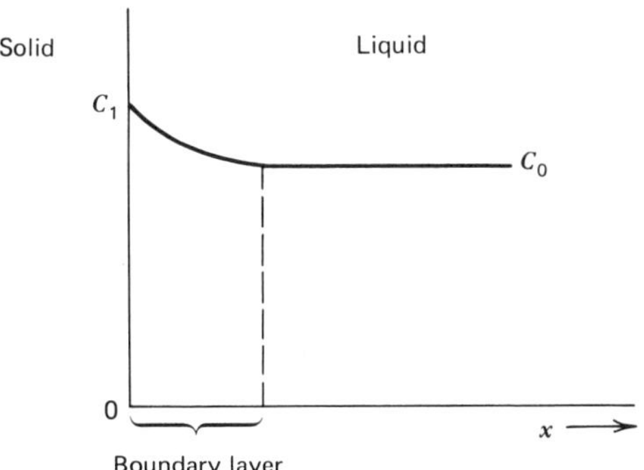

Fig. 5.23 Concentration of solute during liquid-phase epitaxy.

in a practical growth situation is to neglect the last term, so that

$$\frac{\partial C}{\partial t} \simeq D \frac{\partial^2 C}{\partial x^2} \tag{5.42}$$

The amount of solute that is transported through the boundary layer, and deposits on the substrate, is given by M_t per unit area, where

$$M_t = \int_0^t D \left(\frac{\partial C}{\partial x} \right)_{x=0} dt \tag{5.43}$$

Finally, the layer thickness d is given by

$$d = \frac{M_t}{C_s} \tag{5.44}$$

where C_s is the solute concentration in the grown layer ($=$ 0.5 for gallium arsenide). Application to the specific growth processes now follow.

Step Cooling. Here the boundary conditions are

$$C(x, 0) = C_0 \tag{5.45a}$$

$$C(0, t) = C_1 \tag{5.45b}$$

where C_0 and C_1 are shown in Fig. 5.20. For this situation, the amount of supercooling is given by $T_0 - T_1$. The change in temperature is related to the change in composition by the slope of the liquidus curve, $m = dT/dC$, around $T = T_0$.

The diffusion equation can be solved [69] subject to the boundary conditions of Eqs. (5.45a) and (5.45b) so that

$$\frac{C - C_1}{C_0 - C_1} = \text{erf}\left[\frac{x}{2(Dt)^{1/2}}\right] \tag{5.46}$$

Substituting in Eq. (5.43), gives

$$M_t = 2(C_0 - C_1)\left(\frac{Dt}{\pi}\right)^{1/2} \tag{5.47}$$

so that the layer thickness is

$$d = 2\,\frac{T_0 - T_1}{C_s m}\left(\frac{D}{\pi}\right)^{1/2} t^{1/2} \tag{5.48}$$

Thus the layer thickness is proportional to $t^{1/2}$. Moreover, the growth rate of the layer falls with increasing time, and is proportional to $t^{-1/2}$.

Equilibrium Cooling. Consider a cooling rate α, such that

$$C(0, t) = C_0 - \left(\frac{\alpha}{m}\right) t \tag{5.49a}$$

As before,

$$C(x, 0) = C_0 \tag{5.49b}$$

Solution of the diffusion equation, subject to these boundary conditions, gives

$$C(x, t) = C_0 - 4\left(\frac{\alpha t}{m}\right) i^2\,\text{erfc}\left[\frac{x}{2(Dt)^{1/2}}\right] \tag{5.50a}$$

where

$$i^2\,\text{erfc}\,x = \int_x^\infty \int_y^\infty \text{erfc}\,\xi\,d\xi\,dy \tag{5.50b}$$

Combining with Eqs. (5.43) and (5.44), gives the layer thickness as

$$d = \frac{4}{3}\left(\frac{\alpha}{C_s m}\right)\left(\frac{d}{\pi}\right)^{1/2} t^{3/2} \tag{5.51}$$

Here it is seen that the growth rate increases with time, and is proportional to $t^{1/2}$.

A number of different programming sequences have been employed, in an attempt to obtain improved layer morphology. Analysis of the growth rate for these sequences is conveniently done by computer simulation [70].

5.4.3.1 Constitutional Supercooling

The role of constitutional supercooling during crystal growth from the melt was described in Section 3.3.3. Here it was noted that for impurities with $k < 1$, the freezing solid rejects the dopant, resulting in its pileup at the solid–liquid interface. Too great a pileup, as might result from rapid growth, could lead to the condition of constitutional supercooling, accompanied by spurious nucleation and polycrystalline growth. Growth could be stabilized, however, if the temperature gradient in the melt, i. e., the temperature increase with distance from the interface, was sufficiently large (as shown by *XB* in Fig. 3.9*b*) and positive.

A somewhat analogous situation occurs during LPE, except that here we are concerned with a pileup of arsenic (for the case of gallium arsenide growth) at the solid–liquid interface. Here, too, stable steady-state growth requires a sufficient positive temperature gradient to be maintained. Unlike crystal growth from the melt, however, the temperature gradient in an LPE system is often negative, i.e., the solution temperature can be lower than that of the substrate. As a result, the avoidance of constitutional supercooling in LPE is an extremely difficult task. In fact, at first glance it would appear to be impossible. This is not so, however, because of the transient nature of LPE growth.

Consider a gallium arsenide substrate that is immersed in a solution with a finite solute concentration. Initially, the concentration gradient is zero. As the melt temperature is lowered, however, growth of the layer is accompanied by a depletion of the arsenic which gives rise to a concentration gradient. Solutions of this transient process are available in the literature [71], and are not presented here. However, it can be shown that the required temperature gradient must be continually increased during growth in order for stable conditions to be maintained. As the melt cools toward room temperature, however, temperature gradients in the system tend to zero. This latter stage of growth is generally accompanied by surface irregularities and gallium inclusions, which are characteristic of unstable conditions. In practice, therefore, it is customary to avoid this situation by growing the layer during a relatively small temperature change and to terminate growth by removal from the melt, accompanied by wiping.

5.4.4 In Situ Etching

The role of surface preparation has been emphasized previously. In particular, it has been pointed out that an in situ etch step, just prior to epitaxy, is extremely important for the growth of high-quality layers. This etch step can be readily incorporated into LPE systems, and will be described here as an extension of the equilibrium growth method. Referring again to Fig. 5.20, the solution is first saturated to T_0, and its temperature is raised by a small amount to T_2, so that it is now slightly *undersaturated*. Typically, about 10°C undersaturation is used in practice. The substrate, also at T_2, is inserted into the solution, and the temperature is lowered at a controlled rate. Etching takes place as the temperature falls from T_2 to T_0, with growth occurring below T_0. The etch rate for this situation falls off as T_0 is approached, so that the amount that is removed cannot be precisely controlled. However, this is usually not an important process parameter for most devices.

A modified approach can be used to obtain better control of the etched thickness. This consists of holding the solution temperature at T_2 for a fixed time, abruptly lowering it to the equilibrium temperature T_0, and then following with a slow cool for the growth phase. While this approach terminates the etch abruptly, it has the disadvantages of being more complicated, and also of subjecting the system to a sudden temperature drop which can possibly lead to growth instabilities. As a result, the slow, steady cooling technique is generally preferred.

Liquid gallium is a known getter for metallic impurities which may be present in the substrate. Thus an additional advantage of the in-situ etch in LPE comes about because of the gettering action of the gallium solution. This tends to remove deep lying impurities from the substrate just prior to layer growth, and results in a layer–substrate interface that is relatively free of deep levels. This is of prime importance in many electrooptical devices.

5.4.5 Doping

As indicated earlier, it is possible to grow highly doped semiconductor layers by using a solvent that is active. (By way of example, n^+-GaAs layers can be grown in an Sn-GaAs system.) The general requirement, however, is for an intentionally doped layer of finite carrier concentration. This is commonly done by adding a small amount of dopant to the solvent.

Dopants which have been used successfully for gallium arsenide are those belonging to group II (cadmium and zinc) which are *p*-type, and to group VI (sulfur, selenium, and tellurium) which are *n*-type. These dopants generally have a distribution coefficient which is below unity (see Table 3.2) so that the layer becomes more heavily doped as growth proceeds. Dopant enrichment of the melt is usually done in the vapor phase, since these impurities are relatively volatile.*

Elements from group IV can be added directly to the melt for doping purposes [72]. These include tin, germanium, and silicon. These dopants

* An alternate approach, which is commonly used, is to saturate the gallium solution with doped gallium arsenide.

can occupy gallium sites where they behave as donors, as well as arsenic sites where they are acceptor-like. However, the relative attachment to gallium and arsenic sites at the surface is strongly dependent on the nature of impurity–point defect interactions. Unlike VPE however, the incorporation ratio of silicon into gallium and arsenic sites is close to unity. Thus growth at low temperatures (600–700°C) in a silicon-doped solution results in *p*-type material, whereas growth in the 900°C range produces *n*-type layers. Moreover, the actual crossover point is a function of temperature, silicon concentration, and also crystal orientation. Advantage has been taken of this amphoteric doping behavior for the growth of successive *n*- and *p*-layers from a single solution.

Amphoteric behavior has not been observed for other group IV impurities, such as tin or germanium. Tin-doped layers grown by LPE are invariably *n*-type, whereas germanium-doped layers are *p*-type. Interestingly, germanium-doped layers grown by VPE are always *n*-type. The reasons for these differences are not known; in fact this behavior only serves to emphasize the complex nature of impurity–defect interactions at the growing surface.

The growth of semi-insulating, Cr-doped gallium arsenide presents a number of special problems, because of the extreme affinity of this dopant for oxygen. In recent work [73], however, careful attention has been paid to avoiding this problem by subjecting the Cr-doped solvent to a long heat treatment in hydrogen gas prior to growth.

The semi-insulating properties of chromium come about because this dopant compensates residual *n*-type impurities in gallium arsenide. These are invariably present in melt-grown material, due to the high temperatures involved here. This is not true for LPE, however, and it is possible to obtain *p*-type layers of low resistivity, unless the solution is intentionally doped *n*-type (with tin, for example).

Chromium is presumed to occupy gallium sites in the gallium arsenide lattice. In VPE growth, the incorporation of this dopant can be increased by raising the arsenic overpressure, since this increases the gallium vacancy concentration.* The same result can be achieved by increasing the cooling rate in LPE. This has the effect of causing a local arsenic supersaturation at the solid–liquid interface, and a corresponding reduction in the arsenic vacancy concentration.

Gallium arsenide of very high purity can also be grown by LPE, if careful attention is paid to the preparation of the solvent, and to the quality and cleanliness of the equipment. Here the solvent is often formed by saturating ultrapure gallium ($\geq$ 99.9999% purity) with arsenic by passing vapors of $AsCl_3$ over it, in much the same way as was done in the halide process described in Section 5.3.3. This allows the use of ultrapure arsenic, since $AsCl_3$ is also available in purity levels that are comparable to gallium. Doping concentrations in the (1–2) $\times$ 10^{12} cm^{-3} range can be achieved in this way.

* Note that $[V_{As}][V_{Ga}]$ is constant for any given temperature.

5.5 LPE SYSTEMS

The basic LPE process consists of placing the substrate in a solution, programming the system's temperature–time characteristics, removing the substrate, and wiping off the excess solvent. A great variety of systems have been used [5] for carrying out these operations; these include schemes involving tilting furnaces, rotating substrates, and so on. In addition, some systems involve stirring of the solution while others do not.

Early systems [74] utilized a furnace tube which was tilted so that the substrate could be inserted into the solution (and withdrawn from it) by the action of gravity. These systems were difficult to operate and control, and are rarely used today. Systems involving substrate rotation for this purpose are also occasionally encountered. They reduce some of the control problems of the tilting systems, but are relatively complex and prone to malfunction.

Both vertical dipping systems and horizonal sliding systems are in common use today. Most new systems are of the horizontal slide type because of their greater flexibility and convenience. However, dipping systems are especially suited for the rapid growth of thick layers, and so are in continued use.

This section describes the important features of LPE systems. Next, both dipping and sliding systems are considered in some detail, together with their operational advantages and disadvantages.

5.5.1 Hydrodynamic Considerations

The main problems of LPE systems are nonuniformity of layer thickness, and surface perturbations in the form of facets, ridges, and scallops. Some of these nonuniformities are inherent in the growth process. However, the main cause of growth irregularities is uncontrolled fluid flow in these systems [75]. Thus stirring, and systems which stimulate stirring during growth, are not generally favored because they produce unsteady streamlines across the surface. These encourage perturbations and instabilities, some of which can grow as the layer is formed.

Consider an LPE system, using a gallium–arsenic solution in a heated crucible. Here the growth of gallium arsenide is to be accomplished by dipping a substrate vertically into this solution. This growth process causes the formation of a vertical boundary layer which is depleted of arsenic. This layer is more dense* than the surrounding solution, so that it sinks to the bottom. This alters the solutal gradient across the face of the layer as the falling solution becomes increasingly depleted of arsenic, resulting in a wedge-shaped epitaxial growth, which is thinner at the bottom than at the top. In addition, circulating convection currents are created by this process,

* The density of arsenic is 5.727, whereas that of gallium is 6.095.

adding to those already present in the crucible due to the external heat source, and thus leading to further growth nonuniformities.

Many convection-induced problems can be avoided in systems where the substrate is horizontal. Hydrodynamic considerations are important here, also. Thus the substrate can be placed *above* the melt as shown in Fig. 5.24*a*, or *below* it as in Fig. 5.24*b*. Consider first the case of Fig. 5.24*a* for the gallium–arsenic system. Here, as growth proceeds, the boundary layer below the substrate becomes depleted of arsenic and thus more dense. This tends to sink to the bottom, setting up a fluid flow which augments the effects of natural convection. In the system of Fig. 5.24*b*, on the other hand, the dense boundary layer is formed above the substrate, and tends to remain there. Hence it acts to stabilize the growth process.

In some LPE schemes, a source wafer of gallium arsenide, located as shown in Fig. 5.24*c*, is also used. Here, following the above arguments, the region below the source is unstable, whereas that above the substrate is stable. One technique for stablilizing these systems is to apply a temperature gradient so that the source is at a higher temperature than the substrate. This procedure is referred to as *temperature gradient zone melting* (TGZM) and has been the subject of much work [76].

5.5.2 Vertical Systems

These consist of a crucible, usually of high-purity graphite or alumina, mounted in a vertical furnace. The system is enclosed in a quartz chamber with provision for flushing with hydrogen gas. The substrate is held in a quartz or graphite holder attached to the end of a rod, which passes through the top cap. This rod is capable of both rotation and sliding while maintaining hermetic integrity. In some research systems, magnetic support and rotation are used to make the system completely leak-tight. In operation, the substrate is lowered into the solution and its temperature adjusted in a controlled manner. Often substrate rotation is carried out during this process. Growth is terminated by raising the substrate out of the solution.

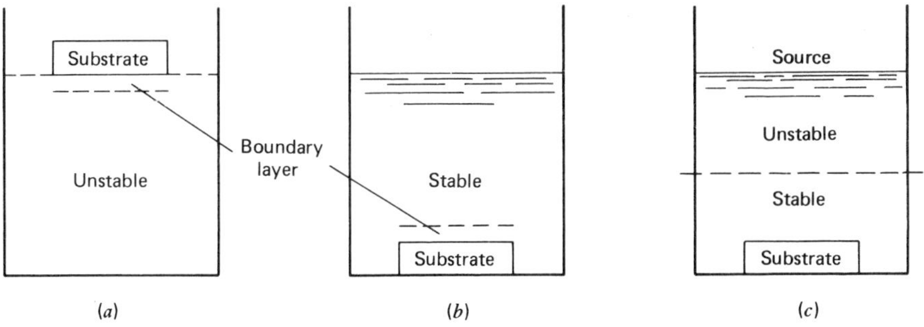

Fig. 5.24 Different horizontal growth considerations.

As has been pointed out, convection currents play an important role in these systems, and lead to tapered growth across the substrate. Moreover, the action of these currents greatly reduces the boundary layer thickness, with a subsequent increase in the growth rate. Typically, for the same cooling rate, films will grow six to eight times faster in vertical systems than in horizontal systems. This feature makes them highly suited for the growth of thick layers, where a slight taper is acceptable. As a result a variety of sophisticated systems have been developed in order to extend the versatility of this technique. These include provisions for source wafers, multiple substrates, and substrate wiping. Elaborate heat shields are also used to minimize thermal gradients, and considerable care has been expended on process optimization [77].

Vertical dipping systems have also been used for the growth of silicon films which must be relatively thick, and have minimal autodoping [61]. Here tin is the usual solvent, although electronically active materials such as gallium and aluminum have also been used [78], to result in heavily doped p^+-layers.

Layers using tin have been grown in the 900–1000°C range. Both doping and saturation of the melt with silicon can be conveniently achieved by the use of doped silicon wafers, which are allowed to saturate the melt at the operating temperature. In one application [79], a step cooling technique was used with a growth temperature of 949°C, and supersaturations of 10–30°C. This technique was found to yield smooth ripple-free layers. Growth under equilibrium cooling conditions, on the other hand, resulted in poor morphology.

5.5.3 Horizontal Systems

By far the most popular LPE systems use horizontal substrates and are of the sliding boat type. These systems are relatively free of morphology problems, which are caused by uncontrolled convection currents. Moreover, their growth rate is much lower than that of dipping systems, so they are better suited for the fabrication of thin layers of the type required in microwave and electrooptical applications.

A variety of such systems are in use; here we describe two schemes in order to outline their basic operational principles. Systems for gallium arsenide are considered for specificity.

Figure 5.25 shows the simplest apparatus for this purpose [5]. Here a well is machined in a high-purity graphite block, and serves to hold the solution. The temperature is monitored by a thermocouple embedded in the block (not shown). A graphite slide is used as a substrate holder, in an arrangement by which it can be moved so as to be located under the well. The entire system is placed in a furnace, in a neutral carrier gas ambient. Commonly hydrogen gas, passed through a palladium purifier, is used because of its availability in highly pure form.

In operation, the system is brought up to temperature with the substrate

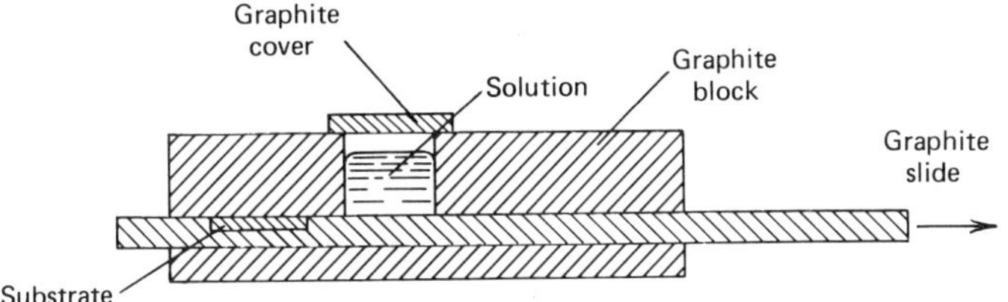

Fig. 5.25 Sliding boat apparatus for LPE.

covered by part of the graphite block. At this point, a push rod is used to position the substrate under the melt and the furnace temperature changed in a programmed manner. A linear reduction of temperature with time is commonly used, with temperature gradients from 1°C/min to 1°C/h. The slide is moved out from under the solution in order to terminate the growth. This also provides a wiping action to remove the solvent, so that often no further substrate cleanup is necessary. The system is now cooled, and the substrate removed. Expansion of the system to multiple slices is accomplished by providing a series of wells in a common graphite block, together with a slide having a corresponding series of indentations for holding the substrates.

There are many modifications of this basic system. Thus, Fig. 5.26 shows an arrangement by which the solution can be saturated with arsenic prior to epitaxial growth [80]. This is done by first sliding a "source" wafer of gallium arsenide under the well; partial dissolution of this wafer is a convenient way of incorporating arsenic into the solution. An important advantage of this approach is that the growth solution can be held in local thermodynamic equilibrium in the vicinity of the eventual growth interface. This results in a high degree of reproducibility in the fabrication of thin LPE layers. This approach can be readily extended to allow the successive growth of multiple layers by the use of multiple source wafers. Systems have been designed to grow more than six different sequential layers in this manner.

Yet another modification consists of placing a source wafer on top of the melt. This is usually held firmly against the solution by means of a graphite block above it, and allows the possibility of continual melt saturation during layer growth.

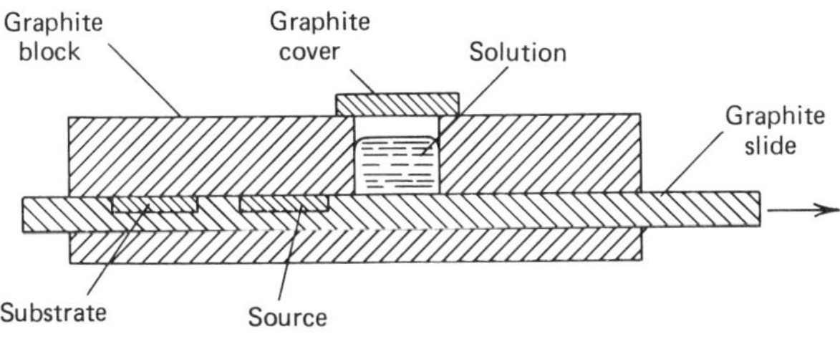

Fig. 5.26 Sliding boat apparatus with arrangement for presaturation with arsenic.

In summary, therefore, the sliding technique, using horizontal substrates, is the most versatile system for LPE. It is extremely flexible and permits the growth of multiple layers in a single operation. Presaturation of the melt just prior to layer growth can be readily accomplished without the necessity of opening up the system, thus avoiding unnecessary contamination problems, and giving a high degree of control of layer thickness. Finally, it can be readily expanded to handle many substrates at a given time.

There are a number of problems associated with systems of this type. First, it is necessary to operate the furnace with a large zone over which the temperature is held extremely constant. Furnace liners, based on the heat-pipe principle, are increasingly used to meet this requirement. Next, some means must be provided to hold the substrates and source slices firmly during movement of the slide. This is because gallium arsenide has a lower density than the solution, so that there is a tendency for the substrates to lift and obstruct movement of the slide. An additional limitation of this system is that the substrate and the solution are at the same temperature prior to contact. Thus some of the operational sequences which are available with the dipping technique are not possible here. However, these disadvantages are minor compared to the many advantages of these systems.

5.5.4 Growth Imperfections

Many different types of growth imperfections can occur during LPE. Some of these are due to the properties of the substrate material, and to imperfections in it. However, the major problems arise from hydrodynamic and thermal conditions during this process.

The nature of any substrate, no matter how carefully prepared, is that it is not flat. On a microscopic scale, a number of different faces are exposed to the solution during growth. In general, the free energy of these faces is a function of the orientation, with the {111} faces having a minimum free energy. During growth, the addition of extra atoms on these faces results in a large change in this energy. Generally, this causes clustering of atoms on these faces, and leads to growth irregularities such as faceting.

Imperfections in the substrate can act as sites for spurious nucleation during early stages of growth. Dopants such as silicon and chromium are particularly bad, since they tend to form inclusions which result in large spiral growth patterns. It goes without saying that substrate selection and surface preparation are particularly important for LPE. In situ etching is a necessity if uniformly flat surfaces are required.

It has been shown that improved morphology can be obtained when there are a large number of nucleation sites on the surface. Thus although it is possible to obtain growth with a supersaturation of as little as 0.25°C, the use of larger values (5–20°C) results in smoother epitaxial layers.

Many problems of layer growth come about from turbulences in the fluid flow, and from the unstable nature of the growth process. These can give rise to growing waves and ripples in the layer surface, when the system is subjected to arbitrarily small mechanical or thermal perturbations. These problems are closely allied to those caused by constitutional supercooling, and are especially severe under conditions of high growth rate. Techniques involving the use of electrodes to control the electric potential gradient at the substrate–solution interface have been considered for solving these problems, but have achieved limited success to date.

In summary, therefore the growth of thin layers by LPE, with a high degree of planarity, is an extremely difficult problem. Unfortunately, this is an important requirement for high-speed VLSI applications. Consequently, VPE is the most favored approach at the present time.

5.6 HETEROEPITAXY

This section concerns itself with two examples of heteroepitaxy, both of which have established advantages for integrated electronics. These are the growth of silicon on sapphire (SOS), and of aluminum gallium arsenide on gallium arsenide (AlGaAs-GaAs). A very brief treatment follows.

The use of silicon films on sapphire allows the formation of MOS transistors with greatly reduced source and drain capacitances. A further advantage is that parasitic latch-up in C-MOS circuits can be avoided by use of this material. Consequently, SOS technology has the potential for high-speed MOS circuits.

The heteroepitaxy of aluminum arsenide on gallium arsenide is unique in that there is an almost exact lattice match between GaAs (5.654 Å) and AlAs (5.661 Å). Consequently, AlAs can be grown on gallium arsenide with a minimum interface dislocation density. Of itself, AlAs is unstable in air; however, $Al_xGa_{1-x}As$ is quite stable for $x \leq 0.8$, so that this ternary alloy is of more practical value. Furthermore, since the tetrahedral radii of aluminum and gallium are nearly identical, a wide spectrum of compounds of the type $Al_xGa_{1-x}As$ can be grown on gallium arsenide with almost no lattice mismatch.

Aluminum arsenide is an indirect gap semiconductor, with a direct gap of 3.1 eV and an indirect gap of 1.26 eV. The admixture of gallium arsenide results in a continuous spectrum of materials, as shown in Fig. 5.27. The value of $x \simeq 0.34$ corresponds to the transition from direct to indirect gap [81], at an energy gap of 1.92 eV. As a result, an appropriate choice of composition allows this material to be used in optical structures as either a window or a light-emitting material. Combined with gallium arsenide, its potential for use in optical integrated circuits is extremely high.

5.6.1 Silicon on Sapphire

The choice of sapphire as an insulating substrate is largely based on its mechanical and chemical stability at processing temperatures, and on its availability in high-grade commercial form. The lattice match between sap-

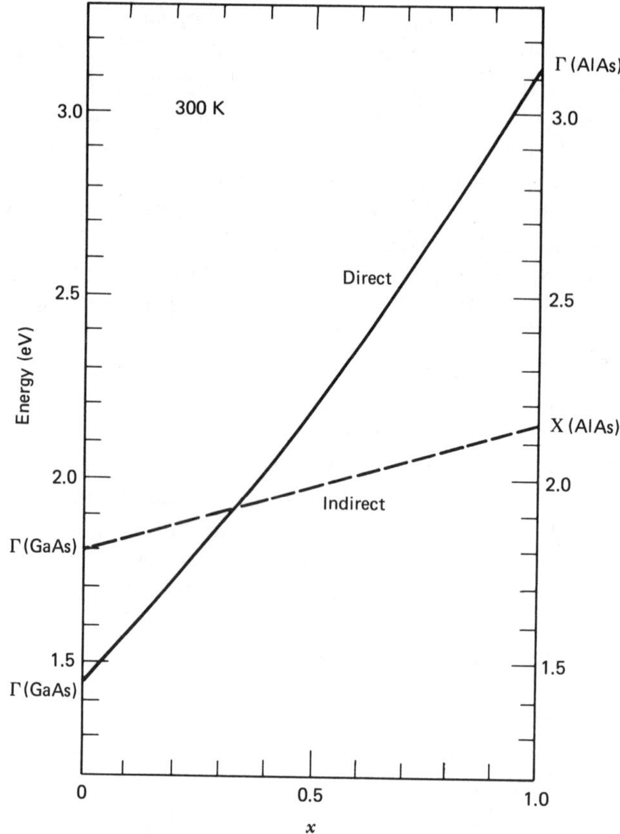

Fig. 5.27 The bandgap of $Al_xGa_{1-x}As$. Adapted from Kressel [81].

phire and silicon is not good on any plane. Nevertheless, long-range order appears to allow reasonable epitaxial growth to be carried out. It is believed that this is achieved by the occupation of aluminum sites by silicon atoms, which are bonded to the oxygen atoms of the substrate. Of greater importance, however, is the built-in stress in the grown films, resulting from differences in the thermal expansion coefficient (about 6×10^{-6} °C^{-1} for sapphire, compared to 3×10^{-6} °C^{-1} for silicon). This produces a compressive stress in the films, so that they are highly defected, and have extremely short minority carrier lifetimes (in the nanosecond region). This is of little consequence for MOS devices. However, the use of SOS films in bipolar transistors and microcircuits is precluded because of this fact.

Many studies have been made of the orientation relationships of silicon and sapphire [1]. From these it is concluded that (111) silicon is closest matched to $(11\bar{2}0)$ sapphire, whereas (100) silicon is best matched to $(01\bar{1}2)$ sapphire. This latter combination is often used for (100)-based MOS microcircuits made on this material. Some work has also been done [82] with

$MgO:Al_2O_3$ (spinel) material, which has the advantage that it is cubic. However, it is not available at the commercial level (and cost) of sapphire substrates. Moreover, the thermal expansion mismatch to silicon is as serious as for the silicon–sapphire system.

Both silicon tetrachloride as well as silane processes have been successfully used for the deposition of silicon on sapphire. Of these, the silane process is preferred because it can be carried out at lower temperatures, thereby minimizing aluminum doping from the substrate. In addition, this process avoids the resultant halogenic etching of Al_2O_3, which occurs at high temperatures and leads to heavily *p*-type layers.

In-situ cleaning of substrates, prior to epitaxy, is an important step in the growth process. Typically, this is done by firing the substrates in hydrogen at 1200–1500°C, for a period of 15–30 min. The silicon growth temperature is a compromise between crystal quality and doping. In general, films grown at higher temperature have better crystal morphology, but are more highly doped with aluminum from the substrate. In addition, silicon growth on sapphire is very sensitive to the presence of even trace amounts of oxygen [83]. Typically, growth in the 950°C range yields films with the highest mobility.

An electron mobility of about 500 cm^2/V s is the best that can be obtained in thin films (0.5–1 μm) of the type used for MOS-based circuits. This disadvantage is outweighed by the fact that extremely small structures, with minimal parasitic capacitance, can be made using this material. Thus it is an attractive candidate for high-speed, densely packed microcircuits.

5.6.2 Aluminum Gallium Arsenide on Gallium Arsenide

Films of AlGaAs are especially useful in optical devices and circuits. Photoluminescent response and optical efficiency are two parameters of great importance in the growth of this material. Both VPE and LPE can be used here. Each process has its special advantages as well as problems, and is now described briefly.

5.6.2.1 VPE Growth

Aluminum gallium arsenide has been grown by an extension of the hydride process described in Section 5.3.4. Here both aluminum and gallium metals were used with transport provided by means of hydrogen chloride. Arsine, together with hydrogen as a carrier gas, were injected so that they entered the deposition zone ahead of the group III reactants.

There are some serious problems associated with this process, one of which is caused by the high reactivity of aluminum chloride with quartz. All-graphite or all-alumina systems must be used to avoid this problem. In addition, the thermodynamics of the gallium arsenide and the aluminum arsenide formation reactions are quite different. Ideally, the formation of

AlAs requires growth in the 1000°C range or higher. On the other hand, gallium arsenide etches very rapidly at these temperatures. Consequently, provision must be made for rapidly moving the substrate into the hot zone just prior to epitaxial growth. As a result of these problems, the hydride process is not practical.

The use of organometallics circumvents many of the problems associated with the chloride process. Here trimethylgallium or triethylgallium, and trimethylaluminum (TMA) can be used as the gallium and aluminum source, respectively, with arsine gas in a hydrogen ambient. Growth is carried out in a cold wall reactor system with a substrate temperature of about 700°C, and excellent morphology can be obtained over the entire useful composition range by this process.

Films of $Ga_xAl_{1-x}As$ have been grown by this process [84] at temperatures of 700–750°C. The highest photoluminescent efficiencies have been obtained for growth temperatures of 750°C and gas-phase arsenic-to-gallium ratios of 5–10. It has been shown that photoluminescent efficiencies comparable to those obtained by LPE can be achieved, if care is taken to remove all traces of oxygen and moisture from the incoming gases.

5.6.2.2 LPE Growth

This is relatively straightforward, and is carried out in the manner described for the LPE growth of gallium arsenide. Multiple-well, horizontal slide structures are commonly used [80, 81], since they allow the successive growth of gallium arsenide and aluminum gallium arsenide layers which are often required for optical circuits.

The growth of AlGaAs layers presents an additional degree of complexity in that, in addition to control of doping level, control of the composition is necessary over the layer thickness, since the energy gap is dependent on this parameter. Moreover, the segregation coefficients of aluminum and gallium are quite dissimilar, so that provision must be made to supply extra aluminum during layer growth, to avoid its depletion. Source wafers of AlGaAs, in conjunction with a temperature gradient, are often used in these systems.

Dopants for AlGaAs are similar to those used for gallium arsenide and all have been used successfully. These include germanium, silicon, tin, tellurium, and zinc. As with gallium arsenide, silicon can be used to grow both *p*- and *n*-type layers, depending on the growth temperature and on the composition. Here, too, *n*-type layers are grown in the 900°C range, with *p*-type at lower temperatures.

The LPE growth of AlGaAs carries with it the same problems that were described for gallium arsenide, in addition to the added problem of composition nonuniformity. These include control of surface flatness and layer thickness, and the incomplete removal of solution from the surface. On the other hand, it is relatively more easy to obtain high photoluminescent efficiencies with LPE than with VPE, since growth takes place in a (primarily) liquid gallium medium which has excellent impurity gettering properties.

5.7 THE EVALUATION OF EPITAXIAL LAYERS

The epitaxial growth of a suitable doped, single-crystal layer forms the starting point for many commonly used microcircuit fabrication techniques. Since the entire microcircuit is eventually fabricated within this layer, it is important that its quality be assessed for this purpose. To do so, it is necessary to evaluate such layer parameters as mobility, carrier concentration, doping profile, and thickness.

The quality of epitaxial silicon is sufficiently good so that the carrier concentration can be determined by measuring the sheet resistance and the junction depth, and assuming an appropriate value of mobility for the resulting concentration. With gallium arsenide, however, a wide range of compensation ratios is encountered. Thus a separate assessment is necessary for the mobility and the carrier concentration.

5.7.1 Sheet Resistance

In silicon microcircuits, a layer of n-type material is usually grown on a p-type substrate. This layer may be electrically characterized in terms of its sheet resistance, as described in Section 4.10.2. Consider a rectangle of the layer, of length l and width w. If its resistance is measured along its length, then

$$R = R_S \frac{l}{w} \tag{5.52}$$

where R_S is defined as the *sheet resistance* of the layer, in ohms.

The sheet resistance of such a layer may be measured by the four-point probe configuration shown in Fig. 5.28*a*. Here four equally spaced collinear probes are placed on the layer. A current I is passed through the outer probes, and the voltage V developed across the inner probes is measured. The value of current is so chosen that V and I are linearly interrelated. Alternating current techniques are sometimes used to facilitate instrumentation.

Since the epitaxial layer is of opposite impurity type to its substrate, it may be assumed that the current flow is restricted within it. Consider a layer of infinite dimensions compared to the probe spacing. For the configuration of Fig. 5.28*b* comprising a positive source and a negative source of current, the potential at any point P is given by

$$\psi_P = \frac{IR_S}{2\pi} \ln \frac{r_2}{r_1} + A \tag{5.53}$$

where r_1 and r_2 are the distances of the point P from the positive and negative source, respectively, and A is a constant of integration. For the four-point probe configuration of Fig. 5.28*c*,

$$\psi_1 = \frac{IR_S}{2\pi} \ln 2 + A \tag{5.54}$$

and

$$\psi_2 = \frac{-IR_S}{2\pi} \ln 2 + A \tag{5.55}$$

where ψ_1 and ψ_2 are the potentials at points 1 and 2, respectively. Thus

$$\psi_1 - \psi_2 = V = \frac{IR_S}{\pi} \ln 2 \tag{5.56}$$

Rearranging Eq. (5.56), gives

$$R_S = \left(\frac{\pi}{\ln 2}\right) \frac{V}{I} \tag{5.57a}$$

$$= 4.5324 \frac{V}{I} \tag{5.57b}$$

Thus the sheet resistance may be directly calculated for the V/I ratio.

Formulas for layers of finite dimensions are available in the literature. By way of example, for probes located centrally on a circular sample of finite diameter, the sheet resistance can be shown to be given by

$$R_S = C \frac{V}{I} \tag{5.58}$$

where C is a function of the probe spacing and the slice diameter, as shown in Fig. 5.29.

Four-point probe measurements are usually made on monitor slices that are placed in the reactor along with those on which the circuits are to be fabricated. This avoids the necessity for interpreting data on slices which already have diffused regions (buried layers) prior to epitaxial growth. These monitor slices are of the same starting material used for the microcircuit. In some situations, the epitaxial layer and the substrate are of the same conductivity type. Four-point probe measurements cannot be made on these layers, necessitating the use of monitor substrates of opposite conductivity type.

The tacit assumption made in taking four-point probe measurements is that the probe–semiconductor contact is highly conductive, and linear in its

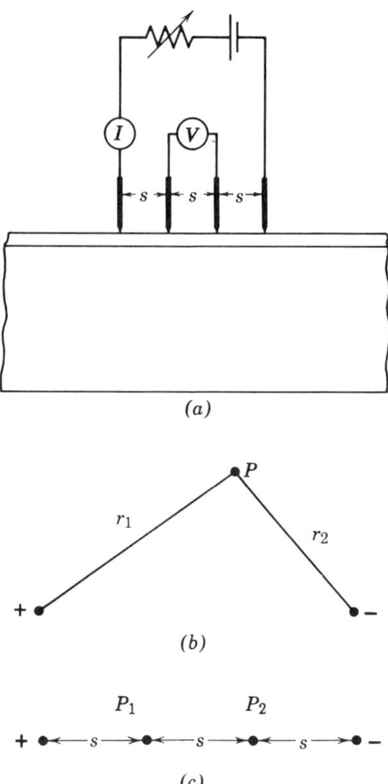

Fig. 5.28 Four-point probe.

V–I characteristic (i.e., ohmic). This assumption is only satisfactory for layers with a sheet resistivity under 10 Ω cm, and the taking of four-point probe data is more difficult beyond this point.

A number of sophisticated techniques have been developed for rapid measurement of film thickness and resistivity. These include the use of

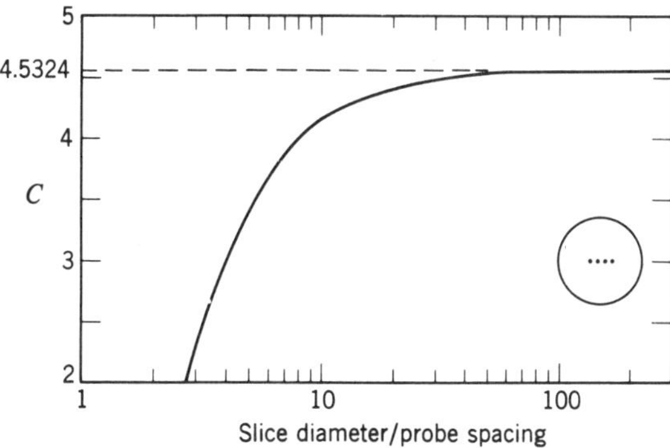

Fig. 5.29 Correction factor for circular slices.

microwave and infrared interferometry, as well as surface-acoustic-wave probing techniques. A description of these is beyond the scope of this book.

5.7.2 Mobility and Carrier Concentration

A measurement of resistivity and junction depth (see Section 4.10.1) is usually sufficient for determining the mobility and carrier concentration of silicon layers. This is because silicon is generally uncompensated. Curves for mobility versus carrier concentration and for resistivity versus carrier concentration in silicon have been given in Figs. 1.10 and 1.11, respectively.

Gallium arsenide is usually compensated. Here the mobility and the carrier concentration in an epitaxial layer may be separately determined by means of the Hall effect. This technique is applicable to layers which are not in communication with the substrate (e.g., n-layers on p-substrates, and vice versa). With gallium arsenide, both p- and n-layers can be evaluated if they are grown on semi-insulating substrates.

Consider a long, uniformly doped layer through which a current flows. A magnetic field is applied across the layer, at right angles to the direction of current flow, as shown in Fig. 5.30. For this condition, there exists an electric field E_H which is mutually perpendicular to both current and the magnetic field; the value of this electric field is given by

$$E_H = \frac{R_H IB}{A} \tag{5.59}$$

where R_H is the Hall coefficient, I the current, A the cross section, and B the magnetic induction. If V_H is the voltage across the bar, and t is its

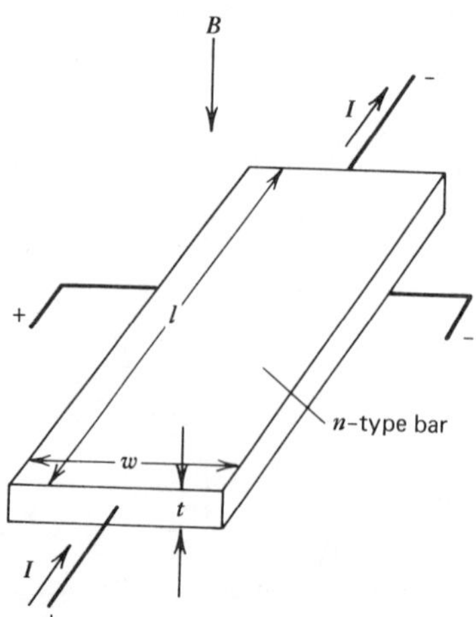

Fig. 5.30 Hall effect.

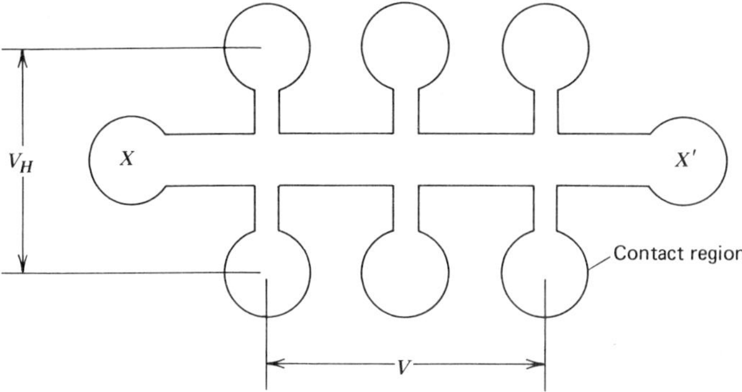

Fig. 5.31 Hall sample.

thickness, then

$$R_H = \frac{t}{IB} V_H \tag{5.60}$$

Finally, the Hall mobility μ_H is given by

$$\mu_H = \frac{R_H}{\rho} \tag{5.61}$$

where ρ is the resistivity of the bar.

Hall measurements are conveniently made on a photolithographically patterned bar structure of the type shown in Fig. 5.31. Here a constant current I is sent through terminals XX'; the other terminals allow a series of voltage measurements to be made on the same bar. These are subsequently averaged to minimize patterning errors. Additional data are taken with both current and field reversed to further smooth the measured data.

Resistivity values can also be obtained by passing a known current through the sample and measuring the voltage across outer pairs of side arms. A combination of these two measurements permits a separation of carrier concentration and mobility values.

Mobility values for gallium arsenide are usually measured at room temperature (300°K) and at liquid nitrogen temperature (77°K). The value of the compensation ratio can be extracted by the technique [86] outlined in Section 1.4.1.4.

Mobility values can also be determined by the technique developed by van der Pauw [87], which is applicable to arbitrarily shaped layers of uniform thickness and composition, with four-point contacts at locations on the sample periphery. This method is sensitive to the placement of these contacts, unless the sample shape is modified by cutting isolating regions between the

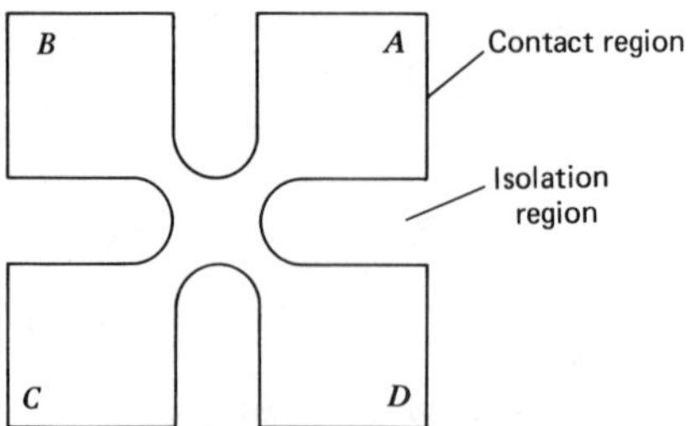

Fig. 5.32 Van der Pauw sample.

contacts to make a "clover-leaf" shape, as shown in Fig. 5.32. This is also done by patterning the epitaxial layer after it is grown on a substrate with which it has no electrical communication.

The specific resistivity of samples of this type can be determined by passing a known current through two adjacent terminals and measuring the voltage developed across the other two. Thus let $R_{AB,CD}$ be the potential difference across CD resulting from unit current flow between the contacts A and B. Then it can be shown that

$$\rho = \left(\frac{\pi t}{\ln 2}\right)\left(\frac{R_{AB,CD} + R_{BC,DA}}{2}\right) F \tag{5.62}$$

where t is the layer thickness, and F is plotted in Fig. 5.33.

The Hall mobility can be determined by passing a known current through a set of nonadjacent contacts, and measuring the voltage drop across the other pair (e.g., $R_{BD,AC}$), both with and without the application of a perpendicular magnetic field. Here

$$\mu_H = \frac{t}{B}\frac{\Delta R_{BD,AC}}{\rho} \tag{5.63}$$

where $\Delta R_{BD,AC}$ is the difference in these voltages.

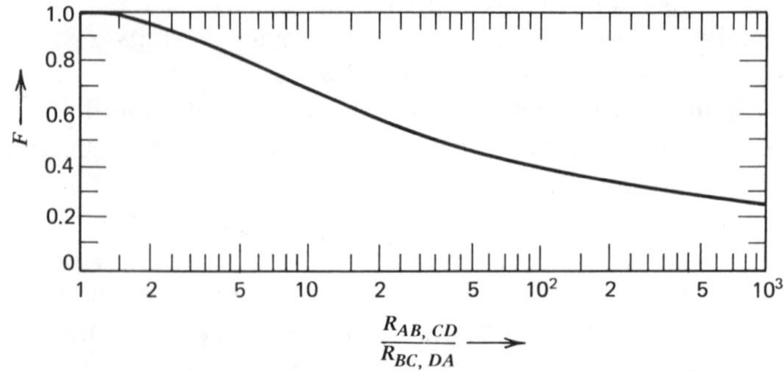

Fig. 5.33 Correction factor for determining specific resistivity. From van der Pauw [87]. With permission from the publisher, Philips Research Reports.

The van der Pauw technique is uniquely suited for small samples of arbitrary shape. Thus it is usually the preferred method for gallium arsenide and other compound semiconductors, where small samples are often used for reasons of economy or availability.

The assumption of uniform carrier concentration is not always true, especially near the substrate–epitaxial layer interface. This presents a serious problem in determining the mobility of very thin expitaxial layers. Techniques involve making a series of measurements as the layer is thinned,* and calculating the mobility and concentration gradient from the small differences in these measurements [88].

Highly conducting inclusions in a sample can often cause serious error in mobility measurements, and lead to artificially high values. Their absence can be ensured if data, taken at different values of magnetic field strength, result in the same values of mobility.

5.7.3 Impurity Profile

Direct techniques such as secondary ion mass spectrometry [89] can be used to measure the impurity profile in epitaxial layers. An alternate approach consists of forming a diode structure and determining the impurity profile of the epitaxial layer by measurements of its capacitance versus reverse voltage characteristic. This necessitates growth of the layer on a heavily doped substrate of the same type, i.e., p on p^+ or n on n^+. Next, a shallow junction is formed on the surface of the layer to obtain this structure. This technique is most readily applicable to the evaluation of n-layers on n^+-substrates. Here a Schottky diode structure can be formed by vacuum evaporation, thus avoiding a high-temperature diffusion step.

The principle behind this technique [90] is described with reference to Fig. 5.34, which shows a diode of this type, having an arbitrary doping profile. Here an increase of bias voltage from V to $V + \Delta V$ results in an increase in the electric field by $\Delta E = \Delta V/x$, and an increase in the depletion layer width from x to $x + \Delta x$. This uncovers an additional charge $qN(x)\,\Delta x$ such that

$$\Delta E = \frac{\Delta V}{x} = \frac{q}{\varepsilon\varepsilon_0} N(x)\,\Delta x \tag{5.64}$$

At this bias, the small-signal capacitance C is given by

$$C = \frac{\varepsilon\varepsilon_0 A}{x} \tag{5.65}$$

* This is done by chemical etching, or by depleting part of the sample by means of a Schottky diode structure.

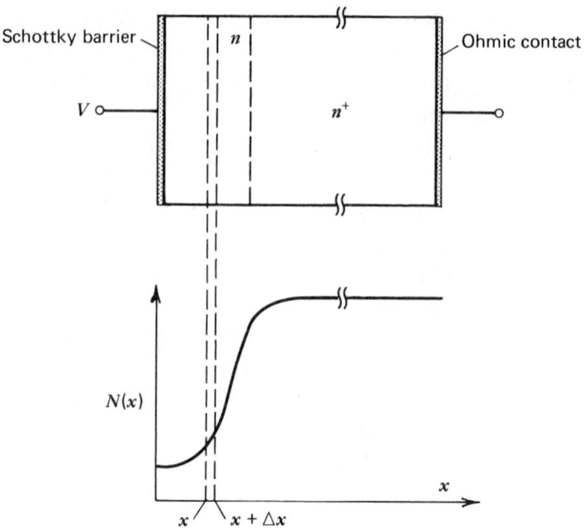

Fig. 5.34 Schottky diode structure.

Eliminating x and Δx from these equations, gives

$$N(x) = -\frac{C^3}{q\varepsilon\varepsilon_0 A^2}\left(\frac{\Delta V}{\Delta C}\right) \quad (5.66)$$

so that measurement of the C–V characteristic of the diode allows calculation of the impurity profile in the epitaxial layer.

Variations of this technique have been developed [91] to provide greater accuracy and/or speed in taking this measurement. Perhaps the most commonly used approach consists of a feedback technique [92] for obtaining a direct indication of this impurity distribution. Details of this and other schemes are outside the scope of this chapter.

REFERENCES

1. H. M. Manasevit, F. M. Erdmann, and A. C. Thorsen, The Preparation and Properties of (111) Si Films Grown on Sapphire by the SiH_4-H_2 Process, *J. Electrochem. Soc.* **123,** 52 (1976).
2. J. E. A. Mauritis, SOS Wafers: Some Comparisons to Silicon Wafers, *IEEE Trans. Electron Dev.* **ED-16,** 859 (1978).
3. K. Pande, D. Reep, A. Srivastava, S. Tiwari, J. M. Borrego, and S. K. Ghandhi, Device Quality Polycrystalline Gallium Arsenide on Germanium/Molybdenum Substrates, *J. Electrochem. Soc.* **126,** 300 (1979).
4. A. Y. Cho and J. R. Arthur, Molecular Beam Epitaxy, *Progr. in Solid State Chem.* **10,** 157 (1973).

5. B. R. Pamplin, *Crystal Growth*, Pergamon, New York, 1980.
6. T. U. M. S. Murthy, N. Miyamoto, M. Shimbo, and J. Nishizawa, Gas-Phase Nucleation During the Thermal Decomposition of Silane in Hydrogen, *J. Crystal Growth* **33,** 1 (1976).
7. R. W. Henke, *Introduction to Fluid Mechanics*, Addison-Wesley, Reading, MA, 1966.
8. P. S. Barna, *Fluid Mechanics for Engineers*, Butterworths, Washington, 1964.
9. C. H. J. van den Brekel, Characterization of the Chemical Vapor Deposition Process, *Philips Research Rep.* **32,** 118 (1977).
10. R. E. Reed-Hill, *Physical Metallurgy Principles*, Van Nostrand, New York, 1973.
11. C. H. L. Goodman, Ed., *Crystal Growth*, Vol. 1, Plenum, New York, 1974.
12. D. C. Gupta, Improved Methods of Depositing Vapor Phase Homoepitaxial Silicon, *Solid State Technol.* p. 33 (Oct. 1971).
13. F. C. Eversteyn, P. J. W. Severin, C. H. J. v. d. Brekel, and H. L. Peek, A Stagnant Layer Model for the Epitaxial Growth of Silicon from Silane in a Horizontal Reactor, *J. Electrochem. Soc.* **117,** 925 (1970).
14. C. W. Menke and L. F. Donaghey, Analysis of Transport Processes in Vertical Cylinder Epitaxy Reactors, *J. Electrochem. Soc.* **124,** 561 (1977).
15. E. Krullman and W. L. Engel, Low Pressure Silicon Epitaxy, IEEE Trans. Electron Dev., **ED-29,** 491 (1982).
16. M. J-P. Duchemin, M. M. Bonnet, and M. F. Koelsch, Kinetics of Silicon Growth Under Low Hydrogen Pressure, *J. Electrochem. Soc.* **125,** 637 (1978).
17. V. S. Ban and S. L. Gilbert, Chemical Processes in Vapor Deposition of Silicon, *J. Electrochem. Soc.* **122,** 1382 (1975).
18. M. E. Jones and D. W. Shaw, Growth from the Vapor, from *Treatise on Solid State Chemistry*, Vol. 5, N. B. Hannay, Ed., Plenum, New York, 1976.
19. L. J. Stinson et al., Sulfur Hexafluoride Etching Effects in Silicon, *J. Electrochem. Soc.* **123,** 551 (1976).
20. C. O. Thomas, D. Kahng, and R. C. Manz, Impurity Distribution in Epitaxial Silicon Films, *J. Electrochem. Soc.* **109,** 1055 (1962).
21. D. Kahng, C. O. Thomas, and R. C. Manz, Epitaxial Silicon Junctions, *J. Electrochem. Soc.* **110,** 394 (1963).
22. A. S. Grove, A. Roder, and C. T. Sah, Impurity Distribution During Epitaxial Growth, *J. Appl. Phys.* **36,** 803 (1965).
23. P. H. Langer and J. I. Goldstein, Impurity Redistribution During Silicon Epitaxial Growth and Semiconductor Device Processing, *J. Electrochem. Soc.* **121,** 563 (1974).
24. P. H. Langer and J. I. Goldstein, Boron Autodoping During Silane Epitaxy, *J. Electrochem. Soc.* **124,** 591 (1977).
25. G. R. Srinivasan, Kinetics of Lateral Autodoping in Silicon Epitaxy, *J. Electrochem. Soc.* **125,** 146 (1978).
26. R. K. Smeltzer, Epitaxial Deposition of Silicon in Deep Grooves, *J. Electrochem. Soc.* **122,** 1666 (1975).
27. C. M. Drum and C. A. Clark, Geometrical Stability of Shallow Surface Depressions During Growth of (111) and (100) Epitaxial Silicon, *J. Electrochem. Soc.* **115,** 664 (1968).
28. C. M. Drum and C. A. Clark, Anisotropy of Macrostep Motion and Pattern Edge-Displacements During Growth of Epitaxial Silicon Near {100}, *J. Electrochem. Soc.* **117,** 1401 (1970).
29. S. P. Weeks, Pattern Shift and Pattern Distortion during CVD Epitaxy on (111) and (100) Silicon *Solid State Technol.*, p. 111 (Nov. 1981).
30. R. H. Finch, H. J. Queisser, G. Thomas, and J. Washburn, Structure and Origin of Stacking Faults in Epitaxial Silicon, *J. Appl. Phys.* **34,** 406 (1963).

31. R. B. Marcus, M. Robinson, T. T. Sheng, S. E. Haszko, and S. P. Murarka, Electrical Activity of Epitaxial Stacking Faults, *J. Electrochem. Soc.* **124,** 425 (1977).

32. H. Kressel, A. Review of the Effect of Imperfections on the Electrical Breakdown of p–n Junctions, *RCA Rev.* **28,** 175 (1967).

33. J. Komeno, S. Ohkawa, A. Miura, K. Dazai, and O. Ryuzan, Variation of GaAs Epitaxial Growth Rate with Distance along Substrate within a Constant Temperature Zone, *J. Electrochem. Soc.* **124,** 1440 (1977).

34. R. G. Sangster, Model Studies of Crystal Growth Phenomena in the III-V Semiconducting Systems, in *Compound Semiconductors*, Vol. 1, R. K. Willardson and H. L. Goering, Eds., Reinhold, New York, 1962.

35. B. J. Baliga and S. K. Ghandhi, Hillocks on Epitaxial GaAs Grown from Trimethylgallium and Arsine, *J. Crystal Growth* **26,** 314 (1974).

36. J. R. Knight, D. Effer, and P. R. Evans, The Preparation of High Purity Gallium Arsenide by Vapor Phase Epitaxial Growth, *Solid State Electron.* **8,** 178 (1965).

37. D. J. Kirwan, Reaction Equilibria in the Growth of GaAs and GaP by the Chloride Transport Process, *J. Electrochem. Soc.* **117,** 1572 (1970).

38. J. J. Tietjen and J. A. Amick, The Preparation and Properties of Vapor-Deposited Epitaxial GaAsP Using Arsine and Phosphine, *J. Electrochem. Soc.* **113,** 724 (1966).

39. G. B. Stringfellow and G. Ham, Hydride VPE Growth of GaAs for FET's, *J. Electrochem. Soc.* **124,** 1806 (1977).

40. H. M. Manasevit and W. I. Simpson, The Use of Metal-Organics in the Preparation of Semiconductor Materials: I. Epitaxial Gallium-V Compounds, *J. Electrochem. Soc.* **116,** 1725 (1969).

41. W-H. Petzke, V. Gottschalch, and E. Butler, Epitaxial Deposition of GaAs in the $Ga(CH_3)_3$-AsH_3-H_2 System (IV). Thermodynamic and Kinetic Considerations, *Kristall und Technik* **9,** 763 (1974).

42. D. H. Reep, Organometallic Chemical Vapor Deposition of Gallium Arsenide: Growth Mechanisms, Ph.D. Dissertation, Rensselaer Polytechnic Institute, Troy, NY, 1982.

43. R. Bhat, B. J. Baliga, and S. K. Ghandhi, Vapor-Phase Etching and Polishing of Gallium Arsenide Using Hydrogen Chloride Gas, *J. Electrochem. Soc.* **122,** 1378 (1975).

44. R. Bhat and S. K. Ghandhi, The Effect of Chloride Etching on GaAs Epitaxy Using TMG and AsH_3, *J. Electrochem. Soc.* **125,** 771 (1978).

45. Y. Seki, K. Tanno, K. Lida, and E. Ichiki, Properties of Epitaxial GaAs Layers from a Triethyl Gallium and Arsine System, *J. Electrochem. Soc.* **122,** 1108 (1975).

46. K. Lindeke, W. Sack, and J. J. Nickl, Gallium Diethyl Chloride: A New Substance in the Preparation of Epitaxial Gallium Arsenide, *J. Electrochem. Soc.*, **117,** 1316 (1970).

47. J. V. DiLorenzo, Vapor Growth of Epitaxial GaAs: A Summary of Parameters which Influence the Purity and Morphology of Epitaxial Layers, *J. Crystal Growth* **17,** 189 (1972).

48. K. P. Pande, D. H. Reep, S. K. Shastry, A. S. Weiner, J. M. Borrego, and S. K. Ghandhi, The Preparation and Properties of Thin Polycrystalline GaAs Solar Cells with Grain Boundary Edge Passivation, *IEEE Trans. Electron Dev.*, **ED-27,** 635 (1980).

49. O. Mizuno, S. Kikuchi, and Y. Seki, Epitaxial Growth of Semi-Insulating Gallium Arsenide, *Jpn. J. Appl. Phys.* **10,** 208 (1971).

50. D. W. Shaw, Influence of Substrate Temperature on GaAs Epitaxial Deposition Rates, *J. Electrochem. Soc.* **115,** 405 (1968).

51. I. A. Sheka, I. S. Chaus, and T. T. Mityureva, *Chemistry of Gallium*, Elsevier, Amsterdam, 1966.

52. J. L. LaPorte, M. Cadoret, and R. Cadoret, Investigation of the Parameters which Control the Growth of $\{111\}$ and $\{\bar{1}\bar{1}\bar{1}\}$ Faces of GaAs by Chemical Vapour Deposit, *J. Crystal Growth* **50,** 663 (1980).

53. D. J. Schlyer and M. A. Ring, An Examination of the Product-Catalyzed Reaction of Trimethylgallium with Arsine, *J. Organomet. Chem.* **114,** 9 (1976).

54. F. W. Williams, The Effect of Orientation on the Electrical Properties of Epitaxial Gallium Arsenide, *J. Electrochem. Soc.* **111,** 886 (1964).

55. J. P. Duchemin, M. Bonnet, F. Koelsche, and D. Huyghe, A New Method for the Growth of GaAs Epilayer at Low H_2 Pressure, *J. Crystal Growth* **45,** 181 (1978).

56. R. Sankaram, S. B. Hyder, and S. G. Bandy, Selective in situ Vapor Etch and Growth of GaAs, *J. Electrochem. Soc.* **126,** 1241 (1979).

57. R. W. McClelland, C. O. Bozler, and J. C. C. Fan, A Technique for Producing Epitaxial Films on Reusable Substrates, *Appl. Phys. Lett.* **37,** 561 (1980).

58. B. A. Joyce, Growth and Perfection of Chemically-Deposited Epitaxial Layers of Si and GaAs, *J. Crystal Growth* **3,4,** 43 (1968).

59. L. R. Dawson, Liquid Phase Epitaxy, in *Progress in Solid State Physics*, Vol. 7, H. Reiss, and J. O. McCaldin, Eds., Pergamon, New York, 1972, Chap. 4, p. 117.

60. H. Morkoc and L. F. Eastman, The Growth of Uniform Submicron GaAs Layers by Liquid Phase Epitaxy, *J. Electrochem. Soc.* **123,** 906 (1976).

61. B. J. Baliga, Buried-Grid Field-Controlled Thyristors Fabricated Using Liquid-Phase Epitaxy, *IEEE Trans. Electron Dev.* **ED-27,** 2141 (1980).

62. C. D. Thurmond and M. Kowalchik, Germanium and Silicon Liquid Curves, *Bell Sys. Tech. J.* **39,** 169 (1960).

63. D. L. Rode, Isothermal Diffusion Theory of LPE: GaAs, GaP, Bubble Garnet, *J. Crystal Growth* **20,** 13 (1973).

64. T. Brys'kiewicz, Investigation of the Mechanism and Kinetics of Growth of LPE GaAs, *J. Crystal Growth* **43,** 101 (1978).

65. R. Ghez, An Exact Solution of Crystal Growth Rates under Conditions of Constant Cooling Rate, *J. Crystal Growth* **19,** 153 (1973).

66. R. Muralidharan and S. C. Jain, Improvements in the Theory of Growth of LPE Layers of GaAs and Interpretation of Recent Experiments, *J. Crystal Growth* **50,** 707 (1980).

67. J. J. Hsieh, Thickness and Surface Morphology of GaAs LPE Layers, *J. Crystal Growth* **27,** 49 (1974).

68. M. B. Small and J. F. Barnes, The Distribution of Solvent in an Unstirred Melt under Conditions of Crystal Growth by Liquid Epitaxy and Its Effect on the Rate of Growth, *J. Crystal Growth* **5,** 9 (1969).

69. J. Crank, *Mathematics of Diffusion*, Oxford University Press, Oxford, 1955.

70. I. Crossley and M. B. Small, Computer Simulations of Liquid Phase Epitaxy of GaAs in Ga Solution, *J. Crystal Growth* **11,** 157 (1971).

71. H. T. Minden, Constitutional Supercooling in GaAs Liquid Phase Epitaxy, *J. Crystal Growth* **6,** 228 (1970).

72. J. S. Harris, Y. Nannichi, and G. L. Pearson, Ohmic Contacts to Solution Grown Gallium Arsenide, *J. Appl. Phys.* **40,** 4575 (1969).

73. M. Otsubo and H. Miki, Liquid Phase Epitaxial Growth of Semi-Insulating GaAs Crystals, *J. Electrochem. Soc.* **124,** 441 (1977).

74. H. Nelson, Epitaxial Growth from the Liquid State and its Applications to the Fabrication of Tunnel and Laser Diodes, *RCA Rev.* **24,** 603 (1963).

75. M. B. Small and I. Crossley, The Physical Processes Occurring During Liquid Phase Epitaxial Growth, *J. Crystal Growth* **27,** 35 (1974).

76. W. G. Pfann, *Zone Melting*, 2nd ed., Wiley, New York, 1966, Chap. 5.

77. J. E. Davies, E. A. D. White, and J. D. C. Wood, A Study of Parameters to Optimize the Design of LPE Dipping Apparatus, *J. Crystal Growth* **27,** 227 (1974).

78. B. Girault, F. Chevrier, A. Joullie, and G. Bougnot, Liquid Phase Epitaxy of Silicon at Very Low Temperatures, *J. Crystal Growth* **37,** 169 (1977).

79. B. J. Baliga, Isothermal Silicon Liquid Phase Epitaxy from Supersaturated Tin, *J. Electrochem. Soc.* **125,** 598 (1978).

80. L. R. Dawson, Near-Equilibrium LPE Growth of GaAs-$Ga_{1-x}Al_xAs$ Double Heterostructures, *J. Crystal Growth* **27,** 86 (1974).

81. H. Kressel, Gallium Arsenide and (AlGa)As Devices Prepared by Liquid Phase Epitaxy, *J. Electron. Mater.* **3,** 747 (1974).

82. G. W. Cullen, The Preparation and Properties of Chemically Vapor Deposited Silicon on Sapphire and Spinel, *J. Crystal Growth* **9,** 107 (1971).

83. C. E. Weitzel and R. T. Smith, Silicon on Sapphire Crystalline Perfection and MOS Transistor Mobility, *J. Electrochem. Soc.* **125,** 792 (1978).

84. E. E. Wagner, G. Hom, and G. B. Stringfellow, Growth of High Quality $Al_xGa_{1-x}As$ by OMVPE for Laser Devices, *J. Electron. Mater.* **10,** 239 (1981).

85. W. R. Runyan, *Semiconductor Measurements and Instrumentation*, McGraw-Hill, New York, 1975.

86. G. E. Stillman and C. M. Wolfe, Electrical Characterization of Epitaxial Layers, *Thin Solid Films* **31,** 69 (1976).

87. L. J. van der Pauw, A Method of Measuring Specific Resistivity and Hall Effect of Discs of Arbitrary Shape, *Philips Res. Rep.* **13,** 1 (1958).

88. I. Hlásnick, Influence of Carrier Concentration Gradients and Mobility Gradients on Galvanomagnetic Effects in Semiconductors, *Solid State Electron.* **8,** 461 (1965).

89. J. I. Goldstein and H. Yakowitz, Eds., *Practical Scanning Electron Microscopy and Ion Microprobe Analyses*, Plenum, New York 1975.

90. J. Hilibrand and R. D. Gold, Determination of the Impurity Distribution in Junction Diodes from Capacitance-Voltage Measurements, *RCA Rev.* **21,** 245 (1960).

91. R. R. Spiwak, Design and Construction of a Direct-Plotting Inverse-Doping Profiler for Semiconductor Evaluation, *IEEE Trans. Instrum. Meas.* **IM-18,** 197 (1969).

92. G. L. Miller, A. Feedback Method for Investigating Carrier Distributions in Semiconductors, *IEEE Trans. Electron Dev.* **ED-19,** 1103 (1972).

PROBLEMS

1. A bubbler, using hydrogen as a carrier gas, has an inlet tube which produces bubbles of 2 mm diameter. Assuming that the diffusivity of the reactant vapor in hydrogen is 0.05 cm^2/s, determine the minimum residence time for the bubbles, in order that they are saturated with the reactant vapor.
2. Given that $D/D_0 = (T/T_0)^{1.75}$, $\rho/\rho_0 = T_0/T$, $v/v_0 = T/T_0$, and $\mu/\mu_0 = (T/T_0)^m$, where m is between 0.6 and 1.0, determine the dependence of the growth rate upon temperature for an endothermic process in the mass-transfer-limited region. Show that this corresponds to an apparent activation energy of 3.5–4.5 kcal/mole.
3. Calculate the width of the stagnant layer at a point 1.0 cm down the susceptor, for a reactor operating with 5 liter/min hydrogen carrier gas, and a reaction tube diameter of 55 cm. Assume that the reactor is operated at 700°C, and at atmospheric pressure.

4. Repeat Problem 3, assuming operation at 0.1 atm. To a first-order approximation, the density varies at P^{-1} and the viscosity varies as P^{-2}. Compare the flux for both cases, when D varies as P^{-1}. Assume identical partial pressures for the reactants.
5. Sketch a stacking fault in an epitaxial layer on a {100} substrate. Indicate the planes along which this fault propagates. Determine the relationship between its height and the length of its sides.
6. The energy of formation of a stacking fault on a {111} surface in silicon is 50 erg/cm^2. Compare this number with the energy of movement of an edge dislocation. Do you think it possible to remove stacking faults by annealing?
7. A phosphorus-doped layer, having a doping concentration of 10^{16} atoms cm^{-3}, is grown on a boron-doped substrate, having a background concentration of 10^{15} atoms cm^{-3}. Determine the junction lag resulting from this growth, assuming that $\phi = 0.5 \times 10^{-4}$.
8. A phosphorus-doped epitaxial layer of concentration N_E is grown on an arsenic-doped buried layer of concentration N_S. An abrupt doping profile results from this process. The slice is subjected to heat treatment at 1200°C for a period of 10 min. Sketch the doping profile resulting from this heat treatment for $N_S = 1000\ N_E$.
9. Repeat Problem 8, assuming a heat treatment period of 100 min. Compare the results for the two cases.

CHAPTER

6

Ion Implantation

CONTENTS

6.1 PENETRATION RANGE 302

6.1.1 Nuclear Stopping, 306
6.1.2 Electronic Stopping, 311
6.1.3 Range, 313
6.1.4 Transverse Effects, 316

6.2 IMPLANTATION DAMAGE 321
6.3 ANNEALING 325

6.3.1 Annealing Characteristics of Silicon, 327
6.3.2 Annealing Characteristics of Gallium Arsenide, 330

6.4 ION IMPLANTATION SYSTEMS 337

6.4.1 Ion Sources, 339
6.4.2 The Accelerator, 340
6.4.3 Mass Separation, 340
6.4.4 Beam Scanning, 342
6.4.5 Beam Current Measurements, 344

6.5 PROCESS CONSIDERATIONS 345

6.5.1 Diffusion Effects during Annealing, 345
6.5.2 Multiple Implants, 346
6.5.3 Masking, 348
6.5.4 Contacts to Implanted Layers, 352
6.5.5 Implantation through an Oxide, 353

6.5.5.1 Junctions, 354
6.5.5.2 Resistors, 355

6.6 APPLICATION TO SILICON 357

6.6.1 Bipolar Devices, 357
6.6.2 MOS Devices, 358

6.7 APPLICATION TO GALLIUM ARSENIDE 362

6.7.1 FET Devices, 363
6.7.2 Integrated Circuits, 365

REFERENCES 366

PROBLEMS 369

Ion implantation, as applied to semiconductor technology, is a process by which energetic impurity atoms can be introduced into a single-crystal substrate in order to change its electronic properties. Implantation is ordinarily carried out with ion energies in the 50–500 keV range. Basic requirements for implantation systems are ion sources and means for their extraction, acceleration, and purification. This is followed by beam deflection and scanning prior to impingement on the substrate.

Ion implantation provides a direct alternative to diffusion as a means for junction fabrication in semiconductor technology. The technique, however, has many unique characteristics which have led to its rapid development from a research tool to an extremely flexible, competitive technology. Some of these unique characteristics are now considered.

1. Mass separation techniques can be used to obtain a monoenergetic, highly pure beam of impurity atoms, free from contamination. Thus a single machine can be used for a wide variety of impurities. Furthermore, the process of implantation is carried out under vacuum conditions, i.e., in an inherently clean environment.
2. A wide range of doses, from 10^{11} to 10^{17} ions/cm^2, can be delivered to the target, and controlled to within $\pm$ 1% over this range. In contrast, control of impurity concentration in diffusion systems is at best 5–10% at high concentrations, and becomes worse at low concentrations. Furthermore, dopant incorporation during diffusion is sensitive to variations in the electronic character of the surface, whereas this is not the case with ion implantation. Thus ion implantation provides inherently more uniform surface coverage than diffusion, particularly when low surface concentrations are required.
3. Ion implantation is usually carried out at room temperature or on sub-

strates held at low temperatures (<400°C). As a result, a wide variety of masks (such as silica, silicon nitride, aluminum, and photoresist) can be used for selective doping. This gives great freedom in the design of unique self-aligned mask techniques for device fabrication, which are not possible with diffusion technology.

4. The penetration of an ion beam increases with the ion energy. In consequence, a wide variety of dopant profiles can be obtained by controlling the energy and dosage of multiple implants of the same or different impurities. Both hyperabrupt and retrograde doping profiles can be made with relative ease in this manner.
5. Ion implantation is a nonequilibrium process, so that the resulting carrier concentration is not limited by thermodynamic considerations, but rather by the ability of the dopant to become active in the host lattice. Thus it is possible to introduce dopants into a semiconductor at concentrations in excess of their equilibrium solid solubility.
6. Ion implantation can also be used for depositing a controlled amount of a charge species in a specific region of a semiconductor. Thus it has important uses in the threshold control of MOS devices.

There are some disadvantages to ion implantation as well. The equipment is highly sophisticated and expensive so that the technology is at an economic disadvantage to diffusion (in those areas where diffusion *can* be used). The competitive disadvantage is worse for gallium arsenide, where energy requirements are in the 200–500 keV range, than for silicon (50–150 keV range). This, however, is offset by the fact that ion implantation techniques lend themselves to a high degree of automation, including in-line process monitoring as well as end-point determination.

A second disadvantage is that ion implantation results in damage to the semiconductor. Annealing at elevated temperatures is necessary to heal some or all of this damage. This does not represent a problem with silicon technology since the vapor pressure of silicon is extremely low at annealing temperatures. Furthermore, silicon is often subjected to later high-temperature processes, where this damage can be completely annealed. With gallium arsenide, however, it is necessary to use a cap, or to conduct the anneal in an arsenic overpressure, in order to avoid dissociation or loss of stoichiometry. Neither approach is completely satisfactory, and this problem has not been solved at the present time.

Another problem comes about because of the relationship of the penetration depth R_p and the $\sqrt{Dt}$ product associated with the anneal process. With silicon, $R_p \gg \sqrt{Dt}$, so that dopant movement during annealing is relatively small.* With gallium arsenide, however, the reverse is often the case, so that dopant motion during annealing can be significant.

* This is not true for gallium in silicon. However, this dopant is rarely used in microcircuit technology.

Both laser and electron beam annealing have the promise of solving these problems for gallium arsenide. These technologies are in a very early stage of research at the present time, and are not considered here.

6.1 PENETRATION RANGE

There are two basic stopping mechanisms by which energetic ions, upon entering a semiconductor, can be brought to rest [1]. The first of these is by energy transfer to the target nuclei. This causes deflection of the projectile ions, and also a dislodging of the target nuclei from their original sites. If E is the energy of the ion at any point x along its path, we can define a *nuclear stopping power* $S_n = (dE/dx)_n$ to characterize this process. Nuclear stopping results in physical damage to the semiconductor, which takes the form of point as well as line defects. Often the semiconductor becomes amorphous as a result of this process.

A second stopping process is by the interaction of the ion with both bound and free electrons in the target. This gives rise to the transient generation of hole–electron pairs as energy is lost by the moving ion. We can define an *electronic stopping power* $S_e = (dE/dx)_e$ to characterize this process.

The average rate of energy loss with distance is given by

$$-\frac{dE}{dx} = N\,[S_n(E) + S_e(E)] \tag{6.1}$$

where N is the number of target atoms per unit volume of the semiconductor. If the total distance traveled by the ion before coming to rest is R, then

$$R = \int_0^R dx = \frac{1}{N}\int_0^{E_0} \frac{dE}{S_n(E) + S_e(E)} \tag{6.2}$$

where E_0 is the initial ion energy. The quantity R is known as the *range*. A more significant parameter, of interest in semiconductor technology, is the projection of this range along the direction of the incident ion, as shown in Fig. 6.1*a*. Because of the statistical nature of this process, this *projected range* is characterized by its mean value R_p, as well as by a standard deviation ΔR_p along the direction of the incident ion. This latter term is sometimes referred to as the *straggle*.

In practice, the ion beam also has a spread at right angles to its incidence, as shown in Fig. 6.1*b*. This transverse term is denoted by ΔR_t, and is of importance in determining the doping distribution near the edge of a mask. It can be ignored if the width of the implantation region is large compared to the depth of the implant. For this case, the statistical distribution for an amorphous target can be described by a one-dimensional gaussian distri-

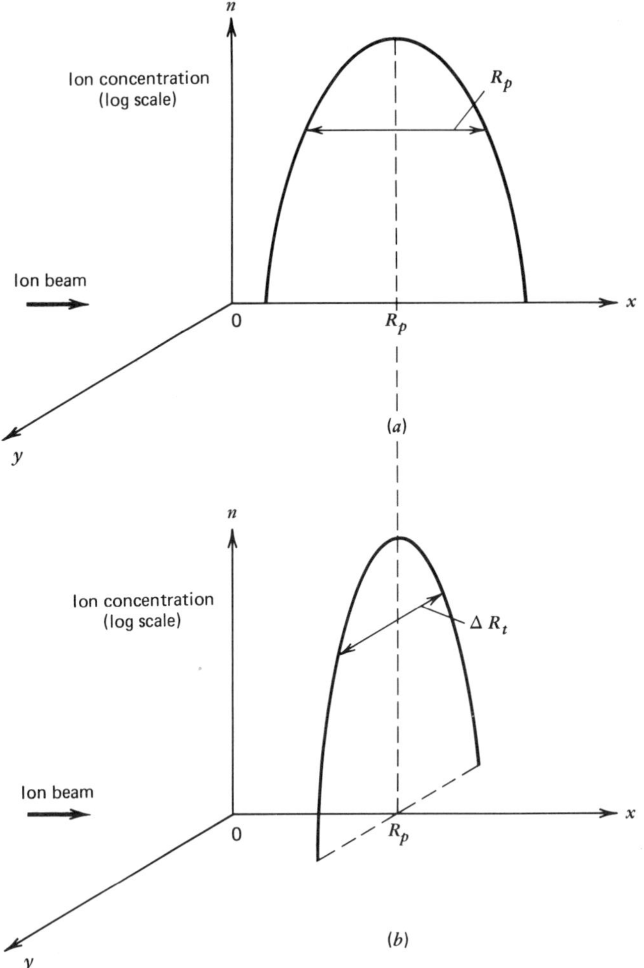

Fig. 6.1 Range and straggles.

bution function of the form

$$\phi(x) = \exp\left[-\frac{1}{2}\left(\frac{x - \bar{x}}{\sigma}\right)^2\right] \tag{6.3}$$

where $\bar{x}$ is the mean value and σ is the standard deviation. Integration of this function, over the limits $\pm\,\infty$ results in the dose. Writing this dose as Q_0 ions/cm^2, and noting that

$$\int_0^\infty e^{-z^2}\,dz = \sqrt{\pi}/2 \tag{6.4}$$

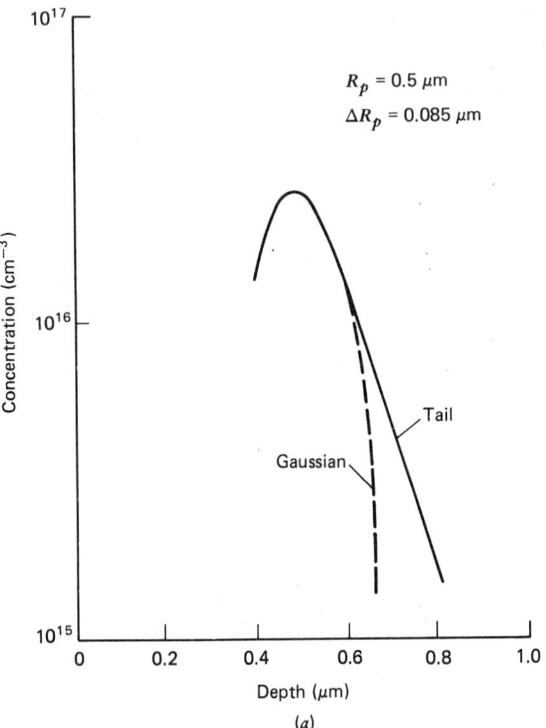

Fig. 6.2 Ion concentration. (*a*) Boron in silicon, 250-keV ions, annealed at 850°C for 30 min. Adapted from Moline [3]. (*b*) Phosphorus in silicon, 300-keV ions, annealed at 800°C for 30 min. Adapted from Dearnaley, Freeman, Gard, and Wilkins [4].

the impurity distribution is given by

$$N(x) = \frac{Q_0}{(2\pi)^{1/2}\,\Delta R_p} \exp\left[-\frac{1}{2}\left(\frac{x - R_p}{\Delta R_p}\right)^2\right] \tag{6.5}$$

This function has a maximum at R_p, and falls off rapidly on either side of this mean value.

Both silicon and gallium arsenide behave as if they were amorphous semiconductors, provided that the ion beam is misoriented from the low-index crystallographic directions by an angle of at least 7–10°. In this situation, the doping profile described by Eq. (6.5) is closely followed for one to two decades below the peak value. Figure 6.2 shows typical doping profiles for boron and phosphorus implants into single-crystal silicon [2, 3] and illustrates this situation.

The ion profile can be dramatically altered if the incident beam is aligned to a major crystallographic axis. This gives rise to a process known as *channeling*, by which ions are steered for a considerable distance through the semiconductor lattice with little energy loss. Such channeled beams

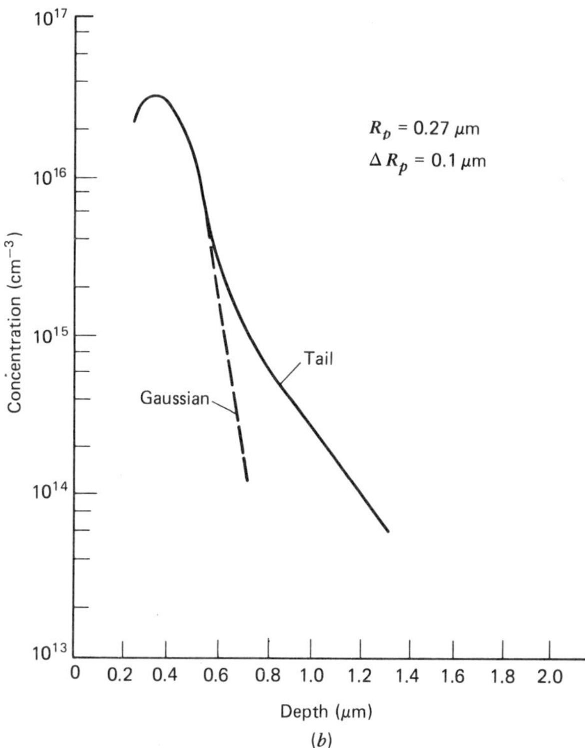

(*b*)

result in greatly increased penetration depth, as well as in reduced lattice disorder. Unfortunately, the range is now critically dependent on the degree of alignment (or misalignment) of the beam and the crystallographic axis, and also on the ion dose.

Figure 6.3 shows an example of normalized doping profiles for phosphorus in silicon [4] for a beam aligned in the ⟨110⟩ direction, and for doses from 1.2×10^{13} to 7.25×10^{14} cm^{-2}. In all cases the position of the gaussian peak is at 0.1 μm, and is only a function of the ion energy. The extent of the channeled region, on the other hand, is a sensitive function of the dose, and results in a penetration depth that is almost a decade larger than would be obtained in the absence of channeling. It is seen here that the higher doses create more lattice disorder, and so the channeling effect is correspondingly reduced.

Ion implantation for commercial microcircuit fabrication is usually carried out so as to avoid any possibility of channeling. The critical misalignment angle is a function of the specific crystal orientation, as well as of the semiconductor material. Typically, a misalignment of 7–10° is used to cover all situations of interest to integrated circuit technology. Even so, it is ex-

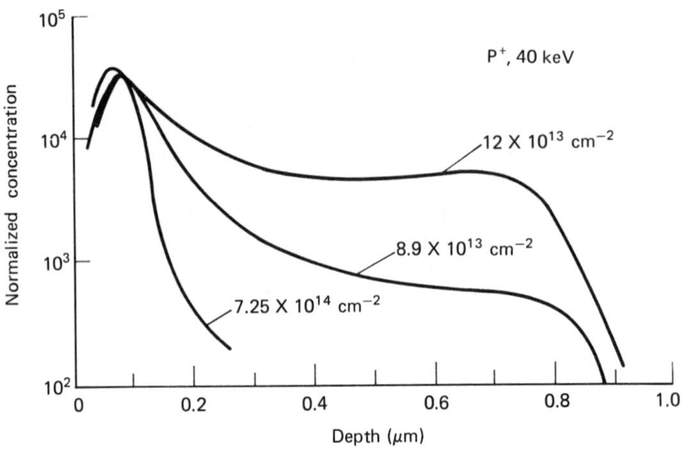

Fig. 6.3 Channeling of phosphorus in silicon, 50-keV ions. Adapted from Dearnaley, Freeman, Gard, and Wilkins [4].

tremely difficult to avoid channeling completely, especially with low-dose implants since some of the ions can be deflected into channeling directions *after* entering the semiconductor. This residual channeling results in the formation of a tail as the dopant penetrates deeper into the semiconductor (see Fig. 6.2). The tail is not observed in heavily damaged material, especially if it has been rendered amorphous, so that the gaussian profile is closely followed in these situations.

The formation of a tail in the doping profile can also be caused by a rapid interstitial diffusion process undergone by the particles once they have lost their incident energy. Such particles continue diffusing, even at room temperature, until they encounter suitable trapping centers such as vacancies or a surface. If τ is an average trapping time and D is the diffusion coefficient associated with this process, it can be shown [5] that the impurity tail will be of the form given by

$$n = n_0 \, e^{-x/\sqrt{D\tau}} \tag{6.6}$$

In some situations, this tail region is enhanced if the semiconductor is annealed (by heat treatment) after the implantation. However, τ can be very short in heavily damaged material, thus suppressing its formation.

6.1.1 Nuclear Stopping

Some of the ions impinging on a semiconductor conductor surface are reflected by collision with the outermost layers of atoms. Some impart sufficient energy to target atoms which are then ejected from the substrate by a process known as *sputtering*. In this section, however, our interest focuses, on the nuclear stopping process, by which the majority of the incident ions penetrate into the semiconductor, and transfer their energy to the lattice atoms by elastic collisions. Detailed calculations for nuclear stopping were first advanced by Linhard, Scharff, and Schiøtt (LSS) and are available in the literature [6, 7]. Here the approach used in developing them is outlined in order to gain some insight into the implantation process.

Consider first the elastic collision of two hard spheres, each of radius R_0, as shown in Fig. 6.4. Let V_0 and E_0 be the velocity and kinetic energy of the moving sphere, of mass M_1; the mass of the stationary sphere is M_2. Upon collision, its velocity and kinetic energy are written as V_2 and E_2, respectively. In like manner, let V_1 and E_1 be the velocity and kinetic energy of the projectile sphere after impact. The distance between spheres is given by an impact parameter p. For this model, energy transfer only occurs if $p \leq 2R_0$. A head-on collision is represented by $p = 0$.

Upon collision, momentum is transferred along the line of centers of the spheres. In addition, the kinetic energy is conserved. Solving for these conditions, the projectile sphere deflects from its original trajectory by an angle θ, such that

$$\cos\theta = \frac{1}{2}\left[\left(1 + \frac{M_2}{M_1}\right)\left(\frac{E_2}{E_0}\right)^{1/2} + \left(1 - \frac{M_2}{M_1}\right)\left(\frac{E_0}{E_2}\right)^{1/2}\right] \tag{6.7}$$

This equation shows that the energy transferred to the stationary particle is related to the scattering angle θ. In addition, the velocity V_1 of the projectile sphere after impact is given by

$$V_1^2 = V_0^2\left[\frac{M_1\cos\theta + (M_2^2 - M_1^2\sin^2\theta)^{1/2}}{M_1 + M_2}\right] \tag{6.8}$$

Maximum energy is lost by the projectile in a head-on collision with a

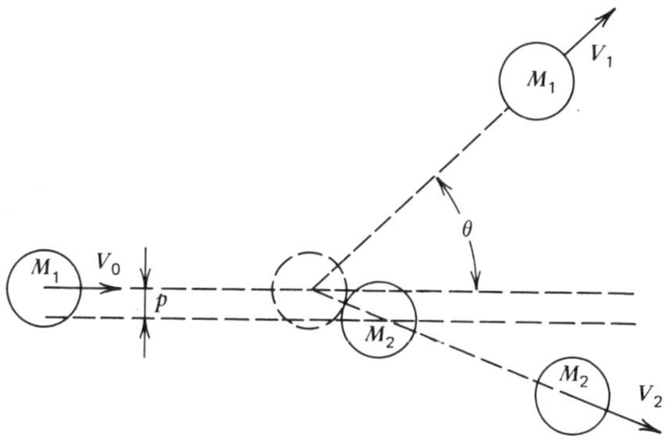

Fig. 6.4 Collision of hard spheres.

target particle that is initially at rest. For this case,

$$V_2 = \left(\frac{2M_1}{M_1 + M_2}\right) V_0 \tag{6.9}$$

so that the energy gained by M_2 is

$$\frac{1}{2} M_2 V_2^2 = \left[\frac{4M_1 M_2}{(M_1 + M_2)^2}\right] E_0 \tag{6.10}$$

This is also the energy lost by the incident particle M_1.

The situation is somewhat more complex when the projectile and the target have an attractive (or repulsive) force between them. Associated with this force is a potential $V(r)$, which usually extends out to infinity. If this is the case, both particles will be continually moving (without physical encounter) as they approach each other, as shown in Fig. 6.5*a*, so that the impact parameter p extends out to infinity. The problem now reduces to a system of point masses moving under the influence of this potential, and has been classically treated [8] by two-particle elastic scattering theory. Here the approach consists of transforming from laboratory coordinates to a set of moving coordinates whose center is located at the center of mass of the system. This renders the problem symmetric, as shown in Fig. 6.5*b*, with a deflection angle ϕ between the particles,

$$\phi = \pi - 2p \int_{-\infty}^{R_M} \frac{dr}{r^2[1 - V(r)/E_r - p^2/r^2]} \tag{6.11}$$

where p is the impact parameter, R_M the minimum distance of separation between the particles, $V(r)$ the interaction potential function, and E_r is given

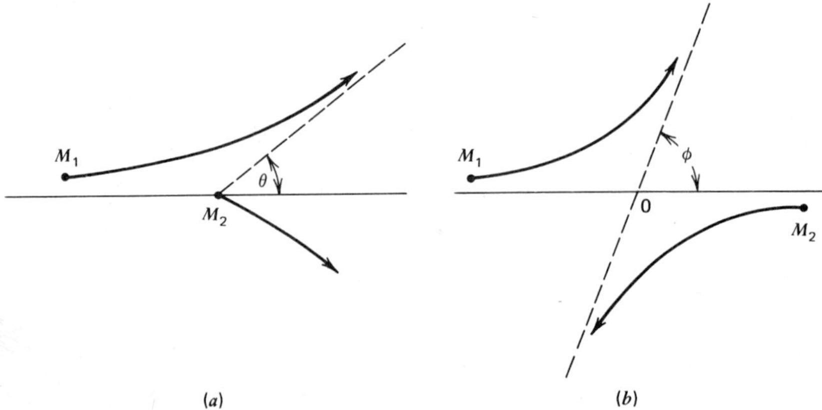

Fig. 6.5 Two-body scattering. (*a*) Laboratory coordinates. (*b*) Center of mass coordinates.

by

$$E_r = \frac{1}{2}\left(\frac{M_1 M_2}{M_1 + M_2}\right) V_0^2 \tag{6.12}$$

The integral of Eq. (6.11) can be evaluated for any given interaction potential function. The magnitude of θ, in laboratory coordinates, is found to be given by

$$\tan\theta = \frac{\sin\phi}{\cos\phi + (M_1/M_2)} \tag{6.13}$$

Consider what happens if two atoms, with atomic numbers Z_1 and Z_2, approach each other at a distance r. The force between these atoms is coulombic in nature, and is given by

$$F(r) = \frac{q^2 Z_1 Z_2}{r^2} \tag{6.14}$$

where $q = 1.6 \times 10^{-19}$ coulombs. The potential is thus

$$V(r) = \frac{q^2 Z_1 Z_2}{r} \tag{6.15}$$

This situation holds only as long as the electrons of each atom are excluded. Their inclusion, however, results in a screening influence on the nuclear repulsion, which causes a modification of the potential function. One general form of such a potential function is

$$V(r) = \frac{q^2 Z_1 Z_2}{r} f\left(\frac{r}{a}\right) \tag{6.16}$$

where a is a screening parameter, and $f(r/a)$ is known as a *screening function*. The Thomas–Fermi value of this screening function, shown in Fig. 6.6, has been found to be useful in calculations of nuclear stopping power. Often it is crudely approximated by a/r, where

$$a = \frac{0.885 a_0}{(Z_1^{2/3} + Z_2^{2/3})^{1/2}} \tag{6.17}$$

and a_0 is the Bohr radius (0.53 Å). For this approximation, the potential is inversely proportional to the square of the distance.

Using either the Thomas–Fermi screening function or its approximation, the deflection angle ϕ (and hence θ) can be computed for any impact pa-

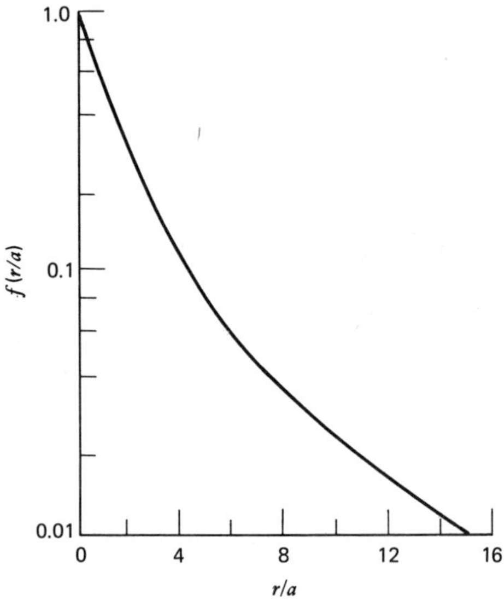

Fig. 6.6 The Thomas–Fermi screening function.

rameter p. This, in turn, allows computation of $T_n(E, p)$, the energy lost by the incident particle in an elastic encounter with a single stationary particle.

In the practical situation of an amorphous target, of thickness Δx and having N atoms/unit volume, energy is transferred to all particles. The total energy transferred is obtained by integrating over all possible values of impact parameter, so that

$$\Delta E = -N \Delta x \int_0^\infty T_n(E, p)\, 2\pi p\, dp \qquad (6.18)$$

The nuclear stopping power is thus given as

$$S_n(E) = \int_0^\infty T_n(E, p)\, 2\pi p\, dp \qquad (6.19)$$

Using this approach and the Thomas–Fermi screening function, LSS have obtained a relationship for nuclear stopping of the form shown in Fig. 6.7. From this figure it is seen that, as an ion enters a semiconductor (i.e., for large values of E), it first transfers energy to the lattice relatively slowly. As the ion slows down, this rate of energy transfer increases, and then eventually decreases to zero as the ion finally comes to rest.

Use of the approximate form of the screening function, where the potential is inversely proportional to the square of the distance, leads to a constant rate of energy loss. This is shown by the dashed line in Fig. 6.7. The ap-

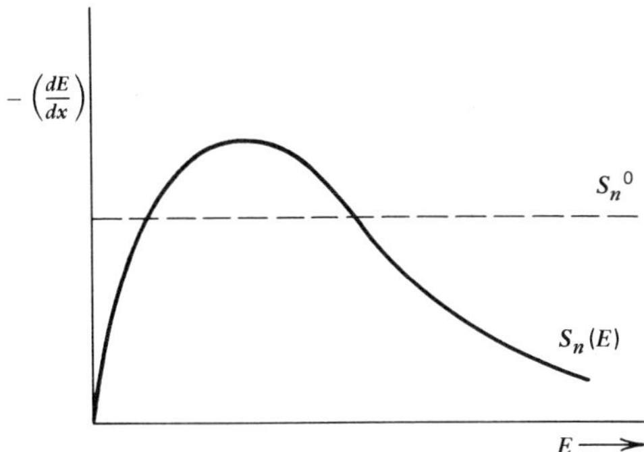

Fig. 6.7 Rate of energy loss as a function of ion energy.

proximate expression for this rate of energy loss turns out to be

$$-\frac{dE}{dx} = 2.8 \times 10^{-15} N \left(\frac{Z_1 Z_2}{Z^{1/3}}\right)\left(\frac{M_1}{M_1 + M_3}\right) \tag{6.20}$$

eV/cm, where

$$Z^{1/3} = (Z_1{}^{2/3} + Z_2{}^{2/3})^{1/2} \tag{6.21}$$

This corresponds to an energy loss rate of about 10–100 eV per angstrom for most ion–target combinations.* The approximate stopping power $S_n{}^0$ is thus given by

$$S_n{}^0 \simeq 2.8 \times 10^{-15} \left(\frac{Z_1 Z_2}{Z^{1/3}}\right)\left(\frac{M_1}{M_1 + M_2}\right) \text{eV cm}^2 \tag{6.22}$$

6.1.2 Electronic Stopping

A comprehensive treatment of inelastic energy exchange processes, which accompany the excitation and ionization of electrons by collisions with incident ions, has not been made at the present time. However, semiclassical approaches can give reasonable estimates for the energy loss rate due to these processes. The fact that electron excitation and ionization can occur is readily seen by considering a head-on elastic collision between an ion of mass M_1 and energy E_0, and an electron of mass m_0. Since $M_1 \gg m_0$, the maximum energy transfer due to this process is given by

$$\frac{4M_1 m_0}{(M_1 + m_0)^2} E_0 \simeq \frac{4M_1}{m_0} E_0 \tag{6.23}$$

* A short listing of Z and M values is given in Table 6.1.

Table 6.1 Properties of Materials Used in Ion Implantation

Element/Compound	Atomic Number	Atomic Weight	Density (g/cm^3)
Al	13	26.98	2.70
Sb	51	121.75	6.62
Ar	33	74.92	5.72
Be	4	9.01	1.85
B	5	10.81	2.34
Cd	48	112.40	8.65
Ga	31	69.72	5.91
Ge	32	72.59	5.32
Mg	12	24.31	1.74
N	7	14.01	1.25[a]
O	8	16	1.43[a]
P	15	30.97	1.83
Se	34	78.96	4.79
Si	14	28.09	2.33
S	16	32.06	2.07
Te	52	127.6	6.24
Sn	50	118.69	7.30
Zn	30	65.37	7.13
GaAs	32[b]	72.32[b]	5.27
SiO_2	10[b]	20.03[b]	2.27
Si_3N_4	10[b]	18.04[b]	3.44

[a] g/liter [b] average

Excitation energies are a few electron volts, whereas $M_1/m_0 \simeq 1000$–2000. Consequently, both electron excitation and ionization are possible since the energy of incident ions during implantation is normally in the 50–500 keV range.

Approximate computations of the inelastic energy loss rate associated with this process have been made by considering the collision of an incident ion with a free electron gas. Using this approach, it has been shown that, if the ion velocity is less than the velocity of an electron having an energy equal to the Fermi energy, then the electronic stopping power is proportional to the ion velocity. In terms of ion energy,

$$S_e(E) = kE^{1/2} \tag{6.24}$$

The value of k depends on both the projectile and the target material. For an amorphous target, however, k is relatively independent of the projectile. For silicon,

$$k_{\mathrm{Si}} \simeq 0.2 \times 10^{-15}\ (\mathrm{eV})^{1/2}\ \mathrm{cm}^2 \tag{6.25a}$$

For gallium arsenide,

$$k_{\mathrm{GaAs}} \simeq 0.52 \times 10^{-15}\ (\mathrm{eV})^{1/2}\ \mathrm{cm}^2 \tag{6.25b}$$

Thus the rate of energy loss due to electronic stopping is several electron volts per angstrom, but is considerably less than the energy loss rate for nuclear stopping.

Although this theory shows a monotonic increase in $S_e(E)$ with energy, experiments have revealed a periodic dependence of this parameter on the atomic number of the ion. Other theories have attempted to treat this problem in more detail, and have resulted in qualitative agreement with this oscillatory behavior.

6.1.3 Range

The LSS calculations [6] have been derived in a system of dimensionless parameters

$$\rho = R\pi a^2 N \frac{M_1 M_2}{(M_1 + M_2)^2} \tag{6.26}$$

and

$$\epsilon = \frac{Ea}{q^2 Z_1 Z_2} \frac{M_2}{M_1 + M_2} \tag{6.27}$$

where a is given by Eq. (6.17).

Using these parameters, it has been shown that

$$-\left(\frac{d\epsilon}{d\rho}\right)_e = k\epsilon^{1/2} \tag{6.28}$$

where

$$k = Z_1^{1/6}\left[\frac{0.0793 Z_1^{1/2} Z_2^{1/2} (M_1 + M_2)^{3/2}}{(Z_1^{2/3} + Z_2^{2/3})^{3/4} M_1^{3/2} M_2^{3/2}}\right] \tag{6.29}$$

Figure 6.8 shows plots of $-(d\epsilon/d\rho)_n$ and $-(d\epsilon/d\rho)_e$ as a function of $\epsilon^{1/2}$ (proportional to the ion velocity). Here the nuclear stopping parameter is represented by a single curve. The curve for constant nuclear stopping power is also shown in this figure by a dashed line. A series of straight lines, one for each value of k, are used to represent the electronic stopping power. In practice, the spread of k values is restricted to those sketched in Fig. 6.8.

A rough estimate of the range can be made by using the approximate

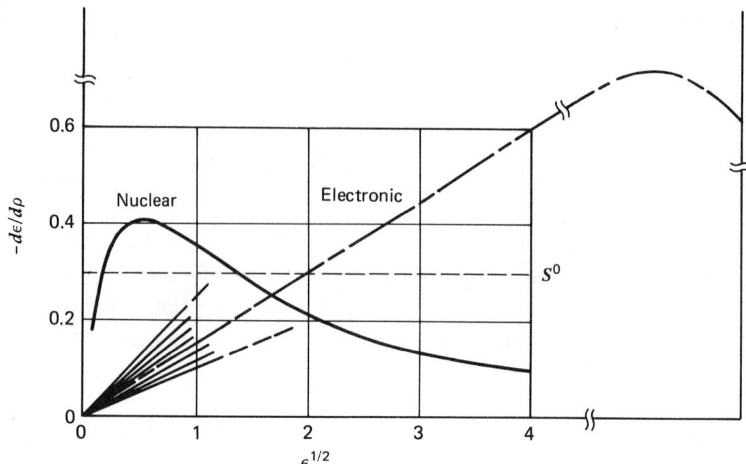

Fig. 6.8 Nuclear and electronic stopping power curves. From J. W. Mayer, L. Eriksson, and J. A. Davies, *Ion Implantation in Semiconductors* [7], 1970. Used with permission of Academic Press, Inc.

values of $S_n{}^0$ and $S_e(E)$ given in Eqs. (6.22) and (6.24), respectively. A sketch of these quantities, shown in Fig. 6.9, identifies a critical energy E_c at which nuclear and electronic stopping powers are equal. For silicon, E_c is about 15 keV for a light projectile such as boron, and as large as 150 keV for a heavy projectile such as phosphorus. Comparable values for gallium arsenide are about 15 keV for beryllium, 150 keV for sulfur, and 800 keV for selenium.

If the initial energy of the projectile is much larger than E_c, the dominant loss mechanism is electronic stopping. Nuclear stopping can thus be ignored in Eq. (6.2), resulting in a range

$$R \simeq K_1 E^{1/2} \tag{6.30}$$

where K_1 is a constant for a particular projectile–target combination.

The range can also be roughly estimated for the situation where nuclear

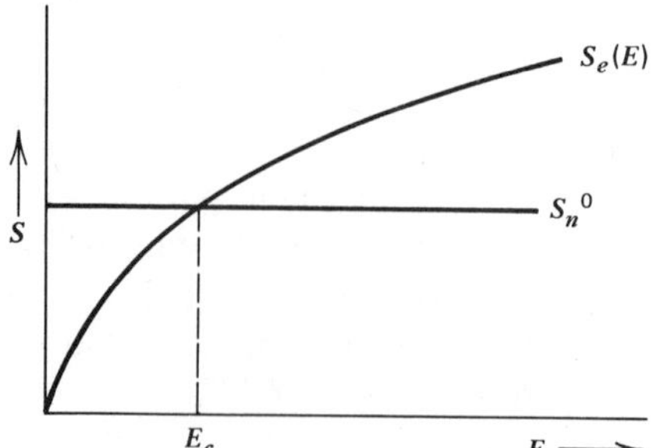

Fig. 6.9 Relative stopping power of an incident ion.

Table 6.2 Nature of the Concentration Profile Due to Ion Implantation

Distance from Mean Value	Magnitude of Ion Concentration (Normalized to Peak Value)
0	1
$\pm 1.18\Delta R_p$	0.5
$\pm 2.14\Delta R_p$	10^{-1}
$\pm 3.04\Delta R_p$	10^{-2}
$\pm 3.72\Delta R_p$	10^{-3}
$\pm 4.29\Delta R_p$	10^{-4}
$\pm 4.80\Delta R_p$	10^{-5}
$\pm 5.25\Delta R_p$	10^{-6}
$\pm 5.67\Delta R_p$	10^{-7}

stopping dominates, i.e., if $E \ll E_c$. For this case,

$$R \simeq K_2 E_0 \tag{6.31}$$

where K_2 is a constant for a particular projectile–target combination.

Differential equations for computing the projected range R_p and straggle have also been given* by LSS. Computer solutions of these equations, together with correction factors, are available in the literature for a variety of ion energies, and ion–target combinations [9, 11].

The doping profile due to ion implantation has been given by Eq. (6.5) as

$$N(x) = \frac{Q_0}{(2\pi)^{1/2}\,\Delta R_p} \exp\left[-\frac{1}{2}\left(\frac{x - R_p}{\Delta R_p}\right)^2\right] \tag{6.5}$$

where Q_0 is the dose in ions/cm^2, R_p the projected range, in cm, and ΔR_p the straggle, in cm. This equation is based on the assumption that transverse straggle effects can be ignored. In practice, these effects are significant in determining the carrier concentration at the edge of a mask, and also in situations where ion implantation is carried out through a narrow slit whose width is comparable to the implantation depth.

The ion profile of Eq. (6.5) is gaussian in character, and falls off symmetrically on either side of R_p. The extent of falloff, as a fraction of the magnitude at the peak value, is given in Table 6.2. With some dopants, considerable skewing of this profile has been observed, with more impurity ions deposited in the region nearest the surface. Higher order moments in range theory can be used to calculate more precise ion profiles for these

* To an approximation, $R_p = R/(1 + M_2/3M_1)$.

situations. However, these ion profiles are often grossly altered by subsequent annealing.

Figures 6.10 and 6.11 give the projected range and straggle for the dopants commonly used in silicon technology. Values for gallium arsenide are given in Figs. 6.12 and 6.13.

6.1.4 Transverse Effects

Transverse straggle is used to characterize the lateral spread of ions upon entering a target. This term is of importance in defining the ion penetration at the edge of a mask, and hence the curvature of a junction that may be formed by this process. It is also important when implantation is to be done through extremely narrow mask cuts, as in the case of microwave devices and in VLSI technology.

Consider an ion beam of infinitely small radius, which is incident on an

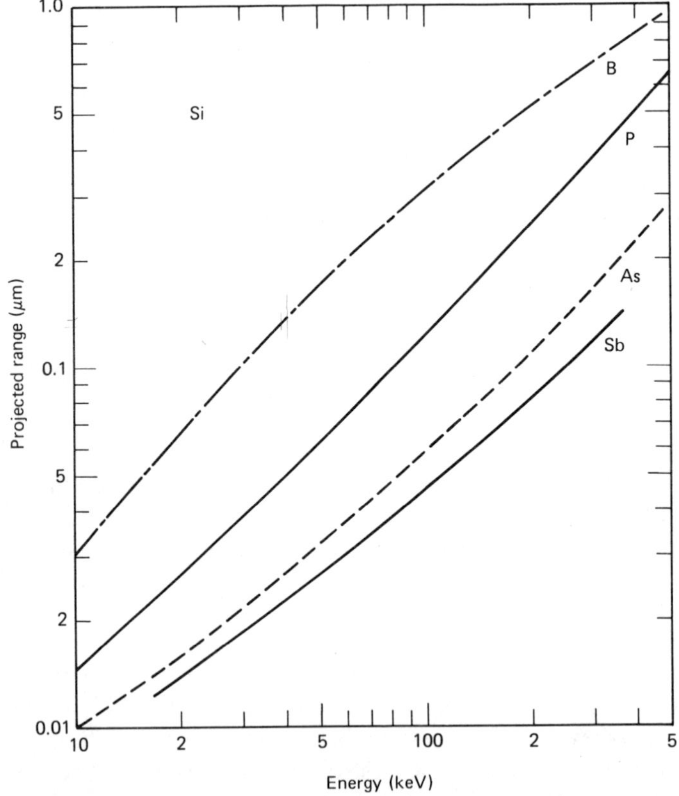

Fig. 6.10 Projected range of boron, phosphorus, arsenic, and antimony in silicon. Adapated from Gibbons, Johnson, and Myrolie [9].

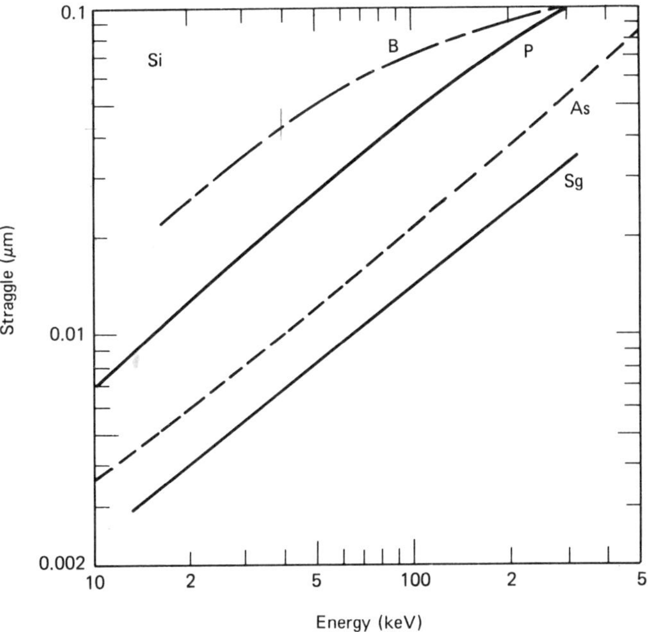

Fig. 6.11 Straggle for boron, phosphorus, arsenic, and antimony in silicon. Adapted from Gibbons, Johnson, and Myrolie [9].

amorphous target. If this beam enters the target in the x direction, then the spatial distribution function $f(x, y, z)$ is given by

$$f(x, y, z) = \frac{1}{(2\pi)^{3/2}\,\Delta R_p \Delta Y \Delta Z} \exp\left\{-\frac{1}{2}\left[\frac{y^2}{\Delta Y^2} + \frac{z^2}{\Delta Z^2} + \frac{(x - R_p)^2}{\Delta R_p^{\,2}}\right]\right\} \tag{6.32}$$

where ΔY and ΔZ are the standard deviations in the y and z directions. The transverse* straggle ΔR_t is then given by

$$\Delta R_t = \Delta Y = \Delta Z \tag{6.33}$$

because of symmetry. Figures 6.14 and 6.15 show the values of ΔR_t as a function of ion energy for implantation into silicon and gallium arsenide, respectively, as calculated from the LSS theory.

Based on the above formulation, it can be shown that the spatial distribution of ions implanted through a narrow mask cut is given, for positive

* Note that the radial straggle is $(\Delta Y^2 + \Delta Z^2)^{1/2} = \sqrt{2}\,\Delta Y$ because of symmetry.

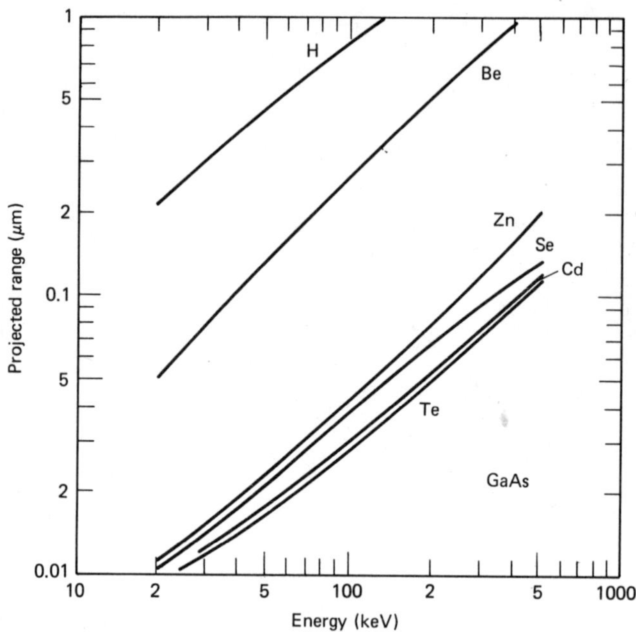

Fig. 6.12 Projected range of hydrogen, beryllium, zinc, selenium, cadmium, and tellurium in gallium arsenide.

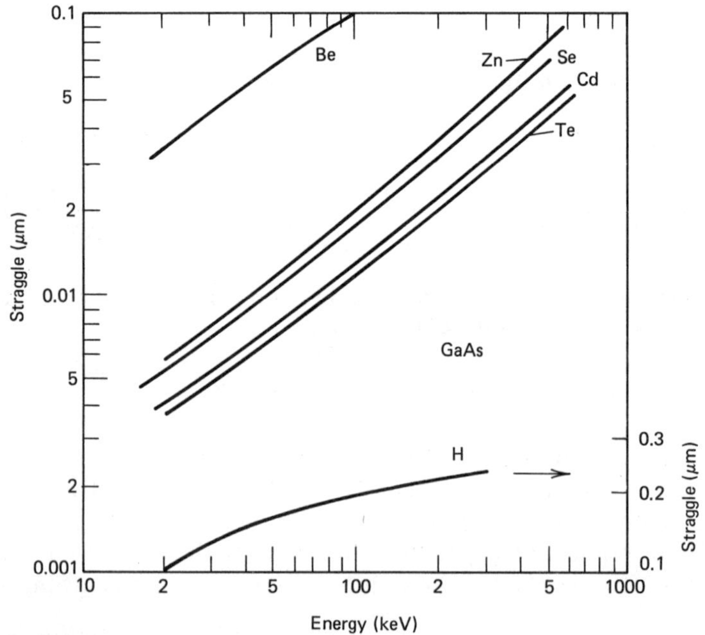

Fig. 6.13 Straggle for hydrogen, beryllium, zinc, selenium, cadmium, and tellurium in gallium arsenide.

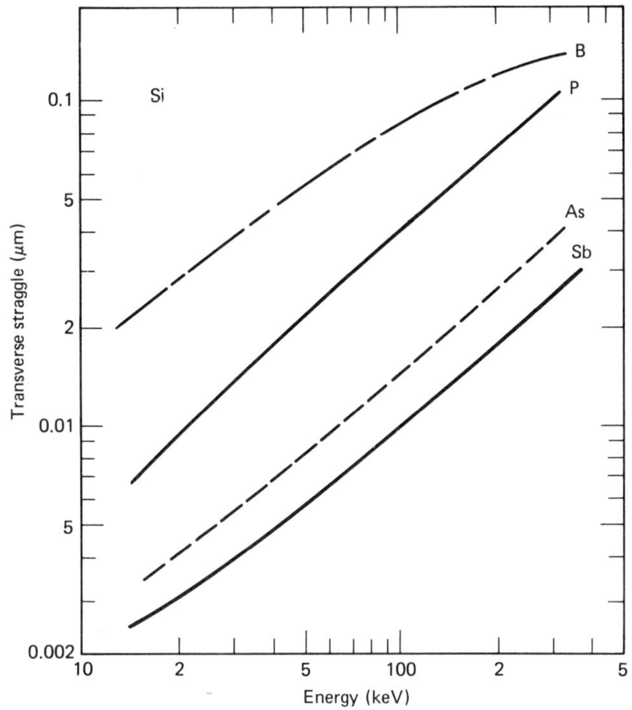

Fig. 6.14 Transverse straggle for boron, phosphorus, arsenic, and antimony in silicon. Adapated from Furakawa, Matsumura, and Ishiwara [11].

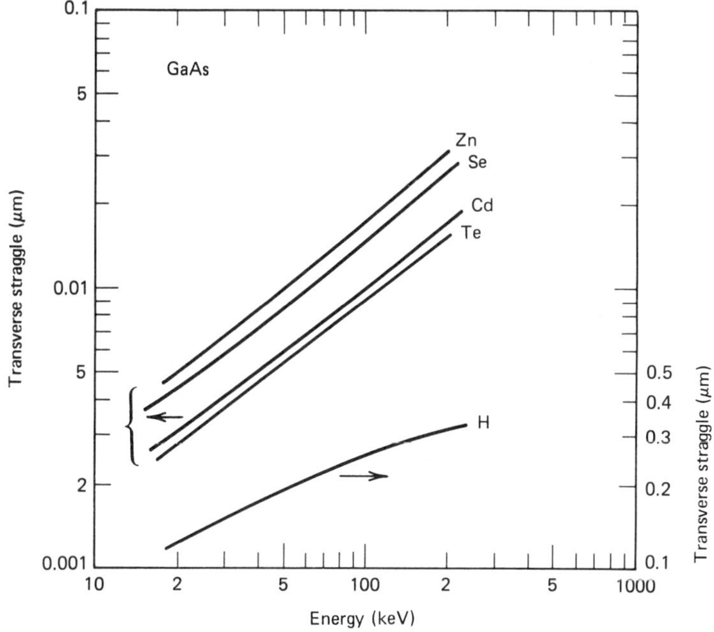

Fig. 6.15 Transverse straggle for hydrogen, zinc, selenium, cadmium, and tellurium in gallium arsenide. Adapted from Furakawa, Matsumura, and Ishiwara [11].

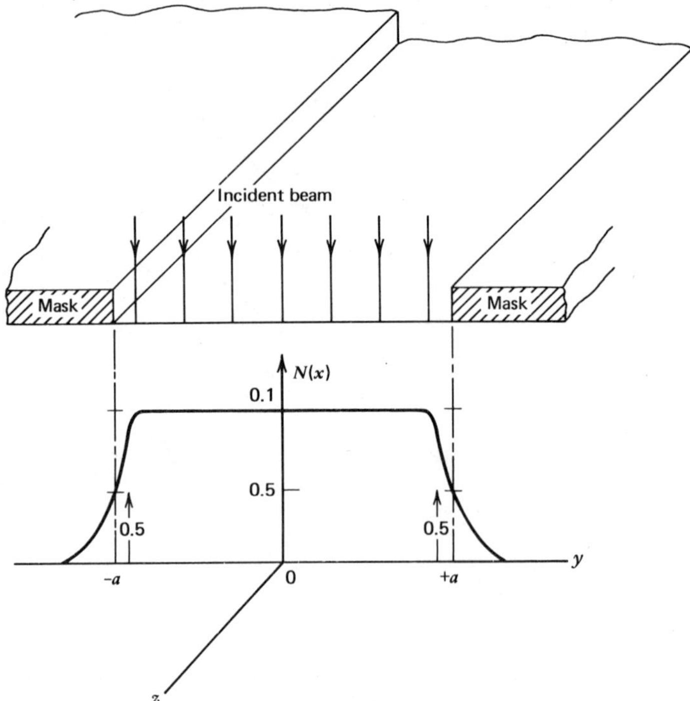

Fig. 6.16 Ion implantation through a slit.

values of y, by [11]

$$N(x, y, z) \simeq \frac{Q_0}{(2\pi)^{1/2}\,\Delta R_p} \exp\left\{\left[-\frac{1}{2}\left(\frac{x - R_p}{\Delta R_p}\right)^2\right]\left[\frac{1}{\sqrt{\pi}}\,\mathrm{erfc}\left(\frac{y - a}{\sqrt{2}\,\Delta R_t}\right)\right]\right\} \tag{6.34}$$

where the width of the slit is $2a$, and the slit is parallel to the y direction,

Table 6.3 Nature of the Concentration Profile at a Window Edge

Distance from Edge	Magnitude of Ion Concentration (Normalized to Peak Value)
0	0.5
$1.28\Delta R_t$	10^{-1}
$2.33\Delta R_t$	10^{-2}
$3.09\Delta R_t$	10^{-3}
$3.72\Delta R_t$	10^{-4}
$4.30\Delta R_t$	10^{-5}
$4.78\Delta R_t$	10^{-6}
$5.22\Delta R_t$	10^{-7}

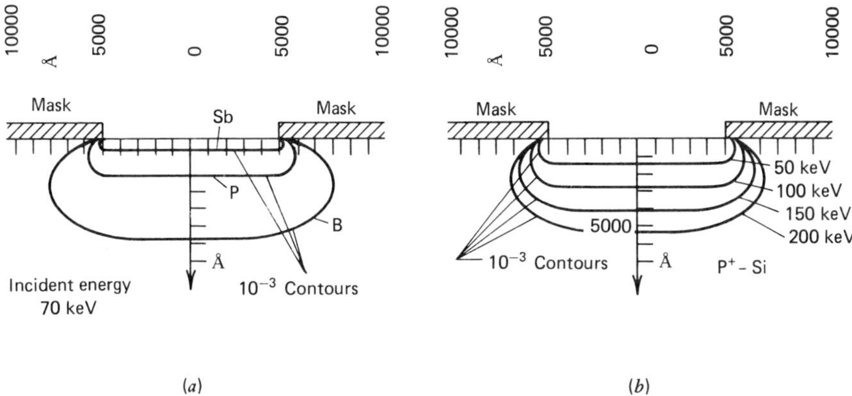

Fig. 6.17 Equi-ion concentration (0.1%) curves for implantation through a 1-μm mask cut. (*a*) 70-keV implants into silicon. (*b*) Variable-energy implants of phosphorus into silicon. Adapted from Furakawa, Matsumura, and Ishiwara [11].

as shown in Fig. 6.16. Thus the ion concentration is 50% at the edge of the mask, and the falloff beyond this point follows an erfc profile. The extent of this falloff is shown in Table 6.3.

Figure 6.17*a* shows equi-ion concentration contours for a variety of 70 keV implants into silicon through a 1 μm mask cut. Each contour represents the 0.1% concentration for this ion species. Figure 6.17*b* shows the 0.1% contours for a phosphorus implantation through the same mask cut, but in different energy ranges. These illustrations give an estimate of the magnitude of this effect, as well as of its implications on device performance.

6.2 IMPLANTATION DAMAGE

As mentioned earlier, the energetic ion makes many collisions with the lattice atoms before finally coming to rest. During this process, sufficient energy is transferred to the lattice so that many of its atoms are displaced, often with considerable energy. These in turn can displace other atoms, which results in a cascade of collisions.

Consider the effect of a single energetic ion as it moves through the semiconductor, making successive collisions with target atoms. Let E_d be the energy required to dislodge one of these atoms from its site, and assume a hard sphere model with elastic collisions. If a lattice atom receives less energy than E_d, it will not be displaced. Similarly, if the incident ion recoils from a collision with an energy less than E_d, it will not dislodge any more atoms. Based on these assumptions, the incident ion must have an energy greater than $2E_d$ if there is to be an *increase* in the net number of displaced atoms. Following this line of argument, the number of displaced atoms per incident ion is roughly equal to $E_0/2E_d$, where E_0 is the initial ion energy. The displacement energy for both silicon and gallium arsenide is about 14–15 eV. Thus about 10^3–10^4 lattice atoms are displaced as a result of each incident ion, and 10^3–10^4 Frenkel pairs are produced by this process.

An understanding of lattice damage during ion implantation is of great importance to the practical exploitation of this technology. Experimental techniques involve direct observation by electron microscopy, backscattering, and channeling techniques [12]. Indirect techniques, based on the effect of damage on the semiconductor, are also used. Thus a crude but effective indicator of large amounts of damage is a visual change in surface reflectivity, accompanied by a milky appearance, with both silicon and gallium arsenide. The sharpness of the optical attenuation characteristic with wavelength is yet another indicator. Mechanical properties, such as surface hardness, have also been used to study ion implantation damage. Perhaps the most sensitive (and directly useful) technique involve studies of changes in the electronic properties resulting from ion implantation. Theoretical computations of damage have also been made [13], based on the assumption that the damage at any given depth is directly proportional to the amount of kinetic energy transferred to the lattice at that depth.

The nature of the damage created by an incident ion will depend upon whether it is light or heavy relative to the lattice atoms. A light ion will generally transfer a small amount of energy during each encounter with the target, and will be deflected through a large scattering angle during this process. The displaced target atom will have a small amount of energy imparted to it, and will probably not create further displacements of its own. The damage from a single, light incident ion will thus take the form of a branching dislocation track, as shown in Fig. 6.18*a*. In addition, much of the energy from a light ion is transmitted to the lattice by electronic stopping, so that there is relatively little crystal damage. Finally, the range is comparatively large, so that the damage will be spread out over an extensive volume of the target.

The effect of a heavy ion is quite different. Here a large amount of energy is transferred with each collision, and the incident ion is deflected through a relatively small scattering angle. Each displaced atom is itself capable of producing a large number of displacements as it is moved away from the path of the incident ion. At the same time, the range is small, and most of the energy is transferred to the lattice by nuclear stopping, so that there is considerable lattice damage within a relatively small volume. This damage takes the form shown in Fig. 6.18*b*, and consists of a somewhat straight dislocation track, surrounded by an ellipsoidal region with a high density of interstitial atoms.

In both situations, the volume of the target in which the incident ion energy is deposited is usually much larger than the volume in which the actual lattice damage occurs. Thus much of the ion energy is transferred to the lattice in the form of a localized thermal spike. There is considerable interaction between the impurity ion and the target atoms during this process, often resulting in some self-annealing and in large deviations from the gaus-

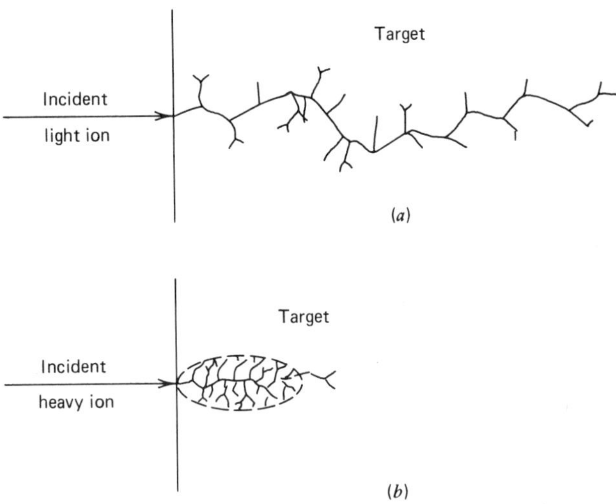

Fig. 6.18 Damage due to (*a*) light ions and (*b*) heavy ions.

sian profile. Such deviations are related to the nature of the charge associated with the ion species, and so differ from one impurity to another.

The disordered region created by each incident ion has a central core [14] in which the defect concentration is extremely large ($\geq$0.02 atom fraction). Otherwise, it consists primarily of vacancies, divacancies, Frenkel pairs, and dislocations, with many of the impurity and target ions occupying both substitutional as well as interstitial sites. As a result, the mobility and minority carrier lifetime in this region are lower than the original bulk values.* In addition, many of the incident ions do not come to rest in substitutional sites, so that the as-implanted carrier concentration is often several decades below the ion concentration. There exists, however, considerable long-range crystallographic order in this structure. Much of the damage can be readily annealed by heat treatment at relatively low temperatures, with some annealing even occurring during the implantation process itself.

With increasing dosage, however, the isolated regions of disorder begin to overlap. In the case of silicon, this results in the formation of an amorphous layer, with no long-range crystallographic order. From Fig. 6.18, and the discussion of the nature of damage, it can be seen that the lighter the ion, the larger the dose that is required to produce this amorphous layer. Typically, for ions of the same energy (for a silicon target), the dose requirements for creating an amorphous layer are about two decades larger for boron ($M \simeq 11$) than for arsenic ($M \simeq 75$).

The formation of an amorphous layer in silicon has been definitely established; for gallium arsenide, however, there is some evidence that the

* Gallium arsenide is often rendered semi-insulating in these regions because of the deep levels associated with these defects.

material becomes heavily polycrystalline under high-dose conditions, but does not transform to a truly amorphous state. It is possible that this comes about because of the tendency of this material to reorder during implantation, resulting in a crystalline material with a dense dislocation network. Often this highly disordered material is referred to as "amorphous" gallium arsenide in the literature.

There are some advantages to achieving an amorphous state during silicon implantation. One is that the ion concentration profile now follows its predicted gaussian shape quite closely, because of the random nature of the host structure. In some low-doping situations, the silicon is implanted with an inert ion (argon, neon, or silicon, for example), to render it amorphous *before* the impurity ion is introduced. The advantage of this procedure is seen in Fig. 6.19, where data [15] are shown for a low-dose 100 keV phosphorus implant into silicon which has been predamaged by 100 keV silicon implants of 10^{14} and 10^{15} ions/cm^2. The more heavily predamaged case corresponds to the situation in which the implant region was rendered amorphous prior to the phosphorus dose. Here the ion concentration profile is seen to closely follow the gaussian shape predicted by the theory.

The heavily damaged, polycrystalline state in gallium arsenide makes this material semi-insulating, with an active carrier concentration less than 10^{11}/cm^3. This characteristic can be used to advantage [16] in order to isolate devices fabricated on the same substrate. In many situations, however, implantation damage in gallium arsenide has been found to create unwanted

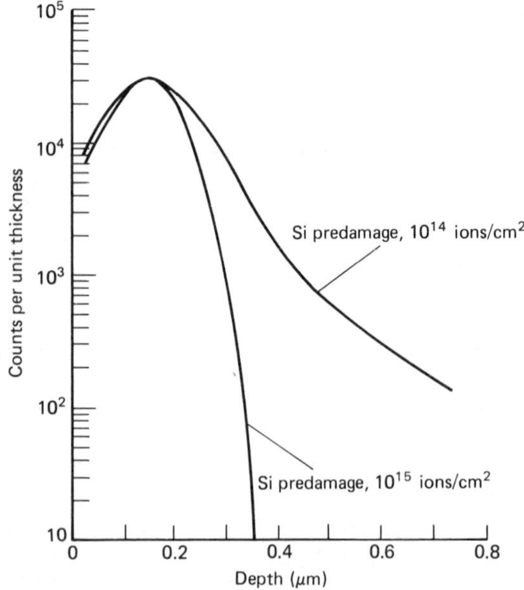

Fig. 6.19 Phosphorus implantation (100 keV, 7 × 10^{12} ions/cm^2) into silicon. Preimplanted. Adapted from Blood, Dearnaley, and Wilkins [5].

layers of semi-insulating material within the substrate [17], resulting in the inadvertent formation of p–i–n or p–n^-–n structures instead of p–n junctions.

The degree of self-annealing that occurs during ion implantation is related to the thermal spike, as well as to the temperature at which the substrate is held. Thus control of this temperature, together with the ion dose, represents a powerful technique for encouraging or delaying the onset of the amorphous state. This approach is often used with both silicon as well as gallium arsenide.

The energy of the incident ion falls off as it moves into the target, so that electronic stopping predominates in the initial penetration region. Consequently, the region of peak damage will be located at some distance from the target surface. In addition, the peak damage will occur further within the target if the incident ion beam has a higher initial energy, as will the range. In all cases, however, computations of the damage distribution show that it precedes the range distribution, and that the damage peak is (very approximately) at 75% of the projected range.

6.3 ANNEALING

Ion implantation damage results in a degradation of material parameters such as mobility and minority carrier lifetime. In addition, only a part of the as-implanted ions are located in substitutional sites, where they are electronically active. It is possible to heal some or all of this damage by annealing the semiconductor at an appropriate combination of time and temperature. Ideally, complete recovery of lifetime, mobility, and carrier activation would occur if the semiconductor could be returned to its original, single-crystalline state. This can only be achieved by melting and epitaxial regrowth.* Practical thermal anneal cycles fall far short of this stage, however, so that all annealing is essentially partial.

The actual details of the anneal treatment depend on the degree to which recovery of mobility, lifetime, and carrier activation are necessary for the satisfactory operation of the devices in a circuit. Thus they are a function of both device and circuit design. By way of example, the operation of silicon-based high-speed bipolar transistor logic circuits places great emphasis on attaining a high emitter carrier concentration, but actually requires a short minority carrier lifetime. On the other hand, the fabrication of a solid-state television camera target requires junctions with extremely low leakage, i.e., a large space charge generation lifetime. Gallium arsenide devices and microcircuits are most often based on majority carrier conduction. Minority carrier lifetime is thus not important.† On the other hand,

* Solid-phase epitaxy is also possible, when the semiconductor is heated to somewhat below its melting point.

† An important exception is the gallium arsenide solar cell.

precise control of carrier activation and depth of the implanted region are primary requirements in devices such as field effect transistors.

Annealing requirements are also related to the physical location of the damage in the semiconductor, and its effect on device operation. Typically, the peak in the ion damage occurs closer to the surface of the semiconductor than the peak in ion concentration. In diode structures of the type illustrated in Fig. 6.20, much of this damage is outside the depletion layer region, thus easing the requirement on maintaining long lifetime. In yet other situations, such as those where it is necessary to maintain an accurate, tail-free gaussian profile, the semiconductor must first be made amorphous by a pre-implant. In order to be successful, both the pre-implant type and the range must be chosen so that its region of damage coincides with the region where the impurity ion is to be deposited.

The anneal cycle is usually 15–30 min in duration. The temperature is dictated by the application, and also by the ability of the semiconductor to tolerate such thermal treatment at this point in the fabrication process. Thus bare silicon can readily be heat treated to 1200°C. On the other hand, a silicon microcircuit with aluminum metallization cannot be heat treated above 500°C if aluminum–silicon alloy formation is to be avoided. Materials such as gallium arsenide will decompose if heated above 600°C for any length of time. Consequently, they must be annealed under an arsenic over-pressure, or covered with a suitable capping layer during heat treatment.

The annealing temperature is also related to the implant dose and to the crystallographic orientation of the substrate. Low-dose implants result in isolated damage trees which can be annealed at low temperatures. High doses, on the other hand, especially those resulting in amorphous layers, are considerably harder to anneal.

The annealing rate of ⟨100⟩ silicon is about ten times more rapid than that of ⟨111⟩ material. No specific variation with orientation has been noted for gallium arsenide.

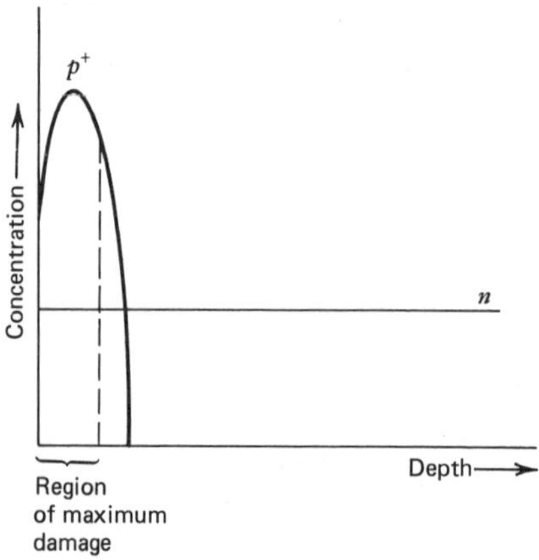

Fig. 6.20 Implanted p^+–n diode structure.

6.3.1 Annealing Characteristics of Silicon

The annealing behavior of implanted silicon is reasonably well understood at the present time. Let us first consider the low-temperature annealing characteristics of a layer which has been made amorphous prior to the implant, or by the implant itself. Individual disorder clusters begin to disappear at temperatures as low as 400°C, together with partial recovery (20–30%) of electronic properties. However, actual recrystallization of this layer occurs within a narrow temperature range between 550 and 600°C, with many of the impurity atoms moving into lattice sites and becoming active. This reordering is accompanied by the formation of dislocation loops, and considerable damage still remains in the lattice, because of the relatively slower rearrangement of silicon atoms on lattice sites.* Thus impurity activation occurs well before recovery of the lattice defect structure.

Lattice damage serves as a sink for heavy metallic impurities such as copper, iron, and gold, all of which degrade the lifetime. Recovery of this parameter is thus difficult to achieve with annealing in the 600°C temperature range. Typically, the 600°C annealing of silicon for these implantation conditions results in an activation† of about 50–90% of the implanted ions, but very little recovery of the lifetime. The degradation of mobility is primarily caused by the unactivated carriers, so that its recovery goes hand in hand with that of the lifetime.

In many situations (MOS devices, for example), carrier activation and mobility are of primary interest. Often the silicon is preimplanted with a heavy, inert ion (argon, for example) in order to make it amorphous before the active impurity implant. Sometimes silicon ions are used for this purpose. An alternate approach is to reduce the tendency to self-annealing by implantation at low substrate temperatures ($\simeq$77°K), but this approach has been confined to laboratory situations.

Dislocation loops, formed during the annealing of amorphous layers, will grow with increasing temperature up to about 800°C. As a result, annealing of high-dose, heavy ion implants at this temperature causes little improvement in lifetime, even though the dopant is 50–70% activated at this point. Recovery of this parameter only takes place at annealing temperatures beyond 950°C, with full recovery by 1000–1100°C.

Low-dose implantation leads to the formation of individual damage trees. Here a general rule is that the higher the dose, the more difficult it is to obtain full carrier activation. Some evidence of annealing occurs by 350°C,

* Note that the activation energy of the usual substitutional impurities in single-crystal silicon is considerably less than the energy of self-diffusion (3.5 eV as compared to 5.5 eV). It is reasonable to expect that, although much lower, the activation energies of movement of these species will follow the same pattern in damaged silicon as well. The activation energies of ion implanted boron and phosphorus in silicon are 0.62 eV and 0.34 eV, respectively.

† The lower value holds for high-dose, heavy ion implants; conversely, the higher value is for low-dose, light ions.

however, as the individual dislocations begin to heal. Mobility values are extremely low at these temperatures. By about 500°C, however, the recovery of this parameter goes hand in hand with the carrier activation. On the other hand, the individual dislocations are readily healed at this temperature, so that lifetime improvement is significant. Typically, 600°C annealing results in a 30–70% recovery of carrier activation. Recovery rates for light ions such as boron are considerably higher than for phosphorus or arsenic, because of the lesser extent of damage that is involved. The mobility typically recovers to 30–70% of the bulk value as well, with lifetime recovery to as much as 50% of its initial value.

With low-dose, light ion implants, almost full recovery of all parameters is obtained by 800–950°C. A further increase in temperature, to 1000°C, results in 100% dopant activation, and full recovery of parameters for heavy ions. By this time, however, there is some change in the as-implanted profile because of diffusion effects.

In summary, therefore, it is relatively easy to anneal low-dose, light ion implants in silicon. However, the difficulty of annealing increases with both the dose as well as the ion mass. In all cases, carrier activation occurs at lower temperatures than the recovery of either mobility or lifetime.

Annealing of implanted silicon is also carried out at 1100–1200°C, primarily as a process step in the formation of p-wells for complementary MOS (C-MOS) technology. Here the final doping profile is essentially one-sided gaussian in shape, and is very similar to that obtained with conventional base diffusions. Full recovery of carrier concentration, mobility, and lifetime also occurs, together with the formation of low-leakage junctions.

Low-dose, low-energy ion implantation is an important step in MOS technology, where it is used to control the threshold voltage. Implantation is conducted into ⟨100⟩ silicon substrates through the gate oxide, with ion doses in the 10^{11}–10^{12} cm^{-2} range, and with ion energies generally below 50 keV. Here the primary effect is to create slow charge states at the Si-SiO_2 interface. For the same dose, the density of these charge states is linearly proportional to the incident ion energy, regardless of the ion type. Thus it is in all probability related to the density of displaced atoms ($\simeq E_0/2E_d$) created by this process. These interface states can be annealed by heat treatment in the 200–400°C range, with complete recovery occurring within 30 min at 400°C. A small fraction of the impurity ion is activated during this process.

There are also many examples of anomalous behavior during the annealing of silicon. Often these include sudden *decreases* in carrier activation with increasing temperature. Behavior of this type is usually caused by the widely different energies of movements of silicon and the impurity ion, as well as of the different types of defects created during implantation. Our knowledge of these processes is far from complete at the present time.

The behavior of some commonly used dopants is now briefly described.

Phosphorus. This is a heavy ion so that its projected range is almost linearly proportional to the implantation energy ($\simeq$1.1 μm/MeV). There is considerable evidence of tails in the implantation profile, particularly at low doses. It has been shown that these are formed during the actual implant process, probably by the steering of ions into channeling directions [5]. Full carrier activation of phosphorus can usually be achieved by annealing at 600°C.

Arsenic. The behavior of arsenic is very similar to that of phosphorus. Again, being a heavy ion, its projected range varies linearly with energy ($\simeq$0.58 μm/MeV). Tails in the arsenic profile have also been observed, prior to annealing, and are caused by partial ion steering into channeling directions. In addition, significant enhanced diffusion effects are also observed upon annealing. About 50% activation can be achieved by annealing to 600°C, and there is some evidence of loss at the surface during this process. This loss can be reduced by using a deeper implant.

It is possible [18] to implant arsenic into silicon in concentrations that are more than its maximum solid solubility, because of the nonequilibrium nature of this process. Furthermore, this arsenic has been found to be located in substitutional sites, and is electronically active. Arsenic concentrations of this magnitude are metastable, and fall in a few minutes to their equilibrium value upon subsequent heat treatment. However, rapid annealing by laser techniques has been shown to preserve this high conductivity state. Thus this technique may become very important for VLSI applications, especially in the fabrication of small-area, high-conductivity, ohmic contacts.

Boron. This is a light ion, with a projected range of approximately 3.1 μm/MeV at energies of 10–100 keV. However, the range variation with energy becomes sublinear above 100 keV, due to the increasing contribution of electronic stopping. Boron implantation into an amorphous target shows considerable asymmetry in its ion profile at energies beyond 100 keV. The complete description of its distribution function requires two additional parameters (skew and kurtosis) in addition to the usual range and standard deviation [19]. This "Pearson IV" type distribution must be further modified by the addition of an exponential tail to include enhanced diffusion effects during annealing. The extent of these effects depends on the quality of the starting material and the nature of the ion damage. As a result, the location of deep, boron-implanted junctions is difficult to control with any degree of accuracy.

Some annealing of boron implants occurs by 300–400°C. However, it is relatively difficult to obtain full activity with low-dose boron implants, unless annealing is carried out at 900–1000°C, or the substrate is made amorphous prior to the implantation.

Boron often exhibits an anomalous annealing characteristic, its activation falling with increasing temperature up to 700°C. Beyond this point activation rapidly increases, up to 100% by 900–1000°C. This has been explained by the fact that a large concentration of interstitial silicon is formed upon implantation. At low annealing temperatures, this silicon competes with the boron atoms for substitutional sites, and can even displace them during the annealing process.

6.3.2 Annealing Characteristics of Gallium Arsenide

Although the basic concepts of damage and its removal in silicon apply to gallium arsenide as well, there are a number of important differences and complications which arise in work with this semiconductor. First, with the notable exceptions of beryllium and magnesium (both *p*-type), almost all dopants are heavy ions. They require considerable energy for implantation (in the 200–500 keV range), have short penetration depths, and create a large amount of lattice damage. Heavily damaged gallium arsenide, even when implanted with an impurity ion, is invariably semi-insulating because of its deep-level defect structure.

Next, a truly amorphous state does not appear to be created under conditions of heavy damage. Rather, the material becomes polycrystalline, with a very dense dislocation network. Reordering of the crystal structure is extremely difficult under these conditions. Annealing temperatures are also higher (≥900°C), and there is no well-defined activation energy* associated with this process, since it requires the movement of two different host species, as well as a variety of defects, into their appropriate sites (e.g., gallium into gallium sites, arsenic into arsenic sites). In addition, there are a greater variety of defect interactions in gallium arsenide than in silicon. Figure 6.21 shows results for a 40 keV tellurium implant [8] in gallium arsenide with a 1×10^{15} ion/cm^2 dose, and illustrates the gradual reduction in residual disorder that comes about with increasing annealing temperature. In contrast, silicon shows a rapid fall in disorder over a narrow (50°C) temperature range, indicating that annealing comes about by means of a single, dominant process.

Gallium arsenide devices and microcircuits are usually based on majority carrier transport, so that minority carrier lifetime is generally not important. Consequently, primary emphasis in this area has focused on carrier activation and on the recovery of bulk mobility, both of which tend to go hand in hand.

As-implanted gallium arsenide generally has no carrier activation. Annealing at temperatures up to 700°C usually has no effect. However, activation of most dopants begins at about this point, and continues to increase with higher temperatures. Often full activity is not achieved even at 900°C, and there appear to be many stable defects beyond this temperature.

As a result of the above difficulties, it is common practice to implant gallium arsenide at elevated temperatures (200–400°C) so that considerable self-annealing can occur during implantation. This prevents the material

* Note that the activation energies of diffusion of gallium and arsenic in single-crystal gallium arsenide are 3.2 and 5.6 eV, respectively.

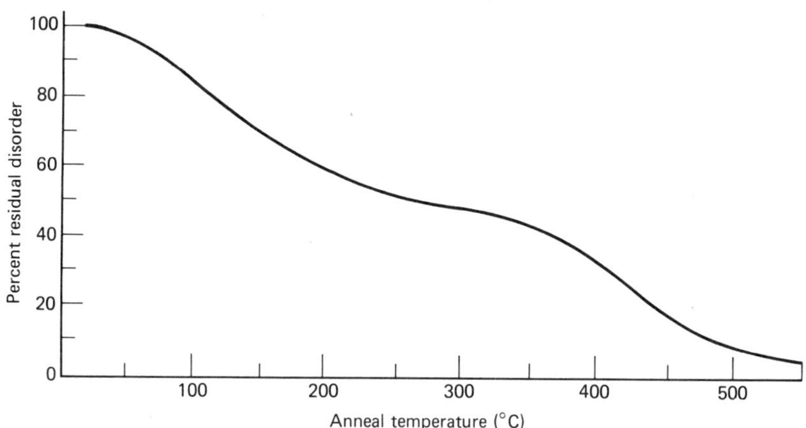

Fig. 6.21 Annealing behavior of lattice disorder in gallium arsenide implanted at room temperature. Adapted from G. Carter and W. A. Grant, *Ion Implantation of Semiconductors* [8], John Wiley and Sons, 1976.

from reaching its "amorphous" state, from which it can only recover at temperatures near its melting point. The effectiveness of this approach is shown in Fig. 6.22, which illustrates the residual disorder for a heavier dose (40 keV tellurium implant at 5×10^{19} ions/cm^2) than shown in Fig. 6.21, as a function of the substrate temperature during implantation, and with *no* subsequent annealing. It goes without saying that further reduction in disorder can be effected by a subsequent anneal step. This approach is in striking contrast to that of silicon technology where reduced temperatures are often used to make the substrate amorphous.

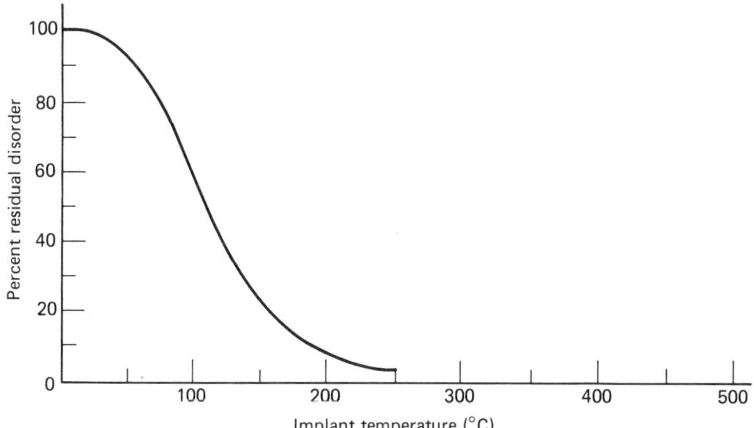

Fig. 6.22 Disorder in unannealed gallium arsenide after implantation at elevated temperatures. Adapted from G. Carter and W. A. Grnt, *Ion Implantation of Semiconductors* [8], John Wiley and Sons, 1976.

Often substantial changes in doping profile have been observed upon annealing. These are usually explained by assuming values for the diffusion constant which are three to four decades larger than those reported in the diffusion literature, and have been attributed to rapid interstitial-diffusion processes. In some cases, a rapid anomalous fall in carrier activation has been noted prior to an increase with annealing temperature. This phenomenon is not fully understood at the present time, and is probably due to competition for substitutional sites at different annealing temperatures, as has been observed with boron in silicon.

Changes in the doping profile from the expected gaussian shape have also been observed *before* annealing, when the implant is made into a substrate held at elevated temperatures. Thus fast diffusion has been noted for selenium implants [20] into substrates which were held at $\geq$ 150°C, even though 1000°C annealing *after* the implantation produced no additional diffusion.

High-temperature anneals carry with them the problem of loss of stoichiometry, since implanted gallium arsenide begins to dissociate by 350°C, with significant arsenic loss by 600°C. One approach to minimizing this problem is to use a capping layer during anneal. Often the implant is made through this cap, as a matter of convenience. This also serves to dope the cap with the impurity, and tends to reduce out-diffusion effects during the anneal step.

A variety of capping materials have been used, with moderate success. Early work with silica and phosphosilicate glass have been superceeded by silicon nitride, aluminum oxide, and aluminum oxynitride. Recently multiple-cap layers of SiO_2 and Si_3N_4 have been used at annealing temperatures as high as 1100°C. Results with capping layers appear to be a strong function of both the capping material as well as the techniques by which they are deposited. These techniques have included sputtering, chemical vapor deposition, and plasma deposition. However, no "optimum" capping technique has emerged at the present time.

Capless annealing, in an arsenic overpressure, has also been used [21]. This approach has the important advantage of process simplification. However, control of the arsenic vapor is important, since this controls the concentration of gallium and arsenic vacancies, and hence their occupation. As expected, dopants such as sulfur, which occupy arsenic sites, have higher activation with lower arsenic vapor pressure. Magnesium, on the other hand, occupies gallium sites and has higher activation at higher arsenic pressures. Silicon, although amphoteric, has a preferred location on gallium sites, and is *n*-type when implanted. Its annealing behavior tends to follow that of magnesium. Carrier activation observed with these implants was as low as 3–6% for high-dose implants (10^{15} cm^{-2}) and as much as 35–60% for low-dose implants (10^{13} cm^{-2}). This may, in part, be caused by loss of dopant during the anneal. More work will have to be done before the role of this method can be definitely established in implantation technology.

Finally, all gallium arsenide is compensated to some degree. This presents an additional problem, both in the implantation process and in interpreting its results. This is especially severe for implantation made into semi-insulating material, and much care must be taken in the incoming selection of substrates if consistent results are to be achieved [22].

Results of implantation with different types of impurities now follow.

***p*-Type.** Work with *p*-type impurities has included zinc, cadmium, beryllium, and magnesium, using room temperature implants [23]. In general, no carrier activation was found upon implantation at room temperature, or with annealing up to 600°C. Above this point, however, there was a steady improvement in carrier activation; by 800°C almost 80–90% activation has achieved for all dopants. Some workers have found that low-dose beryllium implants become almost fully activated at temperatures around 600°C. This result is of particular significance since the decomposition of gallium arsenide becomes increasingly rapid above this temperature. Unfortunately, the validity of this result has not been confirmed at the present time.

The annealing behavior of beryllium at high temperatures has been found to be strongly dependent on the dose [24]. This is seen from Fig. 6.23 for a 250 keV beryllium implant at 1×10^{15} ions/cm^2, followed by a 30 min anneal at a number of different temperatures. Here it is seen that the concentration profile for this high-dose implant undergoes a rapid transition by 800°C, from the gaussian shape to that normally encountered in diffusion situations. In contrast, no such change was observed for a dose of 5×10^{13} ions/cm^2, even after annealing at 900°C for 30 min. Some similar results have been noted for zinc and cadmium. There is some evidence of a sharp fall in carrier activation for zinc upon annealing at 700°C. However, 80–100% carrier activation is obtained by 900°C. There are insufficient data on concentration profiles with these dopants because of the relatively early nature of this work.

***n*-Type.** Considerable work has been done with these dopants because of interest in high-speed gallium arsenide devices, which are usually *n*-type. Dopants under investigation have included sulfur, silicon, selenium, tin, and tellurium. Implantation of these dopants has been found to be more difficult than for *p*-type, and to require critical control of implantation parameters, substrate temperatures, and capping layers. Of these impurities, both silicon and tin are amphoteric; however, their behavior with ion implantation is invariably *n*-type.

Very little work has been done with silicon, because of possible contamination of the silicon ion beam with N_2^+. However, it is potentially useful in device applications because of its high range compared to other, heavier *n*-type dopants. Room temperature implants, with low ion doses (1×10^{14} cm^{-2}), followed by 900°C anneal, have resulted in carrier activation levels of 30% for this impurity [17].

Low-dose tin implants (1×10^{14} cm^{-2}), followed by annealing at 900°C,

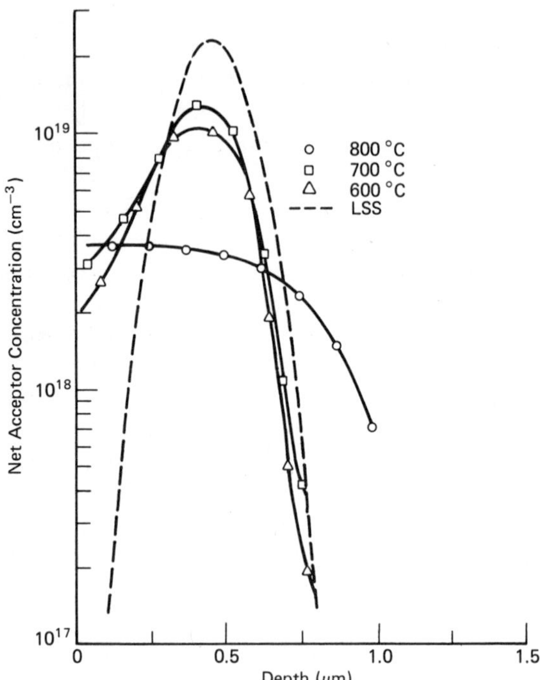

Fig. 6.23 Annealing of beryllium implants in gallium arsenide. From Levige, Helix, Vaidyanathan, and Streetman [24]. With permission from the *Journal of Applied Physics.*

have shown activation levels of 20% with substrate temperatures as low as 125°C. A maximum carrier density of 10^{18} cm^{-3} has been achieved with this dopant [17].

Sulfur is commonly used as an *n*-type dopant in gallium arsenide technology. Results with ion implantation of this impurity have shown carrier activation as low as 20%, accompanied by considerable departure from the predicted gaussian profile. Studies of sulfur annealing have shown considerable out-diffusion into the cap layer, as well as some trapping at the substrate surface. It has also been noted that the doping efficiency is improved with deeper implants, which supports this argument.

Results for a 100 keV sulfur implant, with a dose of 9×10^{12} ions/cm^2, are shown in Fig. 6.24, together with the distribution calculated from LSS theory [25]. Also shown are comparable results for a 400 keV selenium implant, with a dose rate of 2×10^{12} ions/cm^2. Implants were followed by a 30 min anneal at 850°C with a silicon nitride cap. It is interesting to note that both implants resulted in the same peak carrier concentration. However, the selenium profile closely follows LSS theory, whereas that of sulfur is much deeper. Advantage can be taken of these differences in the fabrication of ohmic contact regions to shallow implanted layers, as shown in Fig. 6.46.

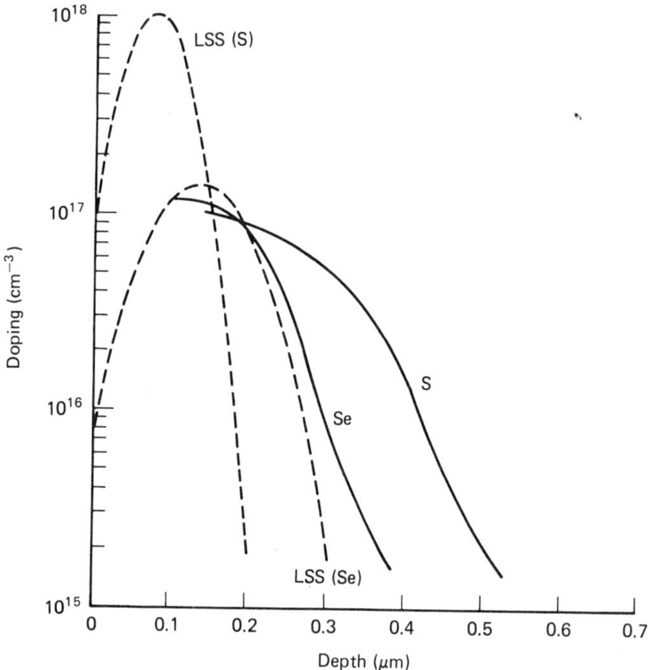

Fig. 6.24 Annealing of selenium and sulfur implants in gallium arsenide. From Eisen [22]. With permission from the publisher, Gordon and Breach, Inc.

The predictable characteristics of selenium make it an extremely useful dopant for the implantation of *n*-type layers into insulating substrates, which serve as a starting point for field effect transistors and integrated circuits. Here control of both implant doping and depth must be precise since they establish the threshold voltage of the transistor. The effect of substrate quality on the doping profile is shown in Fig. 6.25, where results are presented for two different chromium-doped substrates, as well as for implantation into an epitaxial high-resistivity buffer layer. In all cases 1.8×10^{12} cm^{-2}, 400 keV selenium ions were implanted at 350°C and annealed with a silicon nitride cap at 850°C for 30 min. Here implantation into a "good" substrate gave the same results as for an epitaxially grown buffer layer, in close agreement with LSS theory.* Technique for the selection of "good" and "bad" substrates involve the requirement that there be no change in resistivity after capping, implanting with an inert ion such as krypton to simulate implantation disorder, and subjecting to the anneal step.

The use of aluminum oxynitride as a capping material has been investigated in work with this dopant. Annealing data showed a factor of 2–3

* Undoped material, made by in situ compounding techniques described in Chapter 3, behaves as a "good" substrate with selenium implantation.

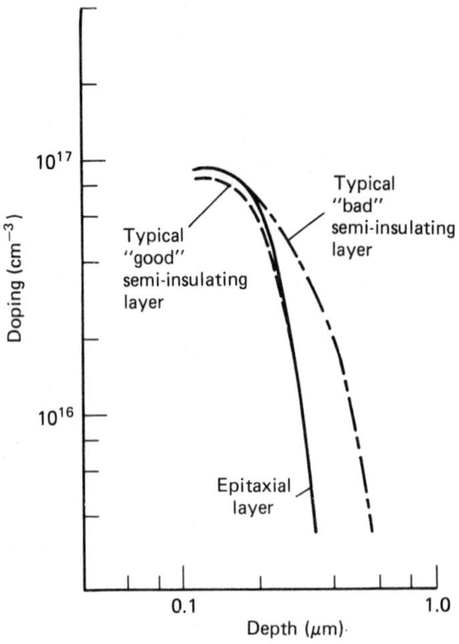

Fig. 6.25 Effect of selenium implantation into semi-insulating gallium arsenide. From Eisen [22]. With permission from the publisher, Gordon and Breach, Inc.

improvement in peak carrier concentration over silicon nitride, for ion doses above 10^{13} cm^{-2}. No advantage of this cap material was seen at lower doses.

The behavior of tellurium [26] is relatively similar to that of selenium. However, it has a higher atomic weight so that implantation to the same depth is accompanied by more damage than for selenium. Doping profiles are found to follow LSS theory at low ion doses, with increasing tail formation at higher levels. Again, the use of aluminum oxynitride caps resulted in an improvement of a factor of 2–3 over silicon nitride at high dose levels.

Semi-insulating. Both *p*- and *n*-type gallium arsenide can be made semi-insulating by proton bombardment. Here 10^{13} H^+ ions/cm^2 in the 100 keV to 3 MeV range [16] have been used to create a disordered layer whose thickness is about 1 μm per 100 keV of implant energy. Layers as thick as 30 μm can be formed in this manner. Free carrier concentration in these layers is under 10^{11} cm^{-3}, so they can be considered to be semi-insulating for all practical purposes. Proton bombardment is useful for isolating devices on a common substrate, and is an important technique for microcircuit fabrication. Both single- and multiple-energy proton bombardments have been used [27, 28] in this application. Deuteron bombardments have also been investigated for this purpose.

One disadvantage of device isolation by proton bombardment is that significant annealing can occur above 350°C. Oxygen implantation [29] has been used to avoid this problem, since it introduces a chemical deep level

(E_c − 0.75 eV) species into the semiconductor. No annealing effects were seen with this dopant up to 800°C; however, implantation depths are much shallower than those obtained with proton bombardment, and are typically 1 μm for implants in the 1 MeV range.

6.4 ION IMPLANTATION SYSTEMS

The development of machines for ion implantation has been very rapid ever since their first application to semiconductor technology. Many recent improvements have been aimed at ease of operation, with sophisticated control electronics and automatic pump-down capability. The performance of these machines has continually improved at the same time, with special emphasis on uniform dose and large wafer handling capacity. In this section we outline some of the basic elements of these systems, without attempting to provide a complete catalog of all these improvements and refinements.

The basic features of an ion implantation machine [30] are a variety of ion sources, means for ion acceleration to energies in the 30–500 keV range, and systems for beam purification and manipulation over the semiconductor target. A number of additional components are used to make the system versatile and suitable for semiconductor applications.

1. Ion sources are generally impure, so that means must be provided for mass separation, and for rejection of undesired species.
2. There is generally a spread of energies attained by the ions upon acceleration. Velocity filtering is sometimes used to provide a monoenergetic beam. In addition, quadrupole lens systems, placed after the accelerator column, are often used for focusing the beam on the target.
3. A wide variety of beam manipulation techniques can be used. These include electrical, magnetic, and mechanical systems. Moreover, hybrid schemes are often used when many slices must be handled at any one time.
4. Techniques for precise monitoring of the integrated ion current must also be provided. Perhaps the key advantage of ion implantation is in this area, where the repeatability of dose can be held to better than 1%.
5. Finally, a wide variety of high-voltage and vacuum techniques are employed to result in a system that is convenient to use. These include load-lock devices which allow substrate loading without breaking vacuum, and auxiliary ports which allow substrate loading on one channel while implantation is being carried out on another.

Figure 6.26 shows a block diagram of one such system in which the ion source is at a high voltage and the rest of the system, beyond the output of

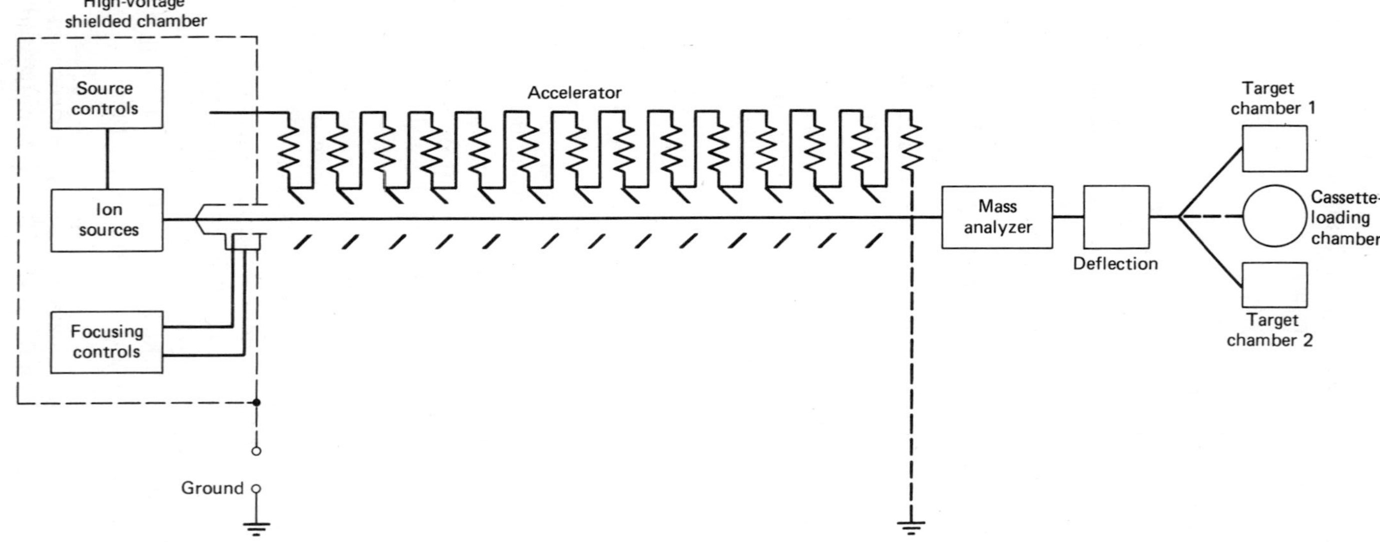

Fig. 6.26 Block diagram of ion implantation system.

the accelerator, is at ground potential. In this system all the ion sources, including their power supplies and control elements, are operated by remote means, as are the ion optical elements of the accelerator column. This lets the rest of the system be kept at ground potential, and allows for ease of access.

Separate valves and pumps are provided so as to allow individual access to the target chambers,* while keeping the rest of the system under high-vacuum conditions at all times. Particular care is taken to operate the target chambers with oil-free pumping systems (e.g., cryogenic, sorption, and turbomolecular), to avoid hydrocarbon cracking of residual oil vapors.

The separate elements of an ion implantation system are now considered.

6.4.1 Ion Sources

Ion sources usually consist of compounds of the desired species, and a means for their ionization prior to delivery to the accelerator column. The choice of materials for ion sources is very wide [31]; almost any ionizable compound can be used here. Gaseous materials are more convenient to use than solid ones since they avoid the necessity of using a vaporization chamber. Consequently, these are the preferred choice in implanation systems today.

Next, the material must be ionized. This is done by passing the vapor through a hot or cold cathode electronic discharge. A magnetic field is provided so as to force the electrons to move in a spiral trajectory, thus increasing the ionizing efficiency of the source. Also provided is a means for extracting the ions from this discharge through an outlet, from which they are fed to the accelerator column. This outlet is usually circular or in the form of a rectangular slit, and defines the geometry of the ion beam. Figure 6.27 shows a sketch of a gas-fed ion source [32], and illustrates many of the features described here.

The effectiveness of an ion source is measured by the magnitude of ion current delivered to the accelerator, and ultimately to the target. If I is the ion current, in amperes, for a species of charge state s, then the rate at which ions arrive at the target is I/qs per second, where q is the electronic charge. Thus a singly ionized beam delivers $6.25 \times 10^{18}\, I$ ions/s to the target. Finally, if A is the target area in cm^2, and t is the implantation time in seconds, the ion dose is given by

$$Q_0 = 6.25 \times 10^{18}\, It/A \text{ ions cm}^2 \tag{6.35}$$

A large beam current is thus highly desirable in commercial systems where many wafers must be handled with a minimum of machine time. In

* In some systems, a cassette loader, with a load-lock system, is used to facilitate slice handling.

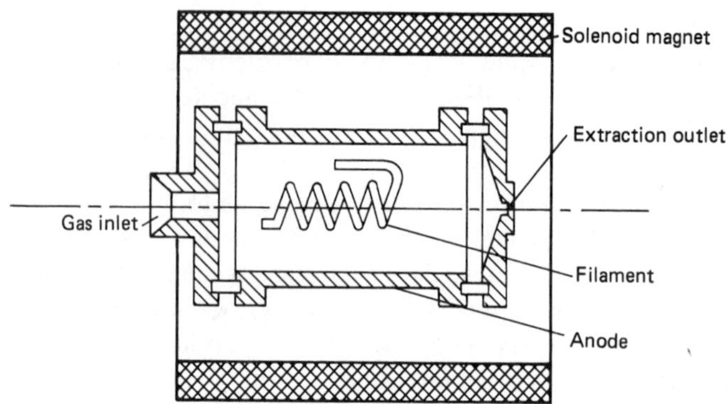

Fig. 6.27 A Nielsen-type gaseous source. Adapted from Nielsen [32].

practice, the beam current is actually a function of the machine design, the ion extraction technique, and the ion source material. By way of example, at the present time high-current arsenic or phosphorus ion beams can be more readily produced than beams of boron. Typical values for a general-purpose, production machine are 800, 800, and 500 μA, respectively, for these ions. These values are for the total beam current, which is reduced by scanning through a circular mask (to cover a semiconductor slice). Typical values of beam current delivered to a 10 cm diameter slice are 500, 500, and 300 μA, respectively, for these ions.

Recently machines have become available with beam currents that are more than an order of magnitude higher than these values. In general, however, the beam energy of these machines is relatively restricted (below 100 keV), so that they are primarily used to replace the predeposition step in diffusion processes.

6.4.2 The Accelerator

Energy is imparted to the ion beam by passing it through a long column across which the accelerating potential is established. The output end of this tube is usually maintained at ground potential for safety reasons.

The beam energy determines the projected range of an ion. However, it is possible to increase this range by using multiply charged ions. This usually results in a lower beam current. For example, a heated elemental arsenic source typically produces As^{+} and As^{2+} ions in a 10:1 ratio. Thus use of an As^{2+} beam results in a large penalty in beam current. For this reason, it is customary to use singly charged ion beams whenever possible.

6.4.3 Mass Separation

Ion source materials are usually compounds of the ion that is required for semiconductor doping (e.g., BF_3, B_2H_6, BCl_3). In addition, they usually produce one or more charged species upon ionization. Moreover, they often contain other impurites as contaminants, even if every effort is made to use chemically pure reagents. As a result, some form of mass separation is usually essential in a practical ion implanter.

Mass separation techniques are highly refined. They provide a unique distinction between ion implantation and diffusion, in that a variety of dopants can be handled in a single machine, with complete freedom from contamination with each other. The most commonly employed technique utilizes a homogeneous field magnetic analyzer. Its principle is based on the dynamics of a charged particle, of mass m and velocity v, moving at right angles to a uniform magnetic field with a flux density B. This particle will experience a force F such that

$$\mathbf{F} = q(\mathbf{v} \times \mathbf{B}) \tag{6.36}$$

This tends to move the particle in a circular path of radius r, thereby creating a centrifugal force mv^2/r. These forces must be equal and opposite. In addition, the velocity of the particle is related to its energy by

$$\tfrac{1}{2}mv^2 = qV \tag{6.37}$$

where V is the accelerating potential. Combining these relations, the radius of the ion path is given as

$$r = \frac{1}{B}\left(\frac{2mV}{q}\right)^{1/2} \tag{6.38}$$

Thus for a given extraction voltage and magnetic flux density, the path radius is directly proportional to the square root of the mass. Trajectories for three different masses are shown in Fig. 6.28. From this figure it is seen that ions of any particular mass can be selected by the appropriate placement of an exit slit.

Velocity filters are sometimes used when extremely precise control of the range is desired. The most common is the $\mathscr{E} \times \mathbf{H}$, or Wien, filter. This filter consists of a chamber through which the beam passes in a direction that is perpendicular to a crossed electric and magnetic field system. Such a beam will travel straight through if the electric and magnetic forces balance each other. If $\mathscr{E}$ and H are the electric and magnetic field intensities, then the condition for zero deflection is given by

$$q\mathscr{E} = Hqv$$

where v is the ion velocity. Thus ions of velocity v pass through such a filter without disturbance, provided that

$$v = \mathscr{E}/H \tag{6.39}$$

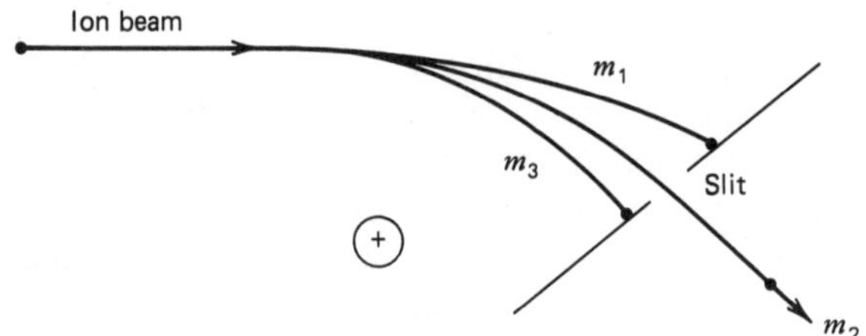

Fig. 6.28 Mass separation through a slit.

The Wien filter can also be used for mass analysis. Thus an ion of mass $M + \Delta M$, traveling at a velocity $v + \Delta v$, would experience a net deflection force F, where

$$F = Hq(v + \Delta v) - q\mathscr{E} \tag{6.40}$$

This would move it along a circular trajectory of radius r, where

$$F = \frac{(M + \Delta M)\,(v + \Delta v)^2}{r} \tag{6.41}$$

Combining Eqs. (6.40) and (6.41), and noting that $v = (2qV/M)^{1/2}$, gives

$$r = \frac{2V}{\mathscr{E}}\left[\left(\frac{M}{M + \Delta M}\right)^{1/2} - 1\right]^{-1} \tag{6.42}$$

resulting in a deflected beam which can be mass separated by means of an exit slit.

Neutral particles are not deflected by any of these techniques. A bend in the ion beam line, of 5–10°, is usually made so that they can be picked off by a collector plate.

6.4.4 Beam Scanning

The primary requirements of commercial implantation systems are uniform coverage and the ability to handle a large number of slices in a single pump-down. This is often accomplished in a hybrid deflection system where the beam is electronically scanned in one direction while the slices are mechanically moved in the other. One such system, shown in Fig. 6.29, consists of electrostatic deflection combined with a rotating carousel which can hold many wafers. Casette loaded implanters, on the other hand, operate on one wafer at a time. Here electronic x and y scanning is more common, with interleaving.

During implantation, practical considerations preclude the possibility of mounting wafers in good thermal contact with the slice carrier; often pres-

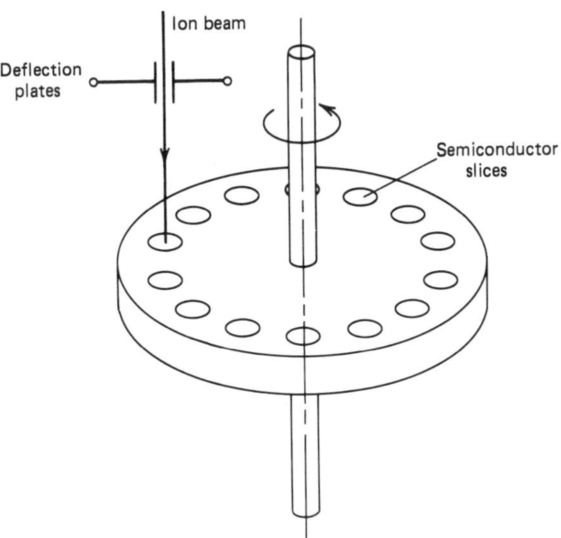

Fig. 6.29 Electromechanical deflection system.

sure fingers or slides are used to hold these slices. In addition, implantation is carried out in a vacuum environment so that the primary mechanism of heat loss from the slice is by radiation. Ion beam heating effects can thus be quite severe, and often limit the rate at which energy can be deposited in a slice.

Figure 6.30 shows the equilibrium temperature that is reached by slices during implantation for a variety of sample conditions, on the assumption that heat loss is only by radiation [30]. Although crude, this calculation allows an estimate of the extent of beam heating effects for a variety of implant conditions. It is seen that temperature excursions of many hundreds of degrees are possible in currently available systems. Furthermore, the temperature across a slice is extremely uneven, since it is usually held at three or four points along its edge. This uneven heating creates a major problem when it is necessary to pre-implant a semiconductor with a heavy ion in order to make it amorphous, or when implantation is to be carried out in cooled substrates for the same purpose.

Thermal effects place a limit on the beam current density that can be used during implantation, and thus on the implantation time required to achieve a desired doping level. A convenient way around these limitations is to reduce the effective beam current density by scanning many slices at any one time, as shown in Fig. 6.29.

Silicon is extremely resistant to thermal shock. Consequently, casette loading, with slices implanted one at a time, is the favored approach for silicon wafer implantation in machines where the beam current is less than 1 mA. However, carousel scanning, of the type shown in Fig. 6.29 must be

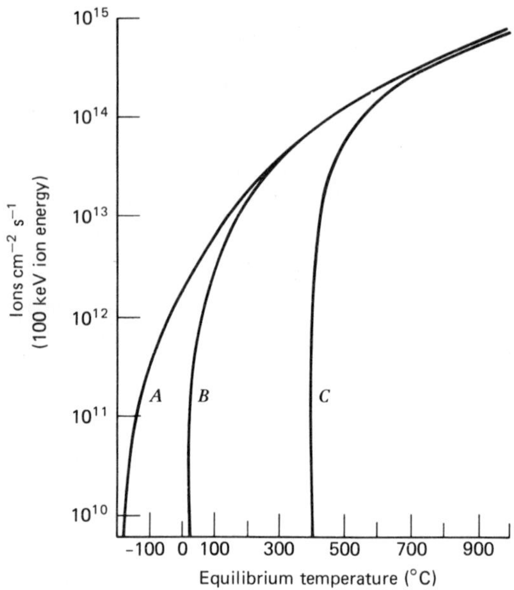

Fig. 6.30 Equilibrium target temperature versus ion dose. *A*—target on 77°K heat sink; *B*—uncooled target; *C*—target heated to 400°C. Adapted from G. Dearnaley, J. H. Freeman, R. S. Nelson, and J. Stephen *Ion Implantation* [30], 1973. North Holland Publishing Company.

used in the higher current "predeposition" machines, and also for the ion implantation of gallium arsenide, which is a considerably more fragile material.

6.4.5 Beam Current Measurements

Dosimetry is an important advantage of ion implantation technology, and is done by measuring the integrated beam current to obtain the total dose. Accurate measurement is a difficult task, and is only achieved if care is taken to eliminate electrons, neutrals, and negative ions from the beam, as well as secondary particles which can be emitted as a result of the ion bombardment of the target. Improvements in machine design have resulted in minimizing the error sources generated from the beam, so that the major problems of measurement are those caused by the secondary particles. Faraday cage arrangements, with both electrostatic and electromagnetic suppression schemes, are used to minimize measurement errors from these secondaries. A schematic of a Faraday cage arrangement is shown in Fig. 6.31. Often a small magnetic suppression flux ($\leq$ 100 gauss) is also incorporated by means of a permanent magnet. Techniques of this type allow dose measurements to be made over six decades of ion flux, with an accuracy of better than 0.5% on all ranges.

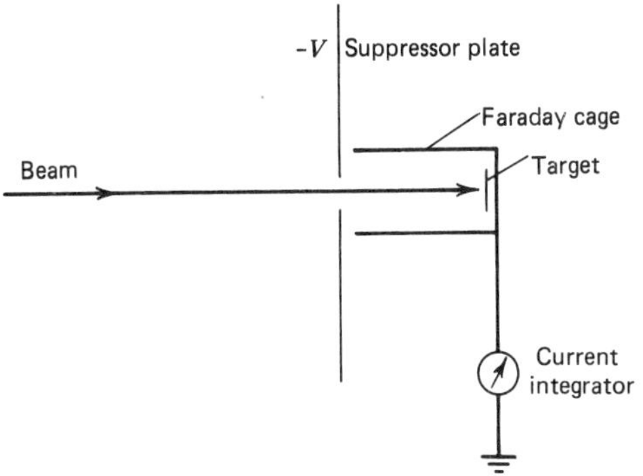

Fig. 6.31 Dosimeter arrangement.

6.5 PROCESS CONSIDERATIONS

Ion implantation is an extremely flexible process. As such, it is not only used as an alternative to diffusion, but also in many ways that are quite unique to this approach. Thus a number of different process considerations must be understood if full advantage is to be taken of this technology. Some of these are now described.

6.5.1 Diffusion Effects During Annealing

We have noted that implantation into an amorphous target (or a crystalline target that is misoriented) results in a gaussian ion profile, at least for a few decades down from the peak ion concentration. Implantation is generally followed by an annealing step so that this profile is modified by diffusion effects. Let D and t be the diffusion coefficient and the time associated with this process, and assume that the substrate is infinitely thick on either side of the implant. For this approximation, the ion concentration will still retain its gaussian shape, with the same mean value. However, the standard deviation will be altered from σ to $(\sigma^2 + 2Dt)^{1/2}$, so that the distribution is now of the form

$$\phi(x) = \exp\left[-\frac{1}{2}\left\{\frac{x - \bar{x}}{(\sigma^2 + 2Dt)^{1/2}}\right\}^2\right] \tag{6.43}$$

The ion concentration is thus given by

$$N(x) = \frac{Q_0}{(2\pi)^{1/2}(\Delta R_p^2 + 2Dt)^{1/2}} \exp\left\{-\frac{1}{2}\left[\frac{x - R_p}{(\Delta R_p^2 + 2Dt)^{1/2}}\right]^2\right\} \tag{6.44}$$

This equation assumes that an appropriate value of D is chosen to characterize diffusion processes during the anneal stage. With silicon, the value of lattice diffusion is usually a reasonable choice. With gallium arsenide, however, it is sometimes necessary to choose a value that is three to four decades larger, depending upon the interaction of the dopant and the lattice defects during the annealing step. Thus use of the diffusion constant gives reasonable values for tellurium and selenium implants, but not for sulfur. (See, for example, Fig. 6.24.)

The presence of the surface, at a finite distance from R_p, can be taken into consideration by assuming perfect reflection at this boundary. An approximate solution for this situation, assuming no out-diffusion effects, has been given [33] as

$$N(x, t) \simeq \frac{Q_0}{(2\pi)^{1/2}(\Delta R_p^{\,2} + 2Dt)^{1/2}} \left(\exp\left\{ -\frac{1}{2}\frac{\Delta R_p}{R_p}\left[\frac{x - R_p}{(\Delta R_p^{\,2} + 2Dt)^{1/2}}\right]^2 \right\} + \exp\left\{ -\frac{1}{2}\frac{\Delta R_p}{R_p}\left[\frac{x + R_p}{(\Delta R_p^{\,2} + 2Dt)^{1/2}}\right]^2 \right\} \right) \tag{6.45}$$

A plot of this equation, for the specific set of values shown, results in the doping profiles of Fig. 6.32. It is interesting to note that, for sufficiently large Dt values, the bell shape is replaced by the type normally expected from diffusion. However, appreciable movement of the peak ion concentration from R_p does not occur until Dt is at least equal to $2.5(\Delta R_p)^2$.

Equations (6.44) and (6.45) can also be used to evaluate the effect of successive processing steps, by replacing Dt with $\sum_{i=1}^{n} D_i t_i$, where D_i and t_i are the appropriate diffusion constant and time for the ith process step.

6.5.2 Multiple Implants

There are a number of situations where it is desirable to make multiple implants. One such case, described earlier, is pre-implantation of silicon with an inert ion in order to make it amorphous, prior to deposition of the impurity. This allows close control of the doping profile to many decades below its peak concentration, and also permits nearly 100% carrier activation at temperatures as low as 600°C. In situations where a deep amorphous region is required, it is necessary to make a series of such implants, at varying energies and doses. One experiment described in [19] used implants of neon at 40, 70, 140, and 380 keV, with doses of 10^{15}, 10^{15}, 10^{15}, and 2.5×10^{15} ions/cm^2, respectively, to create a deep, uniformly damaged region of this type, prior to boron implantation.

Multiple H^+ implants, for the purpose of forming a semi-insulating region in gallium arsenide, have been found [28] to produce significantly more temperature-stable semi-insulating material than is obtained by a single implant. This improvement comes about because the compensation of gallium arsenide by ion bombardment most probably involves many defect levels

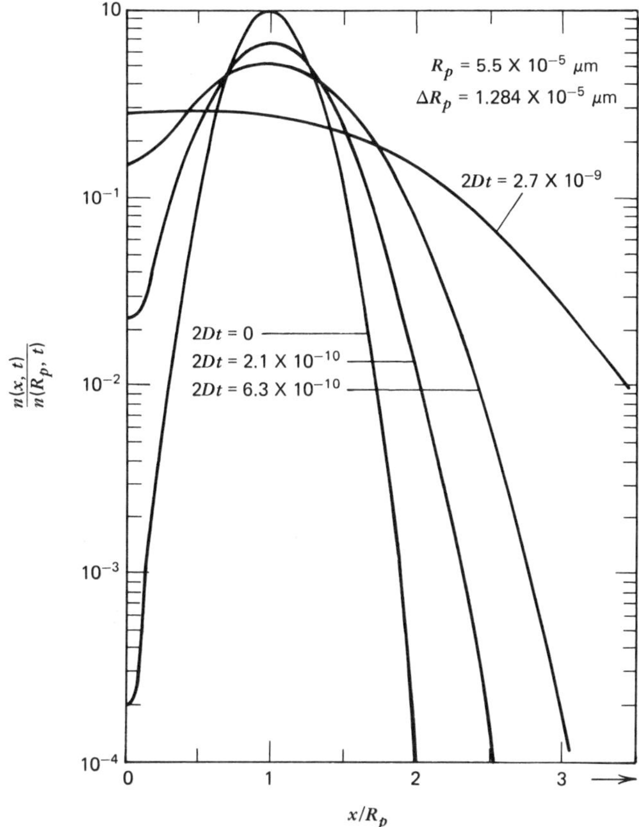

Fig. 6.32 Effect of diffusion on an implant profile, assuming no out-diffusion. From Seidel and MacRae [33]. With permission from the Metallurgical Society of the American Institute of Mechanical Engineers.

which anneal at different temperatures. Consequently, annealing a single implanted device results in many regions of different thermal behavior. On the other hand, uniformity of the defect level structure, as obtained by multiple implants, results in material whose thermal behavior can be better controlled.

A different situation is the case where it is desired [34] to obtain a deep, flat doping profile, as shown in Fig. 6.33. Here four boron implants into silicon were used to provide a composite doping profile, as shown. The actual carrier concentration, as well as that predicted by the LSS theory, are shown in this figure. Unique profiles, unavailable by diffusion technology, can also be obtained by using various combinations of dose and energy for such implantations. One is the profile for a high-efficiency Impatt diode using a double drift region [35].

Multiple implants have been proposed [36] as a means for preserving

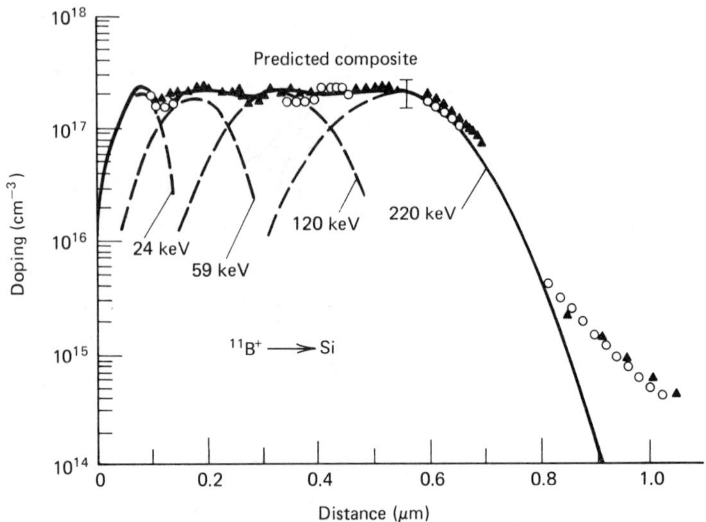

Fig. 6.33 Composite doping profile using multiple implants. From Lee and Mayer [34]. © 1974 by the Institute of Electrical and Electronic Engineers, Inc.

stoichiometry during the implantation and annealing of gallium arsenide. This approach, whereby equal amounts of gallium and an *n*-type dopant (or arsenic and a *p*-type dopant) are implanted prior to annealing, has been suggested as a technique for obtaining higher carrier activation. Results with this technique appear promising. However, there is considerable spread in the data, and further work along these lines is necessary before a definitive evaluation of this approach can be made.

6.5.3 Masking

An unusual feature of ion implantation is that a wide variety of masking materials can be used, since implantation is a low-temperature process. The thickness required for masking is a function of the stopping parameters of the mask material, and can be readily calculated.

Consider the effect of a beam of incident ions on a mask material of infinite thickness, as shown in Fig. 6.34. For this material the dose which is deposited* in the region beyond a depth d (shown shaded) is given by integration of Eq. (6.5) as

$$Q = \frac{Q_0}{(2\pi)^{1/2}\,\Delta R_p} \int_d^{\infty} \exp\left[-\frac{1}{2}\left(\frac{x - R_p}{\Delta R_p}\right)^2\right] dx \qquad (6.46)$$

* It goes without saying that the actual profile of this species depends on its range statistics in the semiconductor, and is not given by the curve of Fig. 6.34.

where Q_0 is the incident ion dose, and R_p and ΔR_p refer to the mask material. Noting that

$$\int_d^\infty e^{-x^2}\,dx = \frac{(\pi)^{1/2}}{2}\,\text{erfc}\,d \tag{6.47}$$

the fraction of the dose that is deposited beyond a depth d is given by

$$\frac{Q}{Q_0} = \frac{1}{2}\,\text{erfc}\left(\frac{d - R_p}{\sqrt{2}\,\Delta R_p}\right) \tag{6.48}$$

To an approximation, this is also the fraction of the incident dose that penetrates a mask of thickness d. From this equation it is seen that a mask of thickness $3.72\Delta R_p + R_p$ is required for a masking effectiveness of 99.99%, or $4.27\Delta R_p + R_p$ for 99.999%. These values should only be used as guidelines since the LSS parameters for most mask materials vary with the manner in which they are deposited or grown. Figures 6.35–6.37 show the minimum thicknesses, based on these considerations, for 99.99% effective masks made of photoresist, silicon dioxide, and silicon nitride, respectively. These materials are most commonly used in device processing, and can also provide the masking function for ion implantation.

Mask thicknesses given in these figures are for boron, phosphorus, and arsenic implants into silicon [30]. However, they can be also used as approximate guidelines for impurity masking in gallium arsenide. Masking data for impurities in gallium arsenide are shown in parentheses in Figs. 6.35–6.37.

There are many practical considerations in the use of masking materials for ion implantation. For instance, the effect of the high-energy beam on an

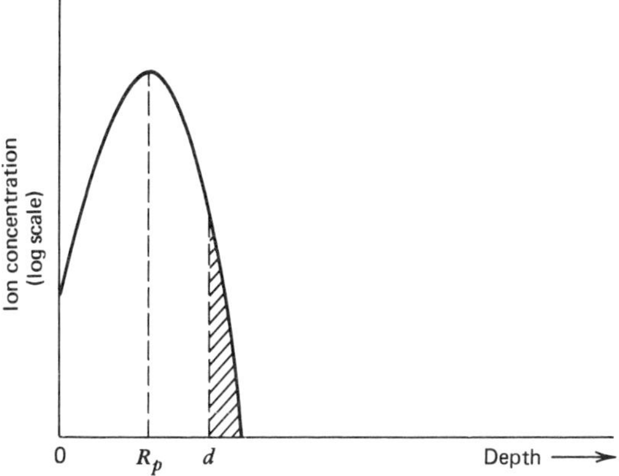

Fig. 6.34 Ion penetration beyond a depth *d*.

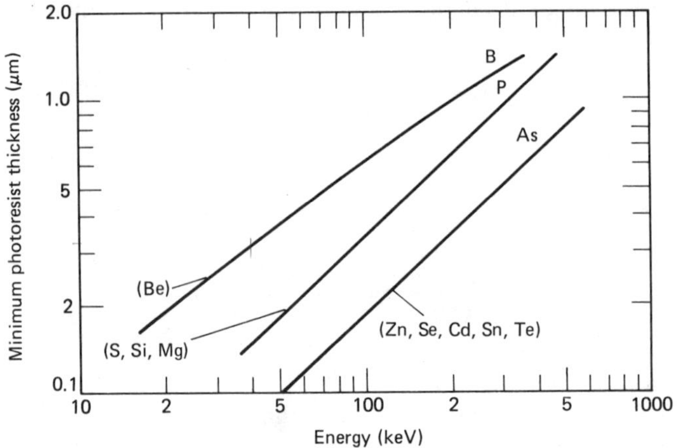

Fig. 6.35 Minimum photoresist thickness for a masking effectiveness of 99.99%.

insulating surface is to cause considerable ionization. This prevents surface charging to high voltages and rupture of the insulator by breakdown, as long as there are conducting paths within 500 μm of this surface. This condition is readily met in microcircuits, so that no special precautions or grounding arrangements have to be provided for implantation of this type.

Ion bombardment of photoresist material results in its becoming heavily

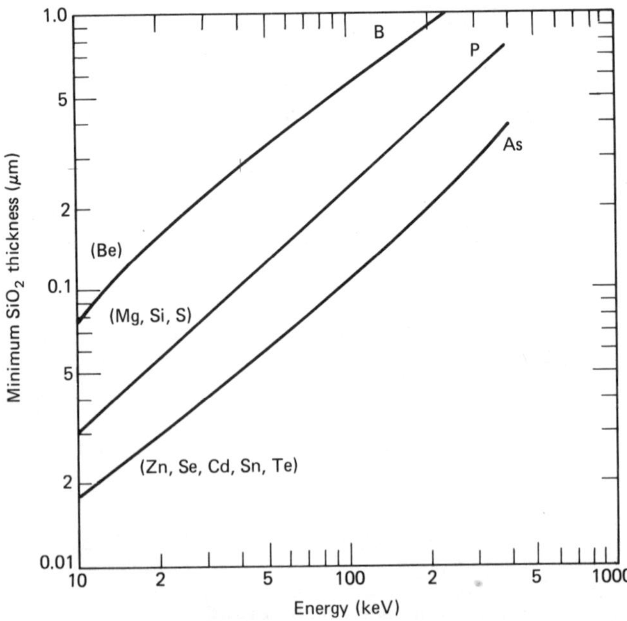

Fig. 6.36 Minimum SiO_2 thickness for a masking effectiveness of 99.99%.

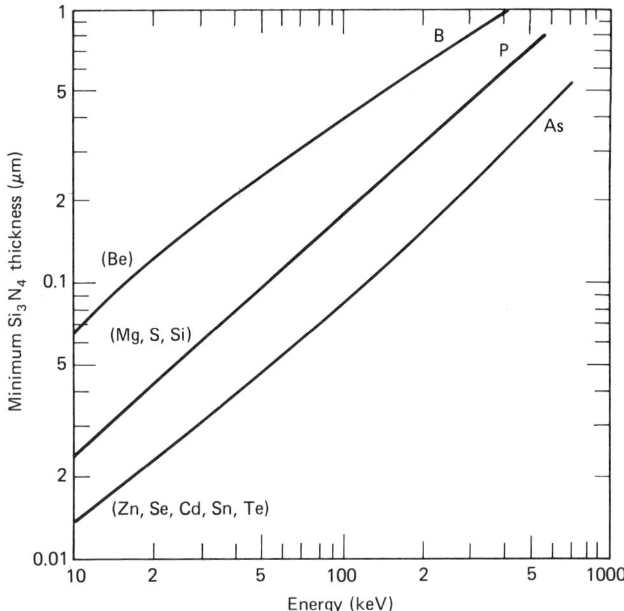

Fig. 6.37 Minimum Si_3N_4 thickness for a masking effectiveness of 99.99%.

cross-linked, and leads to difficulty in its subsequent removal by standard photoresist strippers. One solution here is to make the photoresist film much thicker than is normally dictated by masking requirements. This results in an undamaged layer in direct contact with the semiconductor. This region can be attacked by standard reagents, so that the entire layer can be removed by undercutting and subsequent lift-off. An alternate approach is to use an oxygen plasma for photoresist removal.* In extreme cases, bombardment with high-dose, high-energy ion beams can lead to cracking of the photoresist. It goes without saying that this material cannot be used in these situations.

Ion implantation through a silicon dioxide layer results in creating damage in this material. This has a number of consequences in practical situations. One is that photoresist adhesion is impaired, and there is often lift-off during subsequent masking operations. This lift-off is caused by capillary action of the etchants under the photoresist. Here, too, plasma etching of the oxide is sometimes used to avoid this problem. A second consequence is that the etch rate of the oxide is increased, often by a factor of 2–3. This enhancement is a function of the oxide growth technique, and also of the ion type and energy. It is caused by damage due to the implantation, and can be annealed [37] by heat treatment between 800 and 1000°C.

The increased etch rate of implantation-damaged oxide can be used to

* This technique is described in Chapter 9.

form oxide cuts with sloping sides [38]. This allows good step coverage in the subsequent metallization, and is particularly useful in MOS technology where large changes in oxide thickness are encountered, as is the case in going from the gate oxide (250–1000 Å) to the field oxide (usually > 1 μm).

Ion implantation will sometimes adversely affect the thermal properties of silicon nitride. This material, which is often used as a cap during the annealing of gallium arsenide, has been found to lose its adherence to substrate with subsequent bubbling and lift-off. Again, these problems have only been encountered in high-dose situations, and for high-energy implants. In addition, the severity of the problem is related to the technique by which the nitride layer is grown, and also to its growth rate.

Deposited metal films can also be used as masks for ion implantation. However, some of the mask material can be driven into the substrate during implantation. One approach here, which prevents accidental doping from the residual metal, is to place a thin silicon dioxide or silicon nitride layer between the film and the underlying semiconductor.

It is also possible to implant a semiconductor with its capping layer already in place. In the case of gallium arsenide, this avoids the problems of dopant loss during the deposition of the silicon nitride cap layer, which is usually performed around 700–800°C. With silicon there are additional advantages in the form of surface passivation, so that most silicon implants are performed in this manner. Details of this important technique are discussed in Section 6.5.5.

6.5.4 Contacts to Implanted Layers

The conventional approach for making a contact to a semiconductor region is to deposit the contact metal on it by sputtering or vacuum evaporation, and to heat treat the combination to form a "microalloy" at this interface. This process tends, in practice, to be relatively uneven, with actual contact formation occurring by relatively deep penetration at a number of discrete alloying sites. It is not suitable with ion-implanted layers (or, for that matter, with any type of extremely thin layer), because of the possibility of shorting to the underlying semiconductor. Thus the making of contacts to implanted layers presents a special problem since these layers are extremely shallow.

One approach here consists of forming a deep-diffused region of the same impurity type, connected to the implanted layer, and making the metal contact to this region (see Fig. 6.38*a*). This technique is sometimes not possible, as in the case of a device where this might short out an underlying layer. Here ohmic contact to the shallow implanted region must be made by forming a Schottky barrier contact to the semiconductor, as shown in Fig. 6.38*b*. This requires the choice of a material of suitable work function, which will adhere to the semiconductor. Materials such as titanium and chromium are often used here. An alternate approach with silicon is to use a metal such as platinum or molybdenum, and form its silicides in situ by

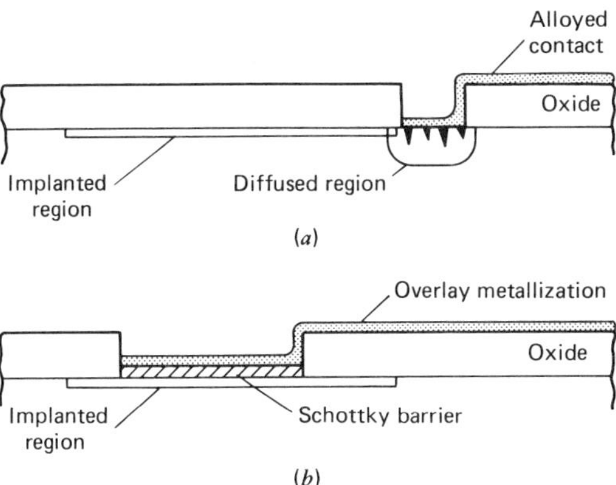

Fig. 6.38 Contacts to implanted layers.

a suitable heat treatment. An overlay metallization is next connected to this Schottky barrier region. The details of both alloy and Schottky-type contacts are outlined in Chapter 8.

6.5.5 Implantation through an Oxide

One of the more significant advantages of silicon for microcircuit fabrication is the fact that its surface can be covered by a dense amorphous layer of grown silicon dioxide. The surface protection afforded by this layer can be maintained by making implantations through this oxide, so that the bare silicon surface is not exposed during processing. Although this results in the formation of charge states in this material, it has been pointed out in Section 6.3.1 that they are readily annealed during subsequent processing.

Implants may be made through an oxide in a number of different ways. This is shown in Fig. 6.39 for the case of silicon covered with a 1200 Å oxide layer, for three different boron implant conditions [34]. Profile *A* represents a situation where the oxide layer allows only a small amount of the total ion dose to be deposited in the silicon. The actual fraction is not only small, but is also a critical function of the oxide quality as well as the implantation parameters. An oxide layer of this type serves essentially as a mask; in a practical application, a slightly thicker film would be used to almost completely prevent boron penetration into the silicon. Profile *B* deposits a large fraction of the dose in the silicon. This is useful in doping the semiconductor while still keeping its surface covered by a passivating film; in a practical situation, a somewhat higher ion energy would result in almost no dopant loss due to the oxide.

Profile *C* illustrates the case where one half of the ion distribution is deposited in the silicon, while the rest is retained in the oxide. Ideally, implantation of group III or V species (such as boron, phosphorus, or arsenic) in this manner results in the placement of a controlled quantity of

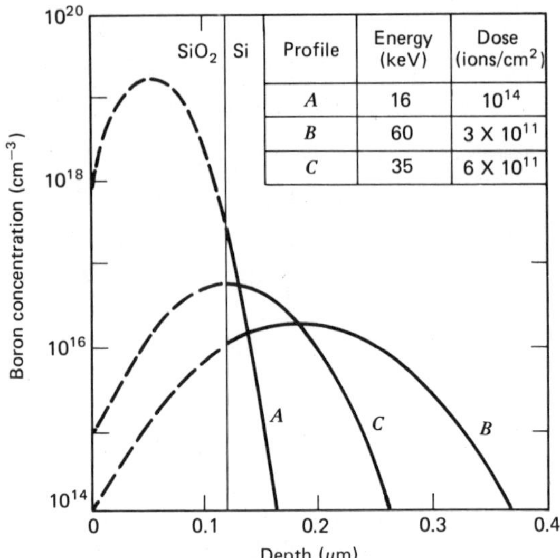

Fig. 6.39 Implantation through an oxide. From Lee and Mayer [34]. © 1974 by the Institute of Electrical and Electronic Engineers, Inc.

sheet charge at the $Si–SiO_2$ interface. This charge, if placed in the channel region, can be used to tailor the threshold voltage of MOS devices.

Specific applications for implantation through an oxide now follow.

6.5.5.1 Junctions

In silicon microcircuit technology it is important that the region where the junction penetrates to the surface be covered by an oxide layer. This is achieved automatically during junction formation by diffusion techniques, since the lateral spread of the impurity beyond the edge of the window is approximately equal ($\simeq 75\%$) to the junction depth. Implantation through an oxide window does indeed result in some junction penetration beyond the oxide edge, due to transverse straggle effects described in Section 6.1.4. However, this penetration is extremely small, and is often not sufficient to protect the junction from contamination during further processing.

One approach to obtaining junction protection is to drive-in the implant with a high-temperature diffusion step. A second technique is to use a double photolithographic process, as shown in Fig. 6.40. Here a mask is used to delineate the implant, which is made through the oxide (see Fig. 6.40*a*). Next the oxide is cut to form the contact region for the implant (see Fig. 6.40*b*). In this manner, the junction edge is far beyond the contact hole and is thus well protected from contamination. Contacts of the type illustrated in Fig. 6.38 are used to make connection to this region.

All oxide-masked junctions have curvature at the window edge. This increases the electric field over that obtained with a parallel-plane structure

[39], and causes premature breakdown. This problem is especially severe with ion-implanted junctions which are shallow and have small radii of curvature.

The radius of curvature of an ion-implanted junction is primarily due to transverse straggle, which is somewhat smaller than the projected range. The actual shape, however, is also determined by the profile of the oxide cut, which is never perfectly vertical. Thus its breakdown voltage cannot be calculated with any precision. In general, however, implanted junctions are found to behave as if they had a radius of curvature of 1–1.5 μm at the edge of the oxide cut. This serves as a good design rule for calculating their breakdown characteristics.

Other approaches have also been used for junction protection. These include direct implant through an oxide window with intentionally tapered edges, and rely upon the variable penetration range provided by this oxide [38]. Guard ring structures can also be used, where a deep diffusion region is provided all around the implant edge. Such a structure, shown in Fig. 6.41, has the additional advantage of producing depletion layer curvature in the opposite direction to that obtained by a single diffusion (or implant). As a consequence, breakdown voltages close to the bulk value can be achieved in this manner [40]. Structures of this type are widely used in avalanche photodiodes which require controlled current multiplication characteristics.

6.5.5.2 Resistors

One very important use for ion implantation through an oxide is in the fabrication of precise resistors for microcircuit technology. Typically, diffused resistors are limited to low sheet resistances (125–180 Ω/square) and

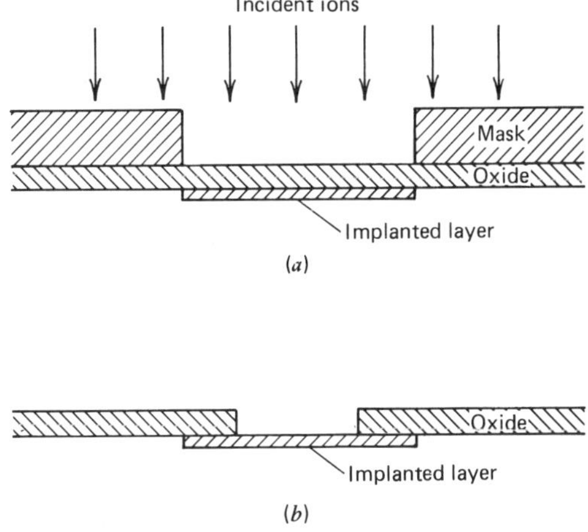

Fig. 6.40 Passivated junction formed by double photolithography.

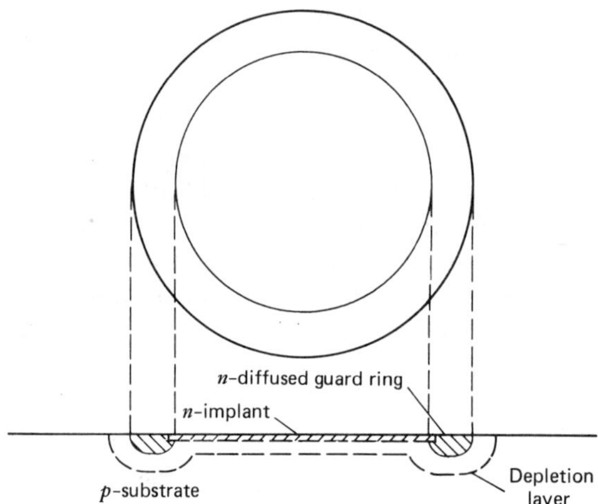

Fig. 6.41 Guard ring structure.

are at best only settable to within ± 10% of their designed values. With implantation, however, almost any sheet resistance is possible, and settable to within ± 1%. Practical considerations limit the values to about 4000 Ω/square, since the resistor becomes extremely sensitive to surface states beyond this value, because of its light doping.

Implanted resistors find wide use in analog circuits where a variety of resistor values, with close settability, provide great freedom to the circuit designer. A typical boron-implanted resistor, which is compatible with existing diffused resistors in the same microcircuit, is shown in Fig. 6.42. Here contact is made to its ends by p^+-regions of the type used to form the diffused base in n–p–n transistors made on the same chip.

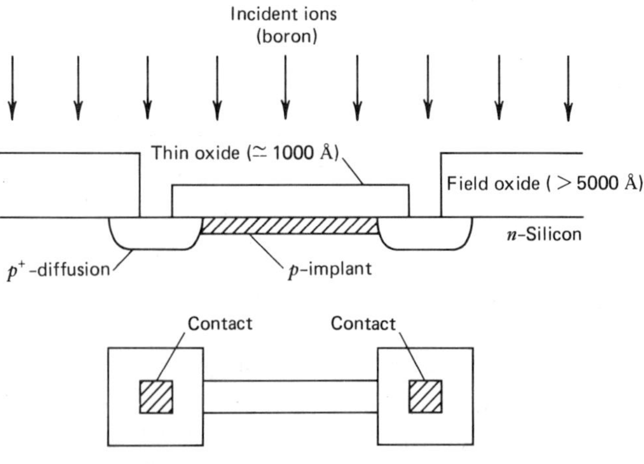

Fig. 6.42 Implanted resistor.

Higher value resistors can also be made by ion implantation. Here it is necessary to cover the resistor implant by a heavily doped shielding layer of opposite conductivity type, in order to make its performance independent of the surface states in the oxide. Values as high as 100 kΩ/square can be obtained in this manner. These resistors are highly nonlinear with voltage, however, because of depletion layer pinching effects during operation.

6.6 APPLICATION TO SILICON

The flexibility and the control available with ion implantation, combined with the ease of annealing, have made it rapidly accepted in the area of silicon technology. This section outlines areas in which this approach has provided a measure of uniqueness, with no attempt at being all-inclusive.

6.6.1 Bipolar Devices

Ion implantation is routinely used as a means for predeposition, prior to base diffusion in bipolar transistors. This is especially true for high-voltage devices, where the base must be lightly doped. In addition, there are a number of unique areas for exploitation of this technology, such as the following:

1. Ion implantation permits selective placement of doped regions on the same slice. Thus it can be used to fabricate high-performance vertical *p–n–p* transistors in the same microcircuit as *n–p–n* devices [41]. Conventional *p–n–p* devices are of the lateral type, and have extremely poor frequency performance as compared to their *n–p–n* counterparts.
2. Transistor fabrication by conventional means necessitates the formation of the emitter *after* the base. This results in the "emitter-push" effect described in Section 4.9.5. Ion implantation allows the formation of these regions in reverse order, so that this problem can be avoided.
3. Ion implantation can be used to make the doping concentration in a semiconductor *increase* from the surface into the bulk. This allows the fabrication of hyperabrupt devices such as the Schottky diode [42] shown in Fig. 6.43. These structures exhibit an anomalously large variation in capacitance with reverse bias, and are used in electronic tuning applications.
4. Extremely small emitter regions can be formed by this technique due to the absence of lateral diffusion effects. Thus ion implantation lends itself to the fabrication of VLSI circuits. Here arsenic is the preferred impurity since its doping profile after heat treatment is relatively abrupt, and free from deep tailing effects.
5. Arsenic is the ideal dopant for buried layers in bipolar microcircuits,

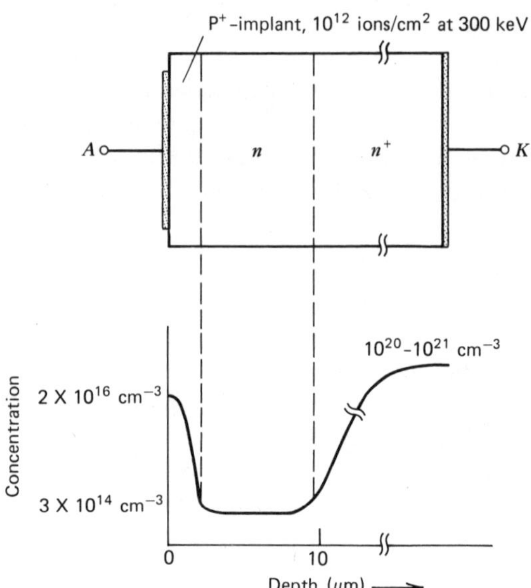

Fig. 6.43 Hyperabrupt Schottky diode.

because its tetrahedral radius is the same as that of silicon. However, buried layer formation requires a long drive-in cycle, which brings with it the serious problem of dopant loss by surface evaporation. Often double deposition and masking are used to replenish this loss. This problem can be greatly reduced by using a deep arsenic implant for predeposition before this drive-in step.

6. Avalanche transit time diodes, consisting of p^+–n–n^+ structures, are conventionally formed by making a p^+-diffusion into an n-epitaxial layer. Precise thickness control of this n-layer is essential for device operation in the 50–100 GHz range. Here the use of implantation for the p^+-region allows the width of the n-layer to be controlled by the epitaxial growth process, and not by a subsequent high-temperature p^+-diffusion step.

In summary, a number of unique characteristics of ion implantation have been exploited in these applications, and serve to indicate different ways in which this approach can be used with bipolar silicon technology. The list is by no means complete, but is indicative of the rapidly expanding use of this process.

6.6.2 MOS Devices

The greatest degree of success enjoyed by ion implantation today is in the area of MOS technology, because it presents a number of situations in which precise control of low-dose levels is required. In addition, the unique masking possibilities with implantation techniques allow considerable flexibility in the process steps, so that device parameters can be tailored to suit the needs of the circuit design.

The most straightforward use of ion implantation is in tailoring the resistivity of the starting silicon. This is done by implantation of an unmasked slice of higher resistivity than the desired value. This sets the resistivity of the surface region to a tight tolerance, even for wide variations in the starting resistivity.

An extremely important use of ion implantation is in the formation of the *p*-well in C-MOS circuits [43]. Here a precisely doped, deep *p*-region is required in *n*-type starting silicon. This region is used to form the *n*-channel transistor portion of the complementary pair. In this application, ion implantation is a direct replacement for the predeposition step in a conventional diffusion process. With predeposition, however, it is extremely difficult to obtain reproducible characteristics from run to run, because of the low carrier concentration requirements for this region.* Ion implantation, on the other hand, accomplishes this with ease, and has led to the successful implementation of C-MOS technology in large scale integration.

The well that is formed by this technique is generally quite deep in order to accommodate a complete MOS device. This requires a long drive-in step (typically 24 h at 1200°C) so that special precautions must be taken to avoid loss of dopant during this process. These include the use of high energies (150–200 keV) in order to deposit the boron ions deep into the semiconductor, as well as drive-in in a nonoxidizing ambient to minimize boron redistribution effects (see Section 4.8).

Ion implantation is ideally suited for controlling the threshold voltage of MOS devices. This is done by implantation through the gate oxide, and results in the deposition of an ionized charge species in the channel region. Use of a boron implant in an *n*-channel device (or a phosphorus implant in a *p*-channel structure) results in a reduction in the threshold voltage. For the ideal case of a sheet charge at the $Si–SiO_2$ interface, which is approximated by profile *C* of Fig. 6.39, this reduction is given by

$$\Delta V_T = \frac{Q}{C_{ox}} \tag{6.49}$$

where ΔV_T is the change in the threshold voltage, Q is the sheet charge at the $Si–SiO_2$ interface, and C_{ox} is the gate capacitance. This linear dependence of the change in threshold voltage with the dose allows for reproducible adjustment of this important MOS device parameter.

The effect of a small change in the gate oxide thickness, due to lack of process control during oxidation, is to create a change in both the gate

* A typical dose for this application is $(1–5) \times 10^{12}$ ions/cm^2.

capacitance as well as the implanted charge. These terms have opposing effects on the threshold voltage, and tend to compensate each other. However, it can be shown [44] that, for each oxide thickness, there is an optimum implant energy which will minimize the effect of these variations.

MOS threshold parameter control can be used in a number of different situations. Thus the direct reduction of V_T with implant dose allows the fabrication of low-threshold devices, which can be operated at low power dissipation levels. Threshold adjustment of *p*- and *n*-channel devices on the same chip allows the fabrication of matched complementary devices [45], and is the cornerstone of practical C-MOS circuits. Finally, it is possible to use this technique to fabricate depletion-type loads on the same substrate as enhancement-type devices. This has resulted in significant improvements in the speed and voltage swing capability of MOS circuits.

High-dose implantation through an oxide results in creating a defect structure which is difficult to anneal, even at temperatures as high as 1000°C. On the other hand, threshold control is usually accomplished at low dose levels. The surface state charge created at these levels can be readily annealed by heat treatment at 300°C. However, significant carrier activation to a reproducible fraction is only accomplished by heat treatment to 525°C or higher, so that it is customary to anneal at this temperature after the threshold-adjustment implant step.

Implantation through an oxide can also be used to increase the threshold voltage outside the channel region. This is desirable to avoid parasitic transistor action by metallization passing over the field oxide, or by surface inversion effects. Figure 6.44 shows how a channel stopper of this type may be formed by means of a single unmasked high-energy implant [46]. In this situation, a 200–300 keV boron implant is used for an *n*-channel MOS device. This implant has a large penetration depth under the thin gate oxide so that it has no effect on the MOS threshold voltage. However, the thick field oxide results in shallow implant penetration in the *p*-type field regions, where it increases the threshold voltage required to cause parasitic action. Multiple

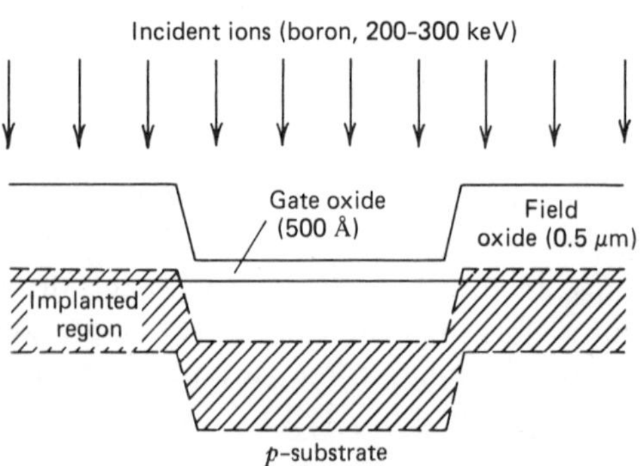

Fig. 6.44 Channel stopper implant.

implants can also be used for channel stopper regions. This represents a more conventional use of ion implantation techniques.

In MOS devices, the overlap of the gate electrode over the drain region creates a parasite feedback capacitance, and deteriorates the frequency response. Precision masking, combined with shallow source and drain diffusions, has been used to minimize this problem. Another approach, which avoids masking problems, is to use self-registered structures where a refractory gate acts as the mask for the source and drain diffusions (see Fig. 6.45*a*). This gate is conventionally made of polycrystalline silicon in order to withstand high diffusion temperatures, although a number of silicides and refractory metals can also be used (see Chapter 8).

Ion implantation offers the alternate approach [47] of building the source and drain region up to the gate edge, after the diffusions have been conducted (see Fig. 6.45*b*). The primary advantage of implantation here is that the gate overlap in this structure is due to transverse straggle effects, which are much less than the overlap due to lateral diffusion of the drain region. This results in a significant improvement in the frequency response.

A second possibility here is that an aluminum gate can be used, if the anneal step is performed at 400–500°C. This advantage is offset, however, by the fact that full carrier activation does not occur at these low temperatures. This leads to a parasitic resistance in the source and drain regions, and deteriorates the frequency response. With the refractory gate, however, annealing can be carried out at 900°C with full carrier activation. Consequently, the refractory gate is commonly used with this implantation technique.

Charge-coupled devices are another area where ion implantation can be used to either simplify the fabrication process, or provide improved device performance. In these devices, charge packets are transported along the

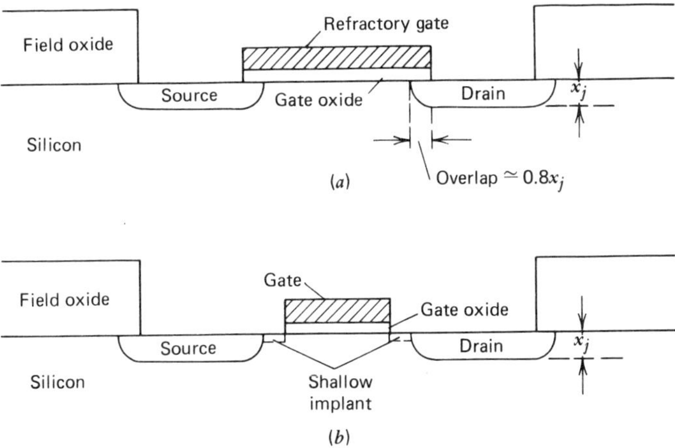

Fig. 6.45 High-speed MOS device using ion implantation.

$Si-SiO_2$ interface and serve to handle data in this manner. The movement of these packets is conventionally achieved by manipulating electrode voltages, and by using different oxide thicknesses with a uniformly doped semiconductor. Ion implantation allows the possibility of introducing an asymmetry into semiconductor doping, by the placement of pockets of varying doping concentration. This allows much simplification in the oxide and metallization processes. In addition, the placement of buried channels [48] by ion implantation permits a reduction of transmission losses in these devices. Details of a wide variety of such processes are available in the literature [49].

6.7 APPLICATION TO GALLIUM ARSENIDE

It was initially predicted that ion implantation would be ideally suited for gallium arsenide and other compound semiconductors, but would find little use in silicon technology. Quite the opposite situation has developed over the past few years, and success with ion implantation of gallium arsenide has been relatively modest during this time. There are a number of reasons for this outcome. First, there is as yet no viable MOS technology for gallium arsenide, so that the needs for well-controlled low doses have not been established. However, all indications are that such a technology is now becoming established [50], and should develop rapidly. Next, the problems of forming heavily doped regions by ion implantation have not as yet been satisfactorily solved [36]. As shown earlier, they necessitate high-dose implants on heated substrates, followed by high-temperature anneals using caps. Each of these technologies has many associated problems. Together they result in a relatively complex sequence of processing steps as compared to those required for processing silicon. Again, this problem area shows promise for being rapidly solved, as evidenced by the recent success of electron beam as well as laser annealing techniques [51]. Nevertheless, there are a number of areas in which ion implantation technology has already proved successful. These are now outlined briefly, with the emphasis on their unique features.

Proton bombardment has been used [52] to produce isolation regions around a number of devices. These include photodiodes and field effect transistors, as well as transferred electron devices and avalanche transit time oscillators. In many of these devices, the unique doping profiles provided by multiple implants have allowed improved performance, and extended both their operating frequencies as well as their power outputs. A unique application of proton bombardment has been in providing optical isolation in couplers and optical waveguides [53]. Oxygen implantation, a somewhat more recent technique, has also been applied to a number of these devices.

Ion implantation has also been used to make ohmic contacts to a number of microwave and optoelectronic devices. Often the uniquely different annealing characteristics of selenium and sulfur implants, as seen in Fig. 6.24, are exploited to form a deep n^+-contact region (sulfur implant) connected to the active n-region (selenium implant). This permits making an alloyed ohmic contact to the shallow implant region with relative ease (see Fig. 6.46).

Hyperabrupt Schottky diodes, suitable for use as electronic tuning elements, have been fabricated [54] using silicon and sulfur implants into gallium arsenide. The doping profile for these structures is of the same type as that for silicon devices, as shown in Fig. 6.43.

6.7.1 Field Effect Transistor Devices

The full exploitation of any electronic technology depends on the development of high-performance three-terminal devices with significant power gain. These devices allow the isolation of input and output signals, and provide the circuit designer with flexibility in designing complex circuits and subsystems. With gallium arsenide, majority carrier devices such as the field effect transistor (FET), based on n-type carrier conduction, can give a speed advantage of 3:1 over silicon, assuming comparable technologies. Although silicon technology is well advanced over that of gallium arsenide

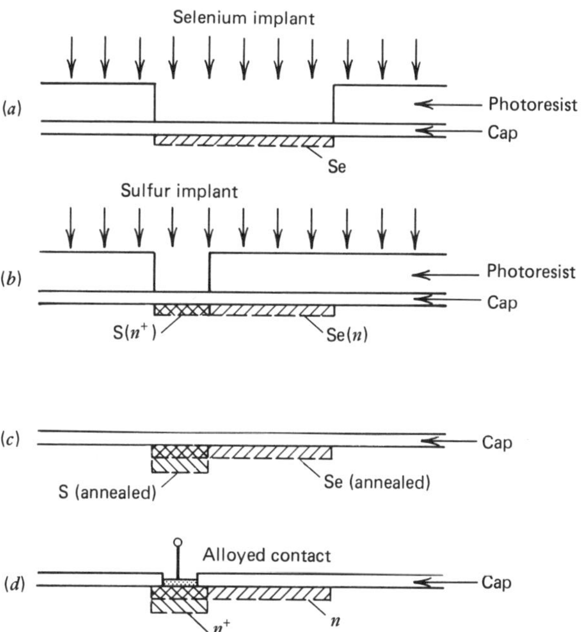

Fig. 6.46 Use of double implantation to form a region for making an alloyed contact to a shallow implant.

today, ion implantation techniques are proving successful in rapidly narrowing this gap, so that this speed advantage has already been realized.

There are three types of field effect transistors that can be made: the Schottky gate device, the junction gate device, and the insulated gate device. Consider first the Schottky gate FET. Here the conventional approach begins* with the epitaxial growth of a thin layer of suitably doped *n*-gallium arsenide on a semi-insulating substrate. Typical parameters for this active region are a carrier concentration of $1–2 \times 10^{17}$ cm^{-3} and a thickness of 0.1–0.3 μm. Often an intervening buffer layer of high impurity material is first grown to prevent out-diffusion effects from the substrate. The Schottky gate FET is fabricated by making ohmic contacts for the source and drain, and using an evaporated metal contact for the gate, as shown in Fig. 6.47*a*.

Problems associated with this fabrication technique are largely due to a lack of sufficient control of the carrier concentration and the layer thickness. These can be greatly reduced by the use of ion implantation. In one approach a high-purity buffer layer is grown in the semi-insulating gallium arsenide and the active layer formed by selenium implantation in it (see Fig. 6.47*b*), followed by annealing under a Si_3N_4 cap [55]. Ion implantation allows the formation of this active layer with better than $\pm$ 3% uniformity over a 2-in. slice diameter.

Double implantation with sulfur and selenium can also be used, and results in the formation of deep source and drain regions, as well as the shallow channel region, after a single anneal step. Here sulfur is used for the source and drain regions, since it penetrates considerably more deeply than selenium during the annealing step (see Fig. 6.47*c*).

Ion implantation can also be used to eliminate the epitaxial growth steps completely, as seen in Fig. 6.47*d*. Here direct implantation is made into semi-insulating substrates to result in a relatively simple process [56]. At the present time, the problem of making suitable high-quality, chromium-doped semi-insulating substrates has not been completely solved (see Fig. 6.25), so that considerable screening has to be done in order to select acceptable material for this application. The introduction of undoped material, made by the in-situ compounding of ultrapure gallium and arsenic (see Chapter 3), has greatly reduced this problem, so that its solution is in clear sight.

Junction gate FET devices can also be made by implantation [57]. For these devices, however, an additional p^+-implant must be made to form the gate region. Both beryllium and magnesium have been used successfully in this application.

Ion implantation technology has only recently been extended to insulated gate FET's, and techniques for forming suitable insulators on the gallium arsenide surface are under active investigation. Eventually it is hoped that all of the techniques that have proved successful with silicon can also be

* A number of process variations are employed in practice to ease the burdens of device fabrication, and are not considered here.

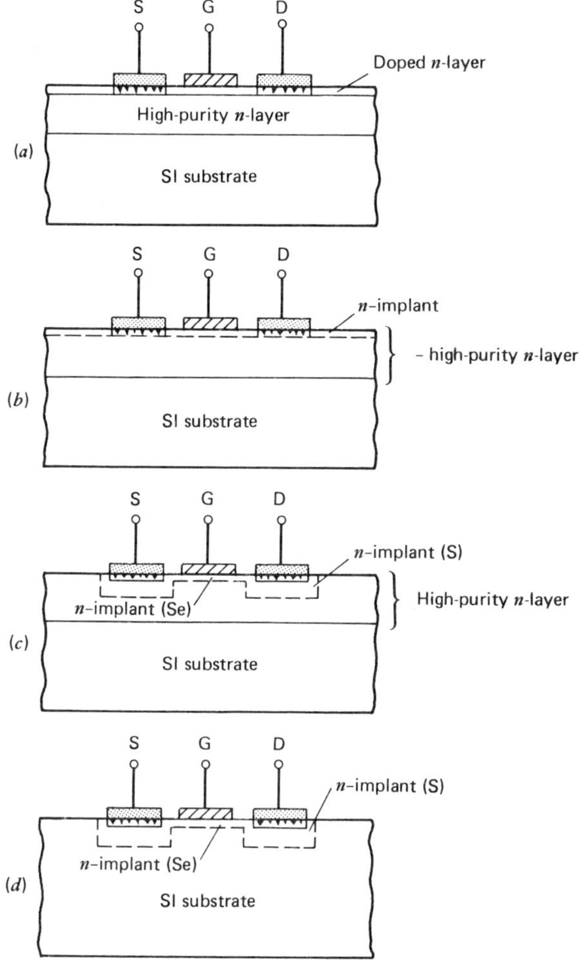

Fig. 6.47 Alternate schemes for junction gate FET devices.

applied here. In these devices, implantation can be used to form the source, drain, and active region [58], and eventually to tailor the threshold voltage.

6.7.2 Integrated Circuits

Integrated circuits in gallium arsenide are based on the use of field effect transistors as active devices, with signal coupling and routing usually done by resistors and by Schottky diodes. The development of these circuits has rapidly followed the development of the FET. Control of doping, provided by ion implantation, has allowed the fabrication of both enhancement mode devices [59] and the more usual depletion type.

Implantation into epitaxially grown *n*-layers has been used to fabricate these integrated circuits. Here sections of the layer are electrically isolated from each other by means of a mesa etch technique. These isolated regions are used for forming the various devices that are required in the circuit [55, 60].

Microcircuits can also be fabricated by direct implantation into semi-insulating gallium arsenide substrates. This approach has the advantage that it results in a planar topology, since mesa etching is not necessary to obtain isolation between components. The resulting planar surface greatly simplifies the task of photolithography at the submicron level, and allows fabrication of ultrahigh-speed circuits [61].

Recently successful integrated circuits using metal-oxide-semiconductor FET's have been demonstrated [62]. Here a double silicon implant is used to form the source, drain, and channel regions. Next a plasma enhanced oxide is grown over the channel, followed by the gate metallization. Circuits using MOS devices of this type are in principle faster than those with junction gate structures. In addition, they offer many possibilities for the use of low-dose implants to tailor their characteristics. Consequently, interest in this area is rapidly increasing.

In all of these techniques, the main impetus for using ion implantation is its ability to provide precise control of the doping and depth of the active channel region. Channel parameters eventually set the threshold voltage of the FET, and thus the signal swing available in linear and digital integrated circuits. The main obstacle to process control in this area is in the reproducible conversion of the ion-implanted impurities into active donors. As noted earlier, the carrier activation in gallium arsenide is far from 100%, so that variations in the damage structure and in the starting material are reflected in run-to-run variations in the doping concentration. Carrier activation is greatly improved by implantation into heated substrates. Unfortunately, this has been found [20] to result in anomalous enhanced diffusion during the implantation, so that control of channel depth is not as tight as can normally be expected from this technique. Nevertheless, practical high-speed integrated circuits are already being successfully fabricated by ion implantation, and developments in the area are extremely rapid [63]. Consequently, there is full expectation that these problems will be solved in the near future.

REFERENCES

1. J. F. Gibbons, Ion Implantation in Semiconductors—Part I. Range Distribution Theory and Experiments, *Proc. IEEE* **56,** 295 (1968).
2. K. A. Pickar, Ion Implantation in Silicon, in *Applied Solid State Science: Advances in Materials and Device Research*, Vol. 5, R. Wolfe, Ed., Academic, New York, 1973, p. 152.
3. R. A. Moline, Ion Implanted Phosphorus in Silicon: Profiles Using C-V Analysis, *J. Appl. Phys.* **42,** 3553 (1971).

4. G. Dearnaley, J. M. Freeman, G. A. Gard, and M. A. Wilkins, Implantation Profiles of ^{32}P Channeled into Silicon Crystals, *Can. J. Phys.* **46,** 587 (1968).

5. B. L. Crowder, Ed., *Ion Implantation in Semiconductors and Other Materials* (International Symposium on Ion Implantation, 1972), Plenum, New York, 1973.

6. J. Linhard, M. Scharff, and Schiøtt, Range Concepts and Heavy Ion Ranges, *Mat. Fys. Medd. Dan. Vidensk. Selsk.* **33,** 1 (1963).

7. J. W. Mayer, L. Eriksson, and J. A. Davies, *Ion Implantation in Semiconductors*, Academic, New York, 1970.

8. G. Carter and W. A. Grant, *Ion Implantation of Semiconductors*, Wiley, New York, 1976.

9. J. F. Gibbons, W. S. Johnson, and S. W. Mylroie, *Projected Range Statistics*, Dowden, Hutchinson, and Ross, Stroudsburg, 1975.

10. D. K. Brice, *Ion Implantation Range and Energy Deposition Distributions*, Plenum, New York, 1975.

11. S. Furukawa, H. Matsumura, and H. Ishiwara, Theoretical Considerations on Lateral Spread of Implanted Ions, *Jpn. J. Appl. Phys.* **11,** 134 (1972).

12. J. F. Zeigler, Determination of Lattice Disorder Profiles in Crystals by Nuclear Backscattering, *J. Appl. Phys.* **43,** 2973 (1972).

13. F. F. Morehead, Jr., and B. L. Crowder, *Ion Implantation*, F. H. Eisen and L. T. Chadderton, Eds., Gordon and Breach, London, 1971, p. 25.

14. J. F. Gibbons, Ion Implantation in Semiconductors—Part II. Damage Production and Annealing, *Proc. IEEE* **60,** 9, 1062 (1972).

15. P. Blood, G. Dearnaley, and M. A. Wilkins, The Depth Distribution of Phosphorus Ions Implanted into Silicon Crystals, in *Ion Implantation in Semiconductor and Other Materials*, (International Symposium on Ion Implantation, 1972), Plenum, New York, 1973.

16. A. G. Foyt, W. T. Lindley, C. M. Wolfe, and J. P. Donnelly, Isolation of Junction Devices in GaAs Using Proton Bombardment, *Solid State Electron.* **12,** 209 (1969).

17. J. M. Woodcock and D. J. Clark, The Ion Implantation of Donors for n^+–p Junctions in GaAs, in *Gallium Arsenide and Related Compounds-1974*, Inst. of Physics, London and Bristol, Publ. 24, 1974, p. 331.

18. A. Lietolla, J. F. Gibbons, T. J. Magee, J. Peng, and J. D. Hong, Solid Solubility of As in Si as Determined by Ion Implantation and CW Laser Annealing, *Appl. Phys. Lett.* **35,** 532 (1979).

19. W. K. Hofker, D. P. Oosthoek, N. J. Koeman, and H. A. M. DeGrefte, Concentration Profiles of Boron Implantations in Amorphous and Crystalline Silicon, *Rad. Eff.* **24,** 223 (1975).

20. A. Lidow, J. F. Gibbons, V. R. Deline, and C. A. Evans, Jr., Fast Diffusion of Elevated Temperature Ion-Implanted Se in GaAs, *Appl. Phys. Lett.* **32,** 149 (1978).

21. J. Kashahara, M. Arai, and T. Watanabe, Effect of Arsenic Partial Pressue on Capless Anneal of Ion-Implanted GaAs, *J. Electrochem. Soc.* **126,** 1997 (1979).

22. F. H. Eisen, Ion Implantation in III-V Compounds, *Rad. Eff.* **47,** 99 (1980).

23. R. G. Hunsperger, R. G. Wilson, and D. M. Jamba, Mg and Be Ion Implanted GaAs, *J. Appl. Phys.* **43,** 1318 (1972).

24. W. W. Levige, M. J. Helix, K. V. Vaidyanathan, and B. G. Streetman, Electrical Profiling and Optical Activation Studies in Be-Implanted GaAs, *J. Appl. Phys.* **48,** 3342 (1977).

25. J. A. Higgins, R. L. Kuvás, F. H. Eisen, and D. R. Chen, Low Noise GaAs FETs Prepared by Ion Implantation, *IEEE Trans. Electron Dev.* **ED-25,** 587 (1978).

26. F. H. Eisen, B. M. Welch, H. Muller, K. Gamo, T. Inada, and J. W. Mayer, Tellurium Implantation in GaAs, *Solid State Electron.* **20,** 219 (1977).

27. J. P. Donnelly, Ion Implantation in GaAs, Institute of Physics Conference Series, Number 33b, 1977, p. 166.

28. J. P. Donnelly and F. J. Leonberger, Multiple Energy Proton Bombardment on n^+-GaAs, *Solid State Electron.* **20,** 183 (1977).
29. P. N. Favennec, G. P. Pelous, M. Binet, and P. Baudet, Compensation of GaAs by Oxygen Implantation, in *Ion Implantation in Semiconductor and Other Materials*, (International Symposium on Ion Implantation, 1972), Plenum, New York, 1973.
30. G. Dearnaley, J. H. Freeman, R. S. Nelson, and J. Stephen, *Ion Implantation*, North Holland, New York, 1973.
31. A. Axmann, Ionizable Materials to Produce Ions for Implantation, *Solid State Technol.* p. 36 (Nov. 1974).
32. K. O. Nielsen, The Development of Magnetic Ion Sources for an Electromagnetic Isotope Separator, *Nucl. Instrum. and Methods* **1,** 289 (1957).
33. T. E. Seidel and A. U. MacRae, Some Properties of Ion Implanted Boron in Silicon, *Trans. Met. Soc. AIME* **245,** 491 (1969).
34. D. H. Lee and J. W. Mayer, Ion-Implanted Semiconductor Devices, *Proc. IEEE* **62,** 1241 (1974).
35. T. E. Seidel and D. L. Sharfetter, High Power Millimeter Wave IMPATT Oscillators Made by Ion Implantation, *Proc. IEEE* **58,** 1135 (1970).
36. R. Heckingbottom and T. Ambridge, Ion Implantation in Compound Semiconductors—An Approach Based on Solid State Theory, *Rad. Eff.* **17,** 31 (1973).
37. T. R. Cass and V. G. K. Reddi, Anomalous Residual Damage in Si after Annealing of "Through-Oxide" Arsenic Implantations, *Appl. Phys. Lett.* **23,** 268 (1973).
38. J. C. North, T. E. McGahan, D. W. Rice, and A. C. Adams, Tapered Windows in Phosphorus-Doped SiO_2 by Ion Implantation, *IEEE Trans. Electron Dev.* **ED-25,** 809 (1978).
39. S. M. Sze and G. Gibbons, Effect of Junction Curvature on Breakdown Voltage in Semiconductors, *Solid State Electron.* **9,** 831 (1966).
40. S. K. Ghandhi, *Semiconductor Power Devices*, Wiley, New York, 1977.
41. P. C. Davis, J. F. Graczyk, and W. A. Griffin, Design of an Integrated Circuit for the T1C Low-Power Line Repeater, *IEEE Trans. Commun.* **COM-27,** 367 (1979).
42. R. A. Moline and G. Foxhall, Ion-Implanted Hyperabrupt Junction Voltage Variable Capacitors, *IEEE Trans. Electron Dev.* **ED-12,** 267 (1972).
43. P. J. Coppen, K. A. Aubuchon, L. O. Bauer, and N. E. Moyer, A Complementary MOS 1.2 Volt Watch Circuit Using Ion Implantation, *Solid State Electron.* **15,** 155 (1972).
44. R. B. Palmer, C. C. Mai, and M. Hswe, The Effect of Oxide Thickness on Threshold Voltages of Boron Ion Implanted MOSFET, *J. Electrochem. Soc.* **120,** 999 (1973).
45. E. C. Douglas and A. G. F. Dingwall, Ion Implantation for Threshold Control in COSMOS Circuits, *IEEE Trans. Electron Dev.* **ED-21,** 324 (1974).
46. H. J. Sansbury, Applications of Ion Implantation in Semiconductor Processing, *Solid State Technol.*, p. 31 (Nov. 1976).
47. J. M. Shannon, Ion Implanted High Frequency Transistors, *Philips Tech. Rev.* **31,** 268 (1970).
48. A. Mohsen and M. F. Tompsett, The Effect of Bulk Traps on the Performance of Bulk Channel Charge Coupled Devices, *IEEE Trans. Electron Dev.* **ED-21,** 701 (1974).
49. C. O. Séquin and M. F. Tompsett, Charge Transfer Devices, Suppl. 8, *Advances in Electronics and Electron Physics*, Academic, New York, 1975.
50. T. Mimura and M. Fukuta, Status of GaAs Metal-Oxide-Semiconductor Technology, *IEEE Trans. Electron Dev.* **ED-27,** 1147 (1980).
51. J. A. Golovchenko and T. N. C. Venkatesan, Annealing of Te-Implanted GaAs by Ruby Laser Irradiation, *Appl. Phys. Lett.* **32,** 147 (1978).

52. G. E. Stillman, C. M. Wolfe, J. A. Rossi, and J. P. Donnelly, Electro-absorption Avalanche Photodiodes, *Appl. Phys. Lett.* **25,** 671 (1974).

53. E. Garmire, H. Stoll, A. Yariv, and R. G. Hunsperger, Optical Waveguide in Proton-Implanted GaAs, *Appl. Phys. Lett.* **21,** 87 (1972).

54. N. Toyada, I. Niikura, Y. Shimura, T. Hozuki, H. Sugibuchi, M. Mihara, and T. Hara, Ion Implanted GaAs Varactor Diodes: Capacitance Uniformity, *Electron. Lett.* **14,** 152 (1978).

55. R. L. van Tuyl, C. H. Liechti, R. E. Lee, and E. Gowen, GaAs MESFET with 4 GHz Clock Rates, *IEEE J. Solid State Circuits* **SC-12,** 485 (1977).

56. B. G. Bosch, Gigabit Electronics—A Review, *Proc. IEEE* **67,** 340 (1979).

57. V. W. Vodicka and R. Zuleeg, Ion Implanted GaAs Enhancement Mode JFETs, 1975 Int. Electron Dev. Meeting, Tech. Digest, 1975, p. 625.

58. T. Nimura, K. Odani, N. Yokoyama, Y. Nakayama, and M. Fukuta, GaAs Microwave MOSFET's, *IEEE Trans. Electron Dev.* **ED-25,** 573 (1978).

59. R. E. Lunggren, C. F. Krumm, and R. P. Pierson, Fast Enhancement-Mode GaAs MESFET Logic, *IEEE Trans. Electron Dev.* **ED-26,** 1827 (1979).

60. R. L. Van Tuyl and C. A. Liechti, High Speed Integrated Logic Using GaAs MESFETs, *IEEE J. Solid State Circuits* **SC-9,** 269 (1974).

61. R. C. Eden, B. M. Welch, and R. Zucca, Planar GaAs IC Technology: Applications for Digital LSI, *IEEE J. Solid State Circuits* **SC-13,** 419 (1978).

62. N. Yokoyama, T. Nimura, and M. Fukuta, Planar GaAs MOSFET Integrated Logic, *IEEE Trans. Electron Dev.* **ED-27,** 1124 (1980).

63. R. Zucca, B. M. Welch, R. C. Eden, and S. I. Long, GaAs Digital IC Technology/Statistical Analysis of Device Performance, *IEEE Trans. Electron Dev.* **ED-27,** 1109 (1980).

PROBLEMS

1. A resistor is made by boron implantation into silicon at 200 keV. Assuming an ion dose of 5×10^{12} cm^{-2}, calculate: (a) the peak concentration; (b) the sheet resistance of the layer; (c) the specific resistivity. Assume $\mu = 250$ cm^2/V s. In addition, provide an approximate algebraic expression for the resistivity.
2. A junction is formed by implanting boron at 100 keV through a window in an oxide. Assume a dose of 10^{15} ions/cm^2 and a background concentration of 5×10^{14} cm^{-3}. Calculate the location of the p–n junction so formed, and make a sketch of this junction. Also, sketch the p–n junction that would be formed by a diffusion to the same depth. Comment on the nature of your answer.
3. Calculate the selenium ion implantation parameters that could be used to provide a channel that is suitable for a GaAs field effect transistor (0.1 μm thick with an average concentration of 10^{17} cm^{-3}). What is the average sheet resistance of this channel if $\mu_n = 4000$ cm^2/V s?
4. An ion implanter has a beam current of 1 mA. Implantation is carried out in a carousel system of the type described, having a 30 wafer capacity (4 in. diameter). Assume 100 keV ions and a total implantation time of

2 min. What is the dose? What is the slice equilibrium temperature for this dose?

5. A silicon slice is implanted with 100 keV As^+ ions. Subsequent annealing is carried out at values of $\sqrt{2Dt}$ given by 0.01 μm and 0.1 μm. Calculate the resulting doping profiles assuming no dopant loss at the surface.
6. It is required to fabricate a *p–n* diode by ion implantation. A *p*-silicon substrate, impurity doped to 10^{16} cm^{-3} is to be used. A window of 15 × 15 μm is cut in the mask. What is the total number of phosphorous ions (at 100 keV) so that the maximum doping concentration does not exceed 10^{19} cm^{-3}. At what depth can you expect the junction to be formed? Draw the impurity core profile (both ideal and realistic).
7. A boron channel implant must be made through a 500 Å thick silicon dioxide gate oxide. What is the masking power of a field oxide of 5000 Å for this implant condition? Assume that $\Delta R_p \simeq 0.3R_p$.

CHAPTER

7

Native Oxide Films

CONTENTS

7.1 THERMAL OXIDATION OF SILICON 373

7.1.1 Intrinsic Silica Glass, 373
7.1.2 Extrinsic Silica Glass, 374

7.1.2.1 Substitutional Impurities, 375
7.1.2.2 Interstitial Impurities, 376
7.1.2.3 Water Vapor, 376

7.1.3 Oxide Formation, 377
7.1.4 Kinetics of Oxide Growth, 378

7.1.4.1 The Initial Growth Phase, 383
7.1.4.2 Doping Dependence Effects, 383
7.1.4.3 Orientation Dependence Effects, 384

7.1.5 Oxidation Systems, 385
7.1.6 Halogenic Oxidation Systems, 388
7.1.7 Oxidation-Induced Stacking Faults, 391
7.1.8 Properties of Thermal Oxides of Silicon, 392

7.1.8.1 Masking Properties, 392
7.1.8.2 Charge States, 394
7.1.8.3 Hot Electron Effects, 399

7.2 THERMAL OXIDATION OF GALLIUM ARSENIDE 400
7.3 ANODIC OXIDATION 401

7.3.1 Oxide Growth, 405

7.3.2 Anodic Oxidation Systems, 408
7.3.3 Properties of Anodic Oxides, 409

7.4 PLASMA ANODIZATION 410
7.5 EVALUATION OF OXIDE LAYERS 411
REFERENCES 414
PROBLEMS 417

The term *native oxide* is given to films which are formed by oxidation of the original semiconductor material. The main advantage of these oxide films is that they are relatively free from contamination, as is the semiconductor from which they are made. Inherently, the chemical composition of native oxides is dictated by the starting semiconductor, and one must resort to deposition techniques if other film compositions are desired. Nevertheless, the native oxide film is widely used in semiconductor processing because of its ease of formation and its comparative freedom from contamination.

The native oxides of both silicon and gallium arsenide form adherent, insulating films on the surface. These films can be used in a number of fabrication processes, depending upon their quality, stability, and electronic properties. Uses include diffusion masks for impurities, surfaces upon which electrical connections can be made between devices, gate oxides, and passivation layers, and antireflective coatings for photo devices.

Thermal oxidation, electrochemical anodization, and plasma anodization can be used to form native oxides on either silicon or gallium arsenide. The properties of the resulting oxides are quite different, however, and each finds use in specific applications. With silicon, thermal oxidation results in the formation of relatively dense, trap-free films, which can be used to protect the slice during its various high-temperature processing steps. Anodic oxides, on the other hand, are relatively porous. To date, their primary use has been in diagnostic situations where thin layers of silicon must be removed (by repeated anodization and oxide dissolution) without subjecting the slice to elevated temperatures.

Electrochemically anodized films, grown on gallium arsenide, are also somewhat porous and have been used in diagnostic situations. With this semiconductor, however, thermal oxidation results in films which contain the oxides of arsenic and gallium as well as free arsenic, and are invariably nonstoichiometric. These thermal oxides are poor electrical insulators, and also poor in their ability to provide protection to the semiconductor surface. Consequently, they are rarely used in gallium arsenide technology. On the other hand, oxides grown by plasma enhanced anodization, as well as electrochemically anodized films, have shown much promise for MOS devices and for MOS-based integrated circuits.

This chapter discusses the physics and technology of native oxides, with the emphasis placed on thermal oxides for silicon and anodic oxides for gallium arsenide. Deposited films, including deposited oxides, are discussed in the following chapter.

7.1 THERMAL OXIDATION OF SILICON

The surface of silicon is covered at all times with a layer of silicon dioxide. This is even true for freshly cleaved silicon, which becomes covered with a few monolayers ($\simeq$ 15–20 Å) of oxide upon exposure to air, and gradually thickens with time to an upper limit of about 40 Å. Considerably thicker films are required in silicon microcircuit fabrication. Here an oxide film is grown on the slice by maintaining it in an elevated temperature in an oxidizing ambient, usually dry oxygen or water vapor. The properties of this oxide are now described, together with the way in which it is modified by the introduction of impurities. This is followed by a discussion of the kinetics of growth, and the manner in which this growth is carried out. Masking and other important properties of this layer are also outlined.

An important property of thermally grown silicon dioxide is its ability to reduce the surface state density of silicon by tying up some of its dangling bonds. In addition, silicon oxides can be grown with good control over interface traps and fixed charge. Their use in controlling the leakage current of junction devices [1], and in the formation of a stable gate oxide for field effect devices, stems from these properties and makes them the cornerstone of modern silicon integrated circuit technology.

7.1.1 Intrinsic Silica Glass

Intrinsic silica glass consists of fused silicon dioxide, having a melting point of 1732°C. It is thermodynamically unstable below 1710°C and tends to return to its crystalline form at temperatures below this value. However, the rate of this devitrification process is usually considered negligible at 1000°C and lower.

The model for pure silica in the vitreous state [2] consists of a random three-dimensional network of silicon dioxide, constructed from polyhedra (tetrahedra or triangles) of oxygen ions. The centers of these polyhedra are occupied with Si^{+2} ions. The tetrahedral distance between the silicon and oxygen ions is 1.62 Å, while the distance between oxygen ions is 2.27 Å.

Silica polyhedra are joined to one another by *bridging* oxygen ions, each of which is common to two such polyhedra. In crystalline silica all such oxygen ions play this role, and all vertices of the polyhedra are tied to their nearest neighbors by these ions. In fused silica or silica glass, however, some of the vertices have *nonbridging* oxygen ions which belong to only one polyhedron. The degree of cohesion between the polyhedra and, hence, of the network as a whole is thus a function of the ratio of bridging to nonbridging oxygen ions.

In pure silica glass, movement of the silicon atom is accomplished by the rupture of four Si—O bonds, whereas the movement of a bridging oxygen atom requires the rupture of only two Si—O bonds. As a consequence, oxygen is freer to move in this glass than is silicon. The movement of this oxygen from its polyhedral site gives rise to the formation of an *oxygen ion vacancy*. Both bridging and nonbridging vacancies may be formed, although the latter is more probable from binding energy considerations. Vacancies of this type represent positively charged defects in the structure.

Silica films, as grown by oxidation of the silicon surface, are amorphous in nature, and consist of a random network of such polyhedra. Typically they have a density of 2.15–2.25 as compared to 2.65 for single-crystal quartz. Because of their open nature, it is possible for a number of impurities to diffuse through them interstitially.

Diffusion processes in intrinsic silica glass are similar to those described in Chapter 4, even though a random structure is involved. As may be expected, the process can be described by a diffusion coefficient D such that

$$D = D_0 e^{-E_A/kT} \tag{7.1}$$

where D_0 is the diffusion constant and E_A is the activation energy of the diffusing species in eV/molecule.* Figure 7.1 shows the diffusion constants of hydrogen, oxygen, and water in silica glass as a function of temperature. These species are important since they are involved in the oxidation process. Also shown in the diffusion coefficient of sodium, which is an important contaminant in thermal oxidation systems. Note that these data represent approximate values only, since the diffusion constant is dependent on the nature of the glass network, that is, on the ratio of bridging to nonbridging oxygen ions in it.

There is much uncertainty about the nature of the species involved in the thermal oxidation of silicon. Some work seems to indicate that the species is a charged form of oxygen, either O_2^- or O_2^{2-}. In any event, it has been established that oxidation involves the transport of the oxidizing species *through* the silicon and to the Si–SiO_2 interface where oxidation occurs [3]. As a result, thermal oxidation of silicon leaves a clean interface, with ionic contaminants transported to the exposed oxide surface. It is this characteristic which results in greatly improving the stability of planar junctions on which this oxide is grown.

7.1.2 Extrinsic Silica Glass

The properties of silica glass are greatly modified [4] by the introduction of impurities. These impurities are primarily of two types: substitutional and

* These values are often given in units of kcal/mole. Note that 1 kcal/mole = 0.0434 eV/molecule.

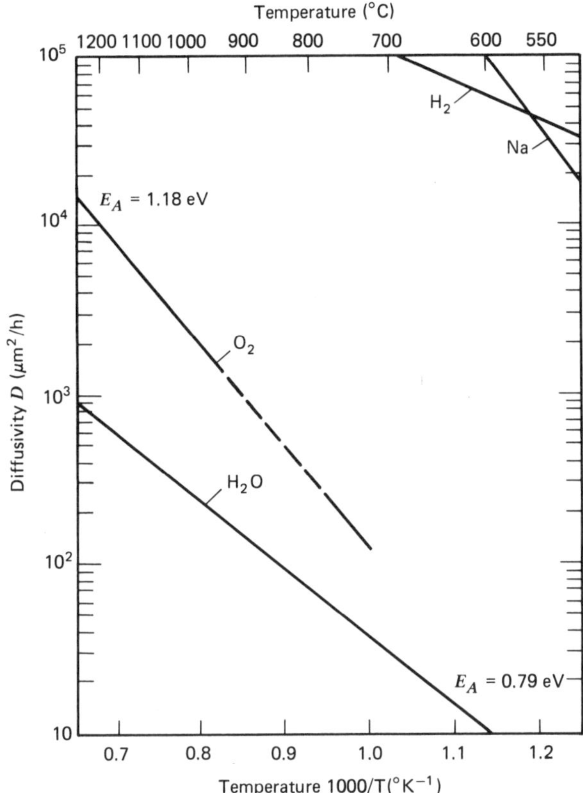

Fig. 7.1 Diffusivities of hydrogen, oxygen, sodium, and water vapor in silica glass.

interstitial. In addition, water vapor can also be considered as an important impurity that is always present in silica glass.

7.1.2.1 Substitutional Impurities

A substitutional impurity is one that replaces the silicon in a silica polyhedron. The most common impurities of this type are B^{3+} and P^{5+}. These impurities are called *network formers*, since it is possible to build vitreous structures using them instead of SiO_2 (i.e., glasses which are totally free of silicon dioxide). In microcircuit fabrication, however, our interest is centered on the effect of small amounts of these impurities in silica glass.

The valence of a substitutional cation is usually either 3 or 5. In the silica lattice, such cations result in charge defects. The presence of column V impurities give rise to the formation of an excess of nonbridging ions, while column III impurities usually (but not always) reduce the nonbridging ion concentration.

7.1.2.2 Interstitial Impurities

These are usually the oxides of large metal ions of low positive charge which enter into the network interstitially between the polyhedra. In so doing, they give up their oxygen to it, thus producing two nonbridging oxygen ions in place of the original bridging ion. This results in weakening the structure and rendering it more porous to other diffusing species. The reaction is shown schematically for Na_2O in the silica lattice as follows:

$$Na_2O + Si—O—Si = Si—O + O—Si + 2Na^+ \qquad (7.2)$$

Impurity oxides of this type are called *network modifiers*, since they are not capable of forming glasses by themselves. Ions such as Na^+, K^+, Pb^{2+}, and Ba^{2+} fall into this class. Aluminum sometimes plays a dual role of network modifier as well as network former.

Sodium is particularly important because it is widely present as part of the environment, especially around humans. It is also an important impurity in the firebrick that is used for making furnace heater cores, and it is even present in the fused quartz tubes that are used in diffusion and oxidation equipment. Since it is a charged, mobile species, its avoidance, removal, or immobilization is important in the fabrication of all types of stable devices, and especially for MOS-based microcircuits. Excessive amounts of sodium in an oxide can accelerate its ability to crystallize, and even result in oxide cracking.

7.1.2.3 Water Vapor

Water vapor is also widely present in the environment. It can reside in molecular form at interstitial sites in oxidized silicon slices to a depth of several hundred angstroms, if they are left unpackaged for as little as a week. Its presence in this form leads to problems of poor adhesion of photoresist in subsequent masking operations, and to instability in the magnitude of the reverse breakdown voltage of diodes which are covered with this oxide. Prolonged bake out (48–72 hs) at low temperatures (200–250°C) is often done just before packaging to remove this water.*

Water vapor is also incorporated during a "wet" oxidation process, and as a contaminant in a "dry" process. On entering, it combines with bridging oxygen ions to form pairs of stable nonbridging hydroxyl groups. The reaction is described schematically by

$$H_2O + Si—O—Si = Si—OH + OH—Si \qquad (7.3)$$

The presence of these hydroxyl groups also tends to weaken the silicon

* Sometimes a short, surface etch is used for this purpose.

network and render it more porous to diffusing species. Thus their behavior is similar to that of interstitial impurities.

Figure 7.2 shows a schematic representation [5] of fused silica glass together with the various types of defect structures that may be present in it.

7.1.3 Oxide Formation

Oxidation of the silicon slice is conveniently carried out by subjecting it to dry oxygen or water vapor while the wafer is maintained at an elevated temperature. In water vapor, or wet oxidation schemes, an inert gas (or oxygen) is bubbled through water which is usually held at 95°C, corresponding to a vapor pressure of H_2O of 640 torr (0.842 atm). Direct oxidation of the silicon surface results in forming an oxide layer which is about 2.27 times the thickness of the consumed silicon. The silica layer so formed contains about 2.2×10^{22} molecules/cm^3 of SiO_2.

The mechanisms of oxide formation are based on the fact that the oxidizing species must move through the growing oxide layer in order to reach the silicon surface. Thus growth proceeds at an ever-decreasing rate as the thickness of the intervening oxide layer increases. This diffusion-controlled regime is observed for layers greater than 40 Å thickness for dry oxidation, and 1000 Å for wet oxidation.

The chemistry of oxidation with dry oxygen is relatively straightforward. It is assumed that the species diffusing through the growing layer are oxygen ions. The chemical reaction at the silicon surface is

$$Si + O_2 \rightarrow SiO_2 \tag{7.4}$$

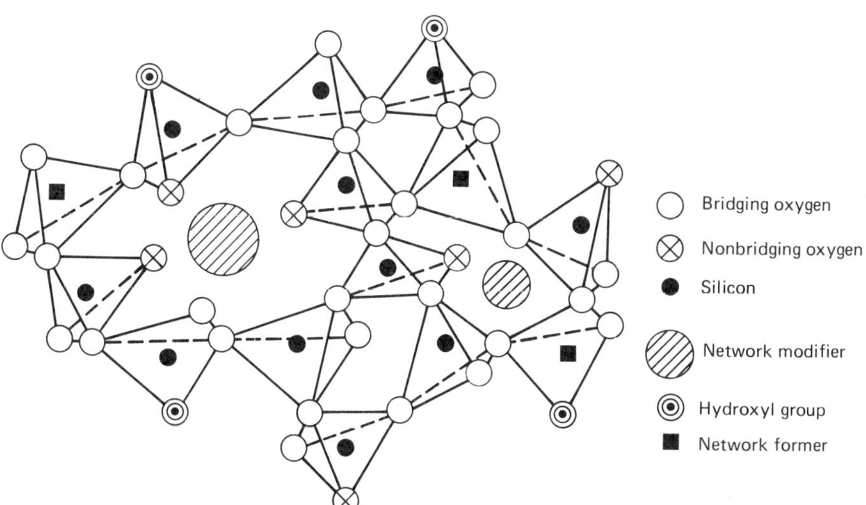

Fig. 7.2 Structure of silica glass. Adapted from Revesz [5].

Here *one* molecule of oxygen results in the formation of *one* molecule of silicon dioxide.

The overall process of oxidation of silicon with water vapor may also be considered as one in which the oxidizing species diffuses through the oxide and reacts with the silicon surface, so that

$$\mathrm{Si} + 2\mathrm{H_2O} \rightarrow \mathrm{SiO_2} + 2\mathrm{H_2} \tag{7.5}$$

Here *two* molecules of water vapor are used to form *one* molecule of SiO_2. The hydrogen evolved by this reaction diffuses rapidly through the growing oxide and leaves the system at the gas–oxide interface.

The detailed nature of this reaction is somewhat more complex [6], and is assumed to proceed in the following manner:

1. Water vapor reacts with the bridging oxygen ions in the silica structure to form nonbridging hydroxyl groups. This reaction, which results in greatly weakening the silica structure, may be written as

$$\mathrm{H_2O} + \mathrm{Si{-}O{-}Si} \rightarrow \mathrm{Si{-}OH} + \mathrm{OH{-}Si} \tag{7.6}$$

2. At the oxide–silicon interface, the hydroxyl groups react with the silicon lattice to form silica polyhedra and hydrogen. The reaction is

$$\begin{matrix}\mathrm{Si{-}OH} \\ \\ \mathrm{Si{-}OH}\end{matrix} + \mathrm{Si{-}Si} \rightarrow \begin{matrix}\mathrm{Si{-}O{-}Si} \\ \\ \mathrm{Si{-}O{-}Si}\end{matrix} + \mathrm{H_2} \tag{7.7}$$

3. Hydrogen leaves the oxide layer by rapid diffusion. In addition, some of the hydrogen reacts with bridging oxygen ions in the silica structure to form hydroxyl groups, as shown by

$$\tfrac{1}{2}\mathrm{H_2} + \mathrm{O{-}Si} \rightleftharpoons \mathrm{OH{-}Si} \tag{7.8}$$

thus further weakening the silica structure.

A somewhat more detailed picture of the role of hydrogen in silica is given elsewhere [7]. However, the relatively simple model, outlined above, is suitable for describing the basic growth mechanism.

7.1.4 Kinetics of Oxide Growth

The kinetics of oxide growth on silicon may be determined [8] with reference to the model of Fig. 7.3. Assume that a silicon slice is brought in contact with the oxidant, resulting in a surface concentration of N_0 molecules/cm^3 for this species. In typical oxidation systems, the mass-transfer coefficient

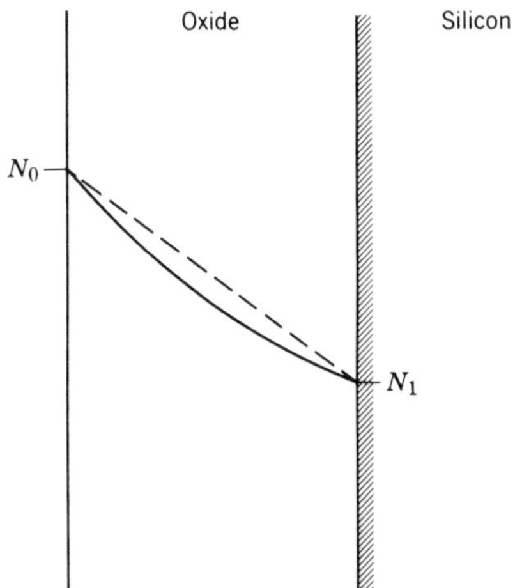

Fig. 7.3 Model for oxidation flux.

is extremely high, so that the magnitude of N_0 is essentially the solid solubility of the species at the oxidation temperature. At 1000°C the solid solubility of these species is 5.2×10^{16} molecules/cm^3 for dry oxygen and 3×10^{19} molecules/cm^3 for water vapor at a pressure of 1 atm.

The oxidizing species diffuses through the silicon dioxide layer, resulting in a concentration N_1 at the surface of the silicon. Transport of the species occurs by both drift and diffusion. Writing D as the diffusion coefficient and ignoring the effects of drift, the flux density of oxidizing species arriving at the gas–oxide interface is given by j, where

$$j = D\frac{\partial N}{\partial x} \simeq \frac{D(N_0 - N_1)}{x} \tag{7.9}$$

and x is the thickness of the oxide at a given point in time.

On arrival at the silicon surface the species enters into chemical reaction with it. If it is assumed that this reaction proceeds at a rate proportional to the concentration of the oxidizing species, then

$$j = kN_1 \tag{7.10}$$

where k is the interfacial reaction rate constant. These fluxes must be equal under steady-state diffusion conditions. Combining Eqs. (7.9) and (7.10), gives

$$j \simeq \frac{DN_0}{x + D/k} \tag{7.11}$$

The reaction of the oxidizing species with the silicon results in the formation of silicon dioxide. Writing n as the number of molecules of the oxidizing impurity that are incorporated into unit volume of the oxide, the rate of change of the oxide layer thickness is given by

$$\frac{dx}{dt} = \frac{j}{n} = \frac{DN_0/n}{x + D/k} \tag{7.12}$$

Solving this equation, subject to the boundary value that $x = 0$ at $t = 0$, gives

$$x^2 + \frac{2D}{k}x = \frac{2DN_0}{n}t \tag{7.13}$$

so that

$$x = \frac{D}{k}\left[\left(1 + \frac{2N_0k^2t}{Dn}\right)^{1/2} - 1\right] \tag{7.14}$$

Equation (7.14) reduces to

$$x = \frac{N_0k}{n}t \tag{7.15}$$

for small values of t, and to

$$x = \left(\frac{2DN_0}{n}\right)^{1/2} t^{1/2} \tag{7.16}$$

for large values. Thus in the early stages of growth, in which the reaction is the rate limiter, the oxide thickness varies linearly with time. In later stages the reaction is diffusion limited, and the oxide thickness is directly proportional to the square root of time.

Equation (7.13) is often written in the more compact form

$$x^2 + Ax = Bt \tag{7.17}$$

Using this form, Eqs (7.15) and (7.16) can be written as

$$x = \frac{B}{A}t \tag{7.18}$$

for the linear region, and

$$x = B^{1/2}t^{1/2} \tag{7.19}$$

for the parabolic region. For this reason, the term B is referred to as the *parabolic rate constant*, whereas the *linear rate constant* is given by B/A. Measured values for these rate constants as a function of temperature are shown [9] in Figs. 7.4 and 7.5 for both dry and wet oxidation (p_{H_2O} = 640 torr). The data of Fig. 7.4 are shown for both (111) and (100) silicon. This orientation dependence effect is considered in Section 7.1.4.2. Mathematical relations, describing the behavior of these rate constants, are given in Table 7.1.

Insight into the physical processes involved during the growth of silica layers can be gleaned by a study of these figures. Thus Fig. 7.4 shows that the logarithm of the linear rate constant falls with $1/T$ at a slope of 2 eV/molecule for dry oxygen, and at 2.05 eV/molecule for wet oxidation. This is in close agreement with the energy required to break Si—Si bonds, which is 1.83 eV/molecule. The logarithm of the parabolic rate constant (Fig. 7.5) also falls with $1/T$, but at a slope of 1.23 eV/molecule for dry oxidation. The comparable activation energy for the diffusion of oxygen in fused silica is

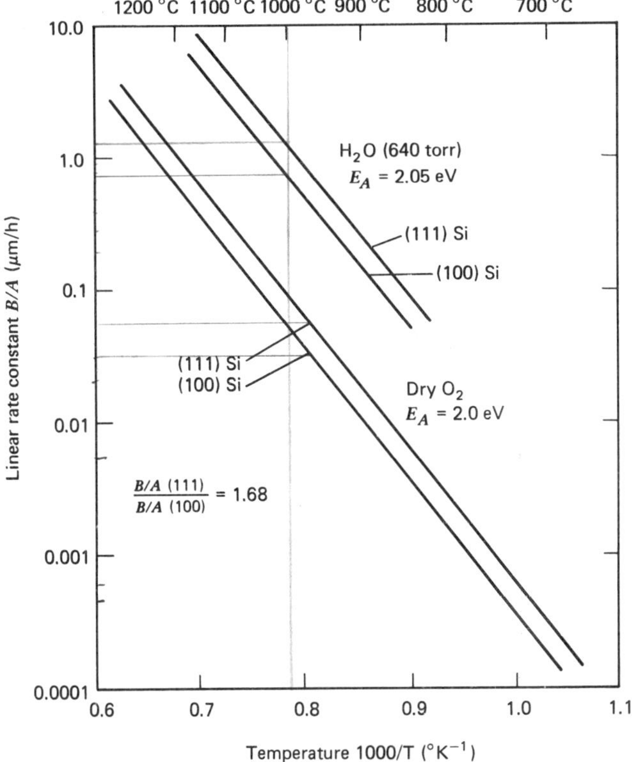

Fig. 7.4 Linear rate constant versus temperature. From Deal [8]. Reprinted with permission of the publisher, The Electrochemical Society, Inc.

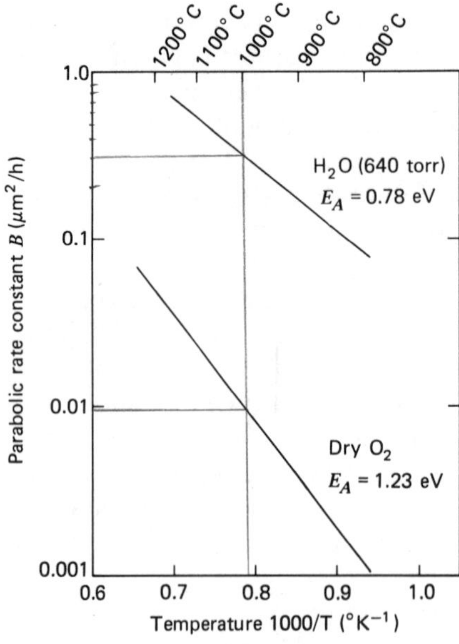

Fig. 7.5 Parabolic rate constant versus temperature. From Deal [8]. Reprinted with permission of the publisher, The Electrochemical Society, Inc.

about 1.18 eV/molecule, as shown in Fig. 7.1. Corresponding values for wet oxidation are 0.78 eV/molecule, which compares favorably with the activation energy of diffusion of H_2O in fused silica (0.79 eV/molecule).

To an approximation, the linear rate constant varies directly with the concentration of oxidizing species at the surface, and thus with its partial pressure. This has been experimentally verified for both wet and dry oxidation conditions. Finally, the parabolic rate constant for wet oxidation is

Table 7.1 Mathematical Relationship for Rate Constants[a]

	Parabolic Linear	$B = C_1 e^{-E_1/kT}$ $B/A = C_2 e^{-E_2/kT}$
	(111) Silicon	(100) Silicon
Dry O_2	$C_1 = 7.72 \times 10^2$ μm²/h $C_2 = 6.23 \times 10^6$ μm/h $E_1 = 1.23$ eV $E_2 = 2.0$ eV	$C_2 = 3.71 \times 10^6$ μm/h All other parameters same as for (111) silicon
H_2O (640 torr)	$C_1 = 3.86 \times 10^2$ μm²/h $C_2 = 1.63 \times 10^8$ μm/h $E_1 = 0.78$ eV $E_2 = 2.05$ eV	$C_2 = 0.97 \times 10^8$ μm/h All other parameters same as for (111) silicon

[a] See reference 9.

found to be much larger than that for dry oxidation. Th' to the significantly greater solid solubility of water ove glass (about three decades larger), which more than compe slightly lower diffusion constant, as shown in Fig. 7.1.

7.1.4.1 The Initial Growth Phase

The theory for the kinetics of oxide formation has been found to apply very well to growth in wet oxygen and steam. There is consistent evidence, however, to indicate the presence of an extremely rapid initial growth phase with dry oxygen. This process can be explained by postulating that the molecular oxygen, on entering the oxide, dissociates to form a negatively charged O_2^- or O_2^{2-} and one or two holes, respectively. The holes have considerably higher mobility than the oxygen ion, and run ahead of it. The result is the formation of a space-charge region. The resulting field enhances the diffusion of the oxygen in the layer by providing an additional drift component. It can be shown that the region of high space-charge density is near the gas–oxide interface, the rest of the oxide layer being almost space-charge neutral. The thickness of this region is on the order of the extrinsic Debye length and depends inversely on the square root of the concentration of the oxidizing species in the layer. This Debye length is about 150–200 Å in dry oxygen, but only 5 Å for water vapor.* In practice, an accelerated oxidation rate is seen for a depth of about 230 ± 30 Å in dry oxygen processing, but unnoticed for wet oxygen or steam processing.

7.1.4.2 Doping Dependence Effects

Heavily doped silicon oxidizes at a faster rate than lightly doped material. However, detailed studies of boron- and phosphorus-doped material have shown considerable differences in oxide growth behavior. During oxidation, boron is preferentially incorporated into the silicon dioxide because of its relatively large segregation coefficient ($\simeq$ 3). This results in a weakening of the bond structure of the silica film, and an increase in the diffusivity of the oxidizing species through it. Consequently, there is an increase in the parabolic rate constant with boron-doping concentration, but little change in the linear rate constant [10].

Phosphorus, on the other hand, has a low segregation coefficient ($\simeq$ 0.1), is only slightly incorporated into the growing oxide, and piles up at the Si–SiO_2 interface. This has been found to result in an increase in the reaction rate, with a corresponding increase in the linear rate constant. On the other hand, the lack of phosphorus incorporation into the oxide results in the parabolic rate constant being relatively insensitive to doping. Experimental

* This is because the solid solubility of water in silica is three orders of magnitude larger than that of oxygen.

data [11], indicating the magnitude of these rate constants, are shown in Fig. 7.6 for a wet oxidation temperature of 900°C. As a result of the above, we can expect that the rate of oxide growth on boron- and phosphorus-doped silicon will depend upon whether the oxidation process is operative in the linear or in the parabolic regime.

With the increased packing density requirements of VLSI, there is an increased need for thin oxides grown at relatively low temperatures, i.e., in the linear regime. This presents a special problem with devices such as n^+–p–n transistors where the oxide growth rate over the heavily phosphorus-doped emitter region can be as much as two to five times faster than the oxidation rate over neighboring regions where the doping is light. This can result in large steps in the oxide, with the attendant possibility of breaks in metal interconnections placed over them.

Oxide growth variations are generally not a problem in situations where the oxidation is diffusion limited. This is partly because the resulting oxide is relatively thick over all regions of the semiconductor. In addition, the change of oxidation rate with doping concentration is only significant with boron-doped silicon. This cannot be doped as heavily as phosphorus, because of the large misfit factor of this impurity atom. As a result, the oxidation rate enhancement with boron-doped silicon is generally no more than 20% over the undoped silicon material.

7.1.4.3 Orientation Dependence Effects

The parabolic growth rate has been found to be independent of crystal orientation during oxidation. This is reasonable, since this parameter is a

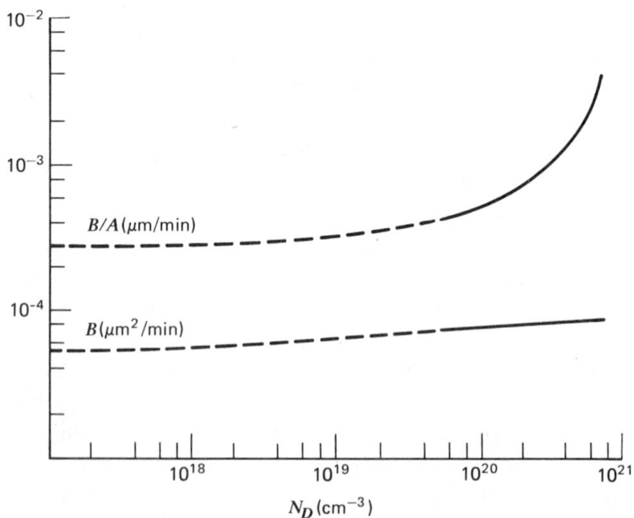

Fig. 7.6 Rate constant versus substrate phosphorus doping level for oxidation at 900°C. From Ho, Plummer, Meindl and Deal [11]. Reprinted with permission of the publisher, The Electrochemical Society, Inc.

measure of the diffusivity of the oxidizing species through an amorphous silica layer. The linear reaction rate, on the other hand, is related to the rate of incorporation of silicon atoms into the silica network. This, in turn, is a function of the atom concentration on the silicon surface, so that it is orientation dependent. The density of silicon atoms on the (111) plane is larger than that on the (100) plane, so we can expect the linear rate constant on (111) silicon to be larger than that on (100) silicon. This is indeed the case, as is shown in Fig. 7.4.

The above argument is an oversimplification, however, since it does not take into consideration the three-dimensional nature of the silicon atomic structure, where atoms in a lower plane are partly shadowed by adjacent atoms in the upper plane, or the relative sizes of the silicon and oxygen atoms. These combined effects can be used to explain the somewhat larger change in the linear rate (111 : 100 = 1.68 : 1) than is given by atom density considerations alone (1.16 : 1).

7.1.5 Oxidation Systems

Oxide growth is carried out in a quartz diffusion tube, in which the silicon slice is maintained at a temperature between 900 and 1200°C. Often a high-density ceramic liner is used to serve as a diffusion barrier to sodium, which is present in the furnace heating elements. Recently, stabilized fused quartz tubes have become available. These tubes have an outer layer of cristobalite quartz which impedes the flow of sodium by a factor of 10. Some systems use diffusion tubes of pure silicon, which provides the highest degree of cleanliness and freedom from sodium contamination.

Wet oxidation processes are quite rapid, but result in relatively porous silica films. These oxides are used for most general-purpose applications such as surface coverage and diffusion masking. Experience has shown that the use of live steam for this purpose leads to poor grades of oxide because of the etching and pitting action of the excess water. Consequently, wet oxidation is usually accomplished by flowing a carrier gas through a water bubbler whose temperature is maintained below the boiling point to prevent its undue depletion. A temperature of 95°C is commonly used and corresponds to a vapor pressure of about 640 torr (0.842 atm). The carrier gas may be either oxygen or an inert species (nitrogen or argon), since the oxidation is almost entirely due to the water vapor.

Pyrogenic water systems are also used for wet oxidation, and are well suited to a manufacturing environment. In these systems, pure hydrogen and oxygen are directly fed to the diffusion tube where they react to form water vapor. These systems allow a wide variation in the partial pressure of H_2O; in addition they avoid the use of a bubbler and all the nuisance problems associated with its continual refilling and cleaning.

Oxides grown in dry oxygen are extremely dense, and have a relatively low concentration of traps and interface states. Consequently, gate oxides

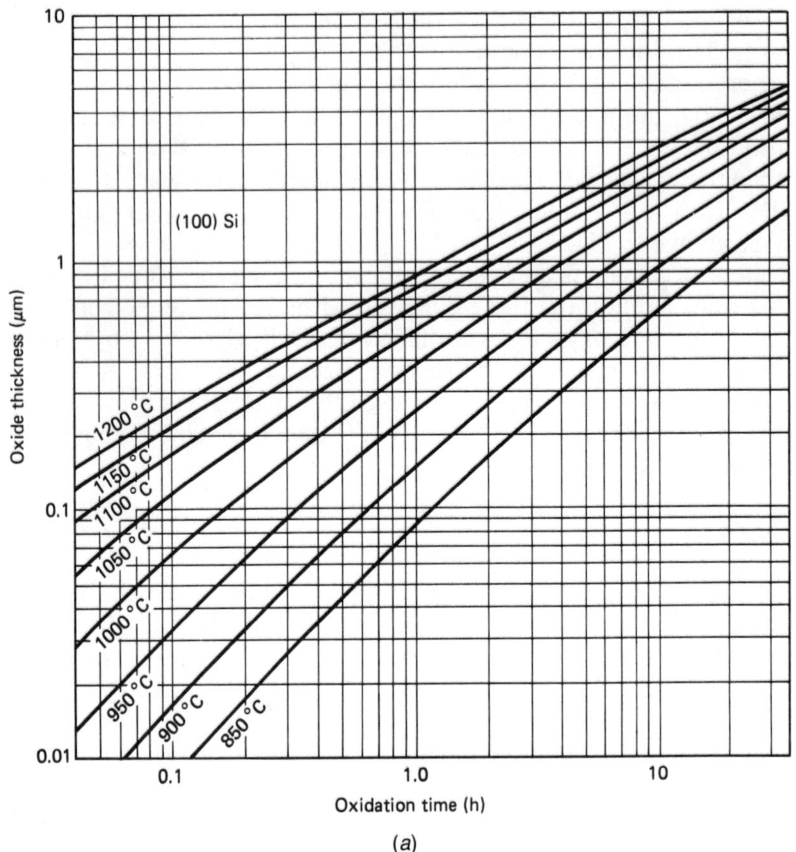

(*a*)

Fig. 7.7 Oxide growth rate for wet oxygen: (*a*) (100) silicon. (*b*) (111) silicon.

for MOS-based circuits are exclusively made by this process, with elaborate precautions taken to ensure a clean, sodium-free system. Also it is extremely important that the oxygen be truly dry, since as little as 25 ppm of water will significantly alter the growth rate, as well as the subsequent properties of the oxide. For this reason, a complete separate system is usually dedicated to growing gate oxides.

Figure 7.7 shows the growth rates for oxide layers in 95°C H_2O for (100) and (111) silicon. Corresponding growth rates for dry oxygen are shown in Fig. 7.8. It is observed that the reaction is diffusion limited over most of the oxidation range, and that oxide growth with wet oxygen is considerably more rapid than with dry oxygen. This is primarily due to the considerably higher solid solubility of water vapor in silicon dioxide, as pointed out in Section 7.1.4.

Extremely thick oxide layers (> 1 μm) are often required in microcircuit fabrication. Growth of these oxides at atmospheric pressure requires long growth times at elevated temperatures. This leads to a poor quality, crystalline oxide film which tends to crack during further processing. The use

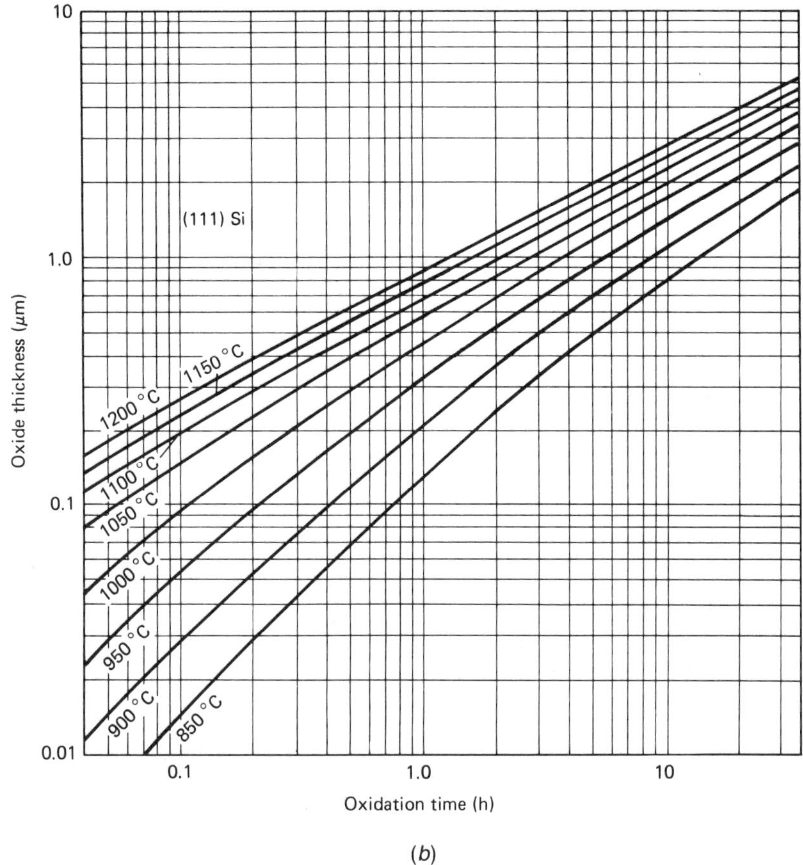

(*b*)

of high-pressure oxidation techniques can eliminate these problems since they allow film growth to be carried out at lower temperatures where devitrification effects are not important. Growth at low temperatures is also desirable in a number of recently proposed VLSI schemes,* which require the selective oxidation of exposed regions of an otherwise masked silicon slice. The movement of diffused impurities during low-temperature oxidation is greatly reduced, so that these techniques are ideal for small geometries. Finally, many oxidation-induced problems can be greatly reduced if oxide growth is accomplished at low temperatures. Again, oxidation at high pressure can be done to alleviate this problem.

Figure 7.9 shows growth rates which can be achieved during a 1-h oxidation in wet H_2O, and indicates the extent of enhancement at increased pressures [12]. This is mainly due to an increase in the parabolic rate constant, which varies almost linearly with the partial pressure of H_2O.

Thermal oxidation of silicon has also been conducted with dry oxygen

* Fabrication aspects of this important technology are described in Chapter 11.

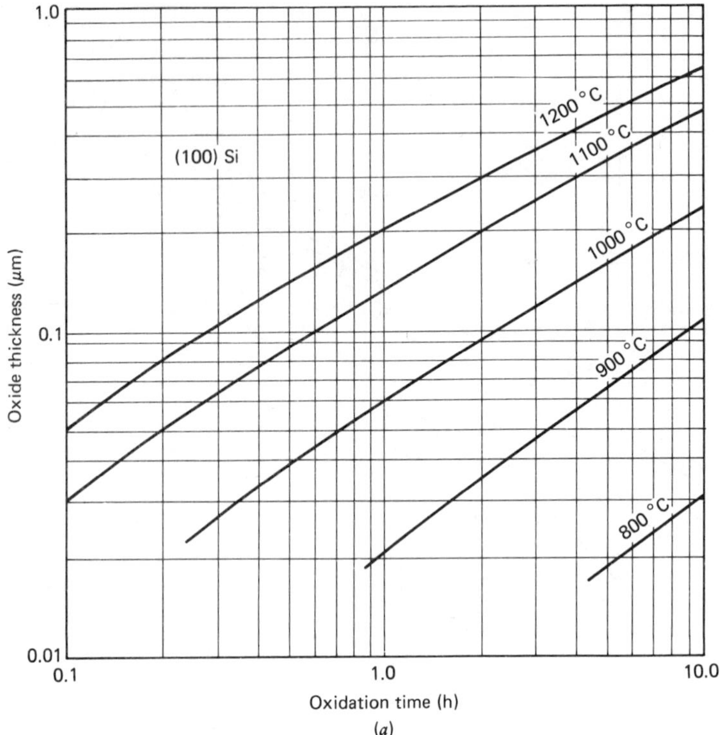

Fig. 7.8 Oxide growth rate for dry oxygen: (*a*) (100) silicon. (*b*) (111) silicon.

pressures as high as 140 atm [13]. Here growth rates at 800°C have been found comparable to those for wet oxidations at 1200°C.

The density of dry oxides is typically 2.25 g/cm^3, whereas that for wet oxides is usually about 2.15 g/cm^3. The dielectric strength of wet oxide films is correspondingly lower. In many oxidation schemes, it is common practice to use a combination of these processes, the oxidation procedure being both initiated and concluded in dry oxygen with an intervening wet oxygen step.

Dry oxidation is always used for MOS gate oxides. The breakdown voltage for such oxides, of thickness 500 Å, is typically 50 V, provided they are free from pinholes. The pinhole problem becomes increasingly severe when thinner oxides are used for these devices.

7.1.6 Halogenic Oxidation Systems

The addition of a halogenic species during dry oxidation results in significant improvements in the electronic properties of the oxide and of the underlying silicon. Typically, a chlorine species is used for this purpose. Oxides with incorporated chlorine have been found to have increased dielectric strength, and to result in MOS devices with improved threshold stability. In addition,

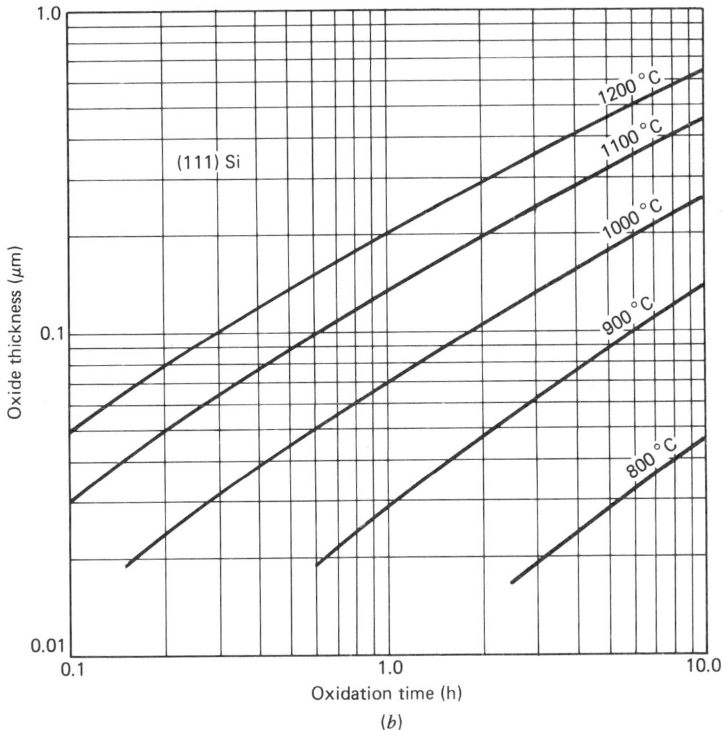

(*b*)

an improvement in the lifetime of the silicon results from the use of these oxides, as well as a reduction in the number of stacking faults on the silicon surface.*

A variety of chlorine-bearing species can be used for this purpose, the most common being chlorine gas and anhydrous hydrogen chloride. Lately there has been considerable interest in trichloroethylene (C_2HCl_3) and its many variants, since they are less corrosive and can be readily handled in a bubbler arrangement. All of the above materials are available in semiconductor grade purity.

Chlorine has been found to concentrate as a neutral species at or near the Si–SiO_2 interface during thermal oxidation when one of these species is involved. This has a number of consequences. First, it interacts with rapidly moving deep lying impurities in the silicon, and removes them by conversion to their chlorides. This results in an improved lifetime in the material on which this halogenic oxide is grown. Next, it appears to capture Na^+ ions to form a neutral system. Thus it effectively removes this highly mobile charge species, and so greatly improves the threshold stability in a

* This topic is discussed in the following section.

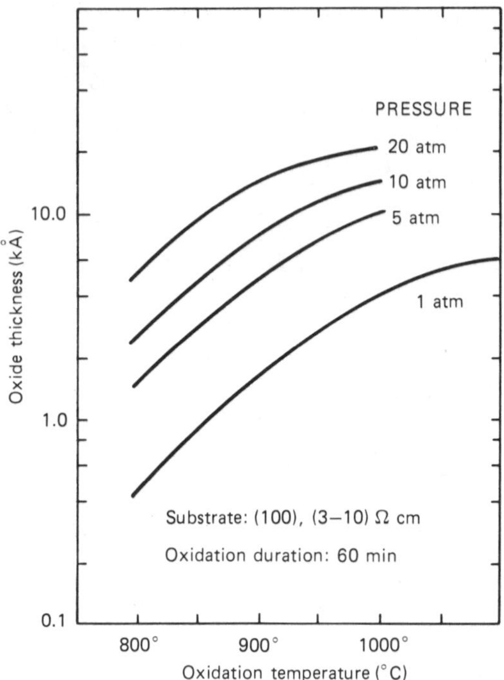

Fig. 7.9 Wet oxide growth at increased pressures. From Su [12]. Reprinted with permission from *Solid State Technology*.

MOS device. It has been shown that only a small percentage of the total chlorine at the interface is involved in this relatively complete capture process [14]. The introduction of a chlorine species during oxidation is also believed to create vacancies at the silicon surface, leading to a reduction in the stacking fault formation during oxidation.

Up to 10^{15} atoms/cm^2 of chlorine can be incorporated into an oxide by any of the above means. Beyond this level, however, a liquid phase (probably a chlorosilane) is found to segregate at the interface, resulting in surface roughness and bubbles in the oxide. This renders the oxide useless for subsequent photolithography, and ruins its masking capability.

The introduction of chlorine during dry oxidation results in an increase in the oxidation rate [15]. Experiments with this system have shown that both the linear and the parabolic rate constants increase with chlorine in-
vever, the mechanisms involved in the growth rate have
ned at the present time.
in the growth rate using O_2-HCl mixtures comes about
oxidation, these species react to form water and chlorine.
ite is thus a result of the combined effects of the addition
orine to the dry oxygen, and HCl does not appear to enter
s process [9].

Oxidation in O_2-C_2HCl_3 mixtures can be represented by the reaction

$$C_2HCl_3 + 2O_2 \rightarrow HCl + Cl_2 + 2CO_2 \tag{7.20}$$

accompanied by the reaction

$$4HCl + O_2 \rightleftharpoons 2Cl_2 + 2H_2O \tag{7.21}$$

Thus the behavior is also governed by the combined effects of water and chlorine. However, this system approaches the O_2-Cl_2 system in its behavior because it produces more chlorine and less water than the O_2-HCl system [16].

Oxidation in H_2O-HCl mixtures proceeds at the same rate as oxidation in pure H_2O alone, and apparently no chlorine is incorporated in the oxide. This is because there is no reaction between the H_2O and HCl. Here, too, the results indicate that HCl does not enter into the oxidation process and that the growth rate is consequently unchanged.

7.1.7 Oxidation-Induced Stacking Faults

Stacking faults can be generated during the thermal oxidation of silicon at high temperatures. It is believed that they are caused by incomplete oxidation at the Si–SiO_2 interface, which results in the formation of interstitial silicon in this region. This interstitial silicon causes fault formation by nucleation at strain centers in the bulk or at the surface [17]. These centers are primarily associated with oxygen precipitates in the silicon and with mechanical damage.

Both wet and dry oxidation result in the formation of oxidation-induced stacking faults (OSF), which can be as much as 40–50 μm long at their point of penetration to the silicon surface. They can be shrunk, and even totally annihilated, by heat treatment at high temperatures in a nonoxidizing atmosphere such as argon.* The logarithm of the rate of shrinkage of these OSF as a function of reciprocal temperature results in a straight line with a slope of 5.2 eV [19]. This is very nearly equal to the activation energy of self-diffusion in silicon (5.13 eV), so that shrinkage is probably caused by the extra plane of atoms diffusing out toward vacancies in the silicon. Support for this argument is provided in studies of impurity diffusion. Thus boron diffuses by an interstitialcy mechanism, so that its diffusion rate is enhanced in situations which encourage the formation of interstitials. In addition, both enhanced diffusion and stacking fault formation are more strongly affected by steam oxidation than by dry oxidation. Finally, both are dependent on crystal orientation.

Phosphorus diffusion occurs partly by the interstitialcy mechanism, so

* Heat treatment in N_2-HCl-O_2 mixtures has also been found effective for this purpose [18].

that these effects are also observed with this dopant, but to a smaller extent than for boron. Arsenic, on the other hand, diffuses in silicon by a purely substitutional mechanism, so that enhanced diffusion behavior is not observed.

The shrinkage of OSF can also be brought about by heavy impurity diffusion into the semiconductor. This is because the vacancy concentration in a semiconductor is increased by heavy doping, partly due to the increased stress in the material, and partly (in the case of donor impurities) by the interaction of the extrinsic donor with the charged vacancy during diffusion (see Section 4.2.3).

The effect of a chlorine species during oxidation can also be explained in terms of the mechanisms outlined here. Thus chlorine appears to promote vacancy formation at the silicon surface, and thus provides a sink by which interstitial silicon atoms can be removed. As a result, OSF formation is greatly reduced by oxidation in a chlorine species. Additional support for this mechanism comes from the observation that both boron and phosphorus diffusion in silicon are retarded if the surface oxide has chlorine incorporated in it.

7.1.8 Properties of Thermal Oxides of Silicon

The thermally grown native oxide of silicon has been shown to be suited for a large number of processing situations. Properties of this oxide, which are relevant to these situations, are now discussed.

7.1.8.1 Masking Properties

An important property of a silicon dioxide layer is its ability to mask against those impurities commonly used in microcircuit fabrication. This masking property may be explained in terms of the ability of the impurity to behave as a network former in the silica structure. Thus both B_2O_3 and P_2O_5 form mixed (borosilicate and phosphosilicate) glasses with the silica layer with which they come into contact. The boundary between the mixed glass phase and the silica phase is quite sharp. The masking properties of the layer are excellent until this boundary extends down to the silicon–oxide interface.

The diffusion process for network formers is concentration dependent, since their presence leads to the creation of charge defects and alters the properties of the oxide layer. Consequently, values for the diffusivity of impurities in silica films are a function both of their impurity concentration and also of the defect structure of the film. A few of these are listed in Table 7.2 [20] and must be taken as very approximate. However, some conclusions can be drawn from these data. Thus the diffusivity of boron, phosphorus, arsenic, and antimony are all orders of magnitude lower than the diffusivities of these impurities in silicon. (By way of comparison, the diffusivity of boron and phosphorus in silicon is 2×10^{-12} cm^2/s at 1200°C.) The reverse situation

Table 7.2 Diffusivities of Elements in SiO_2[a]

Element	D at 1100°C (cm^2/s)	D at 1200°C (cm^2s)
B	3×10^{-17} to 2×10^{-14}	2×10^{-16} to 5×10^{-14}
Ga	5.3×10^{-11}	5×10^{-8}
P	2.9×10^{-16} to 2×10^{-13}	2×10^{-15} to 7.6×10^{-13}
Sb	9.9×10^{-17}	1.5×10^{-14}
Ar	1.2×10^{-16} to 3.5×10^{-15}	2×10^{-15} to 2.4×10^{-14}

[a] See reference 21.

is true for gallium which cannot be masked by the use of silica layers. Aluminum not only moves rapidly through the oxide, but vigorously attacks it, converting to aluminum oxide in the process. Silicon nitride can be used as a mask for these dopants (see Section 8.1.3).

Masking data have been obtained empirically for the more commonly used diffusion systems by a number of workers. Representative results for boron and phosphorus are presented in curves of the type shown in Figs. 7.10 and 7.11 and are typical of commercial practice in silicon device and microcircuit fabrication. An interesting point, shown by these data, is that the masking properties of silica films are about 10 times greater for boron diffusion than for phosphorus diffusion. This is primarily due to the lower temperatures associated with the respective glass compositions, as is seen by a comparison of Figs. 2.17 and 2.18. Typically, a 0.5–0.6 μm oxide thickness is adequate for most conventional diffusion steps, and is used as the starting point in the microcircuit fabrication process.

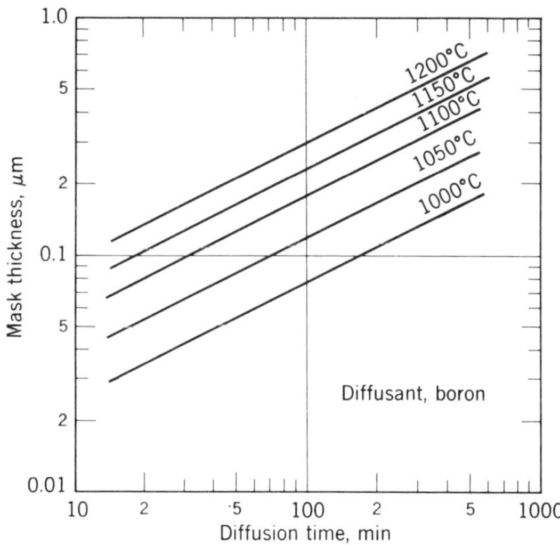

Fig. 7.10 Mask thickness for boron.

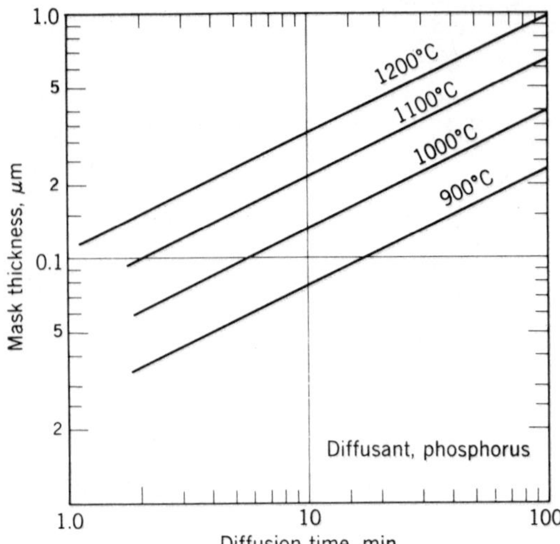

Fig. 7.11 Mask thickness for phosphorus.

7.1.8.2 Charge States

The electronic properties of the oxide, and of the oxide–silicon interface, have a profound effect on the properties of devices in the underlying semiconductor. In some cases, these effects must be accounted for in device design; in others (such as for MOS transistors) they form the very basis of device operation. In all instances, however, mobile ionic charge states give rise to time-varying phenomena and must be minimized if a device is to have long-term reliability [21].

Charge states in the silicon dioxide are intimately associated with the nature of the oxide growth process, and with interaction between the oxide and the silicon surface. This silicon surface presents a major discontinuity in an otherwise periodic crystal lattice. Its electronic properties are largely determined by the defect nature of this discontinuity. Even if atomically clean surfaces were possible, their behavior would be governed by the large number of dangling bonds at the silicon–oxide interface. From purely quantum-mechanical considerations, it can be shown that the discontinuity in the periodic potential at a clean semiconductor surface gives rise to a number of allowed states within the forbidden gap. These states, the so-called *Tamm* or *Shockley states*, are associated with unsaturated covalent bonds at the surface discontinuity and are acceptor-like in their behavior. In freshly cleaved silicon, their density is roughly equal to the density of dangling bonds on the silicon surface ($\simeq 10^{15}$ cm^{-2}).

In microcircuits, in which an oxide layer is grown out of this silicon surface, a number of surface atoms are bound to the oxygen in the form of silica polyhedra. The net result is a partial coherence of the silicon lattice, accompanied by a reduction in the density of *Interface Trapped Charge* to about 10^{11}–10^{12} cm^{-2}. The time constant associated with these states is on the order of 1 μs or less. Consequently, they are referred to as fast states. They reside within the first 25 Å of the silicon surface, and play an important role in altering the electronic properties of the semiconductor surface regions.

Fast states have energy levels which can be represented by a spectrum of surface states, located throughout the energy gap, as shown in Fig. 7.12 [22] for a thermally oxidized silicon surface. Note that the fast surface state density has a peak at about 0.1 eV from the valence band, and also at 0.1 eV from the conduction band. Both of these peaks result from the same lattice distortion effects, as evidenced by the fact that both "good" and "bad" surface treatments have an effect of altering these peaks simultaneously.

Being deep, these fast interface states are responsible for both generation and recombination effects at the surface. Thus their presence results in increased leakage currents in the region where the junction penetrates to the surface, shorter minority carrier lifetimes, and early falloff in transistor current gain at low levels.

The density of interface trapped charge can be reduced by a wet hydrogen heat treatment of the surface. Typically, this process is carried out at 450°C for 15 min and reduces the fast state density by the formation of some satisfied Si—H bonds at this interface.

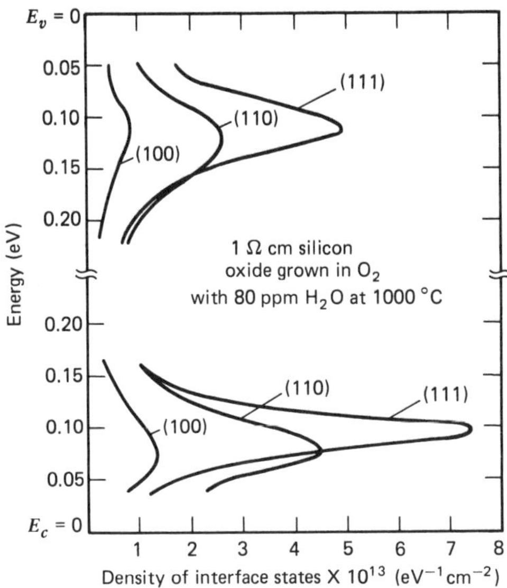

Fig. 7.12 Interface state densities for silicon. From Gray [22]. © 1969 by the Institute of Electrical and Electronic Engineers.

Slow states are also present on a semiconductor surface. With chemically etched silicon, they are present on the surface of the adsorbed layer and may have a positive or negative charge state. In thermally oxidized surfaces, the first 100 Å of surface layer represents a transition between the silicon and the oxide. Consequently, there is a net immobile positive charge associated with the excess silicon ions in this layer. The density of this *Fixed Oxide Charge* is on the order of 10^{11} cm^{-2} in thermally grown oxides on (111) silicon, and about 10^{12} cm^{-2} for chemically treated silicon surfaces.* Values for (100) silicon are typically lower by a factor of 3–4.

Slow surface states primarily behave as traps with trapping times on the order of a few seconds to many months. Consequently, they do not enter directly into electronic processes with the semiconductor, but do so indirectly by establishing the surface potential, by pinning the Fermi level at the surface trap level. This results in an electric field normal to the semiconductor surface, a lowering of the surface mobility in this high-field region, and a change in the surface conductance [23]. With thermally oxidized silicon, the positive charge in the oxide results in an *n*-shift in the semiconductor, i.e., an *n*-type semiconductor behaves more *n*-type whereas *p*-type material behaves less *p*-type, and may even invert to *n*-type. Thus the role of slow surface states is relatively slight on heavily doped silicon, but becomes increasingly significant with more lightly doped material. It is especially important in high-voltage transistors and MOS-based microcircuits.

Slow surface states in silicon can be reduced by processing the samples in such a manner as to reduce the excess ionic silicon in the oxide layer. High-temperature oxidation, in dry O_2, is effective in this situation if carried out at 1100–1200°C. A dry nitrogen treatment, at relatively low temperatures (700°C and higher) is equally effective, and possibly converts the ionic silicon to its nitride. Both treatments result in a reduction of the slow surface state density by a factor of about 5–10. Treatments are equally effective on (111) and (100) silicon.

The remaining bulk of the oxide layer is occupied by oxygen-ion vacancies and by alkali ions. Both of these species are highly mobile, so that their presence in significant amounts affects the reliability of device operation. A schematic view of these various surface states is shown in Fig. 7.13.

Oxygen-ion vacancies can be created in the oxide by two processes. First, we note that the O_2^- or O_2^{2-} ions, which are considered to be the oxidizing species in thermally grown SiO_2, are free to move in the silica layer under the influence of an electric field. Layers of this type have a significant oxygen-ion vacancy concentration, and movement of the O_2^- and O_2^{2-} species is accomplished by jumping from one such vacancy to the next. Thus it is the equivalent of vacancy migration in the silica layer, in the direction of the electric field. Next, the use of aluminum metallization also results

* All values are referenced to the silicon surface.

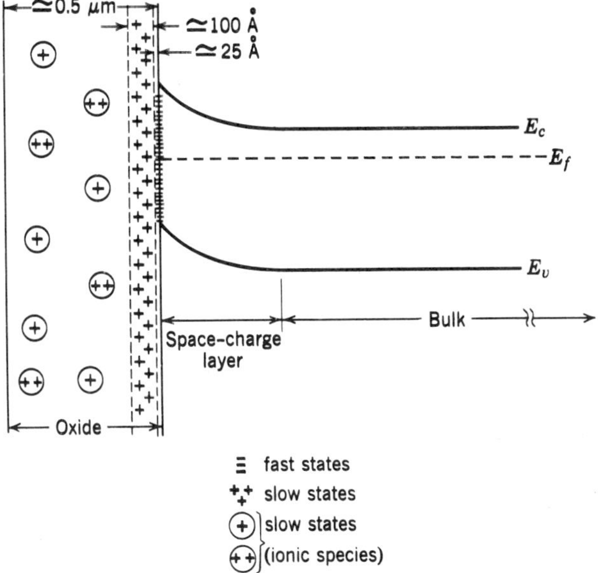

Fig. 7.13 The Si–SiO$_2$ Interface.

in vacancy creation because of its strong affinity for oxygen.* During heat treatment, aluminum substitutes for the silicon in a silica tetrahedron by replacing tetravalent silicon atoms by trivalent aluminum atoms. This necessitates the removal of oxygen ions from some part of the neighboring silica lattice, resulting in vacant sites which are effectively charged positive. The reaction can be written as

$$2Al + 2SiO_2 \rightarrow 2Si + 2AlO_2^- + V^{2+} \quad (7.22)$$

and results in the generation of positively charged vacancies.

Alkali ions such as sodium, potassium, and lithium are also present throughout the bulk of the oxide. Their presence is undesirable since they enter the silica in ionized form (Na^+, K^+, Li^+) and are highly mobile at temperatures as low as 125°C. Consider what happens if these ions are present in the gate oxide of a MOS transistor. The net result of applying a field across the oxide in a direction pointing toward the silicon surface results in an accumulation of a positive space-charge layer close to the silicon–oxide interface and a corresponding *n*-type shift in the surface potential. This will make the threshold voltage more negative in a *p*-channel device, i.e., the

* Indeed, the choice of aluminum for interconnections is primarily based on its excellent bonding properties with SiO$_2$, as a result of this affinity.

magnitude of this voltage increases. On the other hand, movement of this charge away from the silicon surface makes the magnitude of the threshold voltage lower in this device. Similar arguments can be applied to explain the threshold instability in *n*-channel devices.

The amount of positive space-charge accumulation is a function of the magnitude and time duration of the field impressed across the oxide and of the ambient temperature at which the device is operated. On removal of the stress field and reduction of the temperature, the space charge becomes immobilized, resulting in a semipermanent change in the properties of the underlying material. Shifts in the surface potential are reversible, that is, the space charge layer (hence the surface potential) can be altered by varying the stress and temperature conditions. As a result, they impair the long-term stability of these devices.

There are a number of correctives for this problem, which are often used singly or together in microcircuit fabrication. First, gate oxide diffusions must be carried out with dry oxygen, and in ultraclean, dedicated systems having ceramic liners to prevent in-diffusion of sodium from the furnace walls. The use of silicon diffusion tubes has been found to be excellent for this situation. However, their excessive cost has restricted their use to only a few commercial installations at the present time.

A second approach is to use a P_2O_5-doped glass for the outer surface of this oxide. This can be readily accomplished by introducing P_2O_5 during the oxidation process by procedures commonly used for making phosphorus diffusions. The effectiveness of this approach depends on the gettering action of sodium by the phosphosilicate glass [24]. In addition, the introduction of a pentavalent network former such as P_2O_5 provides an excess of oxygen to the silica structure and greatly reduces the number of oxygen-ion vacancies in its vicinity. This makes the glass more dense, and more resistant to the transport of ionic impurities through it. In line with the above arguments it is reasonable to expect that a surface layer of B_2O_3 is not useful in this application because it does not reduce the oxygen deficiency in the oxide. This has indeed been found to be the case in practice.

Excessive amounts of P_2O_5 cannot be used for this purpose since they cause large, fixed threshold shifts due to polarization effects. In addition, they tend to make the gate oxide hygroscopic and hence even more unstable. One approach, often used in the formation of MOS oxides, is to grow this oxide thermally, incorporate a heavy surface layer of P_2O_5, remove this layer in a selective oxide etching solution, and replace it with a lightly doped layer of phosphosilicate glass.

A third approach is to use a chlorine bearing species during gate oxidation. As pointed out in Section 7.1.6, this results in a pileup of chlorine at the $Si–SiO_2$ interface. This chlorine appears to capture the Na^+ ions and form a neutral system.

Finally, a cover layer of silicon nitride is often used. This material is very dense and is an effective barrier to the transport of ionic species, such as sodium, through it. Its growth and technology are described in the following chapter.

7.1.8.3 Hot Electron Effects

In thermally oxidized silicon, the barrier height between the silicon and the silicon dioxide at both the valence- and the conduction-band edges is about 3.2 eV. This is sufficiently low so that it is possible for electrons (or holes), generated in the silicon by device action, to occasionally have enough energy to surmount it, and get trapped into the silica layer. This phenomenon of "hot carrier" trapping is encountered in any high-field semiconductor region which is in contact with the silicon dioxide. One situation where this presents a serious problem is during avalanche breakdown of a *p–n* junction, since the energy of avalanching carriers has a mean value of about 3 eV.

Junctions made by microcircuit technology have curved sides. During avalanche breakdown, the region of high field is very near the surface. In addition, there is a strong fringing field which extends out into the oxide and aids in the trapping of carriers in it. Once trapped, they create a space charge which affects the behavior of the junction. Typically, the value of the breakdown voltage changes by an amount which depends on the extent of avalanching, and consequent carrier trapping. Furthermore, the device recovers over a period of days, at a rate which depends very critically on the quality of the surface oxide and especially on its water content. This problem, known as *junction walkout*, cannot be eliminated, but it can be reduced by careful processing [25]. In particular, a long low-temperature bake out (48–72 h at 250°C), which reduces the water content in the oxide, has been found beneficial.

Hot electron effects also result in the breakage of some of the satisfied silicon bonds (Si—O, Si—H, or Si—OH) which are present on the silicon surface, so that the interface trap density increases. This gives rise to deterioration in the low-level current gain of bipolar transistors whose emitter–base junctions have been subjected to avalanche breakdown [26].

Hot electron effects can also occur, but to a lesser extent, in high-field regions which are not avalanching. These include the channel region of MOS transistors as well as the depletion layer at the drain regions. Both of these effects are of increasing importance in VLSI devices, which require the use of ultrathin gate oxides and short channel lengths [27]. Hot electron effects in these devices result in a slow, long-term change in their threshold voltage. This long-term stability problem is as yet unsolved, although it can be minimized by careful attention to gate oxide formation. However, techniques for its assessment and prediction are well advanced, so that circuit designs can take this into account [28].

Hot electron effects can also be used to advantage in the design of programmable read-only memory (PROM) elements. One such structure consists of a MOS transistor with a floating gate which is buried in the oxide to prevent leakage [29]. Information is entered by causing a momentary avalanche breakdown between the source or drain, and the substrate. This injects hot electrons into the oxide, which causes the floating gate to become charged, and turn the device into its ON state. The leakage rate of such a gate is very long, typically many years, so that data are essentially stored permanently in this manner.

7.2 THERMAL OXIDATION OF GALLIUM ARSENIDE

Gallium arsenide can be thermally oxidized by heat treatment in oxygen. However, the oxidation process is very slow at temperatures below 400°C. Between this temperature and 530°C, it is possible to grow amorphous oxide films with good uniformity [30]. At higher temperature, however, the oxide tends to crystallize and become porous.

Oxidation proceeds by the breaking of Ga–As bonds, reaction with chemisorbed oxygen at the surface, and subsequent nucleation of gallium and arsenic oxides. These nuclei form oxide islands which grow outward until the surface is covered. Further growth of the oxide proceeds in all probability by the diffusion of both gallium and arsenic *through* the oxide layer to the surface, where they are converted to the appropriate oxides. This is quite different to what occurs during the oxidation of silicon, where oxygen diffuses *inward* through the growing oxide layer. One important consequence of this difference is that thermal oxidation of silicon results in a fresh interface region, with surface contaminants on the original silicon ending up on the oxide surface. With gallium arsenide, however, it is highly probably that contaminants on the original semiconductor surface will remain at the interface, even after the oxidation process.

A similarity with silicon oxidation, however, lies in the fact that further oxidation beyond the original surface layer requires the diffusion of various species through the growing oxide. As a result, the growth rate can also be described here by Eq. (7.17) where B/A is a linear rate constant, and B is a parabolic rate constant.

The quality of the thermal oxide of gallium arsenide is dominated [31] by the widely different properties of the elements comprising this semiconductor, and of their oxides. Thus the oxidation potential of gallium is +0.65 V, whereas that of arsenic is −0.25 V. As a result, gallium will oxidize more readily than arsenic. Next, the free energy of formation of Ga_2O_3 is −214 kcal/mole, whereas that of As_2O_3 is −151 kcal/mole, so that the oxide of gallium is thermodynamically more stable than that of arsenic. For a typical oxidation temperature of 450°C, the vapor pressure of gallium is 10^{-11} torr, whereas that of arsenic is 10 torr; thus evaporation of arsenic from the surface of the oxide is rapid, whereas that of gallium is negligible. Finally, the vapor pressure of As_2O_3 at 450°C is almost 760 torr, whereas that of Ga_2O_3 is negligible at this temperature.

As a consequence of the above factors, the arsenic content of a thermally grown oxide of gallium arsenide rapidly falls with distance from the oxide–semiconductor interface, and most of the oxide layer consists of Ga_2O_3, with as little as 5–10% of As_2O_3. In addition, any subsequent heat treatment results in further loss of As_2O_3 from the oxide, and in conversion of the surface oxide to β-Ga_2O_3 or to $GaAsO_4$. These films are brittle and crystalline, and extremely porous to the diffusion of impurities such as zinc. They are unsatisfactory as cover layers for gallium arsenide, either for device processing or for device passivation.

Superior quality thermal oxides can be grown under an overpressure of As_2O_3 to prevent loss of this component [32]. This is done by using a sealed tube with two zones. The gallium arsenide to be oxidized is placed at one end, at 500°C, whereas an excess of As_2O_3 is placed at the other, at 450°C. This arrangement establishes a vapor pressure of As_2O_3 over the gallium arsenide, which is oxidized at a rate of about 0.1 μm/h. Films grown by this technique have high breakdown field strengths [$(5–7) \times 10^6$ V/cm], and consist of both As_2O_3 and Ga_2O_3. The composition of these films tend to be graded, however, becoming slightly depleted of As_2O_3 as the surface is approached. Nevertheless, their quality is sufficiently high so as to allow their use in MOS-based gallium arsenide integrated circuits [33].

7.3 ANODIC OXIDATION

The principal characteristic of anodic processes is that they are usually carried out at room temperature. Thus impurity concentrations present in the semiconductor are not altered during this process. As a result, anodic oxidation is a useful means for the controlled removal of layers of silicon and gallium arsenide at room temperature, and is often used as a diagnostic tool.

This technique can also be used to grow reasonably high-quality oxides on gallium arsenide, since problems associated with the widely different characteristics of Ga_2O_3 and As_2O_3 are minimized when gallium arsenide is oxidized by these techniques. Here the dielectric breakdown strength and the electrical resistivity of these oxides are superior to that obtained by thermal oxidation. As a result, they are useful in technological processes for gallium arsenide, such as masking and coating. In addition, they have been found useful for gate oxides in MOS transistors, and in gallium arsenide integrated circuits based on these devices [34].

Anodic oxidation, or *anodization*, is carried out by placing the semiconductor in an electrolytic cell, as shown in Fig. 7.14, where it is connected to the positive terminal of a power supply so that it serves as the anode. A noble metal such as platinum is connected to the negative terminal of the supply, and serves as a cathode. The source impedance of the supply can be kept low, so that anodization occurs at constant voltage; alternately, a

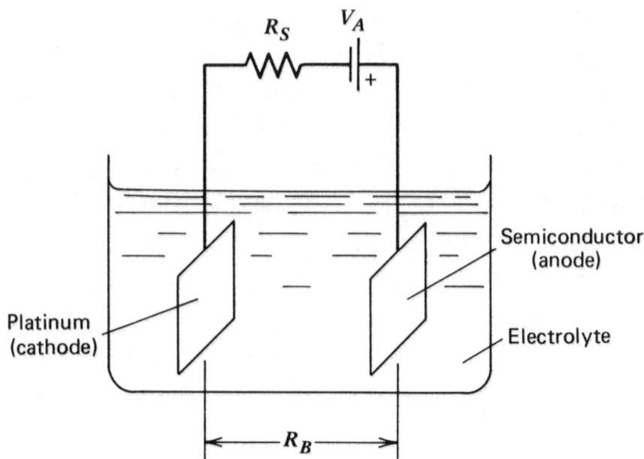

Fig. 7.14 Circuit diagram of an anodization cell.

high source impedance allows anodization to proceed under constant current conditions.

A large variety of electrolyte formulations can be used. The primary oxidizing component of all of these is water, which dissociates into H^+ and $(OH)^-$. The dissociation reaction can be written as

$$H_2O \rightleftharpoons H^+ + (OH)^- \tag{7.23}$$

so that the equilibrium constant is given by

$$K = [H^+][OH^-] \tag{7.24}$$

The equilibrium constant for deionized water at 24°C is 10^{-14}. For this case, since $[H^+] = [OH^-]$, the concentration of $[H^+]$ is 10^{-7}, resulting* in a pH of 7. Often a conductivity/pH modifier is added to this system to vary the resistance of the electrolyte, as well as the dissolution rate of the oxide in it. Typical modifiers are H_3PO_4 (acidic), NH_4OH (basic), and $(NH_4)_2HPO_4$ (almost neutral). Usually very small amounts of these are required for pH control.

It is also possible to carry out an anodization process in a nonaqueous solution. Here the pH is set before addition of the nonaqueous component. Again, the requirement for this electrolyte is that one of its dissociation products be the $(OH)^-$ ion. Ethylene glycol or propylene glycol are often used as the nonaqueous medium. Additional components of the electrolyte serve to control its viscosity and dielectric constant and effect the reproducibility of the process as well as its sensitivity to ambient conditions and to contaminants.

* By definition $\text{pH} = -\log_{10}[H^+]$.

Oxides formed by anodization are generally porous in character, and have water incorporated in them. However, they can be used to advantage in a number of situations. By way of example, anodic oxidation, followed by chemical dissolution, provides an excellent method for removing a controlled amount of semiconductor by a low-temperature, damage-free process. The technique is reproducible and nearly independent of doping concentration or crystallographic orientation. In addition, it can be carried out many times without significant loss of surface planarity, so that it is an ideal means for profiling the impurity concentration in semiconductor layers [35, 36].

Anodic oxides of silicon are generally much poorer than their thermally grown counterparts, so that their use is restricted to singular applications of the type just described. With gallium arsenide, however, the thermal oxide is extremely poor, and anodic oxides are the only native-grown choice for all device and processing applications.

Reactions leading to the anodic oxidation of silicon are as follows:

1. Water in the electrolyte medium dissociates into H^+ and $(OH)^-$

$$2H_2O \rightleftharpoons 2H^+ + 2(OH)^- \tag{7.25}$$

2. The difference in electrochemical potentials between the silicon and the electrolyte results in charge transfer from the silicon until equilibrium is established. This leaves the surface layer partially depleted of electrons. During anodization holes are supplied from the bulk of the semiconductor to the semiconductor–electrolyte interface, thus promoting the silicon surface atoms to a higher oxidation state,

$$Si + 2h^+ \rightarrow Si^{2+} \tag{7.26}$$

3. The Si^{2+} combines with $(OH)^-$ to form the hydroxide

$$Si^{2+} + 2(OH)^- \rightarrow Si(OH)_2 \tag{7.27}$$

4. $Si(OH)_2$ subsequently forms SiO_2 liberating hydrogen in the process,

$$Si(OH)_2 \rightarrow SiO_2 + H_2 \tag{7.28}$$

The overall reaction is given by

$$Si + 2h^+ + 2H_2O \rightarrow SiO_2 + 2H^+ + H_2 \tag{7.29}$$

In the absence of an external battery (or illumination), the continuous supply of holes to the reaction interface ceases. The concentration of thermally generated holes is insufficient for the formation of SiO_2 beyond a few monolayers. However, the battery provides the necessary holes for this anode reaction and thus sustains the process of oxide growth. The H^+ drifts to the cathode, where it is evolved as molecular hydrogen by the addition of electrons

$$2H^+ + 2e^- \rightarrow H_2 \tag{7.30}$$

The growth of anodic oxides of gallium arsenide proceed along somewhat similar lines. Here the ionic states for gallium and arsenic are Ga^{3+} and As^{3+} and are formed by the anodic reaction

$$GaAs + 6h^+ \rightarrow Ga^{3+} + As^{3+} \tag{7.31}$$

The separate reactions for the gallium components are as follows [37]:

1. $3H_2O \rightleftharpoons 3H^+ + 3(OH)^-$ (7.32a)
2. $Ga^{3+} + 3(OH)^- \rightarrow Ga(OH)_3$ (7.32b)
3. $2Ga(OH)_3 \rightarrow Ga_2O_3 + 3H_2O$ (7.32c)

The rections for arsenic are:

1. $2H_2O \rightleftharpoons 2H^+ + 2(OH)^-$ (7.33a)
2. $As^{3+} + 2(OH)^- \rightarrow AsO_2^- + 2H^+$ (7.33b)
3. $2AsO_2^- + 2H^+ \rightarrow As_2O_3 + H_2O$ (7.33c)

so that the overall reaction is

$$GaAs + 12h^+ + 10H_2O \rightarrow Ga_2O_3 + As_2O_3 + 4H_2O + 12H^+ \tag{7.34}$$

Once again, water plays the primary role in forming the oxidation products, and continuous oxidation is sustained by the supply of holes to the anode surface via the external battery. Finally, the delivery of electrons to the cathode (from the battery) is accompanied by hydrogen evolution at this electrode.

The anodization of p-type material is relatively straightforward since there is no problem with delivering holes to the semiconductor surface by means of a battery. The situation is somewhat more complex for an n-type semiconductor, however. Here the initial charge transfer from the semiconductor into the electrolyte (in order to establish thermal equilibrium) creates a depletion layer in the semiconductor, and a barrier to the flow of holes. In effect, then, the electrolyte–semiconductor system behaves much like a Schottky diode with the electrolyte serving the role of the "metal." As mentioned earlier, oxidation will not proceed beyond a few monolayers, unless provision is made to supply holes to the semiconductor surface. One approach is to illuminate the sample to provide these holes by photo generation. Alternately, the anodization cell can be operated at a voltage which exceeds the Schottky diode breakdown voltage, so that avalanche-generated holes can allow the oxidation to proceed. The buildup of this growing oxide reduces the available voltage across the depletion layer; anodization eventually ceases when this voltage falls below the breakdown voltage of the electrolyte–semiconductor system. This effect has been exploited for the selective growth of oxide over different parts of an inhomogeneous semiconductor material such as polycrystalline gallium arsenide [38].

The description of the anodization process provided here is at best elementary. The effect of early stages of anodic growth, as well as the role of double layers at the electrolyte–oxide interface, are considered elsewhere in the literature [39] and are not treated here.

7.3.1 Oxide Growth

Consider an anodization cell as shown in Fig. 7.14, under conditions of no illumination. In this system, initial oxide growth is by the formation of islands on the semiconductor surface. Once these nucleate to form a continuous film, further oxidation proceeds either by the transport of the $(OH)^-$ ion through the oxide to the semiconductor, or by the transport of semiconductor material through the oxide to the electrolyte–oxide interface. As was the case with thermal oxidation, the anodization of gallium arsenide most probably proceeds by the transport of Ga^{3+} and As^{3+} through the oxide, rather than by the motion of $(OH)^-$ through the oxide.

The consumption of semiconductor material (with its conversion to the oxide) is governed by Faraday's law of electrolysis. This law states that the number of grams W of material consumed per coulomb (C) of charge is given by

$$W = W_e/qN \tag{7.35}$$

where W_e is the electrochemical equivalent weight in grams of the material involved, q the electronic charge, and N Avogadro's number ($qN = 96{,}483$). The electrochemical weight of an element or compound is its atomic or molecular weight, respectively, divided by the change in valence involved in the reaction. Thus gallium arsenide undergoes a valence change of 6+ during anodization. From Faraday's law, the amount of gallium arsenide consumed per coulomb of charge is thus 2.5×10^{-4} g. A typical experimental value for this quantity is about 2.4×10^{-4} g/C. The ratio of experimental to theoretical values, quoted as a percentage, is referred to as the *current efficiency*. For gallium arsenide, this is about 95%, whereas the current efficiency for silicon is only 1–3% [40].

The formation rate α of an oxide is defined as the increase in film thickness per unit charge, passing through unit area. This term is closely related to the current efficiency defined above. A second term, of importance during anodization, is the oxide dissolution rate. This is the change in film thickness as a function of time, if left in the electrolyte without any applied voltage. The dissolution rate for oxides of gallium arsenide depends on the electrolyte, and ranges from 2 Å/s to as low as 4.2×10^{-2} Å/s at room temperature.

Consider again the anodization circuit of Fig. 7.14 where V_A is the applied voltage. Assume unit area for the electrodes, and that R_S and R_B are the resistance of the circuit and the bath, respectively. The initial current flow I_0 is given by

$$I_0 = \frac{V_A - V_r}{R_S + R_B} \tag{7.36}$$

Here V_r is the rest potential of the anodization cell (i.e., the reverse e.m.f. generated when the cell is used as a battery) for the case of a p-type semiconductor. With an n-type semiconductor, however, V_r is the sum of the rest potential and the breakdown voltage associated with its depletion layer. Let α be the formation rate, and let f_d be the dissolution rate of the film. Then the equivalent dissolution current I_d is given by

$$I_d = \frac{1}{\alpha} f_d \tag{7.37}$$

During anodization, current flow through the oxide is primarily ionic in nature, and can be described by an equation of the form $J \propto e^{KE_{\text{ox}}}$. For a high-field situation, as in the case of the anodization of gallium arsenide (where $E_{\text{ox}} \simeq 5 \times 10^6$ V/cm), it is possible to assume that there is a constant electric field across the oxide during film growth. For this approximation, the final thickness of oxide which is formed for an applied voltage V_A is given by

$$x_\infty = \frac{V_A - V_r}{E_{\text{ox}}} \left(1 - \frac{I_d}{I_0}\right) \tag{7.38}$$

The growth rate of the film can be determined by writing Kirchhoff's law for the anodization circuit. Thus

$$V_A - V_r = i(R_S + R_B) + \alpha E_{\text{ox}} \int_0^t i \, dt \tag{7.39}$$

Solution of this equation, and insertion of the final value of oxide thickness found from Eq. (7.38), gives

$$x = x_\infty (1 - e^{-t/\tau_f}) \tag{7.40}$$

where the time constant for growth τ_f is given by

$$\tau_f = \frac{R_S + R_B}{\alpha E_{ox}} \tag{7.41}$$

As seen from the above, oxide thickness builds up exponentially until a final value, dictated by the applied voltage, is reached. Typically, the final oxide thickness is 20 Å/V for gallium arsenide and about 3 Å/V for silicon.

Anodization can also be carried out from a constant current source. Here Faraday's law requires that the oxide thickness vary linearly with time. In addition, since the field across the oxide is constant, the voltage drop across it will also increase linearly with time. In many practical anodization systems, it is common practice to exploit this behavior to monitor the film thickness. This is done by using a constant current source, monitoring the linear increase of voltage with time, and then stopping the process at some preset voltage (i.e., at some preset oxide thickness).

The pinch-off voltage of junction gate field effect transistors is determined by the product of thickness and doping concentration of the epitaxial layer. It is important that this parameter be controlled across the wafer, for a practical device fabrication process. This can be done by anodization of the slice in the dark [41], or under uniform controlled lighting [42] if a lower value of $t_{epi} N_d$ is required. In the latter situation, this product is set by this illumination. In either case, anodization of the *n*-GaAs film (on the semi-insulating substrate) proceeds by avalanche breakdown of the substrate. As the layer is consumed, the depletion layer moves closer to the substrate–epitaxial layer interface, and eventually sweeps rapidly into the semi-insulating substrate. As a result, insufficient voltage is left to sustain breakdown so that anodization ceases. Thus semiconductor removal is locally achieved until the pinch-off voltage is constant across the slice.

Table 7.3 lists quantities of interest to the anodization of silicon and gallium arsenide.

Table 7.3 Parameters of Interest for Anodization of Silicon and Gallium Arsenide

Parameter	Si	GaAs
Atomic or molecular weight	28.09	144.64
Valence change	2+	6+
Current efficiency	1–3%	90–95%
Electric field during anodization, oxide (V/cm)	$(1-2) \times 10^7$	5×10^6
Density, semiconductor (g/cm^3)	2.33	5.32
Density, oxide (g/cm^3)	1.025	4.2
Dielectric constant, oxide	3.2	5.4
Refractive index, oxide	1.46	1.80

7.3.2 Anodic Oxidation Systems

Systems for the anodization of silicon have been primarily developed for use in the controlled removal of layers for diagnostic purposes. A nonaqueous solution of *N*-methylacetamide (NMA) with 0.4*N* KNO_3 has been found useful for this application [43]. With this system the oxide thickness is in the 3–5-Å/V range, with measured values of ionic current efficiency of 1.6–3%. It should be noted that this electrolyte does have a small amount of water incorporated in it because of the hygroscopic nature of NMA; sometimes additional water (up to 9%) is added to modify its anodization characteristics.

A solution of ethylene glycol in 0.04*N* KNO_3, with 2.5% water, has also been used [36]. The addition of about 1–2 g of $Al(NO_3)_3 \cdot 9H_2O$ per liter of this solution has been found to improve the uniformity of this electrolyte with successive anodizations [44]. Oxide thickness for this system is typically 2.2 Å /V.

There has been little recent interest in the development of new anodization systems for silicon, because of the overriding advantages of thermal oxidation. The situation is quite different for gallium arsenide, however, since anodic oxides are superior to those formed thermally. Moreover, recent progress in this area has opened up the possibility of obtaining oxides which are suitable for channel conductance control in MOS devices. Consequently, considerable effort has been placed on their development.

Anodization systems for gallium arsenide are of three forms: aqueous, nonaqueous, and mixed. The aqueous solutions are primarily based on water, or on water–H_20_2 mixtures [45]. Water has an extremely high resistivity ($\simeq 16 \times 10^6\ \Omega$ cm) in pure form, so it is necessary to use small amounts of additives for the purpose of pH/conductivity control [46]. These additives may be acidic such as H_3PO_4, which gives good oxides in the 2.5–3.5 pH range. They may also be basic such as NH_4OH; here excessive etching is usually observed, and it is difficult, if not impossible, to obtain reproducible results. Neutral additives can also be used; thus the use of $(NH_4)_2HPO_4$ shifts the pH very slightly. For example, a 0.1*N* solution of this reagent will have a pH of 7.8 as compared to a pH of 7 for pure water.

All of the above electrolytes operate successfully over a very narrow pH range, which must be tightly controlled to avoid excessive dissolution of the oxide while it is being formed. Thus the water–$(NH_4)_2HPO_4$ solution produces better quality oxides, with less dissolution in the electrolyte, if a small amount of H_3PO_4 is added to change its pH from 7.8 to 7.15. Aqueous electrolytes have also been found to be very sensitive to reagent contamination, with consequent nonreproducibility of oxide formation.

Nonaqueous solutions have also been used for anodization of gallium arsenide. Here the pH value loses its meaning in the classical sense; however, it is often used to monitor the process. For one such electrolyte system, a 10 g solution of ammonium pentaborate in 100 cm^3 of ethylene glycol produces a pH reading of 2.2, and has been found to give highly reproducible results with relative insensitivity to bath contamination [47].

Recently considerable work has been done on a system [40] using a mixed solution of water and propylene glycol containing tartaric acid. This system has been found to be relatively insensitive to the presence of contaminants in it, and to consistently give oxides with a resistivity that is 10^3–10^5 better than that obtained with aqueous solutions. In this system, water is the oxidant, and tartaric acid is used for pH/conductivity control. Typically, a 3% aqueous solution of tartaric acid is used, with ammonia to adjust its pH to a suitable value. This solution, which is aqueous, can be used by itself, if its pH is in the 5–7 range. However, it is very sensitive to contamination, and gives results that are difficult to reproduce. The same solution, however, if added to glycol in a 1:2 parts-by-volume ratio (aqueous/glycol), is useful for starting pH values from 2 to 9, and results in highly reproducible anodizations. This mixture, known as AGW, has been used over an aqueous/glycol range from 1:1 to 1:4 with little variation in anodization quality.

The oxide dissolution rate for AGW solutions is very low, typically 4–6 × 10^{-2} Å/s, as compared to dissolution rates of 1–3 Å/s for aqueous oxides. Furthermore, AGW oxides have been found to have a higher As_2O_3 content than those made with aqueous electrolytes.

Electron microprobe studies of the AGW oxide have shown that it approaches stoichiometry more closely than those formed by most of the other methods. As a consequence, it has a high specific resistivity (10^{14} Ω cm) and also a high breakdown field strength.

7.3.3 Properties of Anodic Oxides

Anodic oxidation of silicon results in a relative porous oxide, which is not suited for use as a passivation coating. Its increased porosity also makes it more readily etched than thermal oxides, typically as much as five times faster. Densification of the oxide, by holding it at 450–500°C for a period of time, results in some improvement in these properties, and also serves to remove trapped water from the oxide.

Anodic silicon dioxide has an interface state charge density of about a decade larger than that of conventional thermal oxide. As a result, it cannot be used as a gate oxide in MOS applications. Because of all these disadvantageous features, the anodization of silicon is generally restricted to its use in diagnostic situations, requiring the repeated removal of thin layers.

Anodic oxides of gallium arsenide have considerably better electrical properties than the thermal oxides. Here anodization has the advantage over thermal oxidation, since it is conducted at room temperature. Even so, some departure from stoichiometry is present, because of the significant vapor pressure difference between As_2O_3 and Ga_2O_3.

To date, the closest approach to stoichiometry has been achieved by the AGW formulation, which results in an oxide having a specific resistivity

(10^{14} Ω cm) and also a high breakdown field strength. The surface state trap density at the oxide–gallium arsenide interface is so large that this oxide cannot be used as a gate in a MOS device. In addition, this density is critically dependent on the preanodization treatment given to the gallium arsenide [48]. However, a thermal anneal at 300°C in hydrogen for 30–60 min has been found to greatly reduce this trap density, and MOS action has been demonstrated in microwave devices made from this oxide.

Anodic oxides of gallium arsenide have also been used as antireflection coatings in solar cells, since their refractive index provides a reasonable match between air and gallium arsenide. In addition, they have been used as masks for low-temperature diffusions in gallium arsenide.

7.4 PLASMA ANODIZATION

Plasma anodization of both silicon and gallium arsenide can be accomplished by using activated oxygen, created by an electrical discharge, as the oxidizing species. This is usually carried out in a low-pressure (0.1–0.5 torr) system, where a plasma can be sustained by direct-current or radio frequency excitation [49]. Both inductive and capacitive coupling have been used in radio frequency plasma oxidation systems. In addition, a solenoidal magnetic field is sometimes used to confine the plasma to the region around the semiconductor [50]. Figure 7.15 shows a schematic diagram of one such system, used for the plasma oxidation of gallium arsenide.

A large number of active oxygen species, as well as free radicals and electrons, are formed during this process. However, only a few percent of the oxygen molecules are active, with the primary species being O^-. One

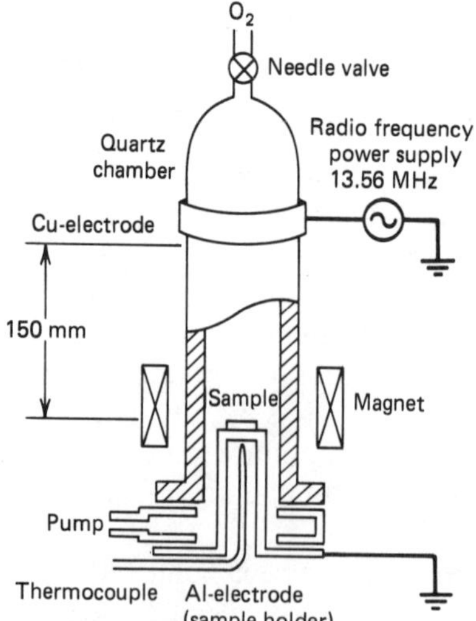

Fig. 7.15 Typical plasma anodization apparatus. From Yokoyama, Mimura, Odani, and Fukuta [50]. Reprinted with permission from the *Japanese Journal of Applied Physics*.

possibility is that this species enters into the oxidation process. An alternate possibility, however, is that oxygen molecules interact with electrons at the semiconductor surface to result in the active oxygen species. In either event, for anodization to proceed, a closed direct-current path must be provided, with the substrate held at positive bias so that it serves as an anode. Typically, the oxide thickness for gallium arsenide is about 40 Å /V of applied bias. However, this high value, compared to wet anodization techniques, is most probably due to the self-bias associated with the plasma potential [51]. Its magnitude is a complex function of both system and plasma parameters, but is 20–30 V in a typical system.

The potential advantage of plasma anodization methods lies in the fact that the electron temperature of the ionized gas is about 10,000°K. As a result, oxidation can be carried out [52] at low thermal temperatures ($\leq$100°C), thus avoiding the problems associated with loss of As_2O_3 during the high-temperature thermal oxidation of gallium arsenide. In addition, films made by plasma techniques have a more abrupt semiconductor–oxide interface, and are cleaner and less porous than those grown by a wet chemical process. Recently high-speed MOS devices and integrated circuits, based on this type of gate oxide, have been successfully operated [53]. These circuits are based on depletion-mode devices, since inversion of the substrate has not been achieved at the present time. Enhancement-mode devices, based on this property, have great promise for the future, so that there is considerable research interest in this area.

Plasma anodization of silicon results in an oxide of poorer quality than that obtained by thermal oxidation [54]. However, this technique allows oxides to be grown at temperatures as low as 500°C, so that it is attractive in VLSI schemes where dopant diffusion must be kept at a minimum during the oxide growth step.

7.5 EVALUATION OF OXIDE LAYERS

The most commonly used methods of evaluation are visual in nature. Thus a check is made to ensure that the oxide is smooth and free from pinholes, crystallites, and other surfacc blemishes. Often the oxidized slice is immersed for a few minutes in a 10% copper sulfate solution which penetrates the pinholes. This results in displacement plating of copper around them, and aids in their visual observation [55].

Oxide layer thickness can be measured by the angle-lap and interferometric technique described in Section 4.10.1. This technique has largely been replaced by ellipsometric methods which are both rapid and nondestructive. Here a monochromatic beam of plane-polarized light is reflected off the oxide layer and also off the oxide–semiconductor interface. These two components will usually experience different amounts of phase shift upon reflection, and also have different reflection coefficients.

The basic ellipsometer system is shown in Fig. 7.16. Here the light source–filter–polarizer combination is often replaced by a laser, which greatly simplifies the source optics. Details of system operation and analysis of the results have been discussed elsewhere [56]. Here it suffices to point out that the data can be processed to obtain both the film thickness as well as its refractive index. In addition, the technique can be extended to determine the thicknesses and refractive indices of multiple layers (such as silica followed by silicon nitride) on a semiconductor substrate.

Oxide thickness may also be determined to a fair degree of accuracy by visual means. If a thin film with a reflecting back surface is viewed by monochromatic light in an almost perpendicular direction, it can be shown that intensity enhancement occurs at wavelengths λ_k, such that

$$\lambda_k = \frac{2nd}{k} \tag{7.42}$$

where $k = 1, 2, \ldots$

$d =$ film thickness, in Å

$n =$ index of refraction of film (1.46 for SiO_2)

If the film is viewed in white light, it will exhibit a brightly colored appearance at one of the wavelengths given by Eq. (7.42). Once the film color is known, however, its thickness can be determined from Table 7.4 which has been derived for SiO_2 layers [57]. The uncertainty of the value of k is generally resolved on the basis of known growth conditions, which allow a rough estimate of film thickness.

Some care is required to make visual measurements of this kind. Thus the chart shown in Table 7.4 is for SiO_2 films, which are observed normal to their surface. If the viewing angle is θ, the true thickness of the film is given by $t_0/\cos\theta$, where t_0 is the value read off the color chart. In addition, it must be emphasized that this chart is for fluorescent light, and that color quality will be different under other conditions of illumination.

Finally, the color chart can also be used for materials of different refractive index. Thus the film thickness t_f for some film other than silica is

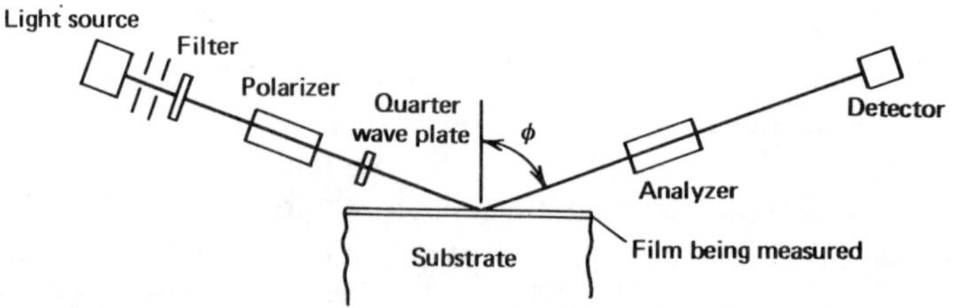

Fig. 7.16 Ellipsometer optics.

Table 7.4 Color Chart for Thermally Grown SiO_2 Films Observed Perpendicularly Under Daylight Fluorescent Lighting[a]

Film Thickness (μm)	Color and Comments	Film Thickness (μm)	Color and Comments
0.05	Tan	0.63	Violet red
0.07	Brown	0.68	"Bluish" (Not blue but borderline between violet and blue green. It appears more like a mixture between violet red and blue green and looks grayish)
0.10	Dark violet to red violet	0.72	Blue green to green (quite broad)
0.12	Royal blue	0.77	"Yellowish"
0.15	Light blue to metallic blue	0.80	Orange (rather broad for orange)
0.17	Metallic to very light yellow green	0.82	Salmon
0.20	Light gold or yellow—slightly metallic	0.85	Dull, light red violet
0.22	Gold with slight yellow orange	0.86	Violet
0.25	Orange to melon	0.87	Blue violet
0.27	Red violet	0.89	Blue
0.30	Blue to violet blue	0.92	Blue green
0.31	Blue	0.95	Dull yellow green
0.32	Blue to blue green	0.97	Yellow to "yellowish"
0.34	Light green	0.99	Orange
0.35	Green to yellow green	1.00	Carnation pink
0.36	Yellow green	1.02	Violet red
0.37	Green yellow	1.05	Red violet
0.39	Yellow	1.06	Violet
0.41	Light orange	1.07	Blue violet
0.42	Carnation pink	1.10	Green
0.44	Violet red	1.11	Yellow green
0.46	Red violet	1.12	Green
0.47	Violet	1.18	Violet
0.48	Blue violet	1.19	Red violet
0.49	Blue	1.21	Violet red
0.50	Blue green	1.24	Carnation pink to salmon
0.52	Green (broad)	1.25	Orange
0.54	Yellow green	1.28	"Yellowish"
0.56	Green yellow	1.32	Sky blue to green blue
0.57	Yellow to "yellowish" (not yellow but is in the position where yellow is to be expected. At times it appears to be light creamy gray or metallic)	1.40	Orange
0.58	Light orange or yellow to pink borderline	1.45	Violet
0.60	Carnation pink	1.46	Blue violet
		1.50	Blue
		1.54	Dull yellow green

[a] See reference 57.

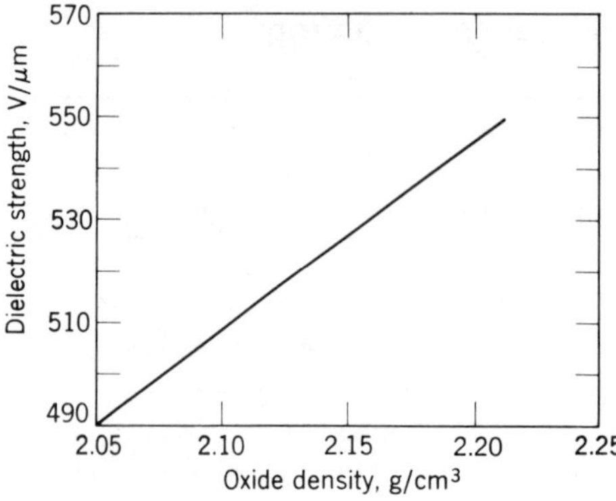

Fig. 7.17 Oxide density versus dielectric strength.

given by

$$t_f = \frac{t_0 n_0}{n_f} \tag{7.43}$$

where t_0 is the thickness as read from the chart for silica films (of refractive index n_0), and n_f is the refractive index of the new film. For anodic oxides of gallium arsenide, $n_f \simeq 1.8$. The color chart of Table 7.4 can thus be applied to oxide films on gallium arsenide as well, provided all listed film thicknesses are multiplied by $1.46/n_f$.

Another measure of oxide quality is its dielectric breakdown strength. In the absence of pinholes in the film, this is usually linearly related to the film density, as shown in Fig. 7.17 for silica films.

Finally, the charge states in the oxide can be evaluated by evaporating a metal dot (aluminum or gold) on the surface to form a MOS capacitor on which electrical measurements can be conducted. No attempt is made to treat this important topic here, since it has been covered thoroughly in many papers as well as in texts on device physics [58].

REFERENCES

1. C. J. Frosh and L. Derick, Surface Protection and Selective Masking During Diffusion in Silicon, *J. Electrochem. Soc.* **104,** 547 (1957).
2. J. M. Stevels and A. Kats, The Systematics of Imperfections in Silicon-Oxygen Networks, *Philips Res. Rep.* **11,** 103 (1956).
3. W. A. Pliskin and R. P. Gnall, Evidence for Oxidation Growth at the Oxide–Silicon Interface from Controlled Etch Studies, *J. Electrochem. Soc.* **111,** 872 (1964).
4. J. M. Stevels, New Light on the Structure of Glass, *Phillips Tech. Rev.* **22,** 300 (1960/61).
5. A. G. Revesz, The Defect Structure of Grown Silicon Dioxide Films, *IEEE Trans. Electron Dev.* **ED-12,** 97 (1965).
6. J. R. Ligenza, Oxidation of Silicon by High Pressure Steam, *J. Electrochem. Soc.* **109,** 73 (1962).

7. A. G. Revesz, The Role of Hydrogen in SiO_2 Films on Silicon, *J. Electrochem. Soc.* **126,** 122 (1979).
8. B. E. Deal, The Oxidation of Silicon in Dry Oxygen, Wet Oxygen, and Steam, *J. Electrochem. Soc.* **110,** 527 (1963).
9. B. E. Deal, Thermal Oxidation Kinetics of Silicon in Pyrogenic H_2O and 5% HCl/H_2O Mixtures, *J. Electrochem. Soc.* **125,** 576 (1978).
10. E. A. Irene and D. W. Dong, Silicon Oxidation Studies: The Oxidation of Heavily B- and P-Doped Single Crystal Silicon, *J. Electrochem. Soc.* **125,** 1146 (1978).
11. C. P. Ho, J. D. Plummer, J. D. Meindl, and B. E. Deal, Thermal Oxidation of Heavily Phosphorus-Doped Silicon, *J. Electrochem. Soc.* **125,** 665 (1978).
12. S. C. Su, Low Temperature Silicon Processing Techniques for VLSIC Fabrication, *Solid State Technol.*, p. 72 (Mar. 1981).
13. R. J. Zeto, C. G. Thornton, E. Hryckowian, and C. D. Bosco, Low Temperature Thermal Oxidation of Silicon by Dry Oxygen Pressure Above 1 Atm, *J. Electrochem. Soc.* **122,** 1409 (1975).
14. A. Rohatgi, S. R. Butler, and F. J. Feigl, Mobile Sodium Ion Passivation in HCl Oxides, *J. Electrochem. Soc.* **126,** 149 (1979).
15. B. E. Deal, D. W. Hess, J. D. Plummer, and C. P. Ho, Kinetics of Thermal Oxidation of Silicon in O_2/H_2O and O_2/Cl_2 Mixtures, *J. Electrochem. Soc.* **125,** 339 (1978).
16. B. R. Singh and P. Balk, Thermal Oxidation of Silicon in O_2-Trichloroethylene, *J. Electrochem. Soc.* **126,** 1288 (1979).
17. S. M. Hu, Formation of Stacking Faults and Enhanced Diffusion in the Oxidation of Silicon, *J. Appl. Phys.* **45,** 1567 (1974).
18. T. Hattori and T. Suzuki, Elimination of Stacking-Fault Formation in Silicon by Preoxidation Annealing in $N_2/HCl/O_2$ Mixtures, *Appl. Phys. Lett.* **33,** 347 (1978).
19. T. Hattori, Elimination of Stacking Faults in Silicon by Trichloroethylene Oxidation, *J. Electrochem. Soc.* **123,** 945 (1976).
20. M. Ghezzo and D. M. Brown, Diffusivity Summary of B, Ga, P, As, and Sb in SiO_2, *J. Electrochem. Soc.* **120,** 146 (1973).
21. E. H. Nicollian and J. R. Brews, *MOS (Metal Oxide Semiconductor) Physics and Technology*. Wiley, New York, 1982.
22. P. V. Gray, The Silicon-Silicon Dioxide System, *Proc. IEEE* **57,** 1543 (1969).
23. C. E. Young, Extended Curves of Space Charge, Electric Field, and Free Carrier Concentration at the Surface of a Semiconductor, *J. Appl. Phys.* **32,** 329 (1961).
24. E. Yon, W. H. Ko, and A. B. Kuper, Sodium Distribution in Thermal Oxide on Silicon by Radiochemical and MOS Analysis, *IEEE Trans. Electron Dev.* **ED-13,** 276 (1966).
25. S. K. Ghandhi, *Semiconductor Power Devices*, Wiley, New York, 1977.
26. J. F. Verwey, On the Emitter Degradation by Avalanche Breakdown in Planar Transistors, *Solid State Electron.* **14,** 775 (1971).
27. S. A. Abbas and R. C. Dockerty, Hot Carrier Instability in IGFET's, *Appl. Phys. Lett.* **27,** 147 (1975).
28. T. H. Ning, P. W. Cook, R. H. Dennard, C. M. Osburn, S. E. Schuster, and H-N. Yu, 1 μm MOSFET VLSI Technology, Part IV: Hot-Electron Design Constraints, *IEEE Trans. Electron Dev.* **ED-26,** 346 (1979).
29. D. Frohman-Bentchkowsky, FAMOS—A New Semiconductor Charge Storage Device, *Solid State Electron.* **17,** 517 (1974).
30. B. Schwartz, GaAs Surface Chemistry—A Review, *Crit. Rev. Solid State Sci.* **5,** 609 (1975).
31. C. W. Wilmsen, Oxide Layers on III-V Compound Semiconductors, *Thin Solid Films* **39,** 105 (1976).

32. H. Takagi, G. Kano, and I. Termoto, Thermal Oxidation of GaAs in Arsenic Trioxide Vapor, *J. Electrochem. Soc.* **125,** 579 (1978).

33. H. Tokuda, Y. Adachi, and T. Ikoma, Microwave Capability of 1.5 μm-Gate GaAs MOSFET, *Electron. Lett.* **13,** 761 (1977).

34. T. Mimura, K. Odani, N. Yokoyama, and M. Fukuta, New Structure of Enhancement-Mode GaAs Microwave MOSFET, *Electron. Lett.* **14,** 500 (1978).

35. H. D. Barber, H. B. Lo, and J. E. Jones, Repeated Removal of Thin Layers of Silicon by Anodic Oxidation, *J. Electrochem. Soc.* **123,** 1404 (1976).

36. H. Muller, F. H. Eisen, and J. W. Mayer, Anodic Oxidation of GaAs as a Technique to Evaluate Carrier Concentration Profiles, *J. Electrochem. Soc.* **122,** 651 (1975).

37. S. K. Shastry, Electrochemically Processed GaAs Solar Cells, M.S. Thesis, Rensselaer Polytechnic Institute, Troy, NY, 1980.

38. K. P. Pande, D. H. Reep, S. K. Shastry, A. S. Weiner, J. M. Borrego, and S. K. Ghandhi, Grain Boundary Edge Passivation of GaAs Films by Selective Anodization, *Appl. Phys. Lett.* **33,** 717 (1978).

39. S. R. Morrison, *The Chemical Physics of Surfaces*, Plenum, New York, 1977.

40. H. Hasegawa and H. L. Hartnagel, Anodic Oxidation of GaAs in Mixed Solutions of Glycol and Water, *J. Electrochem. Soc.* **123,** 713 (1976).

41. D. L. Rode, B. Schwartz, and J. V. DiLorenzo, Electrolytic Etching and Electron Mobility of GaAs for FET's, *Solid State Electron.* **17,** 1119 (1974).

42. A. Shimano, H. Takagi, and G. Kano, Light Controlled Anodic Oxidation of *n*-GaAs and its Application to Preparation of Specific Active Layers of MESFET's, *IEEE Trans. Electron Dev.* **ED-26,** 1690 (1979).

43. E. F. Duffek, C. Mylorie, and E. A. Benjamin, Electrode-Reactions and Mechanism of Silicon Anodization in *N*-Methylacetamide, *J. Electrochem. Soc.* **111,** 1042 (1964).

44. H. D. Barber, H. B. Lo, and J. E. Jones, Repeated Removal of Thin Layers of Silicon by Anodic Oxidation, *J. Electrochem. Soc.* **123,** 140 (1976).

45. R. A. Logan, B. Schwartz, and W. J. Sundberg, The Anodic Oxidation of GaAs in Aqueous H_2O_2 Solution, *J. Electrochem. Soc.* **120,** 1385 (1973).

46. B. Schwartz, F. Ermanis, and M. H. Brastad, The Anodization of GaAs and GaP in Aqueous Solution, *J. Electrochem. Soc.* **123,** 1089 (1976).

47. B. N. Arora and M. G. Bidnukar, Anodic Oxidation of Gallium Arsenide, *Solid State Electron.* **19,** 657 (1976).

48. P. A. Breeze and H. L. Hartnagel, An Assessment of the Quality of Anodic Native Oxides of GaAs for MOS Devices, *Thin Solid Films* **56,** 51 (1979).

49. R. P. H. Chang and A. K. Sinha, Plasma Oxidation of GaAs, *Appl. Phys. Lett.* **29,** 56 (1976).

50. N. Yokoyama, T. Mimura, K. Odani, and M. Fukuta, Low-Temperature Plasma Oxidation of GaAs, *Appl. Phys. Lett.* **32**(1), 58 (1978).

51. K. Yamasaki and T. Sugano, Anodic Oxidation of GaAs Using Oxygen Plasma, *Jpn. J. Appl. Phys.* **17,** 321 (1978).

52. R. P. H. Chang, A. J. Polak, D. C. Allara, and C. C. Chang, Physical Properties of Plasma-Grown GaAs Oxides, *J. Vac. Sci. Technol.* **116,** 888 (1979).

53. N. Yokoyama, T. Nimura, and M. Fukuta, Planar GaAs MOSFET Integrated Logic, *IEEE Trans. Electron Dev.* **ED-22,** 1124 (1980).

54. D. L. Pulfrey, F. G. M. Hathorn, and L. Young, The Anodization of Si in a R.F. Plasma, *J. Electrochem. Soc.* **120,** 1529 (1973).

55. R. M. Burger and R. P. Donovan, *Fundamentals of Silicon Integrated Circuit Device Technology*, Vol. I, Prentice-Hall, Englewood Cliffs, NJ, 1967.

56. W. R. Runyan, *Semiconductor Measurements and Instrumentation*, McGraw-Hill, New York, 1975.
57. W. A. Pliskin and E. E. Conrad, Nondestructure Determination of Thickness and Refractive Index of Transparent Films, *IBM J. Res. Dev.* **8,** 43 (1964).
58. S. M. Sze, *Physics of Semiconductor Devices*, Wiley, New York, 1981.

PROBLEMS

1. A slice of silicon has a 5000 Å thick oxide on its surface. A window is cut in this oxide, and a diffusion made through it by the two-step process. The resulting oxide over this window is 1500 Å thick. Sketch the cross section of the silicon, indicating clearly the location of the oxide relative to the substrate. How long was the drive-in step? Assume (111) silicon and a 1100°C process in wet oxygen.
2. Calculate the oxide thickness which would be obtained by the following sequence: 20 min dry O_2, followed by 20 min in wet O_2, both at 1100°C. Use the graphs of Figs. 7.7 and 7.8 and assume (100) silicon.
3. A slice of (100) silicon has small phosphorus-doped islands, each with a surface concentration of 6×10^{20} cm^{-3}. The slice is given a 60 min wet oxidation at 900°C. Calculate the thickness of the oxide over doped and undoped regions, respectively.
4. Using Fig. 7.9, determine a mathematical expression for the pressure dependence of the oxide growth rate at 800°C, 900°C, and 1000°C.
5. A *n–p–n* transistor is made on (111) silicon by sequential boron and phosphorus diffusions. The boron drive-in step is 30 min at 1100°C in steam. The resulting oxide has a window cut in it for the emitter diffusion, which is carried out at 1000°C for 15 min in dry oxygen. Will the oxide resulting from the base diffusion be sufficiently thick to serve as a mask for the emitter diffusion?
6. It is required to produce a 4400 Å thick layer of silicon dioxide by anodization. Assuming a constant current source of 20 mA/cm^2, and a current efficiency of 2%, determine the anodization time.
7. Repeat Problem 6 for gallium arsenide. Use an appropriate value for current efficiency.

CHAPTER

8

Deposited Films

CONTENTS

8.1 FILMS FOR PROTECTION AND MASKING 421

8.1.1 Silicon Dioxide, 422
8.1.2 Phosphosilicate Glass, 424
8.1.3 Silicon Nitride, 427

8.1.3.1 Silicon Oxynitride, 429

8.1.4 Amorphous Silicon, 430
8.1.5 Self-Aligned Masks, 430

8.1.5.1 Polysilicon, 432
8.1.5.2 Silicides, 435
8.1.5.3 Refractory Metals, 437

8.2 FILMS FOR DOPING 439

8.2.1 Dopant Sources for Silicon, 440
8.2.2 Dopant Sources for Gallium Arsenide, 441

8.3 FILMS FOR INTERCONNECTIONS 443

8.3.1 Single-Metal Interconnections, 446
8.3.2 Electromigration Effects, 448
8.3.3 Multimetal Interconnections, 449

8.4 FILMS FOR OHMIC CONTACTS 451

8.4.1 Single-Layer Contacts, 453
8.4.2 Kirkendall Effects, 455

8.4.3 Multilayer Contacts, 458
8.4.4 Die Bonds, 461

8.5 FILMS FOR SCHOTTKY DIODES 463

REFERENCES 470

A large number of different kinds of deposited films are used in microcircuit technology today; moreover, the functions they perform are continually increasing, as are their varieties. In recent years, this list has greatly expanded because of the new requirements imposed by different VLSI schemes. Both insulating and conducting films are used. The division is not rigid here, since some films are deposited in insulating form, and later made conductive by processing. Even the term "deposited" is not a rigid one, since some films are deposited in elemental form and later converted to their compounds by chemical reaction with the semiconductor.

Deposited films are perhaps most conveniently categorized by the primary functions they are required to perform. Some of these functions are electronic in nature, yet others are process oriented. A number of different types of films are used to provide an insulating, protective cover on which interconnections can be made. Some of these can be patterned for use as diffusion or ion implantation masks, while others can be used to provide the doping function itself. Yet other films make possible the use of greatly simplified processing schemes, such as self-aligned gate technology for MOS microcircuits.

Many films and film combinations can be used to form ohmic contacts, as well as interconnections between devices in a microcircuit. Some of these can also be used to perform device functions as in Schottky diodes and in Schottky gate field effect transistors.

The films to be described in this chapter can be deposited by a wide variety of technologies; indeed, most films can be formed by more than one method. Here the emphasis will be on those technologies that are favored in semiconductor usage.

Physical deposition techniques of practical interest include vacuum evaporation, as well as direct-current and rf sputtering. The recent developments of magnetron sputtering and ion beam deposition both represent improvements over rf sputtering methods, and provide increased growth rates, relative freedom from contamination, and independent adjustment of system parameters such as substrate temperature, beam energy, and current density. In addition, reactive sputtering is also used in some applications where the desired film is not elemental in nature, but is a compound. All of these sputtering techniques share the advantage that substrates can be cleaned (by back sputtering) prior to film deposition. Equipments for physical deposition films form a highly specialized branch of manufacturing technology and have been considered in detail elsewhere [1].

The primary emphasis on chemical techniques has been on chemical vapor deposition (CVD) processes, which were considered in detail in Chapter 5. Many of the systems and techniques developed for vapor-phase epitaxy can be directly used (or modified for use), with a wide variety of insulating and conducting films. Consequently, it is only natural that these have been readily adopted by the semiconductor industry.

Hot wall systems for CVD are resistance heated, whereas cold wall systems use either rf or internal resistance heated susceptors. This latter design is the economical choice over rf if the deposition is to be carried out at low temperatures ($< 600°C$) and the purity requirements are not as severe as in the case of epitaxy.

Systems can be operated at atmospheric as well as low pressure (≥ 0.1 torr). Low-pressure operation results in increasing the diffusivity of the reactant species, and leads to more uniformity and greater throughput than operation at atmospheric pressure. Consequently, recent emphasis on production techniques is in this area. Finally, the use of plasma enhancement allows reactions to be carried out by energetic species with an electron temperature of around 10,000°K. As a result, growth can be accomplished at low temperatures, and sometimes even at room temperature, by these techniques. Thus the choice of CVD techniques for film deposition is a very wide one.

8.1 FILMS FOR PROTECTION AND MASKING

Insulating films are usually deposited on top of a semiconductor in order to provide protection to its surface, or to serve as masks through which selective diffusions and implants can be made. They also serve as the base for electrical connections between semiconductor devices in a microcircuit. Often they provide the role of an interlayer dielectric between two levels of metallization. In all these situations, it is highly desirable that they be free from pinholes and cracks, both when grown, and also if subjected to heat treatments during subsequent processing. Thus their built-in stress, as well as the stress during thermal cycling, must be sufficiently low to maintain their integrity. These requirements become increasingly important in VLSI technology, as wafer size increases and devices become more densely packed.

Films that are useful in masking applications must be capable of preventing the transport of dopant species through them, in addition to maintaining their integrity at diffusion temperatures. Furthermore, they must be capable of being etched into fine line patterns by photolithographic techniques. Often these films are left in place after having provided the masking function. Thus they should be either highly insulating if used as cover layers, or highly conducting if used in the subsequent metallization scheme.

Deposited films can be used for the protection of microcircuits during manufacture, and also for improving their reliability in use. In addition, they can be used to block the movement of light alkali ions such as sodium, or else getter them so as to render them immobile. These films are usually placed over the metallization, to prevent its damage during handling. At the present time, their quality is sufficiently high so that microcircuits with protective films can be used without further packaging in a wide variety of consumer applications.

8.1.1 Silicon Dioxide

Silica films can be grown by the pyrolytic oxidation of a variety of alkoxysilanes [2], in the 700–800°C temperature range. The most commonly used compound is tetraethylorthosilane (TEOS) which is a liquid at room temperature (boiling point = 167°C), and must be transported to the reaction chamber by means of a bubbler arrangement. The oxidation reaction is commonly carried out in a cold wall CVD system at 800°C, and proceeds as follows:

$$Si(C_2H_5O)_4 + 12O_2 = SiO_2 + 8CO_2 + 10H_2O \tag{8.1}$$

As seen from this reaction, a large amount of water is produced as a by-product for each molecule of SiO_2. Consequently, films are of relatively poor quality because of water incorporation. In addition, secondary reaction products such as silicon monoxide, carbon, and organic radicals are also present in these films.

Growth of silica films from TEOS has largely been superseded by deposition involving the oxidation of silane [3]. The silane reaction, when carried out at 600–1000°C, also results in the formation of water as a by-product. However, hydrogen formation is favored at lower temperatures (300–500°C) where the reaction proceeds as follows:

$$SiH_4 + O_2 \rightarrow SiO_2 + 2H_2 \tag{8.2}$$

resulting in high-quality silica films. Typically, this reaction is carried out at atmospheric pressure in a cold wall CVD system of the resistance heated type, because of the low temperatures involved. The resulting film has a built-in tensile stress of about 3×10^9 dyn/cm^2 for a 450°C growth temperature.

Pure silane is a highly pyrophoric gas which burns on exposure to air. Consequently silane is commonly supplied in a low dilution (typically 5–10% by volume) in argon or nitrogen. It is safer to handle in this form and is stable in these dilutions. The growth of silica films from silane proceeds by the strong adsorption of oxygen on the silicon surface, and its subsequent reaction with silane to form silicon dioxide. This can lead to a retardation in the growth rate when oxygen is present in high concentrations [4]. Typically, the $O_2 : SiH_4$ mole ratio must not exceed 8–10 to avoid this retardation effect.

The growth rate of films by the silane process is quite high, typically 500–1000 Å/min. Consequently, these films can be used in many applications where rapid, low-temperature growth is essential. These include the following:

1. Thick field oxides for MOS microcircuits as well as for high-voltage devices. These are usually deposited over a base layer of thermally grown oxide to avoid a high trap density at the silicon surface.
2. Films where the previously grown layer has been removed; for example, deep diffusions of the type used for buried layers and isolation walls. These often utilize fresh masking and redoping partway through the diffusion process.
3. Insulating layers over a metallization layer, to form a base for the next layer of metal.
4. Cover layers to protect the microcircuit from physical abuse during mounting and packaging.
5. Diffusion masks for gallium arsenide. As stated in Chapter 7, the native oxide rapidly deteriorates at high temperatures ($> 700°C$) and cannot be used in this application.
6. Cap layers for regions of gallium arsenide which must not be exposed during processing (the back side of a slice, for example).

Low-pressure CVD systems of the hot wall type are also used, and result in better film uniformity from slice to slice, as well as in increased throughput. In addition, film quality is generally superior to that obtained in an atmospheric pressure system, with a reduced pinhole density. However, the growth rate is significantly slower (100–150 Å/min), so that the technique is suited for the growth of relatively thin oxides.

Silicon dioxide can also be grown at low pressure ($\simeq$ 0.1–0.5 torr) in a plasma-enhanced system of the type shown in Fig. 8.1 [5]. Here the basic reactions that can be used involve SiH_4-O_2, SiH_4-CO_2, and SiH_4-N_2O mixtures. The SiH_4-N_2O system can be operated at low temperatures (250°C) with growth rates that are comparable to those obtained with the SiH_4-O_2 system at atmospheric pressure ($\simeq$ 600 Å/min). A small amount of nitrogen, $\simeq$ 3%, is incorporated in the films as a result of this reaction, but has no deleterious effect on their resulting properties.

Plasma-enhanced CVD techniques result in a built-in compressive stress in the deposited silica films. This greatly reduces the tendency to cracking during subsequent thermal cycling. As a result, films grown by this method can be much thicker than those grown at atmospheric pressure. Finally,

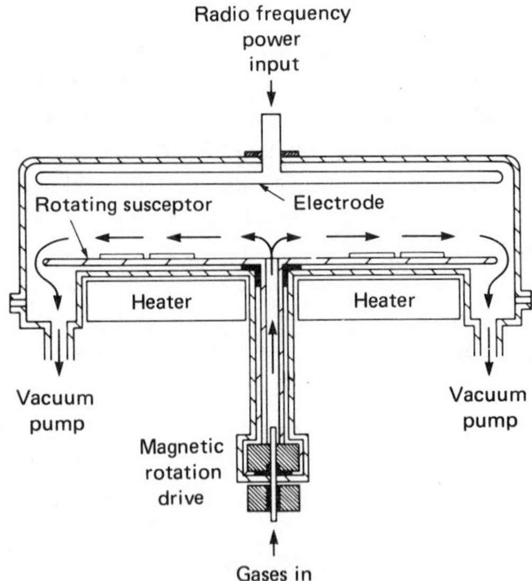

Fig. 8.1 Plasma-enhanced CVD system. From Mattson [5]. Reprinted with permission from *Solid State Technology*.

films grown by this technique are almost completely free of pinholes, so that they are suited for cap layers in VLSI applications.

Notwithstanding their high quality, deposited oxide films have a much higher contamination level (and an associated trap density) than thermally grown oxide films. Thus they are not suitable for use as gate oxides in MOS microcircuits. Usually they are grown over an initial thin (100–200 Å) native oxide to avoid direct contact with the silicon surface. This is especially true when coverage is required over lightly doped silicon, or over regions where a junction is exposed at the $Si–SiO_2$ interface.

8.1.2 Phosphosilicate Glass

Phosphosilicate glass (PSG) films can be grown by the simultaneous pyrolysis of silane and phosphine in oxygen [4, 6]. The temperature range is the same as for the growth of silica films (300–500°C), so that this deposition technique is a natural extension of the silane process, and can be carried out in the same system. Phosphine gas is usually provided in a 5–10% dilution in argon or nitrogen, for this application.

The reaction of phosphine gas with oxygen is usually carried out in the 350–450°C range, and results in the formation of P_2O_5 which is incorporated as a network former in the resulting glass. The phase diagram for the $P_2O_5 \cdot SiO_2$ system is shown in Fig. 2.18. The phosphine oxidation reaction is as follows:

$$2PH_3 + 4O_2 \rightarrow P_2O_5 + 3H_2O \tag{8.3}$$

so that a small amount of water is produced as a by-product of this process. Here, too, growth of the film proceeds by the strong adsorption of oxygen on its surface, so that retardation effects become dominant [7] when the $O_2:(SiH_4 + PH_3)$ mole ratio exceeds 8–10.

Almost any amount of P_2O_5 can be incorporated by this technique. However, the films become increasingly hygroscopic, so that their P_2O_5 content is limited to about 2–8% by weight for films which are left permanently in place on the finished product, and as high as 20% for films which are used only during device processing. The weight percentage of P_2O_5 in the glass is approximately equal to 1.5 times the $PH_3:SiH_4$ mole ratio in the gas phase, over this range of compositions.

The incorporation of P_2O_5 into silica films causes a reduction in the built-in tensile stress from 3×10^9 dyn/cm^2 for the undoped film, to about 2×10^9 dyn/cm^2 for a film with 13% P_2O_5 by weight, and zero for films with 20% P_2O_5 by weight. This reduction in stress improves their integrity during thermal cycling. A further advantage comes about because the thermal expansion coefficient of silica rapidly increases with the incorporation of P_2O_5. Consequently, PSG films can be tailored to provide a more suitable thermal match to the underlying semiconductor. By way of example, silica films, deposited on silicon, will crack upon thermal cyling to 1200°C if they are thicker than 1.5 μm. Silica films, deposited on gallium arsenide, will break this semiconductor upon thermal cycling to 800°C if they are thicker than 2000 Å. In both situations, PSG films can be used to avoid this problem.

The thermal expansion coefficient of PSG is shown in Fig. 8.2 as a function of P_2O_5 content, over the composition range of practical interest [8]. Note that the expansion coefficient* of gallium arsenide (5.9×10^{-6}/°C) can be readily matched at a P_2O_5 concentration of about 20–24% by weight. Films with 15% P_2O_5 concentration by weight have been used in thicknesses as large as 6000 Å, with no tendency toward cracking when cycled to temperatures as high as 1100°C.

PSG is more dense and void-free than silica. Consequently, it can be used as a mask to such dopants as zinc and tin, which are used for diffusion in gallium arsenide [8–10], whereas silica films are relatively transparent to these dopants. For the same reasons, PSG films make better cap layers for gallium arsenide than silica films in this application as well [8].

The etch rate of PSG films increases with the incorporation of P_2O_5. This fact is utilized† [11] to form two-layer films which, upon etching, result in

* Values for undoped silica (6×10^{-7}/°C) and for silicon (2.6×10^{-6}/°C) are also shown in this figure.

† See Section 11.2.

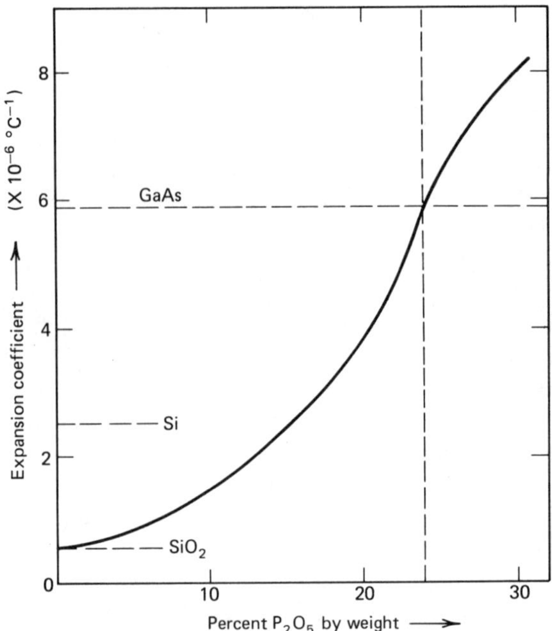

Fig. 8.2 Thermal expansion coefficient of PSG films. Adapted from Baliga and Ghandhi [8].

a tapered oxide cut. This technique finds use in applications where breaks in metallization can occur because of large steps in the oxide (the field oxide for MOS circuits, by way of example). A second application area is in high-voltage devices where a tapered oxide provides control of the depletion layer curvature.

PSG layers have been found to be effective in immobilizing sodium ions in MOS technology (see Section 7.1.8.2) and are commonly used in this important application area. They are also used to enhance the stability of bipolar devices and microcircuits [12]. Often these layers are automatically present, as occurs with *n–p–n* transistors, where the last process is a masked phosphorus diffusion. In those cases, however, when the last diffusion is not phosphorus (*p–n–p* transistors, for example), it is customary to strip the surface oxide which consists of $B_2O_3 \cdot SiO_2$, and grow a PSG film prior to packaging.

PSG layers are often used as coatings on finished microcircuits, to protect the aluminum metallization from scratches during the final bonding operation, and provide a permanent protection against alkali ion migration. These passivation layers must contain no more than 6% by weight of P_2O_5 to prevent corrosion reactions with the aluminum metallization in the presence of moisture [13]. Often a SiO_2-PSG-SiO_2 sandwich is used instead of a single PSG layer, since this avoids direct contact between the aluminum and the

PSG. Moreover, the upper layer of SiO_2 provides better adhesion to the photoresist that is used in cutting the contact holes in this composite layer.

8.1.3 Silicon Nitride

Silicon nitride films are extensively used in both silicon and gallium arsenide device technology. They require a higher deposition temperature than films of silica or doped silica, so that their use is dictated in situations where they provide improved properties. Often they are used in conjunction with silica films to obtain a combination of characteristics which neither can provide alone.

Silicon nitride, in its stoichiometric form, has a composition given by Si_3N_4. Considerable departure from stoichiometry is often encountered during its deposition by the various methods to be outlined here, with Si/N atom ratios from 0.7 to 1.1 being commonly encountered.* It is a dense, wide-gap insulator with a dielectric constant of 5.8–6.1, a refractive index of 1.98–2.05, and a density between 2.3 and 2.8 g/cm^3. Unlike silica, it is an excellent barrier to alkali ion migration, so it is used extensively as a cover layer in MOS technology for this reason. It is superior to PSG since it is not hygroscopic. Its use has allowed unencapsulated circuits to be practical in many consumer applications.

SiN is also an excellent diffusion mask for gallium, and is used for making junctions with this dopant in power device applications. This ability to restrict the diffusion of gallium results in its use as a capping material for gallium arsenide during the high-temperature (900°C) anneal process which must be carried out after ion implantation. It is about 100 times more resistant to thermal oxidation than silicon, so that it can be used as a mask in a number of silicon-based VLSI schemes which require selective oxidation of the semiconductor.† Finally, it is used in radiation hardened devices because of its superior radiation resistance over silicon dioxide.

Although impervious to alkali ion migration, SiN films are poorer than silica in their ability to block electron movement through them. Consequently, they are often used in conjunction with silica films to provide a barrier to both electron and ion transport.

A true native nitride of silicon can be grown by direct nitridation [14]. This process must be carried out at high temperatures (1000–1300°C), and is extremely sensitive to even trace amounts of water or oxygen. Furthermore, film thickness is greatly restricted by the low diffusivity of nitrogen through it. Consequently, all SiN films used in silicon and gallium arsenide technology are deposited today.

Amorphous films of silicon nitride can be successfully deposited by rf sputtering using a silicon target with a nitrogen discharge. Substrate tem-

* Films in this broad range are often referred to as SiN films.

† This important process technology is described in Chapter 11.

peratures of 200–300°C have been used in this process. Direct sputtering from a silicon nitride target with an argon–nitrogen background can also be accomplished [15]. In all cases, the film quality is a sensitive function of the nitrogen partial pressure and the rf power, which together determine its stoichiometry. Films that are rich in either silicon or nitrogen can be obtained; in general their properties (density, resistance to ion migration, etc.) improve as they become more stoichiometric, and also when they have less built-in stress.

The most commonly used process for the growth of SiN films is by chemical vapor deposition involving the reaction of silane gas and ammonia, with nitrogen as the diluent. This reaction is usually carried out at 700°C, and proceeds along the following lines:

$$3SiH_4 + 4NH_3 \rightarrow Si_3N_4 + 12H_2 \tag{8.4}$$

Both hot and cold wall systems are used for this purpose, and deposition can be accomplished at atmospheric as well as at reduced pressures. In all cases, film composition and properties are controlled by the ratio of ammonia to silane in the gas stream. Typical mole ratios of $NH_3:SiH_4$ are 150 or higher for a hot wall system, with growth rates in the 100–200 Å/min range [16].

The present trend in SiN film deposition for large commercial applications is to use low-pressure hot wall systems [17], which allow close stacking of slices while still providing uniform coverage. Operation at low pressure is achieved by greatly reducing or eliminating the use of carrier gases, so that these systems can be run with reactant partial pressures (and growth rates) which are comparable to those achieved with atmospheric systems. A major problem encountered here is the high degree of risk involved in handling concentrated reactants such as silane. Some systems use chlorosilanes which are safer to handle.

Plasma-enhanced cold wall systems can also be used for the growth of silicon nitride [18–20]. Here it is possible to operate the reaction chamber at a low pressure (0.1–1 torr), and use a rf plasma to obtain one or more active species of the reactants. A major advantage of this approach lies in the fact that the deposition reaction can be conducted at relatively low thermal temperatures (275–300°C), because of the high electron temperature ($\simeq$ 10,000°K) associated with the plasma. This allows films to be grown directly on finished microcircuits. Growth rates for these systems are in the 200 Å/min range.

Plasma-enhanced SiN films, grown by the SiH_4-NH_4 reaction in this manner, have large quantities of hydrogen incorporated in them, with the H/Si atom ratio often exceeding unity. This greatly affects properties such as their etch rate. Growth at elevated temperatures (300–400°C) results in a reduction in this ratio to about 0.5, with a resulting improvement in film properties. Another approach is to densify the grown film by heat treatment at 700°C. Unfortunately, this step obviates the initial advantage of low-temperature growth for this approach. Growth can also be carried out using pure nitrogen and silane, since active nitrogen is formed by the rf system.

Deposited films of SiN have a large amount of built-in tensile stress (5×10^9 dyn/cm^2) when grown at 700°C. As a result, they are usually deposited in thicknesses below 1000 Å to avoid breakage or peeling, or damage to the underlying semiconductor. Typically, films with a Si/N ratio of 0.75 have been found to have minimum stress, and are favored for this reason.

The high built-in stress of SiN films has been used to advantage in the fabrication of large-area silicon devices [21]. Here a layer of silicon nitride is deposited on the back surface of the semiconductor slice as the first step in fabrication. This causes a high interfacial stress, resulting in the formation of a dislocation network. Gettering of impurities by this dislocation network, i.e., damage gettering, occurs during subsequent high-temperature processing of microcircuits which are located on the opposite face of the slice. The formation of oxygen-induced stacking faults is also greatly reduced by this technique.

Films grown by plasma-enhanced techniques generally have a lower stress ($\simeq 2 \times 10^9$ dyn/cm^2) than those grown in hot wall systems; in addition, this stress may be either tensile or compressive, depending on the conditions of film growth. Relatively thick (0.5–1 μm) films can be grown with a compressive stress, with excellent adhesion to the semiconductor surface.

The etching properties of SiN films are highly variable, and are related to their Si/N ratio, as well as to the manner in which they are deposited. Pure Si_3N_4 etches at about 68 Å/min in HF (49%), whereas CVD and plasma-grown films usually etch much faster (250–500 Å/min). Boiling phosphoric acid is often used to selectively etch these films in the presence of silica layers. Details of these etch processes are given in Chapter 9.

8.1.3.1 Silicon Oxynitride

Films of silicon oxynitride can also be grown, and have physical properties that are intermediate between silica and silicon nitride [22]. Growth is carried out in a SiN deposition system, with the introduction of nitric oxide in addition to ammonia gas. This results in the partial oxidation of the silane, and produces films which are essentially glassy mixtures of silicon, oxygen, and nitrogen. The chemical formula $Si_xO_yN_z$ is used to designate these films.

Typically, the introduction of about 10% of nitric oxide results in films with $SiO_2:Si_3N_4$ in a 1:1 mixture. The etch rate (in buffered HF) of these films is about 35 times that for silicon nitride.

There are some indications that films of this type may be more suited as passivating layers for gallium arsenide than either silica or silicon nitride. In addition, it is worth noting that the initial 20–30 Å layer of silicon nitride on either a silicon or a gallium arsenide surface has most probably a graded silicon oxynitride composition, because of the native oxide that is always present on these semiconductors.

8.1.4 Amorphous Silicon

Amorphous silicon films have received attention because of their potential use in solar cells. Recently it has been shown [23] that suitably prepared films of this type can be used as an excellent surface passivant for silicon junctions, and are superior to thermally grown silicon dioxide in this regard. A further advantage is that they can be patterned by well-established silicon processing techniques, and require no new technology for their use.

Films of this type can be deposited by the glow discharge decomposition of silane in a rf heated reactor [24]. As-deposited films have hydrogen incorporated in them, since it is a constituent of the silane. Moreover, the hydrogen content can be controlled by intentional doping during the deposition process. Dense, hard layers are obtained as long as this hydrogen content is kept below 50% (atomic). The presence of hydrogen serves to tie up dangling bonds on the crystalline silicon, and thus passivates its surface. Typically, this results in the reduction of the reverse leakage by about two to three decades. Films of this type are commonly designated as a-Si:H.

Films of a-Si:H evolve hydrogen when held at elevated temperatures [25], so that there are still many questions about their stability during subsequent processing, as well as their reliability in service. Present indications are that this is a problem if the device is heated above 400°C during subsequent processing. However, the technique is sufficiently attractive to warrant further research on this materal.

8.1.5 Self-Aligned Masks

Refractory materials, capable of withstanding diffusion temperatures, can often be used for masking purposes in microcircuit fabrication. In some cases they provide unique advantages, as in MOS technology, where they are extensively used for this purpose.

A major factor that limits the high-frequency gain of MOS transistors is the parasitic capacitance caused by overlapping of the gate electrode and the drain. This overlap is necessary to accommodate the tolerances in mask alignment during device fabrication; some overlap is essential to satisfactory device operation.

Consider the MOS transistor shown in schematic form in Fig. 8.3*a*. Here the overlap L is governed by the gate length, the distance between the oxide cuts for the source and drain, and the lateral diffusion of these regions. All of these must be carefully controlled if this distance is to be kept to a minimum. Often the source and drain diffusions are made quite deep ($\simeq 2$ μm) in this structure, to ensure that there will be a finite overlap due to

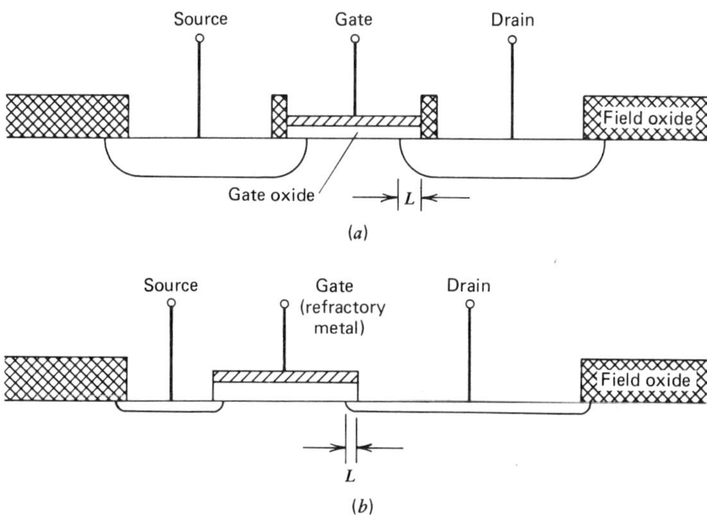

Fig. 8.3 Comparison of Masking Techniques. (*a*) Conventional. (*b*) Self-Aligned.

lateral diffusion effects. This increases the parasitic drain capacitance with a further deterioration in device performance.

Many of these problems can be alleviated if the gate material can be used as a mask during the source and drain diffusions. The use of this refractory material ensures that there will *always* be some overlap between the gate and the diffused regions, regardless of the amount of misalignment or lateral diffusion. Extremely shallow diffusions (0.2–0.5 μm) can be used here, with a concurrent reduction in the overlap. In addition, such junctions have lower parasitic capacitances and result in smaller area devices, which produce additional speed advantages.

Another advantage of using the gate material as the diffusion mask is that the overlap is not dependent on accurate placement of the mask with respect to the field oxide. This is shown in Fig. 8.3*b* where a grossly displaced mask still yields a satisfactory device. Masks of this type are referred to as *self-aligned* for this reason.

Refractory gate technology is ideally suited for VLSI structures, and can be readily implemented by a number of different materials. First, polycrystalline silicon films can be used for the gate material. These have the advantage that they can be formed with a high degree of purity. In addition, all the processes developed for silicon technology can be used with them. Doping of these films allows control of their Fermi level, and hence control of the MOS threshold voltage. In addition, both *p*- and *n*-type doping can be carried out simultaneously in the same microcircuit.

Recently a number of silicide films, with sheet resistances between that of polysilicon and metals, have been considered for this technology. A unique characteristic of these films is that a silica film can be subsequently *grown* on their surface, allowing considerable simplification in forming a second layer of metallization.

Refractory metals, such as molybdenum and tungsten, can also be used directly in a self-aligned masking technology. These have the lowest resistivities and so present an attractive approach for use in interconnections as well.

This section describes and compares these approaches, and outlines the considerations involved in the selection of this masking technology.

8.1.5.1 Polysilicon

Polycrystalline silicon (polysilicon) films can be formed by sputtering or by evaporation. However, CVD methods have the advantage of uniform deposition over oxide steps, and so are most commonly used. These films are grown in a cold wall reactor, by the decomposition of silane, over a wide range of temperatures. Both growth rate and crystallite size increase with increasing growth temperature. A suitable compromise, which produces mirror finish, fine grained films, is to grow them at 650–700°C. A 5% SiH_4 mixture in hydrogen results in a typical growth rate of about 0.5 μm/min.*

Films grown in undoped form have a resistivity in excess of 500 Ω cm, and so are essentially semi-insulating ($\simeq 5 \times 10^6$ Ω/square for a 1 μm thick film). These films can be directly used to provide resistive and field shaping layers as well as surface passivation in high-voltage devices. However, they must be doped for use as gates in MOS transistors. This doping can be carried out in the gas phase, by means of diborane, arsine, or phosphine gas, which can be introduced in dilute form (200–300 ppm) in a hydrogen carrier gas. Use of these dopants alters both the resistivity as well as the growth rate. Results obtained with them, for a typical set of experimental conditions [26], are now summarized:

1. Introduction of B_2H_6 causes a monotonic increase in the growth rate, which doubles as the B/Si atom ratio increases from zero to 2.5×10^{-3}. At the same time, the resistivity falls from 500 to about 0.005 Ω cm. Both the growth rate as well as the resistivity level off for doping ratios in excess of this value. Films remain mirror smooth until the B/Si atom ratio exceeds 3.5×10^{-3}, at which point signs of surface roughness become apparent.
2. The introduction of AsH_3 results in a precipitous fall in the growth rate (from 0.5 to 0.075 μm/min) as well as the resistivity (500–0.01 Ω cm) as the As/Si atom ratio is increased from zero to 2.5×10^{-5}. Both the growth rate and the resistivity level off beyond this point.

* Polysilicon films can also be grown by the decomposition of silicon tetrachloride. However, this method is not favored because of its comparatively high deposition temperatures (900°C).

3. Results with PH_3 doping are qualitatively similar to those for AsH_3, although changes are not as rapid. Here, a P/Si atom ratio of 2×10^{-3} results in a fall in growth rate from 0.5 to 0.2 μm/min. This is accompanied by a fall in resistivity from 500 to 0.02 Ω cm. Both resistivity and growth rate level off beyond this point, and are unchanged with further doping.

The use of polysilicon in MOS fabrication results in its becoming doped by the impurity which it is masking. However, from the above results it is seen that the polycrystalline nature of these films makes it almost impossible to control their doping concentration. This is not a disadvantage, however, since these films are subsequently used for the gate material, where a low sheet resistance is highly desirable. Consequently, films of this type are intentionally used in heavily doped p^+- or n^+-form. Heavy boron incorporation can be satisfactorily obtained by gaseous doping; however, sufficient n-type doping cannot be achieved by this method, as outlined earlier. In addition, it results in greatly reduced growth rates, probably caused by catalytic poisoning of the surface due to the strong adsorption of column V impurities.

A practical alternative for n^+-doping is to grow an undoped film, use it as a diffusion mask (during which process it becomes doped), and subsequently dope it further by a predeposition process of the type used in diffusion technology [27]. Typically, $POCl_3$ can accomplish this purpose, with the polysilicon maintained at 900–1000°C for 30–60 min. This results in sheet resistances of about 10 Ω/square (for a 0.5 μm thick film), whereas the sheet resistance is about 75 Ω/square if this second doping step is omitted. It should be noted that this is still an order of magnitude higher than what would be obtained for a $POCl_3$ predeposition into single-crystal silicon for the same time and temperature. The reasons for this difference are primarily due to the nature of conduction processes in polycrystalline materials. Here conduction between grains is dominated by the trap density at their interface, which results in the creation of an energy barrier [28]. This barrier can be quite low under conditions of heavy doping. Nevertheless, it serves to limit the effective conductivity of the material to a value below that obtained with single-crystal material.

The conductivity of polysilicon films can be further increased by performing the predeposition at a higher temperature (≃ 1150°C), or by using growth techniques which result in large grain sizes. Unfortunately, the photo etching of these films becomes difficult to control, and results in ragged edges due to the uneven etching of the grains and the grain boundaries. Thus they are not suited for VLSI applications.

Polysilicon films can also be grown in low-pressure CVD systems using a hot wall reactor at 625–650°C. The thermal decomposition of silane at these temperatures results in extremely fine-grained films with a growth rate of about 100 Å/min. The use of a low system pressure (≃ 0.5 torr) results in considerably higher wafer handling capacity [29], and makes this technique extremely attractive for commercial applications.

Films made by LPCVD can also be doped with both *p*- and *n*-type impurities using conventional predeposition techniques. The conductivity of heavily doped films is comparable to that obtained in conventional cold wall reactors operating at atomspheric pressure. However, lighter doped films have been found to have an anomalously higher resistivity than that obtained for films made in atmospheric pressure systems [30]. In all probability this is due to the greater number of grain boundaries associated with growth at a lower temperature.

Films can also be doped by ion implantation, and are usually made amorphous by this process, with sheet resistances of about 10^{10} Ω/square. Annealing of these films causes recrystallization to occur between 600 and 700°C, with an extremely rapid fall in sheet resistance, to about 10^2 Ω/square. Annealing above 700°C results in a further slight reduction in the radiation damage, accompanied by a small change in the sheet resistance.

Polysilicon films can be partially oxidized to render them semi-insulating by the addition of controlled amounts of nitrous oxide gas during deposition at 650–700°C.* LPCVD growth, at 0.1–0.2 torr and with a SiH_4/N_2O mole ratio of 5, results [31] in films with an oxygen content of about 20%, and a growth rate of 100–200 Å/min. These films have been used for the passivation of high-voltage silicon devices, where they have significant advantage over grown silica films. The thermal oxide of silicon is particularly unsuited for this application, since its use gives rise to a number of types of instability (see Section 7.1.8), including the following:

1. Instabilities due to the movement in the oxide of sodium ions, which modulate the conductivity of the underlying material and often lead to surface inversion effects
2. Trapping of charge in the oxide during avalanche breakdown, resulting in a temporary increase in the breakdown voltage (the junction walk-out effect)
3. Degradation of the low-level current gain in transistors after avalanche breakdown of the emitter–base junction, caused by an increase in the density of fast surface states

On the other hand, semi-insulating polysilicon films [32] are almost electrically neutral. Furthermore, their electronic properties can be changed by oxygen doping until they are completely converted to SiO_2 (66.7% atomic concentration of oxygen). Experimental data on deep planar junctions reveal that the breakdown voltage falls with increasing oxygen concentration, up to about 8% (atomic). Concurrently, the resistivity of the films increases as the material shifts toward SiO_2, and the reverse leakage current falls.

The use of a semi-insulating polysilicon (SIPOS) film with transistors

* Nitrogen is used as a carrier gas here and as a diluent for the silane.

results in a degradation of their low-level current gain, since it acts as a resistive emitter–base shunt. It has been demonstrated experimentally that this current gain is unaffected for oxygen concentrations above 20%. Thus it is possible to obtain a trade-off on these three device parameters by judicious adjustment of the oxygen content.

Polysilicon films can also be nitrogen doped by the addition of ammonia, until they become silicon nitride, which can be used as a cover layer because of its well-known ability to inhibit the transport of both sodium ions and water. Here, however, no balance need be struck on the film composition, since best protection is achieved with full nitridation. Stable planar junctions with breakdown voltage in excess of 10 kV have been achieved by this process.

8.1.5.2 Silicides

Although polysilicon is well suited for self-aligned masking technology, it suffers from the disadvantage that, even when heavily doped, it has a significantly high resistivity ($\geq 300\ \mu\Omega$ cm). With the trend to finer lines, this results in an increasingly large resistive contribution to the RC delay associated with active devices. Consequently, alternate approaches are being considered for VLSI technology.

The refractory metal silicides are immediately attractive, since many of these have resistivities that are almost ten times lower than that of heavily doped polysilicon. As a result, it is possible to define much thinner lines in these materials without paying a speed penalty in the operating device. In addition, the resistivity of many silicides is sufficiently low so that it is possible to consider their use for interconnections as well, resulting in an overall simplification of the technology.

The possible choices for this technology are many, as seen from Table 8.1, which lists the highly conductive silicides whose lowest binary eutectic temperature is sufficiently high for device processing [33]. The optimum choice, however, must include other considerations as well. Thus a useful silicide should be stable on silica and in oxidizing ambients. It should not be so highly stressed as to crack, or cause cracking of the underlying material, during thermal processing. It should be capable of being etched into fine line patterns by readily available techniques. It should be capable of oxidation, so as to provide an insulating layer over which the second level of metallization can be placed. Finally, it should be capable of being joined to the contact metal (aluminum or gold) which connects to the package terminations, or to the metallization.

Refractory metal silicides can be deposited directly by rf sputtering. However, cosputtering from separate silicon and metal targets allows control of these individual components. Thermal conversion of this layer to the silicide can be subsequently achieved by heat treatment at 900–1000°C for a few minutes.

Silicide formation can also be accomplished by the *sequential* depo-

Table 8.1 Lowest Eutectic Temperatures and Resistivities of Some Silicides of Interest[a]

Silicide	Lowest Binary Eutectic Temperature (°C)	Specific Resistivity (μΩ cm)
$CoSi_2$	1195	18–25
$HfSi_2$	1300	45–50
$MoSi_2$	1410	100
$NbSi_2$	1295	50
$NiSi_2$	966	50–60
Pd_2Si	720	30–35
PtSi	830	28–35
$TaSi_2$	1385	35–45
$TiSi_2$	1330	13–25
VSi_2	1385	50–55
WSi_2	1440	70
$ZrSi_2$	1355	35–40

[a] See reference 33.

sition of the separate components, followed by their thermal conversion. This approach is attractive since it allows patterning of these films (which can often be readily etched) prior to conversion to the considerably more inert silicide form. Etching of silicides is usually done by reactive ion milling, using an aluminum mask, or by plasma methods.

An important requirement on these films is a convenient means for covering them with an insulating film over which the second metal layer can be run. Some silicides, such as $TiSi_2$, oxidize congruently, so that a stable film can be formed by the oxidation of its components. Others, such as $TaSi_2$, do not oxidize in dry oxygen, but form a thin ($\approx$ 100 Å) oxide layer on their surface. However, they oxidize rapidly in steam environments. Most of the silicides have varying oxidation rates for their separate components, so that this process either results in their breakup, or in the formation of cover films which are nonstoichiometric and often leaky.

The oxidation properties of silicide films which are formed on a polysilicon base are uniquely different, however [34]. Here it is found that oxidation takes place by the rapid diffusion of silicon *through* the silicide, followed by conversion of this silicon to silica, during which process the refractory silicide layer remains essentially unchanged. The driving force for this diffusion of silicon appears to be the free energy of formation of SiO_2 (-192 kcal/mole).

A typical process sequence [35] for a tungsten silicide gate system is as follows: a 2500 Å mixture of silicon and tungsten, in a 2:1 atom ratio, is codeposited on a 1500 Å layer of n^+-polysilicon. The silicide is formed by heat treatment at 900–1100°C in an inert gas ambient for 10–30 min, resulting in the formation of a layer with a sheet resistance of 15 Ω/square.

Oxidation is next performed in wet H_2O (95°C) and proceeds at approximately the same rate as for conventional heavily doped silicon. This results in a silicide film, with approximately 1000 Å silicon dioxide over its surface. The sheet resistance of this silicide film is unchanged from that of the original silicide layer, as long as the underlying polysilicon layer is not fully depleted by this process. Etching of this composite film is accomplished by first removing the silicide layer by plasma etching methods, followed by removal of the underlying polysilicon by conventional etching techniques.

The use of silicide films is still in an early phase, so that considerable work needs to be done to determine the optimum choice for any specific application. At the present time, much of the research has focused on the silicides of hafnium, molybdenum, tantalum, titanium, and tungsten, since these materials have a long history of past work.

8.1.5.3 Refractory Metals

Table 8.2 shows the melting points and resistivities for refractory metals which are potentially useful in self-aligned masking technology. The choice is not quite as wide, since the specific resistivity of some of these materials (hafnium, vanadium, titanium, and zirconium) is almost as large as for their silicides. In addition, it is somewhat difficult to achieve the quoted resistivities in thin-film form if the metal is highly reactive (such as chromium, for example), because of oxygen contamination during the deposition step. Most

Table 8.2 Melting Points and Resistivities for Refractory Metals of Interest

Metal	Melting Point (°C)	Specific Resistivity (μΩ cm)
Co	1497	9.8
Cr	1878	12
Hf	2226	35?
Mo	2615	5.7
Ni	1455	7.8
Nb	2472	12
Pd	1554	11
Pt	1772	10
Ta	2996	15.5
Ti	1670	41?
W	3417	5.6
V	1903	25
Zr	1855	41?

work of the present time has been centered around the use of molybdenum, and tungsten to a lesser extent, since their materials technology has been well established for many years.

Both molybdenum and tungsten can be deposited by electron beam evaporation or by sputtering methods. They do not enter into any significant reaction with silicon dioxide, so that they adhere somewhat poorly to layers of this material. Molybdenum is considerably better than tungsten in this respect, so that device research has emphasized this material [36, 37]. The adhesion of molybdenum can be significantly increased if it is deposited on heated substrates (200–300°C). CVD approaches [38] also result in adherent films, and are well suited for device fabrication because of the requirement for relatively thick films (0.5–1 μm).

The most convenient technique for deposition of molybdenum films [39] is by the hydrogen reduction of molybdenum hexafluoride (MoF_6). MoF_6 is a highly volatile liquid, with a vapor pressure of 600 torr at 30°C, and is conveniently handled in a refrigerated bubbler arrangement operated at −10 to −15°C. Using this source, the deposition reaction takes place at temperatures of 400–1000°C, along the following lines:

$$MoF_6 + 3H_2 \rightarrow Mo + 6HF \tag{8.5}$$

Growth at high temperatures (800–1000°C) results in strong pitting of the quartz reaction chamber and of the silica substrate layer, since hydrogen fluoride is one of the reaction products. In addition, the crystal structure tends to become columnar in nature with large crystallites, preferentially oriented in the ⟨110⟩ directions. This results in relatively porous films, which are inadequate as diffusion masks. Low-temperature growth (400–450°C) results in dense films with extremely small grain crystallites. At these temperatures, reaction with the walls and with the substrate is relatively unimportant.

An alternate approach is to use molybdenum pentachloride [40], from which molybdenum films can be grown by the following reaction:

$$2MoCl_5 + 5H_2 \rightarrow 2Mo + 10HCl \tag{8.6}$$

This reaction is typically conducted at 600°C with a growth rate of about 0.4 μm/min, and results in dense films with small grain crystallites. Growth at elevated temperatures again leads to columnar films which are not useful as diffusion masks.

The use of $MoCl_5$ results in HCl as a reaction product, and so avoids the problems of fluoride pitting that occur with MoF_6. On the other hand, the system is considerably less convenient to use. This is because $MoCl_5$ is a powder at room temperature, with a relatively low vapor pressure. This necessitates a heated source (≃ 200°C), with vapor transport achieved by passing an inert gas over a large exposed surface of this powder. In addition, all lines to the reaction chamber must be heated to prevent condensation of this compound, and the formation of particulates in the reactor.

Molybdenum can also be formed [41] by the decomposition of molybdenum carbonyl [$Mo(Co)_6$]. This reaction usually results in films with a large amount of carbon incorporated in them (as much as 30%). These are generally not useful in device technology.

Tungsten is conveniently prepared [38] by the hydrogen reduction of tungsten hexafluoride gas, at a temperature around 600–700°C. The reaction proceeds along the following lines:

$$WF_6 + 3H_2 \rightarrow W + 6HF \tag{8.7}$$

It is also possible to deposit tungsten [42] directly on silicon surfaces at a much lower temperature (400°C), as a result of the following reaction:

$$2WF_6 + 3Si \rightarrow 2W + 3SiF_4 \tag{8.8}$$

This reaction can be used to selectively deposit this metal into holes cut in a masked silicon surface, and provides a simple technique for the formation of W-Si Schottky diodes.

8.2 FILMS FOR DOPING

A number of different films, incorporating dopant impurities, can be used in microcircuit technology. These generally consist of the dopant oxide in an inert binder such as silica, and are formed by CVD techniques. The simplest approach here involves the simultaneous oxidation of silane and a dopant hydride or alkyl, in a process that is very similar to that used for the deposition of silica films.

Doped oxides have many advantages over conventional predeposition methods. First, the surface concentration obtained by making diffusions from a doped oxide is a function of the concentration of the doping source, and also of the relative diffusion rates of the dopant in the oxide and in the semiconductor [43, 44]. Thus use of these sources allows high diffusion temperatures and relatively low surface concentrations at the same time. This is only possible by means of a two-step process when conventional sources are used, or by ion implantation techniques.

A second advantage is that very uniform diffusions can be made using these sources. Furthermore, these diffusions are carried out in an inert gas ambient, so that the diffusion equipment involves no complex gas handling or effluent disposal features, and is extremely simple.

Finally, the doped oxide serves as a capping layer during the diffusion. This is especially significant for gallium arsenide diffusion technology, where the widely disparate partial pressures of gallium and arsenic lead to decomposition of this semiconductor at diffusion temperatures. The conventional approach necessitates the use of sealed ampuls in which a carefully controlled partial pressure of arsenic is established to prevent this decomposition; close control of the dopant partial pressure is necessary as well, so that these techniques are quite impractical outside of the laboratory. On the other hand, the doped oxide source allows precise control of diffusions in gallium arsenide, with retention of the mirror-like surface quality that is essential for subsequent photolithographic processes.

Doped oxide sources usually incorporate only a few percent of the dopant oxide in the silica. As a result, they can be easily removed by conventional etchants, or patterned by fine-line photolithographic techniques. They can be used in situations where simultaneous *p*- and *n*-type diffusions are required in a microcircuit, as in the fabrication of complementary devices on the same chip. Moreover, the patterned placement of doped and undoped oxide regions on a microcircuit allows the fabrication of relatively simple self-aligned structures in MOS technology.

Doped oxide sources can also be formed by spin-on techniques. These have been described in Chapter 5, and are not considered here.

8.2.1 Dopant Sources for Silicon

Borosilicate glass is used as a *p*-type dopant source for silicon. Films of this material can be grown [45] in a cold wall reactor by the simultaneous oxidation of trimethylborate (TMB) and tetraethylorthosilane (TEOS), at a temperature of 750°C or higher. The reactions associated with this process are:

$$2(CH_3O)_3B + 9O_2 \rightarrow B_2O_3 + 6CO_2 + 9H_2O \qquad (8.9)$$

and

$$Si(C_2H_5O)_4 + 12O_2 \rightarrow SiO_2 + 8CO_2 + 10H_2O \qquad (8.10)$$

with the boron trioxide incorporated as a network former in the silica.

Both TMB and TEOS are liquids, so that vapor transport to the reaction chamber is carried out by means of bubbler arrangements. A mixture of these liquids can be used in a single bubbler for simplicity when only one doping concentration is required. However, this gives poor control of dopant concentration with bubbler use, because of their different vapor pressures. A second problem is that the relatively high deposition temperature ($\geq$ 750°C) necessitates the use of rf heating, which is relatively expensive. Consequently, this system has been largely replaced by one involving the oxidation of diborane and silane [46]. Both of these are gases and are used in dilute form in argon or nitrogen, and are reasonably convenient to handle. Typical concentrations are 5–10% for the silane and 0.1–1% for the diborane.

The diborane reaction proceeds as follows:

$$B_2H_6 + 3O_2 \rightarrow B_2O_3 + H_2O \qquad (8.11)$$

in the temperature range of 250–500°C. A growth temperature of 300–500°C is used, and the reaction is carried out in a relatively inexpensive resistance heated, cold wall reactor. Subsequent drive-in at 1200°C results in diffusions whose surface concentration varies roughly linearly with the mole ratio of $B_2H_6:SiH_4$ in the gas phase. Typically, for diffusions at 1200°C, surface concentrations from 10^{17} cm^{-3} (for a gas stream mole ratio of 0.0001) to 10^{20} cm^{-3} (for a gas stream mole ratio of 0.01) are readily achievable by this method.

Similar approaches have been used for the formation of *n*-type doped oxide sources for silicon. Here TEOS and trimethylphosphate mixtures have been used around 750°C, but this technique has been replaced by the oxidation of phosphine and silane gases. Again, dilutions of 0.1–1% for the phosphine and 5–10% for the silane are used, with the reaction carried out at 300–350°C. Here, too, the surface concentration after diffusion varies roughly linearly with the mole ratio of $PH_3:SiH_4$ in the gas stream [46]. Surface concentrations that can be achieved by this method (for 1200°C diffusion) range from 2×10^{17} cm^{-3} (gas stream mole ratio of 0.0001) to 2×10^{20} cm^{-3} (gas stream mole ratio of 0.1). High surface concentrations, up to 10^{21} cm^{-3}, can also be obtained, but these require a significant increase in the gas stream mole ratio.

It has been observed that the surface concentration achieved by all of these doped oxide surfaces is a function of diffusion temperature, but is independent of the diffusion time. This implies that these oxides, usually 0.1–0.2 μm thick, serve essentially as infinite sources for the dopant. The resulting diffusion profiles, which are of the complementary error function type, confirm this observation. More complex, double-dilution methods must be used to obtain low surface concentrations.

8.2.2 Dopant Sources for Gallium Arsenide

p-type diffusions into gallium arsenide can be made by using zinc oxide as the dopant source. During diffusion, this dopant is reduced to elemental zinc, which is then incorporated into the gallium arsenide.

Zinc oxide films can be formed by direct sputtering from a zinc oxide target, by reactive sputtering of zinc in an oxygen discharge, or by chemical vapor deposition. This last technique is most conveniently performed by the pyrolysis of diethylzinc (DEZ) in oxygen [47], at temperatures of 250–500°C. A cold wall reactor with a resistance heated substrate can be used for this purpose.

DEZ is a liquid with a vapor pressure of 15 torr at room temperature, so that it must be transported to the reactor by a bubbler arrangement. It reacts with oxygen even at room temperature, resulting in the formation of zinc oxide and a number of organic by-products. However, growth at temperatures above 250°C, in the presence of excess oxygen, results in zinc oxide films that are free from carbon contamination. The oxidation reaction for these conditions proceeds along the following lines:

$$(C_2H_5)Zn + 7O_2 \rightarrow ZnO + 4CO_2 + 5H_2O \tag{8.12}$$

Zinc oxide films cannot be used directly on gallium arsenide for diffusion purposes. This is because their reduction to zinc is excessive at these temperatures, leading to the formation of zinc–gallium and zinc–arsenic alloys and compounds, with a resultant loss in surface quality. Two approaches can be used to eliminate this problem. First, a thin barrier layer of silica can be deposited between the ZnO and the GaAs. This moderates the delivery of zinc to the gallium arsenide, and thus eliminates surface damage. This technique has been used successfully [48] for open-tube diffusions at temperatures up to 800°C and junction depths of 1–25 μm.

An alternate approach is to use a mixed film of ZnO and SiO_2, grown by the simultaneous oxidation of silane with the DEZ. Growth of a 0.1 μm thick layer of this type is conveniently carried out around 350°C in a resistance heated, cold wall reactor. Diffusion from a $ZnO{\cdot}SiO_2$ source results in excellent surface characteristics, and has been used [49] at temperatures up to 700°C, for the formation of shallow (0.03–1.5 μm) junctions with high surface concentration ($\simeq 10^{20}$ cm^{-3}) and excellent control of junction depth (± 0.01 μm).

In all of these open-tube diffusions, it is important to prevent the out-diffusion of zinc from the top surface of the dopant source. This is done by using a cap layer of PSG, 0.2 μm thick, on its surface,* since this material is impervious to the diffusion of zinc through it [10]. Use of this cap makes the diffusion relatively independent of thickness and composition of the dopant source, and also of the gas ambient in which the diffusions are conducted.

A mixture of tin oxide and silica can be used as a dopant source for *n*-type diffusions into gallium arsenide. Doped oxides of this type have been deposited by the simultaneous oxidation of tetraethylorthosilane (TEOS) and tetramethyltin (TMT) in the 750–800°C temperature range [50, 51], resulting in significant damage to the gallium arsenide surface. This problem has been eliminated, however, by growing these oxides at lower temperatures (350–450°C) by the use of silane and TMT [52]. Film growth is carried out in a resistance heated, cold wall reactor because of the low temperatures involved.

TMT is relatively stable in air and moisture; it is a liquid at room tem-

* This PSG layer is also grown on the back surface of the gallium arsenide slice to prevent its decomposition during the diffusion process.

perature and can be handled without hazard. Its relatively high vapor pressure (100 torr at room temperature) allows its ready transport to the reaction chamber by means of a bubbler arrangement. The oxidation reaction of TMT proceeds along the following lines:

$$(CH_3)_4Sn + 4O_2 \rightarrow SnO_2 + CO_2 + 6H_2O \tag{8.13}$$

The simultaneous oxidation of silane results in the formation of SiO_2 which is doped by the SnO_2.

These films are suitable for use as doped oxide sources for gallium arsenide [53], and are stable at diffusion temperatures up to 1100°C. Films of this type have also been used for open-tube *n*-type diffusions into gallium arsenidc at temperatures up to 900°C, provided they are capped with a layer of PSG as described earlier. Here, too, a mirror-like surface is preserved because of the low concentration of dopant in the source.

8.3 FILMS FOR INTERCONNECTIONS

An important application area for conductive films is to provide interconnections between contacts which are made to devices in a microcircuit. Both single- and multi-metal films are used for this purpose. Examples of these interconnections are shown in Fig. 8.4.

Films used for interconnections must adhere firmly to the semiconductor contact and also to the insulating layer that is placed over the semiconductor surface. They should be readily deposited by a relatively low-temperature process, since metallization is one of the last steps in microcircuit fabrication. They should be easily patterned with high resolution, without etching the insulating layer on which they are placed. They should be relatively soft and ductile, so that they can withstand cyclic temperature variations in service without failure. They must be highly conductive, and capable of handling high current densities while still maintaining their electrical integrity. Finally, they should be easily connected to external terminations.

Most of these requirements are met by gold and aluminum films. In

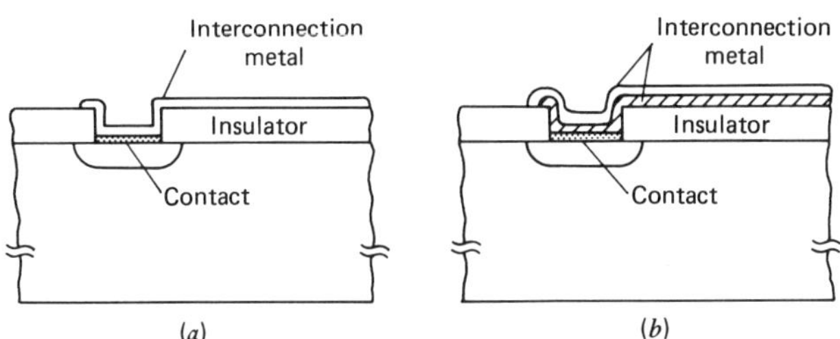

Fig. 8.4 Interconnections. (*a*) Single metal. (*b*) Multi-metal.

addition, both gold and aluminum wires, ranging from 0.5 to 2 mils in diameter, can be used for connecting these films to the terminal posts. Gold wires are usually drawn from 99.999% pure gold and are work-hardened to provide enough stiffness for handling purposes. With aluminum, however, it is necessary to use a 99% Al–1% Si alloy to provide this stiffness.

Gold welds readily to both aluminum and gold bond pads by the simultaneous application of heat and pressure. This process, known as *thermocompression bonding*, is commonly carried out with the aid of a nail head bonder. Figure 8.5 shows the various steps in its operation [54]. A gold wire with a sphere at one end is fed through a capillary from a spool. The sphere is aligned over the bonding pad and then lowered. During this process the semiconductor die, mounted on its header, is maintained at 280–300°C in a nitrogen gas ambient. The gold sphere is pressed against this pad for a few seconds until a weld is formed. This also results in plastic flow of the sphere into the form of a nail head. On raising the capillary, gold wire is drawn out from the spool, resulting in a lead. The capillary is now aligned over the terminal post and the procedure repeated. Finally, a hydrogen torch is used to break the wire, resulting in the formation of spheres at its ends. Thus the wire is ready for the next bonding operation.

A disadvantage of this type of bonding is the presence of the cantilever "tail" formed with each bond. Under conditions of extreme vibration this

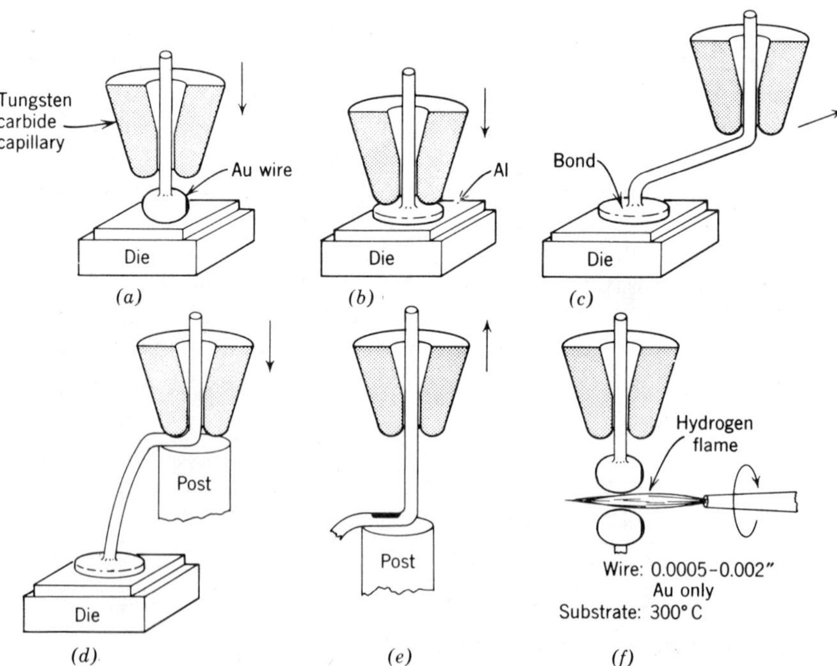

Fig. 8.5 Sequence of nail head bonding steps. Adapted from [54].

tail can be broken off within the package, and is thus an incipient cause for circuit failure. A stitch bonder is more commonly used since it avoids this problem. Here steps in the bonding operation are as shown in Fig. 8.6, and cutters are used to prevent the formation of the dangling lead.

The problem associated with the purple plague has been described in Section 2.2.6. Of the many systems that have been tried to avoid this problem, the simplest and best appears to be one in which aluminum leads are bonded to aluminum pads, or gold leads are bonded to gold pads. In this way the binary Au–Al system is avoided at the lead–chip interface. However, if an aluminum lead is used, it must eventually be connected to the gold-plated terminal post. This tends to alleviate the problem to some extent, since compound formation is accelerated in the presence of a third component such as silicon [55].

Ultrasonic methods have been found suitable for bonding aluminum leads to microcircuit pads. The bonding operation is accomplished as shown in Fig. 8.7. Here the wire is placed in contact with the aluminum pad, and pressure is brought to bear on this combination by an ultrasonically driven tool. Welding occurs as the tough aluminum oxide layer is broken by the ultrasonic vibrations.

Ultrasonic bonding is done at room temperature. It is commonly used with VLSI circuits which have a large number of bonding pads, since this avoids subjecting the chip to elevated temperatures during this process.

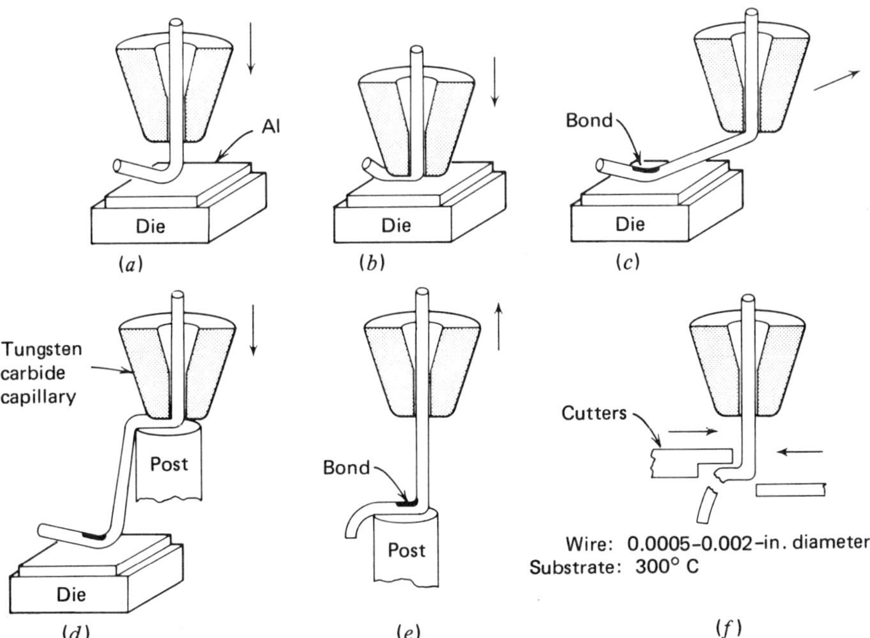

Fig. 8.6 Sequence of stitch bonding steps. Adapted from [54].

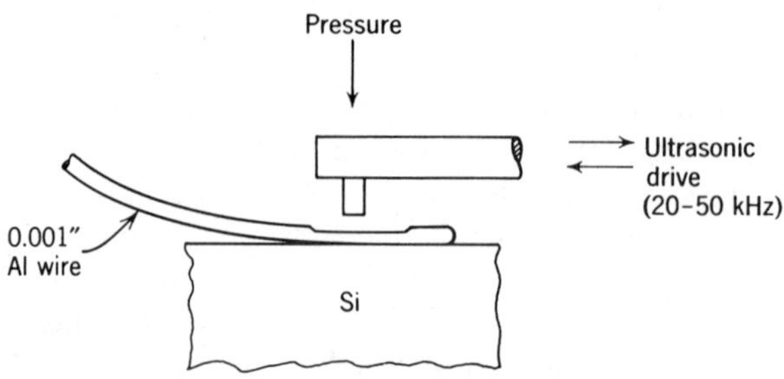

Fig. 8.7 Ultrasonic bonding.

However, its success is a sensitive function of the pressure as well as of the ultrasonic power level. By comparison, the thermocompression bonding of gold leads is a considerably easier technique, and requires less operator skill.

A number of techniques have been developed to avoid the expensive process of individually bonding leads to a microcircuit. These include gang bonding, as well as solder bump and mutliple-weld methods, and involve a high degree of automation. These techniques are of a specialized nature, and are not considered here.

8.3.1 Single-Metal Interconnections

Aluminum is a viable choice for single-metal interconnections, since it bonds well to silica as well as to silicon nitride insulating layers by a relatively short heat treatment. During this process, the aluminum reacts with the surface layer on which it is placed, at a number of localized points. Thus it becomes firmly attached to this surface, while still providing for stress relief during subsequent thermal cycling.

The bonding of aluminum to silica comes about by chemical reaction, with the formation of aluminum oxide. The reason for the bonding of aluminum to silicon nitride is less clear, however. In all probability, actual bonding occurs via a thin interfacial layer of silicon dioxide on the silicon nitride film.

There are many additional advantages to the use of this metal [56]. It can be readily deposited by well-developed vacuum evaporation techniques. Both resistance heated sources as well as electron beam evaporation methods are commonly used. More recently, magnetron sputtering techniques have been developed for this purpose. These are especially useful when more than one level of interconnects are required, since they allow the use of back sputtering for cleaning purposes, prior to deposition of the next interconnection level.

Evaporated aluminum is generally free from contamination, because of the high rate at which it can be deposited, relative to the impingement rate of impurities in the vacuum system (as governed by the background pressure). As a result, its electrical resistivity in thin-film form (3.0 μΩ cm) is not very different from its bulk resistivity (2.7 μΩ cm). A typical aluminum interconnection film, 1 μm thick, has a sheet resistance of about 0.0125 Ω/square.

Aluminum films are readily etched into fine-line patterns by chemicals which do not attack the insulating layers on which they are deposited. A variety of phosphoric acid based etches can be used for this purpose. One such, with an etch rate of about 200 Å/min, consists of phosphoric acid, nitric acid, acetic acid, and water in a 15:0.6:3:1 ratio by volume, and is commonly used with silicon microcircuits. This etch will attack gallium arsenide; however, an alternate etch, consisting of $HCl:H_2O = 1:4$ by volume, is often used at 80°C for this purpose. A number of proprietary etches are also in common use for both these applications.

Finally, aluminum welds readily to gold leads by thermocompression bonding, and to aluminum leads by ultrasonic welding. Both of these techniques result in strong, reliable, low-resistance connections to the package terminals.

Perhaps the biggest advantage of aluminum interconnections is in silicon-based microcircuits, where, as will be shown later, this metal can also be used to form the ohmic contacts. This allows the deposition of both the contact and the interconnection metal in a single step. In addition, a single heat treatment serves to form this contact, and also to bond the metal to the silica surface layer. Thus the overall process is a very simple one, and results in a low-cost technology which is widely used today.

The use of aluminum interconnect metal has some disadvantages, however. Although aluminum has a natural protective coating of native oxide, this metal is electronegative so that it is prone to corrosion in both acidic and basic environments [57]. In addition, it is soft, and relatively easily scratched during wafer and chip handling. Both of these problems can be greatly reduced by coating the complete microcircuit slice with a film of either PSG or silicon nitride. This additional step involves an extra masking operation to cut holes for the bonding pads, but greatly improves the processing yield and also the long-term reliability of the packaged microcircuit.

Another problem with aluminum is that of compound formation where it interfaces with the gold lead, during storage at temperatures in excess of 300°C. This phenomenon has been described in Section 2.2.6, together with methods by which its effects may be minimized.

A problem area with aluminum metallization is the difficulty of obtaining complete coverage at steps or irregularities on the semiconductor surface. This is especially severe in MOS circuits where oxide thicknesses can vary from 500 Å for the gate to over 15,000 Å for the field oxide. Microcracks in the metal film occur at these steps, even if considerable care is used to provide distributed sources for the evaporating aluminum. The most direct method of eliminating this problem is to taper* the oxide cut in order to

* Tapering and other planarizing techniques are described in Chapter 11.

avoid sharp corners. In addition, evaporation of aluminum on heated substrates ($\simeq$ 300°C) tends to promote increased coverage; this method can be used when the step profile is not too severe. More recently, sputtering techniques have been used for aluminum metallization, because they provide excellent step coverage.

Finally, an important problem comes about because of the requirements for VLSI technology. Here the emphasis on high-speed, densely packed structures has resulted in smaller dimensions, and an increase in the operating current densities of these films. This has given rise to an important failure mode known as electromigration.

8.3.2 Electromigration Effects

The term *electromigration* refers to the transport of mass in metals when stressed to high current densities [58]. It occurs during the passage of direct current through thin metal conductors in integrated circuits, and results in the piling up of metal in some regions, and void formation in others. The pileup can occasionally cause shorting to adjacent conductors; the voids, however, result in locally increasing the current density, and thus aggravating this problem.

Electromigration is only one of several transport phenomena that occur in solids [59]. However, it is the most significant failure mode in interconnection technology for microcircuits where current densities can often exceed 10^6 Å/cm^2 without catastrophic effects such as melting.

Figure 8.8 shows a metal conductor in which the electric field $\mathscr{E}$ is directed from left to right, as is the current flow. Here there are two forces on the (positive) metal ions in the conductor. The first, a field-ion force F_1, is directed toward the negative terminal, and is proportional to the electric field strength and the valence of the metal ion. A second force F_2 comes about because of the exchange of momentum between the electrons which comprise the current flow, and the metal ions. These electrons move from right to left, toward the positive terminal, and cause metal ion motion in this direction by an electron wind effect. In general, this second effect dominates, so that the net result is a movement of metal ions toward the positive terminal, i.e., with the electron flow. Ideally, this cannot cause void formation since all the metal is moved simultaneously. In practice, however,

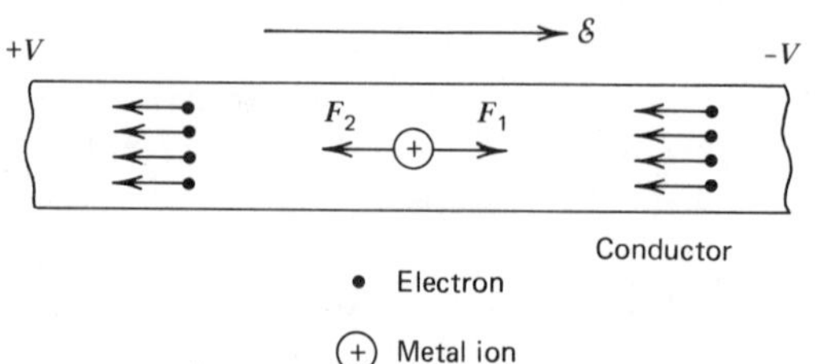

Fig. 8.8 Forces on the metal ion during current transport.

minor variations in mobility along the length of the conductor will cause the metal to be moved at different rates, with the eventual formation of voids.

Ignoring the field-ion effect, the rate of metal transport R is directly proportional to the electron momentum and to the electron flux. It is also directly proportional to the target cross section of the metal. Finally, metal motion is governed by its activation energy of self-diffusion E_d. The electron momentum and the electron flux are, in turn, each proportional to the current density, J. It follows that

$$R \propto J^2 e^{-E_d/kT} \tag{8.14}$$

The mean time to failure (MTF) of a conductor is commonly used [60] as a measure of R. Thus a curve of log $\{(\text{MTF})\ J^2\}$ versus $1/T$ results in a straight line from which the value of E_d can be extracted.

Metal interconnections are polycrystalline in character, so that diffusion is primarily via grain boundaries. As a consequence, the value of E_d is well below that which would be obtained for single-crystal material, and is highly dependent on the grain size. For single-crystal aluminum [61], $E_d = 1.4$ eV, whereas E_d varies from 0.4 for small-grain films to 0.5 for large grain evaporated films. The effect of this lower activation energy of diffusion is seen from experiments with both single-crystal and polycrystalline aluminum films. Single-crystal films, grown epitaxially on MgO substrates, have been operated at 175°C with a current density of 5×10^6 A/cm^2, for a period exceeding 26,500 h. Under these same conditions, evaporated polycrystalline aluminum films have failed in less than 30 h.

A number of approaches can be used to reduce electromigration effects in aluminum films. Primarily, they include the control of deposition parameters and the seeding with impurities such as copper and silicon, since these techniques critically affect the grain size. Thus aluminum doped with 4% by weight of copper results [62] in films with an improvement in the MTF by a factor of 10.

Impurities, such as aluminum oxide, have also been introduced for improving the electromigration characteristics of aluminum films. However, improvements claimed for this latter approach have not been corroborated by other workers. An alternate approach is to restrain conductor motion by means of an insulating cap layer, and some success has been reported along these lines. Here, too, corroboration of this improvement has not been obtained by other workers so that the results are questionable. In the final analysis, however, the best solution at the present time is to avoid the problem entirely by restricting the current density to about 10^5 Å/cm^2. In VLSI technology, where space for the metallization is restricted, this places a premium on the use of devices which can operate at low current levels.

8.3.3 Multimetal Interconnections

The problems of electromigration can be greatly reduced by use of a metal having a larger activation energy of diffusion than aluminum. One such metal, which meets many of the requirements of an interconnect layer, is gold. This is a ductile, weldable metal which can be deposited by vacuum evaporation techniques, with a resistivity in thin-film form that is almost identical to its bulk value (2.44 $\mu\Omega$ cm). It can be readily patterned into fine lines by a number of different etchants. A solution of potassium iodide in iodine, consisting of 4 g KI, 1 g I_2, and 40 ml H_2O, can be used for this purpose, and has an etch rate of about 0.5 μm/min. Another formulation, consisting of 2 g NH_4I, 0.5 g I_2, 15 ml C_2H_5OH, and 10 ml H_2O, has a considerably slower etch rate ($\simeq$ 700 Å/min) and is more suited for VLSI technology. In addition to the above, a number of cyanide-based etches are available, as well as a variety of prepackaged proprietary formulations.

Evaporated gold films have an activation energy of self-diffusion of about 0.85–1.0 eV, so that they can be operated at considerably higher current densities than aluminum films without suffering from electromigration effects. Typically, for the same current density and operating temperature (10^6 Å/cm^2 at 170°C), the MTF for gold-based interconnection systems is about 25–40 times higher than that for aluminum films [63]. At higher operating temperatures, the advantages of gold over aluminum are even more significant.

Gold is highly electropositive, and is not subject to the corrosion problems of aluminum. Unfortunately, its inert nature prevents adhesion to the insulating layer by chemical bonding. As a consequence, it can only be used in a multimetal system, where one or more additional layers are used for adhesion to the insulator, as well as to the gold. Transition metals such as titanium, chromium, and tantalum have all been used for this purpose. These films are somewhat difficult to deposit because of their reactive nature, and care must be taken to prevent their conversion to the oxides during this process. The use of these materials to promote adhesion to the gold film results in the possibility of reactions with it to form intermetallic compounds [64], during subsequent processing at 400–500°C. This necessitates the use of a barrier metal such as tungsten or molybdenum between the gold and the transition metal film. One popular approach is to combine these functions in a film of titanium–tungsten (20% Ti by weight), which can be easily deposited by rf sputtering. This alloy has excellent adhesion to both the gold and the insulating layer, and also serves as an interdiffusion barrier between the gold and the semiconductor.

Gold films can also be used over aluminum metallization, and can be plated up to form thick beam leads by which the chip can be directly bonded to the circuit board. Compound formation can be avoided, if interfacial barriers of platinum and titanium are used [65]. The resulting system consists of Au–Pt–Ti–Al, where the platinum prevents Au–Ti reactions, and the titanium prevents Pt–Al reactions. Layers of platinum and titanium, each 2000 Å thick, result in a system with no purple plague formation at temperatures as high as 450–500°C.

A problem with schemes of this type is that the barrier does not provide protection at its edge where it is delineated. This necessitates the use of wide conductors, so that barriers can still be used for beam leads, however, which are typically 20–25 μm wide.

8.4 FILMS FOR OHMIC CONTACTS

An ohmic contact to a semiconductor region must have good mechanical properties. It must adhere firmly, during both formation and subsequent processing, and also in service. It must not cause excessive stress in the underlying semiconductor, since this can result in a change in its electronic character. It must have low electric resistance, and be "ohmic," that is, its voltage–current characteristic should approximate a straight line going through the origin and extending over the entire range of voltages and currents to which the contact is subjected. It must serve purely as a means for getting current into and out of the semiconductor, but play no part in the active processes occurring within the device, i.e., it must not be an injecting contact. Finally, it must be compatible with the metal system used for the interconnection technology, since the top layer (or layers) of both the contact and the interconnection system are the same.

There are many ways for forming an ohmic contact that are suitable for use in a laboratory environment. Thus almost any contact to a highly damaged semiconductor region will be noninjecting because of the extremely high recombination rate associated with this region. Contacts which are made far from the active region, e.g., contacts to a Hall bar, are also "ohmic" in the sense that the region of device operation is many diffusion lengths away from the contact, so that injection effects can be neglected.

The needs for small dimensions and smooth surfaces preclude the use of techniques of this type for VLSI applications. Consequently, other approaches must be used to form a contact which meets the necessary mechanical requirements, and also offers a minimum barrier to the flow of electrons and holes into the device. One such approach consists of a two-step process, which can often be performed in a single operation. First, a heavily doped p^+- or n^+-region is formed on (or in) the semiconductor. Next, a metallic contact is made to this region. This two-step process results in great latitude in the choice of dopants and contact metals, and is commonly used for both silicon and gallium arsenide devices.

The behavior of an ohmic contact of this type can be studied by considering each step separately. Thus the first step results in the following possibilities.

***n^+*-Contact on *p^+*-Semiconductor or *p^+*-Contact on *n^+*-Semiconductor.** A contact of this type forms a symmetric tunneling junction whose built-in contact potential is sufficient to cause breakdown at zero volts. It results in a low-resistance "ohmic" contact with the voltage–current characteristics of Fig. 8.9*a*. Moreover, a contact of this type is made between highly damaged, low-lifetime regions so that minority carrier injection into the semiconductor is negligible.

p^+-Contact on p-Semiconductor or n^+-Contact on n-Semiconductor. This also results in "ohmic" behavior. In effect, the "high–low" junction [66] formed by this combination has an extremely large leakage current which completely masks the usual diode-like characteristics. Its voltage–current characteristic is of the form shown in Fig. 8.9*b*, and approximates a straight line, representative of ohmic behavior.

p^+-Contact on n-Semiconductor or n^+-Contact on p-Semiconductor. These situations result in the formation of a well-defined p–n junction diode whose characteristics are quite nonohmic. However, ohmic contact to a lightly doped region can be achieved by the series connection of the two schemes outlined earlier. Thus contact to an n-region is made by first forming an n^+-region on it, followed by a p^+-region, resulting in a sandwich of p^+–n^+ (tunneling junction) and n^+–n (high–low junction) regions, which is ohmic in character. (In like manner, an n^+–p^+ junction, followed by a p^+–p junction, results in an n^+-contact to a lightly doped p-region.)

From a device physics viewpoint, the second step, which consists of the placement of a metal in intimate contact with a semiconductor, creates a Schottky barrier. The properties of this barrier are a function of the choice of metal and the semiconductor doping. It is shown in Section 8.5 that these characteristics can be exploited in a variety of diode structures. Here, however, it suffices to point out that the voltage–current characteristic of any Schottky barrier will approximate ohmic behavior if the metal layer is in contact with a region of very high doping, $> 10^{19}$ cm^{-3}, regardless of the barrier height [67]. This is because field-emission dominated conduction,

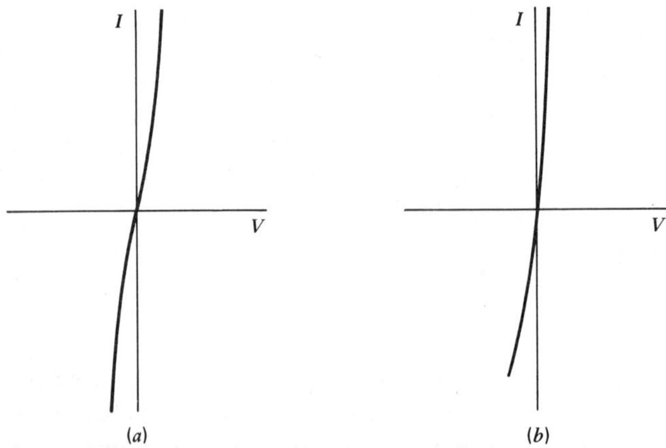

Fig. 8.9 Ohmic contact characteristics. (*a*) High–low junction. (*b*) Tunneling junction.

i.e., tunneling through the barrier, prevails so that it is almost transparent to the carrier flow. Thus a tunneling junction is formed in this manner. The use of this principle allows great latitude in the choice of contact metal, which can be now selected on metallurgical considerations alone.

8.4.1 Single-Layer Contacts

In silicon technology, the use of aluminum metal, which belongs to group III of the Periodic Table, allows the combination of these steps in a single process. The use of aluminum for the interconnection as well results in a simple, low-cost process that is both highly reproducible as well as highly reliable. In order to form an ohmic contact, aluminum is deposited on the silicon and subjected to a heat treatment at a temperature that is somewhat below the melting point of the combination. Consider the metallurgical processes that occur when an evaporated aluminum film is alloyed on a silicon substrate in the neighborhood of the eutectic temperature. Figure 8.10*a* shows an aluminum–silicon combination of the type considered here. On raising its temperature, some of the aluminum combines with the silicon to form a melt (see Fig. 8.10*b*). On cooling, a solid solution of aluminum in silicon freezes out epitaxially as an extension of the single-crystal substrate. This is followed by a polycrystalline Al–Si alloy and, finally, by pure aluminum, as shown in Fig. 8.10*c*. Thus a highly aluminum-doped, p^+-region is formed on the silicon. Connection is made to it by a polycrystalline Al–Si alloy, and eventually by the aluminum metal.

It would at first appear necessary to perform this alloying operation above the eutectic temperature. This is not the case because, on a microscopic scale, contact between the aluminum and the silicon only occurs at a discrete number of points. Here melting occurs below the eutectic temperature because of considerable localized pressure at these points.* With a micro-alloy contact of this type, bonding occurs at a large number of isolated points on the film surface. Stresses created during circuit operation are relieved by plastic flow of the metal, thus greatly reducing the possibility of failure due to thermal cycling. It can be shown [68] that a contact of this type, if made over 10% of the surface, has as low an electrical resistance as one made over the full area of the contact region.

In this contact technology, the aluminum and p^+-silicon form a tunneling Schottky contact, without the necessity of further steps or the deposition of additional layers. One particularly important advantage of this system lies in the fact that the trace oxide (10–20 Å), which is normally present on all silicon surfaces, is readily dissolved by the aluminum during contact formation. Thus elaborate cleaning procedures, involving back sputtering or ion milling, are not necessary prior to aluminum deposition.

* Typically, this heat treatment is carried out at 450–550°C, whereas the eutectic temperature of the Al–Si system is 577°C.

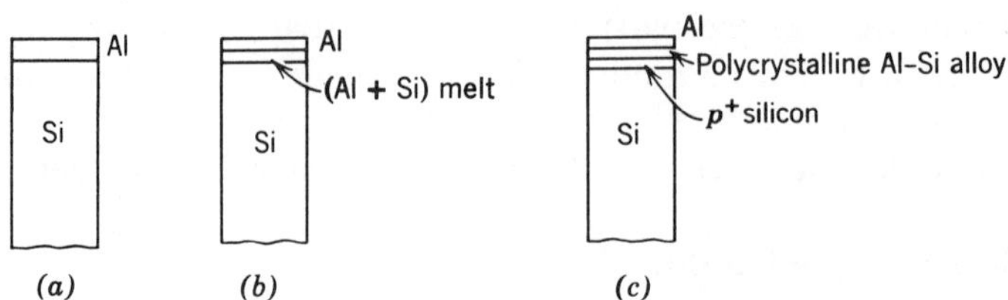

Fig. 8.10 Steps in the Formation of an Aluminum–Silicon Alloyed Contact.

Figure 8.11*a* illustrates the types of ohmic contact that are made to an n^+–p–n silicon transistor using this contact technology. Here the emitter contact consists of a metal–p^+–n^+ combination. The base contact is a metal–p^+–p structure. The collector contact consists of a metal–p^+–n^+ in series with an n^+–n junction. The n^+-region is formed by diffusion into the collector. However, it is made at the same time as the emitter diffusion, so that it requires no additional process steps. In contrast, connection to a p^+–n–p silicon transistor, as shown in Fig. 8.11*b*, requires an additional process step for the n^+-diffusion which is necessary for ohmic contact to the *n*-base.

Single-layer systems can also be used in simple contacting schemes for gallium arsenide [69]. Here a successful contact material for *n*-type gallium arsenide is a gold–germanium alloy, of eutectic composition (12% by weight of germanium). Deposition of this material, followed by a 2 min heat treatment at 450–500°C, serves to form an n^+–n contact, with the gold alloy acting as a Schottky barrier connection to it.

The exact mechanism of contact formation with this system is not known. In fact, its success is somewhat surprising since germanium can be incorporated on both gallium and arsenic sites. One argument is that the presence of the gold serves as a getter for gallium during the heat treatment. This results in the increased formation of gallium vacancies in the semiconductor, accompanied by their preferential occupation by the germanium so that the material becomes heavily *n*-type. Sometimes a little nickel ($\simeq$ 1%) is added to this material to enhance the diffusion of germanium into the lattice.

An alternate argument [70] is that heat treatment of the Au–GaAs system

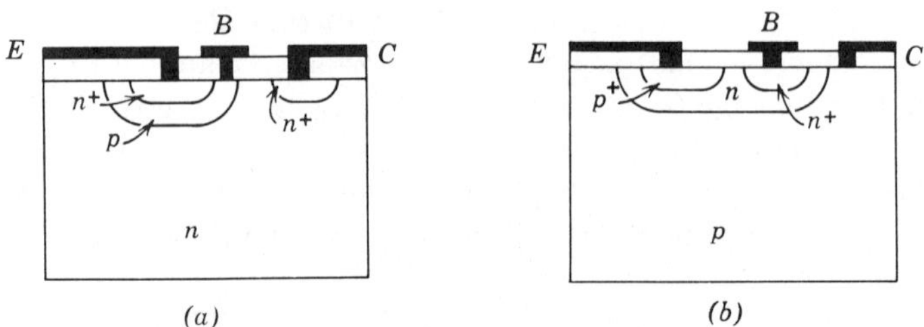

Fig. 8.11 Ohmic Contacts. (*a*) *n–p–n* transistors. (*b*) *p–n–p* transistors.

lowers the barrier height, and the in-diffusion of germanium seems to further enhance this effect. In addition, the germanium does substitute for both gallium and arsenic atoms, but appears heavily n-type because of the strain in the lattice which is created by the indiffusion of both these elements (gold and germanium).

Gold, doped with 5–10% tin by weight, or with 1–2% selenium by weight, has also been used successfully in the formation of ohmic contacts to n-gallium arsenide. Contacts to p-gallium arsenide are often made by means of a Au–Zn alloy (1% zinc by weight). In both cases, the gold forms a Schottky barrier having suitable metallurgical properties for use as a contact metal.

8.4.2 Kirkendall Effects

The single-layer contacting situations described above are based on the microalloying of the contact metal to the semiconductor, by means of heat treatment. This results in the dissolution of the semiconductor, which tends to proceed more slowly along the ⟨111⟩ directions, resulting in the formation of voids. These eventually become backfilled with the contact metal or its alloy, to form conducting spikes as shown in Fig. 8.12. These spikes are of little consequence in deep junction structures. With shallower structures, however, they often lead to leaky or to shorted junctions. They represent a special problem in VLSI technology where the requirement for densely packed structures mandates the use of shallow junction devices.

The tendency to spike formation comes about because of the differences in the solid solubilities of the contact metal in the semiconductor, and vice versa. Upon heat treatment, this gives rise to the differential transport of materials, known as the *Kirkendall effect*. This can result in the formation of voids in the region in which there is a net loss of material.

Kirkendall effects present a serious problem in the formation of ohmic contacts to silicon. Here connection is made by heat treatment at 450–550°C as described previously, resulting in alloying at discrete points. From Fig. 8.13 it is seen that the solid solubility of silicon in aluminum is quite significant at these alloying temperatures [56]. On the other hand, the solid solubility of aluminum in silicon is much lower, $\simeq$ 0.0011% by weight at 600°C (see Fig. 2.22). This differential solid solubility results in the net dissolution of silicon into the aluminum, with the resultant spike formation.

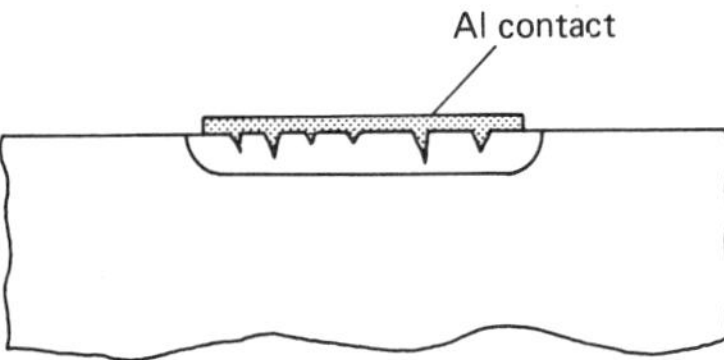

Fig. 8.12 Alloy spiking effects.

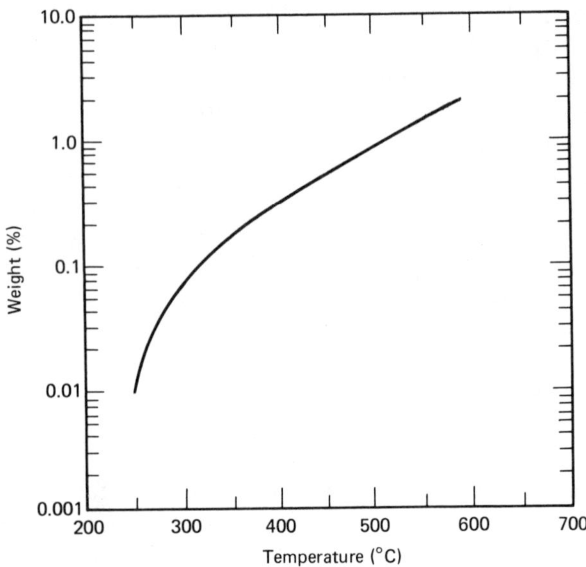

Fig. 8.13 Solid solubility of silicon in aluminum. From Schnable and Keen [56]. © 1971 by the Institute of Electrical and Electronic Engineers.

A second mechanism, which greatly enhances the Kirkendall effect, is the high diffusivity [71] of silicon in aluminum films, as shown in Fig. 8.14. This mechanism allows the silicon to be transported a considerable distance into the aluminum during heat treatment. Furthermore, void formation will be especially severe if the contact is made in a region where there is a large aluminum sink, such as a bonding pad.

One approach to reducing the penetration depth of these spikes is to carry out the microalloying operation at 400°C or lower. Unfortunately, this often results in high-resistivity ohmic contacts. An additional approach is to use aluminum metal which is already presaturated with silicon. Typically, about 2–3% by weight of silicon is commonly used for this purpose. The use of Al–Cu–Si alloys, with about 3% by weight of copper and silicon, reduces problems of electromigration as well. These alloys are commonly used for metallizations in VLSI technology. Other approaches, using multilayer contact schemes, are described in the next section.

Kirkendall effects are also present in the gold–gallium arsenide system when it is heated to such a temperature that decomposition of the semiconductor takes place. Here the solubility of gallium in gold is considerably higher than that of gold in gallium arsenide, resulting in the formation of gallium vacancies. This process eventually leads to void formation and alloy spikes. Here, too, multilayer contacting schemes are used to minimize these problems in VLSI technology.

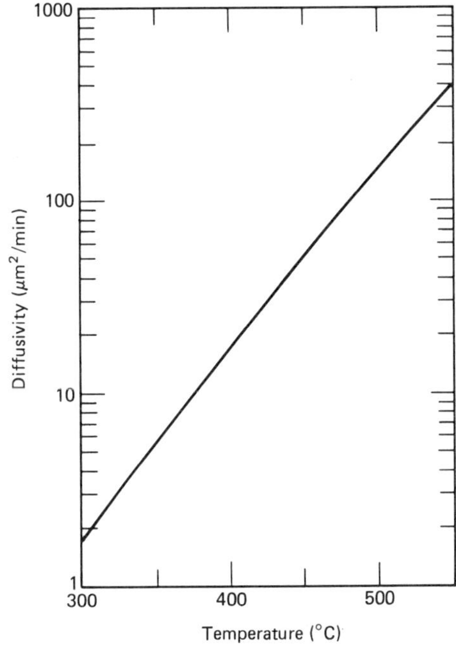

Fig. 8.14 Diffusivity of silicon in deposited aluminum films. Adapted from McCaldin and Sankur [71].

Both electromigration and Kirkendall effects can cooperate in a number of situations, as shown in Fig. 8.15, which depicts a diffused resistor with aluminum connections. The bias conditions for this resistor are also shown, with the path of electron flow from interface A to interface B. Here both electromigration and Kirkendall effects aid in transporting silicon into the aluminum at interface B. At A, electromigration effects tend to transport aluminum into the silicon. However, very little actual material transfer is involved, because of the low solid solubility associated with this process. As a result, the interface at B will steadily degrade during device operation, as the voids and alloy spikes continue to grow. On the other hand, the interface at A will be relatively unaffected.

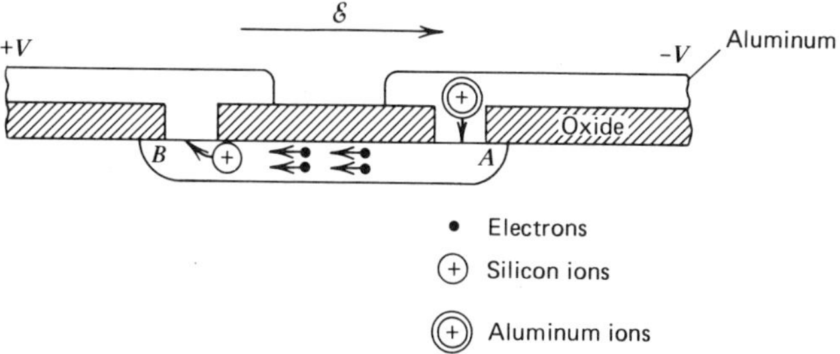

Fig. 8.15 Cooperation of electromigration and Kirkendall effects.

8.4.3 Multilayer Contacts

Contact systems of this type are more complex than those described earlier. Consequently, they are used when single-layer contacts have poor electrical performance or are unreliable. One problem area, described in Section 8.4.1, is that of making an aluminum contact to *n*-silicon. This structure resulted in nonohmic behavior and necessitated the use of an additional n^+-layer between the aluminum and the contact. Thus it was, in effect, a multilayer contact.*

A second problem area arises in VLSI technology, where single-layer contacts, formed to shallow junctions by alloying, result in poor yield due to localized penetration of alloys of the contact metal into the semiconductor. Here one approach is to introduce a barrier layer between the top contact metal and the semiconductor. The function of this layer in the aluminum–silicon system is to prevent the net flow of silicon into the aluminum during contact formation by heat treatment.

Platinum silicide is a suitable barrier material for this purpose. In practice, a layer of platinum, about 500–1000 Å thick, is deposited on the silicon surface which is masked except for openings where the contacts must be made. Next, the combination is heat treated at 600–650°C for about 10 min, in a nonoxidizing atmosphere such as argon. This serves to convert the platinum to PtSi, which makes a dense, adherent contact to the silicon. A unique advantage of this contact is that, unlike alloying, this conversion results in planar penetration into the silicon and avoids alloy spiking [72]. In addition, the conversion process consumes silicon, and produces a clean interface with the semiconductor. Typically, 500 Å of platinum consumes 665 Å of silicon to form 990 Å of PtSi. Eventually, aluminum or a multimetal system is placed over the platinum silicide film.

The formation of a platinum silicide film on silicon is a difficult task because of the refractory nature of this material. Platinum must be deposited by E-beam evaporation or by rf sputtering. Back sputtering or ion milling must be used to clean the contact regions prior to platinum deposition, and also for etching away the unwanted platinum from the rest of the surface. Aqua regia is also effective in removing the unreacted platinum. However, this chemical is extremely difficult to use in an industrial environment. The presence of even trace amounts of oxygen or water during the silicide formation process must be avoided here, so that it is sometimes carried out in a vacuum.

Many of these processing difficulties can be avoided by the use of palladium [73] which can be more easily deposited than platinum. In addition, it can be readily etched by chemical techniques, using the same KI + I etch

* In the example of Fig. 8.11*a*, this n^+-layer was formed by diffusion. It goes without saying that epitaxy can be used in some situations as well.

formulations that are suitable for gold. Deposited palladium is fully converted to Pd_2Si by heat treatment at 200°C, and there is some evidence of partial conversion even at room temperature. Often the deposition and conversion to silicide are accomplished in a single step, by using a heated substrate during vacuum evaporation. Typically, 500 Å of palladium is deposited, with a consumption of about 500 Å of silicon during this process; the resulting Pd_2Si film is about 720 Å thick.

Aluminum contact layers, deposited on these silicides, bond readily to them by forming the respective aluminides, but eventually result in the penetration of the aluminum to the silicon interface during subsequent processing [64]. Consequently, a 2000 Å thick titanium–tungsten (20% Ti by weight) layer is used as a barrier between the silicide film and the aluminum.

Silicide films can also be used in systems where the upper metal is gold. Gold does not adhere to these silicides, so that an intermediate layer must be used for this purpose. Here additional complexities arise because of the incompatibility of these materials. Thus in one system, referred to as *sealed junction technology* [74], a layer of titanium is placed over the PtSi to form an adherent seal over the junction region, as well as to provide a film that bonds to the insulating surface over which the interconnections are to be made. Next this layer is covered with 2000 Å of platinum metal, since the titanium–gold contact can lead to intermetallic compound formation. Gold is placed over this layer to form the complete ohmic contact and interconnection system. In addition, the gold film can be selectively plated up to form built-in beam leads by which the chip can be directly attached to the circuit board. The resulting system, consisting of Au–Pt–Ti–PtSi–Si, allows subsequent processing to 500°C.

It is also possible to make a Schottky barrier contact directly to the semiconductor, and avoid the alloying step completely. This technique allows contact to extremely shallow junctions; however, it greatly restricts the choice of both the contact metal and the doping concentration (and type) of the underlying semiconductor. For example, both PtSi and Pd_2Si have large barrier heights to *n*-silicon, and thus make excellent ohmic contacts to *p*-type material. Chromium and titanium have a low barrier height ($\simeq$ 0.5 eV) to both *n*- and *p*-silicon, and can be used to make ohmic contacts directly to moderately doped silicon, in addition to adhering to the surface insulator. These reactive metals are generally deposited by vacuum evaporation techniques and are subsequently followed by evaporated layers of gold to which they make strongly adherent contact. Systems of this type, Au–Ti–Si or Au–Cr–Si, while not as reliable as the Au–Pt–Ti–PtSi–Si system described previously, are considerably simpler, and are in more common use at the present time [76].

A second approach consists of building up the thin semiconductor region, prior to making an alloyed contact to it. Both doped and undoped polycrystalline silicon layers can be used for this purpose. This system is particularly convenient if aluminum is used as the top contact metal. In this case, heat treatment of the combination results in some epitaxial regrowth of the silicon, as well as alloying of the aluminum contact material.*

Figure 8.16 indicates experimental data on the results of postmetallization heat treatment for a number of such contacts to n^+-silicon [77]. The behavior of the single-layer (Al/n^+-Si) system is also shown for comparison. It is seen that the specific resistivity of the Al/n^+-Si system is initially high, but falls rapidly with heat treatment between 350 and 450°C. This fall is, unfortunately, accompanied by spike formation during alloying. The resistivity of the Al/n^+-poly Si/n^+-Si system is low, and can be attained by heat treatment at 200°C, with a resulting reduction in spike formation. In addition, the extra layer of silicon that builds up the underlying semiconductor serves to further reduce spike penetration, and improves the yield of the resulting contact.

Undoped polysilicon can also be used for this layer. Here, however, the specific resistivity is significantly higher than that obtained if the layer is heavily doped.

Multilayer contact schemes are also used to form low-resistivity contacts to n-gallium arsenide. Here the most common approaches are based on the formation of an n^+–n contact by germanium doping, followed by Schottky contact to the metal. Some of these schemes use additional layers of other materials in order to improve the wettability of the system, to enhance the incorporation of germanium into the gallium arsenide, or to act as barriers between the separate films that comprise the system.

One approach [78] consists of the deposition of a 1000 Å layer of germanium on the gallium arsenide, followed by a 1000 Å layer of nickel, with a subsequent heat treatment at 450–550°C for 30 min, followed by a gold metal contact. In this system, use of the nickel greatly enhances the diffusion of germanium and gold, presumably by the creation of defects in the gallium arsenide.

A second approach [79] involves the deposition of germanium followed by a film of gold of sufficient thickness to allow significant indiffusion during heat treatment. In this system the excess gettering of gallium by the gold must be prevented, since this leads to loss of planarity of the interface [80]. Consequently, a barrier layer of titanium–tungsten (20% Ti by weight) is often placed between the contact system and the thick gold film (0.5–1 μm) to which a lead can be attached. This combination, upon heat treatment, provides both the n^+-region as well as a soft metallic connection to it. Deposition of the barrier layer also serves to provide an adherent film over the microcircuit surface, so that the gold contact can be extended to the interconnections and the bonding pads.

Refractory metal ohmic contact systems have recently been developed

* Note that epitaxial silicon cannot be directly grown for this purpose, because of the high growth temperature ($\simeq$ 1200°C). In contrast, the growth temperature for polysilicon can be as low as 600°C, or 400°C if plasma-enhanced techniques are used.

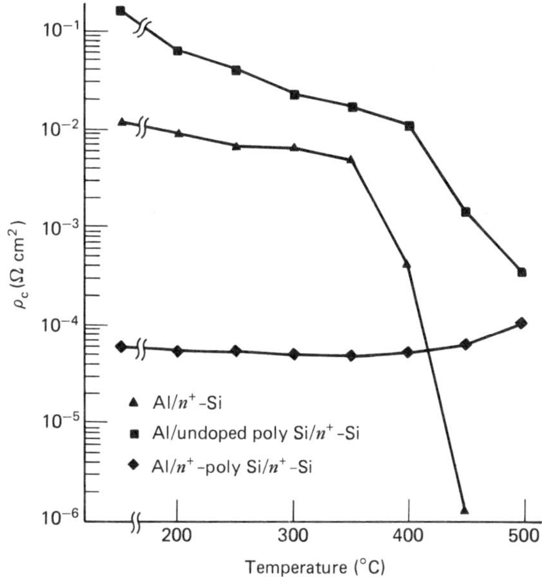

Fig. 8.16 Contact resistivity of a variety of aluminum–silicon systems. From Finetti, Ostaja, Solmi, and Soncini [77]. Reprinted with permission from the Pergamon Press, Ltd.

[81] for *n*-gallium arsenide. These are free from alloy spiking, and are potentially more reliable than the previous approaches. These systems are based on the fact that an n^+-Ge/n^+-GaAs combination forms a heterojunction tunneling contact. In addition, a Schottky metal contact can be readily made to the germanium because of its low energy gap.

Contacts of this type are quite different in their physics of operation compared to those described earlier. Thus they require no heat treatment for indiffusion. In fact the ideal contact metals are molybdenum, titanium, or tungsten, which are relatively inert to the gallium arsenide. Alloy spiking effects are avoided in this manner. The complete contact thus consists of Mo–n^+-Ge–n^+-GaAs–*n*-GaAs. A cover metal of gold is also used to prevent oxidation of the molybdenum, and to serve in a low-resistance interconnection scheme. No heat treatment is used other than during the epitaxial growth of heavily doped germanium on the gallium arsenide. Laser annealing of an evaporated germanium film has also been proposed for making this type of contact [82].

8.4.4 Die Bonds

Contact schemes are also required for forming a strong bond between the chip and the header.* Here the problems of spike formation are not impor-

* Note that microcircuits are one sided, and do not require an "ohmic" contact to the header; epoxy bonding is used in many consumer applications.

tant, since the active region of the device or microcircuit is 100–200 μm away from the contact region. Consequently, single-layer contacts are invariably used here. Dies are usually bonded by means of a preform which is placed between a plated header and the chip, as shown in Fig. 8.17. The combination is raised to the bonding temperature and pressure is applied to the die in conjunction with a vibratory scrubbing motion. The process is carried out in an inert or a slightly reducing N_2/H_2 gas ambient.

Gold is often used to make a strong bond to the chip in silicon technology. Vacuum evaporation of this gold, in a film about 500 Å thick, avoids both the cost as well as the handling complexities of a separate preform. In some cases, as in discrete devices, this bond must make an ohmic contact as well. Typically, this is obtained by using gold that is doped with about 1% gallium or arsenic. Upon alloying, this results in a Schottky connection to p^+- or n^+-doped silicon, respectively. Often a predoped gold–germanium alloy, which is of eutectic composition (12% Ge by weight) is used. This provides a better wetting action with the silicon, and results in a Au—Ge—Si bond at a slightly reduced bonding temperature.

Bonds of this type are commonly made at about 390–420°C, which is well above the eutectic temperature for these systems. As a result, they form strong, large area bonds, with considerable damage to the silicon surface. As mentioned earlier, this is of little consequence since the damage is far from the active regions.

Soft solder preforms, consisting of lead–tin alloys, are used for bonding large-area power devices where thermal cycling effects are important. These soft solders provide stress relief during device operation, and prevent fatigue by plastic deformation of the interface layer between the silicon chip and the header. Typical soft solders are 95% Pb, 5% Sn by weight, and have a melting point of 310–314°C. Alloys of Pb–In–Ag are also used because of their improved thermal cycling characteristics. One example is an alloy of 92.5% Pb, 5% In, and 2.5% Ag by weight.

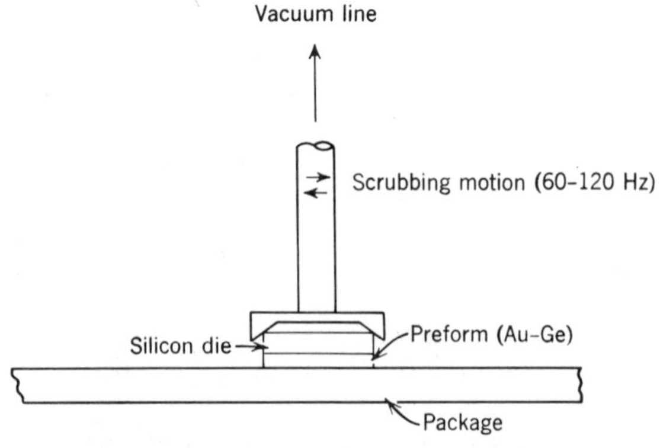

Fig. 8.17 Die bonding arrangement.

Nickel-plated steel headers are commonly used in this application. The silicon is also nickel plated, the deposit being commonly formed by electroless plating [83, 84]. This technique, sometimes known as autocatalytic plating, depends on the action of a reducing agent in the bath to convert the metallic ions to the metal. The most commonly used reducing agent for electroless nickel baths is NaH_2PO_2, which results in the deposition of nickel films with as much as 3–15% phosphorus. A short heat treatment after deposition greatly improves the adherence of these films, and forms an n^+-contact to silicon, which becomes phosphorus doped during this process.

Boron-doped electroless nickel films can also be formed, with dimethylamine borane as the chemical reducing agent. These films have about 0.3–10% dissolved boron, and can be used for contacts to *p*-silicon.

A single-layer scheme, consisting of Au–Ge eutectic, Au–1% Se, or Au–1% Zn, can be used to form an ohmic contact to the reverse side of the gallium arsenide chip. Bonding is usually done by heating the combination to 450°C in a flowing N_2–H_2 gas mixture (10–20% H_2 by volume) for 1 min. An alternate approach is to evaporate indium metal on this surface. Bonding is accomplished by heat treatment at 350°C for 5 min in a flowing N_2–H_2 gas mixture. It has been proposed that this results in the formation of a thin regrowth layer of gallium–indium–arsenide. This material has a lower band gap than gallium arsenide, and thus provides a heterojunction contact to both *p*- and *n*-type materials. The exact details of this system have not been worked out at the present time.

8.5 FILMS FOR SCHOTTKY DIODES

The Schottky diode is firmly established in both silicon and gallium arsenide technology, since it performs functions which cannot be achieved by conventional junction devices. Thus, it has a higher conductance than is possible with *p–n* structures. In addition, it is a majority carrier device so that it has extremely fast recovery. This combination of characteristics allows its use in the circuit arrangement of Fig. 8.18 [85]. Here the turn-on voltage of the diode is lower than that of the collector–base junction, so that its conduction prevents the transistor from going into saturation. This allows fast switching of the transistor, but still preserves the clamping action that is desirable in saturated logic circuits. As a result, this transistor–diode configuration has a twofold advantage over a conventional gold-doped silicon transistor. First, it is much faster. Second, it avoids the need for gold doping and thus eliminates one of the most poorly controlled steps in silicon microcircuit technology [86].

An extremely important advantage of Schottky diode technology with gallium arsenide is that it avoids the high-temperature steps that are associated with junction formation by diffusion. Furthermore, since this structure requires only the deposition of one or more contact materials followed by

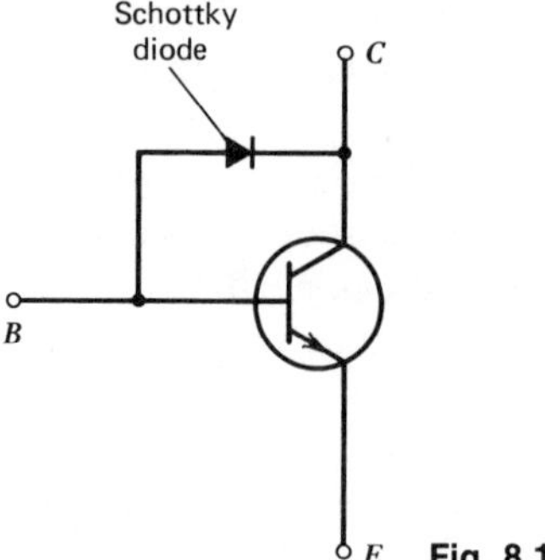

Fig. 8.18 Schottky clamp circuit.

photolithographic delineation, it is possible to precisely control its physical dimensions to within fractions of a micron. Thus it can be used as the gate in short-channel, high-speed field effect transistors. These transistors, with Schottky gates, are the almost universal choice for VLSI applications in gallium arsenide.

The factors which determine diode characteristics can be evaluated by noting that the Fermi levels of a metal and a semiconductor will line up if they are placed in intimate contact under conditions of thermal equilibrium. This gives rise to band bending in the semiconductor, and a resultant barrier ϕ_{Bn} or ϕ_{Bp}, as shown in Fig. 8.19 for both *p*- and *n*-type material. The height of this barrier is related to the work function of the metal, the resistivity and electron affinity of the semiconductor, the Schottky barrier reduction due to image force lowering of the barrier height, and the nature of the semiconductor surface states [87]. Experimental values for ϕ_{Bn} are given in Table 8.3 for a variety of metals and silicides on *n*-Si, and are relatively insensitive to the doping level, provided it is below 10^{17} cm^{-3}.

Experimentally, it is found that $\phi_{Bn} = C_1\phi_m + C_2$, where ϕ_m is the work function, and C_1 and C_2 are constants for the semiconductor. Figures 8.20 and 8.21 show this relationship for Schottky barriers on *n*-type silicon and

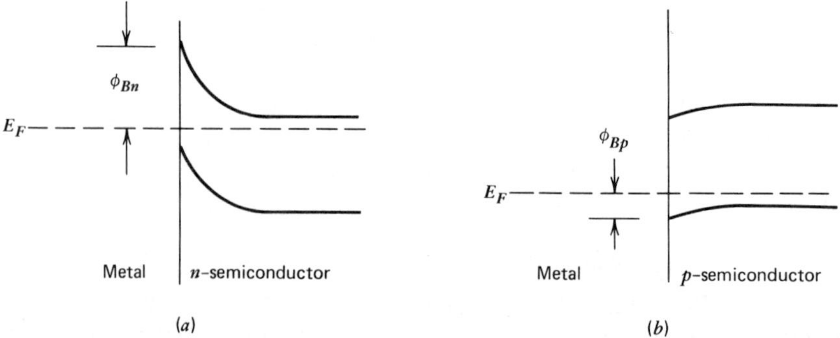

Fig. 8.19 The metal–semiconductor system.

Table 8.3 Barrier Heights of Some Metals and Silicides on *n*-Si

Metal	ϕ_{Bn} (eV)
Al	0.7–0.75
Au	0.79–0.80
$CrSi_2$	0.57
Mo	0.57
$MoSi_2$	0.55
NiSi	0.66
Pd_2Si	0.74
PtSi	0.85–0.87
Ti	0.50
RhSi	0.68–0.70
WSi_2	0.65
$ZrSi_2$	0.55

gallium arsenide, respectively. It is seen that values of ϕ_{Bn} for silicon are considerably less than the contact potential of a p–n junction ($\simeq$ 1 eV), so that the "threshold voltage" for significant current conduction (1 A/cm^2) will be considerably lower for these diodes than for junction diodes ($\simeq$ 0.7 V). Typical values for this threshold range from 0.2 V for Al–n-Si to 0.35 V for PtSi–n-Si devices.

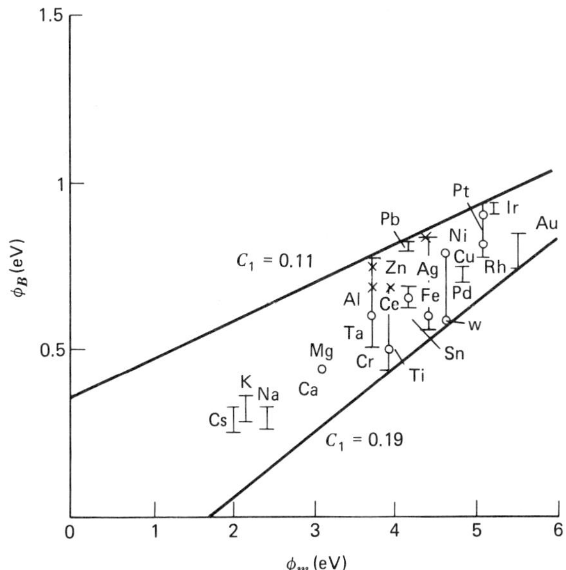

Fig. 8.20 Experimental values of ϕ_{Bn} for metals on *n*-silicon. From Sharma and Gupta [88]. Reprinted with permission from *Solid State Technology*.

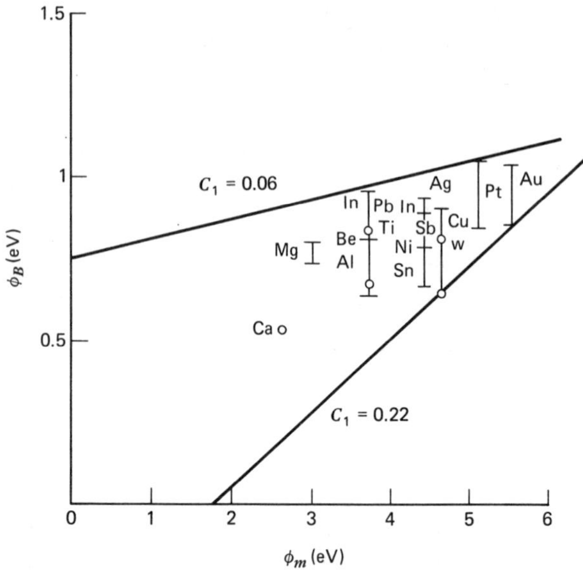

Fig. 8.21 Experimental values of ϕ_{Bn} for metals on *n*-gallium arsenide. From Sharma and Gupta [88]. Reprinted with permission from *Solid State Technology*.

The room temperature value of reverse leakage current of a Schottky diode increases by one decade for each 0.06 eV reduction in the barrier height. It follows, therefore, that this leakage current will be many decades higher than that of a *p–n* junction structure. Consequently, there are only a few metals which result in acceptable reverse current values for silicon devices, from the point of view of circuit design.

Barrier heights to *p*-type silicon are considerably lower than to *n*-type, since $\phi_{Bn} + \phi_{Bp} = E_g = 1.11$ eV. Thus good rectifiers are usually made to *n*-type silicon, whereas ohmic contacts are more easily made to *p*-type silicon. Note that materials such as molybdenum and titanium have relatively low values of ϕ_{Bn}. These metals find use in Schottky ohmic contacts to silicon that is not heavily doped (see Section 8.4.3).

It is worth remembering at this point that a Schottky diode made on a highly doped semiconductor ($> 10^{19}$ cm^{-3}) will have an extremely narrow depletion width. Here conduction is primarily by field-assisted tunneling *through* the barrier, rather than *over* it. As a result, these barriers are relatively transparent to carrier flow, regardless of their height, and are used for ohmic contacts, as described in Section 8.4.1.

Gallium arsenide has a relatively large energy gap (1.38 eV) compared to silicon (1.11 eV) so that barrier heights to this semiconductor are comparably larger. In consequence, the choice of metal–semiconductor combinations which result in diodes with acceptable leakage current is much greater. An extensive catalog of such combinations is available in the literature [88], but the practical choice is actually quite small, because of metallurgical incompatibility with the gallium arsenide. A few of the more common systems are listed in Table 8.4.

A silicon Schottky diode will have a high electric field at its corners, due to depletion layer curvature (see Fig. 8.22*a*). This causes premature avalanche breakdown when reverse biased, with a "soft" characteristic which varies from device to device, depending on the exact contour of the metal edge. The breakdown voltage of these devices is usually found to be about one-third of what would be obtained for a parallel-plane diode made on the same bulk semiconductor.

Two approaches are used for improving this reverse characteristic, and near-ideal diode behavior can be obtained by these means. The first consists of using a metal field plate which extends over the surface insulator. This reduces the curvature in the depletion layer edge as shown in Fig. 8.22*b*, and thus delays the onset of premature breakdown. This technique requires control of the surface insulator, which must be tapered to provide suitable depletion layer shaping. In practice, slight softening of the junction characteristic near breakdown is observed with this structure.

A second approach is to use a deep-diffused guard ring. This curves the depletion layer in the opposite direction (see Fig. 8.20*c*) to what is normally obtained, and reduces the electric field at the edge to a value below that in the parallel-plane region. In consequence, bulk breakdown prevails for this structure, with a near-ideal reverse characteristic. The diffused guard ring approach is conventionally integrated into the transistor-clamp circuit of Fig. 8.18 by means of the configuration of Fig. 8.23. This configuration is commonly used in silicon-based Schottky T^2L logic circuits.

The breakdown voltage of a gallium arsenide Schottky diode is anomalously high, and often approaches the bulk breakdown of a *p–n* junction. One reason advanced for this behavior is that the Fermi level in this semiconductor is pinned at the surface, to a value close to its intrinsic level. Thus depletion layer curvature is controlled by this pinning effect, and not by the contour of the metal edge.

At first glance, it would appear that the Schottky diode is an extremely simple structure to make, since it can be formed by physical deposition methods, with no high-temperature steps. However, it has been found that these devices need great care in their fabrication, with particular emphasis

Table 8.4 Barrier Heights of Metals on *n*-GaAs

Metal	ϕ_{Bn} (eV)
Ag	0.88–0.93
Al	0.8
Au	0.9–0.95
Pt	0.86–0.94

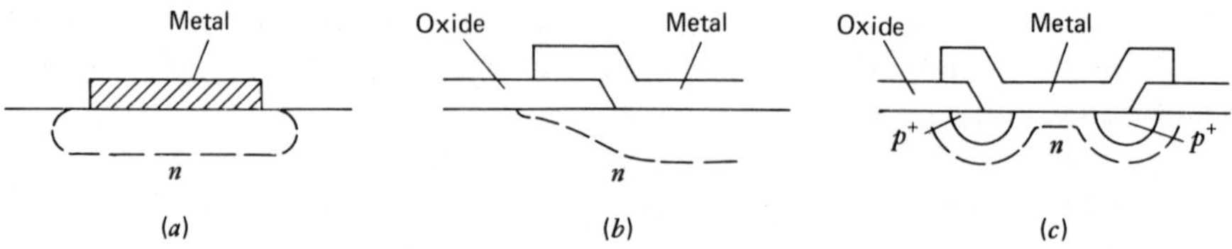

Fig. 8.22 Depletion layer curvature for a Schottky diode.

placed on cleaning the interface prior to deposition of the metal contact material. In addition, many combinations are precluded because of their low barrier heights, or for reasons of metallurgical incompatibility, as seen by changes during subsequent heat treatment, or in circuit operation. Thus the fabrication of a reliable Schottky diode can often place severe restrictions on the subsequent processing steps that can be employed.

Aluminum has been considered for silicon diodes, since it can also be used for the ohmic contacts and the metallization. Its barrier height to n-silicon is about 0.7–0.75 eV, so that it results in diodes with a turn-on voltage of about 0.2 V. The fabrication of these devices requires care in the preparation of the silicon surface; this is usually done by back sputtering or ion etching just prior to metal deposition. An important problem with this structure is that it is extremely sensitive to subsequent heat treatment [89]. This results in microalloying, thus partially converting the Schottky diode to a p^+–n junction structure. As a result, the forward voltage increases if the device is subjected to heat treatment in excess of 350°C.

Both PtSi and Pd_2Si are preferred contact materials for Schottky diodes

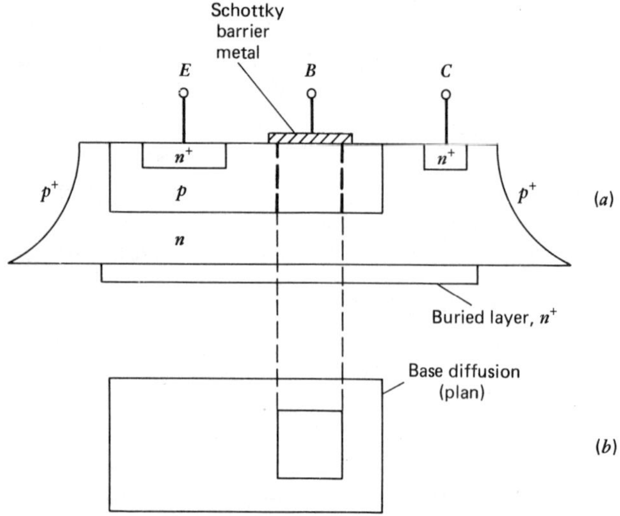

Fig. 8.23 Schottky diode realization in T^2L logic gates.

on silicon, and are formed by the same method as that described in Section 8.4.3. An important advantage is using these materials stems from the fact that they are grown by reaction of the metal with the silicon. Thus their interface is clean and relatively stable, in addition to being free from alloy spiking. The use of these materials on *n*-silicon results in diodes with threshold voltages of 0.35 V (for PtSi) and 0.25 V (for Pd_2Si), respectively.

A cover metal of aluminum can be used here, and forms an adherent contact to the silicide, in addition to serving as the interconnect metal. However, rapid diffusion of aluminum through this silicide results in a change in diode properties upon heat treatment above 350°C. A barrier layer of titanium–tungsten (20% Ti by weight) is commonly used to prevent this interdiffusion effect, and allows heat treatment to 500°C without degradation [90]. Chromium can also be used as a barrier layer in this application [91].

Schottky diodes, using PtSi and Pd_2Si, can also be used with a cover metal of gold. Here, the Ti–W layer is necessary to provide an adherent contact between the gold and the silicide, and also to allow the gold interconnection to adhere to the surface insulating layer.

Reliable Schottky diodes, which can tolerate subsequent heat treatment, are considerably more difficult to make with gallium arsenide, since metallurgical compatibility with both gallium and arsenic is a requirement that must be met simultaneously. This problem is further complicated by the fact that the chemical characteristics of gallium and arsenic are very different.

Materials such as gold and silver make excellent Schottky barriers to gallium arsenide. However, heat treatment to 260°C results in dissolution of the gallium into the contact metal [80]. This causes morphologically irregular contacts, with considerable alloy spiking, and changes in the barrier height. Often some of the gallium reaches the metal–air surface to form a thin layer of Ga_2O_3 which makes bonding difficult.

Metals such as aluminum, tungsten, molybdenum, and titanium have relatively inert interfaces with gallium arsenide, and can be used for Schottky diodes. Aluminum devices can be heat treated to 450°C for a few minutes without any change in their diode characteristics. Beyond this temperature, however, diode properties have been found to be adversely affected; in the case of aluminum metal, definite signs of AlAs formation have been noted, with a fall in the barrier height by as much as 1 eV.

Titanium and tungsten can be incorporated into multilayer metal systems for gallium arsenide. Thus a system consisting of Au–Mo–Ti–gallium arsenide, with 2000 Å thick layers of Mo and Ti [92], can tolerate heat treatment to 550°C. Beyond this point, the formation of some β-TiAs takes place. Although it has been shown [93] that this alters the barrier height by a negligible amount (from 0.82 eV for Ti to 0.84 eV for TiAs), this compound formation impairs the ability of the titanium to serve as a diffusion barrier, allowing gallium diffusion through the molybdenum film by Mo–Ga eutectic formation at 580°C, and into the gold.

The use of tungsten instead of molybdenum improves the situation, since the W–Ga eutectic temperature is around 860°C; multilayer Schottky diodes of Au–W–Ti–gallium arsenide have been found to be stable with heat treatment up to 650°C. Finally, diodes in which the titanium and tungsten are cosputtered from a Ti–W (10–20% Ti) target can also be used with an upper film of gold. This fabrication technique is somewhat more convenient. However, separate layers of these metals have been found to be slightly more stable with subsequent thermal processing.

Metals such as platinum and palladium, which form stable compounds to both gallium and arsenic, can also be used. Heat treatment of the resulting devices, in the range of 250–350°C, produces a layer of $PtAs_2$ (or $PdAs_2$), followed by a distinct layer of Pt_3Ga (or Pd_3Ga), with the arsenide layer serving as a barrier to the further rapid out-diffusion of the gallium [64]. Once these reactions are complete, these devices are relatively temperature stable up to 350°C.

The primary advantage of these metallurgically complex Schottky diode structures lies in the formation of a clean interface with the gallium arsenide; however, systems based on the more inert metals (titanium, tungsten, molybdenum, aluminum) are considerably simpler, and also superior in their thermal stability.

Cleaning procedures, prior to Schottky diode formation, are extremely important since device performance is highly sensitive to the nature of the metal–semiconductor interface. This is especially true for gallium arsenide, where the commonly used gate metals (titanium, tungsten, molybdenum, and aluminum) do not react with the semiconductor. Here sputter etching procedures cannot be used since they result in selective loss of surface arsenic, and hence in arsenic vacancies. Moreover, such damaged surfaces cannot be annealed by subsequent heat treatment; rather, the vacancy distribution becomes spatially altered during annealing. As a result, wet chemical etching procedures are commonly used for this purpose. Details of some of the more successful of these are given in Section 9.4.1.

Up to the present time, most research has been aimed at developing gallium arsenide Schottky diodes with the ability to withstand high temperatures without metallurgical degradation. Recent work with deep-level transient spectroscopy [94] has shown that a number of the materials used with these devices introduce multiple deep levels into the semiconductor, with a consequent degradation of its electrical properties. Thus an "ideal" Schottky diode technology for gallium arsenide has not been achieved at the present time.

REFERENCES

1. J. L. Vossen and W. Kern, Eds., *Thin Film Processes*, Academic, New York, 1978.
2. C. R. Barnes and C. R. Geesner, Pyrolytic Deposition of Silicon Dioxide for 600°C Thin Film Capacitors, *J. Electrochem. Soc.* **110**, 361 (1963).

3. N. Goldsmith and W. Kern, The Deposition of Vitreous Silicon Dioxide from Silane, *RCA Rev.* **28,** 153 (1967).
4. B. J. Baliga and S. K. Ghandhi, Growth of Silica and Phosphosilicate Films, *J. Appl. Phys.* **44,** 990 (1973).
5. B. Mattson, CVD Films for Interlayer Dielectrics, *Solid State Technol.*, p. 60 (Jan. 1980).
6. M. M. Schlacter, E. S. Schlegel, R. S. Kan, R. A. Lathlaen, and G. L. Schnable, Advantages of Vapor-Plated Phosphosilicate Glass Films in Large Scale Integrated Circuit Arrays, *IEEE Trans. Electron Dev.* **ED-17,** 1077 (1970).
7. W. Kern, G. L. Schnable, and A. W. Fisher, CVD Glass Films for Passivation of Silicon Devices: Preparation, Composition, and Stress Properties, *RCA Rev.* **37,** 3 (1976).
8. B. J. Baliga and S. K. Ghandhi, Lateral Diffusion of Zinc and Tin in Gallium Arsenide, *IEEE Trans. Electron Dev.* **ED-21,** 410 (1974).
9. S. R. Shortes, J. A. Kanz, and E. L. Wurst, Zinc Diffusion in GaAs through SiO_2 Films, *Trans. AIME* **230,** 300 (1964).
10. B. J. Baliga and S. K. Ghandhi, PSG Masks for Diffusion in Gallium Arsenide, *IEEE Trans. Electron Dev.* **ED-19,** 761 (1972).
11. L. K. White, Bilayer Etching of Field Oxides and Passivation Layers, *J. Electrochem. Soc.* **127,** 2687 (1980).
12. M. Yamin, Observations on Phosphorus Stabilized SiO_2 Films, *IEEE Trans. Electron Dev.* **ED-13,** 256 (1966).
13. R. B. Comizzoli, Aluminum Corrosion in the Presence of Phosphosilicate Glass and Moisture, *RCA Rev.* **37,** 483 (1976).
14. T. Ito, S. Hijiya, T. Nozaki, H. Arakawa, M. Shinoda, and Y. Fukukawa, Very Thin Silicon Nitride Films Grown by Direct Thermal Reaction with Nitrogen, *J. Electrochem. Soc.* **125,** 448 (1978).
15. G. J. Kominiak, Silicon Nitride Films by Direct RF Sputter Deposition, *J. Electrochem. Soc.* **122,** 1271 (1975).
16. R. Ginsburgh, D. L. Heald, and R. C. Neville, Silicon Nitride Chemical Vapor Deposition in a Hot Wall Diffusion System, *J. Electrochem. Soc.* **125,** 1557 (1978).
17. W. Kern and G. L. Schnable, Low-Pressure Chemical Vapor Deposition for Very-Large-Scale Integration Processing—A Review, *IEEE Trans. Electron Dev.* **ED-26,** 647 (1979).
18. H. F. Sterling and R. C. G. Swann, Chemical Vapour Deposition Promoted by rf Discharge, *Solid State Electron.* **8,** 653 (1965).
19. R. S. Rosler, W. C. Bensing, and J. Baldo, A Production Reactor for Low Temperature Plasma-Enhanced Silicon-Nitride Deposition, *Solid State Technol.*, p. 45 (June 1976).
20. A. K. Sinha, H. J. Levinstein, T. E. Smith, G. Quintana, and S. E. Haszko, Reactive Plasma Deposited Si–N Films for MOS-LSI Passivation, *J. Electrochem. Soc.* **125,** 601 (1978).
21. P. M. Petroff, G. A. Rozgonyi, and T. T. Sheng, Elimination of Process-Induced Stacking Faults by Preoxidation Gettering of Si Wafers, *J. Electrochem. Soc.* **123,** 565 (1976).
22. D. M. Brown, P. V. Gray, F. K. Heumann, H. R. Philipp, and E. A. Taft, Properties of $Si_xO_yN_z$ Films on Si, *J. Electrochem. Soc.* **115,** 311 (1968).
23. M. L. Tarng and J. I. Pankove, Passivation of *p–n* Junction in Crystalline Silicon by Amorphous Silicon, *IEEE Trans. Electron Dev.* **ED-26,** 1728 (1979).
24. J. I. Pankove and D. E. Carlson, Photoluminescence of Hydrogenated Amorphous Silicon, *Appl. Phys. Lett.* **31,** 450 (1977).
25. J. I. Pankove, M. A. Lampert, and M. L. Tarng, Hyodrogenation and Dehydrogenation of Amorphous and Crystalline Silicon, *Appl. Phys. Lett.* **32,** 439 (1978).
26. F. C. Eversteyn and B. H. Put, Influence of AsH_3, PH_3, and B_2H_6 on the Growth Rate

and Resistivity of Polycrystalline Films Deposited from a SiH_4-H_2 Mixture, *J. Electrochem. Soc.* **120,** 106 (1973).

27. H. Yamamoto, T. Wada, O. Kudoh, and M. Sakamoto, Polysilicon Interconnection Technology for IC Device, *J. Electrochem. Soc.* **126,** 1415 (1979).
28. J. Y. W. Seto, The Electrical Properties of Polycrystalline Silicon Films, *J. Appl. Phys.* **46,** 5247 (1975).
29. R. S. Rosler, Low Pressure CVD Production Processes for Poly, Nitride, and Oxide, *Solid State Technol.,* p. 63 (Apr. 1977).
30. T. I. Kamins, Resistivity of LPCVD Polycrystalline Silicon Films, *J. Electrochem. Soc.* **126,** 833 (1979).
31. H. R. Maxwell, Jr., and W. R. Knolle, Densification of SIPOS, *J. Electrochem. Soc.* **128,** 576 (1981).
32. T. Matsushita, T. Aoki, T. Ohtsu, H. Yamoto, H. Hayashi, and M. Okayama, Highly Reliable High-Voltage Transistors by Use of the SIPOS Process, *IEEE Trans. Electron Dev.* **ED-23,** 826 (1976).
33. S. P. Murarka, Refractory Silicides for Integrated Circuits, *J. Vac. Sci. Technol.* **17,** 775 (1980).
34. S. Zermsky, W. Hammer, F. d'Heurle, and J. Baglin, Oxidation Mechanisms in WSi_2 Thin Films, *Appl. Phys. Lett.* **33,** 76 (July 1, 1978).
35. B. L. Crowder and S. Zirinsky, 1 μm MOSFET VLSI Technology: Part VII, Metal Silicide Interconnection Technology—A Future Perspective, *IEEE Trans. Electron Dev.* **ED-26,** 369 (1979).
36. D. M. Brown, W. E. Engeler, M. Garfinkel, and P. V. Gray, Self-Registered Molybdenum-Gate MOSFET, *J. Electrochem. Soc.* **115,** 874 (1968).
37. F. Yanagawa, K. Kiuchi, T. Hosoya, T. Tsuchiya, T. Amazawa, and T. Mano, A 1 μm Mo-Gate 64 kbit MOS RAM, *Transactions of the IEEE Electron Device Conference,* Washington, D.C., 1979, p. 362.
38. C. F. Powell, J. H. Oxley, and J. M. Blocher, Eds., *Vapor Deposition,* Wiley, New York, 1966.
39. J. G. Donaldson and H. Kenworthy, Vapor Deposition of Molybdenum-Tungsten Alloys, Bureau of Mines Rep. 6853, 1966.
40. T. Sugano, H-K. Chou, M. Yoshida, and T. Nishi, Chemical Deposition of Mo on Si, *Jpn. J. Appl. Phys.* **7,** 1028 (1968).
41. J. L. Lander and L. H. Germer, Plating Molybdenum, Tungsten and Chromium, *Metals Technol.,* p. 648 (Sept. 1947).
42. J. M. Shaw and J. A. Amick, Vapor-Deposited Tungsten as a Metallization and Interconnection Material for Silicon Devices, *RCA Rev.* **30,** 306 (1970).
43. M. L. Barry and P. Olofsen, Doped Oxides as Diffusion Sources—I. Boron into Silicon, *J. Electrochem. Soc.* **116,** 854 (1969).
44. M. L. Barry, Doped Oxides as Diffusion Sources—II. Phosphorus into Silicon, *J. Electrochem. Soc.* **117,** 1405 (1970).
45. J. Scott and J. Olmstead, A Solid-to-Solid Diffusion Technique, *RCA Rev.* **26,** 357 (1965).
46. A. W. Fisher, J. A. Amick, H. Hymann, and J. H. Scott, Jr., Diffusion Characteristics and Applications of Doped Silicon Dioxide Layers Deposited from Silane, *RCA Rev.* **29,** 533 (1968).
47. J. R. Shealy, B. J. Baliga, and S. K. Ghandhi, Preparation and Properties of Zinc Oxide Films Grown by the Oxidation of Diethylzinc, *J. Electrochem. Soc.* **128,** 558 (1981).
48. J. R. Shealy, B. J. Baliga, and S. K. Ghandhi, Open Tube Diffusion of Zinc in Gallium Arsenide, *IEEE Electron Dev. Lett.* **ED-1,** 119 (1980).

49. S. K. Ghandhi and R. J. Field, Precisely Controlled p^+-Diffusion in GaAs, *Appl. Phys. Lett.* **38,** 267 (1981).
50. W. von Muench, Gallium Arsenide Planar Technology, *IBM J. Res. Dev.* **10,** 438 (1966).
51. H. Yamazaki, K. Kawasaki, M. Fugimoto, and K. Kudo, Tin Diffusion into Gallium Arsenide from Doped SiO_2 Films, *J. Appl. Phys.* **14,** 717 (1975).
52. B. J. Baliga and S. K. Ghandhi, The Preparation and Properties of Tin Oxide Films Formed by Oxidation of Tetramethyltin, *J. Electrochem. Soc.* **123,** 941 (1976).
53. B. J. Baliga and S. K. Ghandhi, Planar Diffusion in Gallium Arsenide from Tin-Doped Oxides, *J. Electrochem. Soc.* **126,** 135 (1979).
54. Integrated Circuits Course, Integrated Circuit Engineering Corp., *Electron Eng.* p. 63, (1966).
55. B. Selikson and T. Longo, A Study of Purple Plague and its Role in Integrated Circuits, *Proc. IEEE* **52,** 1638 (1964).
56. G. L. Schnable and R. S. Keen, Aluminum Metallization—Advantages and Limitations for Integrated Circuit Applications, *Proc. IEEE* **57,** 1570 (1971).
57. A. J. Learn, Evolution and Current Status of Aluminum Metallization, *J. Electrochem. Soc.* **123,** 894 (1976).
58. H. B. Huntington and A. R. Grone, Current Induced Marker Motion in Gold Wires, *J. Phys. Chem. Solids* **20,** 76 (1961).
59. F. d'Heurle, Electromigration and Failure in Electronics: An Introduction, *Proc. IEEE* **59,** 1409 (1971).
60. J. R. Black, Electromigration Failure Modes in Aluminum Metallization for Semiconductor Devices, *Proc. IEEE* **57,** 1587 (1969).
61. H.-U. Schreiber, Activation Energies for the Different Electromigration Mechanisms in Aluminum, *Solid State Electron.* **24,** 583 (1981).
62. I. Ames, F. M. d'Heurle, and R. Horstmann, Reduction of Electromigration in Aluminum Films by Copper Doping, *IBM J. Res. Dev.* **14,** 461 (1970).
63. L. E. Terry and R. W. Wilson, Metallization Systems for Silicon Integrated Circuits, *Proc. IEEE* **57,** 1580 (1969).
64. J. M. Poate, K. N. Tu, and J. W. Mayer, *Thin Films-Interdiffusion and Reactions*, Wiley, New York, 1978.
65. S. P. Murarka, H. J. Levenstein, I. Bleck, T. T. Sheng, and M. H. Reed, Investigation of the Ti-Pt Diffusion Barrier for Gold Beam Leads on Aluminum, *J. Electrochem. Soc.* **125,** 156 (1978).
66. R. W. Lade and A. G. Jordan, A Study of Ohmicity and Exclusion in High-Low Semiconductor Devices, *IEEE Trans. Electron Dev.* **ED-10,** 268 (1963).
67. C. Y. Chang, Y. K. Fang, and S. M. Sze, Specific Contact Resistance of Metal Semiconductor Barriers, *Solid State Electron.* **14,** 541 (1971).
68. R. Holm, *Electric Contacts: Theory and Application*, Springer-Verlag, New York, 1967.
69. V. L. Rideout, A Review of the Theory and Technology for Ohmic Contacts to Group III-V Compound Semiconductors, *Solid State Electron.* **18,** 541 (1975).
70. C. R. M. Grovenor, Au/Ge Based Ohmic Contacts to GaAs, *Solid State Electron.* **24,** 792 (1981).
71. J. O. McCaldin and H. Sankur, Diffusivity and Solubility of Si in the Al Metallization of Integrated Circuits, *Appl. Phys. Lett.* **19,** 524 (1971).
72. J. M. Andrews, The Role of the Metal–Semiconductor Interface in Silicon Integrated Circuit Technology, *J. Vac. Sci. Techn.* **11,** 972 (1974).
73. C. J. Kircher, Metallurgical Properties and Electrical Characteristics of Palladium Silicide Contacts, *Solid State Electron.* **14,** 507 (1971).

74. M. P. Lepselter, Beam-Lead Technology, *Bell Sys. Tech. J.* **45,** 233 (1966).

75. J. H. Forster and J. B. Singleton, Beam-Lead Sealed-Junction Integrated Circuits, *Bell. Lab. Rec.* **44,** 312 (Oct.-Nov. 1966).

76. P. H. Holloway, Gold/Chromium Metallizations for Electronic Devices, *Solid State Tech.*, p. 109 (Feb. 1980).

77. M. Finetti, P. Ostoja, S. Solmi, and G. Soncini, Aluminum-Silicon Ohmic Contact on Shallow n^+/p Junctions, *Solid State Electron.* **23,** 255 (1980).

78. W. T. Anderson, Jr., A. Christou, and J. E. Davey, Development of Ohmic Contacts for GaAs Devices Using Epitaxial Ge Films, *IEEE J. Solid State Circuits* **SC-13,** 430 (1978).

79. M. N. Yoder, Ohmic Contacts in GaAs, *Solid State Electron.* **23,** 117 (1980).

80. D. C. Miller, The Alloying of Gold and Gold Alloy Ohmic Contact Metallizations with Gallium Arsenide, *J. Electrochem. Soc.* **127,** 467 (1980).

81. W. J. Delvin, C. E. C. Wood, R. Stall, and L. F. Eastman, A Molybdenum Source, Gate and Drain Metallization System for GaAs MESFET Layers Grown by Molecular Beam Epitaxy, *Solid State Electron.* **23,** 823 (1980).

82. S. Margalit, D. Fekete, D. M. Pepper, C.-P. Lee, and A. Yariv, *Q*-Switched Ruby Laser Alloying of Ohmic Contacts on Gallium Arsenide Epilayers, *Appl. Phys. Lett.* **33,** 346 (1978).

83. F. A. Lowenheinm, Ed., *Modern Electroplating*, Wiley, New York, 1974.

84. A. Brenner and G. Riddell, Deposition of Nickel and Cobalt by Chemical Reduction, *J. Res. NBC,* **39,** 385 (1947).

85. R. H. Baker, Maximum Efficiency Switching Circuits, MIT Lincoln Lab. Rep. TR-110, 1956.

86. A. Tarui, Y. Hayashi, H. Teshima, and T. Sekigawa, Transistor Schottky-Barrier-Diode Integrated Logic Circuit, *IEEE J. Solid State Circuits* **SC-4,** 3 (1969).

87. S. M. Sze, *Physics of Semiconductor Devices*, Wiley, New York, 1981.

88. B. L. Sharma and S. C. Gupta, Metal Semiconductor Schottky Barrier Junctions—Part I, *Solid State Tech.*, p. 97 (May 1980); Part II, p. 90 (June 1980).

89. K. Chino, Behavior of Al–Si Schottky Barrier Diodes under Heat Treatment, *Solid State Electron.* **16,** 119 (1973).

90. P. C. Parekh, R. C. Sirrine, and P. Lemieux, Behavior of Various Silicon Schottky Barrier Diodes Under Heat Treatment, *Solid State Electron.* **19,** 493 (1976).

91. J. O. Olowolafe, M.-A. Nicolet, and J. W. Mayer, Chromium Thin Film as a Barrier to the Interaction of Pd_2Si with Al, *Solid State Electron.* **20,** 413 (1977).

92. E. Kohn, High Temperature Stable Metal-GaAs Contacts, *Proceedings of the 1979 International Electron Devices Conference*, Washington, D.C., p. 469.

93. A. K. Sinha, T. E. Smith, M. H. Read, and J. M. Poate, *n*-GaAs Schottky Diodes Metallized with Ti and Pt/Ti, *Solid State Electron.* **19,** 489 (1976).

94. D. L. Partin, A. G. Milnes, and L. F. Vassamillet, Hole Diffusion Lengths in VPE GaAs and $GaAs_{0.6}P_{0.4}$ Treated with Transition Metals, *J. Electrochem. Soc.* **126,** 1584 (1979).

CHAPTER

9

Etching and Cleaning

CONTENTS

9.1 WET CHEMICAL ETCHING 477

9.1.1 Crystalline Materials, 478
9.1.2 Application to Silicon, 479
9.1.3 Application to Gallium Arsenide, 482
9.1.4 Anisotropic Etches, 487

9.1.4.1 Silicon, 488
9.1.4.2 Gallium Arsenide, 490

9.1.5 Crystallographic Etches, 490
9.1.6 Non-crystalline Films, 492

9.1.6.1 Silica Films, 493
9.1.6.2 Phosphosilicate Glass (PSG), 494
9.1.6.3 Mixed Oxides, 495
9.1.6.4 Silicon Nitride, 495
9.1.6.5 Polysilicon and Semi-Insulating Polysilicon, 496
9.1.6.6 Silicides, 497
9.1.6.7 Metals, 498

9.2 PLASMA ETCHING 499

9.2.1 Resist Removal, 502
9.2.2 Loading Effects [55], 503
9.2.3 Pattern Delineation, 504

9.3 PLASMA-ASSISTED ETCHING 510

9.3.1 Sputter Etching, 512
9.3.2 Ion Milling, 514

9.4 CLEANING 517

9.4.1 Schottky Diodes and Gates, 520

TABLES 522
REFERENCES 527
PROBLEMS 531

Etching and cleaning processes are involved at many points in the microcircuit fabrication process. Thus saw-cut slices of suitably oriented semiconductor material are first mechanically lapped to remove gross damage, and then chemically etched and polished to obtain an optically flat, damage-free surface. Often this involves the removal of many micrometers of surface material. Next, slices are chemically cleaned and scrubbed to remove contaminants produced by handling and storing, before they are covered with an initial protective layer of thermally grown silica (for silicon-based microcircuits), or deposited silicon nitride (for gallium arsenide circuits). Etching processes, in conjunction with patterning, are used to cut openings in this protective film through which implants or diffusions are made to form the semiconductor regions. This process is repeated until all the components are formed.

Further etching and patterning processes are used to delineate one or more layers of metallization. This is often followed by protective coatings of deposited PSG or silicon nitride. Again, both etching and lithographic processes are used to cut holes for access to the bonding pads, to which leads can be attached.

The term etching is used to describe all techniques by which material can be uniformly removed from a wafer as in surface polishing, or locally removed as in the delineation of a pattern for a microcircuit. It also includes chemical machining of a semiconductor as part of the fabrication process, as well as the delineation of surface features such as defects.

Chemical etching includes the simple dissolution of a material in a solvent as well as the conversion of the material into a soluble compound which can be dissolved by the etching medium. This process is sometimes carried out in the gas phase at high temperature, and has been described previously as part of an important deposition process (e.g., halogen etching of the semiconductor surface prior to epitaxy).

Low-temperature gaseous etching processes can also be conducted, and play an important role in fabrication technology. One such technique is *plasma etching*, where the material to be removed undergoes conversion to a volatile state by chemical reaction with one or more energetic species which are produced in a gas discharge.

A number of etching processes involve material removed by momentum transfer from a rapidly moving inert projectile (usually argon). *Ion milling* is an important mechanical etching process of this type, and has been appropriately referred to as "sandblasting on an atomic scale" [1]. Another often used process is *sputtering*, which is similar in its essential features to ion milling. Both of these techniques require the formation of a gas discharge to produce the high-velocity argon ions. However, energetic species are not intentionally involved in etching processes. Thus these are referred to as *plasma-assisted* processes.

The term *reactive ion etching* is used to describe systems which combine the use of energetic species with ion bombardment. The dominant effect of ion bombardment is damage of the surface, which can be chemically removed by the energetic species. Here the atom–substrate reactions of the energetic species are enhanced by the ion bombardment, so that the etch rate is usually greater than the sum of the two separate processes.

This chapter describes the many etching processes that are used in the fabrication of microcircuits. Both wet and dry processes are considered, the latter being especially important for VLSI fabrication technology. Patterning processes, which are often used in conjunction with etching are described in the following chapter.

The cleaning of semiconductor surfaces is treated in this chapter. This cleaning is necessary to remove a variety of organic and inorganic contaminants, often of unknown origin, which are present on the as-purchased semiconductor surface. These include films of native oxide, airborne contaminants such as common salt and bacteria, and chemical contaminants such as oil and plasticizer films from the packages in which wafers are shipped. Cleaning between processing steps is also important in order to avoid the introduction of contaminants.

Many of these cleaning processes are closely related to etching, in that they involve some removal of the semiconductor surface; however, this is not their primary intent. They play an important role in microcircuit fabrication technology.

9.1 WET CHEMICAL ETCHING

The wet chemical etching of a material consists both of transporting etchant and reaction products to and from its surface, respectively, and also of the surface reaction itself. If the first of these is the limiting process, the reaction is said to be diffusion limited. Adsorption and desorption processes also affect the rate of wet chemical etching, and may be the limiting factors in the overall process.

Most etching processes in semiconductor technology are carried out at a relatively slow, controlled rate. Consequently, they tend to be limited by the rate of diffusion of the reactant through a stagnant layer which covers the surface. This stagnant layer is usually a few micrometers thick, and can be broken up if the chemical reaction results in the evolution of a gas.

Wet chemical etching processes are often limited by the rate of dissolution of the reaction products into the solution. Here agitation of the solution helps in increasing the etch rate by enhancing this out-diffusion effect.

The etching of polycrystalline and amorphous materials is isotropic; however, etches for crystalline materials may be isotropic or anisotropic, depending on the nature of the reaction kinetics. Isotropic etches for crystalline materials are often called polishing etches, since they result in a smooth surface. Anisotropic etches often sharply delineate planes or surface defects. They are used in chemical machining applications, and also as crystallographic etches.

A large variety of chemical reagents, and their mixtures, are used for etching purposes. Many of these are now available in "electronic grade" purity and are preferred in order to avoid inadvertent contamination of the semiconductor during processing. Deionized water is invariably used as a diluent, for the same reason. The compositions of commonly used aqueous reagents are given in Table 9.1. This table, as well as others on this topic, are collected at the end of this chapter for ready reference.

9.1.1 Crystalline Materials

Single-crystal silicon and gallium arsenide fall into this category; both exhibit long- as well as short-range order, and their etching can be either isotropic or anisotropic in character.

Etching by direct chemical dissolution can be achieved with these materials. This is most commonly done in the gas phase, at elevated temperatures, as in the preparation of a semiconductor surface prior to epitaxial growth (see Chapter 5). Typical etchants are the halogens and their compounds. Thus HCl and SF_6 are used in the pre-epitaxy etching of silicon, whereas HCl and $AsCl_3$ are used with gallium arsenide.

Low-temperature etching of these semiconductors, using wet chemicals, usually proceeds by their oxidation, followed by the dissolution of the oxide (or oxides) by a chemical reaction. Both of these processes are carried out simultaneously by a mixture of the reagents in the same etching solution. The oxidation chemistry is identical to that of anodic oxidation described in Section 7.3. Here, however, no clearly defined anode or cathode is established by a battery. Rather, points on the surface of the semiconductor behave randomly as localized anodes and cathodes. The oxidation reaction proceeds from the action of these localized electrolytic cells and gives rise to relatively large corrosion currents, often in excess of 100 A/cm^2.

Over a period of time each localized area (which is large compared to atomic dimensions) adopts the role of both anode and cathode. If the proportion of time allocated to each role is roughly equal, uniform etching occurs. Conversely, selective etching occurs if these times are very different. Such factors as the defect nature of the semiconductor surface, the etchant temperature, impurities in the etchant, and adsorption processes at the semiconductor–etchant interface play an important role in determining the degree of selectivity of the etchant as well as its etch rate. A brief review of the oxidation process now follows [1].

Consider a localized anodic site. Here the semiconductor is promoted from its initial oxidation state to some higher oxidation state, as given by

$$M^0 + xh^+ \rightleftharpoons M^{x+} \tag{9.1}$$

where M represents the semiconductor. This oxidation reaction requires holes for its execution.

Reduction occurs simultaneously at a localized cathodic site, and is accompanied by the liberation of holes. Writing N^0 as the oxidizing species,

$$N^0 \rightleftharpoons N^{x-} + xh^+ \tag{9.2}$$

The entire reaction, which is charge neutral, is given by

$$M^0 + N^0 \rightleftharpoons M^{x+} + N^{x-} \tag{9.3}$$

In semiconductor etching, the primary oxidizing species is $(OH)^-$. This is usually formed by the dissociation of water, which is present in the etchant, as given by

$$H_2O \rightleftharpoons (OH)^- + H^+ \tag{9.4}$$

The formation of an oxide presents a barrier to further oxidation of the semiconductor, so that it is necessary to add additional chemicals for its dissolution into water soluble compounds or complexes. Stirring removes these from the semiconductor surface so that further oxidation can proceed. The choice of chemicals for this purpose is quite wide. Both acids and bases can be used, in addition to salts involving CN and NH_4 groupings. In practical systems, the choice is limited by the availability of high-purity reagents, and the desire to avoid metallic ion contamination. Thus hydrofluoric acid is the invariable choice for silicon etching systems. Gallium arsenide systems often use sulfuric, phosphoric, and citric acid, or ammonium hydroxide.

9.1.2 Application to Silicon

The most commonly used etchants for silicon are mixtures of HNO_3 and HF in water or acetic acid. Here the anode reaction is given [2] by

$$Si + 2h^+ \rightarrow Si^{2+} \tag{9.5}$$

This Si^{2+} combines with $(OH)^-$ so that

$$Si^{2+} + 2(OH)^- \rightarrow Si(OH)_2 \tag{9.6}$$

which subsequently liberates hydrogen to form SiO_2,

$$Si(OH)_2 \rightarrow SiO_2 + H_2 \tag{9.7}$$

Hydrofluoric acid can be used to dissolve this SiO_2, the reaction being given by

$$SiO_2 + 6HF \rightarrow H_2SiF_6 + 2H_2O \tag{9.8}$$

Stirring serves to remove the soluble complex H_2SiF_6 from the vicinity of the silicon slice. This reaction is referred to as a *complexing reaction* for this reason.

In anodic oxidation, the supply of holes for reaction (9.5) is permitted by the battery. In this system, these holes are produced by the reduction of NO_2 at a localized cathode. This reaction is *autocatalytic* [3], in that the reaction products promote the reaction itself. It proceeds in the presence of trace impurities of HNO_2, as shown,

$$HNO_2 + HNO_3 \rightarrow N_2O_4 + H_2O \tag{9.9a}$$

$$N_2O_4 \rightleftharpoons 2NO_2 \tag{9.9b}$$

$$2NO_2 \rightleftharpoons 2NO_2^- + 2h^+ \tag{9.9c}$$

$$2NO_2^- + 2H^+ \rightleftharpoons 2HNO_2 \tag{9.9d}$$

The HNO_2 generated in reaction (9.9d) reenters into reaction with HNO_3 in reaction (9.9a), and the process is thus autocatalyzed.* The first of these reactions is the rate-limiting one; in some cases NO_2^- ions are deliberately added (in the form of ammonium nitrite) to induce the reaction. Since the HNO_2 is regenerated in this reaction, the oxidizing power is a function of the amount of undissociated HNO_3.

Water can be used as a diluent for this etchant. However, acetic acid is preferred because of its lower dielectric constant (6.15 as compared to 81 for water). Use of this diluent results in less dissociation of the nitric acid, and hence in a higher concentration of the undissociated species. This preserves the oxidizing power of the HNO_3 for a wider range of dilution than if water were used. Thus the oxidizing power of the etchant tends to remain relatively constant during its operating life, which is an advantage in commercial microcircuit fabrication.

The overall etching process is preceded by an induction period during which the autocatalysis of HNO_3 is initiated. This induction period is followed by the cathodic reduction of HNO_3, resulting in a source of holes

* Note that reaction (9.9c) proceeds in the presence of the silicon surface, and not in the volume of the etchant.

which enter into the oxidation reaction. The oxidation products are reacted in hydrofluoric acid to form the soluble complex H_2SiF_6. All of these processes occur within a single etch mixture, resulting in the overall reaction

$$Si + HNO_3 + 6HF = H_2SiF_6 + HNO_2 + H_2O + H_2 \qquad (9.10)$$

Extensive studies [4–7] of the HF-HNO_3 system have been made using both water and acetic acid as the diluents. Rapid stirring was used to prevent the formation of localized hot spots and to present the silicon surface with fresh etchant at all times. The uncertain induction period (for certain compositions) was avoided by the addition of nitrite ions in the form of NH_4NO_2. Figure 9.1 shows their results in the form of isoetch curves for the various constituents by weight. It should be noted here that normally available concentrated acids are 49.2 wt. % HF and 69.5 wt. % HNO_3, respectively. In addition, the etch rates given in this figure represent silicon removal on both sides of the wafer. Thus they must be divided by 2 to give the thickness removed from each surface.

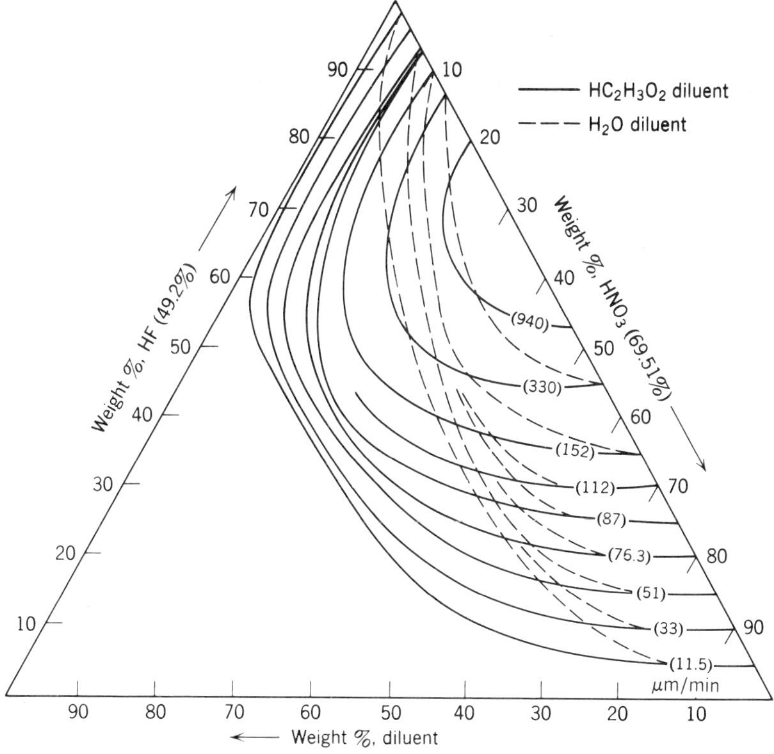

Fig. 9.1 Isoetch curves for silicon (HF:HNO_3:diluent system). From Robbins and Schwartz [5]. Reprinted with permission of the publisher, The Electrochemical Society, Inc.

Either water or acetic acid may be used as the diluent for the system. Qualitatively, both show similar behavior. Common to both systems are the following characteristics:

1. At low HNO_3 and high HF concentrations, corresponding to the region near the upper vertex of Fig. 9.1, the etching contours run parallel to the lines of constant HNO_3. Thus the etch rate is controlled by the HNO_3 concentration in this region. This is due to the fact that there is an excess of HF to dissolve the SiO_2 formed during the reaction. Etching with these formulations tends to be difficult to initiate because of an uncertain induction period. In addition, they result in relatively unstable silicon surfaces, which proceed to slowly grow a layer of SiO_2 over a period of time. Finally, the etch is limited by the rate of the oxidation–reduction reaction, so that it tends to be somewhat orientation dependent. However, this effect is slight, since etching is relatively rapid, and is accompanied by the liberation of heat.
2. In the region of the lower right vertex (low HF and high HNO_3 concentrations), the etch-rate contours are parallel to the lines of constant HF. Here there is an excess of HNO_3 and the etch rate is governed by the ability of the HF to remove the SiO_2 as it is formed. These etches are "self-passivating" in that a freshly etched surface is already covered with a relatively thick layer of SiO_2 (30–50 Å). They are used extensively in the fabrication of nonplanar microwave devices where they are known as "mesa" etches because of the resulting shape of the device. In addition, the primary limit on the etch rate is the rate of removal of the complexes by diffusion. Consequently, etches in this region are not sensitive to crystallographic orientation, and are true polishing etches.
3. Etch formulations in the HF : HNO_3 = 1 : 1 range are initially insensitive to the addition of diluent. Eventually, they fall off very sharply in etch rate, until the system becomes critical with respect to the diluent.

As seen from Fig. 9.1, an almost infinite choice of formulations can be used for silicon etching. A number of these are available in premixed form, and have been well characterized. Some are listed in Table 9.2. Also included in this table are a few etchant formulations which are based on such oxidizing agents as Br_2 [8], I_2 [9], and $KMnO_4$ [10].

9.1.3 Application to Gallium Arsenide

A wide variety of etches have been investigated for gallium arsenide; however, very few of them are truly isotropic. This is because, unlike silicon, the surface activity of the (111) Ga and (111) As faces is very different. The arsenic face, terminated on arsenic, has two unsatisfied bonds per atom. Consequently, although some reordering occurs in the surface layer, it is still considerably more reactive than the gallium face, and so etches at a faster rate. As a result, most etches give a polished surface on the arsenic face. The gallium face etches much more slowly and tends to show up surface features and crystallographic defects. Often the appearance of an etched (111) Ga face appears cloudy or frosted for this reason.

The details of the oxidation reactions are given in Chapter 7, but are summarized here [2]. Immersion into the electrolyte system results in electronic charge transfer from the semiconductor to the electrolyte, as the separate Fermi levels line up, so that

$$GaAs + 6h^{+} = Ga^{3+} + As^{3+} \tag{9.11}$$

Reaction with $(OH)^{-}$ ions in the electrolyte occurs in a sequence of steps, outlined in Eqs. (7.32) and (7.33), to produce Ga_2O_3 and As_2O_3. Dissolution of these oxides occurs in acids or bases which are part of the etchant formulation, to form soluble salts or complexes.

One of the earliest (and still very popular) etching systems for gallium arsenide is based on the use of small concentrations of bromine in methanol [11, 12]. Here the $(OH)^{-}$ ion is provided by the methanol, while bromine serves the dual role of being a strong oxidizing agent, and also dissolving the oxidation products to form soluble bromides.

The Br_2–CH_3OH system can be used over a wide range of concentrations. At low concentrations, its removal rate is linearly proportional to the bromine content. Typically, a removal rate of about 0.075 μm/min can be obtained with 0.05 vol. % Br_2 in CH_3OH. Higher bromine concentrations, up to 10%, can also be used to obtain a high degree of surface polish, and are used in the removal of saw-cut damage.

The etch rate for the Br_2–CH_3OH system is different for the various planes, with the (111) As plane etching at the fastest rate, and the (111) Ga at the slowest [13]. Typically, etch rates are (111) As:(100):(111) Ga = 3:2:1. More uniform etch rates are observed with the faster etching formulations, i.e., those with a high bromine content.

The Br_2–CH_3OH system can be used for polishing all principal crystal faces except the (111) Ga. Here its relatively slow etch rate results in preferential etching and feature delineation. The addition of Syton*, in conjunction with mechanical lapping, allows polishing of this face as well. A 1 ml Br_2, 20 ml CH_3OH, 300 ml Syton formulation has been used for this purpose, with an etch rate of about 0.2 μm/min [14].

Many etch systems for gallium arsenide are based on the use of H_2O_2, which is strongly oxidized, in combination with an acid or base to dissolve the oxidation products. Sulfuric, phosphoric, nitric, hydrochloric, and citric acids, as well as ammonia and sodium hydroxide, have all been used for this purpose. The properties of some of these etching systems are now described.

* Colloidal silica suspension (Remet Chemical Corp., New York).

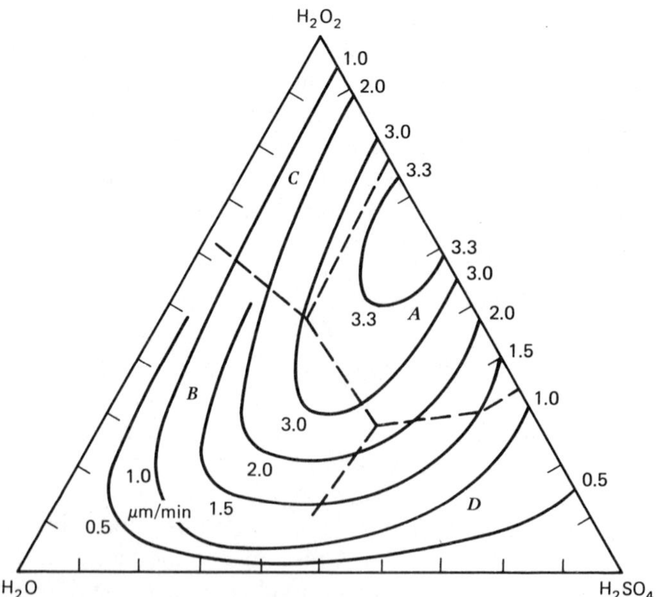

Fig. 9.2 Isoetch curves for gallium arsenide ($H_2SO_4:H_2O_2:H_2O$ system). From Iida and Ito [15]. Reprinted with permission of the publisher, The Electrochemical Society, Inc.

The H_2SO_4–H_2O_2–H_2O System. This system can be used in a wide variety of formulations, as shown by the 0°C isoetch curve of Fig. 9.2 [15]. Etch rates for this system increase exponentially with temperature, so that room temperature values are about three to five times as large as those in this figure. Formulations with either high H_2SO_4 or high H_2O_2 content fall into regions *C* and *D* in this figure, and result in surfaces with mirror finish. Formulations in region *A*, involving large amounts of both these chemicals, etch very rapidly and result in a cloudy appearance. Region *B* results in very slow etch rates, and can be used to delineate surface defects such as etch pits.

Typically, an 8:1:1 volume ratio of $H_2SO_4:H_2O_2:H_2O$ results in an etch rate of 0.8 μm/min for the (111) Ga face and 1.5 μm/min for all other faces.* Etches with a low $H_2SO_4:H_2O_2$ ratio tend to be somewhat more anisotropic. Thus a 1:8:1 volume ratio of $H_2SO_4:H_2O_2:H_2O$ results in etch rates of 3:8:8:12 μm/min for the (111)Ga, (100), (110), and (111)As faces, respectively.

The H_3PO_4–H_2O_2–H_2O System. The use of phosphoric acid in combination with H_2O_2 produces etch formulations which are relatively independent of the doping concentration. Figure 9.3 shows isoetch curves for

* A commonly used 4:1:1 formulation, known as Caro's etch [16], has a room temperature etch rate of about 0.5 μm/min for all faces other than the (111) Ga. Its etch rate at 50°C is 3μm/min.

this system [17], and indicates essentially four regions of interest *A*, *B*, *C*, and *D*. The amount removed varies linearly with time for etch formulations in regions *A*, *C*, and *D*. The removal rate for etches in region *B*, however, is proportional to the square root of time. This is because the etchant has a high concentration of H_3PO_4, and is relatively viscous. Consequently, diffusion of H_2O_2 to the gallium arsenide is the rate-limiting term here.

Etch formulations in regions *A*, *B*, and *C* have approximately equal etch rates on all principal planes except the (111) Ga, which etches at approximately one half the rate. Thus an etch consisting of 3 : 1 : 50 by volume of $H_3PO_4 : H_2O_2 : H_2O$ results in an etch rate of 0.4 μm/min for the (111) Ga plane and 0.8 μm/min for the other principal planes. Etches in region *C* are somewhat more anisotropic.

The $C_3H_4(OH)(COOH)_3H_2O$–H_2O_2–H_2O System. Etches based on citric acid have also been studied. These etches, consisting of a mixture of a 50 weight % citric acid aqueous solution with concentrated H_2O_2 in a volume ratio *k* : 1, have been used over a wide compositional range, with a correspondingly large range of etch rates. Figure 9.4 shows [18] the etch rate as a function of *k*, over the range of 1–100 Å/s. It is interesting to note that

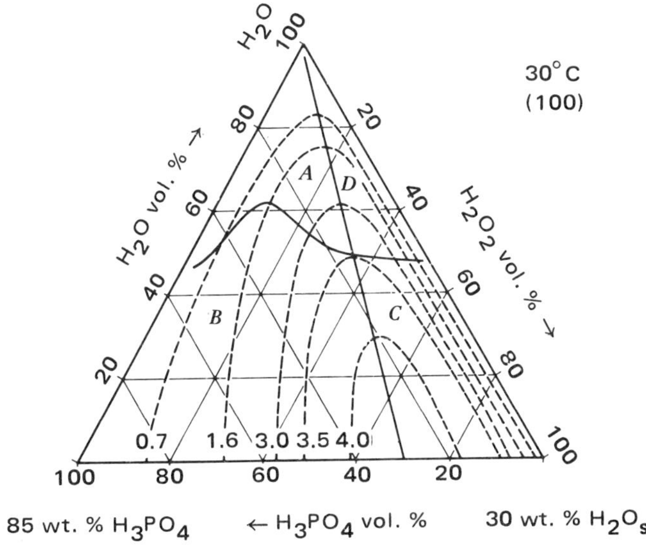

Fig. 9.3 Isoetch curves for gallium arsenide ($H_3PO_4 : H_2O_2 : H_2O$ system). From Mori and Watanabe [17]. Reprinted with permission of the publisher, The Electrochemical Society, Inc.

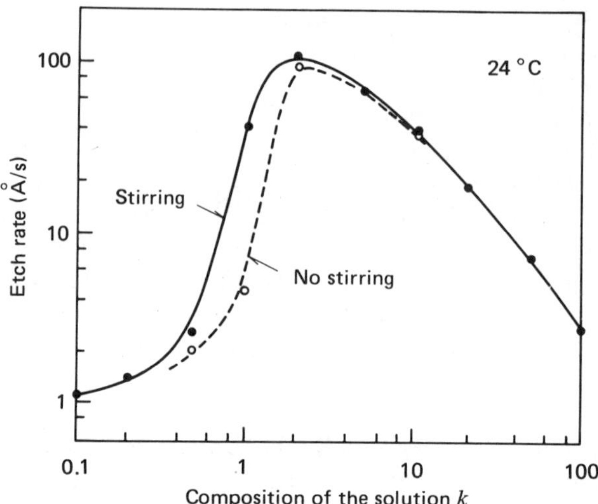

Fig. 9.4 Etch rate versus composition for gallium arsenide (citric acid:H_2O_2:H_2O system). From Otsubo, Oda, Kumaba, and Miki [18]. Reprinted with permission of the publisher, The Electrochemical Society, Inc.

etches with $k < 2$ are relatively sensitive to stirring. Here the rate-limiting step is the availability of citric acid for removing the oxidation products from the surface of the gallium arsenide. Etch rates for the principal crystallographic planes are essentially identical, except for the (111) Ga plane, which etches at a rate of about 60% of the others.

The NH_4OH–H_2O_2–H_2O System. This system can be used to etch the (111) Ga plane nonpreferentially, in addition to the other principal planes [14]. Here a formulation consisting of 1 ml NH_4OH and 700 ml H_2O_2 results in an etch rate of about 0.3 μm/min for the (100), (111) As, and (111) Ga faces, so that the etch can be considered to be truly isotropic for this formulation. Other formulations can be prepared, however, which exhibit anisotropic characteristics. An advantage of this system, of importance to wafer manufacturing, is its considerably superior aging qualities over the Br_2-CH_3OH and the Br_2-CH_3OH-Syton systems.

The NaOH–H_2O_2–H_2O System. Very little work has been done with this system, but indications are that it is similar in etch characteristics to ammonia-based formulations [19]. Typically, a 1*M* NaOH, 0.76*M* H_2O_2 formulation has been shown [15] to result in an etch rate for the (100) plane of about 0.2 μm/min at 30°C, increasing to 1.0 μm/min at 55°C.

Etches using HCl–H_2O_2–H_2O have been investigated, but little is known of their detailed behavior. A variety of etchants, usually modifications of the systems described here, have also been developed [20, 21]. These involve the substitution of CH_3OH instead of H_2O, HNO_3 instead of H_2O_2, and such oxidizing agents as $HClO_4$, NaOCl, and $KMnO_4$. Their use has generally been confined to special-purpose applications.

9.1.4 Anisotropic Etches

The etching of a semiconductor proceeds by the successive dissolution of layers of this material. As a consequence, it is reasonable to expect that this process will be slowest on the (111) planes, since they are the closest packed low-index planes. This has indeed been found to be true for etches which are reaction-rate limited, provided that they are slow so that they do not generate much heat. Etches which are diffusion limited, as well as those which are fast and result in a localized rise in temperature, tend to etch uniformly in all directions.

In recent years, the work with anisotropic etches has focused on their use for the chemical machining of semiconductor materials. They have been used for cutting apart semiconductor chips for beam lead devices [22], and for providing sloping faces on microwave diodes. The etching of both V-shaped [23] and vertical grooves has been accomplished [24] by these methods, and devices as well as microcircuits have been based on this ability. Anisotropic etching has been used in purely nonelectronic applications, such as the development of precision nozzles for ink-jet printers [25, 26].

As pointed out earlier, etching is slowest on the {111} planes because of their close spacing. Following this line of reasoning, etching will be fastest on the {100} planes, whereas etching on the {110} planes will proceed at an intermediate rate. This has been verified for silicon etching, and all anisotropic etches for this material follow this trend.

Unlike silicon, gallium arsenide is a polar semiconductor and has different surface activities in the (111) Ga and (111) As directions. Specifically, the (111) As face, terminated on arsenic atoms, is highly reactive, and will usually etch faster than any other plane. As a consequence, the sequence of etch rates for silicon orientations is invariably (100), (110), and (111) [with (111) being the slowest], whereas the sequence for gallium arsenide is usually (111) As, (100), (110), and (111) Ga. Moreover, a very high degree of anisotropy can be achieved with many silicon etches, whereas this is not the case for gallium arsenide. In fact, some etches for gallium arsenide do not show the preferential ordering listed here.

The effect of anisotropic etching of (100) silicon is shown in Fig. 9.5, where the orientations of a window cut in a mask are as indicated. Here it is assumed that the etch rates for all {111} planes are identical, and negligibly small. This is usually the case, and etch ratios for (100):(111) planes are typically greater than 25. For this situation, the etch profile is trapezoidal after some fixed period of time. Eventually, however, the etchant will delineate a V-groove, as shown by the dashed line in Fig. 9.5. In practice, the etch rate on the {111} planes is small, but finite, so that some degree of undercutting will also occur.

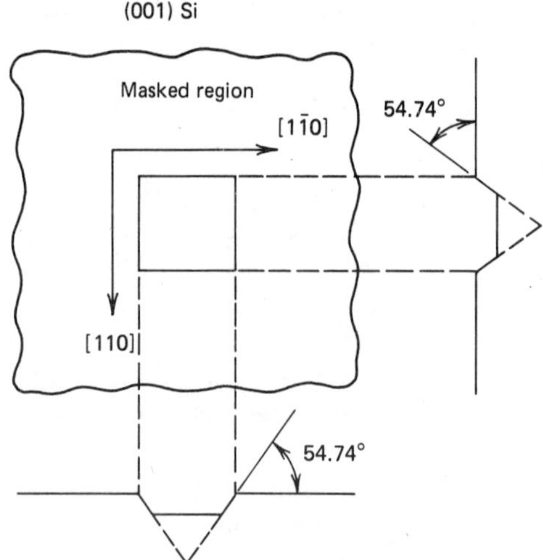

Fig. 9.5 Etch profiles for (001) silicon.

Figure 9.6*a* shows the same situation for gallium arsenide. Here the effect of anisotropic etching is quite different, since the (111) Ga planes etch the slowest, and the (111) As the fastest.* As a result, the etch profile is trapezoidal for one direction, and dove tailed in the other. Furthermore, many of the higher order planes present either gallium or arsenic faces. Consequently, the etch angles are not precisely 54.74°, as for silicon. It has been shown [15], however, that rectangular etch profiles can be obtained if the window is cut at 45° to the [110] and [1$\bar{1}$0] directions, as shown in Fig. 9.6*b*. A wide variety of more complex profiles can be obtained for other orientations of the gallium arsenide and of the windows. These have been detailed elsewhere [13].

Some anisotropic etches are now described, together with their etching characteristics.

9.1.4.1 Silicon

A commonly used anisotropic etch for silicon consists of a mixture of KOH in water or isopropanol alcohol [27]. Typically, etching is carried out in 19 wt. % of KOH in water at 80°C, to result in an etch ratio of 400:1 for (110):(111), with an etch rate of 0.59 μm/min on the (110) plane. Dilution reduces this ratio, but results in a more controllable etch rate. The use of isopropanol alcohol instead of water also results in a better controlled etch;

* The (1$\bar{1}\bar{1}$), ($\bar{1}$1$\bar{1}$), and ($\bar{1}\bar{1}$1) planes all have gallium atoms on their surface and are (111) Ga faces. The ($\bar{1}$11), (1$\bar{1}$1), and (11$\bar{1}$) planes, on the other hand, have arsenic atoms and are thus (111) As faces.

however, a condenser arrangement is needed to avoid evaporation of the alcohol during the etching process.

Etches using hydrazine and water have also been used [28] for the anisotropic etching of V-grooves for microcircuits. Here an etching solution consisting of 100 g N_2H_4 and 50 ml H_2O is used at 100°C to result in an etch rate of 3 μm/min on the (100) plane, with an etch rate of about 0.3 μm/min on the (111) plane.

A series of etchants, consisting of mixtures of ethylenediamine, pyrocatecol, and water (EPW), have been designed for use with silicon [29]. A typical formulation of this type, consisting of 17 ml ethylenediamine, 3 g pyrocatecol, and 8 ml water, results in etch rates of (100):(110):(111) =

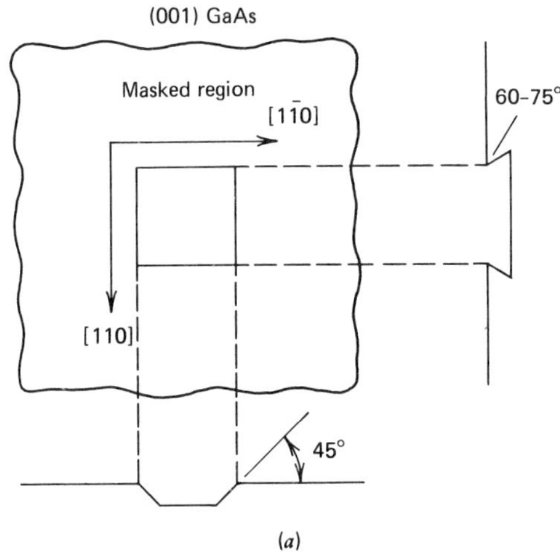

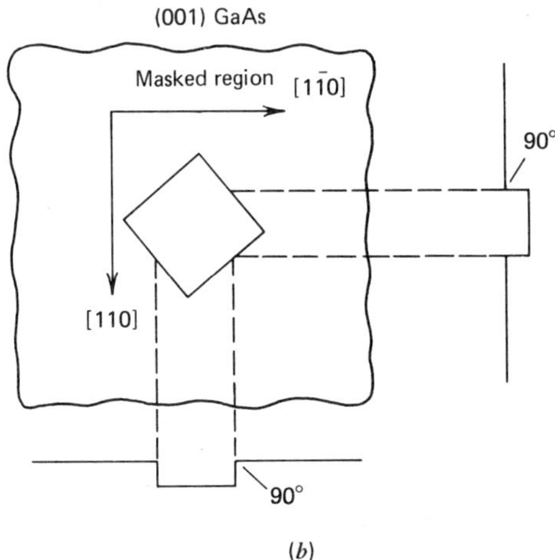

Fig. 9.6 Etch profiles for (001) gallium arsenide. From Tarui, Komiya, and Harada [13]. Reprinted with permission of the publisher, The Electrochemical Society, Inc.

50:30:3 μm/h at 110°C. Lower etch rates can be obtained by reducing the water content of this etch. The use of hydrazine instead of ethylenediamine as the oxidant has also been investigated.

EPW etches are normally used near their boiling point, so that etching must be carried out in a condenser arrangement. Operation at lower temperatures results in lower etch rates, and is sometimes employed for fine-line delineation for polysilicon. However, one problem here is that of residue formation, and modified formulations have been developed where this is not a serious problem [30]. Catalysts such as diazine or pyrazine are sometimes added to EPW solutions. Their use has been shown to double the etch ratios in some EPW formulations.

9.1.4.2 Gallium Arsenide

Almost all of the polishing etches described for gallium arsenide show some degree of anisotropy. At the same time, because of its polar nature, anisotropic etches are usually not as selective as for silicon. By way of example, bromine–methanol formulations, which are normally considered to be polishing, show some degree of anisotropic behavior [13]. A 1% by weight solution of bromine in methanol results in etch ratios of 6:5:4.6:1 for (110):(111) As:(100):(111) Ga. The etch rate for this formulation is about 1 μm/min in the (110) plane.

Anisotropic etching has also been investigated for the NH_4OH–H_2O_2–H_2O system. Here an etch consisting of $NH_4OH:H_2O_2:H_2O$ = 20:7:973 results in etch rates [31] of 0.037:0.12:0.2 μm/min for the (111) Ga, (100), and (111) As planes, respectively.

One disadvantage of both these etchants is that they destroy photo-resist, so that a SiO_2 mask must be used with them. This is not a problem with etches containing phosphoric acid [17], which also show preferential etching behavior. Thus the H_3PO_4–H_2O_2–H_2O system can be used in region *C* of Fig. 9.3 as an anisotropic etch. Here a 1:9:1 formulation by volume results in etch rates of 5:4.2:3:1.5 μm/min for the (110):(111) As:(100):(111) Ga planes, respectively.

The citric acid, H_2O_2–H_2O, system can also be used with a photoresist mask. This etchant exhibits preferential behavior for high values of k (see Fig. 9.4), in excess of 10.

In summary, no really satisfactory polishing etch system has been developed for gallium arsenide. However, most of the formulations show some degree of anisotropy. In general, formulations with low etch rates are more anisotropic than those where the rate is rapid. Finally, etching is often done at low temperatures to increase the anisotropic behavior of these etchants.

9.1.5 Crystallographic Etches

Crystallographic etches can be used to delineate the regions where dislocations intersect with the semiconductor surface. These dislocations, together with their associated strain fields, are present in starting material, and are also created during strain-inducing processes such as dopant incorporation and oxide growth. They result in highly localized shifts in the surface potential, which will etch selectively, if the etch is slow and reaction-rate limited. Fast etches, on the other hand, tend to generate considerable amounts of heat, which obscures these localized variations.

Crystallographic etches are thus generally slow, and are often composed of the same constituents as polishing etches. Frequently one or more heavy metal ions are added; they tend to plate out during etching, and give further visual contrast to the etched regions.

The most commonly used crystallographic etches for silicon, together with their properties, are compared in Table 9.3 at the end of this chapter. Of these, Dash etch [32] has historical value since it was used to definitely establish a correlation between dislocations within a material and the pattern of etch pits created by them at the surface. This etch requires several hours of immersion to be fully effective.

Sirtl [33], Secco [34], and Wright [35] etches all utilize chromium salt oxidizers, which also provide the heavy metal ion. All delineate etch pits in a few minutes time, and are used extensively. Sirtl etch tends to be anisotropic, and has found most use on (111) silicon, where triangular pits are delineated by its action. Secco etch, on the other hand, is isotropic, and results in circular or elliptical etch pits. It is equally effective on (100) as well as on (111) silicon. Wright etch uses copper nitrate as a plating agent so as to more sharply delineate etch pits. It is anisotropic in its etch characteristics, and can be used on (100) and (111) silicon where the geometric shape of these pits gives information about the crystallographic orientation of the defects. Other etch formulations have been developed for special applications such as junction delineation [9]. A few of these are also listed in Table 9.3.

A number of crystallographic etches have been developed for use with gallium arsenide. Here, because of its polar nature, it is relatively easy to distinguish the slow etching (111) Ga face from the fast etching (111) As. Etch pits on these faces tend to be triangular in shape. On the other hand, etch pits on the (001) face are rectangular (instead of square) because of the difference in etch rates for the gallium and arsenic faces which become exposed during etching [36].

Delineation of etch pits in the ⟨100⟩ and ⟨110⟩ directions is particularly useful for gallium arsenide, since these faces are exposed in the formation of lasers and other electrooptical structures. An etch [37] consisting of CrO_3 as the oxidizer, together with $AgNO_3$ and HF, has been found useful for this application. Successful use of this etchant requires 65°C operation, and rapid stirring to remove residues from the semiconductor surface.

All of the crystallographic etches developed for silicon can also be used with gallium arsenide. For example, Sirtl etch is extensively used for delineating etch features in the ⟨100⟩ and ⟨110⟩ directions in gallium arsenide. An unusual feature of this etch is that mounds and hillocks are formed at defect sites, rather than the usual etch pits. A dislocation etch that has found wide use in the delineation of interfaces in gallium arsenide and other compound semiconductors is known as the *A-B* etch [37, 38]. This is a two-part solution with CrO_3 and $AgNO_3$, and is used on cleaved faces to indicate compositional changes resulting from multilayer epitaxy.

Etchant solutions, of the type developed for anisotropic etching, can also be used for etch pit delineation. Thus the dilute bromine–methanol formulations can be used for delineating etch pits in gallium arsenide for this purpose. Etches based on the use of H_3PO_4 and H_2O_2 have also been used here. The properties of a few crystallographic etches for gallium arsenide are compared in Table 9.4 at the end of this chapter.

9.1.6 Non-crystalline Films

The patterned removal of a wide variety of thin film materials is necessary during microcircuit fabrication. These films are formed by the techniques described in Chapters 7 and 8, and are either vitreous, amorphous, or polycrystalline in nature. They lack long-range order, so that etching by wet chemicals is usually isotropic in character, i.e., the etchant spreads out under the mask layer by an amount roughly equal to the etched thickness. In many instances, stresses at the resist–film interface, combined with capillary action of the wet chemical, can cause excessive undercutting at this point, and even lifting or tearing of the resist [39].

Excessive undercutting can also occur because of stresses at the interface between the film and the semiconductor. One such example is the SiO_2–Si interface. Here the process of opening a window in the oxide can lead to undercutting of the SiO_2 near the oxide–silicon interface. In extreme situations this may result in uncovering the oxide protection over a junction edge, as shown in Fig. 9.7. This problem is sometimes encountered in the fabrication of shallow-diffused high-speed transistors in which the emitter–metallization window is as large as the emitter–diffusion window (the so-called *washout* emitter).

Etching is commonly done with the same chemicals which dissolve these materials in bulk form, and usually involves their conversion into soluble salts or complexes. A wide variety of etching chemicals are available for

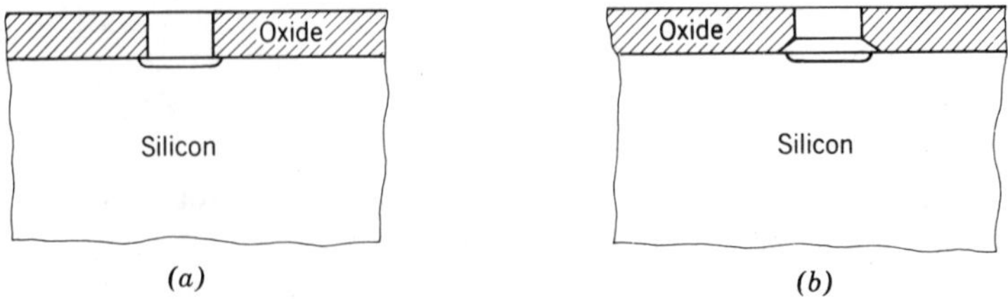

Fig. 9.7 Undercutting effects: (*a*) ideal and (*b*) actual.

each material. Their characteristics will depend upon such film parameters as its microstructure, its porosity, how it is formed, and on the nature of the previous processes to which it has been subjected. In general:

1. Film materials will etch more rapidly than their bulk counterparts, so that reagents must be used in dilute form to reduce the etch rate to manageable proportions.
2. Films which have been irradiated will generally etch rapidly. This includes films which have been ion implanted, those that have been grown by E-beam evaporation, and even those that have been subjected to an E-beam evaporation environment at some previous step. The exception here are certain resists which toughen by polymerization under these conditions. These materials, referred to as *negative* resists, are described in the following chapter.
3. Films which have a high built-in stress will etch rapidly. Often the stress in a film can be controlled by the rate at which it is deposited, the deposition technique, and the substrate temperature.
4. Films with poor microstructure will etch rapidly. This includes films which are porous or loosely structured. Such films can often be densified by heat treatment at a temperature in excess of their growth temperature. These densified films will etch at a slower rate than as-grown films.
5. Films of compounds will etch faster if their preparative technique results in departures from stoichiometry. Silicon nitride is a good example of a material which falls in this category.
6. Mixed films will often etch faster than films made of a single component. This is because, often, the etching of one component causes a rapid increase in the film porosity with a consequent increase in the wetting surface. PSG is a film material which falls in this category.

The etching behavior of materials which are important to microcircuit fabrication are now described. A brief formulary of useful etchants is provided in Table 9.5 at the end of this chapter; an encyclopedic listing of this type is provided in [40].

9.1.6.1 Silica Films

Silica films are widely used as cover layers in both silicon and gallium arsenide technology. They are readily etched by hydrofluoric acid, as shown by the complexing reaction of Eq. (9.8). In practice, this reaction is performed in a dilute solution of HF, buffered with NH_4F to avoid depletion of the fluoride ion. Addition of this NH_4F results in etching characteristics that are consistent from run to run. Furthermore, it has been found that this also lessens attack of the photoresist by the hydrofluoric acid. This etch formulation is commonly referred to as buffered HF (BHF).

Often BHF is used in diluted form in order to slow down its etching rate. The nature of the reaction kinetics is quite complex, and is related [41] to the concentration of HF and HF_2^- ions which are present in this etchant.* The relative concentrations of these species is related to the pH of the solution. Typical BHF solutions, having a pH of around 3, are almost entirely made up of the HF_2^- ion.

Both thermally grown as well as deposited silica films can be etched in BHF. However, etching of deposited films (grown by the silane process at 350–500°C) proceeds much more rapidly than that of thermal oxides, which are normally grown at 1000–1200°C. Typically, the etch rate in BHF is about 3000 Å/s for silane oxides grown at 450°C, but only 900–1000 Å/min for dry thermal oxides.† Densification of these films, by heat treatment at 1000–1200°C for about 15 min, results in a fall in the etch rate to a value that is approximately the same as that for the thermally grown oxide [42].

9.1.6.2 Phosphosilicate Glass (PSG)

PSG has a lower as-deposited stress than silica, and is a diffusion barrier to light alkali ions. Its uses include the passivation of silicon devices and the capping of gallium arsenide wafers before heat treatment. It is grown at 350–500°C by the simultaneous oxidation of SiH_4 and PH_3.

PSG will readily etch both in HF as well as in BHF, the etch rate increasing with the amount of P_2O_5 incorporated in the glass. This is because the HF attacks the SiO_2, rendering the glass porous, with the rapid dissolution of the P_2O_5 in the water. Typically [42], the etch rate of PSG with 8 mole % P_2O_5 content is 5500 Å/min in BHF, whereas that for undoped SiO_2 is about 2500 Å/min. The same films, upon densification at 1100°C, have a greatly reduced etch rate of 3000 Å/min for the PSG and 800 Å/min for the SiO_2.

A wide variety of BHF dilutions can be used, in order to optimize etching characteristics for a given situation. In addition, an etch formulation based on the combination of HF and HNO_3, known as P-etch, has also found much use in the etching of both SiO_2 and PSG films [43].

The etch rate of P-etch increases logarithmically with the phosphorus content of the film. Typically, undoped thermal oxide etches at about 1.8 Å/s. In contrast, the etch rate for a PSG film formed during phosphorus diffusion through a silicon mask ($\simeq$ 16 mol % P_2O_5) is about 500 Å/s for this formulation. Thus P-etch is especially useful in removing layers of PSG which are formed on the surface of SiO_2 films during phosphorus diffusion, and also in some diagnostic situations. Some care is required in its use, however, because of the possibility of dissolution of exposed silicon with which it may come in contact.

* The reaction rate of HF_2^- with SiO_2 is about four to five times that of HF.

† Wet oxides etch slightly faster than dry oxides because they are somewhat more loosely structured.

9.1.6.3 Mixed Oxides

These include* borosilicate glass (BSG) as well as $ZnO \cdot SiO_2$ and $SnO_2 \cdot SiO_2$. All are useful as dopant sources. The BSG is used as a *p*-type doped oxide source for silicon, whereas $ZnO \cdot SiO_2$ and $SnO_2 \cdot SiO_2$ are used as *p*- and *n*-type dopant sources for gallium arsenide, respectively. They are grown at 350–450°C, by the simultaneous oxidation of silane and the appropriate dopant hydride or alkyl (diborane, diethylzinc, and tetramethyltin, respectively). They can be readily etched in HF, or in BHF since SiO_2 is their primary constitutent. Again, etching proceeds by the dissolution of the SiO_2 in the HF, rendering the structure porous and thus more susceptible to dissolution.

Dilute BHF is commonly used with these doped oxides. Its dissolution of BSG has been investigated in some depth [42]. Typically, the etch rate is seen to increase monotonically with the B_2O_3 content. In some instances, however, anomalous characteristics have been observed during the etching of BSG films.

A thin layer of BSG is also formed during the boron doping of masked silicon wafers. Selective etches, known as R-etch and S-etch, can be used [44] to remove this film while leaving the underlying SiO_2 unetched. Both etches are modifications of the P-etch formulation; typically they are about five to six times more rapid in their ability to remove the BSG layer.

The etch characteristics of both $ZnO \cdot SiO_2$ and $SnO_2 \cdot SiO_2$ have not been studied. However, they are comparable to undoped SiO_2 grown by the silane process. Dilute BHF has been found useful for this purpose.

9.1.6.4 Silicon Nitride

Silicon nitride (commonly written as SiN) is an inert dense material which is an excellent diffusion barrier to both sodium and gallium. For this reason, it is widely used as a protective coating for silicon microcircuits, and also as a cap during the annealing of ion-implanted gallium arsenide. Its protective characteristics are generally superior to those of SiO_2 and PSG. However, its patterned removal is more difficult and is highly dependent on the growth technique.

Both HF and BHF can be used to etch these films [45]. However, even at elevated temperatures, the etch rates are extremely slow, so that photoresist films are ruined during this process. Typically, the etch rate of CVD films (SiH_4/NH_3 process) is a function of the growth temperature [46]. The etch rate is concentrated HF is about 1000 Å/min for films grown at 800°C, falling to as low as 140 Å/min for films grown at 1100°C. The etch rate in BHF is considerably lower, about 5–15 Å/min for films grown at 1100°C.

* PSG belongs in this category as well. However, it has been treated separately because of its importance in microcircuit fabrication technology.

The problem of photoresist destruction can be avoided by depositing a molybdenum film between the nitride and the photoresist. This film can be readily etched with excellent edge definition by standard photolithographic techniques. Once patterned, it can be used as a mask for etching the underlying SiN film, after the resist has been stripped.

Silicon nitride is often used as a cover layer over a film of silica. In these situations, neither HF nor BHF can be used, since this would result in deep undercutting of the SiO_2 layer, which is etched rapidly by these solutions. Instead, etching is carried out in boiling* H_3PO_4, and a reflux boiler apparatus is used to avoid changes in the etchant composition during this process. Typically [47], the etch rate for CVD grown SiN films is about 100 Å/min, while the etch rate for thermally grown SiO_2 layers is about 10–20 Å/min. For these same conditions, the etch rate of silicon is under 3 Å/min, so that this technique can be used in the presence of exposed silicon surfaces. A curve of comparative etch rates of these films is shown in Fig. 9.8.

There are many situations in which consecutive layers of SiN and SiO_2 must be etched at the same rate (the gate oxide of a metal–nitride–oxide semiconductor (MNOS) field effect transistor, for example). Here mixtures of HF and glycerol are used at 80–90°C, and experimentally adjusted to provide equal etch rates. Typically, a 1–3 mole/liter concentration of HF etches SiO_2 and CVD SiN films at equal rates of about 100 Å/min at 80°C [48].

The etch rate of SiN films is extremely sensitive to the incorporation of even trace amounts of oxygen in them. In general, the etch rate in HF and BHF increases with oxygen incorporation, whereas the etch rate in H_3PO_4 decreases. Typically, the etch rate in concentrated HF varies [46] from 350 Å/min for $Si_xO_yN_z$ films grown at 1000°C with 7% SiO_2 incorporated in them, to as high as 5000 Å/min for films with 50% SiO_2 incorporation. Also, it should be noted that SiN films, grown at low temperatures by plasma-enhanced CVD (see Chapter 8), contain a large amount of incorporated hydrogen and have considerably faster etch rates than those grown by conventional techniques.

9.1.6.5 Polysilicon and Semi-Insulating Polysilicon

The same chemical systems used for etching silicon can be used for both undoped and doped polysilicon. In general, their etch rate is considerably faster, so that their use results in films with poor edge definition. However, these etches can be modified to be suitable for polysilicon. This usually consists of greatly reducing the amount of HF, using large ratios of HNO_3 to HF, and large amounts of diluent. Typical etch rates for these formulations are about 1500–7500 Å/min for undoped films. Etch rates for doped films are strongly dependent on the crystallite size and the doping concentration.

* The etching temperature is thus related to the concentration of the H_3PO_4.

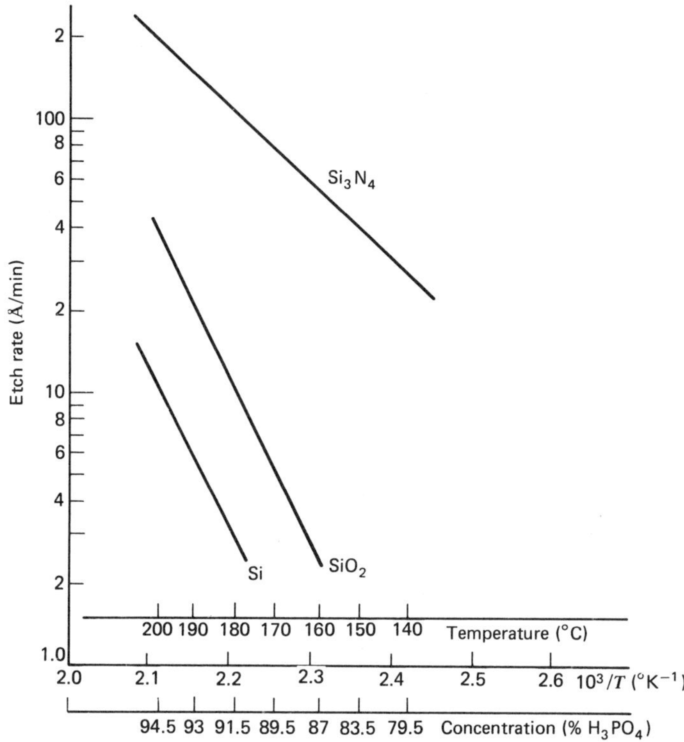

Fig. 9.8 Comparative etch rates of silicon, silicon dioxide, and silicon nitride in phosphoric acid. From van Gelder and Hauser [47].

Formulations, based on the use of ethylenediamine, pyrocatecol, and water can also be used for etching polysilicon. Originally developed as anisotropic silicon etches [29], their slow etch rate makes them attractive for this application.

Semi-insulating polysilicon (SIPOS) is formed by oxygen doping of silicon during growth, so that it can be readily etched in dilute BHF, which dissolves the silica content. The etch rate for SIPOS films is highly dependent on the amount of oxygen incorporated in them. Generally, about 20% O_2 (atomic) is used in films of this type, and dilute BHF can be used to provide a reasonable etch rate of around 2000 Å/min.

9.1.6.6 Silicides

These are extremely difficult, if not impossible, to etch by wet chemical methods. Although etchants are available for a few materials, the etch rate is extremely slow. These etches generally attack resist materials as well, so that auxiliary masks must be used with them. Plasma-assisted etching methods are commonly used for these reasons.

9.1.6.7 Metals

A large variety of metal films are employed in microcircuit fabrication today. Their uses include ohmic contacts, Schottky barriers, and conducting metallization. In addition, they serve as interdiffusion barriers in multilayer metal systems for both silicon and gallium arsenide microcircuits. A brief discussion of chemical etches for the more important metals now follows. Suitable etchants are listed in Table 9.5.

Aluminum. This is certainly the most important metal, and is used for ohmic contacts, Schottky barriers, and interconnections. It can be etched readily by a wide variety of acidic formulations. The most commonly used etchant systems are based on mixtures of H_3PO_4 and HNO_3. Formulations based on HCl and water are used for etching these films in the presence of gallium arsenide, which would otherwise be etched by HNO_3.

Gold and Silver. Gold is used in multilayer interconnection systems, in beam lead microcircuits, and as a Schottky barrier and ohmic contact metal for gallium arsenide microcircuits. It can be etched in aqua regia, but this approach is not practical since it destroys most resist and metal masks. A common etchant for gold consists of a dilution of KI and iodine in water. Photoresist can be used to pattern gold films using this etchant. However, this etchant is opaque, so that visual observation of its progress is only possible by rinsing off the wafer prior to inspection. A number of proprietary etch formulations are also available which do not have this problem. Most of these are cyanide based.

Silver has the same barrier height to *n*-gallium arsenide as gold, and is sometimes used for Schottky devices. Other applications include multilayer metallization systems, where it is used together with titanium. Unlike gold, it can be readily etched in both acidic and basic etchants.

Chromium. Chromium is used in metal systems and ohmic contacts, usually in combination with gold. It is also used in hard surface photomasks. This metal is characterized by a surface that is passivated with a thin oxide film, which must be partially destroyed before etching can take place. This depassivation can be brought about by momentarily touching the film with a wire of some electropositive metal such as aluminum or zinc. This results in the brief evolution of hydrogen gas, followed by ready etching of the film in many acids. Etchants which do not require this depassivation treatment are also available.

Molybdenum and Tungsten. Both these metals are used in refractory gate MOS devices (silicon) and in Schottky diodes for gallium arsenide. A variety of successful etchants are available for each metal.

Platinum and Palladium. These metals are used for the formation of Schottky barriers and ohmic contacts to gallium arsenide. Their deposition, followed by thermal conversion to the silicide, is used in forming Schottky barriers and ohmic contacts to shallow junction silicon devices. In this application, the unreacted metal is subsequently removed after the heat treatment. A "dilute" version of aqua regia can be used for platinum, since the platinum silicide is unaffected by it. Palladium, on the other hand, can be readily etched in a number of different solutions.

9.2 PLASMA ETCHING

Plasma techniques have been known and widely used by the chemical industry for many years. It is only recently, however, that they have been applied to microcircuit fabrication technology. Here their cost and convenience advantages have been so great that they have won ready acceptance in a short period of time.

Plasma chemistry deals with the conduct of reactions in a partially ionized gas composed of ions, electrons, and free radicals. These ionized species are produced most conveniently (and cleanly) by means of a rf discharge at pressures of 0.01–10 torr. In this discharge, free electrons gain energy from the electric field, which they lose by collision with the gas molecules. Energy transfer from the electrons to the gas molecules can occur by elastic as well as by inelastic collisions. A very small amount of energy is transferred by elastic collisions, because of the relatively low ratio of their masses ($\simeq 10^{-5}$). On the other hand, inelastic collisions involve much larger energy transfer, accompanied by excitation of the gas molecule. This results in a wide range of energetic, metastable species. By way of example, as many as 29 different reactions have been identified for a relatively simple oxygen plasma [49].

Plasmas of this type are characterized by the absence of thermal equilibrium between gas molecules and electrons. Typical electron temperatures of 10,000°K are achieved, although the thermal temperatures are only 50–100°C. The electron concentrations in these plasmas are quite low, on the order of 10^9–10^{12} cm^{-3}. A plasma can thus be considered as an ensemble of highly reactive particles in a relatively cool medium. It is these reactive particles, and their characteristics, that can be used effectively in plasma etching.

A typical plasma etching system is shown in schematic form in Fig. 9.9*a*, and consists of a rf excited quartz reaction chamber which can be pumped down to the requisite pressure. A gas, or gas mixture, is fed into this chamber through a controlled leak; the reaction products are exhausted through the pumping apparatus.

Although plasma etching in these systems is primarily due to reaction

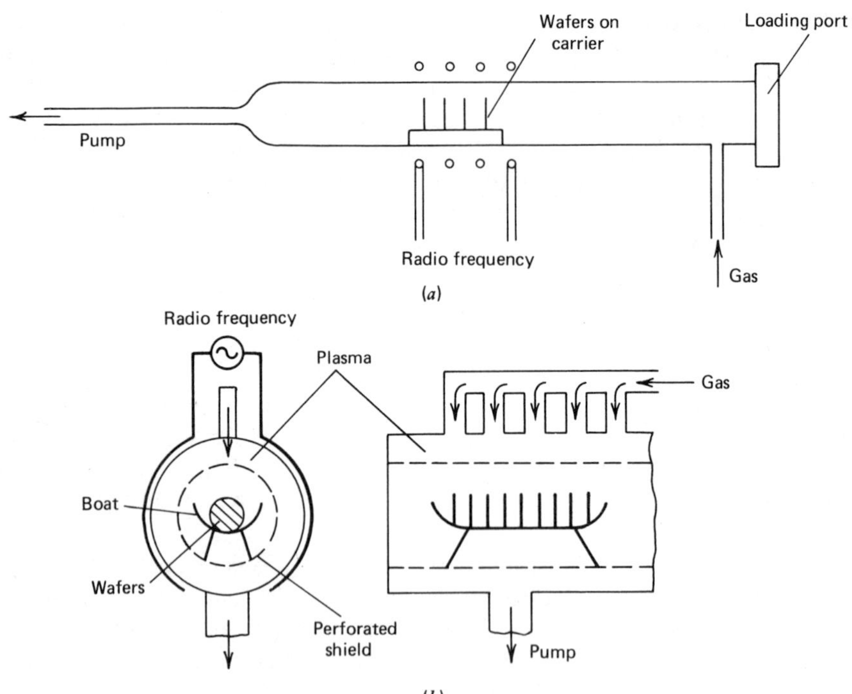

Fig. 9.9 Plasma reactor: (*a*) schematic and (*b*) modern tunnel unit. Adapted from W. Kern and J. L. Vossen, Eds., *Thin Film Processes* [40], 1978.

with the energetic chemical species, a considerable amount of electron and ion bombardment also occurs [50]. This results in sputtering effects, and can lead to surface contamination, radiation damage, and uneven etching across the wafer. In practice, it has been found that the etching characteristics of systems of this type can be greatly improved by enclosing the wafers in a conductive mesh. This tunnel shields the wafers from these species. In addition, it eliminates the effects of fringing fields at the wafer edge which would otherwise produce nonuniform etching due to localized heating, as well as localized bombardment by high-velocity particles [51]. It is particularly effective when the lifetime of the reactant species is long enough to permit its diffusion from the plasma into the tunnel. Tunnels of this type are often used in commercial reactors.

Uniformity of etching is an important consideration when plasma techniques are used to transfer a pattern from one layer (photoresist, for example) to another. This has resulted in new tunnel reactor designs involving multiple gas inlets, as shown in Fig. 9.9*b*. Typically, a lightly loaded reactor of this type, with wafers widely spaced and with a slow etch rate, will have a

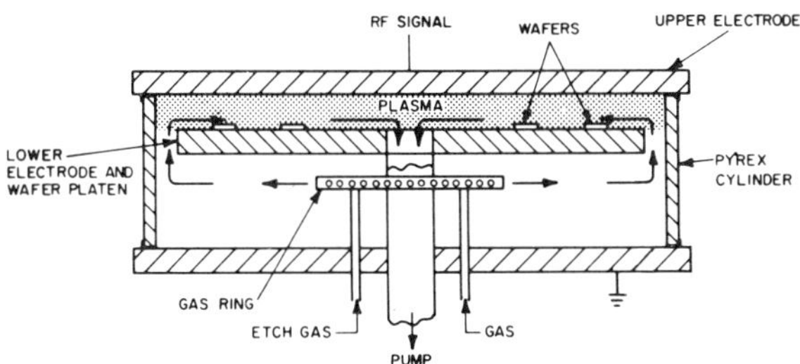

Fig. 9.10 A planar reactor. From Mogab [55]. Reprinted with permission of the publisher, The Electrochemical Society, Inc.

uniformity that is better than ±3% across a slice. With faster etch rates, and with closer wafer spacing, diffusion of the reactant species becomes a limiting factor and can increase the nonuniformity of etch rate by a factor of 10. One approach [52] to obtain better uniformity without a considerable loss in throughput is to control the diffusion of the reactant between wafers by the placement of annular aluminum* rings between them.

A second approach is to use a planar reactor of the type shown in Fig. 9.10. In this type of reactor, the substrates are at normal to the gas flow so that etching is more uniform (but at the price of lower throughput). Wafers in this system are directly in the plasma, so that it allows the use of energetic species which have high recombination rates. In addition, the wafers are at normal to the rf field so that ion movement is both highly directional and rapid. Consequently, a high degree of anisotropic etching can occur, aided by momentum transfer. Here etching comes about because of the enhanced desorption rate of the substrate material in the presence of species which are both active as well as rapidly moving. Plasma etching of this type is more properly called *reactive ion etching* [53].

Plasma etching systems are extremely attractive for a number of reasons. First, the dry etching process avoids problems of undercutting of the resist by capillary action of liquid etchants.† Moreover, highly anisotropic etching can be achieved by use of a planar reactor where momentum transfer is an important mechanism for material removal. Next, the amount of reagent gases, and hence the effluent, is quite small. Although it includes some dangerous and explosive species such as carbon monoxide and active oxygen, they are relatively easy to handle because of the small quantities involved. Finally, one or more of the reactants or the effluent can be monitored by spectrometric methods to signal the end of the etching process. Consequently, these systems can be automated, with cassette loading, automatic pump-down, and load-lock features.

* For reasons to be discussed later, aluminum is a remarkably inert material in a plasma medium.
† An additional advantage is that the etching process is not unintentionally prolonged wherever droplets of wet chemicals cling to the wafer upon removal from the etching solution.

9.2.1 Resist Removal

Plasma techniques achieved their first major acceptance in microcircuit fabrication when they were used for the removal of exposed negative resists. Conventionally, two approaches have been used here. In the first, this tough, polymeric material is "burned" by treatment in hot oxidizing agents, such as H_2SO_4–H_2O_2 mixtures. Alternately, the wafer is soaked in one of many hot chlorinated hydrocarbon mixtures (trichloroethylene, chlorobenzene, etc.) which induce swelling of the polymer and loss of adhesion to the substrate. In either case, mechanical scrubbing is often employed in order to remove remnants of these materials.

Both these approaches require the handling and eventual disposal of large amounts of corrosive chemicals. In addition, the combination of rough treatments described here can result in damage to underlying films, and especially to metallization films which are soft (gold and aluminum) and easily attacked (aluminum) by acids and bases. The requirements of VLSI technology for fine-line structures have made these problems extremely severe so that plasma removal techniques are almost universally employed here.

Plasma removal of resists, or *plasma ashing* as it is often called, consists of placing the resist-covered wafers in an oxygen plasma. A tunnel reactor is commonly used for this purpose. A large number of energetic species are generated in this plasma. Of these species, perhaps the most predominant are formed by

$$e + O_2 \rightarrow 2O + e^- \qquad (9.12a)$$

$$\rightarrow O^{\bullet} + O^{\bullet} + e^- \qquad (9.12b)$$

Temperatures as low as 40–50°C are sufficient to cause oxidation, or "burning" of the resist by these free radicals. The reaction products consist mostly of water, carbon monoxide, and carbon dioxide.

Requirements for such plasmas are relatively modest; about 100–300 W of radio frequency power,* for 5–10 min, can be used to process a batch of wafers coated with 1 μm photoresist in a single operation. The rate of removal of resists depends on the gas pressure, the gas flow rate (which establishes its residence time in the reactor), the rf power, the wafer temperature, and the number of wafers in any given load.

The resist removal process is diffusion controlled, so that the removal rate is a function of the spacing between the wafers and also of their diameter [54]. In addition, the etch rate will be faster at the perimeter and slower at the center of the wafer because of this diffusional process. These problems are of little consequence, however, since almost all of the materials upon which the resist is placed are unaffected (or very slightly affected) by the

* This is the rf power output of the generator. The actual power coupled to the reactants or to the wafers is rarely known, but is somewhat lower.

oxygen plasma. These include silicon, gallium arsenide, aluminum, gold, silicon dioxide, and silicon nitride. As a result, plasma etching can be carried out until all the resist is removed without fear of etching the underlying substrate film. In addition, the use of modern resists leaves almost no residue so that the wafers are ready for the next process step after a minimal amount of cleaning up and rinsing.

A problem area with plasma etching, or with any plasma-assisted process, is the introduction of fast interface states in the gate oxide of low-threshold MOS devices, due to radiation damage. The use of a tunnel shields the wafers from these species, and some of this damage can be annealed by heat treatment at 300–400°C. Nevertheless, some damage remains in the oxide, since this annealing is only partial.

Slow interface states can also be introduced, especially during long plasma ashing processes. These appear to be caused by the movement of light alkali ion contaminants from the resist into the oxide, and can only be removed by high-temperature processes of the type described in Section 7.1.8.2. Consequently, plasma stripping over the gate oxide in MOS circuits is not recommended. However, plasma stripping of photoresist which is placed over the gate *metal* does not present this problem.

Plasma cleaning of silicon is advantageous, since it removes all organic contamination from the surface. Subsequent dry thermal oxidation results in a high-quality gate oxide, and also serves to remove the damaged layer produced by the plasma process.

Plasma cleaning of gallium arsenide prior to gate metal deposition (for Schottky gate FET devices) is not desirable. Here selective sputtering of the gallium arsenide results in the formation of donor-like arsenic vacancies, and leads to enhanced leakage current and a poor ideality factor. Moreover, heat treatment spatially redistributes these vacancies, so that the damage is not completely removed by this approach.

9.2.2 Loading Effects [55]

Most of the process parameters which control the etching rate in plasma reactions can be optimized, once and for all, by experimental runs of the system. Unfortunately, however, it has been found that the etch rate is *also* related to the number of wafers in a load or, more precisely, to the area of the surface to be etched. Specifically, the etch rate increases as this area is reduced, so that an adjustment of the etching time must be made to compensate for each load.

The magnitude of this loading effect can be determined if some simplifying assumptions are made concerning the reactant species and the nature of the reaction. Thus assume that a single reactant species is produced by the plasma, with a generation rate G per unit volume and unit time. Let τ be the recombination rate associated with this process. It is further assumed that the reaction can be characterized by a linear reaction rate constant k

at any given wafer temperature, and that a single reaction product is formed. Let N be the concentration of the reactant species and A the surface area of the material that remains to be etched. If j is the flux density of this species, then the flux is given by

$$jA = kNA \tag{9.13}$$

Let V be the volume of plasma from which this species is delivered to the wafer. Then from considerations of continuity, it follows that

$$\frac{\partial N}{\partial t} = G - \frac{kNA}{V} - \frac{N}{\tau} \tag{9.14}$$

Let N_0 be the steady-state concentration of the active species. Then

$$\frac{\partial N}{\partial t} = 0 = G - \frac{kN_0A}{V} - \frac{N_0}{\tau} \tag{9.15}$$

The etch rate R is given by j/n, where n is the number of atoms per unit volume of the layer being etched. Combining with Eqs. (9.13) and (9.14), gives

$$R\left(\frac{n}{G}\right) = \frac{\tau k}{1 + \tau kA/V} \tag{9.16}$$

From this equation it is seen that if $\tau kA/V \ll 1$, the etch rate is independent of the area being etched. However, the etch rate rises with decreasing A, when $\tau kA/V > 1$. Experimental results, for the etching of silicon and silicon nitride in a CF_4 plasma [56], are shown in Fig. 9.11 and serve to illustrate this point.

The temperature dependence of the etch rate can also be determined by noting that the reaction rate constant takes the form

$$k = k_0 e^{-E_a/kT} \tag{9.17}$$

where E_a is the activation energy for the reaction. In some etching situations, the wafer temperature is controlled in order to vary the etch rate. More often, however, this temperature inadvertently rises during the reaction with an upward shift in the etch rate.

9.2.3 Pattern Delineation

The transference of a photographically produced pattern to an underlying film requires a number of new considerations, making the problem signifi-

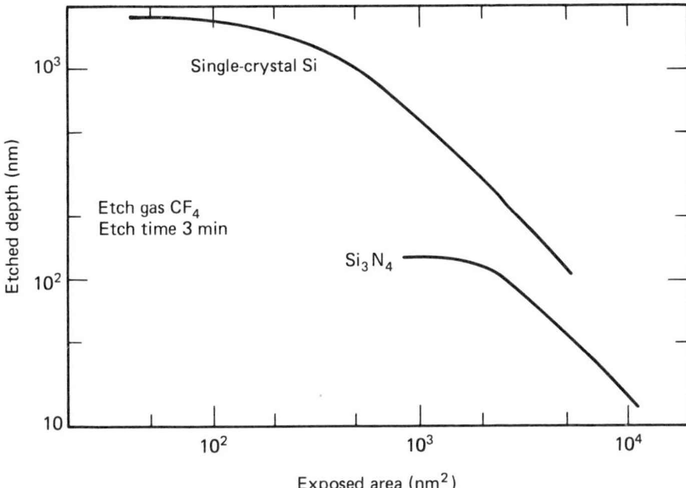

Fig. 9.11 Illustration of the loading effect. From Enomoto [56]. Reprinted with permission from *Solid State Technology*.

cantly more difficult than that of plasma ashing. Here if a photoresist pattern is used, it is necessary that it be unattacked by the plasma, or attacked at a much lower rate than the underlying film. If this is not possible, it is necessary to use an additional layer of a suitable material in which the pattern has been previously delineated by the photoresist. Next, the etching of the film must not be accompanied by inadvertent etching of subsequent layers, so that selectivity is an important consideration. Finally, uniformity of etching is very important in pattern delineation, and undercutting must be controlled. Reactive ion etching, which combines directed ion bombardment with one or more energetic species, is the preferred choice in VLSI applications, since it results in anisotropic etching with almost no undercutting.

Most commonly used films, including both silicon and gallium arsenide, are unetched in an oxygen plasma. Consequently, new reactants, and combinations of reactants, are being actively investigated in order to etch these materials. Again, selectivity of etching characteristics is extremely important for the reasons outlined above.

An important problem that arises in the patterning of fine-line structures for VLSI applications is caused by the loading effect described previously. Consider, for example, the etching of an aluminum metal film using a photoresist stencil. Here operation of the reactor in the region where $\tau kA/V > 1$ results in an etch rate that increases as the aluminum film clears (i.e., as A falls). Consequently, any undercutting that occurs during this phase will proceed at an accelerated rate, and lead to over-etching.

One way of avoiding the problem of over-etching due to loading effects is to monitor by spectrometry the change in the concentration of either the reaction products or the active species and to use these changes to terminate the etching process [53]. An alternate approach is to load the reactor very lightly ($\tau kA/V \ll 1$) so that loading effects are unimportant. A disadvantage of this procedure is the lowered throughput and the difficulty in end point measurement. Nevertheless, cassette loading, combined with the processing of wafers one at a time, is increasingly used to avoid this problem.

The patterning of films of silicon and its compounds is commonly required in both silicon and gallium arsenide technology. These films include polysilicon, SiO_2, PSG, doped oxides, and Si_3N_4. This can be done by carrying out the etch process in a discharge containing fluorine. A number of fluorine compounds, such as CF_4, C_2F_6, and C_3F_8, have been studied for this purpose.* Of these, CF_4 is most commonly used because of its stability. The reaction of CF_4 that is most probable in a plasma environment is

$$e^- + CF_4 \rightarrow CF_3{}^+ + F^{\bullet} + 2e^- \qquad (9.18)$$

The fluorine free radical is the primary active species which etches the silicon by converting it to SiF_4, which is a stable, volatile reaction product. Thus the formation of decomposition products such as silicon, which can deposit downstream from the wafers, is avoided. On the other hand, the strongly reducing character of the CF_3^+ ion is important in etching SiO_2 by the intermediate step of reducing it to SiO. The etching of silicon nitride comes about by the direct reaction of $F^{\bullet}$ with it. Thus [57]

$$12F^{\bullet} + Si_3N_4 \rightarrow 3SiF_4 + 2N_2 \qquad (9.19)$$

Addition of small quantities of oxygen (1–10% by volume) to the CF_4 plasma results in a rapid increase in the amount of atomic fluorine, by reaction between the atomic or molecular oxygen and the CF_3^+ in the plasma. Carbon monoxide is also produced during this process. This is shown in the experimental data [58] of Fig. 9.12, as evidenced by the intensity of emission of fluorine at a wavelength of 704 nm, and of carbon monoxide at 482 nm. Also shown in this figure is the etch rate for silicon which tends to follow the same trend as the fluorine emission intensity.

Silicate glass films, grown by the thermal oxidation of silane and the hydrides of phosphorus, boron, or arsenic, can be readily etched [59] in a CF_4–O_2 plasma. The etch rate in this plasma is relatively independent of the temperature at which the glass is grown (450–650°C) or the amount of oxygen in the CF_4 plasma (0–66%). However, it tends to increase linearly with amount of dopant incorporated in the glass. Typical etch rates in a CF_4–O_2 plasma for heavily doped oxides (10^{21} dopant atoms/cm^3) are from two to ten times the etch rate for CVD-grown layers of undoped silica.

Silicon nitride can also be readily etched in a CF_4–O_2 plasma, and has an etch rate that is slightly higher than that of SiO_2. Typical etch rates [51]

* Carbon-free gases, such as SiF_6, have also been investigated for this purpose.

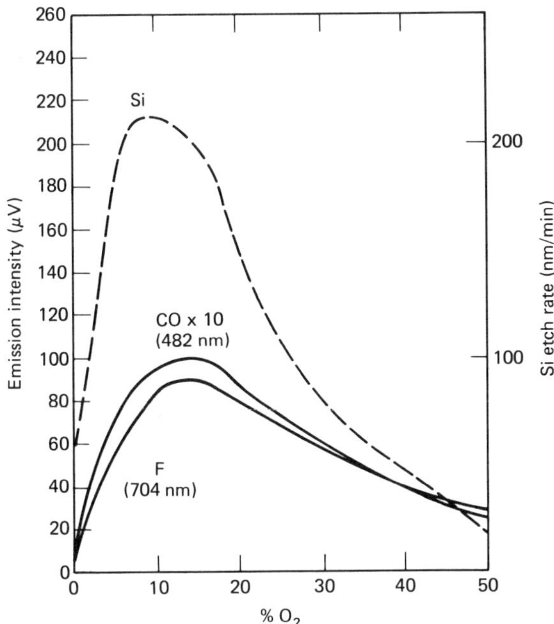

Fig. 9.12 Emission intensity and silicon etch rate as a function of oxygen concentration. Adapted from Harshbarger and Porter [58].

for the CF_4–4% O_2 plasma are Si:Si_3N_4:SiO_2:AZ 1350 photoresist = 17:3:2.5:1.

The use of a CF_4 plasma for etching windows in a protective cover layer of PSG or Si_3N_4 (through which contacts are made to the bonding pads in a microcircuit) is relatively straightforward. This is because aluminum is not etched by fluorine plasmas since they result in the formation of a tough, nonvolatile aluminum fluoride film which prevents further attack of the metal.* Thus the aluminum film effectively stops the etching process.

The etching of windows in an oxide or nitride, prior to diffusion, is considerably more difficult, since the etch rate of silicon is about three to ten times faster than that of the silicon dioxide. Consequently, care must be taken to precisely terminate this etch step in order to avoid undercutting of the exposed silicon. This can be done by monitoring the change in intensity of the fluorine emission line whose magnitude is sensitive to whether the silicon or SiO_2 is being etched.

An alternate approach lies in altering the selectivity of this process by controlling the oxygen content of the etchant gas and thus varying the relative amounts of CF_3^+ and $F^\bullet$ that are produced. An increase in the strongly reducing CF_3^+ ion increases the etch rate of SiO_2, whereas the etch rate of silicon is affected by the concentration of the fluorine free radical. A problem

* Aluminum slice carriers are often used in plasma etching systems for this reason.

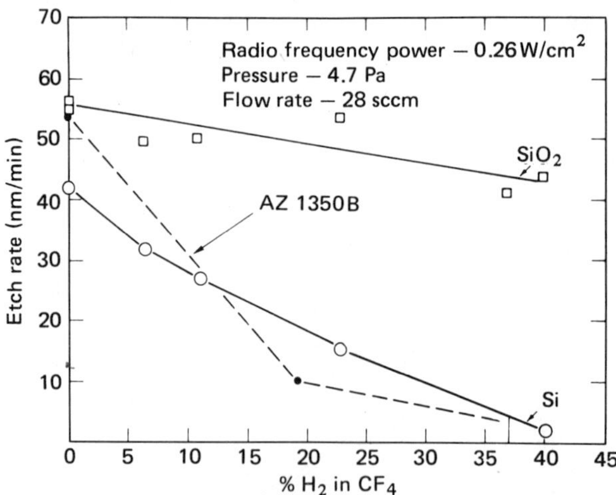

Fig. 9.13 Etch rates for Resist, silicon, and silicon dioxide in a CF_4–H_2 plasma. From Ephrath [60]. Reprinted with permission of the publisher, The Electrochemical Society, Inc.

with this approach is that the etch rate of the photoresist is greatly increased with oxygen concentrations above 8%, so that this method may not always be feasible.

An alternate approach to this problem lies in the use of a CF_4–H_2 plasma for this purpose. The addition of hydrogen to the CF_4 results in the subsidiary reaction

$$2F^{\cdot} + H_2 \rightarrow 2HF \tag{9.20}$$

thus suppressing the fluorine concentration in the system, and hence the etch rate of silicon. Using this approach, etch ratios for SiO_2:Si of as large as 35 have been achieved. An additional advantage here is that this plasma has no oxygen so that photoresists are relatively unattacked by it. Figure 9.13 shows etch rates for one type of resist, SiO_2, and silicon as a function of the concentration of hydrogen in the plasma [60].

The generation of HF as a product of this reaction causes serious problems in the handling of the exhaust gases. An alternate approach is to use a fluorine-deficient molecule, such as C_3F_8, as the etchant. This approach [61] provides SiO_2:Si etch ratios of 5–6, and greatly reduces the problems of effluent disposal. Plasma etching using this hydrocarbon is limited to planar reactors because of the short recombination time of the energetic species.

The anisotropic characteristics of an ion-assisted etching process in a planar reactor are illustrated [62] in the etching of masked layers of SiO_2 and silicon in a CF_4 plasma. Here the etching of SiO_2 is dominated by an ion-assisted mechanism involving CF_3^+, so that highly anisotropic etching can be obtained with a planar reactor. The etching of silicon, on the other hand, is due to the neutral fluorine radical whose transport is diffusion controlled. Consequently, it is etched isotropically and will thus be undercut. A cross section of the resulting structure is shown in Fig. 9.14.

Polysilicon and SIPOS films are very rapidly etched in fluorine plasmas, with etch rates that are three to ten times that for single-crystal silicon. Etching of these films is commonly carried out in a CF_4–4% O_2 mixture that has been diluted with nitrogen for this reason.

It is not possible to etch aluminum in O_2 or CF_4–O_2 plasmas, because of the formation of tough, nonvolatile films of the resulting oxide and fluoride. However, chlorine can be used for this purpose [63]. The active species in these plasmas are usually Cl^+, Cl_2^+, and $Cl^{\bullet}$ [64]. Use of CCl_4 gas leads to the formation of the free radicals $CCl_3^{\bullet}$ and $Cl^{\bullet}$ as follows:

$$CCl_4 \rightarrow CCl_3^{\bullet} + Cl^{\bullet} \tag{9.21}$$

The CCl_3 is strongly reducing and serves to break down the surface layer of aluminum oxide, so that the metal can be etched by the chlorine, with the formation of volatile aluminum chloride.

One of the problems with the plasma generation process for active chlorine species is that its recombination time is very short. As a result, etching must be carried out in a planar reactor of the type shown in Fig. 9.10 where the wafers are in the immediate vicinity of the plasma.

Gallium arsenide is not etched in either an oxygen or a CF_4 plasma, since gallium fluoride is a high melting point solid. Consequently, plasma etching of cover layers of PSG or Si_3N_4 can be readily accomplished in CF_4–O_2 plasmas.

The etching of gallium arsenide can be accomplished in a chlorine plasma, using etch gases that include CCl_4, HCl, or Cl_2 in their formulations [65]. This reaction must be carried out in a planar reactor because of the short lifetimes involved here. Moreover, ion bombardment in this reactor aids in the removal of surface carbon, and thus enhances the formation of the volatile species.

Refractory metal films of molybdenum, tungsten, titanium, and tantalum have also been successfully etched [66] in both CF_4 as well as CF_4–O_2 plasmas. Etch rates for these metals are comparable, and are about 100–150

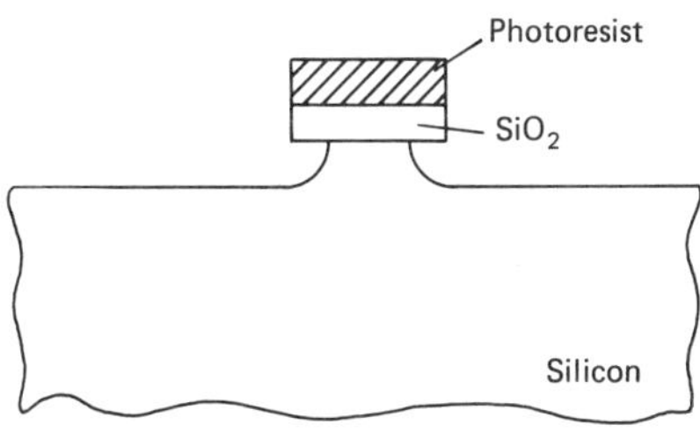

Fig. 9.14 Effect of etching SiO_2 and Si with a CF_4 plasma in a planar reactor.

Å/min for 100 W of input rf power [67]. The silicides of these metals can also be etched in a CF_4–4% O_2 plasma [68].

9.3 PLASMA-ASSISTED ETCHING

An alternate technique for removing material from a substrate is by physically bombarding it with projectiles such as atoms or ions. In this approach, a gas discharge is used for the purpose of imparting energy to the projectile so that it is moving at high velocity when it impinges on the substrate. Upon so doing, momentum is transferred by elastic collision to the substrate atom; dislodging of these substrate atoms will occur when this energy exceeds the binding energy. Figure 9.15 shows the threshold energy for a number of different elements as a function of their heats of sublimation [69], for bombardment with an argon beam.

This process is known as sputtering or ion etching, and a variety of plasma-assisted schemes can be used to bring it about. In all of these, the function of the plasma is to impart energy to an *inert* projectile.* Thus these approaches are distinctly different from plasma etching where energetic species are produced by means of a gas discharge.

Sputtering is accomplished by momentum transfer to the substrate atoms. The rate of sputtering, or ion etching, will therefore be related to the projectile momentum, flux density, and angle of incidence [70]. In addition, it will be related to the sputtering yield of the target. This sputtering yield is a characteristic parameter for a given material and is defined as the number of atoms ejected from the substrate per incident projectile. The sputtering yield for materials of interest is shown in Table 9.6 at the end of this chapter, for bombardment with argon ions in the 0.6 keV range.

The collision process which occurs is essentially an elastic one. Thus the maximum amount of energy T_1 that can be transferred from a projectile of mass M_1 and energy E to a substrate of mass M_2 is given by

$$T_1 = \left[\frac{4M_1M_2}{(M_1 + M_2)^2}\right] E \tag{9.22}$$

Consequently, it is advantageous to use heavy, inert gases for the projectile. Argon is most commonly used, although krypton and xenon have also been investigated.

The sputtering yield of most materials is relatively independent of ion energy. This is because, at high energies, the projectile has an increasing penetration depth and hence increasing energy losses below the surface. Consequently, not all ejected atoms are able to reach the surface in order

* The use of chemically active species, on the other hand, results in *reactive ion etching* by the process described in Section 9.2.

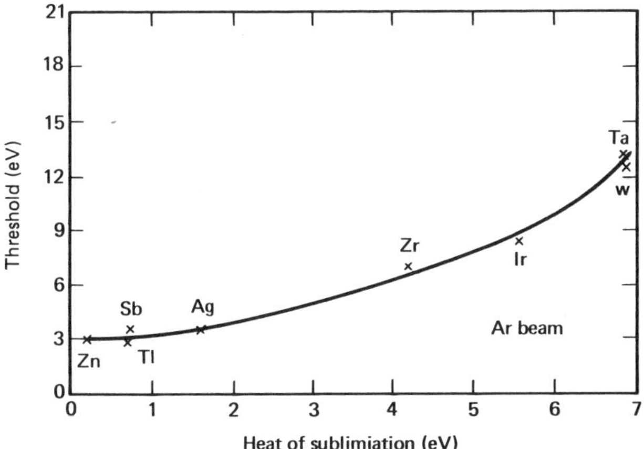

Fig. 9.15 Threshold energy versus heat of sublimation. Adapted from Spencer and Schmidt [69].

to escape. Concurrently, a high-energy projectile can transfer enough momentum to a substrate atom, which can itself cause further dislodgings, resulting in a cascade effect. With sufficiently high energy, this effect can penetrate quite deeply into the substrate, causing considerable damage.* For these reasons, projectile energies are usually kept below 1 keV in order to minimize this problem.

Another parameter of interest is the angle of incidence of the projectile. Typically, for polycrystalline targets, the yield increases as this angle is increased, with a maximum at around 60° (where 90° represents grazing incidence). This is because the penetration depth increases with smaller angles, so that the dislodging process is less efficient. Yield falls off quite rapidly beyond 70°, however, since many of the projectiles simply do not hit the sample at these near-grazing angles.

As seen from the above, this etching technique is strikingly similar to ion implantation, with the main difference being that low-energy ions are used as the projectile in this case, in order to minimize implantation damage. For this reason, the sputtering yield for most projectile–target combinations is around 1–3.

All plasma-assisted techniques share a number of common characteristics. Thus they generally lead to highly anisotropic etching since material removal is primarily by momentum transfer. They are ideally suited for the removal of refractory films such as the silicides or platinum. These materials must otherwise be removed in aqua regia which is so virulent that it destroys the photoresist.

Plasma-assisted etching techniques cause varying amounts of damage to the underlying material. This damage is unavoidable, since it is inherent to

* This process is known as nuclear stopping, and has been described in Chapter 6.

the dislodging process. In some instances, as with MOS devices, this can cause a large shift in the threshold voltage. Schottky-type structures, made by depositing a metal film on an ion-cleaned substrate, have deteriorated characteristics because of this damage [71], even when low ion energies are used (100–150 eV).

Ion bombardment can also cause damage to the mask layer; for example, photoresist becomes extremely difficult to remove by chemical means after bombardment, necessitating the use of plasma ashing techniques. Furthermore, etching techniques of this type are relatively nonselective, so they cannot be used in all situations.

9.3.1 Sputter Etching

Sputter etching is carried out in a self-sustained glow discharge which is created by the breakdown of a heavy inert gas such as argon. The physics of this process is illustrated in Fig. 9.16*a*. Here a dc electric field is impressed across two water-cooled electrodes which are located in this gas. At sufficiently low electric field intensities, a very small current will flow, primarily by the transport of electrons between these electrodes. These electrons may be produced by photoemission or by cosmic ray stimuli, and are always present to some degree in any gaseous medium.

Transport of electrons between these electrodes will result in some collisions with the gas molecules, so that ionization will occur in those encounters where a sufficiently large amount of energy is transferred to the gas. The products of these few ionizing collisions, Ar^+ and electrons, are of themselves accelerated (in opposite directions) because of the electric field, so that they can also enter into collisions with neutral argon molecules, thus resulting in an avalanche multiplication effect. The actual magnitude of this avalanche multiplication* will depend on the electric field, since this determines the acceleration of the electrons and Ar^+, and also on the gas pressure which establishes the mean free path for collisions. It will also depend upon the distance between electrodes, since the number of collisions per unit length traveled is reciprocally related to the mean free path.

Figure 9.16*a* shows this situation, where the multiplied electrons reach the anode and the Ar^+ ions reach the cathode. At sufficiently high applied voltages, some of these Ar^+ ions can eject secondary electrons from the cathode, thus adding to the supply of electrons which contribute to the avalanche multiplication process. Avalanche multiplication can eventually lead to gaseous breakdown with the current being limited by the circuit impedance.

During sputtering, Ar^+ ions bombard the cathode, and hence the wafers which are placed on it, resulting in sputtering by momentum transfer. In

* About 5 argon ions/cm for an electron energy of 1 keV, at a pressure of 1 torr.

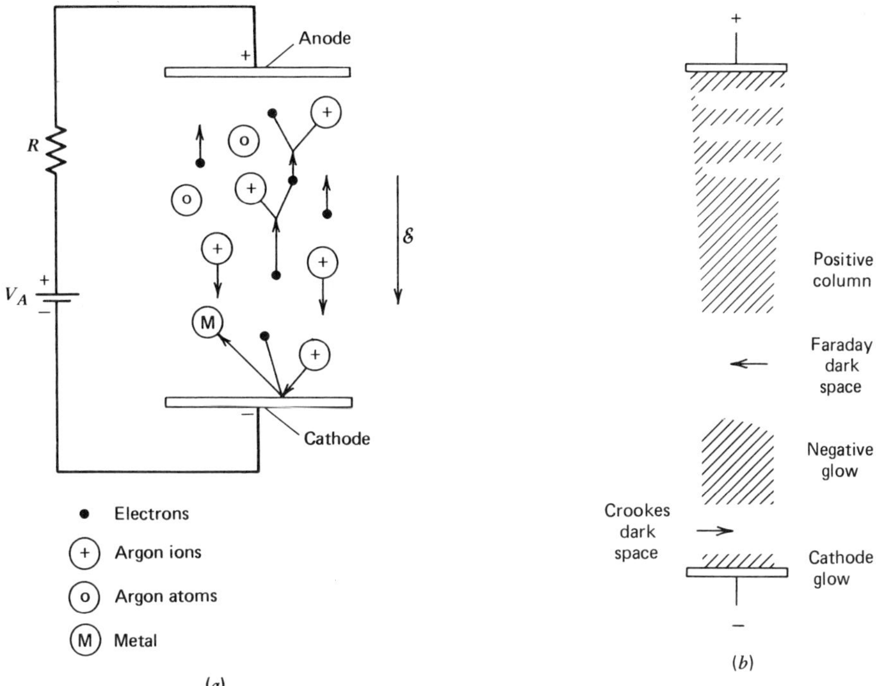

Fig. 9.16 Sputtering in a direct-current excited glow discharge.

addition, secondary electrons* emitted at the cathode participate in sustaining the discharge by ionizing collisions with argon molecules.

Figure 9.16*b* shows the nature of the discharge. The most important region here is the cathode dark space across which nearly all of the applied voltage is dropped. Both ions and electrons created at breakdown are primarily accelerated across this region. Its thickness d is related to the chamber pressure p; for argon the pd product is about 0.3 torr cm.

Atoms ejected from the substrate are mostly neutral, although both positively and negatively charged species are also produced. They have a wide energy spread, typically 0–20 eV, and most of them diffuse to the anode or the chamber walls. Thus the system can be used for deposition purposes as well. In sputter etch applications, however, it is common practice to make the anode large with respect to the cathode, to prevent contamination by back sputtering. In addition, the sputtering of the cathode (on which the substrates are placed) is minimized by covering all exposed parts with a relatively pure, low sputtering-yield material such as quartz.

With dc sputter etching, it is necessary that positive ions impinging on

* About 1 electron per 10 incident argon ions.

the target surface have an opportunity to recombine with electrons to prevent charge accumulation on the surface. While discharging arrangements can be provided to minimize this problem, it can be completely avoided if the plasma is maintained by a rf field. This approach is most commonly used, since it is equally effective with both conducting and insulating substrates.

Radio frequency sputtering is usually carried out with a 1–3 kV peak-to-peak rf potential, and at a pressure of 2–5 mtorr. Radio frequency excitation is commonly provided at 13.56 MHz, which is internationally assigned for equipment usage. A circuit arrangement of this type of system is shown in Fig. 9.17. Here advantage is taken of the diode-like character of the plasma to produce a self-bias by means of the capacitor C, so that the electrodes take on the role of anode and cathode as shown. Sputter etching is carried out by placing substrates on the cathode, as before.

Typically, the etch rate for these systems is about 100–500 Å/min, and is relatively independent of the material used. This lack of selectivity has both good and bad consequences. Thus rf sputtering is highly advantageous in removing refractory materials such as silicides. However, the lack of selectivity in etch rates means that pattern delineation can only be achieved by adjusting the relative thicknesses of the mask and the layer which is to be etched.

Resist masking is a particular problem with rf sputtering, since this material suffers radiation damage and becomes difficult (if not impossible) to remove by chemical means after sputtering. A second problem is that photoresist rapidly deteriorates if its temperature is allowed to exceed 150–200°C during etching over long periods of time. Consequently, a water-cooled cathode, to which good thermal contact is made with the substrates, is essential with photoresist masking. Finally, the inadvertent introduction of any oxygen into a rf sputtering system results in extremely rapid etching of this material.* Thus although rf sputtering systems are commonly operated at relatively high pressures, they must have the same vacuum integrity and pumping capability as systems used for vacuum evaporation, in order to avoid this problem.

Reactive ion etching can be accomplished by the deliberate introduction of small amounts of various chemical species (such as CF_4, C_2F_6, or CCl_4) into the sputtering chamber [73]. Now the etching proceeds partly by the use of energetic species and partly by enhanced surface desorption effects, as described in Section 9.2.

9.3.2 Ion Milling

Here, too, argon is most commonly used. However, a separate ion source [74] is employed to produce the projectiles. In one such system [75], a hot filament provides the source of electrons which travel to an anode. A

* Interestingly, the etch rates of materials such as aluminum, silicon, and silicon dioxide are greatly reduced in the presence of oxygen [72].

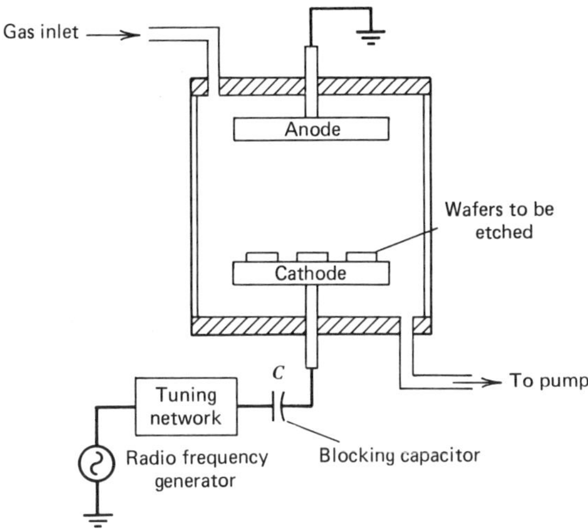

Fig. 9.17 Radio Frequency sputtering equipment.

solenoidal magnetic field increases the electron path, and thus its ability to ionize argon gas which is then accelerated by means of an electric field (see Fig. 9.18). This allows independent control of the ion energy and the ion density. Next, the accelerated Ar^+ ions are made to leave the ion source through aligned, multiapertured grids. A hot filament, outside the ion source, injects electrons into this beam so as to neutralize the overall ion beam space charge. This prevents defocusing, so that a reasonably collimated beam of argon ions (and some neutral argon atoms) are delivered into a working chamber in which the substrates are located on a rotating platform.

Ion milling is very similar to sputtering, since both systems involve momentum transfer as one mechanism for the removal of material from the substrate. Ion milling has a number of advantages over sputtering in that the system is considerably more flexible and can be easily controlled. Thus both conductors as well as insulators can be milled by this technique. Both the ion energy and the ion density can be independently controlled by separate adjustment of thc filament current and the accelerating voltage. In addition, the substrate temperature can be independently controlled, so that heated or chilled substrates are possible. The angle of incidence of the beam can be independently adjusted. Typically, an angle of about 60° results in a maximum etch rate for polycrystalline or amorphous targets. Finally the working chamber can be operated at a considerably lower pressure than the ion source (one to two decades) so that the system is comparatively cleaner.

The milling rate for a number of materials of interest to microcircuit fabrication technology is shown in Table 9.7 at the end of this chapter. This table highlights one of the major problems with ion milling, i.e., the relative lack of selectivity in this process. This selectivity can sometimes be obtained

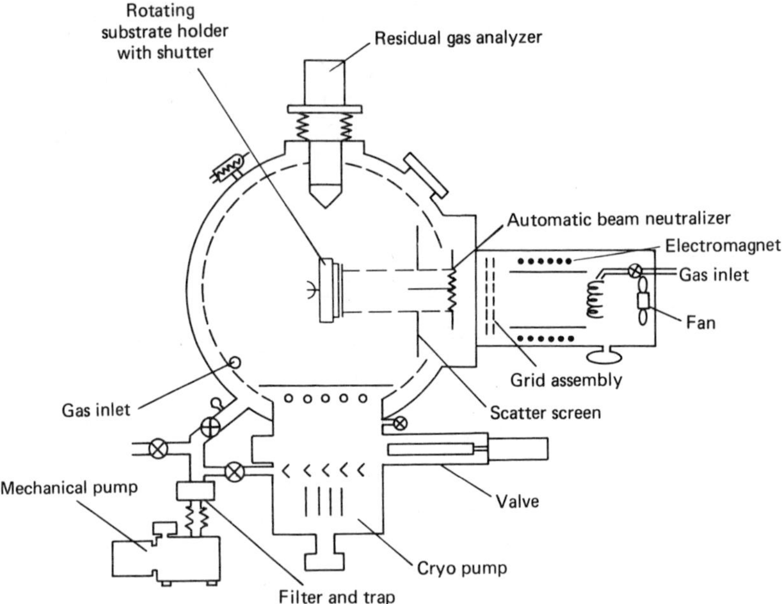

Fig. 9.18 Ion milling equipment. From Thompson [75]. Reprinted with permission from *Solid State Technology*.

by control of the thickness of the mask layer, which should ideally be about twice the etch depth. This is possible if the mask will ultimately be discarded, as in the use of photoresist patterning of aluminum. However, there are many situations in which this mask layer thickness cannot be controlled, so that ion milling is not always feasible.

The high degree of anisotropic etching that can be achieved by this technique makes it extremely attractive for use in VLSI applications. Consequently, a number of studies have been made to determine the role of substrate orientation and of the mask thickness in determining the etch profiles [76]. These studies have also included consideration of such important problems as the redeposition of sputtered material onto the edge of a milled step [40]. A treatment of these topics is beyond the scope of this work.

Reactive ion milling is also being investigated for improving the selectivity in these systems. This can be done by the introduction of reactive gases into the working chamber. Gases such as oxygen are often introduced in this manner if the system is to be used for the deposition of oxides. The use of this gas has been shown [72] to greatly reduce the etch rates of such materials as aluminum, silicon, and silicon dioxide.

Reactive gases can also be introduced directly into the gas feed for the ion gun, so as to produce a much higher concentration of rapidly moving energetic species. This technique has considerable promise; however, much work needs to be done in the design of processes as well as of suitable equipment, before it is accepted outside of the laboratory environment.

9.4 CLEANING

Semiconductor wafers are subjected to physical handling during the processes of cutting, lapping, polishing, and packaging. This leads to large amounts of molecular contamination, much of which is of unknown origin. Included in the list of such contaminants are airborne bacteria, grease and wax from cutting oils and from physical handling, abrasive particulates (usually silicon carbide, alumina, and diamond dust) from the grinding and sawing operations, as well as a variety of plasticizers which come from the containers and wrapping in which the wafers are shipped. Removal of these contaminants is the first step in wafer cleaning, and is usually done by rinsing in hot organic solvents such as trichlorethylene or xylene, accompanied by mechanical scrubbing, ultrasonic agitation, or by compressed gas jets.

Ionic contaminants are also present on these wafers, and result from improper cleaning or from etch steps during processes, such as the opening of a window, or the delineation of a metal interconnection pattern. These ions (usually light ions such as sodium and potassium) are due to trace impurities in the etchants and adhere to the semiconductor surface by physical adsorption or by chemisorption. Light metal ion contamination is particularly undesirable in MOS-based microcircuits, where they can cause instabilities in the threshold voltage. However, they can often be tolerated to a greater extent both in gallium arsenide as well as in bipolar silicon microcircuits.

Desorption of a large fraction of these ions can be accomplished by rinsing in hot deionized water for a few minutes [77]. However, their complete elimination generally requires chemical reaction with the semiconductor surface, accompanied by flushing to remove these soluble reaction products. HCl treatments are especially effective for this purpose.

Heavy metal impurities can also be present as contaminants on the semiconductor surface, and usually come about by deposition out of the etchant solutions during device fabrication. Gold, silver, copper, iron, and nickel are common impurities found in most "electronic" grade chemicals of the type used in microcircuit fabrication technology. Although present in a few parts per million, this represents a significant amount of chemical contamination, especially in circuits where the control of minority carrier lifetime is important. Etching processes will generally leave only small amounts (less than 10^{13} cm^{-2}) of these impurities on the surface, since they involve removal of semiconductor material, with continual flushing away of the surface reaction products. Oxidizing agents, such as H_2O_2 and HNO_3, tend to form a thin protective coating on the semiconductor, so that heavy metal contamination is not severe when these chemicals are used.

Heavy metal contamination primarily comes about during processes which dissolve the semiconductor protective oxides, but leave the surface unattacked. An example of one such process is the opening of a window in an oxide or nitride layer, prior to diffusion. Here uninhibited electrochemical displacement plating can occur, because of the large difference between the oxidation potentials of the semiconductor and the metal ions in solution, resulting in contamination levels of as high as 10^{16} cm^{-2}.

The tolerance to ionic and atomic contamination, and hence the appropriate surface cleaning treatment, is a function of the device operation. Thus a cleaning solution involving the use of sodium-based materials (NaOH, NaOCl, etc.), which may be quite acceptable for devices in gallium arsenide, cannot be tolerated for MOS-based silicon circuits, unless a second cleaning solution is used to remove the sodium contamination. Here, surface cleaning should be carried out by the use of chemicals that only contain volatile species* in their composition. Chemicals such as HCl, HNO_3, NH_4F, H_2O_2, and H_2O are favored for this purpose.

All semiconductor wafers must receive an initial cleaning in hot organic solvents. In addition, further cleaning of wafers must be done after each processing step in the fabrication sequence, and especially before each high-temperature operation. As far as possible, the same cleaning process should be used at each point, and made as routine as possible, in order to avoid operator error.

One highly successful approach [78] to cleaning silicon is to use two solutions in sequence. The first of these consists of 1:1:5 to 1:2:7 volumes of $NH_4OH:H_2O_2:H_2O$. Here the hydrogen peroxide functions to oxidize all remaining organic contaminants on the surface, which are present because of incomplete removal of resists,† and also due to airborne materials and physical handling. The ammonia is effective in removing heavy metals such as silver, cadmium, cobalt, copper, mercury, and nickel, by forming complex -amine groupings with them. Next, a solution consisting of $HCl:H_2O_2:H_2O$ in a 1:1:6 to 1:2:8 volume ratio is used to remove the light alkali ions, and to prevent displacement plating from the solution.

Each of these steps is carried out for 10–20 min at 75–85°C, under conditions of rapid agitation. Nitrogen gas bubbling through the etchants is often used for this purpose. Finally, wafers are blown dry and stored in a clean environment until further processing. Results with this cleaning technique

* This does not preclude the presence of ionic or atomic impurities in them.

† Neither wet chemistry nor plasma stripping does a complete job in this area. Residues from the plasma process also present a problem here.

make it quite suitable for both bipolar as well as MOS microcircuits, so that it is in wide use in industry at the present time.

A large number of special-purpose techniques are also used in individual situations. Many of these are unique to each manufacturer, and are often closely guarded trade secrets. A few of the more well-known cleaning steps are briefly described.

Heavily doped silicon will occasionally become stained by the formation of a suboxide on its surface. A 10–15 s dip in a solution [10] consisting of 2 volumes HF to 1 volume $KMnO_4$ (6% solution) will remove this stain. A small amount of silicon is removed during this process.

An excellent cleanup step, widely used for silicon, consists of $HNO_3:HF:HClO_4:CH_3COOH$ in a 100:14:10:10 volume ratio. This solution is particularly effective in removing heavy metal impurities, and has been used prior to the oxidation of silicon in high-power semiconductor circuits requiring long lifetime [79].

Oxidized silicon wafers that are stored for any length of time will often become contaminated within the top 50 Å of the oxide layer. This contamination is best eliminated by actual removal of this upper layer of oxide in a short etch consisting of $HF:H_2O = 1:50$ by volume. The etch rate of this formulation is about 70 Å/min so that an immersion time of 45–60 s is satisfactory [80].

Water is a major contaminant of oxidized silicon wafers and can be adsorbed in molecular form if slices are stored for any length of time. Its removal, by a 250°C heat treatment for 24–64 h, is an important process step just prior to packaging, or to the deposition of the final cover layer of PSG or Si_3N_4. Junction walkout problems, and other instabilities in diode reverse characteristics, can be greatly reduced by this treatment [81].

In principle, the cleaning of gallium arsenide is very similar to that of silicon, and proceeds along the same lines. The starting point here is the use of strong solvents for removing organic residues which are present on the as-purchased wafer. The successive use of hot trichloroethylene, acetone, and methanol, each for 5–10 min, is very effective for this purpose. Wafers cleaned in this manner are blown dry and can be kept under methanol for a short time until ready for further cleaning or processing.

Most of the cleaning formulations for gallium arsenide are of an etching nature, i.e., they remove some of the gallium arsenide during the cleaning process. Typically, a quick cleanup etch is used and involves one of the many formulations described in Section 9.1.3. Again, heavy metal impurities such as gold, silver, copper, iron, and nickel have a tendency to plate out of the reagents during this process. This displacement plating can be minimized if the etch is rapid, and is accompanied by quick flushing away of the reaction products. In addition, the etch should involve large amounts of deionized water so as to minimize the concentration of these metals. One formulation, consisting of $HCl:H_2O_2:H_2O$ in a 1:10:80 ratio by volume, used at 70°C, has been found highly effective for this application [82]. Its relatively high etch rate, 4.0 μm/min, requires only a short immersion time for cleanup purposes.

9.4.1 Schottky Diodes and Gates

Cleaning procedures of the above type are satisfactory prior to epitaxy, diffusion, alloying, and post-implant annealing. In all of these processes, the primary requirement is to avoid surface contamination with heavy metal impurities, which degrade lifetimes during the subsequent high-temperature step. The fabrication of Schottky diodes and gates, however, poses an additional problem since these devices are highly sensitive to the nature of the metal–semiconductor interface [83].

Schottky diodes on silicon can be made by the sputter deposition of a wide variety of gate metals, some of which are listed in Table 8.3. Back sputtering is commonly used to clean the silicon surface of its residual oxide film (15–20 Å) prior to deposition of this metal. Increasingly, the silicides of platinum and palladium are used for the Schottky barrier, because of their improved stability under subsequent processing conditions. These layers are formed by the deposition of the elemental metal on a sputter-cleaned silicon surface, followed by thermal conversion to the silicide.* This ensures a clean interface with the silicon, and anneals the damage caused by back sputtering. Diodes with near ideal ($n \simeq 1$) behavior have been made by this means.

The problems with cleaning gallium arsenide surfaces prior to Schottky gate deposition are somewhat more severe, since sputter etching causes considerable damage to this material which cannot be removed by subsequent thermal annealing [84]. However, this technique can be used with contact metals such as platinum or palladium, where the metal is subsequently heat treated to form compounds [85] with the gallium arsenide (see Section 8.5), and thus consumes the damaged layer. Metals such as aluminum, gold, molybdenum, and tungsten are used in elemental form on the gallium arsenide. Here wet chemical techniques are more commonly used to remove the residual oxide layer.

Vacuum-cleaved gallium arsenide will rapidly develop a surface oxide of about 8 Å upon exposure to air. This oxide increases in thickness to about 15–20 Å within an hour, and eventually to about 30 Å over a period of days. Cleaning in organic solvents such as trichloroethylene and acetone, followed by a rinse in water or methanol, leaves this oxide layer thickness essentially unchanged [86]. A surface oxide of this type will generally consist of a nonstoichiometric mixture of As_2O_3 and Ga_2O_3, with an excess of As_2O_3.

* Details of this process have been outlined in Section 8.4.3.

This is believed to produce traps in the bandgap of the Ga_2O_3. Consequently, Schottky diodes made by the evaporation of a metal on surfaces of this type will generally have poor ideality factors ($n > 1$) and soft reverse V–I characteristics [87].

A variety of surface treatments can be given prior to evaporation of the Schottky metal. The effect of some of these are now considered. Etchants, of the type used for the removal of semiconductor material (see Section 9.1.3), usually result in the formation of a thick surface oxide (≃ 50 Å) which is As_2O_3 rich, and is unsuitable for Schottky diode formation. Consequently, the etch step must be followed by an additional treatment to reduce its thickness. Both HCl or NH_4OH, in a 1:1 concentration by volume, can be used for this purpose. A 30 s immersion in either of these, followed by a methanol rinse, results in a 8–12 Å residual oxide which is slightly gallium deficient. Gallium-to-arsenic ratios of 0.78–0.82 have been observed [88] with the 1:1 HCl treatment; the oxide with the 1:1 NH_4OH treatment is somewhat superior,* with gallium-to-arsenic ratios of about 0.84–0.94. Diodes made on these surfaces have ideality factors that are close to unity [89]. Etches based on the use of H_2SO_4 result in highly arsenic-rich surfaces [90], and are undesirable for Schottky diode applications.

The bromine methanol etch is somewhat unusual, in that it results in a surface oxide that is deficient in arsenic. This is because the bromine preferentially brominates the gallium, inducing its migration to the surface where the $GaBr_3$ converts to the oxide. This process leaves behind a conducting film of elemental arsenic at the oxide–GaAs interface, so that further treatment is necessary to remove this layer. Both 1:1 HCl and 1:1 NH_4OH can be used for this purpose. These act by dissolving the surface oxide, oxidizing the underlying arsenic, and then removing it. In both cases, the residual oxide is slightly As_2O_3-rich.

In summary, therefore, a successful etch procedure preceding Schottky gate formation is to use a rapid etch (such as H_2SO_4:H_2O_2:H_2O in a 5:1:1 ratio by volume), followed by a 1 min treatment in HCl or NH_4OH in a 1:1 concentration by volume in order to thin the surface oxide to about 10 Å. Often the wafer is immersed under the solution at this point, and subsequent gate formation is carried out by immediate transfer to the deposition apparatus, after it has been blown dry.

The final oxide thinning step results in an almost bare gallium arsenide surface, so that heavy metal plating is a strong possibility during this process. Often, however, this does not present a problem since the Schottky metal is formed by a low-temperature process, and the device is not subjected to further heat treatments. In some situations, however, it is important to

* Treatment in concentrated NH_4OH leads to an increase in the As_2O_3 content of the residual oxide. Thus the role of water is important in removing this compound.

minimize this type of contamination. This is often done by passing the final reagents through a column containing crushed, "sacrificial" gallium arsenide where these impurities can plate out, prior to using it for the gallium arsenide wafer [89]. A two-to-three decade reduction in metal contamination can be achieved by this technique.

Table 9.1 Compositions of Commonly Used Concentrated Aqueous Reagents

Reagent	Weight %
HCl	37
HF	49
H_2SO_4	98
H_3PO_4	85
HNO_3	70
$HClO_4$	70
CH_3COOH	99
H_2O_2	30
NH_4OH	29 (as NH_3)

Table 9.2 Some Polishing Etches for Silicon

Formulation	Remarks
3 ml HF 5 ml HNO_3 3 ml CH_3COOH	CP-4A, 80 μm/min
2 ml HF 15 ml HNO_3 5 ml CH_3COOH	Planar etch, 5 μm/min
10 ml H_2O_2 3.7 g NH_4F	0.7 μm/min, almost neutral etchant
2 ml HF 1 ml $KMnO_4$ (6%)	0.3–0.4 μm/min
50 ml HF 100 ml HNO_3 110 ml CH_3COOH 3 g I_2	

Table 9.3 Some Crystallographic Etches for Silicon

Formulation	Remarks
1 ml HF 3 ml HNO_3 10 ml CH_3COOH	Dash etch, 8 h
1 ml HF 1 ml CrO_3 (5*M* in H_2O)	Sirtl etch, for (111) silicon, 5 min
2 ml HF 1 ml $K_2Cr_2O_7$ (0.15*M* in H_2O)	Secco etch, for (100) and (111) silicon, 5 min
60 ml HF 30 ml HNO_3 60 ml CH_3COOH (glacial) 60 ml H_2O 30 ml solution of 1 g CrO_3 in 2 ml H_2O 2 g $(CuNO_3)_2 \cdot 3H_2O$	Wright etch, for (100) and (111) silicon, 5 min, long shelf life
2 ml HF 1 ml HNO_3 2 ml $AgNO_3$ (0.65*M* in H_2O)	Silver etch, for faults in epitaxial layers
200 ml HF 1 ml HNO_3	For *p–n* junction delineation, 1 min

Table 9.4 Some Crystallographic Etches for Gallium Arsenide

Formulation	Remarks
1 ml Br_2 100 ml CH_3OH	Distinguishes between (111) Ga and (111) As planes
1 ml HF 2 ml H_2O 8 mg $AgNO_3$ 1 g CrO_3	Etch pits on (100) and (110) planes
A: 40 ml HF 40 ml H_2O 0.3 g $AgNO_3$ *B*: 40 ml H_2O 40 g CrO_3	*A–B* dislocation etch, separate parts store indefinitely, used for delineation of epitaxial layers. Mixed in a 1 : 1 ratio before use.
1 g $K_3Fe\ (CN)_6$ in 50 ml H_2O 12 ml NH_4OH in 36 ml H_2O	Used for the delineation of epitaxial layers. Mixed in a 1 : 1 ratio before use.

Table 9.5 Etchants for Noncrystalline Films[a]

Material	Etchant	Remark
SiO_2	28 ml HF 170 ml H_2O 113 g NH_4F	BHF, 1000–2500 Å/min at 25°C
	15 ml HF 10 ml HNO_3 300 ml H_2O	P-etch, 128 Å/min at 25°C
	1 ml BHF 7 ml H_2O	800 Å/min
BSG	1 ml HF 100 ml HNO_3 100 ml H_2O	R-etch, 300 Å/min for 9 mole % B_2O_3, 50 Å/min for SiO_2
	4.4 ml HF 100 ml HNO_3 100 ml H_2O	S-etch, 750 Å/min for 9 mole % B_2O_3, 135 Å/min for SiO_2
PSG	28 ml HF 170 ml H_2O 113 g NH_4F	BHF, 5500 Å/min for 8 mole % P_2O_5
	15 ml HF 10 ml HNO_3 300 ml H_2O	P-etch, 34,000 Å/min for 16 mole % P_2O_5, 110 Å/min for SiO_2
	1 ml BHF 7 ml H_2O	800 Å/min
Si_3N_4	HF	140 Å/min, CVD at 1100°C 750 Å/min, CVD at 900°C 1000 Å/min, CVD at 800°C
	28 ml HF 170 ml H_2O 113 g NH_4F	BHF, 5–10 Å/min
	H_3PO_4	100 Å/min at 180°C
Polysilicon	6 ml HF 100 ml HNO_3 40 ml H_2O	8000 Å/min, smooth edges
	1 ml HF 26 ml HNO_3 33 ml CH_3COOH	1500 Å/min
SIPOS	1 ml HF 6 ml H_2O 10 ml NH_4F (40%)	2000 Å/min for 20% O_2 film

Table 9.5 (*Continued*)

Material	Etchant	Remark
Al	1 ml Hcl 2 ml H_2O	80°C, fine line, can be used with gallium arsenide
	4 ml H_3PO_4 1 ml HNO_3 4 ml CH_3COOH 1 ml H_2O	350 Å/min, fine line, will attack gallium arsenide
	16–19 ml H_3PO_4 1 ml HNO_3 0–4 ml H_2O	1500–2500 Å/min, will attack gallium arsenide
	0.1 *M* $K_2Br_4O_7$ 0.51 *M* KOH 0.6 *M* $K_3Fe(CN)_6$	1 μm/min, pH 13.6, no gas evolved during etching
Au	3 ml HCl 1 ml HNO_3	Aqua regia, 25–50 μm/min
	4 g KI 1 g I_2 40 ml H_2O	0.5–1 μm/min, can be used with resist
Ag	1 ml NH_4OH 1 ml H_2O_2 4 ml CH_3OH	3600 Å/min, can be used with resists, must be rinsed rapidly after etching
Cr	1 ml HCl 1 ml glycerine	800 Å/min, needs depassivation
	1 ml HCl 9 ml saturated $CeSO_4$ solution	800 Å/min, needs depassivation
	1 ml, 1 g NaOH in 2 ml H_2O 3 ml, 1 g $K_3Fe(CN)_6$ in 3 ml H_2O	250–1000 Å/min, no depassivation, resist mask can be used
Mo	5 ml H_3PO_4 2 ml HNO_3 4 ml CH_3COOH 150 ml H_2O	0.5 μm/min, resist mask can be used
	5 ml H_3PO_4 3 ml HNO_3 2 ml H_2O	Polishing etch
	11 g $K_3Fe(CN)_6$ 10 g KOH 150 ml H_2O	1 μm/min

Table 9.5 (*Continued*)

Material	Etchant	Remark
W	34 g KH_2PO_4 13.4 g KOH 33 g $K_3Fe(CN)_6$ H_2O to make 1 liter	1600 Å/min, high resolution, resist mask can be used
Pt	3 ml HCl 1 ml HNO_3	Aqua regia, 20 μm/min, precede by a 30 s immersion in HF
	7 ml HCl 1 ml HNO_3 8 ml H_2O	400–500 Å/min, 85°C
Pd	1 ml HCl 10 ml HNO_3 10 ml CH_3COOH	1000 Å/min
	4 g KI 1 g I_2 40 ml H_2O	1 μm/min, opaque, must be rinsed before visual inspection

[a] Listed in the order in which they are described in Section 9.1.6.

Table 9.6 Sputtering Yields for Materials Bombarded By Argon at 0.6 keV[a]

Target	Sputtering Yield
Al	1.2
Au	2.8
Mo	0.9
Ni	1.5
Pd	2.4
Pt	1.6
Si	0.5
Ta	0.6
Ti	0.6
W	0.6
GaAs	1.2 molecules/ion
SiO_2	0.1 molecules/ion
Si_3N_4	0.05 molecules/ion

[a] See reference 40.

Table 9.7 Ion Milling Rates for Argon[a]

Material	Milling Rate (Å/min)	Energy (eV)
Al	300–700	500
	450–750	1000
Au	1050–1500	500
	1600–2150	1000
Mo	230	500
	400	1000
Ta	130–330	500
Ti	200	500
	200	1000
W	180	500
SiO_2	280–420	500
	380–670	1000
Si	215–500	500
	360–750	1000
GaAs	650	500
	2600	1000
AZ 1350 (Shipley photoresist)	200–420	500
	600	1000
KTFR (Kodak photoresist)	390	1000
PMMA (positive electron resist)	840	1000
COP (negative electron resist)	860	500

[a] See reference 40.

REFERENCES

1. H. C. Gatos and M. C. Lavine, Chemical Behavior of Semiconductors: Etching Characteristics, in *Progress in Semiconductors*, Vol. 9, Temple, London, 1965.
2. H. Gerischer and W. Mindt, The Mechanisms of the Decomposition of Semiconductors by Electrochemical Oxidation and Reduction, *Electrochem. Acta* **13,** 1329 (1968).
3. D. R. Turner, On the Mechanism of Chemically Etching Ge and Si, *J. Electrochem. Soc.* **107,** 810 (1960).
4. H. Robbins and B. Schwartz, Chemical Etching of Silicon, I. The System HF, HNO_3, and H_2O, *J. Electrochem. Soc.* **106,** 505 (1959).
5. H. Robbins and B. Schwartz, Chemical Etching of Silicon, II. The System HF, HNO_3, H_2O and $HC_2H_3O_2$, *J. Electrochem. Soc.* **107,** 108 (1960).
6. B. Schwartz and H. Robbins, Chemical Etching of Silicon, III. A Temperature Study in the Acid System, *J. Electrochem. Soc.* **108,** 365 (1961).
7. B. Schwartz and H. Robbins, Chemical Etching of Silicon, IV. Etching Technology, *J. Electrochem. Soc.* **123,** 1903 (1976).
8. P. J. Holmes, Ed., *The Electrochemistry of Semiconductors*, Academic, London, 1962.

9. W. R. Runyan, *Semiconductor Measurements and Instrumentation*, McGraw-Hill, New York, 1975.

10. D. G. Schimmel and N. J. Elkind, An Examination of the Chemical Staining of Silicon, *J. Electrochem. Soc.* **125,** 152 (1978).

11. C. S. Fuller and H. W. Allison, A Polishing Etchant for III-V Semiconductors, *J. Electrochem. Soc.* **109,** 880 (1962).

12. M. W. Sullivan and G. A. Kolb, The Chemical Polishing of Gallium Arsenide in Bromine-Methanol, *J. Electrochem. Soc.* **110,** 585 (1963).

13. Y. Tarui, Y. Komiya, and Y. Harada, Preferential Etching and Etched Profile of GaAs, *J. Electrochem. Soc.* **118,** 118 (1971).

14. J. C. Dyment and G. A. Rozgonyi, Evaluation of a New Polish for Gallium Arsenide Using a Peroxide-Alkaline Solution, *J. Electrochem. Soc.* **118,** 1346 (1971).

15. S. Iida and K. Ito, Selective Etching of Gallium Arsenide Crystals in H_2SO_4–H_2O_2–H_2O System, *J. Electrochem. Soc.* **118,** 768 (1971).

16. I. Shiota, K. Motoya, T. Ohmi, N. Miyamoto, and J. Nishizawa, Auger Characterization of Chemically Etched GaAs Surface, *J. Electrochem. Soc.* **124,** 155 (1977).

17. Y. Mori and N. Watanabe, A New Etching System, H_3PO_4–H_2O_2–H_2O for GaAs and its Kinetics, *J. Electrochem. Soc.* **125,** 1510 (1978).

18. M. Otsubo, T. Oda, H. Kumabe and H. Miki, Preferential Etching of GaAs Through Photoresist Masks, *J. Electrochem. Soc.* **123,** 676 (1976).

19. D. W. Shaw, Enhanced GaAs Etch Rates Near the Edges of a Protective Mask, *J. Electrochem. Soc.* **113,** 958 (1966).

20. D. J. Stirland and B. W. Straughan, A Review of Etching and Defect Characterization of Gallium Arsenide Substrate Material, *Thin Solid Films* **31,** 139 (1976).

21. W. Kern, Chemical Etching of Silicon, Germanium, and Gallium Arsenide, *RCA Rev.* **39,** 278 (1978).

22. M. P. Lepselter, Beam Lead Technology, *Bell Sys. Tech. J.* **45,** 233 (1966).

23. T. J. Rodgers, W. R. Hiltpold, B. Frederick, J. J. Barnes, F. B. Jenné, and J. D. Trotter, VMOS Memory Technology, *IEEE J. Solid State Circuits* **SC-12,** 515 (1977).

24. B. W. Wessels and B. J. Baliga, Vertical Channel Field Controlled Thyristors with High Gain and Fast Switching Speeds, *IEEE Trans Electron Dev.* **ED-25,** 1261 (1978).

25. E. Bassous, Fabrication of Novel Three-Dimensional Microstructures by the Anisotropic Etching of (100) and (110) Silicon, *IEEE Trans. Electron Dev.* **ED-25,** 1178 (1978).

26. E. Bassous and E. F. Baran, The Fabrication of High Precision Nozzles by the Anisotropic Etching of (100) Silicon, *J. Electrochem. Soc.* **125,** 1321 (1978).

27. D. L. Kendall, On Etching Very Narrow Grooves in Silicon, *Appl. Phys. Lett.* **26,** 195 (1975).

28. M. J. Declercq, L. Gerzberg, and J. D. Meindl, Optimization of the Hydrazine-Water Solution for Anisotropic Etching of Silicon in Integrated Circuit Technology, *J. Electrochem. Soc.* **122,** 545 (1975).

29. R. M. Finne and D. L. Klein, A Water–Amine–Complexing Agent System for Etching Silicon, *J. Electrochem. Soc.* **114,** 965 (1967).

30. A. Reisman, M. Berkenblit, S. A. Chan, F. B. Kaufman, and D. C. Green, The Controlled Etching of Silicon in Catalyzed Ethylenediamine–Pyrocatechol–Water Solutions, *J. Electrochem. Soc.* **126,** 1406 (1979).

31. J. J. Gannon and C. J. Nuese, A Chemical Etchant for the Selective Removal of GaAs through SiO_2 Masks, *J. Electrochem. Soc.* **121,** 1215 (1974).

32. W. C. Dash, Copper Precipitation on Dislocations in Silicon, *J. Appl. Phys.* **27,** 1193 (1956).

33. E. Sirtl and A. Adler, Chromsäure-Flussäure als spezifisches System zur Ätzgrubenentwicklung auf Silizium, *Z. Metallk.* **52,** 529 (1961).

34. F. Secco d'Aragona, Dislocation Etch for (100) Planes in Silicon, *J. Electrochem. Soc.* **119,** 948 (1972).

35. M. W. Jenkins, A New Preferential Etch for Defects in Silicon Crystals, *J. Electrochem. Soc.* **124,** 757 (1977).

36. W. R. Wagner, L. I. Greene, and L. I. Koszi, Defect-Revealing Etches on GaAs: A Comparison of the AHA with the A/B and KOH Etches. *J. Electrochem. Soc.* **128,** 1091 (1981).

37. M. S. Abrahams and C. J. Buicchi, Etching of Dislocations on the Low Index Planes of GaAs, *J. Appl. Phys.* **36,** 2855 (1965).

38. G. H. Olsen and M. Ettenberg, Universal Stain/Etchant for Interfaces in III-V Compounds, *J. Appl. Phys.* **45,** 5112 (1974).

39. L. I. Maisell and R. Glang, Eds., *Handbook of Thin Film Technology*, McGraw Hill, New York, 1970.

40. W. Kern and J. L. Vossen, Eds., *Thin Film Processes*, Academic, New York, 1978.

41. J. S. Judge, A Study of the Dissolution of SiO_2 in Acidic Fluoride Solutions, *J. Electrochem. Soc.* **118,** 1772 (1971).

42. A. S. Tenney and M. Ghezzo, Etch Rates of Doped Oxides in Solutions of Buffered HF, *J. Electrochem. Soc.* **120,** 1091 (1973).

43. W. A. Pliskin and R. P. Gnall, Evidence for Oxidation Growth at the Oxide-Silicon Interface from Controlled Etch Studies, *J. Electrochem. Soc.* **113,** 263 (1966).

44. L. R. Plauger, Etching Studies of Diffusion Source Boron Glass, *J. Electrochem. Soc.* **120,** 1428 (1973).

45. C. A. Deckert, Etching of CVD Si_3N_4 in Acidic Fluoride Media, *J. Electrochem. Soc.* **125,** 320 (1978).

46. D. M. Brown, P. V. Gray, F. K. Heumann, H. R. Philipp, and E. A. Taft, Properties of $Si_xO_yN_z$ Films on Si, *J. Electrochem. Soc.* **115,** 311 (1968).

47. W. van Gelder and V. E. Hauser, The Etching of Silicon Nitride in Phosphoric Acid with Silicon Dioxide as a Mask, *J. Electrochem. Soc.* **144,** 869 (1967).

48. C. A. Deckert, Pattern Etching of CVD Si_3N_4/SiO_2 Composites in HF/Glycerol Mixtures, *J. Electrochem. Soc.* **127,** 2433 (1980).

49. J. R. Hollahan and A. T. Bell, Eds., *Techniques and Applications of Plasma Chemistry*, Wiley, New York, 1974.

50. B. Chapman, *Glow Discharge Processes*, Wiley, New York, 1980.

51. R. J. Poulsen, Plasma Etching in Integrated Circuit Manufacture—A Review, *J. Vac. Sci. Tech.* **14,** 266 (1977).

52. M. Doken and I. Miyata, Etching Uniformities in a CF_4 + 4% O_2 Plasma, *J. Electrochem. Soc.* **126,** 2235 (1979).

53. D. F. Downey, W. R. Bottoms, and P. R. Hanley, Introduction to Reactive Ion Beam Etching, *Solid State Technol.*, p. 121 (Feb. 1981).

54. J. F. Battey, The Effects of Geometry on Diffusion-Controlled Chemical Reaction Rates in a Plasma, *J. Electrochem. Soc.* **124,** 437 (1977).

55. C. J. Mogab, The Loading Effect in Plasma Etching, *J. Electrochem. Soc.* **124,** 1263 (1977).

56. T. Enomoto, Loading Effect and Temperature Dependence of Etch Rate of Silicon Materials in a CF_4 Plasma, *Solid State Technol.*, p. 117 (Apr. 1980).

57. H. Abe, Y. Sonobe, and T. Enomoto, Etching Characteristics of Silicon and its Compounds in a Gas Plasma, *Jpn. J. Appl. Phys.* **12,** 154 (1973).

58. W. H. Harshbarger and R. A. Porter, Spectroscopic Analysis of RF Plasmas, *Solid State Technol.*, p. 99 (Apr. 1978).
59. K. Jinno, H. Knoshita, and Y. Matsumoto, Etching Characteristics of Silicate Glass Films in CF_4 Plasma, *J. Electrochem. Soc.* **124,** 1258 (1977).
60. L. M. Ephrath, Selective Etching of Silicon Dioxide Using Reactive Ion Etching with CF_4–H_2, *J. Electrochem. Soc.* **126,** 1419 (1979).
61. R. A. H. Heinecke, Control of Relative Etch Rates of SiO_2 and Si in Plasma Etching, *Solid State Electron.* **18,** 1146 (1975).
62. G. C. Schwartz, L. B. Rothman, and T. J. Schopen, Competitive Mechanisms in Reactive Ion Etching in a CF_4 Plasma, *J. Electrochem. Soc.* **126,** 464 (1979).
63. K. Tokunaga and D. W. Hess, Aluminum Etching in Carbon Tetrachloride Plasmas, *J. Electrochem. Soc.* **127,** 928 (1980).
64. V. M. Donnelly and D. L. Flamm, Anisotropic Etching in Chlorine-Containing Plasmas, *Solid State Technol.*, p. 161 (Apr. 1981).
65. G. Smolinsky, R. P. Chang, and T. M. Mayer, Plasma Etching of III-V Compound Semiconductor Materials and their Oxides, *J. Vac. Sci. Technol.* **18,** 12 (1981).
66. R. L. Bersin, A Survey of Plasma Etching Processes, *Solid State Technol.*, p. 31 (May 1976).
67. K. Maeda and K. Fujino, The Patterning of Metal Films by Gas Plasma Technique, *Denki Kagaku* **43,** 22 (1975).
68. T. P. Chow and A. J. Steckl, Plasma Etching Characteristics of Sputtered Mo_2Si Films, *Appl. Phys. Lett.* **37,** 466 (1980).
69. E. G. Spencer and P. H. Schmidt, Ion-Beam Techniques for Device Fabrication, *J. Vac. Sci. Technol.* **8,** S52 (1971).
70. K. L. Chopra, *Thin Film Phenomena*, McGraw Hill, New York, 1969.
71. S. K. Ghandhi, P. Kwan, K. N. Bhat, and J. M. Borrego, Ion Beam Damage Effects During the Low Energy Cleaning of GaAs, *IEEE Electron Dev. Lett.* **EDL-3,** 48 (1982).
72. M. Cantragel and M. Marchal, Argon Ion Etching in a Reactive Gas, *J. Material Sci.* **8,** 1711 (1973).
73. C. G. Schwartz and P. M. Schaible, Reactive Ion Etching of Silicon, *J. Vac. Sci. Technol.* **16,** 410 (1979).
74. P. D. Reader and H. R. Kaufman, Optimization of an Electron-Bombardment Ion Source for Ion Machining Applications, *J. Vac. Sci. Technol.* **12,** 1344 (1975).
75. G. R. Thompson, Ion Beam Coating—A New Deposition Method, *Solid State Technol.*, p. 73 (Dec. 1978).
76. L. Maeder and J. Hoepfner, Ion Beam Etching of Silicon Dioxide on Silicon, *J. Electrochem. Soc.* **123,** 1893 (1976).
77. W. Kern, Radiochemical Study of Semiconductor Surface Contamination, II. Deposition of Trace Impurities on Silicon and Silica, *RCA Rev.* **31,** 234 (1970).
78. W. Kern and D. A. Puotinen, Cleaning Solutions Based on Hydrogen Peroxide for Use in Silicon Semiconductor Technology, *RCA Rev.* **31,** 187 (1970).
79. R. E. Blaha and W. R. Fahrner, Passivation of High Breakdown Voltage *p–n–p* Structures by Thermal Oxidation, *J. Electrochem. Soc.* **123,** 515 (1976).
80. M. Polinsky and S. Graf, MOS-Bipolar Monolithic Integrated Circuit Technology, *IEEE Trans. Electron Dev.* **ED-20,** 239 (1973).
81. S. K. Ghandhi, *Semiconductor Power Devices*, Wiley, New York, 1977.
82. D. L Partin, A. G. Milnes, and L. F. Vassamillet, Effect of Surface Preparation and Heat Treatment on Hole Diffusion Lengths in VPE GaAs and $GaAs_{0.6}P_{0.4}$, *J. Electrochem. Soc.* **126,** 1581 (1979).

83. G. Goldfinger, Ed., *Clean Surfaces: Their Preparation and Characterization for Interfacial Studies*, Dekker, New York, 1970.
84. P. Kwan, Characterization of the N.I.B. System and the Effect of Ion Damage Due to Ion Milling of *n*-GaAs, Master's thesis, Rensselaer Polytechnic Institute, Troy, NY, 1981.
85. J. M. Poate, K. N. Tu, and J. W. Mayer, *Thin Films—Interdiffusion and Reactions*, Wiley, New York, 1978.
86. W. Kern, Radiochemical Study of Semiconductor Surface Contamination, III. Deposition of Trace Impurities on Germanium and Gallium Arsenide, *RCA Rev.* **32,** 64 (1971).
87. A. C. Adams and B. R. Pruniaux, Gallium Arsenide Surface Evaluation by Elliposmetry and its Effect on Schottky Barriers, *J. Electrochem. Soc.* **120,** 408 (1973).
88. C. C. Chang, P. H. Citrin, and B. Schwartz, Chemical Preparation of GaAs Surfaces and Their Characterization by Auger Electron and X-ray Photoemission Spectroscopies, *J. Vac. Sci. Technol.* **14,** 943 (1977).
89. C. M. Garner, C. Y. Su, A. Saperstein, K. G. Jew, C. S. Lee, G. L. Pearson, and W. E. Spicer, Effect of GaAs or $Ga_xAl_{1-x}As$ Oxide Composition on Schottky Barrier Behavior, *J. Appl. Phys.* **50,** 3376 (1979).
90. A. Aydinli and R. J. Mattauch, The Effects of Surface Treatments on the Pt/*n*-GaAs Schottky Interface, *Solid State Electron.* **25,** 551 (1982).

PROBLEMS

1. Silicon is to be etched in $HF:HNO_3:CH_3COOH = 1:x:1$. Sketch the etch rate as x is varied from 0 to 20. Describe the characteristic of etches at $x = 0.01$, 1.0, and 10.
2. A 100 cm diameter silicon wafer has 800 μm × 800 μm microcircuits on it. It is required to chemically cut this wafer into chips by anisotropic etching. Outline a scheme for doing this, and calculate the fraction of the surface area that is lost in the process. Assume a 300-μm thick wafer of ⟨100⟩ orientation.

   Suggest a scheme with less waste, assuming the same wafer thickness.
3. A slice of (111) silicon is anisotropically etched through a triangular window, whose sides are in the ⟨110⟩ directions. Sketch this window, the shape of the hole, and the planes delineated by the etch.
4. Repeat Problem 3 for (111) As gallium arsenide and identify the surface atoms on the etch planes.
5. What are the directions of the window edges in Fig. 9.6*b*. Show that the planes delineated by this etch are the {100}.
6. Sketch the (322) plane of gallium arsenide, and identify the atoms in it.
7. A window must be cut in (110) silicon so that all four sides will etch vertically in an anisotropic etch. Sketch this window, identify its directions, and also the etch planes.

CHAPTER

10

Lithographic Processes

CONTENTS

10.1 PATTERN GENERATION AND MASK MAKING 535

10.1.1 Optical Techniques, 535
10.1.2 Electron-Beam Techniques [2, 3], 538

10.2 PRINTING AND ENGRAVING 541

10.2.1 Optical Printing [7], 542

10.2.1.1 Lift-Off Techniques, 548

10.2.2 E-Beam Printing, 550
10.2.3 Ion-Beam Printing, 552
10.2.4 Photoelectron Printing [19], 554
10.2.5 X-Ray Printing [21], 555
10.2.6 Resists, 556

10.3 PROBLEM AREAS 560

10.3.1 Mask Defects, 561
10.3.2 Printing and Engraving Defects, 562

REFERENCES 564

Microcircuit fabrication requires the precise positioning of a number of appropriately doped regions in a slice of semiconductor, followed by one or more interconnection patterns. These regions include a variety of implants and diffusions, cuts for gates and metallizations, and windows in protective cover layers through which connections can be made to the bonding pads. A sequence of steps is required, together with a specific layout pattern, for each of these regions.

Lithographic processes are used to perform these operations, and are carried out in succession during circuit fabrication. Typically, five to ten complete lithographic steps are required on each wafer. By way of example, a conventional silicon-based microcircuit, using bipolar transistors, requires seven separate lithographic processes to define openings for the buried layer, the isolation wall, the base, the emitter/collector, the ohmic contact, the metallization, and the cover layer. Thus lithographic processes play a central role in microcircuit technology.

The placement of a pattern on a wafer can be done in a single step, and some of the research in this area is along these lines. Typically, however, the process is broken into two separate operations: the generation of a mask, and its utilization to define cuts in a large number of wafers. The mask pattern is usually produced by serial exposure, one spatial element at a time. Its design involves circuit and device designers, who together produce the layout of the circuit and of its individual masks. This layout is usually in the form of a set of drawings for circuits of low complexity, and computer tapes for the more complex VLSI circuits. Next, these are converted into masks by the photo-reduction of large drawings, or by the use of computer-controlled pattern generators using optical or electron beams. The sophistication involved here increases rapidly with circuit complexity. As a result, masks for circuits of low complexity are often made with in-house capability, whereas masks for VLSI circuits are generally made by specialty houses, except in the case of large manufacturers. This is warranted on economic grounds, since one master mask set serves for the entire production run of a particular circuit for which it is designed.

The placement of mask patterns on the wafer is an in-house activity, since this must be done on each wafer. It is usually a parallel process, involving many (if not all) of the spatial elements at a time. Optical printing methods are commonly used, but it does not appear that these will meet the requirements for submicron structures over the long term. Electron, X-ray, and ion techniques are under active consideration for this reason. Some schemes, which have the greatest promise for the long term, combine pattern generation and wafer masking into a single operation. The cost of these direct writing schemes will be warranted when the performance of the separate processes proves to be inadequate, and the processed wafer throughput is acceptable.

This chapter describes both photolithographic and electron-beam lithographic processes in some detail, together with the new directions for these technologies. Photoelectron, ion-beam, and X-ray lithography are briefly considered as well.

10.1 PATTERN GENERATION AND MASK MAKING

The most critical part of the lithographic process is conversion of the layout pattern into a *master mask*. This mask is often used directly in projection printing. Alternately, *working masks* are made from this master by contact printing and are used for defining the pattern on each wafer to be processed. It goes without saying that the quality of the master is important in determining the process yield for the microcircuit over its entire production life. Consequently, extreme care is taken to ensure that it will be free from defects.

The complexity of a microcircuit is limited by three factors. The first is the ingenuity of the circuit designer in reducing the number of devices required to perform a given electronic function. Although there have been many dramatic advances along these lines, it appears likely that further progress will be slight, and only in specialized areas of circuit development. The second is the maximum size of the chip that can be made with a reasonable processing yield. Materials and process technologies, outlined in previous chapters, have a strong bearing on this size. The third limit is the size of the *minimum element* which can be placed on the chip. This is determined by lithographic techniques, which are used in conjunction with pattern transfer processes to delineate the various regions in a microcircuit. The minimum element, in turn, defines the size of a *minimum feature* (such as a source region) which is made up of a number of such elements.

Optical techniques for pattern generation, in combination with conventional optical printing methods, can be routinely used to fabricate circuits with a minimum element size of 1.5–3 μm. On the other hand, VLSI technology already requires a 1 μm minimum element size. Over the next 10–15 years, it is hoped that this requirement will be further tightened to 0.1 μm.

Electron-beam techniques show greatest promise for approaching these requirements at the present time. Sophisticated equipment of this type is already being used in conjunction with optical printing methods for the fabrication of masks with 1 μm elements. The extension of these techniques to ion-beam lithography presents one approach to meeting the 0.1 μm goal at the wafer level as well. An alternate approach, direct writing on a chip by electron and ion beams, is also under active consideration.

10.1.1 Optical Techniques

A series of large drawings, each 250 times the final pattern size (× 250), is the usual starting point for many microcircuits of low complexity [1]. One drawing is required for each of the regions (the source/drain diffusions, for example) in the microcircuit. This drawing is made on a plastic laminate, consisting of a dimensionally stable mylar sheet bonded to a thin veneer of red plastic. This plastic is cut on a table known as a *coordinatograph*, and peeled off to form the desired pattern of clear and red regions.

Next, this pattern is illuminated and reduced in size to form a glass *reticle mask* which is usually × 10. This mask consists of a polished* glass plate, coated with a high-resolution, orthochromatic emulsion, about 4 μm thick. This emulsion is insensitive to red light, so that a high contrast pattern can be replicated in it. Quartz plates are used in those situations where subsequent optical processing will be done with deep UV light. Thin plastic or metal films are necessary for masks which are used with X-ray printing techniques.

The pattern formed on these reticle masks can take the form of a gelatin emulsion. Increasingly, hard surface reticle masks are coming into use. These consist of a thin (1000–2000 Å) film of chromium or iron oxide† which is covered by a photosensitive resist. After exposure, this resist is used to delineate the pattern in this film, which provides better edge resolution since it is extremely thin. In addition, it is more hardwearing than the soft gelatin-based emulsion, so that it is less prone to wear and tear during use.

Figure 10.1 shows a flow diagram for this approach. Also shown is an alternate method, in which a computer tape is used to drive an optical pen which directly writes the pattern on the reticle mask. This approach has two advantages over the last. First, mask errors can be readily corrected by altering the computer tape instead of the large drawing. Second, the drawing size becomes unmanageably large with increases in chip size. By way of example, a × 250 drawing for a 0.5 cm × 0.5 cm VLSI circuit would be almost 5 ft × 5 ft in size. Almost all VLSI masks are made from computer-generated patterns for these reasons.

Smaller reticle masks, of × 4 to × 10 can also be made. These are usually of the hard surface type, and are used in some applications, to be outlined later.

Figure 10.2 shows subsequent steps in this process, up to the eventual printing and photoengraving step on the wafer surface. Starting with a × 10 reticle mask, the conventional approach is to use a *step-and-repeat* camera to form the master mask. This camera is essentially an inverted microscope, and projects a × 1 image on an emulsion-coated glass plate, or on an emulsion/hard-surfaced glass plate if this type of master is required. This plate is mounted on a mechanical stage which is programmed to move after each pattern is exposed, in a step-and-repeat manner. In this way, multiple images can be formed on a photographic plate of almost any size that is required to accommodate the silicon wafer, while still permitting the

* Although optically flat over small areas, these plates may have as much as 5-10 μm warpage over their entire working surfaces.

† Borosilicate glass films can also be used if subsequent printing is to be done with deep UV.

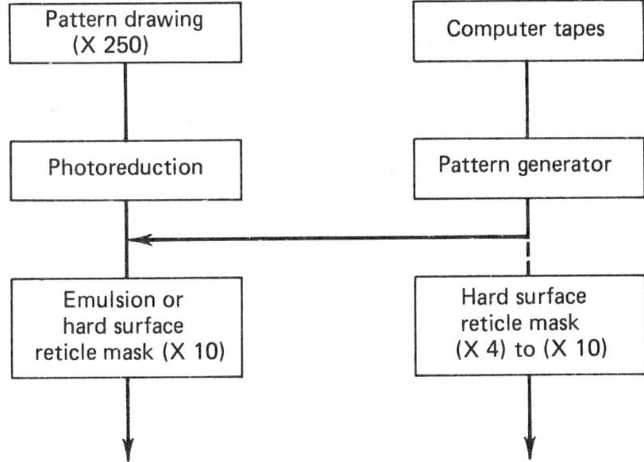

Fig. 10.1 Flow chart for reticle mask making.

optics to cover a relatively small field. Typically, 10 cm × 10 cm ultrahigh-resolution plates are used for work with 4 in diameter wafers.

The step-and-repeat process can be speeded up by the use of a multi-barrel camera which projects four images of the reticle at one time. Multi-barrel systems are in common use today.

Contact printing of the master, or of an intermediate submaster, is used

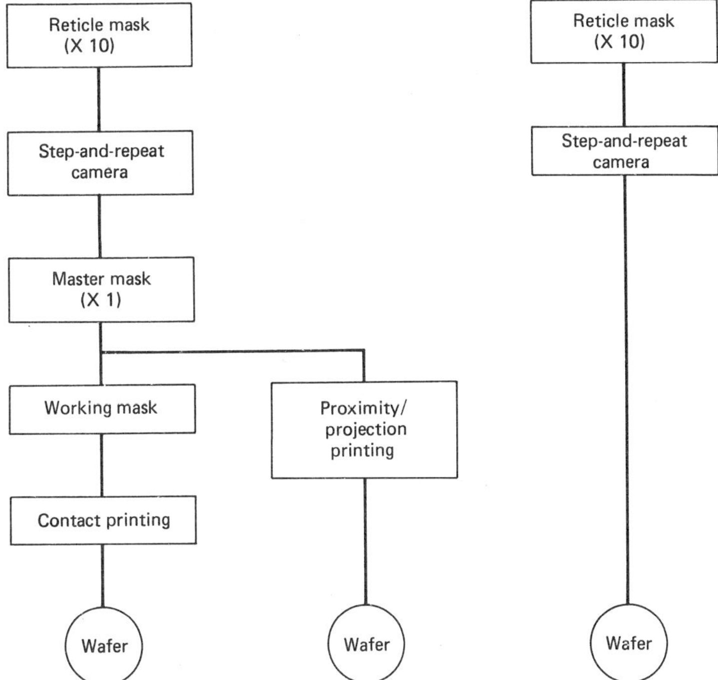

Fig. 10.2 Subsequent steps from reticle mask to printing.

to make multiple working masks of each pattern. These masks are used to define the pattern on each microcircuit wafer, and are thus subjected to considerable wear and tear. Typically, the life of each working mask ranges from 5 to 25 wafers, depending on the resolution requirements of the microcircuit.

Patterns can also be placed on the wafer by projection printing, or by printing with a small proximity gap between the photographic plate and the wafer (see Fig. 10.2). Both schemes result in essentially infinite mask life, so that a master mask can be directly used here. Typically, a hard surface mask is dictated for this application.

Improved resolution can be obtained by directly imaging the reticle mask on to the wafer in a step-and-repeat sequence. Hard surface reticle masks, $\times$ 10, are commonly used here, with a single-barrel camera, as shown in Fig. 10.2. The main disadvantage to this approach is the greatly increased time required for processing each wafer. However, improvements in photoresist sensitivity may reduce this problem in the future.

10.1.2 Electron-Beam Techniques [2, 3]

The fundamental limit to optical techniques, i.e., the diffraction limit, is set by the wavelength of light. This limit can be slightly extended by the use of deep UV ($\lambda \simeq 2000$ Å). Nevertheless, the minimum element size that is obtained by optical pattern generation and printing techniques in a manufacturing environment is about 1.5 μm. On the other hand, VLSI technology over the next 10–15 years will necessitate minimum element sizes as small as 0.1 μm.

Electron-beam writing does not suffer from a practical diffraction limitation. At the present time, the use of this technique for mask making, in combination with optical printing techniques, allows the fabrication of circuits with a 1 μm element size. It should be capable of providing a workable technology in the submicron range when coupled with nonoptical printing methods such as those employing electron and X-ray techniques.

The flow chart for a lithographic system using E-beam mask making equipment, in combination with optical printing, is shown in Fig. 10.3. Here the layout information is provided in the form of computer tapes, since most circuits using this approach are of VLSI complexity. These data are used to control an electron beam which can be driven to directly produce a hard surface reticle mask at a $\times$ 1 to $\times$ 10 magnification. The reticle mask for this purpose is of the same type as for optical patterning described in Section 10.1.1, except that an electron-sensitive resist is placed on the hard surface. This mask is eventually used in printing schemes of the type outlined earlier. Thus a step-and-repeat camera is used to produce the $\times$ 1 master from which working masks are made by contact printing. Alternately, a step-and-repeat system is used for direct writing on the wafer.

The basic components of an E-beam pattern generator are shown in Fig.

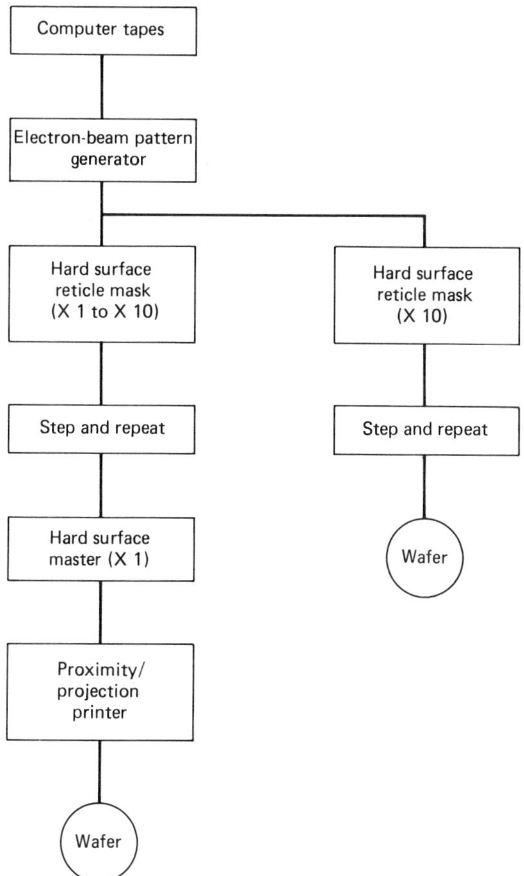

Fig. 10.3 Flow chart for E-beam mask making and optical printing.

10.4. The system resembles, in many ways, a scanning electron microscope with the addition of beam blanking and computer-controlled deflection. Important additional features are the use of a laser-driven stage and fiducial mark detectors [4]. This combination, together with fiducial marks which are printed during the first pattern writing, allows thc system to be precisely positioned for each successive step-and-repeat operation. Typically, the beam has a deflection field of about 2 mm × 2 mm. Consequently, a complete VLSI mask usually requires "stitching" together a number of such fields to form the entire pattern [5].

Two types of scan systems are in use today—the raster scan and the vector scan [6]. In the raster scan system, rectangular strips of the circuit are scanned by a series of lines in order to form the complete chip pattern. In a vector scan system, on the other hand, the E-beam is controlled to scan a feature, move directly to the next feature, and so on. While this often requires wider scan deflection and considerably more complex data handling

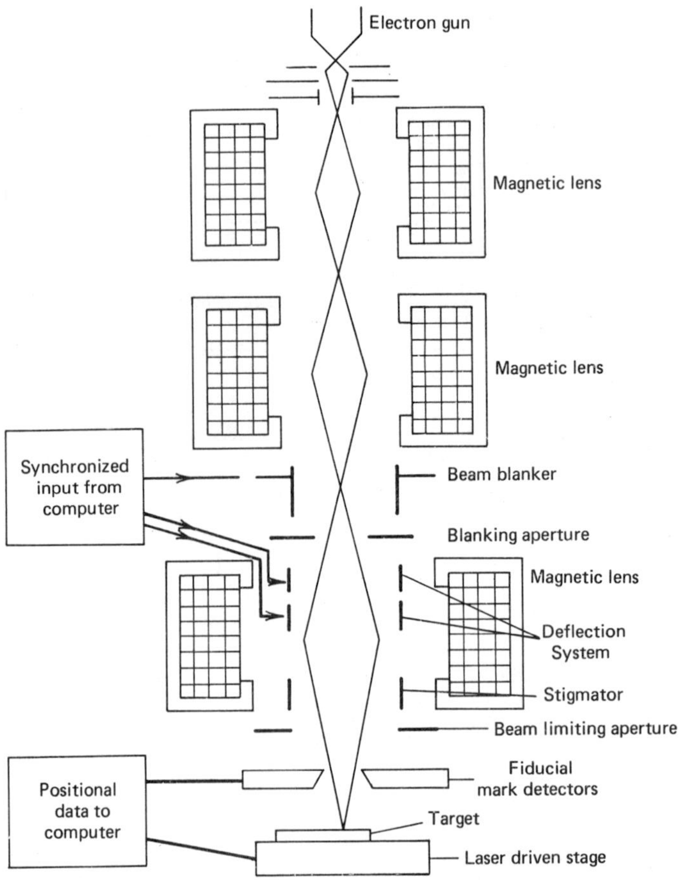

Fig. 10.4 Schematic of an E-beam pattern generation system. From Thornton [2]. Reprinted with permission from *Advances in Electronics and Electron Physics.*

and beam blanking techniques, it results in a much faster system since the beam does not spend time scanning featureless regions. This is an important advantage, since one of the main limitations of E-beam systems is their low throughput.

Electron beam machines are used to make the master mask. Here a writing time of about 1 h for a 10 cm × 10 cm mask size is considered acceptable. This is due to the fact that a single master is used to define one particular set of regions (base diffusions, for example) for an entire microcircuit production run. On the other hand, multiple working masks are required for the many printers that are used to perform the engraving operations at the wafer level.

The writing time of mask making equipment of this type is set primarily by limitations in the intensity of the electron beam, and the sensitivity of the E-beam resist. Factors such as mask loading and pump-down must also be considered, but are relatively unimportant, due to the availability of load-lock techniques and high-speed vacuum systems.

Direct E-beam writing on the wafer promises the ultimate in resolution as well as flexibility. Thus the ability to compose the pattern at the wafer level will allow the possibility of stitching a large number of different fields on the wafer, and will ultimately result in fabricating the entire computer on a single slice. An important advantage of this technique is the ability to stretch the writing field electronically in order to compensate for departures from wafer flatness due to processing. This automatic correction of distortion will become increasingly important as the minimum element size becomes comparable to such circuit features as steps in the oxide.

The practical realization of these capabilities will, however, necessitate a total exposure time of no more than 5–10 min for a 10 cm diameter.* This will require improvements in high-intensity guns, new systems for exposure such as matrix lens techniques which can be used to form multiple circuit patterns simultaneously, improved drive electronics, and new scanning systems which minimize the dead time between active semiconductor regions. Last, but most important, new electron resists, with greatly improved sensitivity, will be required for these systems.

10.2 PRINTING AND ENGRAVING

The printing process consists of taking a suitably patterned mask, and imaging it on the surface of a wafer which is precoated with a resist. This is followed by the engraving process, where the exposed resist pattern is used for opening windows in a protective underlying mask layer to define semiconductor regions, or to remove metal from a coated wafer in order to delineate the interconnection pattern. These processes must be carried out on each wafer, and for each of the five to ten masking operations that are necessary for any particular microcircuit. Thus the amount of time that this step takes is an important consideration. At the present, it is generally accepted that the time required to insert, align, and expose the wafer should be 1 min or less. This requires resists of high sensitivity, as well as automatic (and accurate) means for alignment of the image on the wafer.

This section describes the basic processes for printing and pattern engraving. Optical printing processes are described in some detail since they are the most commonly used. This is followed by brief descriptions of other methods which are under active development, and appear as promising replacements for optical printing in the future. A description of resists, suitable for the different printing techniques, concludes this section.

The emphasis in this section is on processes which handle the entire wafer in a single operation. An important problem, common to all such systems,

* Present optical printers can perform this operation, at the 1 μm level, in 1 min.

is the lack of wafer planarity, combined with its process-induced curvature which varies as the wafer progresses from the as-bought state to the finished product. These in-plane distortions are a function of the specific wafer process, and can be as large as 1–2 μm, so that the success of printing schemes of this type will eventually depend on the development of processes which maintain wafer planarity throughout its fabrication steps.

10.2.1 Optical Printing [7]

This process can be described by considering first the problem of cutting windows in a film of the type that is used for masking purposes. Typical films are silica, polysilicon, silicon nitride, silicides, or refractory metals. Steps in this operation are detailed in the flow chart of Fig. 10.5; some of the results of these steps are shown in Fig. 10.6.

Coating with Photoresist. This consists of laying a film of a photoresist material on the surface of the wafer (see Fig. 10.6*b*). Ideally, such a film should be thin, highly adherent, uniform, and completely free from dust or

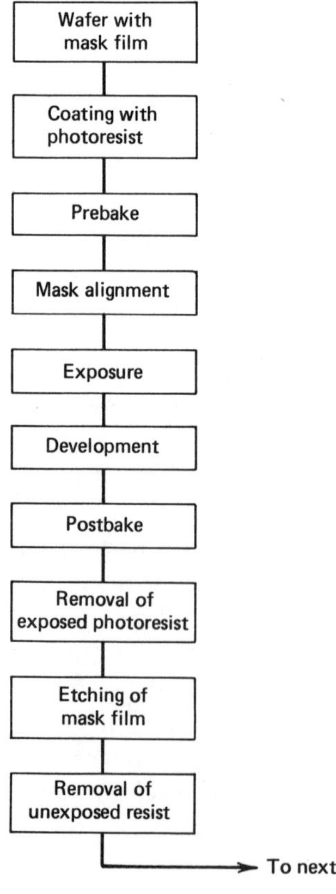

Fig. 10.5 Flow chart for the printing and engraving process.

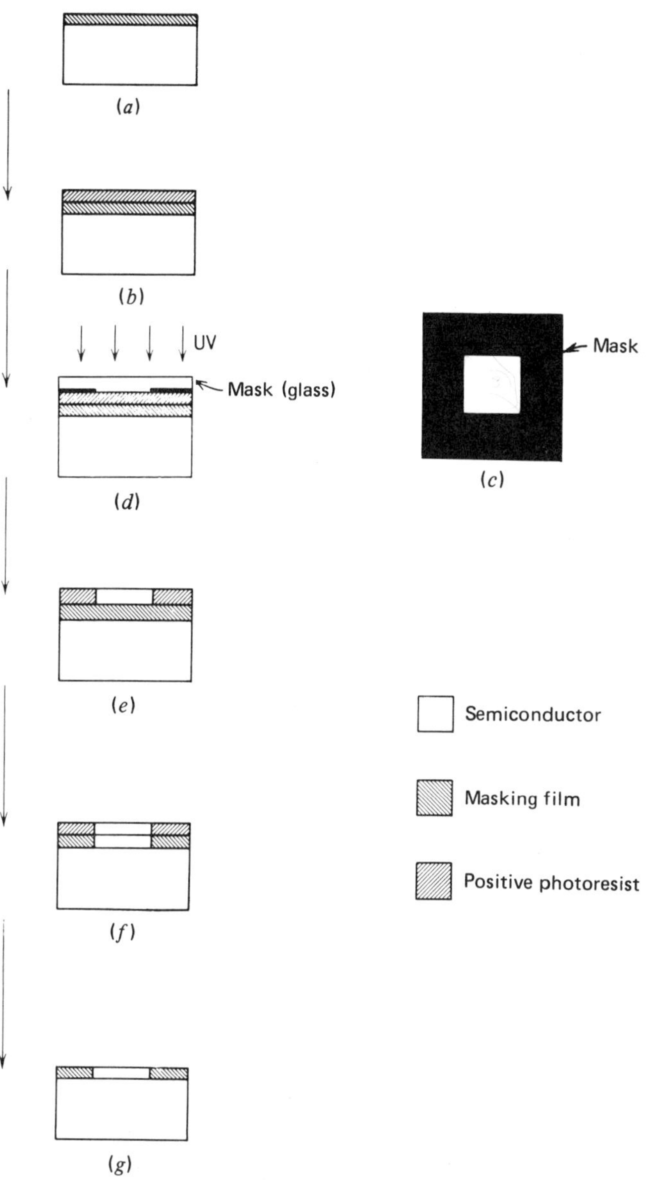

Fig. 10.6 Steps in opening a window in a masking film.

pinholes. A positive photoresist is used in the example of Fig. 10.6, for illustrative purposes.* Photoresists of this type become degraded upon exposure to ultraviolet light [8]. Negative photoresists, on the other hand, become toughened by this irradiation.

The coating process consists of spinning the wafer at high speed after a small quantity of prefiltered photoresist has been placed on it. The film thickness is inversely proportional to the square root of the spin rate; typically, spinning speeds range from 1000 to 5000 rpm and result in films that are about 0.5–2.5 μm thick. It is necessary to bring the spinner rapidly up to full speed in order to obtain a uniform coating. Consistent results are obtained only if the viscosity of the photoresist is maintained constant on a routine basis.

Extreme care must be taken to use clean, dry slices to obtain good adhesion of the photoresist. Freshly prepared wafers may be coated directly; however, slices that have been stored must be subjected to cleaning and drying procedures before coating. Adhesion to some surfaces, such as phosphosilicate glass, often presents a serious problem. A dip in a coupling agent, just prior to photoresist application, is sometimes used in these problem situations. Hexamethyldisilizane, $(CH_3)_3SiNHSi(CH_3)_3$, commonly abbreviated to HMDS, is often used for this purpose [9].

Although ultraclean conditions should be maintained during the entire operation, the coating step is the most critical one from the point of view of dust contamination. This is because the spinning action creates an air suction along the axis of the slice and promotes the delivery of any airborne particles of dust to its surface. Moreover, the photoresist is sticky at this point.

Drying. After coating, the slice is heated for about 10–20 min at 80–90°C to drive out all traces of solvent from the photoresist. With positive photoresists, the drying time and temperature are critical, and care must be taken to follow the manufacturer's recommendations closely in order to avoid changes in sensitivity to subsequent exposure.

Mask Alignment. There are many variations here. Conventionally, a mask consisting of a gelatin photographic emulsion on a glass plate is placed over the slice, brought into contact with it, and then backed off slightly to produce an air gap. Next it is manipulated into its desired position by micrometer adjustment. This alignment process is performed with the aid of a microscope. Finally, physical contact is reestablished between the mask and the slice, and the exposure is made. This is known as contact printing (see Fig. 10.6*c* and *d*).

Most of the variations in this approach have come about because of the needs for improved resolultion, increased throughput, or economy. For ex-

* The subject of photoresists is considered in some detail in Section 10.2.6.

ample, physical contact between the mask and the wafer causes damage to the soft gelatin emulsion. In addition, this damage results in defects which are then transferred to *all* successive slices using this mask. As a result, mask life is very short. Depending upon the density and resolution requirements, these "working masks" must be replaced after every 5–25 operations.

A hard surface mask is sometimes used instead of the gelatin photographic plate. This consists of a patterned thin film (1000–2000 Å) of some hard material such as iron oxide or chromium on a glass substrate. Masks of this type were originally introduced because they were three to four decades more abrasion resistant than gelatin masks, and were dimensionally stable during wet chemical processing. In VLSI applications, their primary advantage lies in their superior edge resolution characteristics and absence of shadowing effects, because of their extreme thinness. (Typically, gelatin masks must be 4 μm thick in order to have the same opcity as these materials.)

Masks with an iron oxide coating have a number of advantages over chromium masks. They are about five times more abrasion resistant and provide better adhesion to the photoresist than does the chromium film [10]. Although opaque to ultraviolet light, they are transparent in visible light. This see-through characteristic simplifies both visual and automatic alignment of these masks. Unfortunately, they are not suitable for E-beam patterning, since iron oxide is an insulator.

An alternate approach, which is becoming increasingly popular, is to use proximity gap printing instead of contact printing. Here a small gap, 2.5–25 μm wide, is always maintained between the mask and the slice. This necessitates a collimated light source. A large-area flood source can be used; however, a line source, under which the mask–wafer combination is mechanically scanned, is an alternate for systems where flood sources are not available. With collimated sources, it can be shown that the minimum resolution increases with the square root of the gap, so that there is a slight loss of sharpness in this approach. The width of this gap is thus dictated by mask and wafer flatness, and by the precision of the mechanical system. Its incorporation avoids damage to the mask, and is thus a great advantage in VLSI applications.

An alternate to proximity gap printing is projection printing, where the mask pattern, which may be anywhere from one to ten times the actual size, is imaged on the plane of the slice. This technique is comparable in resolution to that obtained by proximity gap printing for chip sizes up to 0.5 cm × 0.5 cm, and allows greater flexibility in wafer handling. However, it may have difficulty in meeting the resolution and size requirements of future systems.

Printing of VLSI circuits requires rapid, accurate mask alignment in addition to high resolution. A rule of thumb in the industry is that alignment accuracy should be about one-fourth the minimum element size. Thus a 0.25 μm alignment accuracy is required in 1 μm systems. These specifications can only be met by automatic mask alignment systems.

Automatic mask alignment can be built into systems in which the mask does not come into contact with the wafer. Typically, a series of fiduciary marks are placed on the mask, and illuminated by means of a He-Ne laser. The diffraction pattern produced by the edge of these marks is measured by photosensors, and the mask movement automatically adjusted for equal response. Moiré patterns, reproduced on the mask, can also be used in automatic alignment schemes.

Exposure. Photoresists are exposed by means of collimated ultraviolet light. Some filtering of the source is necessary to prevent undue heating of the mask during exposure. On the other hand, monochromatic light sources are undesirable because they promote the formation of standing wave patterns during printing, which can produce variations in the defined resist pattern. This problem is of increasing importance in VLSI technology, as the resolution requirements approach the diffraction limit. Finally, improvements in optical resolution can be achieved by using shorter wavelengths [11], in the deep UV region (2000–3000 Å). This is true for both contact and proximity-gap printing systems.

A high intensity (1–3 kW) Xe-Hg lamp serves as the light source in deep UV systems. Plates of synthetic quartz are used for the masks, since they are transparent at these wavelengths. A thin film of borosilicate glass, iron oxide, or chromium can be used to form the masking pattern on these plates, since these materials are opaque to deep UV.

Development. The slice is now rinsed in an appropriate developer, specified by the photoresist manufacturer. This dissolves the exposed positive photoresist but does not affect the unexposed regions (see Fig. 10.6*e*). A short postbake period follows at a temperature specified by the manufacturer (about 70–150°C), and toughens the remaining photoresist.

Etching. The slice is now etched in a chemical which can dissolve those parts of the underlying film that are not covered by the photoresist. This results in the formation of windows in the mask film, as shown in Fig. 10.6*f*. To avoid unnecessary undercutting, this process is monitored and arrested as soon as full etching is accomplished. Increasingly, plasma etching, reactive ion etching, or ion milling are used to remove this unwanted material. The advantages of these dry etching processes for VLSI applications have been outlined in Chapter 9.

Stripping. The final step consists of removal of the exposed photoresist. Positive photoresists are usually removed by means of a chemical solvent such as acetone or methylethylketone. Negative photoresists are considerably harder to remove. Here one approach is to boil the slices in concentrated sulfuric acid for about 20 min, followed by mechanical agitation. Often a 1:1 mixture of hot sulfuric acid and hydrogen peroxide is used, at a somewhat lower working temperature.

The most common removal technique for negative photoresists consists of using hot chlorinated hydrocarbons to swell the polymer, together with acids to loosen its adhesion to the substrate. Solvent mixtures of trichlorethylene, methylene chloride, and dichlorobenzene, combined with formic acid or phenol, are used in the form of proprietary mixtures for this purpose.

The most recent technique for photoresist removal is by plasma oxidation (commonly called *plasma ashing*), which has many advantages over wet chemical methods. Thus undercutting by capillary action of the liquids is avoided, so that photoresist adhesion is not critical. Waste disposal problems are eliminated, since only a small amount of plasma reaction products (water, carbon dioxide, and carbon monoxide) are produced. Perhaps the most important advantage is that a number of monitoring methods are available to signal the end of the process, without the need for visual inspection. Because of this, plasma ashing processes can be automated. These factors add up to a major cost advantage for this approach, so that it has received rapid acceptance by industry.

Plasma methods present a problem in the area of silicon MOS circuits. Here radiation damage to the gate oxide can cause significant threshold shifts, especially for short-channel, low-threshold devices. However, subsequent annealing at about 400°C has been found to reduce the effects of this damage to within acceptable design limits.

The engraving process for the contacts and the metallization is only slightly different from the above procedure. Here aluminum and gold–Ti/W are the most commonly used materials, as outlined in Chapter 8. The insulating layer covering silicon microcircuits usually consists of silicon dioxide, whereas silicon nitride is commonly used with gallium arsenide circuits. Contacts and interconnections are made as shown in Fig. 10.7.

By use of a mask with a contact pattern, windows are cut in the insulating layer over the appropriate regions. This is shown in Fig. 10.7*a* for an n^+–p diode. The metallization film (or films) are deposited over this insulating layer, and make contact to the semiconductor, as in Fig. 10.7*b*. The wafer is now covered with photoresist and exposed through a mask carrying the interconnection pattern (Fig. 10.7*c*). Positive photoresists are almost universally used in this step. After development, the metallization that is not protected by the photoresist is removed by suitable etching techniques, as described in Chapter 9. The resulting metal connects to the contact regions, and runs over the insulating layer, as shown in Fig. 10.7*d*.

Interconnection lines may be widened once contact is made to the device. In addition, it is possible to widen these lines further to form large bonding pads to which leads may be attached. Both situations are shown in the plan view of Fig. 10.7*e*.

The slice is now heated in an inert gas ambient to a temperature of about

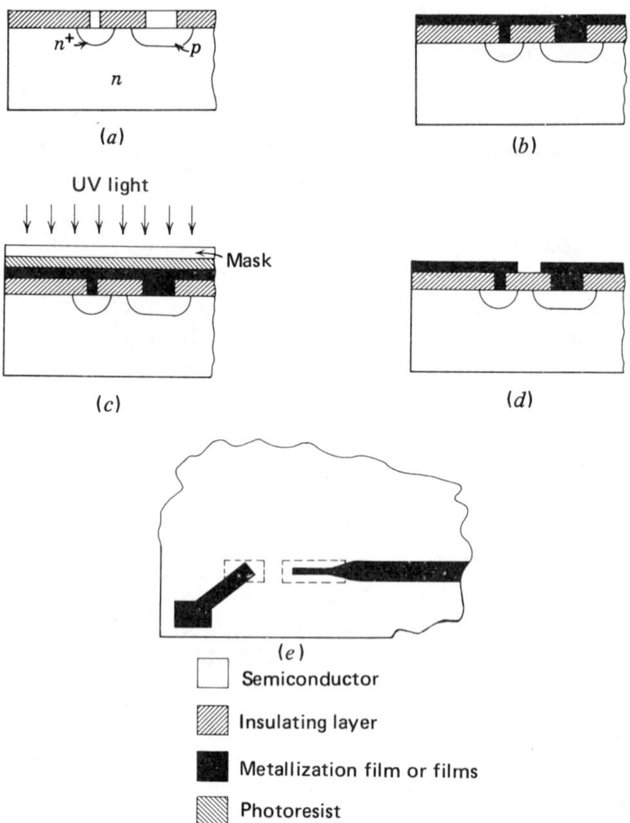

Fig. 10.7 Steps in making contacts and metallization.

450–500°C. This serves to make the contact to the semiconductor, and also to bond the metallization to the insulating surface. Cracking or wrinkling of the photoresist during this process is relatively unimportant with conventional devices and microcircuits. With VLSI circuits, however, it is customary to remove the remaining photoresist prior to the annealing step, in order to avoid damage to the fine-line metallization. This is done by means of organic solvents, or by plasma ashing. Strong oxidizing agents or mechanical scrubbing cannot be used at this point, because of potential damage to the delicate metal film.

10.2.1.1 Lift-Off Techniques

The *lift-off technique* is an important variation of the above process [12]. Here the placement of, say, a metal pattern on an insulating substrate proceeds along the following lines. First a photoresist film is placed on the substrate, and patterned so that it covers all those regions where *no* metal

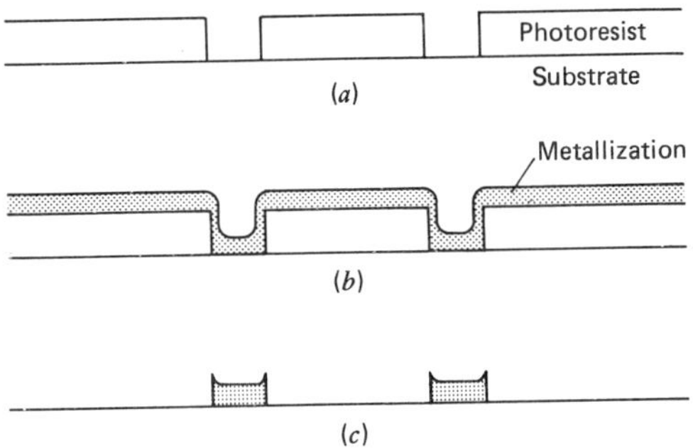

Fig. 10.8 The lift-off process.

film is desired, as in Fig. 10.8*a*. Next, the metal film is deposited over the substrate-resist combination, as shown in Fig. 10.8*b*. Thus the metal film contacts the substrate *only* in those regions where it is required. Finally, the photoresist is removed by a solvent which does not attack the metal film. In so doing, it "lifts off" the material which is on its surface, leaving behind the patterned metal film, as in Fig. 10.8*c*. The success of this approach requires the use of a relatively thick photoresist film so that the deposited metal film is very thin at the sides of the step, or even discontinuous. This allows rapid dissolution of the photoresist mask, and makes it easier to lift off the raised portions of the film without breakage.

Two advantages are realized by this process. First, it can be used with films such as platinum, gold, and the silicides, which are very difficult to etch by conventional means. Next, the use of a thick photoresist avoids problems due to pinhole formation. Unfortunately, however, the definition of fine-line patterns for VLSI applications requires the use of thin photoresist layers. These conflicting requirements can be accommodated by a double-layer process, where the upper layer is made very thin [13]. At the same time, the lower layer is made relatively thick, and is selected to be more easily soluble. Now the patterned photoresist film takes the overhang form shown in Fig. 10.9*b*. Vacuum evaporation of the metallization results in a discontinuous film as shown in Fig. 10.9*c*, whose subsequent lift-off can be accomplished without tearing or breaking.

A variety of thin films, both organic and inorganic, can be used for these multilayer structures. Moreover, different techniques can be used (vacuum deposition, sputtering, chemical vapor deposition, etc.) for placing the more readily etchable layer. The simplest approaches, however, use photoresist for both layers. One scheme consists of two layers of different photoresist material. A simpler technique, however, is to spin on a thick layer of a single photoresist, and toughen its surface by a pretreatment [14]. Positive photoresists such as AZ1350-J* can be surface toughened by immersion in an aromatic solvent such as chlorobenzene for a fixed period of time after the

* Shipley Company, Newton, MA.

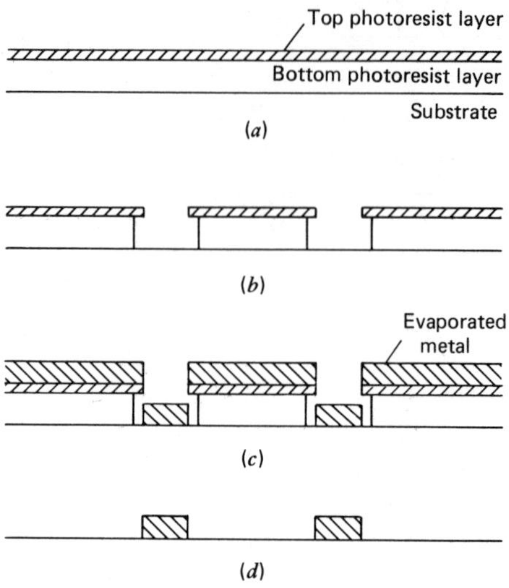

Fig. 10.9 The lift-off process, using a double-layer mask.

prebake period, and before or after pattern exposure. Control of the soak time and prebake temperature permits photoresist cuts with the overhang pattern shown in Fig. 10.10. Here the AZ1350-J photoresist was modified to a depth of 2000 Å by the use of chlorobenzene prior to development. This thin upper layer allows high-resolution patterning. At the same time, the thick photoresist (1–2 μm) permits the use of metal films of sufficiently large cross section, so as to handle the current requirements of high-speed circuits. Thus the approach represents a practical alternative to the anisotropic plasma etching methods described in Chapter 9. Furthermore, it avoids the problems of damage and oxide charge associated with plasma processes, and is extensively used in VLSI applications for this reason.

10.2.2 E-Beam Printing

We have seen that E-beam techniques can be used to circumvent the optical diffraction limit. Masks made by this method, in combination with optical printing, are in use today for producing microcircuits with a 1 μm minimum element size. However, it will be necessary to remove the optical diffraction limit during printing as well, if the eventual goal of a 1000 Å element size is to be achieved. Thus there is a considerable impetus to extend the use of electron beams to printing at the wafer level.

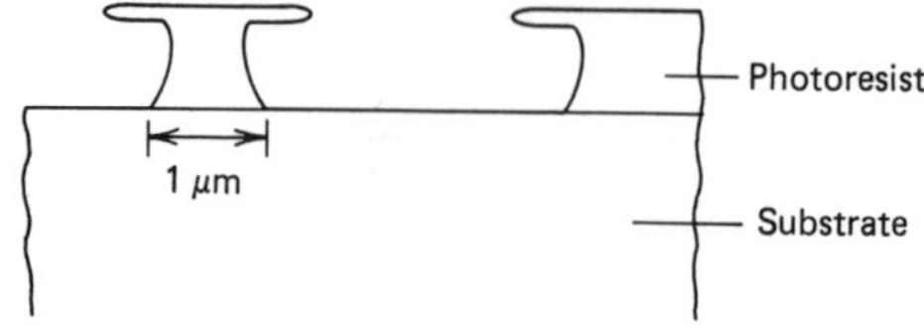

Fig. 10.10 Profile of photoresist using a chlorobenzene-modified film prior to development. Adapted from Hatzakis, Canavello, and Shaw [14].

One approach is to use an E-beam system that combines pattern generation and printing in a single machine [2]. An alternate approach is to separate these functions into two distinct systems. Since five to ten masking operations are involved on *each* wafer during microcircuit fabrication, there is a considerable advantage to a separate projection system which can expose a complete chip, or even an entire wafer, in a single operation.

The viability of a printing system also depends on the time taken to perform a single printing (plus alignment) operation. Thus a 1 h process time for fabricating a master mask is entirely acceptable. However, the printing time for a wafer must be held below 5 min for the process to be practical.* In addition, most microcircuit fabrications require a number of in-house printers to meet their production needs, so that the cost of each printer is also an important consideration.

E-beam projection printing is one approach that is presently under consideration [15]. A schematic of a demagnifying projection system for this purpose is shown in Fig. 10.11. Here a self-supporting foil mask is used with a flood electron gun source which provides a collimated beam of electrons. The reticle mask, usually × 10, is imaged directly on a photoresist-coated wafer which is located on a laser-controlled table for positioning purposes.

Systems of this type are considerably simpler than E-beam systems used for pattern generation, since the beam is not required to maintain its precise convergence under scan conditions. This greatly simplifies the electro-optical system that can be used, in addition to the digital electronics required for data handling and beam deflection. Moreover, the system has the potential of handling a complete wafer in one exposure, although most of the present research involves step-and-repeat techniques for printing one chip at a time. Even with this limitation, electron projection printing is considerably faster than are direct writing systems where the pattern is written serially, one element at a time. This can be a major advantage in a development situation, where it can provide quick turnaround for prototype circuits.

Problems of mask fabrication for this system are severe, since the pattern consists of a large number of isolated structures. Moreover, they must be physically connected for a mask to exist. One approach is to use a fine grid or mesh on which the pattern is placed. The resulting fine bars which connect isolated structures produce a small distortion at the edge of each structure; thus they can be kept within acceptable limits if a × 10 mask is used.

Problems of mask alignment in these systems have already been solved, as have mechanical problems such as wafer loading and automatic pump-down. Thus the eventual success of E-beam projection printing systems will depend on advances in electron gun sources and in resist technology. Together these hold the key to the design of economical systems with fast throughput capability.

* By way of comparison, a printing time of 1 min per wafer or less is achieved by optical methods.

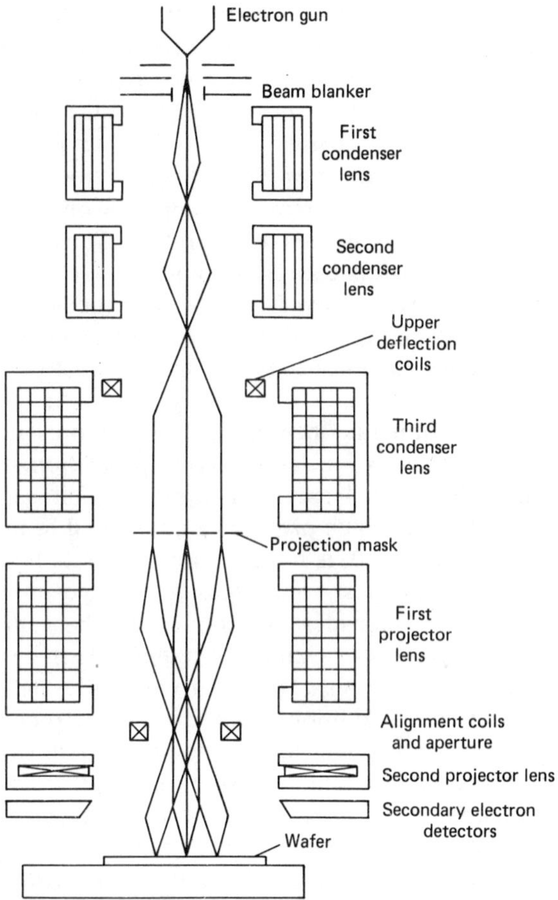

Fig. 10.11 Schematic of a demagnifying electron projection printing system. From Thornton [2]. Reprinted with permission from *Advances in Electronics and Electron Physics.*

10.2.3 Ion-Beam Printing

The use of ions instead of electrons in the lithographic process has two potential advantages [16]. First, ion beams do not suffer from as much scattering as electron beams, when they penetrate a resist. In addition, secondary electrons produced by ion bombardment have very low energy so that scattering caused by them is limited. The ultimate limit to the minimum element size is set by these scattering processes.

Next, the ion sensitivity of resists is a function of the ion energy and mass, and is many orders of magnitude greater than that for electrons. Thus ion-beam printing is potentially more rapid than electron-beam printing.

Ion projection printing systems, ultilizing a rf ion source and focusing optics, have been developed [17] with ion sources using H^+, He^+, and Ar^+ in the 100 keV range, to image a field 5 mm × 5 mm in a single exposure. Important problems with this system are its large size, so that it is prone to vibration. Moreover, considerable work needs to be done to improve the resolution of the ion-optical system in order for it to meet the 1000 Å minimum element requirement.

An alternate approach, shown in Fig. 10.12, uses the channeling action of ⟨110⟩ silicon in a proximity printer [18]. Here a metal mask pattern is placed on a thin, suitably oriented, single-crystal silicon sheet. A spacing of about 20 μm is held between the silicon and the resist coated substrate, into which the pattern is to be transferred. Only channeled ions are transmitted through the mask, resulting in a collimated beam which arrives at the substrate. A 180 keV H^+ beam is used in this system.

Problems with this system are associated with the difficulty of forming a large-area membrane of single-crystal silicon. In addition, the beam spread from such a system will eventually limit the minimum achievable element size.

Direct writing by ion beams is also under consideration. Here the critical elements are finely focused ion sources of the desired impurity. Although much progress has been made in their development, the writing speed of these systems has confined their use to special applications.

Ion-beam systems are in their very early stages of development. However, there are strong indications are that they will be suitable for 1000 Å features when their many problems have been resolved.

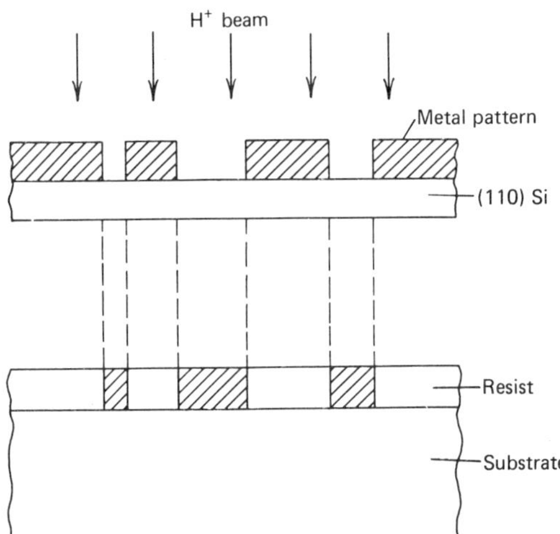

Fig. 10.12 Schematic of a channeled ion projection printing system.

10.2.4 Photoelectron Printing [19]

This is an alternate approach to the use of electrons in the printing process. Thus it shares all of the advantages of E-beam printing, and the same considerations of cost and throughput must be applied in order to evaluate its potential use in this application.

A photoelectron projection printing system is shown [20] in schematic form in Fig. 10.13. Here the mask consists of a quartz plate, coated with an UV-opaque film. A pattern is delineated in this mask by E-beam techniques, as described in Section 10.1.2. Next, the plate is covered with a thin film (50–100 Å) of photoemissive material such as palladium. This is placed in a vacuum environment, about 1 cm away from a wafer which is coated with an electron-sensitive resist. A high voltage (10–20 kV) is applied across these, with the mask held negative. The entire sandwich is placed between a system of external focusing and deflection coils.

In operation, the mask is illuminated with deep UV light (2000–2500 Å) from an external source, as shown. This results in the generation of a photoelectron emission pattern which is imaged on the electron resist by means of a homogeneous magnetic field.

Deflection coils are used for automatic mask alignment in this system. This is done by the placement of heavy metal markers on the wafer, which generate X-ray signals upon bombardment by the electron beam. Alternate systems, requiring the drilling of holes in the wafer, have also been proposed. Alignment accuracies of 0.2 μm have been achieved using both these techniques.

Photoelectron printing is attractive since there is no limit on the size of the wafer that can be processed in a single printing operation. However, there are many problems which will have to be solved before this approach

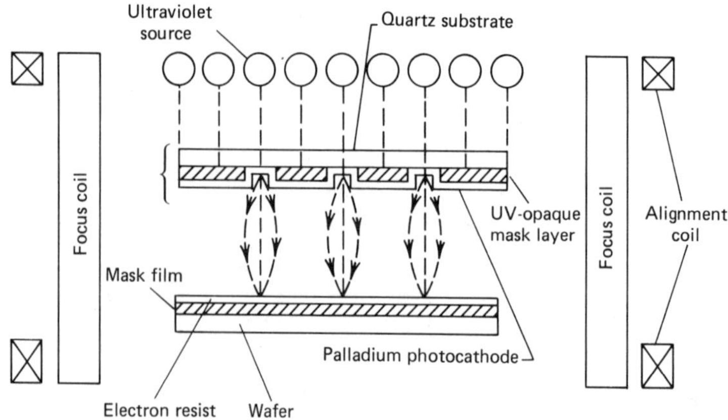

Fig. 10.13 Schematic of a photoelectron printing system. From *Livesay* [20]. Reprinted with permission from Solid State Technology.

becomes commercially viable. These include problems with charge retention on the resist surface, the chromatic spread of photoemitted electrons, and interactions between the magnetic fields and the backscattered electrons during the printing operation. In addition, further improvements in electron resists will reduce the printing time per wafer.

10.2.5 X-Ray Printing [21]

The attractiveness of this approach stems from the extremely short wavelengths that are possible with soft X-ray sources (< 50 Å). This eliminates the diffraction limitation of optical and ultraviolet lithographic techniques, and has the potential of providing a means for fabricating submicron structures at a cost that is competitive with optical projection systems. There are, however, major differences in the design of these systems. Thus optical systems can use collimated light, whereas X-ray systems are generally restricted to a point source. Proximity gap printing is highly desirable in both approaches, since this allows for rapid and precise alignment by automatic techniques. In an X-ray system this results in a slight magnification of the pattern. While this is usually small enough to be neglected, the magnitude of the gap (and more important, its spatial variation), must be kept constant to prevent distortion effects. This necessitates a somewhat larger gap than is used in optical proximity gap printers. Typically, systems of this type can be operated with a gap of about 25–40 μm and still give satisfactory performance at the submicron level.

The ideal source for X-ray lithography is a point source. This is produced by a high-intensity electron beam impinging on a target, which is water cooled and rapidly rotated to prevent overheating. The choice of X-ray wavelengths is relatively wide, ranging from palladium (4.36 Å) and upward. By way of example, an aluminum target and a 10–20 keV beam can be used to produce X-rays from a wavelength of 8.34 Å (the aluminum K_α line). These are filtered to remove stray radiation, and are fed out of the vacuum through a beryllium window [22], as shown in Fig. 10.14. All processing of slices in these systems can be done outside the vacuum for X-ray wavelengths below 9 Å, which is a distinct advantage over E-beam lithography. The beam is usually confined to a helium gas environment to prevent unnecessary absorption. However, X-ray lithography is relatively insensitive to airborne particles, so that cleanliness requirements can be readily met.

Mask substrates for X-ray lithography must be extremely thin so as to be transparent at these wavelengths. Membranes of both inorganic (Si, SiC, Si_3N_4, Al_2O_3, for example) and organic materials have been investigated. Organic materials have proved to be the most easily used for this purpose. These consist of tightly stretched films of mylar or polyimide, about 1–3 μm thick. These materials are both transparent to X-rays and also optically transparent, so that alignment of the mask to the wafer can be done by conventional means.

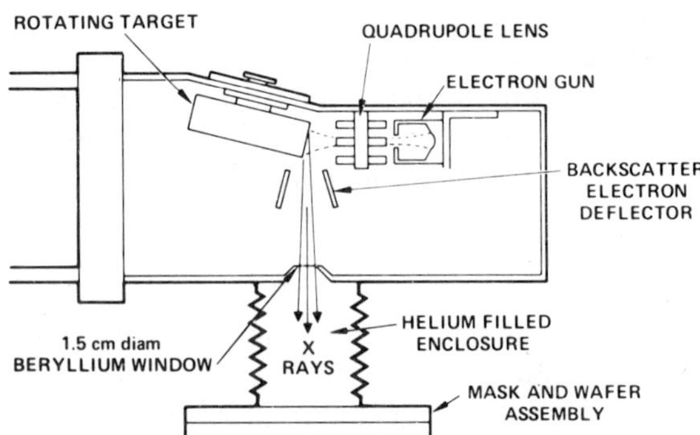

Fig. 10.14 Schematic of an X-ray printing system. From *Stover, Hause, and McGreevy* [22]. Reprinted from Solid State Technology.

A gold film, in which the pattern is defined, is used to absorb the X-rays. Typically, this is vacuum evaporated on the substrate, to a thickness of about 5000 Å. The mask pattern is made in this film by conventional E-beam techniques, and the excess gold is removed by ion milling.

Here, as in E-beam printing, the major problem is in the availability of suitable resists, in order to keep the exposure time to an acceptable value. Thus improved photoresist materials and high-power X-ray sources are crucial to the commercial realization of this technique. Nevertheless, a number of practical systems [23] have already been developed, and microcircuits fabricated using this approach.

10.2.6 Resists

There are many requirements that a resist must satisfy before it can be useful for a specific technology [24]. Above all, it must be capable of resolving the required minimum element size. Other factors, over which there is some degree of flexibility, are as follows.

Sensitivity. A high sensitivity is desirable since this reduces the time for exposure, and thus the operating cost.

Adhesion. A resist must adhere sufficiently well to the underlying film so that it does not lift off during subsequent processing. This is usually not a problem, provided appropriate procedures (the use of HMDS, for example) are used for surface preparation. This requirement has been somewhat relaxed with VLSI technology because of the increasing use of dry processes such as plasma etching and ion milling.

Etch Resistance. This requirement has become harder to meet since many resist materials degrade in a plasma or ion-beam environment. In fact, a tenuous balance between sensitivity to radiation during imaging, and insensitivity to radiation during subsequent processing, must be struck by the resist designer in order to produce a useful product.

Other features of a resist that are highly desirable are stability in storage, uniformity from batch to batch, and the absence of large particulates. It is fair to say that the optimum combination of all of these features is not available in any resist at the present time.

Resist materials are sensitive to radiation, which alters their chemical properties sufficiently so that a pattern can be delineated in them. Those used for optical printing respond to radiation of 3300–4300 Å, and are designed for maximum absorption (and hence sensitivity) over this UV range. Recently interest has focused on the deep UV region (1500–3000 Å) because of the higher resolution that can be obtained at these shorter wavelengths. The region of practical interest is $\geq$ 2000 Å, since a vacuum environment is necessary for operation in the 1500–2000 Å range. This range is further restricted by the availability of high-intensity deep UV sources. The most practical source today is a Xe-Hg lamp, which has a peak output of 2500–2900 Å.

The use of E-beam techniques for reticle mask generation has resulted in the need for resists which are sensitive to electron irradiation in the 10–30 kV range. Finally, resists are also required for use in printing systems which operate in the soft X-ray region (5–50 Å).

Resists are classified as negative or positive. *Negative* resists have the property that they become more durable upon exposure to radiation [25]. On the other hand, *positive* resists degrade and become more easy to remove under these conditions.

Most negative resists consist of polymeric organic materials, with long-chain molecules having a molecular weight range of 10^4–10^6. Irradiation of these molecules by an energetic beam results in bonding or cross-linking of adjacent polymer chains, and thus in an increase in the average molecular weight. These materials become more durable after they have been subjected to energetic radiation, and are sometimes referred to as *cross-linking* resists.

Negative photoresists are usually chemically inert synthetic rubbers, typically cyclized cis-poly isoprene, in combination with a radiation-sensitive cross-linking component. They are commonly applied in layers of 0.3–1 μm thickness, and are sensitive to UV light in the 3300–4300 Å range. A common rule of thumb here is that the resist thickness should be one-third the minimum element size. They are readily soluble in aromatic solvents such as benzene, toluene, and xylene, *prior* to cross-linking by exposure to UV light. These chemicals (as well as proprietary formulations) are used as developers in the photoengraving process [26].

Negative photoresists are well established in conventional microcircuit fabrication, and there is a large body of experience in their use. However, they are not suited for resolving elements under 2 μm because of a solvent-induced swelling that takes place during development. This results in ragged edges or a loss of pattern fidelity that makes them unsuited for VLSI applications. Moreover, they can only be removed with great difficulty once they have been exposed to UV light. Consequently, optical printing techniques have turned almost exclusively to positive photoresists for this purpose.

Positive photoresist materials are designed so that the effect of radiation is to make them increasingly capable of dissolution by chemical solvents. Such materials are sometimes known as *degrading* photoresists for this reason. These photoresists do not suffer from solvent-induced swelling problems, so that they can readily define elements as small as 0.5 μm in size. They can also be used in much thicker coatings than negative photoresists (by a factor of 3). As a result, they are more resistant to chemical etches, and are more reliable in problem areas such as steps, which lead to resist thinning. They are also relatively free from pinhole formation, since they can be applied in thick layers.

Positive resists can be made up of polymeric materials with weak links, which degrade upon irradiation by the process of *scission*. However, the most commonly used photoresists do not operate on this principle. Rather, they consist of three basic components: a low molecular weight, alkali-soluble resin such as phenol formaldehyde novolac, a photoactive dissolution inhibiter which usually consists of orthoquinone diazide, and a solvent such as xylene. They are usually applied in one or two layers, 1–3 μm thick, and are sensitive to the UV range of 3300–4300 Å. Exposure to light degrades the photoactive component, so that the resin becomes readily soluble* in sodium or potassium hydroxide, both of which can be used as developers [8]. The recent emphasis on MOS microcircuits has resulted in the introduction of proprietary nonionic developers, which are free from these contaminants. The development process is one of direct dissolution, without any swelling of unexposed material, and accounts in large measure for their resolution capability. At the present time, these resists can resolve elements as small as 0.25 μm.

Positive resists are more difficult to use than the negative type, and tight control of manufacturing processes must be maintained to obtain consistent adhesion to the mask film. Often the wafer is given a bake in an oxidizing environment such as air, and is spin-coated with HMDS to promote this adhesion. In addition, the pre-exposure bake is a critical step since the photoactive component is subject to thermal decomposition. Thus excessive baking temperatures result in a loss of sensitivity during subsequent exposure. However, some form of prebake step is necessary in order to densify the resist and make it more resistant to subsequent chemical etching processes.

The unexposed resist can be removed by chemicals such as acetone or methyl ethyl ketone. Proprietary chemicals are available for removing these

* The dissolution rate of exposed positive resists is about three decades faster than that of unexposed materials.

photoresists as well. In rare situations, an alternative approach consists of flooding the slice with UV irradiation to degrade the remaining resist, prior to its removal.

Deep UV-sensitive photoresists are increasingly used in VLSI applications in order to meet the 1 μm minimum element requirement. Most of the resists that are useful in this wavelength range are positive resists involving photoactive dissolution inhibitors.

Thermally degrading resists are increasingly finding use in this application, even though originally developed for use with E-beam systems. Materials such as polymethyl methacrylate (PMMA) and its derivatives have a maximum absorption at about 2150 Å so that they are not well matched to the range of available deep UV light sources. Typically, a 1 min exposure is necessary for a 1 μm thick film, using a Xe-Hg arc lamp. Materials such as poly methyl isopropenyl ketone can extend this absorption range out to 2900 Å, resulting in improved sensitivity by a factor of 5. This can be further improved by the addition of sensitizing agents, such as *p*-terbutyl benzoic acid.

Very few negative photoresists are available for the deep UV range, although a number are under active development. One proprietary formulation, whose principal ingredient is cyclized rubber, is commercially available for use in this region. This resist is about ten times more sensitive than the best positive resists, and does not appear to have the swelling problems associated with other cross-linking materials [11].

The technology of electron resists has been the subject of intense research in the last 10 years. Here the goal is to achieve resists with submicron resolution. Furthermore a high sensitivity (10^{-6} C/cm^2 or less) to electrons in the 10–30 kV range is also required in order to obtain reasonable exposure times. Finally, since most of these resists will be used in conjunction with plasma etching or ion milling of the underlying films, the ability to withstand these processes is extremely important.

A large number of electron resists have been investigated. However, most of the work has been done with positive resists such as PMMA and its derivatives [27]. This material, while having excellent resolution capability, has a low sensitivity (5×10^{-5} C/cm^2 for a 15 kV beam). It also has a poor resistance to ion milling, and a tendency to flow at elevated temperatures.

Techniques for improving the sensitivity of PMMA include the introduction of chemical configurations which tend to weaken the chain stability of the polymer by copolymerization. One such cross-linked PMMA has been reported to have a sensitivity of 8×10^{-6} C/cm^2.

Lift-off techniques have been successfully applied to PMMA resists, using a double-layer technique [13]. The lower layer consists of a 1–2 μm thick film of PMMA dissolved in chlorobenzene. After baking, a 4000 Å film of MMA dissolved in ethyl cellosolve acetate is spun on the wafer. This upper layer defines the pattern, while the lower layer is undercut during development so that easy lift-off is achieved.

Electron resist systems, based on the poly (olefin sulfones) have also been developed. One such, poly butane sulfone (PBS), is commercially available, and has a sensitivity of 7×10^{-7} C/cm^2. Its unique characteristic is that it can be developed by gaseous means. This allows an all dry process, which greatly reduces the adhesion requirements of the resist and the underlying layer.

Very little work has been done with negative electron resists in recent years. This is because they are generally prone to solvent-induced swelling effects, and cannot resolve submicron features. One such, based on a copolymer of glycidyl methacrylate and ethyl acrylate (COP), has a sensitivity of 4×10^{-7} C/cm^2 at 10 kV, but is limited to resolving a minimum element size of 1.5 μm. In contrast, the resolution capability of PMMA is ten times better. However, much research has been aimed at reducing these swelling effects, while still preserving the generally superior sensitivity of negative resists.

Resists developed for E-beam lithography have all been found to be relatively insensitive in their X-ray absorption capability. Attempts to increase this sensitivity have involved the inclusion of heavy metal ions such as cesium and thallium [28], in order to increase the absorption capability of the resist at the wavelength of interest. Typical resists that have been used to date include both PMMA (a positive resist) and COP (a negative resist). COP is about 20–30 times more sensitive than PMMA, but has much poorer resolution and chemical resistance properties. Typical exposure times are quite large, as much as 1–5 min for the more sensitive negative resists. Research in this area is very active, so that considerable improvements in sensitivity should be forthcoming.

Recently resist improvements have focused on the development of ultrathin layers which can be used with E-beam and X-ray lithography, and are capable of development by dry processes [29]. Some are based on inorganic resist systems. One such system [30] uses a 2000 Å thick film of $GeSe_2$ on which is placed a 100 Å film of $AgSe_2$. Here the upper layer is photoactive, and extremely thin so as to provide high resolution. The $GeSe_2$ layer is normally quite soluble in a number of alkaline solutions. Irradiation of the top layer produces silver which rapidly migrates into the $GeSe_2$ and renders it almost insoluble in these solutions. This photo-induced process is highly anisotropic in character, with migration almost entirely in the direction of the beam, and results in a high contrast image.

Inorganic resist systems of this type are compatible with dry processing and have extremely high resolution (200 Å). Their speed is about a factor of 10 lower than that of conventional organic resists, so much work is being done to improve their sensitivity [31].

10.3 PROBLEM AREAS

A variety of problems can be encountered at all points in the lithographic process. One is that of unwanted particulate matter. This ranges from glass particles left on the plates by the manufacturer to airborne particles which are always present to some extent in a clean room. Additional problems come about because of operator-induced errors during processing. In general, these result in random defects in the microcircuit.

Consider a microcircuit chip of area A. To an approximation, the number of such chips on a slice of diameter d is given by N, where

$$N \simeq \frac{\pi}{4A}(d - \sqrt{A})^2 \tag{10.1}$$

Let N_G be the number of good chips on the wafer, so that N_G/N is the fractional yield.

Most defects from the lithographic process are extremely small and of random nature. Let N_D be the total number of such defects per wafer.

If a random defect is added to a wafer, its probability of destroying a good chip is N_G/N. Thus,

$$dN_G = -N \Big/ d\left(\frac{N_G}{N}\right) = \left(\frac{N_G}{N}\right) dN_D \tag{10.2}$$

so that the fractional yield is

$$\frac{N_G}{N} = e^{-N_D/N} \tag{10.3}$$

Minimizing the number of defects on a wafer is thus increasingly important for VLSI, where chip area is correspondingly large (i.e., where N is small).

A discussion of the different types of defects that can arise during the lithographic process now follows.

10.3.1 Mask Defects

Mask defects of the visual type include pinholes, spots, intrusions, and protusions. Such defects are a function of the quality control exercised by the manufacturer, as well as of the care taken in using the starting materials. Often these masks are supplied precoated with a strippable laquer film for surface protection. This film is removed immediately prior to loading in the mask equipment.

Extreme cleanliness is required during the fabrication of reticle masks, and a detailed examination of each exposed mask is essential before it can be replicated. A number of different methods are used for mask inspection today [32]. The simplest approach is to directly compare one die pattern with another on the same mask. This is done by optically overlaying these patterns, followed by visual inspection. A second approach is to compare a die with a defect-free pattern that is stored in digital form in a computer. This approach has the advantage that it can be carried out automatically.

A somewhat similar approach is to scan two dies by a dual-beam flying spot scanner and compare the signals by electronic means. Spatial filtering techniques have also been used for this inspection purpose.

Dimensional errors can also occur during mask making. These can only be determined by both line-width and run-out measurements on the mask.

Visual defects on reticle masks at × 10 magnification can also be repaired if damaged. This painstaking operation consists of applying minute droplets of ink to fill the pinholes, after which they are laser trimmed if necessary. On the other hand, dimensional errors require that the mask be redone.

10.3.2 Printing and Engraving Defects

A number of problems may be encountered during the opening of windows in the masking film. Some result in defective circuits that are rejected. More serious, however, are those problems that result in an inadvertent spread of device parameters which may impair the reliability of the microcircuit and degrade its performance [33]. Some of these defects are now considered for a few practical situations.

Undercutting of the Resist. If the resist adheres poorly to the masking film, undercutting may occur during the etching process, resulting in a larger window (see Fig. 10.15*a*) and changes in the associated device parameters after diffusion. In extreme cases a defective circuit can result by short-circuiting out adjacent regions. Undercutting of the photoresist comes about because of capillary action of the (wet) chemicals that are used during the etching process. This problem is greatly reduced when plasma etching techniques are used.

Dimensional Variations. Dimensional variations may occur in the width of cuts in the masking film (as in Fig. 10.15*b*) because of limitations of the photographic process. These, too, result in changing the effective area of the masked regions. The problem is especially severe when long, very narrow cuts are required and may lead to breaks in the metallization, as shown in Fig. 10.15*c*. These variations can also occur because of swelling of the resist during development. Positive resists are less susceptible to this problem than negative ones, so that they are preferred in VLSI applications.

Drafting errors should be considered in those cases where the mask is drawn × 250 prior to photoreduction. Here errors of 25 μm on the drawing are translated into 0.1 μm variations in the final circuit. Thus dimensional variations due to these errors can usually be neglected.

Lack of Registration. This occurs because of mask alignment errors, or because of cumulative errors in the step-and-repeat process. It can lead to altered device configurations, as shown in Fig. 10.15*d* for the emitter and base regions of a transistor. Here, too, defective circuits can result in ex-

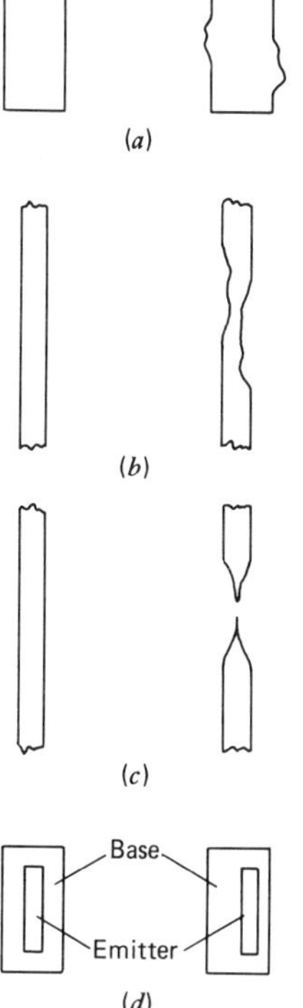

Fig. 10.15 Defects and their results.

treme cases. In some situations, misregistration may result in exposing a projected junction edge, thus degrading its characteristics.

Registration problems have become increasingly important with the advent of the high-density structures that are required for VLSI technology. At the present time a misalignment error of about 0.25 μm is generally achieved in printing systems operating at the 1 μm level.

Pinholes. These may be present in the resist and result in the inadvertent opening of a small window in the underlying mask film. Their effect is a function of their specific location. As may be expected, the severity of this problem increases with component packing density in a microcircuit.

Dust Particles. These present a serious problem during the delineation of patterns in mask films. Some dust, such as carbon or metallic particles, is opaque. Its presence on the clear parts of a mask will prevent exposure of the underlying resist. This will cause holes in a negative resist after exposure and development, and thus holes in the mask film. The effect with positive resists is to not completely clear an opening in the underlying mask. This latter effect is usually less serious than the former.

Most dust particles are transparent and cause diffraction effects. This can have the same result as opaque particles; often, however, the effect is over a wider area because of the diffraction pattern. In either case, the problems of dust are minimized by the use of good clean-room technique and by careful visual inspection of the masks prior to alignment and exposure.

Scratches and Tears. The printing of metallization patterns is a particularly difficult task because of the delicate nature of the film. Here scratches are by far the most common type of defect and are caused by rough handling during the fabrication process. Scratches greatly increase the current density in their vicinity, giving rise to the formation of hot spots, and the premature onset of electromigration. Thus although the circuit operation is unchanged, its reliability may be greatly impaired in this manner. In extreme cases, scratches result in defective circuits by breaking the metallization path.

Tearing of the metallization may occur during the removal of the photoresist. This is often due to dimensional changes and damage to the photoresist during the alloying step. This defect also results in reducing the effective area of the current path or in an open circuit in extreme cases. The two-level lift-off technique is extremely important in controlling this problem.

Step Coverage. This occurs when an excessively thin layer of resist goes over a step in an oxide. This can give rise to a failure of the resist during the subsequent etching process. The problem is especially severe with negative resists, since they are applied in much thinner layers than positive resists. Here, too, the use of two (or more) levels of resist can alleviate this problem.

Problems of step coverage can also be reduced by the use of planarization techniques. This important topic is discussed in Section 11.4.

REFERENCES

1. Techniques for Photolithography, Eastman Kodak Company, Rochester, NY, 1966.
2. P. R. Thornton, Electron Physics in Device Microfabrication—I. General Background and Scanning Systems, *Adv. Electron. Electron Phys.* **48,** 271 (1979).
3. P. R. Thornton, Electron Physics in Device Microfabrication—II. Electron Resists, X-Ray Lithography and Electron Beam Lithography Update, *Adv. Electron. Electron Phys.* **54,** 69 (1980).
4. D. E. Davis, R. D. Moore, M. C. Williams, and O. C. Woodard, Automatic Registration in an Electron-Beam Lithographic System, *Solid State Technol.*, p. 61 (Aug. 1978).
5. S. Okazaki, K. Mochiji, E. Takeda, and Y. Maruyama, Electron Beam Mask Fabrication for MOSLSI's with 1.5 μm Design Rule, *Jpn. J. Appl. Phys.* **19,** 51 (1980).
6. G. I. Varnell, D. V. Spicer, and A. C. Rodger, E-Beam Writing Techniques for Semiconductor Device Fabrication, *J. Vac. Sci. Technol.* **10,** 1048 (1973).
7. D. A. Doane, Optical Lithography in the 1 μm Limit, *Solid State Technol.*, p. 101 (Aug. 1980).

8. *Shipley AZ Photoresists*, The Shipley Company, Newton, MA, 1977.
9. D. L. Flowers, Lubrication in Photolithography, *J. Electrochem. Soc.* **124,** 1608 (1977).
10. W. R. Sinclair, M. V. Sullivan, and R. A. Fastnacht, Materials for Use in a Durable Selectively Semitransparent Photomask, *J. Electrochem. Soc.* **118,** 341 (1971).
11. S. Iwamatsu and K. Asanami, Deep UV Projection Systems, *Solid State Technol.*, p. 81 (May 1980).
12. L. I. Maisell and R. Glang, Eds., *Handbook of Thin Film Technology*, McGraw-Hill, New York, 1970.
13. M. Hatzakis, Multilayer Resist Systems for Lithography, *Solid State Technol.*, p. 74 (Aug. 1981).
14. M. Hatzakis, B. J. Canavello, and J. M. Shaw, Single Step Optical Lift-Off Process, *IBM J. Res. Dev.* **24,** 452 (1980).
15. M. B. Heritage, Electron-Projection Microfabrication System, *J. Vac. Sci. Technol.* **12,** 1135 (1975).
16. W. L. Brown, T. Venkatesan, and A. Wagner, Ion Beam Lithography, *Solid State Technol.*, p. 60 (Aug. 1981).
17. G. Stengl, R. Kaitna, H. Loschner, P. Wolf, and R. Sacher, Ion Projection System for IC Production, *J. Vac. Sci. Technol.* **16,** 1883 (1979).
18. D. B. Rensch, R. L. Seliger, G. Csonsky, R. D. Olney, and H. L. Stover, Ion Beam Lithography for IC Fabrication with Submicrometer Features, *J. Vac. Sci. Technol.* **16,** 1897 (1979).
19. T. W. O'Keefe, Fabrication of Integrated Circuits Using the Electron Image Projection System (ELIPS), *IEEE Trans. Electron Dev.* **ED-17,** 465 (1970).
20. W. R. Livesay, Computer Controlled Electron-Beam Projection Mask Aligner, *Solid State Technol.*, p. 21 (July 1974).
21. D. L. Spears and H. I. Smith, High Resolution Pattern Replication Using Soft X-Rays, *Electron. Lett.* **8,** 102 (1972).
22. H. L. Stover, F. L. Hause, and D. McGreevy, X-Ray Lithography for One Micron LSI, *Solid State Technol.*, p. 95 (Aug. 1979).
23. W. P. Buckley and G. P. Hughes, An X-Ray Lithography System, *J. Electrochem. Soc.* **128,** 1106 (1981).
24. M. J. Bowden and L. F. Thompson, Resist Materials for Fine Line Lithography, *Solid State Technol.*, p. 72 (May 1979).
25. W. S. Deforest, *Photoresist*, McGraw-Hill, New York, 1975.
26. Photosensitive Resists for Industry, Eastman Kodak Company, Rochester, NY, 1966.
27. M. J. Hatzakis, Electron Resists for Microcircuit and Mask Production, *J. Electrochem. Soc.* **116,** 1033 (1969).
28. N. Taylor, X-Ray Resist Materials, *Solid State Technol.*, p. 73 (May 1980).
29. K. Harada, O. Kugure, and K. Murase, Poly (Phenyl Methacrylate-Co-Methacrylic Acid) as a Dry Etching Durable Positive Electron Resist, *Trans. IEEE Electron. Dev.*, **ED-29,** 518 (1982).
30. K. L. Tai, W. R. Sinclair, R. G. Vadimsky, J. M. Moran, and M. J. Rand, Bilevel High Resolution Photolithographic Technique for Use with Wafers or Stepped and/or Reflecting Surfaces, *J. Vac. Sci. Technol.* **16,** 1977 (1979).
31. M. J. Bowden, Forefront of Research on Resists, *Solid State Technol.*, p. 73 (June 1981).
32. D. B. Novotny and D. R. Ciarlo, Automated Photomask Inspection, *Solid State Technol.*, Part 1, p. 51 (May 1978); Part 2, p. 59 (June 1978).
33. J. R. Brauer, Poor QC Yields Bad IC's, *Electron. Eng.*, p. 78 (Aug. 1966).

CHAPTER

11

Device and Circuit Fabrication

CONTENTS

11.1 ISOLATION 569

11.1.1 Mesa Isolation, 570
11.1.2 Oxide Isolation, 570
11.1.3 *p–n* Junction Isolation, 572

11.2 SELF-ALIGNMENT 575
11.3 LOCAL OXIDATION 576
11.4 PLANARIZATION 582
11.5 METALLIZATION 585
11.6 GETTERING 587
11.7 MOS-BASED MICROCIRCUITS 589

11.7.1 The *p*-Channel Transistor, 589
11.7.2 The *n*-Channel Transistor, 591
11.7.3 Complementary Transistors, 593
11.7.4 Nonvolatile Memory Devices, 595
11.7.5 Silicon-on-Sapphire Devices, 596

11.8 MESFET-BASED MICROCIRCUITS 598

11.8.1 The Depletion-Mode Transistor, 599
11.8.2 The Enhancement-Mode Transistor, 602
11.8.3 Digital Circuit Approaches, 603

11.9 BJT-BASED MICROCIRCUITS 605

11.9.1 The Buried Layer, 606

11.9.2 Choice of Transistor Type, 609
11.9.3 Transistor Properties, 610

11.9.3.1 Breakdown Voltage, 610
11.9.3.2 Gain–Bandwidth Product, 611
11.9.3.3 Active Parasitics, 612

11.9.4 *p–n–p* Transistors, 613
11.9.5 Special-Purpose Bipolar Transistors, 615
11.9.6 Field Effect Transistors, 616
11.9.7 Self-Isolated VLSI Transistors, 618
11.9.8 Diodes, 622

11.9.8.1 Schottky Diodes, 626

11.9.9 Resistors, 626

11.9.9.1 Diffused Resistors, 626
11.9.9.2 Ion-Implanted Resistors, 630
11.9.9.3 Thin-Film Resistors, 631

11.9.10 Capacitors, 631

REFERENCES 633

The preceding chapters have described a number of basic processes that are used in semiconductor device fabrication technology. The emphasis has been on the theoretical and practical aspects of specific processes which are in use today, as well as on their limitations. In this chapter, the fabrication of complete microcircuits is treated with the emphasis on VLSI schemes. This requires a description of some of the ways in which these processes can be combined to form devices and components, which can be interconnected to make a complete microcircuit. These combinations of processes are by no means unique, although a few basic systems have evolved over the last 15 years. However, the requirements of VLSI circuits have resulted in a shift of emphasis to schemes which allow the use of smaller active device regions, the delineation of fine-line multi-level patterns, and the preservation of relatively planar surfaces during the fabrication process. Technologies such as ion implantation, plasma etching, and microlithography [1] are extensively used in these schemes. New sequences of fabrication steps have developed, as each of these technologies has been incorporated into the fabrication process. As a consequence, there is much room for creativity as well as for change, as these different fabrication sequences are evaluated in terms of performance and cost.

Some of the basic integrated circuit fabrication processes [2–4] are described in this chapter together with their unique features. The electrical characteristics of the components so formed are also described in order to highlight the limitations of these processes. No attempt has been made to be all-inclusive here; instead, emphasis is placed on well-accepted approaches, and also on those that appear most promising for VLSI circuits. An extensive reference list provides further examples of the variety of fabrication processes that are currently under investigation.

Many different kinds of active and passive components are used in microcircuits today. Of these, the primary active devices are the Metal-Oxide Semiconductor Field Effect Transistor (MOSFET, often shortened to MOS), the Metal Semiconductor Field Effect Transistor (MESFET), and the Bipolar Junction Transistor (BJT). Many other active devices are also used. Nevertheless, it is convenient to classify microcircuits in terms of their primary active elements. This chapter consequently focuses on the special features of circuits based on these active devices.

11.1 ISOLATION

A microcircuit can be described as an "ensemble of active and passive components, interconnected within a monolithic block of semiconductor material." Thus the first requirement is to devise schemes for fabricating components that are electrically isolated from each other in order to allow design flexibility. A number of different approaches are in use today. Of these, the one that is conceptually the most straightforward is to fabricate devices in an active film which is grown on an insulating or semi-insulating substrate, and to obtain isolation by etching moats around each device. The resulting islands are known as *mesas*, and this process is referred to as *mesa isolation*. Circuits made in silicon films on sapphire, as well as those made in epitaxial gallium arsenide on semi-insulating gallium arsenide, are examples of this approach.

An important extension of this technique is to fabricate devices in pockets of active semiconductors which are directly formed *in* the insulating substrate. This technique is unique to gallium arsenide, where both undoped and chromium-doped single-crystal substrates can be used as the starting material, and ion implantation or diffusion used to form pockets in which active devices can be made.

Oxide isolation techniques can also be used to form insulating tubs in order to isolate a number of pockets of single-crystal semiconductor [5]. This approach has been widely adopted for the fabrication of microcircuits which must be radiation hardened, as well as in some special high-voltage applications where the degree of isolation between components must be high.

A third approach makes use of the fact that a reverse-biased p–n junction in either silicon or gallium arsenide has an extremely low leakage current (in the picoampere range at room temperature). Thus two regions of a semiconductor are effectively isolated to direct current if they are of the opposite conductivity type and are suitably reverse biased. The significant coupling between regions is primarily capacitive in nature so that its effect on the microcircuit need only be considered at high frequencies.

Finally, isolation can be achieved by a combination of these processes. Thus *p–n* junction isolation, combined with isolation produced by the localized oxidation of silicon, results in greatly reducing the physical dimensions of some active devices, and aids in retaining the planarity of the silicon surface. This combination represents one of the more important approaches to the fabrication of VLSI circuits.

11.1.1 Mesa Isolation

This approach can be used for circuits which are fabricated on insulating or semi-insulating substrates. These include circuits in silicon-on-sapphire, and *n*-gallium arsenide on semi-insulating gallium arsenide. Here isolation of active regions is obtained by masking them with photoresist, and etching the exposed semiconductor.

Wet chemical etching is often employed for this purpose. Here (100) starting material is used, in conjunction with an anisotropic etch of the type described in Chapter 9. Windows are delineated along ⟨110⟩ directions; next V-grooves are etched deep enough so as to produce electrical separation of the components. Typically, this etch process is carried out after device fabrication, and just before the metallization step.

V-groove isolation processes result in nonplanarity of the semiconductor surface, and excessive waste of active chip area. However, their sloping walls (54.74°) allow excellent step coverage for the subsequent metallization. Ion milling, followed by a light plasma etch, has been used [7] to reduce this wasted area. Both approaches create problems with doping concentration along island-edge surfaces, and give rise to enhanced leakage.

The trend in VLSI circuits is toward schemes which preserve the surface planarity, and also provide a protective oxide cover. One such approach consists of covering the active regions with a refractory material such as silicon nitride, and oxidizing the exposed silicon to provide silicon dioxide isolation. This local oxidation approach is rapidly proving to be one of the more important VLSI techniques, and is discussed more fully in Section 11.3.

11.1.2 Oxide Isolation

One method for obtaining oxide isolation in silicon is described here in order to illustrate the approach to the problem. Most practical techniques are based on this process, with minor modifications. The process sequence is as follows:

1. An *n*-type (100) silicon slice is masked with silicon dioxide, as shown

in Fig. 11.1*a*, and etched so as to result in the structure shown in Fig. 11.1*b*. Anisotropic etching, with windows oriented along ⟨110⟩ directions, are used to delineate the V-shaped grooves.

2. After removal of the mask, an n^+-layer is diffused across the entire slice (see Fig. 11.1*c*). This layer is heavily doped so as to provide a low-resistance ohmic contact to this region.
3. A thermal oxide is now grown across this wafer. This oxide becomes the isolation between the single-crystal and the subsequent polycrystalline silicon, which is next deposited (Fig. 11.1*d*) to a thickness of 250–500 μm. The thermal reduction of $SiHCl_3$ in hydrogen at 1150°C provides growth rates as high as 10 μm/min for this purpose. An alternate approach is to grow this layer by the thermal decomposition of silane.

   Considerable stress, accompanied by deformation of the slice, can result from this step. One approach to its control is to build up the

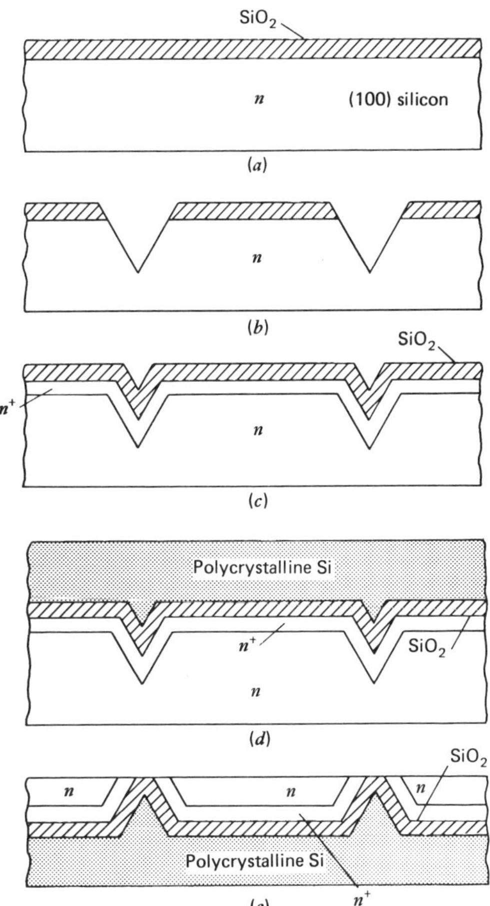

Fig. 11.1 Oxide isolation scheme.

insulating support with a sandwich of alternating layers of polysilicon and silicon dioxide [8].

4. The single-crystal side of the slice is now lapped down, resulting in the structure shown in Fig. 11.1*e*. The resulting slice consists of the desired tubs of single-crystal silicon, isolated from each other by a layer of silicon dioxide. Various types of components may be fabricated within these pockets as desired. Each tub is lined with an n^+-layer which provides a collector connection with a low parasitic resistance, and is essential for high-performance bipolar transistors fabricated in these tubs. This connection is also employed if the tub is to be used for resistor placement.

The oxide isolation process results in near-perfect isolation between these single-crystal tubs. Although the thermal conductivity of the SiO_2 is 1/50 times that of silicon, this does not present a serious problem since the layer is thin (0.5–1 μm).

The main disadvantage of oxide isolation is that it requires careful mechanical alignment of the slice during the lapping step. Since the depth of the isolation pocket is only 5–10 μm, any misalignment results in single-crystal tubs of variable depths across the slice. As a consequence, a number of modifications have been made to improve this process. One is to locate a series of variable-depth V-markers during the etch step by using oxide cuts of different widths. These can serve as indicators to mark the end of the lapping step, and also to give early warning so that adjustments of the lapping process can be carried out just before termination.

An alternate approach, which is very promising, is outlined [5] in the sequence of Fig. 11.2. Here a heavily doped starting substrate of (100) orientation is used. The active layer is epitaxially grown on it, oxide masked, and then anisotropically etched as shown in Fig. 11.2*b*. Next, the thermal oxide is grown over this layer (Fig. 11.2*c*), followed by a deposit of polycrystalline silicon, as in Fig. 11.2*d*. Finally the heavily doped starting silicon is removed by use of a concentration dependent etch, which stops at the higher resistivity epitaxial layer. Yet another variation is to use an electrochemical etch which will stop at the appropriately doped epitaxial layer [9].

Other approaches make use of the fact that the ethylene diamine, pyrocatechol, and water etch (EPW) described in Chapter 9 will not etch a p^+-region. Thus a region of this type serves as an automatic etch stop.

11.1.3 *p–n* Junction Isolation

Junction isolation can be used for many different types of active and passive components. Thus a resistor can be formed by making a diffusion into a semiconductor substrate of opposite impurity type, and is isolated from it as long as a reverse (or zero) bias is maintained across the junction formed in this way. Moreover, any number of resistors, placed side by side, will

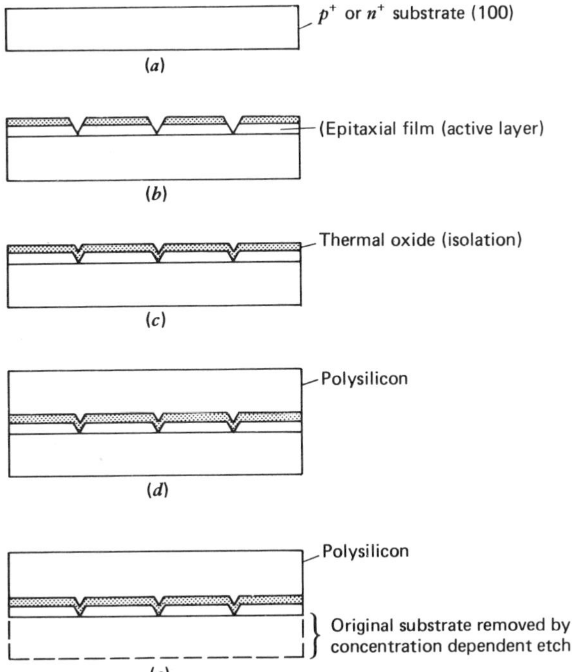

Fig. 11.2 Alternate oxide isolation scheme. From Bean and Runyan [5]. Reprinted with permission of the publisher, The Electrochemical Society, Inc.

be isolated from the substrate as well as from each other, provided that this junction bias is maintained. An additional condition is that, during operation, their depletion layers do not touch. This requirement is easily met in conventional (low-voltage) microcircuits; however, it must be taken into account for high-voltage applications.

The condition of self-isolation is also met by insulated gate, field effect devices since all of their junctions (source, drain, and channel) are either reverse or zero biased with respect to the substrate. Thus no special arrangement has to be made for these devices.

Bipolar junction transistors can be formed by the successive fabrication of a base and an emitter in a semiconductor substrate. All of these transistors will share a common collector, so that these devices are *not* self-isolating. Consequently, each* transistor must be put in an appropriately biased separate tub, of opposite conductivity type to the substrate. This can be done in a number of different ways.

1. A series of n-type tubs can be diffused into a p-type substrate. A transistor is made in each tub by the successive formation of its active

* The exception is transistor groups which share a common collector.

regions. Again, circuit arrangements must be made to ensure that these tubs are reverse or zero biased. This scheme is shown in Fig. 11.3*a*, together with typical dimensions for a diffused structure in silicon.

Transistors formed in tubs of this type have relatively poor performance since their collector series resistance is excessively large. Moreover, they are difficult to make, since the base region is delineated by the precise placement of three diffusions. Thus relatively widebase (2–3 μm) structures are necessitated for this process. The use of ion implantation has greatly reduced this problem. Nevertheless, these devices are still relatively poor in speed performance compared to conventional transistors.

An additional problem with this structure is that it results in a *p–n–p–n* sandwich. Such a sandwich is electrically equivalent to an *n–p–n* and a *p–n–p* transistor, connected in a positive feedback loop. This structure can become regenerative if, during circuit operation, the sum of the common-base current gains of the two transistors becomes equal to or greater than unity. The current gain of the *n–p–n* transistor is typicaly 0.98–0.99, so that extreme care must be taken in biasing the circuit to ensure that the parasitic *p–n–p* transistor is cut off at all times.

2. A second approach is to form isolated tubs by diffusing completely

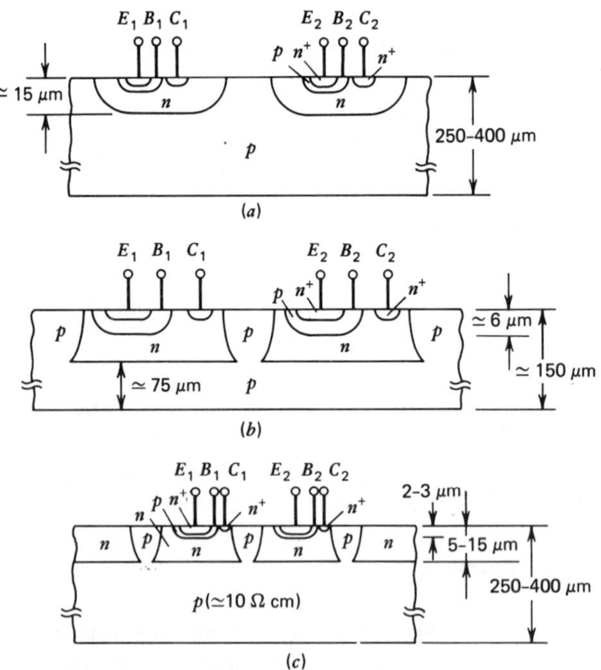

Fig. 11.3 Junction isolation schemes. (*a*) Triply diffused. (*b*) Two-sided process. (*c*) Doubly diffused epitaxial process.

through the slice from both sides. This approach has an advantage over the last process in that transistors fabricated in these tubs can be of the conventional type, since each tub consists of uniformly doped starting material. However, even with relatively thin substrates, this diffusion step requires at least 40–50 h at 1250°C, and results in excessive contamination and consequently soft junctions.

Figure 11.3*b* shows a schematic cross section for this process. The first bipolar microcircuits were fabricated in this manner, so that the process is of historical significance. Interestingly, this approach was the precursor to the modern bipolar microcircuit which is fabricated by the double diffused epitaxial (DDE) process.

3. In the DDE process, a thin *n*-layer is epitaxially grown on a *p*-type substrate, and isolated tubs are formed in this layer by *p*-diffusions from the *top* surface. Subsequent base and emitter diffusions result in the transistors as shown in Fig. 11.2*c*. Note the similarity to the configuration resulting from the two-sided diffused process of Fig. 11.3*c*. The single, but all important, difference is in the width of the active *n*-layer.

Isolated tubs, formed by the DDE process, can be used for fabricating both active and passive components, many of which are not self-isolating. Consequently, this process represents the conventional approach to integrated circuits which are based on the use of bipolar transistors.

11.2 SELF-ALIGNMENT

Self-alignment techniques are important in VLSI fabrication technology since they reduce the difficulties of precise alignment, and thus allow a considerable shrinkage of device size. Many of these techniques are based on the use of masking materials which can withstand high-temperature processing steps.

An important example of the use of these techniques is in the fabrication of polysilicon gate MOS devices, as outlined in Section 8.1.5. Here a polysilicon mask is used to define the gate region during the source and drain diffusions. This serves to form these regions so that the gate oxide overlaps them (Fig. 8.3), even when extremely shallow diffusions are used. In addition, the gate becomes heavily doped during this process, thus avoiding an additional process step.

Yet another example of the self-aligned technique is the use of a washout emitter to form an ohmic contact to this region. This is shown in Fig. 11.4, where an emitter contact of this type is compared to one of the same contact area, made by a separate alignment step. The advantages of the washout technique are that it reduces the area of the emitter region by approximately a factor of 3 and eliminates an additional mask and alignment step. Thus it is extensively used in high-speed device fabrication technology. Note, how-

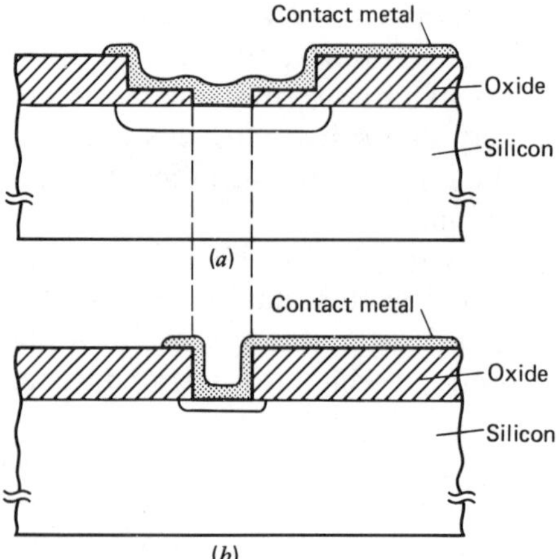

Fig. 11.4 Emitter contact schemes. (*a*) Conventional. (*b*) Self-aligned washout emitter.

ever, that considerable care must be taken to prevent exposing the edge of the junction to avoid excessive leakage, or failure due to short circuits.

One of the unique characteristics of ion implantation is that it is a low-temperature process, so that a number of different mask materials can be used. This feature has been exploited in many self-aligned masking schemes. In one such scheme, aluminum gate MOS circuits were fabricated by ion implantation followed by laser anneal [10], with the gate used as a mask.

Silicon nitride can also be used as a mask region in self-aligned structures, since it can withstand subsequent high-temperature steps. A unique feature of this material, in addition to its excellent masking quality, is that it can withstand an oxidizing environment. It forms the basis for the local oxidation process which is outlined in the following section.

Doped oxides, which can be used as diffusion sources, have also been used in self-alignment schemes for the fabrication of MOS devices [11]. A unique feature of this approach is that both phosphorus- and boron-doped oxide patterns can be defined in the same circuit, so that complementary transistors can be fabricated in a single operation.

11.3 LOCAL OXIDATION

This technique is specifically applicable to silicon microcircuits. Here the localized oxidation of silicon (LOCOS) serves as the starting point in a technology that shrinks device size and improves device performance [12]. In addition, variations of this approach result in preserving the planarity of the overall microcircuit, thus greatly reducing problems associated with the step coverage of metal. It is understandable, therefore, that this approach (and its variations) are widely used in VLSI fabrication schemes.

The LOCOS process is based on the fact that silicon nitride can be used as a mask against thermal oxidation. Typically, its oxidation rate in 95°C H_2O is about 30 times slower than that of silicon. In addition, H_3PO_4 etches are available which will remove silicon nitride but not attack silicon dioxide.

Figure 11.5 shows two basic LOCOS structures. In both, a layer of silicon nitride is deposited on the silicon substrate and patterned. The approaches differ in the pre-etch of the silicon that is done prior to the local oxidation. The resulting structures show that, in each case, the oxide is countersunk in the silicon. As a result, the area of unoxidized silicon is *smaller* than the original nitride mask. This unoxidized silicon is used to form an active device. Its reduced area is a significant advantage over regions formed by diffusion, which are *larger* than the window that is cut in the mask.

The displacement of the plane of the oxide from the silicon surface is controlled by the amount of pre-etch prior to oxidation. (Note that the thickness of the grown silicon dioxide layer is 2.2 times that of the consumed silicon.) Ideally, the most planar situation, the one with the fully recessed oxide (Fig. 11.5*b*), is the best for the subsequent placement of metallization patterns.

This process can be directly applied to form isolated islands in silicon microcircuits on sapphire substrates [13]. In addition to surface planarity, an important advantage of this process is passivation of the edge of the islands, with a concurrent reduction in the leakage current. An important

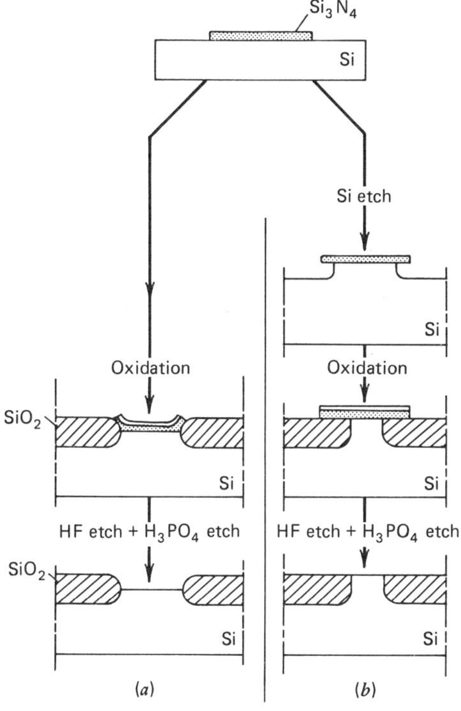

Fig. 11.5 Locally oxidized structures. From Appels, Kooi, Paffen, Schatorjé, and Verkuylen [12]. With permission from Philips Research Reports. (*a*) Partially recessed. (*b*) Fully recessed.

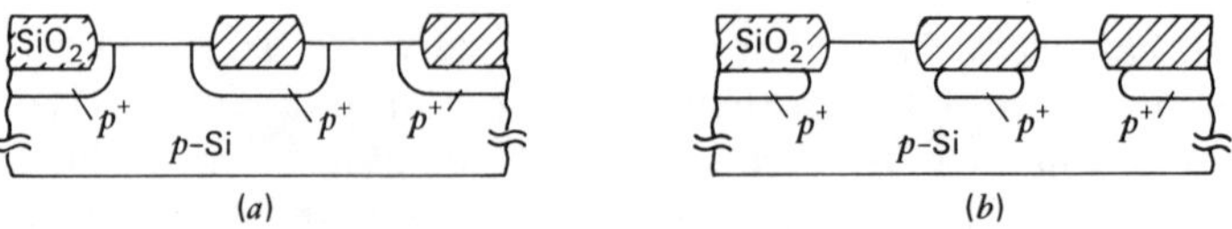

Fig. 11.6 Combination of impurity diffusion with local oxidation.

feature of the LOCOS process is that the nitride film can be used as a mask against impurity diffusion, as well as oxidation. This is seen in Fig. 11.6*a*, where a p^+-diffusion is first made using the nitride film as a diffusion mask, and followed by a deep local oxidation. A variation of this process results in the impurity configuration of Fig. 11.6*b*. Both of these schemes are useful in the placement of field inversion channel stops between devices in a MOS-based microcircuit.

Figure 11.7 compares transistors of equal emitter area, built by conventional processes and by local oxidation. The main differences in these structures are seen in the collector–base junction. The latter device has a smaller area resulting in smaller size and less parasitic capacitance. Moreover, it has less junction curvature so that its breakdown voltage is higher. Thus the locally oxidized structure is not only smaller, but also superior in its electrical characteristics.

Variations of this technique are possible. Thus double nitride masking and oxidation steps can be used to form a bipolar transistor where both the emitter–base and the collector–base junctions can be made between the countersunk oxide, with a reduction of the emitter transition capacitance as well.

An important advantage of local oxidation is that any region, bounded by locally oxidized regions, can have contact made to it without the need for an additional mask and alignment step. This self-alignment feature greatly reduces the minimum size of any exposed region. It provides the same

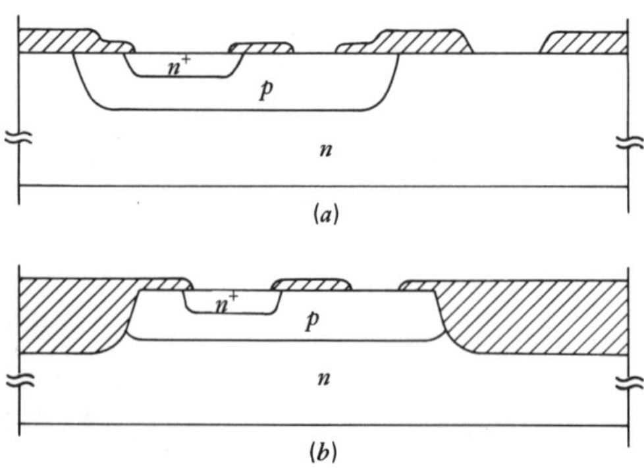

Fig. 11.7 Bipolar transistor. From Appels, Kooi, Paffen, Schatorjé, and Verkuylen [12]. With permission from Philips Research Reports. (*a*) Conventional. (*b*) LOCOS.

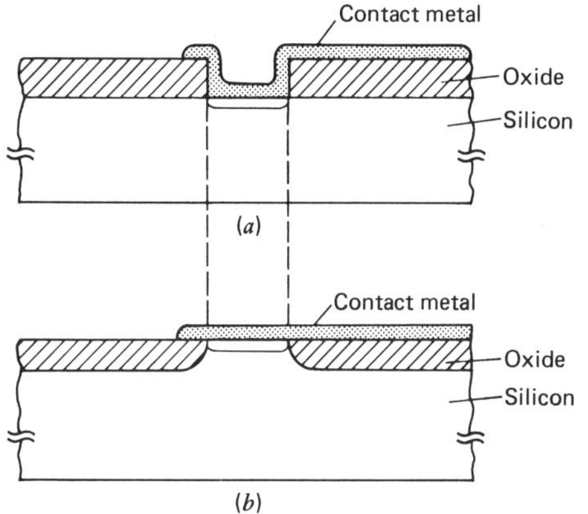

Fig. 11.8 Self-aligned contacting schemes. (*a*) Washout emitter. (*b*) Locally oxidized emitter.

reduction of size as the washout emitter of Fig. 11.4*b*, but is less prone to leakage, since the junction edge is fully protected by the recessed local oxide. Figure 11.8 shows this structure, as well as the washout emitter, by way of comparison. This technique, combined with the use of doped polysilicon for resistors and interconnections, has been successfully exploited [14] in the fabrication of high-speed emitter-coupled logic gates with 0.6 ns delay time. Integrated Schottky logic with comparable delay times has also been developed using local oxidation techniques [15].

Isolation between MOS devices in a microcircuit is conventionally obtained by the growth of a field oxide, which is usually ≥ 1 μm in thickness. Thus step coverage of the interconnection metal is a particularly severe problem. The use of recessed oxides presents an especially powerful approach in this situation. As a result, extensive use of this approach has been made in MOS-based VLSI circuits [16, 17]. Figure 11.9 shows one version of a MOS device formed by these techniques.

There are a number of technological problems which arise during the application of local oxidation technology. The first of these is that the oxidation of silicon proceeds slightly under the nitride as well. Although oxide penetration can be significant, the projecting oxide lip can be easily removed causing little problem.

A second problem comes about because of the extremely poor thermal expansion coefficient match between the silicon nitride and the underlying silicon. This causes the formation of a massive dislocation network in the

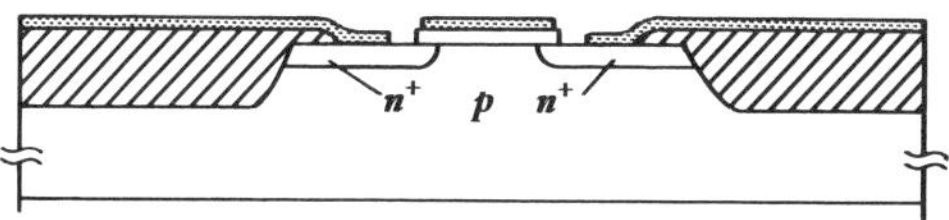

Fig. 11.9 MOS device made by local oxidation. From Appels, Kooi, Paffen, Schatorjé, and Verkuylen [12]. With permission from Philips Research Reports.

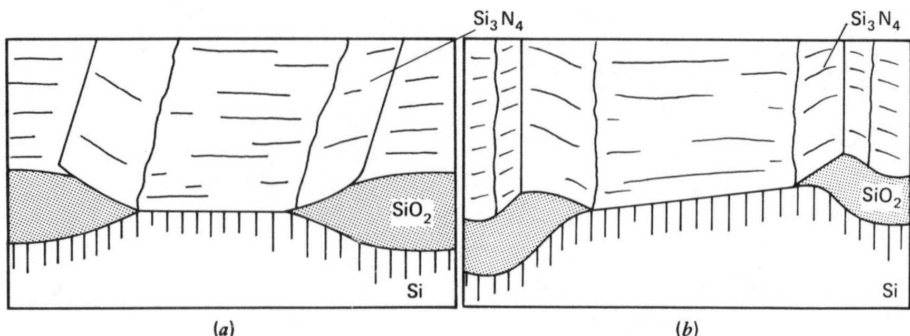

Fig. 11.10 Bird beak and bird crest configurations. From Bassous, Yu, and Maniscalco [19]. Reprinted with permission of the publisher, The Electrochemical Society, Inc.

unoxidized silicon regions. It has been shown that this can be greatly suppressed by growing a thin layer of silicon dioxide prior to placement of the silicon nitride mask [18]. Typically, a 100–200 Å thickness has been found sufficient for this purpose. Unfortunately, this greatly enhances the penetration of oxide under the nitride-masked region, resulting in structures of the type shown in Fig. 11.10. These oxide configurations, referred to as "bird beaks" and "bird crests," are undesirable, and studies have been made to minimize these effects [19]. In particular this effect gets worse as the oxide structure is recessed further, i.e., a fully recessed local oxide has a more prominent feature than the unrecessed oxide. This is shown in Fig. 11.11 where cross sections are shown for three different conditions of etching prior to local oxidation. For all cases, a 4500 Å oxide was grown with a pad thickness of 100 Å and a nitride thickness of 4500 Å.

It has been shown that a thicker (i.e., more rigid) nitride film tends to suppress these effects, but results in greater damage to the underlying silicon. Consequently, a trade-off is involved in the selection of these two layer thicknesses.

The thermal oxidation temperature is also an important process parameter, with less beak formation at lower temperatures. Thus there is consid-

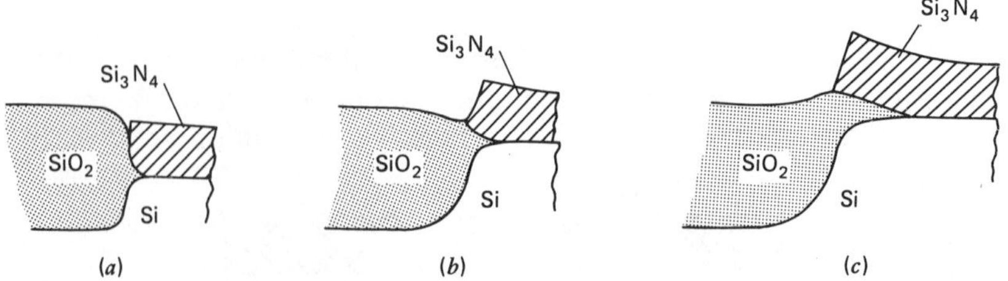

Fig. 11.11 Development of bird beak and crest. From Bassous, Yu, and Maniscalco [19]. Reprinted with permission of the publisher, The Electrochemical Society, Inc. (*a*) No pre-etch. (*b*) 1000-Å pre-etch. (*c*) 2000-Å pre-etch (fully recessed).

erable emphasis on the use of low-temperature processes for the local oxidation of silicon. Of these, oxidation at high pressure has been extensively studied [20, 21]. Here it has been shown that silicon nitride can be used as a mask up to at least 90 atm. Some work with wet chemicals as well with plasma techniques for silicon anodization has also been undertaken.

A typical sequence of steps, used in the preparation of a recessed oxide surface for MOS-based microcircuits, now follows. It is assumed that low-threshold *n*-channel devices are required, necessitating a *p*-type starting substrate of (100) orientation. Intermediate cleaning steps in this sequence are omitted for simplicity.

1. A 100–200 Å thick oxide is thermally grown on the silicon slice to serve as the pad oxide. This is covered with a 2000 Å thick layer of silicon nitride, grown by the reaction of silane in ammonia as described in Section 8.1.3.
2. A cover layer of SiO_2, typically 500–750 Å thick, is grown by CVD (see Section 8.1.1), followed by a positive photoresist film. At this point, the wafer is as shown in Fig. 11.12*a*.
3. The photoresist is now patterned along ⟨110⟩ directions, and the exposed parts are removed in acetone. The unexposed regions serve as a mask for the CVD oxide layer, which is delineated by means of BHF. After removal of the photoresist, this CVD oxide serves as a mask to delineate the underlying nitride, which is etched in hot (180°C) phosphoric acid. Note that photoresist masking cannot be used in this process. The resulting structure is shown in Fig. 11.12*b*.
4. The wafer is now ion implanted with boron in the unmasked regions, to increase the field inversion threshold between devices. A deep implant (150–200 keV) is necessary to avoid loss of dopant during the subsequent etch and local oxidation steps.
5. Both the CVD oxide as well as the exposed pad oxide are now removed in BHF.
6. Anisotropic etching of the silicon is now carried out in one of the many etch systems described in Chapter 9. Typically, about 2000 Å is removed, as shown in Fig. 11.12*c*.
7. Next follows a 4400 Å field oxide growth, which results in a fully recessed structure. Steam oxidation is commonly used at 1100°C. Alternately, high-pressure steam oxidation, at 900°C, is increasingly chosen for this step. The resulting structure is shown in Fig. 11.12*d*. A thin film of oxide ($\simeq$ 150 Å), which grows on the nitride during this field oxidation step, is removed in BHF.
8. Next, the nitride pattern is removed in hot H_3PO_4. Removal of the pad oxide in BHF results in a wafer which is now ready for MOS device fabrication (see Fig. 11.12*e*).

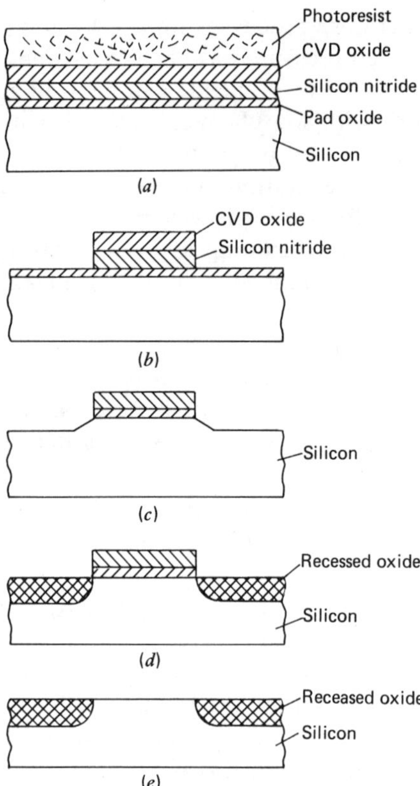

Fig. 11.12 Processing sequence for a fully recessed local oxide.

The above process is by no means exclusive, and many modifications can be made by way of improvement. One such uses sidewall oxidation to control the profile of the oxide features [22]. Yet another scheme involves the use of a second nitride layer after substrate etching to result in two thicknesses of local oxide [23].

The greatest impact of local oxidation technology has been in VLSI applications of MOS-based microcircuits. This is because all of the advantages of this technology can be simultaneously exploited here. These include: recessed oxides to provide a planar surface; field inversion channel stops by diffusion or ion implantation, followed by local oxidation; and extensive use of self-aligned techniques, using silicon nitride as a mask for ion implantation, diffusion, and local oxidation.

Details of these approaches are provided later in this chapter.

11.4 PLANARIZATION

The starting material for a microcircuit is ideally flat. However, the process of photolithographic patterning, combined with the growth or deposition of both insulating and conducting films, results in an increasingly nonplanar structure as the wafer proceeds to the metallization stage. Typically, the gate oxide of a MOS transistor may be 300–500 Å thick, whereas the thickness of a neighboring field oxide may be as much as 10,000–20,000 Å. This presents a severe problem in maintaining step coverage of metallization patterns, often resulting in a break in continuity at the step. This problem is especially serious in VLSI schemes where the lateral dimensions are further shrunk, but the metallization and oxide thicknesses are relatively unchanged. In these applications, techniques for flattening these steps greatly improve the process yield. These are commonly referred to as *planarization* techniques.

The local oxidation technique, described in the previous section, results in considerable planarization of itself, especially if the silicon is pre-etched so that the final oxide is fully recessed. Yet other approaches, aimed at removing the abrupt step when a cut is made in an oxide, are also desirable in VLSI applications. By way of example, Fig. 11.13*a* shows the etch profile obtained with an isotropic etch in an amorphous material such as silicon dioxide. Here equal etch rates in the lateral and vertical directions result in a circular contour. One approach to planarization, sometimes referred to as taper control, is to make the step in a cut more gradual by making the etch rate faster at the surface, by means of ion implantation, followed by sub-

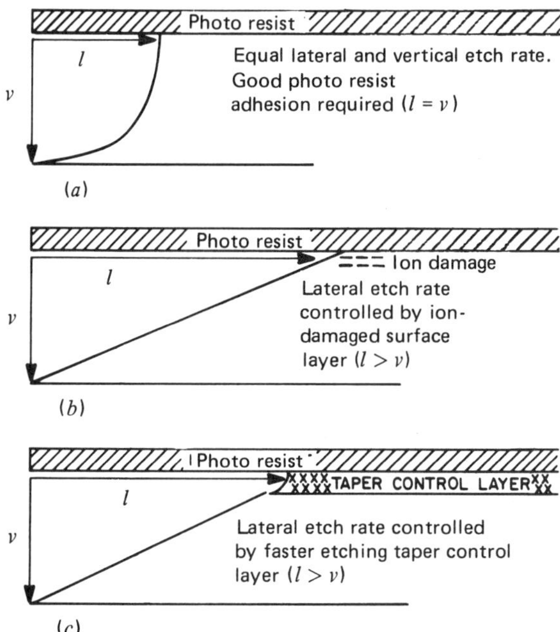

Fig. 11.13 Contours obtained with chemical etch. From White [25]. Reprinted with permission of the publisher, The Electrochemical Society, Inc. (*a*) Isotropic. (*b, c*) Tapered.

sequent etching [24]. An inactive ion such as argon is commonly used to damage the surface prior to the photoresist step. The dose is established by trial and error so that the resultant cut has a tapered wall as in Fig. 11.13*b*. This technique has proved to be quite controllable, and has been used successfully for tapering sidewalls in SiO_2, Si_3N_4, and polysilicon windows. One problem here is caused by the fact that photoresist adhesion to an ion-damaged surface is relatively poor. Consequently, an adhesion promoter such as hexamethyldisilazane (HMDS) must be used before photoresist application. Moreover, plasma etching is preferable to wet chemical methods, where the capillary action of the liquid tends to promote lifting of the photoresist.

Taper control by ion implantation damage cannot be used with MOS-based microcircuits, since this process produces both fixed positive charge states in the oxide as well as interface traps at the Si–SiO_2 interface. Annealing in forming gas can reduce this problem, but not eliminate it completely. An alternate approach involves the use of a two-layer sandwich whose upper layer is more readily etched [25]. This technique is most suited for tapering cuts in a thermally grown oxide. Here a thin layer of CVD silicon dioxide, which etches more rapidly, is used as the taper control layer on which the photoresist is placed. The precise taper angle is a function of both the oxide growth parameters as well as the etching solution. Figure 11.13*c* shows details of the tapering process using this layer.

Taper control can also be achieved by the use of a surface film of a low-melting-point glass, which can be reflowed by heating to elevated temperatures. PSG films, grown by the simultaneous oxidation of silane and phosphine at low temperatures (350–500°C) and with about 8% phosphorus by weight, can be reflowed at 850–900°C to give smooth window edges [26]. Typically, a 1000 Å layer of this glass is used in this process.

Polymide films can also be used for planarization [27]. These films, which consist of resins such as dimethylformamide (DMF) in acetone or methanol, have excellent thermal stability and chemical resistance, as well as good dielectric properties. Thus they can be left permanently in place, and are

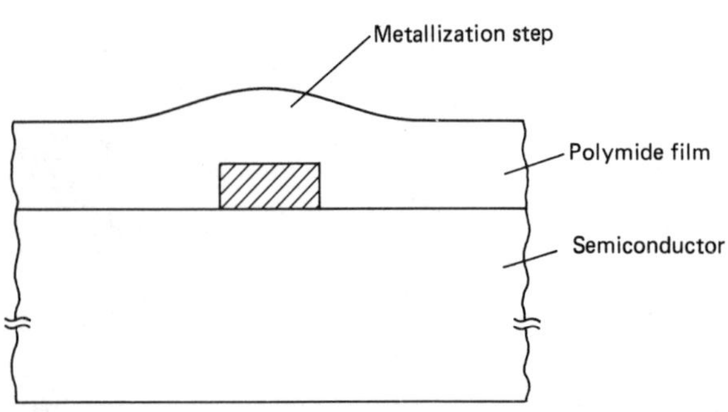

Fig. 11.14 Planarization effect with polymide film.

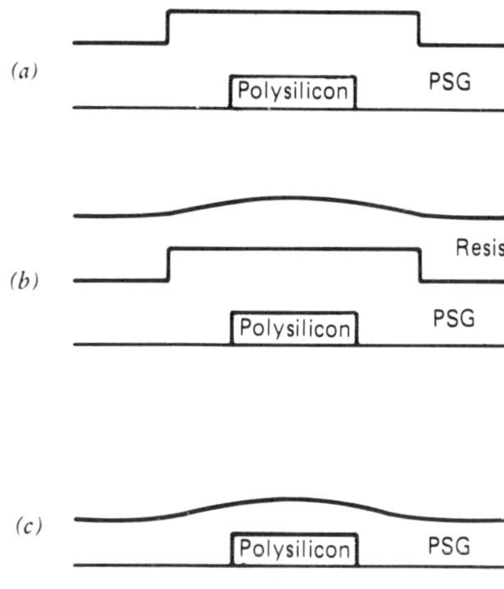

Fig. 11.15 Planarization by plasma etching. From Adams and Capio [28]. Reprinted with permission of the publisher, The Electrochemical Society, Inc.

often used instead of SiO_2 or PSG layers in microcircuits. They can be applied in layers, 1–3 μm thick, by conventional photoresist spinning techniques. Upon application, they flow around projections and steps on the surface, resulting in the composite structure of Fig. 11.14. The film shape becomes permanent upon curing at around 300°C. In use, the degree of planarization is controlled by the viscosity of the material, i.e., by the ratio of DMF to solvent, and by the film thickness.

The planarization of PSG glass films can also be accomplished by coating them with a resist, and subjecting the sandwich to gaseous etching in a planar reactor, with a CF_4–O_2 plasma [28]. This gas mixture can be controlled to remove the resist and the underlying PSG at the same rate, resulting in a smooth glass surface on which further layers can be formed. Although considerably more complex than the previous process, this approach is potentially more reliable since it leaves no residual organic layer on the microcircuit.

Steps in this process sequence are illustrated in Fig. 11.15 for a polysilicon projection on an otherwise flat surface. Note that the CVD growth of PSG results in a surface which follows the abrupt contour of the projection. The resist, on the other hand, planarizes the structure because of its viscosity. Here, too, the degree of planarization depends on the choice of resist and its thickness. A 1.0 μm layer results in relatively smooth surfaces. These are made smoother by flowing the resist at temperatures above 150°C before plasma etching in a planar reactor.

11.5 METALLIZATION

The process technology for depositing metal films has been described in Section 8.3, and photographic processes for forming interconnection patterns have been covered in Chapter 10. Here, however, we emphasize the special problems of VLSI circuits, and the approaches used for their solution.

First, area utilization of the silicon surface, as well as the ability to lay out a microcircuit, become increasingly difficult as circuits become more complex. Both of these problems can be alleviated by the use of multilevel metallization schemes. This not only provides additional surface area, but it also greatly aids in solving many of the topological problems of interconnection, since crossovers can be made by connecting between layers.

Next, planarization techniques must be used to minimize the problems of step coverage. These have been outlined in the last section. Finally, special photolithographic techniques, which allow the delineation of fine-line patterns, must be used in VLSI circuits. Some of the ways of addressing these problems are now described.

Multilayer metallization is quite commonplace in VLSI fabrication. Three layers of metal are often used, and some schemes call for even more. All these schemes rely upon the ability to deposit insulating films at low temperature. Silicon dioxide and PSG are commonly used for these layers; however, PSG is preferred since it can be deposited with less stress than undoped films.

Films of polymide have also been used for intermediate insulating layers. These are laid by spinning, and can be readily etched by conventional photoetching techniques, using hydrazine as an etchant. After curing, these films are stable to temperatures as high as 350°C. Recently new materials have been developed [29] which are unaffected by heat treatment to 450°C. Electrically, films of this type are comparable to silicon dioxide in their insulating capability. Moreover, they can be made 2–4 μm thick with much lower residual stress, and can provide a considerable degree of planarization.

Sputtered films of SiO_2 and Si_3N_4 have also been used for these intermediate layers. It has been shown that excellent quality films can be deposited if careful attention is paid to ensure a clean sputtering system, using oil-free pumping [30].

Contacting between aluminum metal layers and silicon is relatively easy. Here the contact system consists of Al–SiO_2–Si. Heat treatment results in breaking down the residual oxides of silicon to form an excellent microalloyed contact. With interconnections between layers, however, the problem is considerably more difficult. Here the contact system consists of Al–Al_2O_3–Al so that great care must be taken to remove the native oxide before the upper layer is deposited. Sputtering techniques are finding increasing use here, because of the ability to clean the metal area to be contacted by back-sputtering prior to sputter deposition of the upper layer. Here, too, a sputtering system that is oil-free is essential for the formation of these contacts with high reliability.

The lift-off technique, described in Section 10.2.1.1, is a powerful method for defining high-resolution metallization patterns for VLSI circuits. Here the use of two-layer photoresist films to provide an overhang is common practice [31]. An intermediate dielectric layer has also been used in some of these schemes.

Lift-off techniques rely upon the undercutting effect of wet chemical etches. Consequently, they are difficult to implement with dry etching processes which are becoming accepted in VLSI fabrication. Polymide films have been successfully used in this application [32]. Here a molybdenum pattern delineation mask is used, since it allows side etching of this film during sputter etching in an oxygen plasma.

11.6 GETTERING

The role of defects in reducing the circuit yield becomes more important with increasing device density. Thus successful VLSI fabrication requires the use of starting wafers of optimum quality. Nevertheless, even if fabrication is carried out under completely contamination-free conditions, a number of process-induced defects will eventually limit the circuit yield. The aim of gettering techniques is to remove or reduce these defects (as well as any defects caused by contamination), especially those that are formed in the vicinity of the active devices. Some of these processes remove defects completely; yet others move the defects away from the active region to other areas (the substrate backface, for example) where they have no effect.

Metallic impurities are often electronically active in both silicon and gallium arsenide. As a result, they reduce device lifetime and increase the leakage current of reverse biased junctions. Often they degrade the reverse breakdown characteristic, resulting in "soft" junctions with excess reverse currents at low voltages below their avalanche voltage. Elimination or reduction of these impurities is thus important in VLSI applications [33].

Oxygen and carbon are present as impurities in both starting silicon and gallium arsenide. These impurities usually form complexes in the lattice with the host atoms (silicon, gallium, and arsenic) as well as with impurities in the semiconductor. This, in turn, increases the possibility of forming other crystal defects by pinning dislocations at these sites.

Dislocations and stacking faults are also present in the semiconductor. Dislocations are primarily caused by the effect of mechanical stress which is induced during high concentration diffusions, and by rapid cool-down of slices after high-temperature processing. Stacking faults on the other hand come about because of imperfections during epitaxial layer growth, especially at the epitaxial layer–substrate interface. In silicon, these stacking faults can also be created during oxidation (i.e., the oxidation-induced stacking fault). The primary effect of stacking faults is that they act as precipitation sites for metallic impurities, which cluster around them in order to relieve the stress in the lattice. This in turn results in degraded junction characteristics, as well as in reduced lifetime.

A number of different gettering techniques are used for the removal or reduction of these defects [34]. Some of these are considered here.

Intrinsic Gettering. Silicon wafers, especially those which are grown by the Czochralski process, contain large amounts of dissolved oxygen. During device processing, this oxygen tends to precipitate in regions of stress, in the form of large clumps 0.1–1 μm in diameter, and about 1 μm apart. The intrinsic gettering process consists of heat treatments to silicon wafers prior to processing [35]. Its function is to precipitate out the dissolved oxygen in a region that is several microns below the silicon surface. Heat treatment in the 900–1100°C range for 2–4 h is commonly used for this purpose, and results in removing the dissolved oxygen in a 5–10 μm layer from the surface.

Halogenic Dopant Sources. Use of these sources during diffusion serves to convert metallic impurities (which are usually fast diffusers) into their volatile halides, which then leave the system by incorporation into the gas stream. These dopants must be used with care, however, since excessive halogen usage can result in pitting of the semiconductor surface.

Chlorine Oxidation. This process, described in Chapter 7, is highly effective in preventing the formation of oxygen-induced stacking faults in silicon. It is also effective in removing metallic impurities from the bulk crystal in much the same way as halogenic dopant sources during diffusion.

Mechanical Damage. Both sandblasting and physical abrasion can be used to create a layer of damage on the back surface of the slice. During subsequent high-temperature processing, this damaged region acts as a sink for the removal of both metallic impurities as well as stacking faults.

Controlled amounts of damage can also be introduced by depositing a layer of silicon nitride on the back surface of the slice [36]. Here the high interfacial stress created by the deposition process results in the formation of a dislocation network which acts as a getter during subsequent processing. Recently considerable success has been obtained by the use of ion implantation [37] as well as laser irradiation [38] for this purpose.

Glassy Layers. Both BSG and PSG in contact with silicon at elevated temperatures, have been found to be extremely effective in reducing metallic impurities, stacking faults, and dislocations in silicon slices [39]. The effectiveness of these layers comes about because, during heat treatment, they create a diffusion-induced stress which is well ahead of the impurity diffusion front [40]. In addition, they provide a viscous layer in which it is possible for some of the metallic impurities to become entrapped. Often these glass layers are deposited from a halogenic source, and so have some chlorine or bromine incorporated in them. This further provides a gettering role by chemical conversion of the impurity to its halide.

All of the above methods can be used singly during VLSI circuit fabrication. Often, however, combinations of these techniques are found to give improved results. Moreover, many of these techniques are used at different points in the process cycle.

11.7 MOS-BASED MICROCIRCUITS

The MOS transistor is the most promising active component for silicon VLSI circuits at the present time. There are a number of reasons for this choice. First, MOS transistors are self-isolating, so that devices can be placed side by side on a chip without the need for providing an isolation tub. As a result, they are considerably smaller than their bipolar counterparts. The use of local oxidation processes, however, has reduced somewhat this size advantage. Next, MOS transistors can be made in bulk silicon, thus avoiding the costly step of epitaxial growth. Recently, however, expitaxy has been shown to be necessary in some very-high-density microcircuits, in order to reduce noise problems caused by substrate interactions.

MOS transistors require less process steps (and masking operations) than bipolar transistors. Here, too, many special devices, necessitated by VLSI considerations, have tended to increase the process complexity, and thus reduced this advantage.

The MOS transistor is a high-impedance device, so that its power dissipation is low. In addition, its dynamic storage capability can be exploited to minimize the number of devices per memory cell in digital applications. On the other hand, the bipolar transistor is faster, and can be made in a variety of special configurations. As a consequence, its application areas tend to be more varied.

A few different basic MOS structures are now described in order to outline their fabrication principles. Many modifications to these processes are continually being developed, so that any attempt to be all inclusive would be fruitless. Structures based on D-MOS and V-MOS devices are not considered here. The D-MOS transistor is ideally suited for discrete power devices, but has highly asymmetrical characteristics which make it less useful in integrated circuit applications. Use of the V-MOS transistor results in exceptionally good area utilization, but suffers from the disadvantage that all devices must share a common source terminal. Fabrication sequences for these devices have been adequately described elsewhere [4].

11.7.1 The *p*-Channel Transistor

This is the earliest practical form of MOS device structure. Although not suited for VLSI applications, it is in wide use at the present time since it

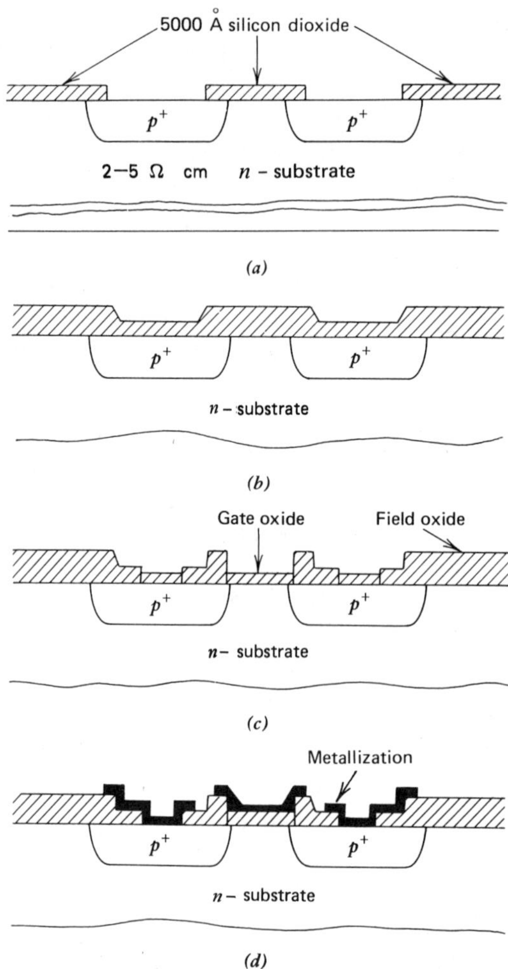

Fig. 11.16 Processing sequence for a *p*-channel aluminum gate MOS transistor. From R. A. Colclaser, *Microelectronics: Processing and Device Design* 1980. With permission from John Wiley and Sons.

involves relatively conventional processing steps to make an enhancement-mode device.* Its process sequence begins with a slice of *n*-type silicon [4]. Typical starting resistivities of 2–5 Ω cm are commonly used; alternately, 10 Ω cm slices, with a substrate-adjust implant, have been used to obtain more precise control of the threshold voltage. The slice is oxidized to about 5000 Å; source and drain diffusions are made through windows in this oxide, as shown in Fig. 11.16*a*. Next the slice is oxidized further to build up the field oxide (see Fig. 11.16*b*).

* The enhancement-mode device, which is normally off in the absence of an input signal, is the preferred choice for digital applications since it minimizes the standby power dissipation.

Cuts are made to define the source and drain contacts as well as the gate region. A dry thermal oxide is next grown, about 500 Å thick (Fig. 11.16*c*). The source and drain contact windows are opened for a second time by etching through the thin oxide grown during the thermal oxidation step. Finally an aluminum metal film is deposited on the slice and patterned, resulting in the MOS transistor shown in Fig. 11.16*d*.

Resistive loads are also used in circuits involving MOS devices. These are essentially transistors, with the drain connected to the gate. Thus they operate in the ON state, with a nonlinear V–I characteristic. This is generally not a problem, since they are used primarily in digital applications.

As mentioned earlier, the MOS transistor of Fig. 11.16 is not suited for VLSI applications. It requires precise definition of the gate so that it overlaps the source and drain. This requirement necessitates that the source and drain regions be deep, to take advantage of the increased lateral diffusion effect. Unfortunately, this leads to devices which are relatively slow because of the increased overlap and drain capacitances, both of which are parasitic terms.* Finally, the *p*-channel device is inherently about three times slower than its *n*-channel counterpart, since $\mu_n/\mu_p \simeq 3$. For these reasons, the *n*-channel enhancement-mode device is the primary vehicle for high-performance, high-density applications.

11.7.2 The *n*-Channel Transistor

The *n*-channel transistor, fabricated by self-alignment techniques, avoids many of the problems (and disadvantages) of the last device. This structure is inherently more difficult to make as an enhancement-mode device, and process controls are generally more stringent than for the previous structure.

Here the starting silicon is a *p*-type substrate, with a starting resistivity of about 5 Ω cm. An oxide layer, about 5000 Å thick, is grown over the slice and opened in regions where the source, drain, and gate electrode are to be placed, as shown in Fig. 11.17*a*. A 500 Å thick gate oxide is now grown on the slice by dry thermal oxidation. A layer of polycrystalline silicon is next deposited by a CVD process of the type outlined in Section 8.1.5. This is etched in order to define the gate. This polysilicon stripe is used as a self-aligned mask to selectively remove the exposed thin oxide, resulting in the structure of Fig. 11.17*b*.

An *n*-type diffusion is made for the source and drain regions. Again, the polysilicon gate serves as a self-aligned mask for this step. The gate is doped heavily *n*-type during the diffusion; this provides the appropriate work function for an *n*-channel transistor (see Fig. 11.17*c*).

Next, a layer of PSG is deposited by CVD and reflowed. This increases the thickness of the oxide over the field regions between devices, and thus

* This problem can be avoided by the use of ion implantation after gate metallization, followed by laser annealing [10].

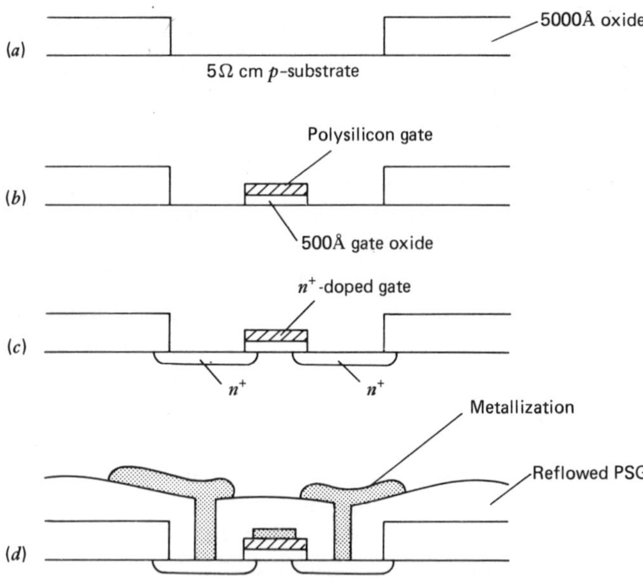

Fig. 11.17 Processing sequence for an *n*-channel polysilicon gate MOS transistor.

avoids parasitic MOS action. Moreover, it provides an insulator over the gate metallization, on which a second metal pattern can be placed. Contact windows for the gate, source, and drain connections are cut in this PSG layer. The final device, after metallization and pattern definition, is shown schematically in Fig. 11.17*d*.

Variations can be made to this sequence for the purpose of meeting the process requirements, and also to obtain improved device performance. These include the use of a partly or fully recessed local oxidation process to reduce the parasitic capacitance of the drain, and to planarize the structure; the use of ion implantation for substrate adjust or for adjusting the channel resistivity; and ion implantation under the field oxide to prevent channeling effects. A structure, incorporating these various features [17], is shown in Fig. 11.18. Here selective ion implantation is used to provide a depletion-mode load device as well. This provides a more desirable nonlinear load line and results in faster circuit operation.

Improvements in device performance can also be made by the use of silicide gate and interconnection materials instead of polysilicon. These materials have been described in Chapter 8; their unique advantage results in their lower resistance, as well as in their ability to be surface passivated, so that the next level of metallization can be placed over them [41].

Yet another area is the development of process sequences which lend themselves to the extensive use of self-alignment methods. These reduce (or eliminate) errors in the masking steps, reduce the number of masks that are required, and allow the tight physical tolerances to be maintained with less difficulty [42].

11.7.3 Complementary Transistors

Complementary MOS (C-MOS) circuits require a balanced pair of *n*- and *p*-channel enhancement-mode devices on the same chip. A typical fabrication sequence for these devices, using metal gate technology, is shown in Fig. 11.19. Again, many of the intermediate steps have been omitted in order to emphasize the process flow.

The starting material for these circuits is usually *n*-type silicon of about 5 Ω cm resistivity. The *p*-channel device is made directly in this substrate, using the process sequence outlined in Fig. 11.16. The *n*-type device, on the other hand, must be made in a *p*-type substrate. This is provided by means of a *p*-type well for each device, as shown in Fig. 11.19*a*, which must be diffused into the substrate. This diffusion must have a low surface concentration, and be precisely controlled, if the threshold voltages of the two devices are to match. This combination, resulting in a *p*-well with a resistivity of about 5 Ω cm, is extremely difficult to achieve by conventional diffusion methods. One approach here is to make a sealed-tube diffusion, using a dilute doped oxide as the impurity source [43]. Increasingly, however, ion implantation is used as a precise predeposition step. This implantation is made through an oxidized region of the slice; the drive step is conducted in a nonoxidizing atmosphere in order to avoid boron depletion effects. Subsequent steps in the formation of the *n*-channel transistor are carried out within this well, and follow the process sequence of Fig. 11.16.

Parasitic *p–n–p* and *n–p–n* transistors are present in C-MOS pairs of the type shown here. Thus a vertical *n–p–n* is formed by the n^+-source/drain regions, the *p*-well, and the *n*-substrate. At the same time, a lateral *p–n–p*

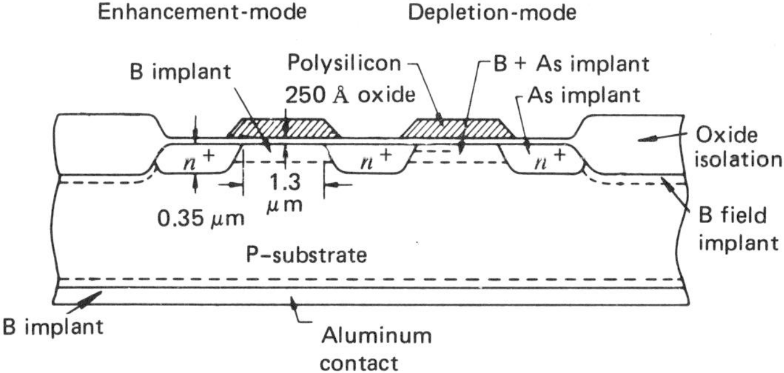

Fig. 11.18 A modern *n*-channel polysilicon gate MOS with a depletion-type load device. From Dennard, Gaensslen, Walker, and Cook [17], © 1979. The Institute of Electrical and Electronic Engineers.

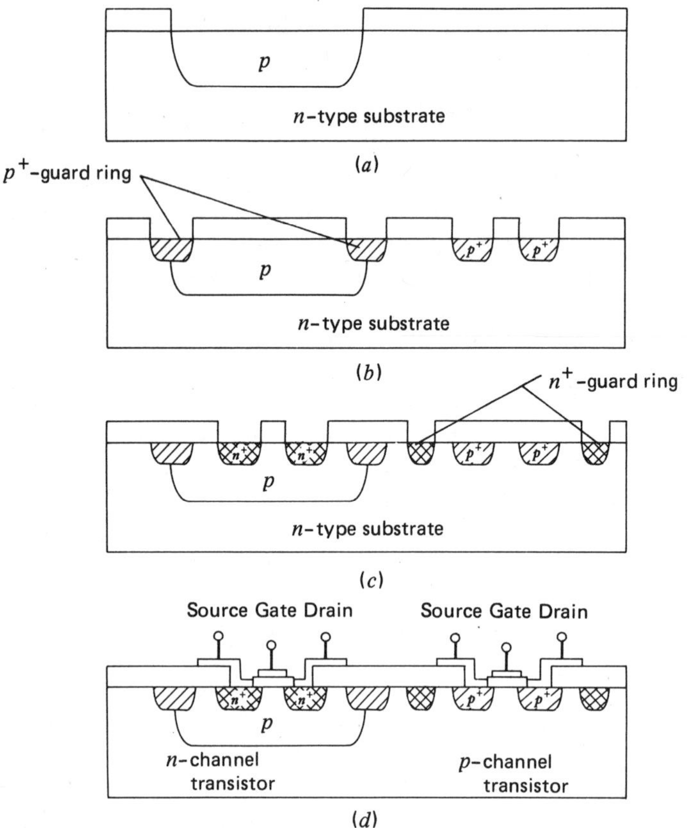

Fig. 11.19 Process sequence for metal gate C-MOS devices. From R. A. Colclaser, *Microelectronics: Processing and Device Design* [4], 1980. With permission from John Wiley and Sons.

transistor is present due to the p^+-source/drain regions, the n-substrate, and the p-well. Together, these constitute a p–n–p–n transistor which can latch up under certain bias conditions, and thus present a potential reliability problem. One approach to removing this problem is to surround the n-channel device with a p^+-guard ring which is grounded to the p-well. In addition, an n^+-guard ring, which is grounded to the n-substrate, must be used to surround the p-channel transistor. Both of these guard rings are shown in Fig. 11.19. In some situations, suitably placed field implants have been used to avoid this problem.

Silicon gate devices can also be used with C-MOS circuits. Here, too, the first step is the fabrication of the p-well in an n-type substrate. Subsequent steps follow the process sequence for polysilicon gate devices shown in Fig. 11.17. The resulting schematic cross section of these devices is shown in Fig. 11.20 [44]. Note the use of field implants to avoid parasitic *p–n–p–n* action and local oxidation for reducing the device size.

An alternate approach to C-MOS fabrication begins with a *p*-type substrate in which an *n*-well is diffused. This type of circuit has the advantage that it is compatible with *n*-channel MOS transistors which can be made on the same substrate [45]. This concept has also been extended to the incorporation of bipolar transistors in the same chip, resulting in an extremely flexible technology that is suitable for combined analog–digital circuits [46].

C-MOS circuits are inherently difficult to make because of the requirements for matched threshold voltages. Ion implantation has been extensively used for adjustment of this parameter as well as for the diffusions and field regions [47]. In addition, local oxidation and self-alignment techniques have greatly eased the fabrication problems of this process. Thus many modifications have been made to these processes in order to produce microcircuits with acceptable process yields.

11.7.4 Nonvolatile Memory Devices

A nonvolatile memory device is one that retains its information even when power is removed from the circuit. Thus it is a direct replacement for magnetic core memories, which also have this characteristic. MOS transistors can be used for this purpose since their gate has a high impedance to ground, and is thus capable of long-term charge retention. One such structure consists of a *p*-channel MOS transistor with a floating polysilicon gate [48]; erasure is by UV or X-ray irradiation, which increases temporarily the conductivity of the gate oxide, causing the gate to return to an uncharged state. Cells in a microcircuit array can thus be individually programmed, but must be collectively erased. The schematic configuration for a cell of this type is shown in Fig. 11.21*a*.

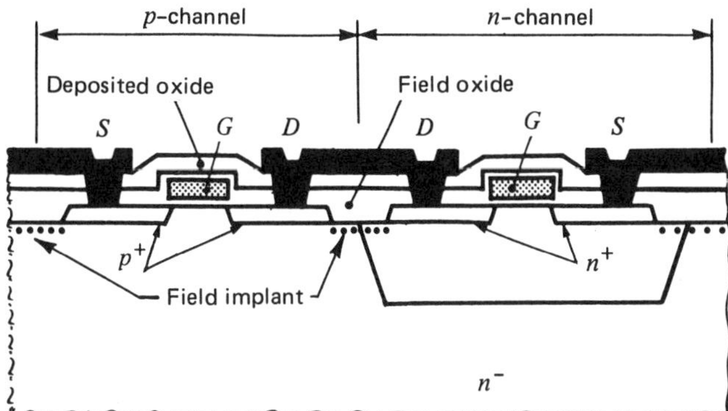

Fig. 11.20 Silicon gate C-MOS devices. From Salama [44], © 1981. The Institute of Electrical and Electronic Engineers.

From the point of view of circuit design flexibility, however, cells which are individually erasible are much more desirable and have been the subject of intensive research. One approach here consists of using a stacked polysilicon gate structure, where an upper control gate can be used to extract charge stored on the lower floating gate. This device can use the same photomask for both gates [49] which allows the fabrication of a self-aligned structure, as shown in Fig. 11.21*b*. Other configurations of this type utilize a dual gate structure, with variations in the shape and positioning of the gates over each other. Their device structures (and their fabrication steps) are somewhat more complex than the structure shown here.

Individually erasible cells can also be made with a control gate in a dual-dielectric structure. Figure 11.22 shows the schematic of a cell of this type [50]. This cell operates on the principle of charge storage on traps at an interface. This interface is established by the use of a double-layer dielectric, consisting of an extremely thin ($\simeq$ 20 Å) film of grown silicon dioxide on which is deposited a thicker film ($\simeq$ 500 Å) of silicon nitride. Here charging of the interface traps, as well as traps in the bulk silicon nitride, is accomplished by breakdown of the drain–substrate junction, and tunneling through the thin silicon dioxide layer [51]. The resulting structure is thus a metal–nitride–oxide semiconductor (MNOS) transistor.

It has been found that the write–erase characteristics of MNOS cells depends critically on the properties of the interface, and of the bulk silicon nitride. One improvement is to introduce a thin layer of a refractory metal such as tungsten at this interface, so that charge storage is concentrated in this region. Layer thickness must be one monolayer or less, so that specialized vacuum evaporation techniques have been employed for this purpose [52].

11.7.5 Silicon-on-Sapphire Devices

Microcircuits made in silicon films on sapphire (SOS) comprise an important area for VLSI technology, since they are more radiation resistant than those

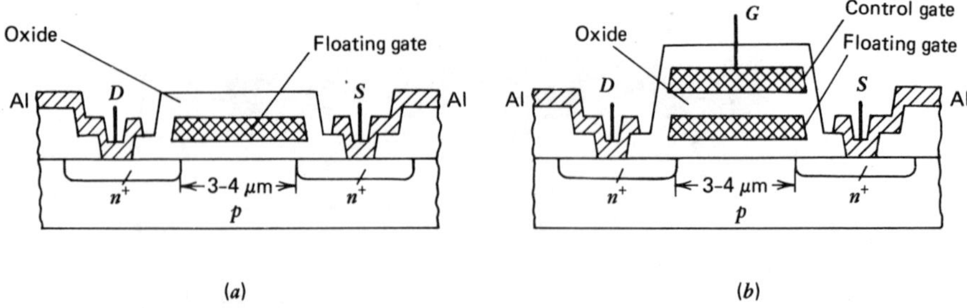

Fig. 11.21 Nonvolatile memory devices. From Scheibe and Schalte [49], © 1977. The Institute of Electrical and Electronic Engineers: (*a*) floating gate and (*b*) floating and control gate.

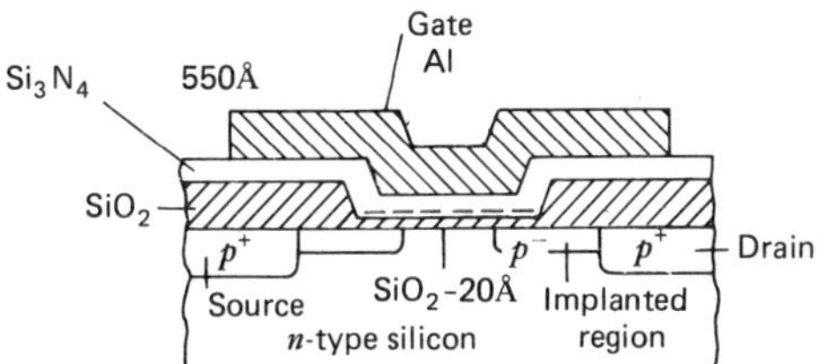

Fig. 11.22 Schematic of a metal–nitride–oxide semiconductor (MNOS) transistor.

using bulk silicon. Moreover, the insulating substrate greatly reduces the parasitic drain capacitance, so that these circuits are about 50% faster than conventional circuits made on bulk silicon. Deep depletion devices are commonly made with *n*-channel. In addition, *p*-channel inversion devices are used as the complementary device in C-MOS circuits.

A number of special factors must be taken into account in the fabrication of these circuits. Thus island definition leads to exposing the edges of the source and drain junctions. This results in a parasitic edge transistor which is made on a (111) face* whose threshold voltage is lower than that of the bulk *n*-channel device. This parasitic effect can be reduced if local oxidation techniques are used to form the islands [53]. Alternately, stops placed on these edges can completely avoid [7] this problem, which creates anomalous kinks in the device characteristics. (It is worth noting that *p*-channel devices do not suffer from premature turn-on since the threshold voltage of the parasitic device is higher than that of the bulk structure.)

Films of 4000–5000 Å are commonly used for these circuits, since their mobility degrades very rapidly if the film thickness is reduced. Even so, typical bulk mobilities are only 40–50% of that obtained in single-crystal silicon, because of the defect structure of these films. Thus their primary speed advantage comes from elimination of the parasitic capacitance of the drain to the substrate.

Ion implantation into SOS films can be difficult because of charging effects due to the insulating substrate. Metal films are sometimes used for masking these circuits during this step, so as to avoid this problem.

Figure 11.23 shows both *p*- and *n*-channel devices, made by a C-MOS process [6]. Here a p^+-doped silicon gate is used for a conventional enhancement-mode *p*-transistor, whereas the enhancement-mode *n*-transistor is a deep depletion device. In this circuit, a combination of ion milling followed by plasma etching was used in the various delination steps to minimize under-cutting problems, while reducing problems due to ion damage. Glass reflow techniques using a CVD film of PSG, which was flowed at 875°C, served to minimize step problems in the metallization.

* This assumes that a conventional anisotropic etch process is used for edge definition.

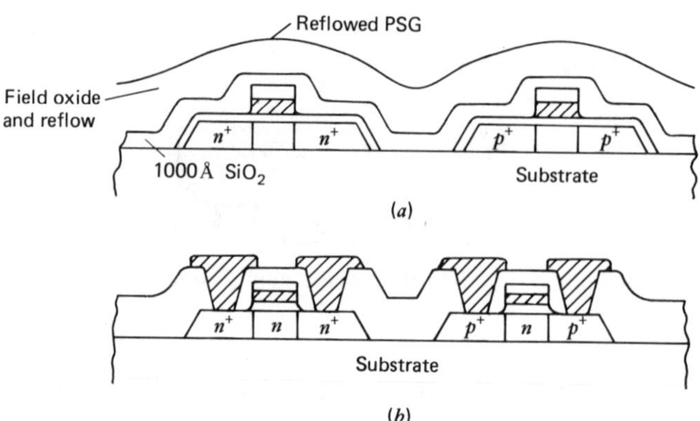

Fig. 11.23 Silicon-on-sapphire devices. From Splinter [6], © 1976. The Institute of Electrical and Electronic Engineers: (*a*) after reflow and (*b*) after metallization.

11.8 MESFET-BASED MICROCIRCUITS

The metal-semiconductor field effect transistor (MESFET) is the active component used in these circuits. Here a Schottky gate serves to control the flow of majority carriers between source and drain. Gallium arsenide microcircuits are based on the use of these devices, and *n*-type material is used because of its high electron mobility.

There are a number of additional factors that make the MESFET device ideal for use with gallium arsenide.

1. The wide energy gap of gallium arsenide results in a relatively high barrier height ($\simeq$ 0.8–0.9 eV) with many metallurgically compatible gate materials, so that the leakage current of the input diode is low.
2. The gate is formed by vacuum evaporation, which is a low-temperature process. The source and drain regions are usually fabricated by a 30–60 s heat treatment at 450°C.
3. The use of photo-patterned metal films for the gate as well as for the source/drain regions results in a structure in which extremely tight tolerances can be held, so that these devices are capable of high speed. At the present time, the gallium arsenide MESFET has a speed advantage over silicon devices (of the same physical dimensions) by a factor of 3.
4. Finally, a viable technology for MOS-type enhancement-mode structures in gallium arsenide does not exist at the present time. Moreover, diffused junctions are difficult to form precisely in gallium arsenide, since lateral diffusion effects are not fully understood.

Both linear and digital circuits can be fabricated using these devices. Linear circuits include microwave amplifiers, couplers, mixers, and receivers, and have been operated at frequencies from approximately 100 MHz to over 100 GHz [54]. The prospects for these circuits are excellent.

The prospects for high-density digital circuits using gallium arsenide are also excellent. However, no clearly optimum approach has been developed at the present time, and there is much room for choice of device type and circuit configurations [55, 56]. This is because the gallium arsenide MESFET is inherently a depletion-mode device. Thus relatively complex circuit configurations, involving dual power supplies and voltage level shifting, are necessary to minimize power dissipation in digital logic circuits. In contrast, enhancement-mode devices can be used in extremely simple circuit configurations, with a single power supply. MESFETs of this type can be made by using a narrow channel region which is lightly doped so that it is pinched off by the contact potential of the Schottky junction. However, this structure is more difficult to fabricate, and has a lower voltage swing than its depletion-mode counterpart.

The use of a junction gate allows an increase in voltage swing as well as in channel width. Unfortunately, junction formation is a considerably more difficult process than the fabrication of a Schottky gate at the present time. However, the recent development of techniques for making precisely controlled p^+-type diffusions by an open-tube process [57] indicates that a solution to this problem may be forthcoming.

11.8.1 The Depletion-Mode Transistor

This device, which is most commonly used, is fabricated in an n-type layer, epitaxially grown on a $\langle 100 \rangle$ oriented semi-insulating substrate. Typically this layer is 0.1–0.2 μm thick, doped to about $(1\text{–}3) \times 10^{17}$ carriers/cm^3.

The interface between the epitaxial layer and the substrate is a somewhat defected region because of instabilities during the initial growth phase. These result in deep acceptors and traps which can interact with majority carriers in the channel. In addition, both chromium and other impurities in the bulk substrate have been shown to out-diffuse during epitaxial growth. For this reason, a lightly doped ($n < 10^{15}$ cm^{-3}) buffer region, 4–10 μm thick, is often grown prior to the active layer. This buffer layer is especially important in low-noise applications; however, it can usually be dispensed with in digital circuits.

The MESFET is not self-isolating. Here isolation of the device is obtained by surrounding it with an etched moat which penetrates through to the semi-insulating substrate. This results in a slight loss of surface planarity. Selective proton implantation, described in Chapter 6, can also be used for isolation between devices, to result in planar circuits.

Variations in the pinch voltage of MESFET devices in a microcircuit can be caused by local nonuniformity in epitaxial layer doping and thickness. These variations can be eliminated by selective anodic thinning of the slice [58] so as to obtain a constant doping–thickness product across the wafer. This technique has been outlined in Chapter 7.

Ion implantation can be used to form the active layer in a semi-insulating substrate, or in the buffer layer. The doping and penetration depth of this process is sufficiently well controlled so that circuits can be made without the use of anodic thinning techniques. After implantation, the layer is capped with silicon nitride and annealed and then patterned and etched as before.

Isolated regions can also be formed directly into the semi-insulating substrate by ion implantation [59] through a photomask and a cap layer of silicon nitride, which is used in the subsequent anneal step. Individual isolated islands are formed in a planar structure by this means. Here, too, outdiffusion of chromium and other impurities from the substrate can deteriorate the quality of the channel region. However, many of these problems have been greatly reduced by the use of undoped material, grown by the liquid encapsulated Czochralski (LEC) process, in conjunction with in-situ compounding [60]. In addition, the anodic thinning step is not required with this process. Details of the implantation step have been given in Section 6.3.2.

A wide choice of metals is available for the Schottky gate (see Section 8.5). Of these, the most generally used is aluminum. This metal can be vacuum evaporated, and etched in a fine-line pattern with relative ease. Moreover, it can tolerate the subsequent heat treatment which is required for forming the ohmic contacts for the source and drain. Molybdenum, tungsten, and titanium are popular choices as well, and can operate at higher temperatures than aluminum. These must be deposited by E-beam evaporation, sputtering, or ion-beam deposition.

The basic approach for gate definition is by means of the lift-off process. Optical patterning, without the use of undercutting, can provide gate lengths in the 1–2 μm range with some care. Increasingly, however, this lift-off process is combined with a two-layer system so as to allow undercutting of the gate metal. Gate widths of 0.5 μm can be obtained readily by this method.

The gain–bandwidth product of a MESFET is critically dependent on the parasitic resistance between the gate and the source (including that of the source contact). Contacts with low specific resistance ($\simeq 10^{-5}$–$10^{-6}\ \Omega\ \text{cm}^2$) are thus essential for high-performance devices. The formation of these contacts has been described in detail in Section 8.4. At the present time, lowest contact resistance values are obtained by the use of evaporated films of Au–Ge eutectic, or Au–Ge–Ni. After deposition and patterning, these films are heat treated to about 450°C for 30–60 s to form the contact [61]. Laser annealing has been found to be effective in reducing the contact resistance, and is under active consideration at the present time [62].

Much work has been done to minimize the deterioration of surface flatness which occurs during heat treatment. Here, too, laser annealing is one approach that is under consideration. Another is the use of a cap layer of SiO_2 to contain the film during its alloying phase. Note, however, that this layer must be grown by an extremely low temperature process ($\leq$ 150°C) such as ion-beam deposition, to prevent excessive interface reaction between the gold and the semiconductor [63].

Yet another approach to reducing the parasitic gate–source resistance is to use etching to form a recessed structure [64]. This is shown in Fig. 11.24*a*, together with a conventional plane structure in Fig. 11.24*b* for comparison. It should be emphasized that techniques of this type are more suited for discrete devices and analog integrated circuits, than for high-density digital applications.

A process sequence for a MESFET using a nonrecessed structure is now outlined [65]. The starting material here consists of *n*-gallium arsenide on a semi-insulating substrate, in which active regions have been defined by mesa etching.

A 0.5 μm thick aluminum layer, which eventually serves as the gate metal, is evaporated on the slice. A positive photoresist is spun on the wafer, and its surface layers are toughened by a soak in chlorobenzene (see Section 10.2.1.1). Thc gate region is defined, and developed to result in an overhang, as shown in Fig. 11.25. The aluminum is now etched away from source and drain areas. At the same time, it is undercut so that the gate length is much smaller than the defined pattern. Typically, a 3 μm defined pattern results in a 0.5 μm gate by this process.

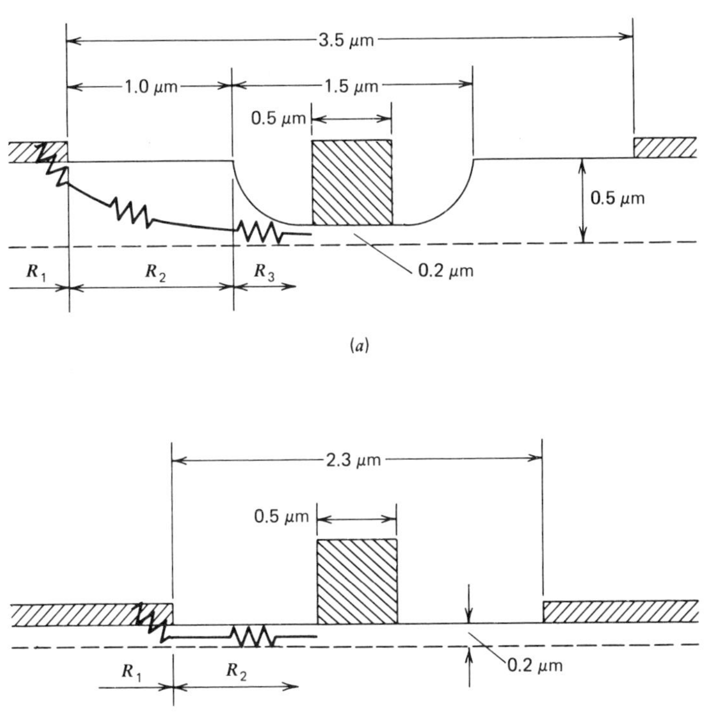

Fig. 11.24 MESFET structures. From Ohata, Itoh, Hasegawa, and Fujiki [64], © 1980. The Institute of Electrical and Electronic Engineers.

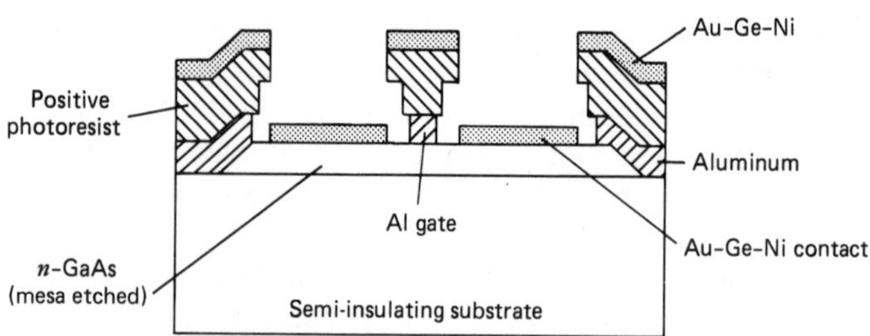

Fig. 11.25 Schematic of a depletion-mode transistor. From Donizelli [65]. Reprinted with permission from *Electronic Letters.*

A layer of Au–Ge–Ni is next evaporated for the source and drain contacts. After lift-off from unwanted regions, a 450°C heat treatment for 30 s serves to alloy these contacts. Note that the aluminum gate is unaffected by this thermal process. Finally, the exposed aluminum is etched away from unwanted areas.

Variations of this process usually involve alternate gate metals such as molybdenum and tungsten, alternate means for achieving the overhang required to facilitate the lift-off process, and alternate processes for controllably undercutting the gate resist. One of these [66] uses a molybdenum gate on which is deposited a gold upper layer. Photoresist patterning, followed by ion milling, is used to define the gold region. Next the molybdenum gate is undercut in a CF_4-O_2 plasma, using the gold layer as a mask to provide the necessary overhang. The use of plasma etching at this stage serves to more controllably transfer the resist pattern, and avoids irregular etching caused by capillary action and bubble formation during wet chemical processing.

11.8.2 The Enhancement-Mode Transistor

As mentioned earlier, circuits based on enhancement-mode devices are highly suited for VLSI applications because of their simplicity and low power dissipation. One approach to fabricating devices for these circuits follows the sequence outlined in Section 11.8.1 for the depletion-mode MESFET [67]. In this case, however, the *n*-layer is more lightly doped ($\simeq 10^{17}$ carriers/cm^3), and must be precisely thinned by anodic means to ensure that the channel is pinched off with zero gate bias. A 0.5 μm mesa etch serves to define the isolated islands on which the device is made. In this work, a titanium gate was used, followed by an evaporated gold layer. This Ti–Au system also provided the metallization which was placed over a CVD layer of SiO_2. Source and drain contacts were made by means of a Au–Ge eutectic film, as described earlier.

Enhancement-mode field effect transistor can also be made using a diffused junction gate. Although not MESFET devices, these can be used in gallium arsenide digital circuits as well, and are considered here for convenience.

Figure 11.26 shows the schematic of an enhancement-mode junction gate FET (E-JFET) with a series-connected resistor which serves as a load in a typical inverter configuration [68]. Here multiple ion implants into a semi-insulating substrate were used to form the deep, heavily doped regions over which the contacts were made, and the somewhat shallower (but less heavily doped) regions for the channel. Next, the implants were capped with a silicon nitride layer and annealed for 20 min at 750–925°C.

The gate region was next implanted with zinc, and diffused at 600°C to provide the necessary junction depth required to pinch off the device under conditions of zero bias. The thickness of the channel under the gate was 1000 Å in this application.

11.8.3 Digital Circuit Approaches

This section describes two VLSI digital circuit schemes for the purpose of illustrating differences in approach. The first of these [69] utilizes ⟨100⟩ oriented semi-insulating substrates in which a 2 μm thick buffer layer is grown by liquid-phase epitaxy. Selenium ions are implanted over the entire substrate which is maintained at 350 °C during this process. Next, the slice is capped with silicon nitride and annealed at 850 °C. This forms the active n-layer with a sheet resistance of 500 Ω/square and a ± 3% uniformity over the wafer.

Au–Ge contact material is evaporated, defined, and alloyed at 460 °C, followed by defining and etching mesas to isolate the devices. The lift-off process is next used to form a gate of 1 μm length. Cr–Au–Pt layers are used for this gate, as well as for the interconnection metal. An SiO_2 layer grown by CVD is used for the substrate insulation. Figure 11.27 shows a MESFET, a diode, and an interconnection formed using this approach.

Figure 11.28 shows a scheme for the fabrication of a planar device and two types of Schottky diodes in semi-insulating substrates, formed by direct implantation without the use of a buffer layer [59]. This approach begins by

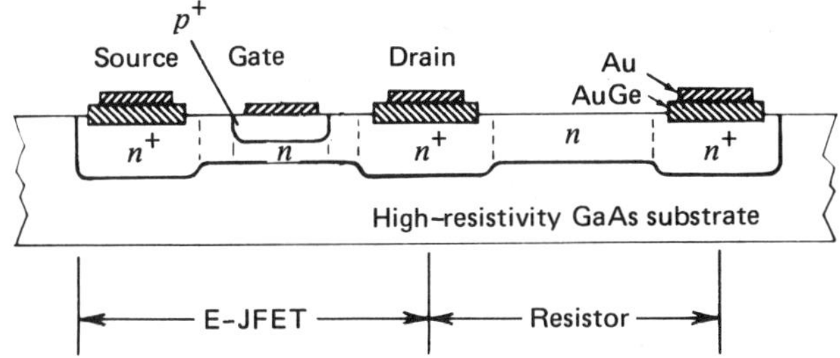

Fig. 11.26 Schematic of an enhancement-mode JFET inverter circuit. From Zuleeg, Nothoff, and Lehovec [68], © 1968. The Institute of Electrical and Electronic Engineers.

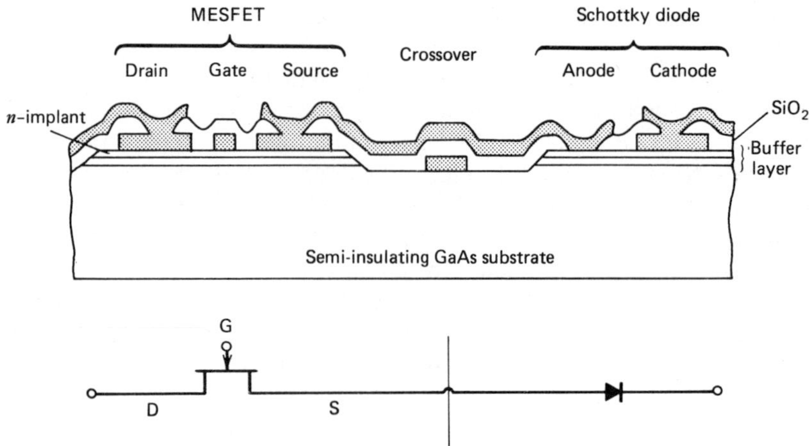

Fig. 11.27 Digital circuit scheme using mesa isolation. From Van Tuyl, Liechti, Lee, and Gowen [69], © 1977. *The Institute of Electrical and Electronic Engineers.*

coating the ⟨100⟩ substrates with a thin layer of Si_3N_4 by CVD techniques. Two implants are made through this layer, using a photoresist film as a mask. A selenium implant defines the active channel ($\simeq$ 0.15 μm deep), whereas a sulfur implant defines somewhat deeper ($\simeq$ 4 μm) source and drain regions (see Section 6.3.2). Next, additional Si_3N_4 is deposited before an annealing step which is required to activate these implants.

Windows are opened in the deep implanted regions, and Au–Ge eutectic is deposited for the source and drain contacts. These are formed by heat treatment at 450°C for 30 s. The gate metal consists of Ti–Pt–Au layers, and is patterned by a double-layer lift-off process. The Ti–Au layers are also carried over to produce the first and second levels of interconnect metal. Routing diodes, of the Schottky type, are formed by using the deeper sulfur implants as the active region.

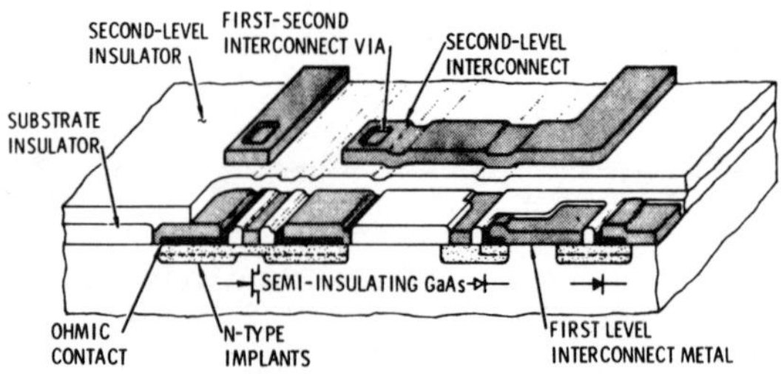

Fig. 11.28 Digital circuit scheme using self-isolation. From Welch, Shen, Zucca, Eden, and Long [59], © 1980. The Institute of Electrical and Electronic Engineers.

11.9 BJT-BASED MICROCIRCUITS

Here the active device is the *n–p–n* bipolar junction transistor (BJT), which is used in conjunction with a variety of components, both active as well as passive. Passive elements include many types of diodes, resistors, and capacitors. Active devices include one or more types of *p–n–p* transistors, special-purpose devices such as low-inverse-beta, superbeta, multi-emitter, and multi-collector transistors, and often junction gate or insulated gate field effect transistors. Bipolar transistors are minority carrier devices, so that there is no advantage to building these in gallium arsenide. As a result, BJT-based microcircuit technology is only used with silicon starting material.

The double diffused expitaxial (DDE) process, outlined in Section 11.1.3, is commonly used today for BJT-based microcircuits, and results in isolated tubs of the type shown in Fig. 11.3*c*. In this process the starting material is a lightly doped slice of *p*-type silicon. A broad range of starting resistivity (3–10 Ω cm) is often specified; in many instances, no upper limit is set on this parameter. A layer of *n*-type silicon, usually 3–15 μm thick and of the appropriate resistivity type, is epitaxially grown on this slice. Isolation is accomplished by diffusing highly doped *p*-type moats from the upper surface (the epitaxial layer side); transistors and other components are fabricated in the resulting pockets of *n*-type material. This process has the advantage that the epitaxial layer needs only to be thin enough to fabricate a device. On the other hand, the substrate can be of arbitrary thickness, so that breakage problems during microcircuit fabrication are minimized. Thicknesses of 250–400 μm are typical of modern practice. A slice diameter of 75–100 mm is commonly used. Finally, lateral penetration of the isolation diffusion is approximately equal to its penetration normal to the silicon surface. Consequently, the relatively shallow diffusions required by the DDE process result in a minimum of waste area on the silicon chip and lead to greatly improved yields in processing.

Bipolar transistors of this type have a high parasitic collector resistance, as is seen from Fig. 11.29*a*, where the current path from the emitter to the n^+-collector has been delineated. Figure 11.29*b* shows the same device in which a low-resistivity *buried layer* has been incorporated prior to epitaxial growth. The epitaxial layer is grown over this buried layer, and the microcircuit is fabricated in the resulting slice. The collector-current flow path is also shown for this transistor. Here much of this path is traversed through the low-resistivity buried layer, resulting in a reduction of the collector parasitic resistance. For a typical high-speed transistor of small geometry, the use of the buried layer may reduce the collector resistance by as much as a factor of 20, which is a significant improvement in many circuit applications. In this manner, performance comparable to that of epitaxial transistors may be obtained.

Details of the process sequence for these devices are shown in Fig.

11.30*a–g*. Also shown in this figure, for illustrative purposes, is a second isolation tub with a diffused resistor. Process variations include the use of implantation for one or more regions, and the use of double masking and predeposition for regions such as the buried layer and the isolation wall, which require high conductivity as well as long drive-in times. Some specific process and device considerations now follow.

11.9.1 The Buried Layer

The buried layer is the first region to be introduced into the silicon slice. In addition, it is a relatively highly doped region. This causes three problems. First, it is subject to considerable movement during subsequent high-temperature processing. The slow diffusers (antimony and arsenic) are used for this application to minimize this movement. Next, the buried layer should be highly doped to minimize its sheet resistance. Unfortunately this leads to strain-induced damage, on top of which the epitaxial layer must be grown. In addition, severe redistribution takes place during subsequent epitaxial growth and diffusion steps. Partially to offset these problems, this layer is diffused with a long drive-in step, typically 8–10 h at 1200°C, to reduce the

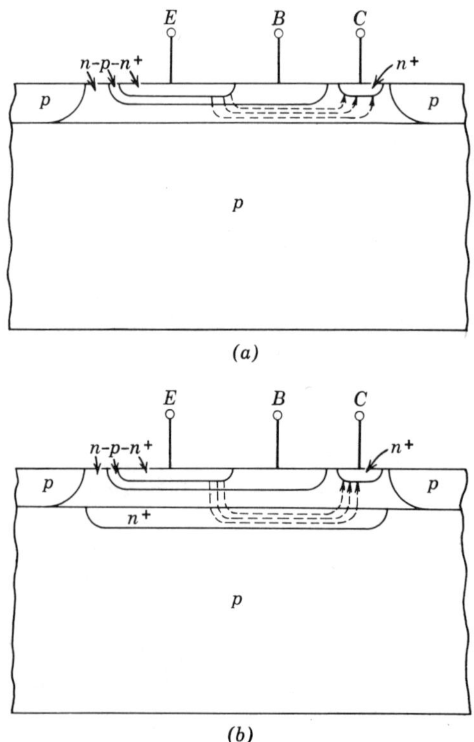

Fig. 11.29 The buried layer.

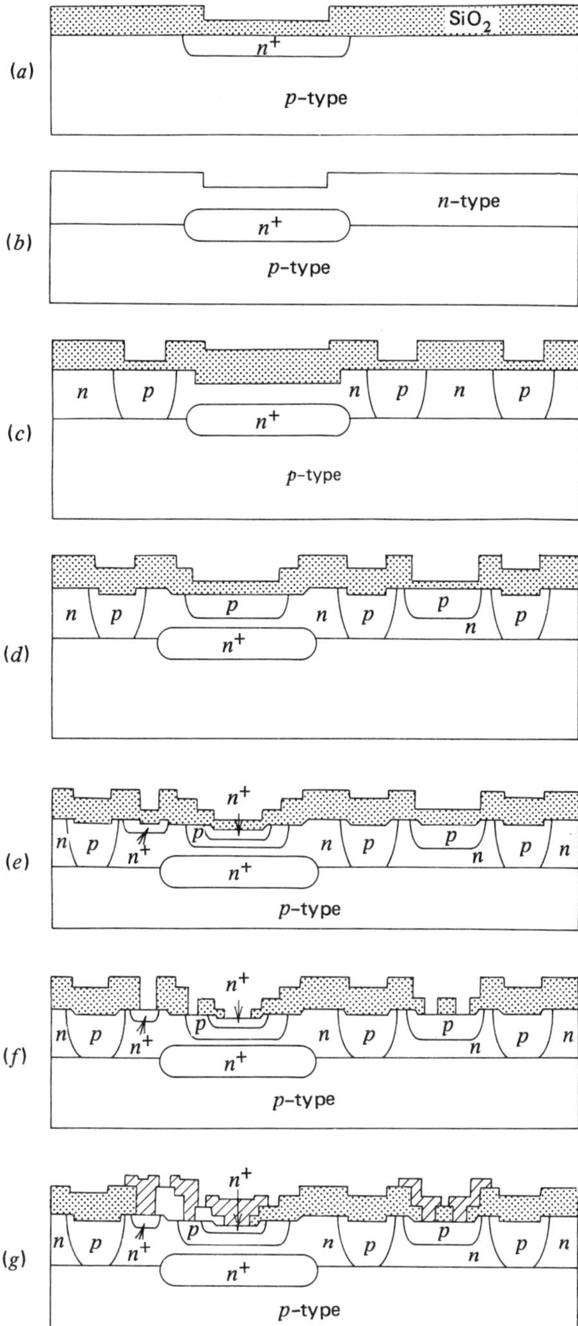

Fig. 11.30 The junction-isolated bipolar integrated circuit process. From R. A. Colclaser, *Microelectronics: Processing and Device Design* [4], 1980. With permission from John Wiley and Sons. (*a*) *p*-type substrate with n^+-buried layer diffusion. (*b*) Epitaxial layer growth. (*c*) p^+-Isolation diffusion. (*d*) *p*-Base diffusion. (*e*) n^+-Emitter and collector contact diffusions. (*f*) Contact pattern. (*g*) Metallization and interconnect pattern.

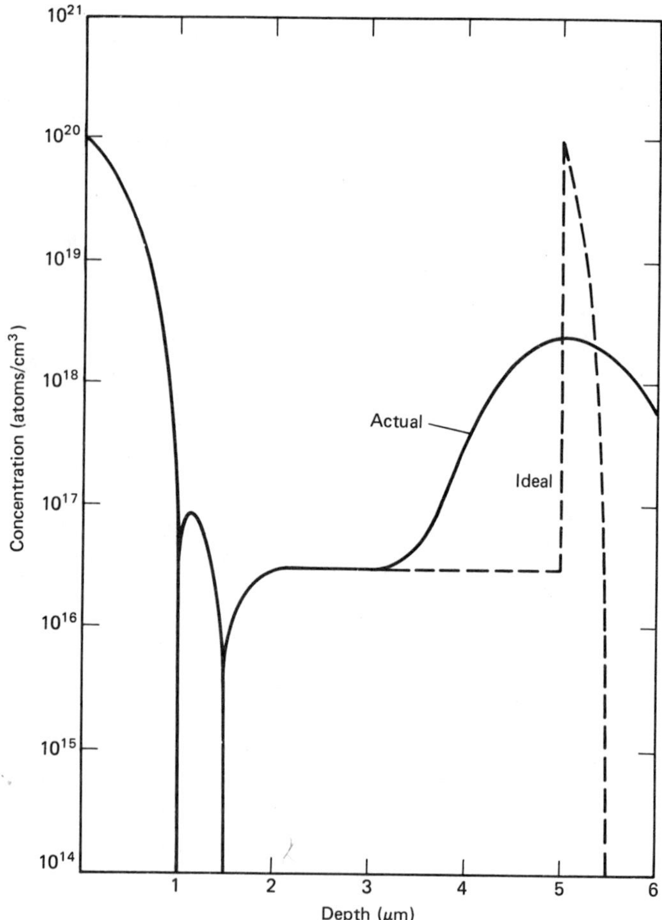

Fig. 11.31 Ideal and actual doping profiles with a buried layer.

surface concentration and still maintain a low sheet resistance. Finally, as a result of out-diffusion and etch-back effects, the ideal abrupt n^+–n interface is not obtained. For thin epitaxial layers (where the collector–base junction is close to the substrate–epitaxial layer interface) these effects can significantly alter the doping profile of the collector region, as shown in Fig. 11.31. For comparison the ideal doping profile of a buried layer transistor is also shown. Typically, the combined effects of out-diffusion and etch-back alter the doping profile of the epitaxial layer for about 2 μm from the epitaxial layer–substrate interface.* This assumes that the epitaxial layer is grown by the silicon tetrachloride process. Use of the silane process

* Note that a long drive-in of the buried layer prior to epitaxy reduces the severity of this problem.

greatly reduces these effects. However, this process is considerably more difficult so that its use is confined to structures with ultrathin (1–2 μm) epitaxial layers.

Antimony is sometimes used for the buried layer. Its large misfit factor ($\epsilon = 0.153$) and consequent low solid solubility limits its use to layers with a sheet resistance of about 15–20 Ω/square.

In view of the perfect fit of arsenic to the silicon lattice and its consequent high solid solubility, it should be possible to incorporate a substantially higher concentration of this dopant without excessive lattice damage. Assuming that solid solubility is the only limitation here, a sheet resistance of less than 5 Ω/square can be achieved. Unfortunately, arsenic has a high vapor pressure, and considerable loss of this dopant occurs during a conventional predeposition and drive-in cycle.

Approaches around this problem include the use of sealed-tube diffusion techniques, as well as doped oxide sources. An alternate, and considerably more convenient, method is to predeposit the arsenic by means of high-energy ion implantation. This allows placement of the dopant *below* the silicon surface, and thus greatly reduces this problem. A sheet resistance of 5–10 Ω/square can be achieved in this manner.

11.9.2 Choice of Transistor Type

The DDE process results in the possibility of only one type of transistor for a specific type of substrate. Thus only n^+–p–n transistors can be fabricated in the examples shown in Fig. 11.3*c*. Conversely, for a choice of *n*-type starting material, only p^+–n–p devices would be possible.

In general, the use of n^+–p–n devices is preferred over p^+–n–p devices in microcircuit fabrication for the following reasons:

1. The emitter region of a transistor must be as highly doped as possible for efficient operation. This favors phosphorus over boron (i.e., an n^+-type emitter over a p^+-type emitter) because it has a lower misfit factor (0.068 versus 0.254).
2. The drive-in step causes significant impurity depletion in regions doped with boron but not with phosphorus. In a p^+–n–p transistor, the region that is most lightly doped (the collector) is *p*-type; hence the depletion effect is most serious for this region. For an n^+–p–n transistor, boron depletion occurs in the more heavily doped base region, and the problem is easier to handle.
3. The ohmic contact to the collector of an n^+–p–n transistor is made by an n^+-region which is diffused simultaneously with the emitter. For a p^+–n–p device, however, it is necessary to make contact to the base by means of an n^+-region, as seen in Fig. 8.11. Such a region requires an additional masking and diffusion step.

4. It has been noted that the charge state of a thermally grown oxide layer tends to induce an n-type shift in the surface regions beneath it, that is, n-type surface layers become more n-type while p-type layers become less p-type. Thus a lightly doped p-region is most likely to be affected and can actually invert to n-type under certain conditions. The collector of a p^+–n–p transistor is prone to this surface inversion effect because it is the most lightly doped region in this device.

   Surface inversion in p^+–n–p transistors leads to short circuits between adjacent base and isolation regions. This can be prevented by diffusing a narrow p^+-guard ring through the middle of the collector surface at the same time as the emitter diffusion. Being heavily doped, this region remains p-type, and prevents the inverted layer from extending over the entire collector surface. However, its presence necessitates the use of a larger collector pocket.
5. The last diffusion step for n^+–p–n based microcircuits involves the n^+-emitter. This is usually formed by means of a phosphorus diffusion, resulting in a surface coating of PSG. A glass of this type greatly inhibits sodium ion transport through the oxide, and thus prevents long-term degradation of the underlying device. In contrast, a p^+–n–p based microcircuit has a final layer of BSG. This must be removed and replaced by PSG, in order to stabilize these circuits against sodium migration.

These extra processing steps are not required for circuits based on n^+–p–n transistors.

As a consequence of the foregoing, almost all bipolar microcircuits are made with n^+–p–n transistors. The starting materials shown in Fig. 11.3 are thus indicative of current practice.

11.9.3 Transistor Properties

The physics of operation of discrete transistors has been described in many texts. Here, however, we consider those areas of device operation that have been altered because of the microcircuit fabrication process [3, 4].

11.9.3.1 Breakdown Voltage

The microcircuit junction is parallel plane in character, except at the edge of the window cut in the oxide, where it takes on a cylindrical shape. This results in distortion of the space-charge lines, and a reduced breakdown voltage in this region.

Exact computations of the breakdown voltage of a cylindrical emitter–base junction, formed by an erfc diffusion into a gaussian background, are extremely difficult to make. However, calculations have been made by the use of a semi-empirical approach [70] with results that are in reasonable agreement with those observed in practice. Experimentally, it has been found that, for practical bipolar transistors, the breakdown voltage is dominated by junction curvature, so that it is relatively independent of the emitter and base diffusion profiles. Typically, devices with emitter–base junction depths of 2.5 μm, which are commonly used in linear applications, have a breakdown voltage of $BV_{EB0} \simeq 6.5$ V. This voltage falls roughly linearly with junction depth. In modern high-speed digital transistors, where the emitter diffusion is 0.5–1 μm deep, the breakdown voltage is about 3–4 V.

The breakdown voltage of the collector–base junction is more readily determined, since the base diffusion is made into a region of uniform background concentration. Numerical solutions of this problem are available in the literature [71].

11.9.3.2 Gain–Bandwidth Product

Figure 11.32 shows a microcircuit transistor made by the DDE process, together with its discrete counterpart. Differences in both parasitic resistances and capacitances between these devices result in a reduced value of the gain–bandwidth product for the microcircuit transistor. This occurs for the following reasons:

1. Both the emitter–base junction and the collector–base junction have curved sidewall areas in addition to their plane floor areas.
2. The collector tub also has a floor and a sidewall which result in a substrate capacitance that is not present in the discrete transistor.
3. The collector contact is made on the same side as the emitter and base contacts. Thus the current flow path in this region is significantly longer than in a discrete transistor, resulting in a greatly increased collector parasitic resistance.

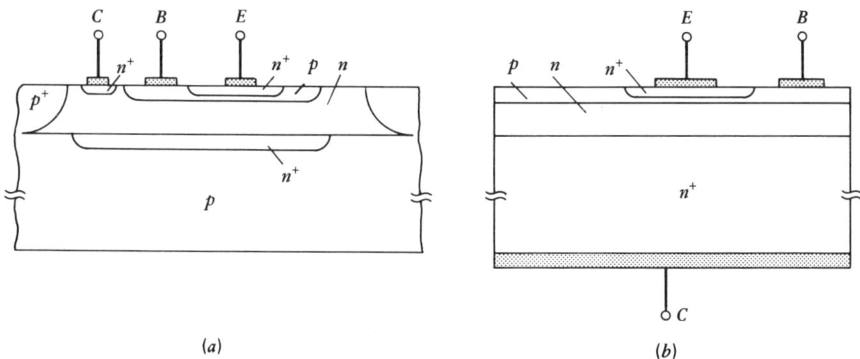

Fig. 11.32 Microcircuit and discrete transistors.

Calculations of the gain–bandwidth product of a transistor have been made, including the effect of these parasitic terms [3]. It has been shown that the net effect is that microcircuit devices will have a lower gain–bandwidth product than their discrete counterparts. Reduction of the epitaxial layer thickness, and the use of shallow junctions, are thus important approaches to improving device performance. Concurrently, both approaches result in smaller devices, so that this represents the trend in VLSI circuit fabrication.

11.9.3.3 Active Parasitics

In microcircuit structures, isolated tubs are formed by tying the *p*-type substrate to the most negative point in the circuit, since this ensures that the collector–substrate junction is reverse biased. Under certain conditions of circuit operation, this allows the possibility of obtaining transistor action in the *p–n–p* device comprising the base, collector, and substrate regions of Fig. 11.33*a*.

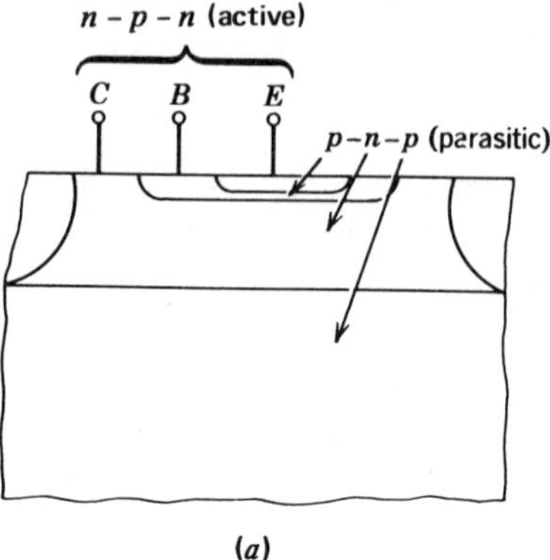

(*a*)

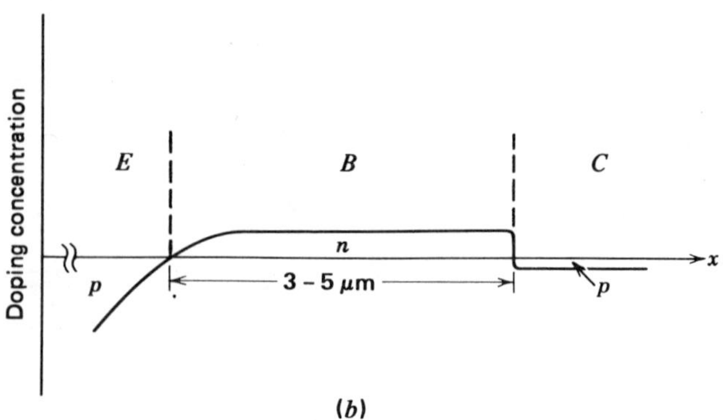

(*b*)

Fig. 11.33 Parasitic *p–n–p* transistor with no buried layer.

The doping profile of this *vertical* parasitic *p–n–p* transistor is shown in Fig. 11.33*b*. Its emitter–base junction has an injection efficiency of about 0.5–0.7, so that the device has a maximum common-emitter current gain of about 2.

In circuit operation the presence of the *p–n–p* parasitic transistor may give rise to a number of different types of effects. Thus as an active device, it may introduce a variable parasitic impedance in the microcircuit of which it is a part. It may provide a "sneak" path by which unwanted coupling occurs between two otherwise isolated components. It may provide a shunt path which diverts part of the current in the circuit. Finally, latch up of the resulting *p–n–p–n* transistor is possible under some circumstances.

In gold-doped devices the diffusion length of minority carriers in the collector region* is short compared to its width. Consequently, the current gain of the *p–n–p* transistor becomes vanishingly small because of its degraded base transport factor. The presence of a buried layer also serves to reduce the base transport factor to zero by incorporating a retarding field into this region. Thus devices which are gold doped or have buried layers (or both) are not prone to active *p–n–p* parasitic effects.

Parasitic *p–n–p* transistor action can also occur laterally to the sidewall, which can serve as the collector. Here the absence of a retarding field can sometimes cause this problem. However, it can be avoided by widening the effective "base" width, or by placing a p^+-channel stopper in this region.

11.9.4 *p–n–p* Transistors

There is considerable interest in providing both *p–n–p* and *n–p–n* transistors on the same substrate, with no additional processing steps. Two approaches are reviewed here briefly. Both result in relatively poor transistors, and are used in circuit schemes in which the high performance demands are made only on the *n–p–n* device. Figure 11.34*a* shows one approach by which such a transistor can be formed at the same time as its *n–p–n* counterpart. This technique results in a wide-base device with poor frequency performance. Moreover, the substrate of this structure serves as its collector, so that it can only be used in circuit applications where this electrode is at the common ground potential.

This vertical *p–n–p* transistor is used in the complementary-pair output stage of operational amplifiers, in the emitter–follower circuit configuration of Fig. 11.34*b*. Its incorporation into a microcircuit necessitates the use of a relatively low resistivity starting *p*-type layer. Typically, a value of 0.25 Ω cm ± 10% is used for this application. Even so, its performance is extremely poor compared to that of the *n–p–n* device, in terms of both speed as well as current-carrying capability.

Figure 11.35 shows a lateral *p–n–p* transistor which is completely isolated

* That is, in the base region of the parasitic *p–n–p* transistor.

from the substrate. A conventional n–p–n transistor, with which it is process compatible, is also shown. Here the base width is set by geometrical placement of the emitter and collector so that the device is relatively slow ($f_t \simeq$ 10–20 MHz). Moreover, emitter current injection is limited to the top edge so that its current-carrying capacity is much less than that of the vertical n–p–n transistor shown in this figure. In addition, since transistor action occurs near the surface, surface phenomena are important in determining device characteristics.

An important problem associated with the lateral p–n–p transistor is that part of its emitter current is lost by parasitic collection at the isolation wall and at the substrate. This effectively reduces the current gain of the useful transistor to ≤ 5. Lateral injection to the isolation wall can be minimized by making the collector wrap around the emitter as shown. Vertical injection to the substrate is greatly reduced by placement of a buried layer, which

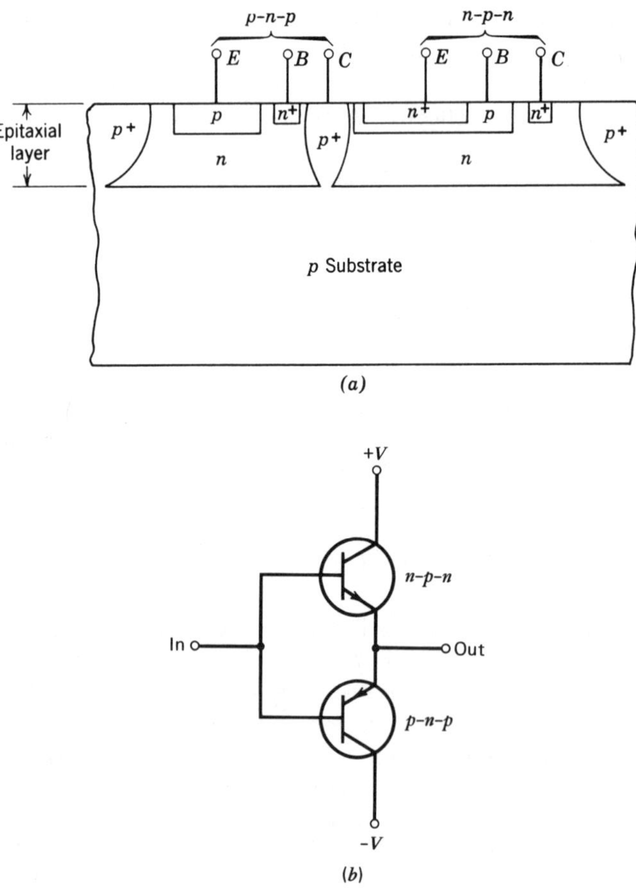

Fig. 11.34 Vertical *p–n–p* transistor.

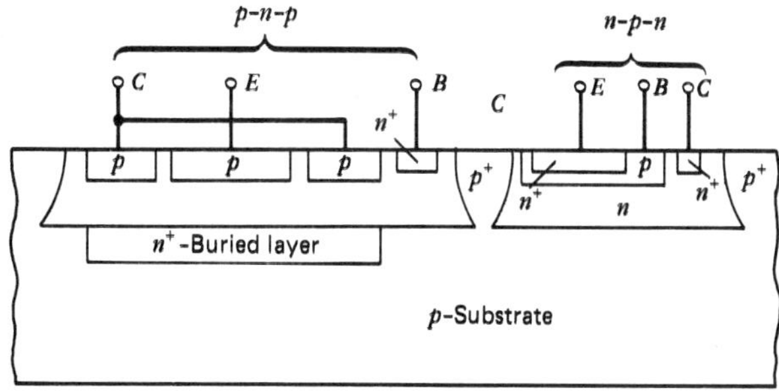

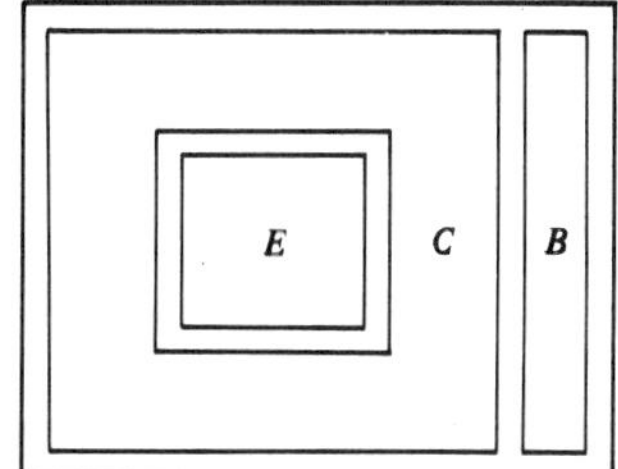

Fig. 11.35 Lateral *p–n–p* transistor.

provides a retarding field* and suppresses this term. The combined use of these techniques allows common emitter current gains of 50 to be achieved in practice.

11.9.5 Special-Purpose Bipolar Transistors

A number of devices, tailored for special applications, are also used in BJT-based microcircuits. A few of these are considered here.

Transistors with multiple emitters are often used in analog–digital conversion operations, and also at the input to I^2L logic gates. These require minor masking changes from the conventional device. Transistors with multiple collectors are also used in some current steering applications.

Superbeta transistors find use in the input stage of some operational amplifier circuits. These devices, with current gains of 1000 or more, are made by a double masking and diffusion operation, which serves to drive the emitter region deeper than that of other transistors in the circuit. An alternate approach, which is sometimes employed, is to use double masking at the base diffusion level, so as to form a shallower base for these devices. This approach is somewhat more difficult to control than the previous one. In both cases, however, these transistors have a very low collector–base

* An incidental improvement in base resistance is also accomplished in this manner.

punch through* voltage, typically about 2–5 V, because of their extremely narrow base width.

An important requirement for satisfactory operation of transistor–transistor logic gates is the need for making an input transistor with an extremely low value of reverse current gain (≤ 0.001). Typically, this device is used in a multi-emitter configuration. Figure 11.36*a* shows a means whereby this may be accomplished in an *n–p–n* compatible process. Here the base region is shaped so as to have a relatively high lateral resistance. In the forward direction the device behaves like a normal transistor with a high extrinsic base resistance. In the reverse direction, however, with the collector acting as an emitter, the voltage drop due to the lateral resistance of the base region causes almost all injection to occur near the base contact. Thus there is negligible transistor action in this direction. Devices of this type are commonly used with multiple emitters, as shown in Fig. 11.36*b*. Here the base regions are usually folded to minimize the device area.

11.9.6 Field Effect Transistors

Both MOS and junction gate field effect transistors can also be made on the same chip as bipolar n^+–*p*–*n* transistors by means of a compatible process. Typically, these devices are used in the input stages of linear circuits where an extremely high input impedance is required, as is the case in some operational amplifiers. Other applications include active loads for bipolar transistors, as well as level shifters.

Figure 11.37 shows the configuration for a process compatible, MOS field effect transistor. A BJT device is also shown side by side for comparison. Here the *p*-channel device is made by using the base diffusion to form its source and drain regions. This is followed by the n^+-diffusion which forms the emitter of the BJT as well as its collector contact region. Next, an additional masking and oxidation step is used to form the gate oxide, over which the gate is formed during the final metallization.

It is important that the emitter region not move excessively during this oxide growth step. Consequently, a wet oxidation process is used, at a relatively low temperature (875°C). The resulting oxide has a large density of both fast and slow surface states. However it has been shown [72] that these can be reduced by a 10-min high-temperature (1000°C) forming gas treatment immediately following the oxidation step. Enhancement-mode devices made by this process result in threshold voltages in the 1.6–2.1 V range, without any degradation of the current gain of associated n^+–*p*–*n* transistors on the same chip. It goes without saying that ion implantation can also be used if necessary to adjust the threshold voltage after the entire circuit has been fabricated, but prior to the deposition of the interconnect metal.

* It is worth noting that transistors made by the DDE process usually breakdown by avalanching of the collector–base junction.

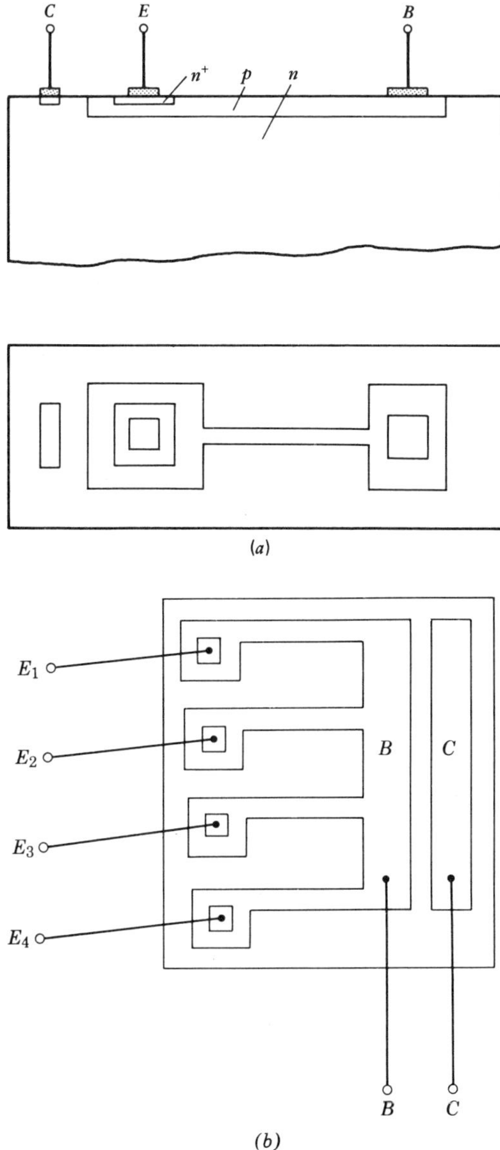

Fig. 11.36 Transistors with low inverse gain. (*a*) Single emitter. (*b*) Multiple emitters.

Compatible processes have also been developed [46] for fabricating C-MOS, *n*-channel MOS, and bipolar transistors on the same chip. These circuits are especially useful in VLSI schemes requiring both analog and digital signal conditioning.

Junction gate field effect transistors can also be made compatible with the DDE process. Here a *p*-base diffusion serves as the channel region in

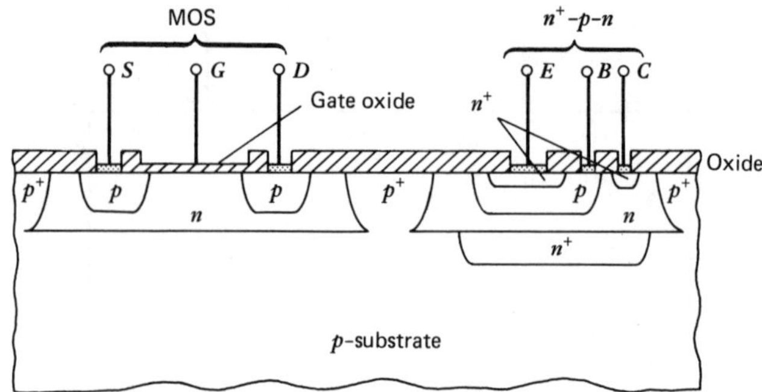

Fig. 11.37 Process-compatible MOS and BJT devices.

which ohmic contacts are made for the source and drain. An n^+-emitter diffusion forms the gate on the top side of the resulting p-channel. The isolation tub can be used as a gate on its bottom side; generally, however, this region is tied to a fixed reverse potential with respect to the channel.

An important layout requirement of the junction gate field effect transistor is that the source region be topologically isolated from the drain, in order to avoid sneak conduction paths. Concentric structures can be used in order to meet this requirement. On the other hand, the layout of Fig. 11.38 is equally satisfactory and takes up less space. Thus it is commonly used for these devices.

11.9.7 Self-Isolated VLSI Transistors

The junction isolated transistor, in its conventional form, has found use in a wide variety of both digital as well as linear circuits. However, its size and processing complexity, as compared to MOS devices, places it at a competitive disadvantage in VLSI applications. In recent years, a number of new techniques have been developed for removing this disadvantage. All are aimed at reducing the physical dimensions of the device, as well as the number of processing steps. Concurrently, device performance has been improved, so that these approaches are preferred in very-high-speed integrated circuit (VHSIC) applications.

All these approaches share the following characteristics:

1. Use of self-isolated devices to shrink overall area and reduce the number of processing steps.
2. Use of self-aligned techniques to simplify the masking requirements, and thus allow further shrinkage of device dimensions.
3. Use of extremely thin epitaxial layers (1–2 μm), combined with local oxidation, to provide improved high-frequency performance.

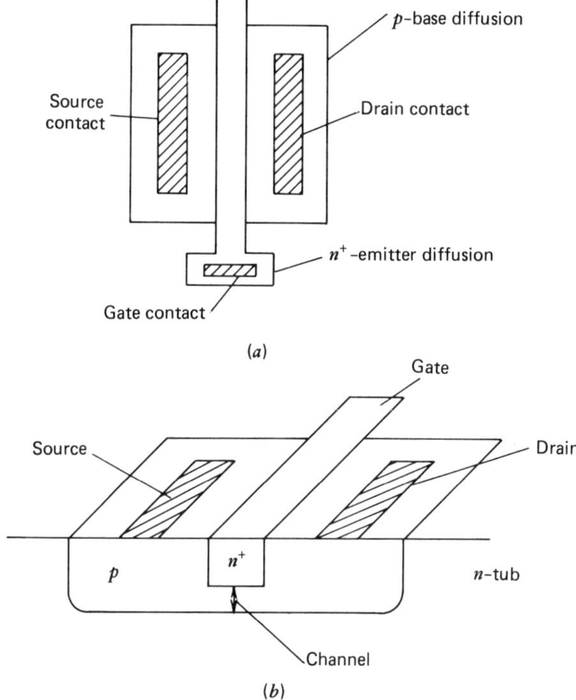

Fig. 11.38 Process-compatible J-FET: (*a*) plan and (*b*) isometric view.

4. Use of washout contacts to reduce the device size and improve its high-frequency performance.
5. Extensive use of local oxidation methods, which allow many of these device-shrinking techniques to be exploited. Here high-pressure oxidation at low temperatures provides a further degree of process flexibility, since it minimizes the movement of previously diffused regions.

One approach is to use the collector region to isolate* the device from the substrate [73]. The Collector Diffusion Isolation (CDI) process uses high-resistivity p-type starting silicon in which an n^+-buried layer diffusion is first made. This is followed by a thin (1–2 μm) p-type epitaxial layer of 0.2 Ω cm resistivity. Deep n^+-diffusions are next made so as to contact the buried layer; a final n^+-emitter diffusion completes the device, as shown in Fig. 11.39.

Here considerable area saving is achieved since the device is self-isolated. Furthermore, the use of an extremely thin epitaxial layer (grown by the

* Base diffusion isolation [74] has also been proposed, but is not used since it results in poor device performance.

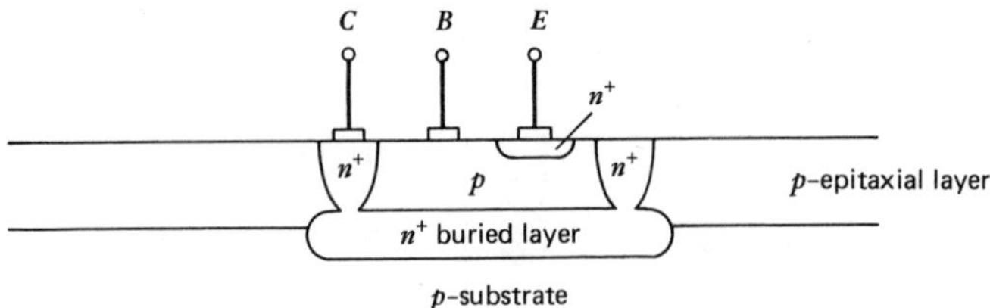

Fig. 11.39 Collector diffusion isolated transistor.

silane process) results in improved high-frequency performance over conventional devices. An additional advantage of this approach is that the number of masking steps is reduced, as is the process complexity.

It is worth noting that the device of Fig. 11.39 has a uniform base. Consequently, it is prone to surface inversion, resulting from redistribution effects during diffusion. Moreover, the uniform base results in a device with a high inverse beta, so that it is not suitable in many digital circuit applications.

Both of these problems can be avoided if the surface doping is enriched by a shallow boron implant, to result in a pseudograded base device. An alternate approach is to grow the epitaxial layer with a p^{-}–p doping sequence for this purpose.

The CDI process can be combined with local oxidation [2] to result in a greatly improved device structure, as shown in Fig. 11.40. Here local oxidation, together with washout contacts, is used to reduce physical dimensions and improve the breakdown voltage. Additional local oxidation within the device serves to further reduce the capacitance of the collector–base junction, with an additional improvement in the gain–bandwidth product. This approach has proved highly successful and is used in many commercial VLSI circuits.

Ion implantation techniques are used extensively in these devices, and often permit unique device processing steps. For example, it is possible to up-diffuse a base by making a selective implant on top of the buried layer, prior to epitaxy. Devices resulting from this process are highly desirable for digital circuits based on the use of I^2L logic.

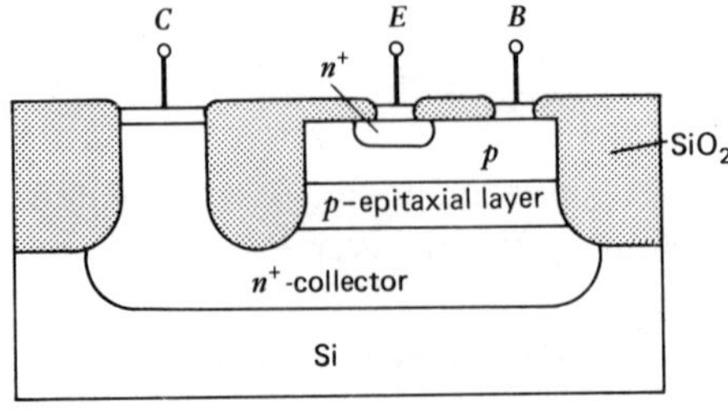

Fig. 11.40 Collector diffusion isolated device with local oxidation. From Deal and Early [2]. Reprinted with permission from the publisher, The Electrochemical Society, Inc.

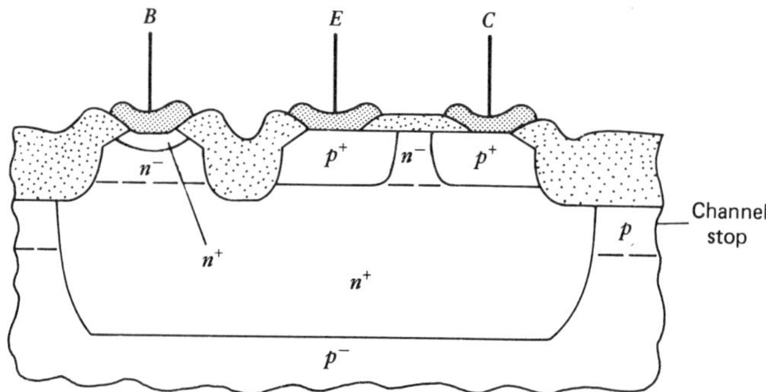

Fig. 11.41 Lateral *p–n–p* transistor using local oxidation process. From Agraz-Güereña, Panousis, and Morris [75]. © 1975. The Institute of Electrical and Electronic Engineers.

Ion implantation can also be used to fabricate high-performance lateral *p–n–p* devices on the same chip as *n–p–n* transistors [75] as shown in Fig. 11.41. Here local wall oxidation is used to avoid the necessity of a wrap-around collector, resulting in small devices with improved frequency response over the structure shown in Fig. 11.35.

The self-alignment capabilities of both local oxidation and polysilicon technology can be combined to take further advantage of the CDI process. Figure 11.42 outlines the processing sequence for one such approach [14], resulting in an n^+–p–n transistor and an integrated polysilicon resistor. Here the starting material consists of a high-resistivity *p*-substrate in which an n^+-diffused collector region is formed, followed by a p^+-channel stopper diffusion. Next the slice is locally oxidized, using a silicon nitride mask, as shown in Fig. 11.42*a*.

A boron implant, followed by diffusion, forms the *p*-base (Fig. 11.42*b*). After a second local oxidation step, a layer of polysilicon is deposited by CVD, and capped with Si_3N_4, as in Fig. 11.42*c*. Next the mask is selectively removed, and a phosphorus diffusion made to form the collector contact region as well as the emitter. The polysilicon covering these regions becomes highly doped during this process, and serves as the device metallization. The undoped polysilicon (i.e., the part covered with Si_3N_4) is used as a resistor in the circuit of Fig. 11.42*e*.

In summary, therefore, many new combinations of processing sequences are being employed to fabricate bipolar transistors for VLSI and VHSIC applications. These processes make extensive use of local oxidation. High-pressure oxidation is increasingly used, to allow this step to be carried out at low temperatures. Ion implantation and polysilicon technology, in combination with local oxidation, allow the use of self-aligned mask technology. Thus they play an important role in simplifying the device process, and also in improving device performance. As a result, prospects for the use of these devices in VHSIC applications are excellent.

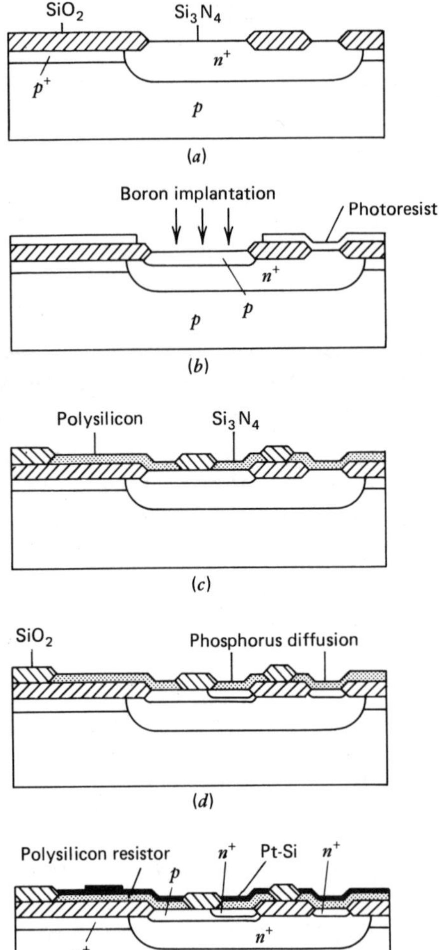

Fig. 11.42 Transistor and resistor formed by local oxidation and polysilicon technology. From Okada, Aomura, Nakamura, and Shida [14], © 1979. The Institute of Electrical and Electronic Engineers.

11.9.8 Diodes

Diodes in bipolar integrated circuits are of the diffused junction type, and can be fabricated by making a single p-type base diffusion into the n-type epitaxial layer. Alternately, the diode may be obtained by connecting together the various regions of a transistor. Figure 11.43 shows the different possible combinations, together with their equivalent circuits. Their characteristics may be evaluated by setting up the equations for the direct-current terminal voltages and currents of a transistor under any bias condition. Analysis of this type lead to the following conclusions concerning their

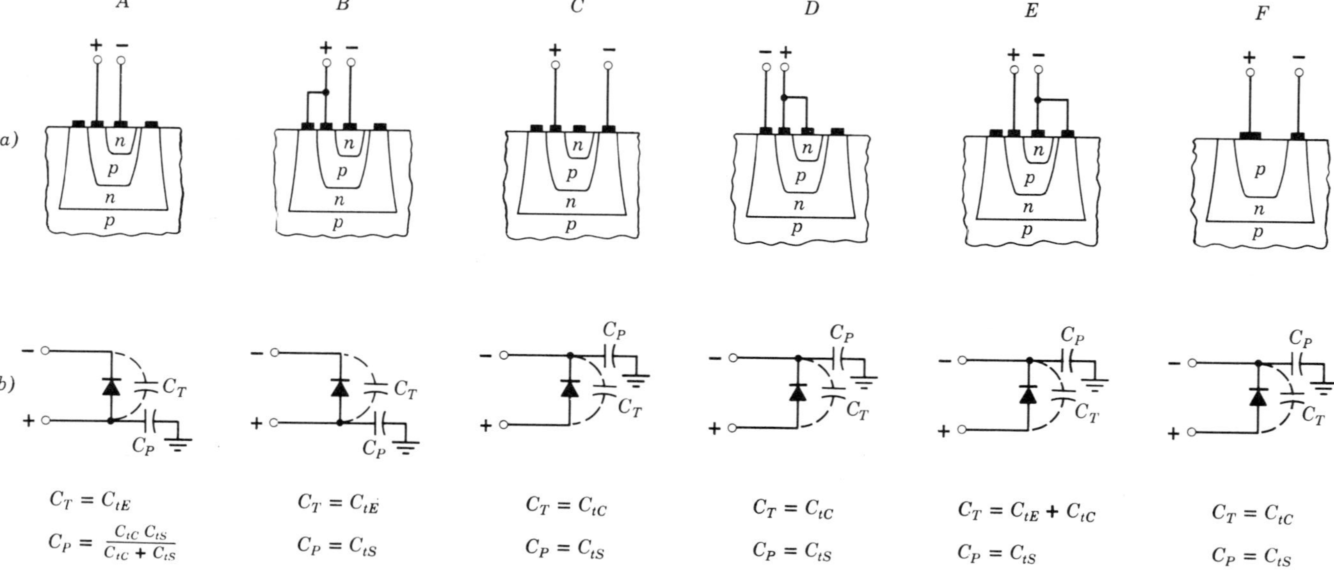

Fig. 11.43 Diode connections and equivalent circuits.

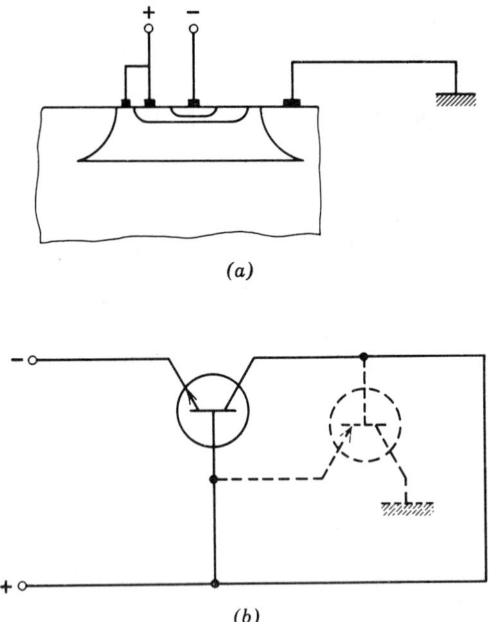

Fig. 11.44 A diode configuration with its parasitics.

relative device performances:

1. Under conditions of forward current flow, the collector–base junction is reverse biased for case *B*, slightly forward biased for case *A*, and heavily forward biased for cases *C*, *D*, and *E*. Thus the amount of stored charge in the transistor configurations is the least for case *B*, and increases in the following order: *B*, *A*, *D*, *C*, *E*. The charge stored in the single diffused diode configuration of case *F* is the highest since it is stored in both the *n*- as well as the *p*-region. The recovery times of the various diodes are in this same order.
2. Under high current conditions, the forward conductance of these diodes is in the order *B*, *E*, *D*, *A*, *C*, *F*, with device *B* having the highest value. This is because *B* is a "short" diode.*
3. From Fig. 11.44 it is seen that the emitter–base diode of the parasitic *p–n–p* transistor is short-circuited for diode *B*. This is not the case for the other configurations. Consequently, there is no possibility of active *p–n–p* effects with this structure.

In view of the above it is concluded that the diode configuration of case *B* is the best one for general circuit applications because it has the fastest

* Diode *D* is also "short", but has a high series resistance due to the collector region.

recovery characteristics, the highest forward conductance at operating current, and complete freedom from parasitic p–n–p effects. Its single disadvantage is the fact that it has a breakdown voltage corresponding to that of the emitter–base diode. In the event that a higher breakdown voltage is required, it is necessary to use the configuration of either case C or D. The configurations of cases E and F have no special advantage and are not used in practice.

The choice of diodes is not restricted to the above, and other structures have been found useful in special applications. Thus diodes fabricated by making an emitter diffusion into the highly doped p-type isolation wall exhibit a large depletion-layer capacitance per unit area. They suffer from the disadvantage of having a slightly lower breakdown voltage than diodes made from emitter–base structures. In addition, they can only be used in circuits where the p-side is grounded.

In some circuit applications, it is necessary to provide a number of diodes with their p-regions connected together, as shown in Fig. 11.45a. This may be done by fabricating all diodes in a common isolation pocket with separate

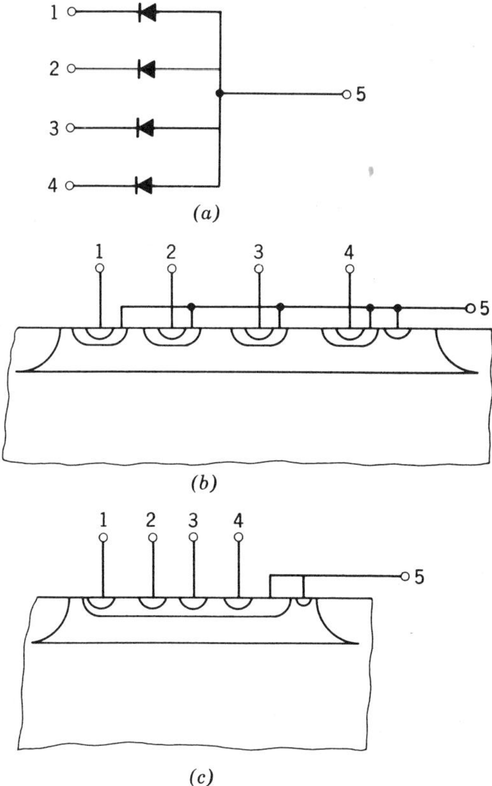

Fig. 11.45 Multiple diodes.

base diffusions, as in Fig. 11.45*b*; a further area saving is obtained by using a common base tub as in Fig. 11.45*c*. In this last configuration, care must be taken to avoid active *n–p–n* transistor action between adjacent emitters. This is done by fabricating the structure with a large lateral spacing between the various emitter regions, and using gold doping to intentionally degrade the minority carrier lifetime.

11.9.8.1 Schottky Diodes

Schottky diodes are also used in BJT-based microcircuits. Here their application is primarily confined to high-speed logic stages, where they are placed in shunt with the collector–base regions of transistors to prevent them from going into saturation [76]. An extremely important end result is that this avoids the necessity for gold doping. This step is one of the most difficult to control in silicon processing, so that its avoidance is often the main reason for using these devices.

The circuit connection for the Schottky diodes, and its integration with a bipolar transistor, has been described in Section 8.5. The commonly used gate metals are platinum silicide, palladium silicide, molybdenum, and tungsten. Aluminum has also been used here; however, its barrier height to silicon is sensitive to subsequent thermal processing, so that the more refractory gate materials are favored.

11.9.9 Resistors

A number of different resistor types, with widely varying properties, can be fabricated within a BJT-based microcircuit chip. These include diffused resistors which are made during the DDE process, as well as ion-implanted and thin-film types which require additional fabrication steps. Their process technology, as well as their electrical characteristics, are described in the following sections.

11.9.9.1 Diffused Resistors

One approach to making these is by means of a *p*-type diffusion into an *n*-type background. Resistors of this type are made simultaneously with the base diffusion. Consequently, many of their characteristics are determined by the requirements of the associated active devices which form the basis of the microcircuit design.

The sheet resistance of a base diffusion is typically between 100 to 200 Ω/square. Thus the base diffusion results in reasonably proportioned resistors for about 50 Ω to 10 kΩ and is useful for the majority of circuit applications. This sheet resistance is monitored during the diffusion process and can be held to a tolerance of ± 5% of its room temperature value. In practice, a tolerance of ± 10% of the initial value is generally assumed as a design rule for resistors of this type.

The temperature characteristics of a diffused resistance are difficult to compute accurately because of the variation of its doping concentration with diffusion depth. To an approximation, however, they are dominated by the region near the surface where the doping concentration is the highest. Curves of normalized resistance, based on the variation of mobility with temperature, are shown in Fig. 11.46 for p-type material with different carrier concentrations. From these curves it is seen that resistors with a surface concentration of around 10^{19} atoms/cm^3 will have a minimum variation with temperature. Fortunately, this is a convenient surface concentration for base diffusions as well, so it is often used in practice. A typical boron-diffused resistor, with a surface concentration of 10^{19} atoms/cm^3, has a temperature variation (from the nominal value at 25°C) of approximately +4% at −55°C up to +8% at +125°C.

A diffused resistor is isolated from its background by the contact potential of the associated p–n junction or by a higher reverse voltage if the tub is appropriately biased. As a consequence, a number of resistors may be placed in the same tub with a physical saving of space on the silicon chip. In practice, it is customary to lump as many resistors as conveniently possible within an individual region of this type.

The breakdown voltage between resistors is equal to the collector–base breakdown voltage of transistors made on the same slice, provided that they are sufficiently far apart so that punch-through does not occur prematurely. In addition, the reverse-leakage current density is also the same as that of the associated collector–base junction.

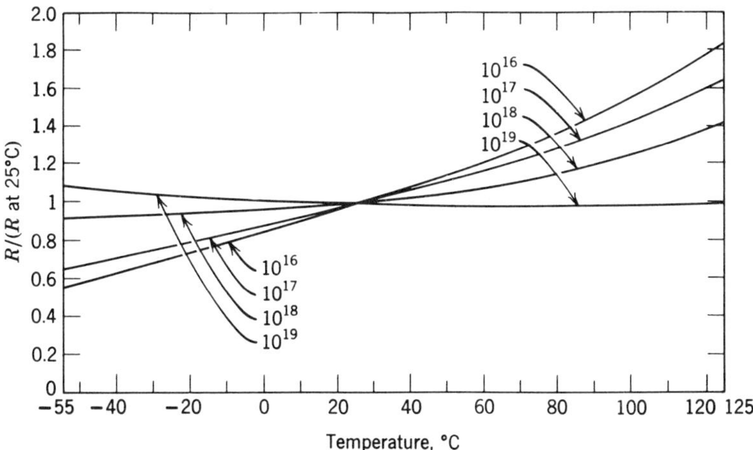

Fig. 11.46 Temperature behavior of doped silicon.

Tubs containing resistors are usually connected to the most positive voltage in the system. This effectively cuts off the parasitic p–n–p transistor formed by the p-resistor, the n-tub, and the p-substrate. Consequently, active parasitics, and all their attendant problems, are avoided by this means.

Ohmic contacts are made to the resistor by aluminum evaporation, followed by subsequent alloying as described in Section 8.4.1. The presence of this contact distorts the current flow lines in its vicinity and alters its effective resistance. Many configurations have been attempted to minimize these effects. However, the practical approach is to use a contact scheme that is reasonably uniform in its current flow characteristics and, at the same time, convenient to implement. Three such schemes are shown in Fig. 11.47, together with experimentally measured values of resistance associated with the end effects.

The effect of a foldover has also been handled empirically. It has been found [77] that a bend with the proportions shown in Fig. 11.48 has an

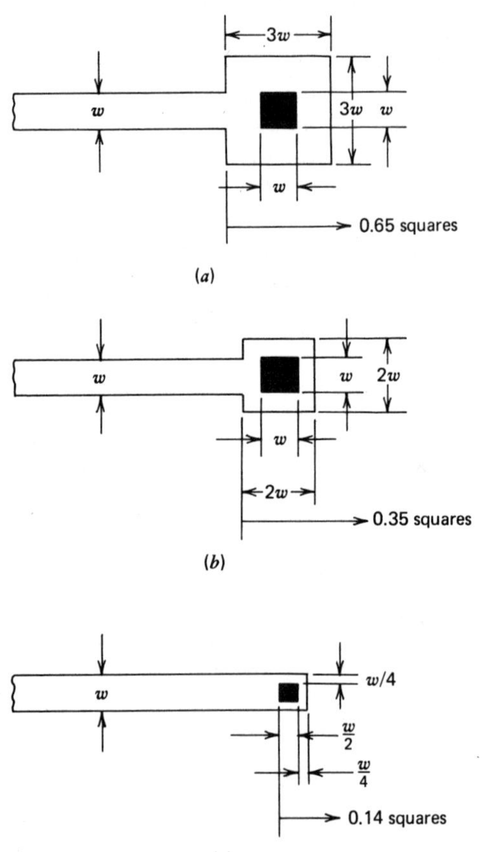

Fig. 11.47 Resistor terminations and end effects.

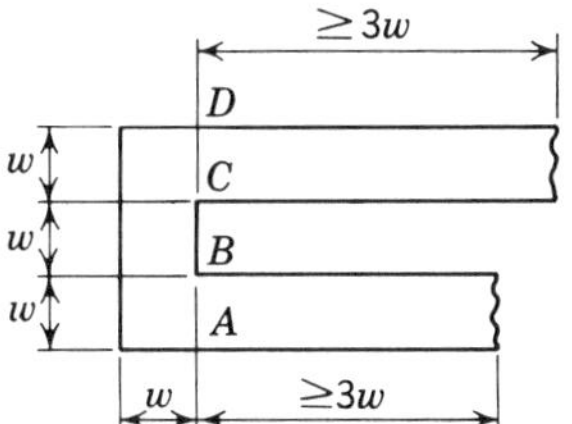

Fig. 11.48 Folded resistors.

effective resistance of 1 square between *AB* and *CD*, provided it is followed on either side by a minimum of 3 squares of resistor length. A sharp, right-angle bend of the type shown here leads to an infinite concentration of current at its inside corner. In spite of this fact, however, it is used in low-level situations because of the ease with which it is implemented in the photomask. If necessary, a small fillet can be added to relieve this current concentration.

Low values of resistors may be obtained by using the properties of the diffused emitter region. Here a sheet resistance of 2–10 Ω/square is typical and results in a reasonable aspect ratio for resistor values as low as 1 Ω. Unfortunately, the value cannot be set with any degree of precision, since the emitter sheet resistance is not monitored during microcircuit fabrication. As a result, this type of resistor is not commonly used. However, it is often integrated within the emitter of a power output stage where it provides a ballasting function. An example of this approach is shown in the transistor layout of Fig. 11.49.

Large-value resistors can be made by utilizing the sheet resistance of the base region *under* the emitter. This is done by making a *p*-diffused resistor and placing an emitter region (*ABCD*) over it as shown in Fig. 11.50. This

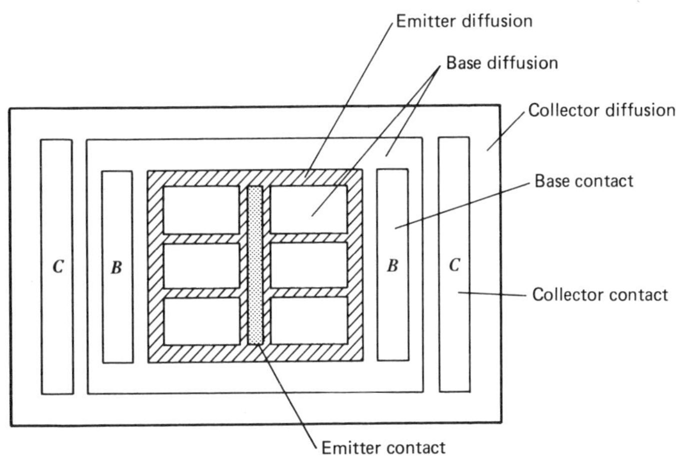

Fig. 11.49 Emitter-ballasted output transistor.

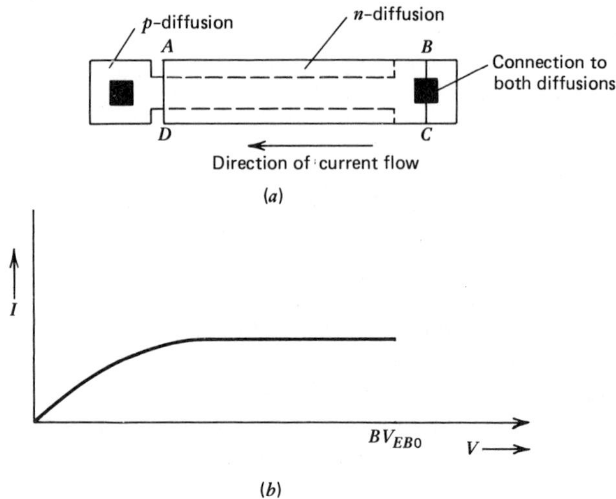

Fig. 11.50 Pinch resistor and its *I–V* characteristic.

emitter region is biased positive with respect to the *p*-diffusion which it covers. Often this is done by tying it to the "high" end of the resistor, provided that the current flow through it is known to be unidirectional. Resistor operation in this manner results in pinching the current flow through the resistor, as in the channel of a function gate FET. The *V–I* characteristic for such a *pinch resistor* is similar to the output curves of a FET so that it is highly nonlinear. Furthermore, resistor operation is limited to voltages under the emitter-base breakdown voltage.

A sheet resistance of 10 kΩ/square is typical for these resistors, with a greater than ±50% tolerance on the value. Furthermore, these devices are highly temperature sensitive, changing by as much as 100% over the operating range. Needless to say, they can only be used in noncritical applications.

11.9.9.2 Ion-Implanted Resistors

Ion-implanted resistors can be made with sheet resistances as high as 5000 Ω/square and with a ± 1% tolerance. Thus they allow considerable reduction in chip area, and are especially suited for low-power digital and linear microcircuits. These resistors are made by first forming two base diffusions to which contact can be made, and implanting the resistor between them. A typical process sequence consists of delineating the windows for the contacts, which are made at the same time as the base diffusion. Next, the oxide over the implant region is stripped and a thin oxide, about 1000 Å, is thermally grown in its place. The implant is made through this thermal oxide. Details of this process have been given in Section 6.5.5.2.

11.9.9.3 Thin-Film Resistors

In some applications, particularly for analog-to-digital conversion, it is necessary to provide ultrastable high-precision components, which are preset to a precise value. These are made [78, 79] by the vacuum evaporation or sputtering of thin films of resistive materials, such as nichrome, chrome-silicon, or a variety of refractory metal silicides, directly on top of the oxide layer of a microcircuit chip which has in it the various active components that make up the entire circuit. All of these materials adhere firmly to the oxide and are used in film thicknesses of about 100–1000 Å. A large amount of effort has been devoted to studying their characteristics and their deposition technology, to the point where they are quite satisfactory for precision applications. Typically, these resistors are preset by laser trimming before the chip is encapsulated. Other techniques, such as anodic oxidation, have also been used for this purpose.

Thin-film technology is a subject in itself and will not be covered here. The interested reader is referred to references 78–80.

11.9.10 Capacitors

A capacitor must have a high cutoff frequency and a capacitance value that is independent of the amplitude of the signals that are impressed across it. The only capacitors satisfactorily meeting these requirements are those made by the vacuum evaporation of thin conducting films on either side of an insulating material. These have been treated extensively in the literature [78–80] and are not considered here.

A reverse-biased diffused p–n junction can also be used for this purpose. Figure 11.51 shows a number of possibilities, each with its equivalent circuit. Here it is seen that all of these structures have parasitic resistances which degrade their cutoff frequency (which varies inversely with the CR product). In addition, some have parasitic capacitances C_P associated with them. All are voltage dependent and introduce distortion in the circuit in which the capacitor is used. The configuration of Fig. 11.51c is restricted to situations where one side of the capacitor is grounded. Its advantage, however, is that it has the largest capacitance per unit area.

Diffused junction capacitors are used in applications where their nonlinear behavior can be tolerated. Thus they are suitable for use as speedup capacitances in digital circuits, as temporary storage devices in triggering schemes, and as bypass capacitances.

Ion-implanted capacitors can also be used. Here hyperabrupt structures, of the type described in Chapter 6, are widely used in electronic tuning applications.

MOS capacitors can be fabricated by using an emitter region as one plate, the aluminum metallization as the second plate, and the intervening oxide layer as the dielectric. The construction of the capacitor is shown in Fig.

11.52. Its fabrication involves removal of the oxide layer that is normally over this region (≃ 5000–6000 Å thick) and its replacement by a grown oxide (≃ 500–1000 Å) in order to obtain a higher capacitance per unit area. Low-temperature steam oxidation is commonly used, to avoid excessive movement of the emitter diffusion during this process.

Although its capacitance per unit area is well below that of a p–n junction, the MOS device is superior in all other respects, as follows:

1. Since the lower plate of this capacitor is made of highly doped material, the capacitance is essentially independent of voltage, and the device does not distort signals applied to it.

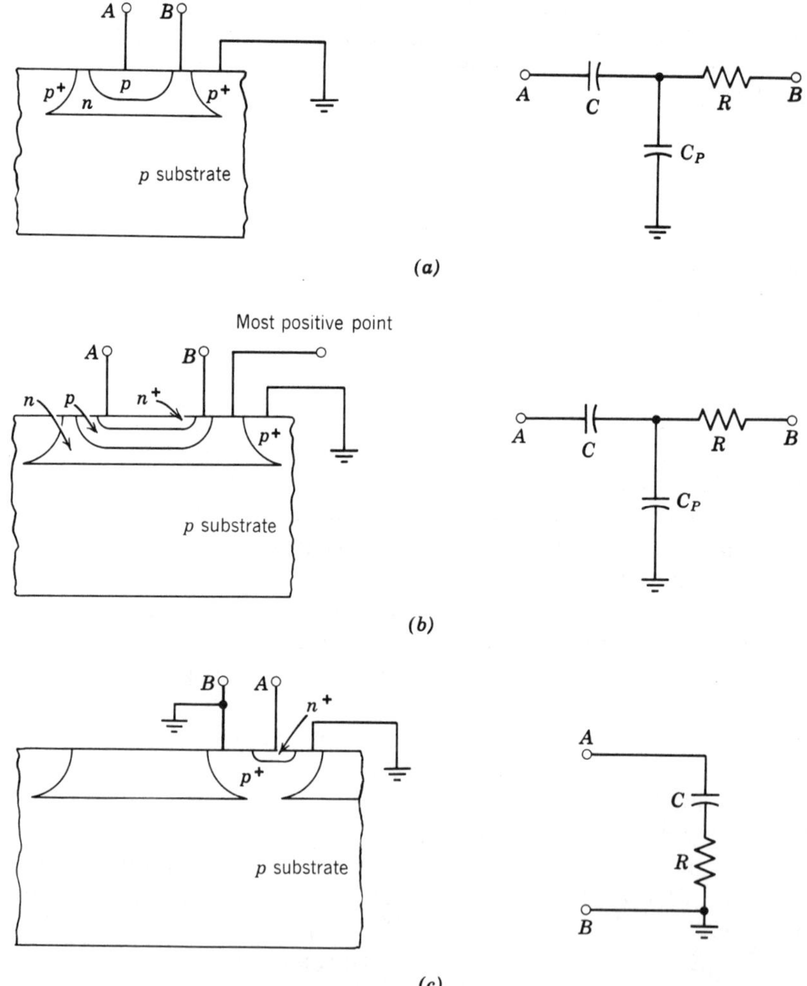

Fig. 11.51 Diffused junction capacitors.

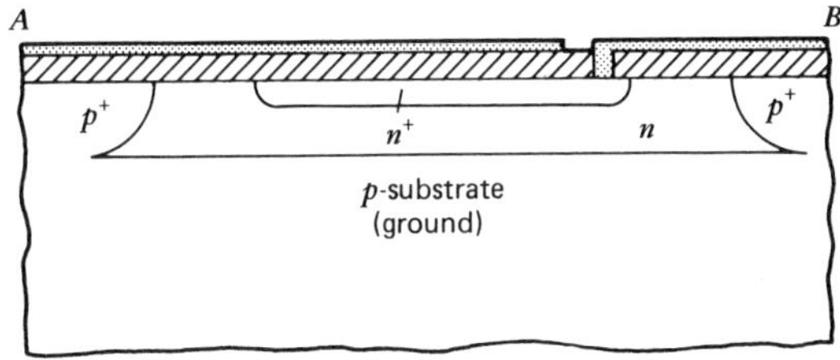

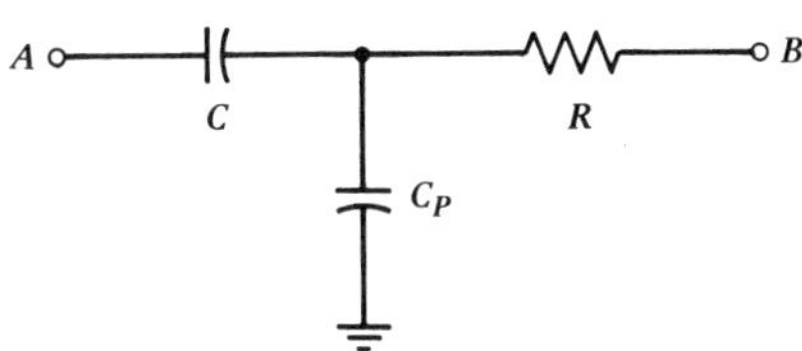

Fig. 11.52 MOS-type capacitor.

2. Since it is larger in area than a p–n junction of equal capacitance, its dimensions can be held to tighter tolerances.
3. The series parasitic resistance associated with it (see Fig. 11.52) is that of a diffused emitter layer, and is considerably lower than that obtained with the p–n junctions of Fig. 11.51. The shunt parasitic capacitance is that of the collector–substrate junction and is lower in magnitude than that of other junctions in the microcircuit.
4. The temperature variation of a MOS capacitance is caused by the change in the dielectric constant of the oxide layer, and is about 20 ppm/°C. This is about 50 times lower than that of a p–n junction.

The MOS capacitor is extensively used in those situations where a tolerance requirement is placed on the value of the capacitance, and where it is important that distortion products are not introduced into the circuit. Thus its main uses are in linear integrated circuits. One important application area is its use as a feedback capacitance for phase shift control in operational amplifiers.

REFERENCES

1. R. K. Watts and J. H. Brunning, A Review of Fine-Line Lithographic Techniques: Present and Future, *Solid State Technol.*, p. 99 (May 1981).
2. B. E. Deal and J. M. Early, The Evolution of Silicon Semiconductor Technology: 1952–1977, *J. Electrochem. Soc.* **126,** 20C (1979).

3. S. K. Ghandhi, *The Theory and Practice of Microelectronics*, Wiley, New York, 1968.
4. R. A. Colclaser, *Microelectronics: Processing and Device Design*, Wiley, New York, 1980.
5. K. E. Bean and W. R. Runyan, Dielectric Isolation: Comprehensive, Current, and Future, *J. Electrochem. Soc.* **124,** 5C (1977).
6. M. R. Splinter, A 2 μm Silicon Gate C-MOS/SOS Technology, *IEEE Trans. Electron Dev.* **ED-25,** 996 (1978).
7. S. N. Lee, R. A. Kjar, and G. Kinoshita, Island Edge Effects in C-MOS/SOS Transistor, *IEEE Trans. Electron Dev.* **ED-25,** 971 (1978).
8. T. Suzuki, A. Mimura, T. Kamei, and T. Ogawa, Deformation in Dielectric Isolated Substrates and its Control By a Multilayer Polysilicon Support Structure, *J. Electrochem. Soc.* **127,** 1537 (1980).
9. W. R. Runyan, *Semiconductor Measurements and Instrumentation*, McGraw-Hill, New York, 1975.
10. S. Iwamatsu and M. Ogawa, Self-Aligned Aluminum Gate MOSFET's Fabricated by Laser Anneal, *J. Electrochem. Soc.* **128,** 384 (1981).
11. W. M. Cosney and L. H. Hall, The Extension of Self-Registered Gate and Doped-Oxide Diffusion Technology to the Fabrication of Complementary MOS Transistors, *IEEE Trans. Electron Dev.* **ED-20,** 469 (1973).
12. J. A. Appels, E. Kooi, M. M. Paffen, J. J. H. Schatorjé, and W. H. C. G. Verkuylen, Local Oxidation of Silicon and its Application in Semiconductor Device Technology, *Philips Res. Rep.* **25,** 118 (1970).
13. M. Isobe, J. Iwamura, M. Ohhashi, H. Koike, K. Maeguchi, T. Sato, and H. Tango, A 7000 Gate Microprocessor on SOS-PULCE, *IEEE Trans. Electron Dev.* **ED-26,** 588 (1979).
14. K. Okada, K. Aomura, T. Nakamura, and H. Shida, A New Polysilicon Process for a Bipolar Device—PSA Technology, *IEEE Trans. Electron Dev.* **ED-26,** 385 (1979).
15. J. Lohstroh, J. D. P. v.d. Crommenacker, and A. J. Linssen, Oxide Isolated ISL Technologies, *IEEE Electron Dev. Lett.* **EDL-2,** 30 (1981).
16. E. Kooi, J. G. van Lierop, W. H. C. G. Verkuijlen, and R. deWerdt, LOCOS Devices, *Philips Res. Rep.* **26,** 166 (1971).
17. R. H. Dennard, F. H. Gaensslen, E. J. Walker, and P. W. Cook, 1 μm MOSFET VLSI Technology: Part II. Device Designs and Characteristics for High Performance Logic Applications, *IEEE Trans. Electron Dev.* **ED-26,** 325 (1979).
18. K. Shibata and K. Taniguchi, Generation Mechanism of Dislocations in Local Oxidation of Silicon, *J. Electrochem. Soc.* **127,** 1383 (1980).
19. E. Bassous, H. N. Yu, and V. Maniscalco, Topology of Silicon Structure with Recessed SiO_2, *J. Electrochem. Soc.* **123,** 1729 (1976).
20. R. J. Powell, J. R. Ligenza, and M. S. Schneider, Selective Oxidation of Silicon in Low Temperature High-Pressure Steam, *IEEE Trans. Electron Dev.* **ED-21,** 636 (1974).
21. S. C. Su, Low Temperature Silicon Processing Techniques for VLSI Fabrication, *Solid State Technol.*, p. 72 (Mar. 1981).
22. K. Y. Chiu, J. H. Moll, and J. Manoliu, A Bird's Beak Free Local Oxidation Technology Feasible for VLSI Circuit Fabrication, *IEEE Trans. Electron. Dev.* **ED-29,** 536 (1982).
23. E. Kooi and J. A. Appels, Selective Oxidation of Silicon and its Device Applications, in *Semiconductor Silicon 1973*, Electrochem. Soc., Princeton, NJ, 1973, p. 860.
24. J. Götzlich and H. Rysell, Tapered Windows in SiO_2, Si_3N_4 and Polysilicon Layers by Ion Implantation, *J. Electrochem. Soc.* **128,** 617 (1981).
25. L. K. White, Bilayer Etching of Field Oxides and Passivation Layers, *J. Electrochem. Soc.* **127,** 2687 (1980).

26. W. Kern and R. S. Rosler, Advances in Deposition Processes for Passivation Films, *J. Vac. Sci. Technol.* **14,** 1082 (1977).
27. L. B. Rothman, Properties of Thin Polymide Films, *J. Electrochem. Soc.* **127,** 2216 (1980).
29. A. C. Adams and C. D. Capio, Planarization of Phosphorus Doped Silicon Dioxide, *J. Electrochem. Soc.* **128,** 423 (1981).
29. A. Saiki, S. Harada, T. Okubo, K. Mukai, and T. Kimura, A New Transistor with Two-Level Metal Electrodes, *J. Electrochem. Soc.* **124,** 1619 (1977).
30. H-V. Schreiber and E Fröschle, High Quality RF-Sputtered Silicon Dioxide Layers, *J. Electrochem. Soc.* **123,** 30 (1976).
31. M. Hatzakis, B. J. Canavello, and J. M. Shaw, Single Step Optical Lift-Off Process, *IBM J. Res. Dev.* **24,** 452 (1980).
32. Y. Homma, H. Nozawa, and S. Harada, Polymide Liftoff Technology for High-Density LSI Metallization, *IEEE Trans. Electron Dev.* **ED-28,** 552 (1981).
33. L. Jastrebski, Origin and Control of Material Defects in Silicon VLSI Technologies: An Overview, *IEEE Trans. Electron. Dev.* **ED-29,** 475 (1982).
34. S. K. Ghandhi, *Semiconductor Power Devices*, Wiley, New York, 1977.
35. R. A. Craven and H. W. Korb, Internal Gettering in Silicon, *Solid State Technol.*, p. 55 (July 1981).
36. P. M. Petroff, G. A. Rozgonyi, and T. T. Sheng, Elimination of Process-Induced Stacking Faults by Pre-oxidation Gettering of Si Wafers, *J. Electrochem. Soc.* **123,** 565 (1976).
37. H. J. Geipel and W. K. Tice, Reduction of Leakage by Implantation Gettering in VLSI Circuits, *IBM J. Res. Dev.* **24,** 310 (1980).
38. Y. Hayafuji, T. Yanada, and Y. Aoki, Laser Damage Gettering and its Application to Lifetime Improvement in Silicon, *J. Electrochem. Soc.* **128,** 1975 (1981).
39. M. Nakamura and N. Oi, A Study of Gettering Effect of Metallic Impurities in Silicon, *Jpn. J. Appl. Phys.* **7,** 512 (1968).
40. S. P. Murarka, A Study of the Phosphorus Gettering of Gold in Silicon by the Use of Neutron Activation Analysis, *J. Electrochem. Soc.* **123,** 765 (1976).
41. H. J. Geipel, Jr., N. Hsieh, M. H. Ishaq, C. W. Koburger, and F. R. White, Composite Silicide Gate Electrodes—Interconnections for VLSI Device Technologies, *IEEE Trans. Electron Dev.* **ED-27,** 1417 (1980).
42. K. Ohta, K. Yamada, M. Saitoh, K. Shimizu, and Y. Tarui, Quadruply Self-Aligned MOS (QSA-MOS)—A New Short-Channel High-Speed High-Density MOSFET for VLSI, *IEEE Trans. Electron Dev.* **ED-27,** 1352 (1980).
43. E. Vittoz, B. Gerber, and F. Leuenberger, Silicon-Gate CMOS Frequency Divider for the Electronic Wrist Watch, *IEEE J. Solid State Circuits* **SC-7,** 100 (1972).
44. C. A. T. Salama, VLSI Technology for Telecommunication IC's, *IEEE J. Solid State Circuits* **SC-16,** 253 (1981).
45. R. L. Maddox, Reverse CMOS Processing, *Solid State Technol.*, p. 128 (Feb. 1981).
46. G. Zimmer, B. Hoefflinger, and J. Schneider, A Fully Implanted N-MOS, CMOS Bipolar Technology for VLSI of Analog-Digital Systems, *IEEE Trans. Electron Dev.* **ED-26,** 390 (1979).
47. D. Ong, An All Implanted CCD/CMOS Process, *IEEE Trans. Electron Dev.* **ED-28,** 6 (1981).
48. D. Frohman-Bentchkowsky, FAMOS:A New Semiconductor Charge Storage Device, *Solid State Electron.* **17,** 517 (1974).
49. A. Scheibe and H. Schalte, Technology of a New *n*-Channel One-Transistor EAROM Cell Called SIMOS, *IEEE Trans. Electron Dev.* **ED-24,** 600 (1977).

50. Y. Uchida S. Saito, M. Nakane, N. Endo, T. Matsuo, and Y. Nishi, 1K-Bit Nonvolatile Semiconductor Read/Write RAM, *IEEE Trans. Electron Dev.* **ED-25,** 1066 (1978).
51. J. J. Chang, Non-Volatile Memory Devices, *Proc. IEEE* **64,** 1039 (1976).
52. J. R. Ligenza, D. Kahng, M. Lepselter, and E. Labate, A Method of Tungsten Dopant Deposition for Dual Dielectric Charge-Storage Cells, *IEEE Trans. Electron Dev.* **ED-24,** 581 (1977).
53. K. Maegushi, M. Okhaski, J. Iwamura, S. Taguchi, E. Sugino, T. Sato, and H. Tango, 4 μm LSI on SOS Using Coplanar-II Process, *IEEE Trans. Electron Dev.* **ED-25,** 945 (1978).
54. W. R. Wisseman and J. G. Oakes, Eds., Special Issue on Advances in Monolithic Microwave III-V Devices and Circuits, *IEEE Trans. Electron Dev.* **ED-28,** 133–228 (1981).
55. R. C. Eden, B. M. Welch, R. Zucca, and S. I. Long, The Prospects for Ultrahigh-Speed VLSI GaAs Digital Logic, *IEEE J. Solid State Circuits* **SC-14,** 221 (1979).
56. K. Lehovec and R. Zuleeg, Analysis of GaAs FET's for Integrated Logic, *IEEE Trans. Electron Dev.* **ED-27,** 1074 (1980).
57. S. K. Ghandhi and R. J. Field, Precisely Controlled p^+-Diffusions in GaAs, *Appl. Phys. Lett.* **38,** 267 (1981).
58. D. L. Rode, B. Schwartz, and J. V. DiLorenzo, Electrolytic Etching and Electron Mobility of GaAs for FET's, *Solid State Electron.* **17,** 1119 (1974).
59. B. M. Welch, Y-D. Shen, R. Zucca, R. C. Eden, and S. I. Long, LSI Processing Technology for Planar GaAs Integrated Circuits, *IEEE Trans. Electron Dev.* **ED-27,** 1116 (1980).
60. R. D. Fairman, R. T. Chen, J. R. Oliver, and D. R. Ch'en, Growth of High Purity Semi-Insulating Bulk GaAs for Integrated Circuit Applications, *IEEE Trans. Electron Dev.* **ED-28,** 135 (1981).
61. M. N. Yoder, Ohmic Contacts in GaAs, *Solid State Electron.* **23,** 117 (1980).
62. J. F. Gibbons, L. D. Hess, and T. W. Sigmon, Eds., Laser and Electron-Beam Solid Interactions and Materials Processing, *Proc. Materials Research Society Annual Meeting*, Boston, MA, 1980.
63. D. C. Miller, The Alloying of Gold and Gold Alloy Ohmic Contact Metallizations with Gallium Arsenide, *J. Electrochem. Soc.* **127,** 467 (1980).
64. K. Ohata, H. Itoh, F. Hasegawa, and Y. Fujiki, Super Low-Noise GaAs MESFET's with a Deep-Recess Structure, *IEEE Trans. Electron Dev.* **ED-27,** 1029 (1980).
65. G. P. Donizelli, High Yield Self-Alignment Method for Submicrometre GaAs M.E.S.F.E.T.S., *Electron. Lett.* **14,** 523 (1978).
66. S. Takahashi, F. Murai, and H. Kodera, Submicrometer Gate Fabrication of GaAs MESFET by Plasma Etching, *IEEE Trans. Electron Dev.* **ED-25,** 1213 (1978).
67. M. Ida, T. Mizutani, K. Asai, M. Uchida, K. Shimada, and S. Ishida, Fabrication Technology for an 80-ps Normally-Off GaAs MESFET Logic, *IEEE Trans. Electron Dev.* **ED-28,** 489 (1981).
68. R. Zuleeg, J. K. Notthoff, and K. Lehovec, Femtojoule High-Speed Planar GaAs E-JFET Logic, *IEEE Trans. Electron Dev.* **ED-25,** 628 (1978).
69. R. L. Van Tuyl, C. A. Liechti, R. E. Lee, and E. Gowen, GaAs MESFET Logic with 4 GHz Clock Rate, *IEEE J. Solid State Circuits* **SC-12,** 485 (1977).
70. P. R. Wilson, The Emitter-Base Breakdown Voltage of Planar Transistors, *Solid State Electron.* **17,** 465 (1974).
71. D. P. Kennedy and R. R. O'Brien, Avalanche Breakdown Calculations for a Planar p–n Junction, *IBM J. Res. Dev.* **10,** 213 (1966).
72. M. Polinsky and S. Graf, MOS-Bipolar Monolithic Integrated Circuit Technology, *IEEE Trans. Electron Dev.* **ED-20,** 239 (1973).

73. B. T. Murphy, V. J. Glinski, P. A. Gary, and R. A. Pederson, Collector-Diffusion Isolated Integrated Circuits, *Proc. IEEE* **57,** 1523 (1969).

74. L. S. Senhouse, Jr., D. L. Kushler, and B. T. Murphy, Base Diffusion Isolated Transistors for Low Power Integrated Circuits, *IEEE Trans. Electron Dev.* **ED-18,** 355 (1971).

75. J. Agraz-Güereña, P. T. Panousis, and B. L. Morris, OXIL, A Versatile Bipolar VLSI Technology, *IEEE Trans. Electron Dev.* **ED-27,** 1397 (1980).

76. Y. Tarui, Y. Hayashi, H. Teshima, and T. Sekigawa, Transistor Schottky-Barrier-Diode Integrated Logic Circuit, *IEEE J. Solid State Circuits* **SC-4,** 3 (1969).

77. R. M. Warner, Jr., and J. N. Fordemwalt, *Integrated Circuits*, McGraw-Hill, New York, 1965.

78. L. I. Maisell and R. Glang, Eds., *Handbook of Thin Film Technology*, McGraw-Hill, New York, 1970.

79. J. L. Vossen and W. Kern, Eds., *Thin Film Processes*, Academic, New York, 1978.

80. L. Holland, Ed., *Thin Film Microelectronics*, Wiley, New York, 1965.

APPENDIX

The Mathematics of Diffusion

CONTENTS

A.1 SOLUTIONS FOR A CONSTANT DIFFUSION COEFFICIENT 640

A.1.1 Reflection and Superposition, 640
A.1.2 Extended Initial Conditions, 641
A.1.3 Diffusion through a Narrow Slot, 642
A.1.4 Miscellaneous Useful Solutions, 643
A.1.5 Some Useful Error Function Relations, 645
A.1.6 General Solution for Diffusion into a Semi-Infinite Body, 645

A.2 SOLUTION FOR A TIME-DEPENDENT DIFFUSION COEFFICIENT 649

A2.1 Ramping of a Diffusion Furnace, 650

A.3 SOLUTION FOR CONCENTRATION-DEPENDENT DIFFUSION COEFFICIENTS 651

A.4 DETERMINATION OF THE DIFFUSION CONSTANT 653

A.4.1 The *p–n* Junction Method, 653
A.4.2 The Boltzmann–Matano Method [5], 654

REFERENCES 655

This appendix considers some important mathematical techniques which can be used for solving problems involving diffusion in seminconductors. Exhaustive treatments of this subject are given elsewhere [1–3].

A.1 SOLUTIONS FOR A CONSTANT DIFFUSION COEFFICIENT

Many diffusion situations can be handled if the diffusion coefficient is assumed to be concentration independent. For this case, Fick's second law, in one dimension, takes the form

$$\frac{\partial N}{\partial t} = D \frac{\partial^2 N}{\partial x^2} \tag{A.1}$$

where N is the volume concentration of the diffusing impurity. It can be shown by direct differentiation that

$$N = \frac{A}{t^{1/2}} e^{-x^2/4Dt} \tag{A.2}$$

is a solution of this equation, where A is an arbitrary constant. This expression is symmetric in x, and tends to zero as x tends to infinity. The total amount of impurity involved in this diffusional process, at any given time, is

$$Q = \int_{-\infty}^{\infty} N \, dx \tag{A.3}$$

Substituting in Eq. (A.2), gives

$$Q = 2A\sqrt{\pi D} \tag{A.4}$$

This is also the initial amount of material at $t = 0$ and $x = 0$. Solving for A, the volume concentration is obtained as

$$N(x, t) = \frac{Q}{2\sqrt{\pi D t}} e^{-(x/2\sqrt{Dt})^2} \tag{A.5}$$

A.1.1 Reflection and Superposition

The symmetric nature of the solution of Eq. (A.5) allows us to build up solutions for other diffusion situations by using the concept of reflection at a boundary, combined with superposition. This is possible because of the linear character of the diffusion equation. Consider, for example, the above situation for diffusion in the positive direction alone, i.e., for diffusion from a sheet of impurities Q (cm^{-2}) into a semi-infinite body ($0 \leq x \leq \infty$). A solution can be obtained from Eq. (A.5) by assuming that the solution for negative values of x is reflected in the plane of $x = 0$, and superposed on the distribution of $x \geq 0$. Thus for this case,

$$N = \frac{Q}{\sqrt{\pi D t}} e^{-(x/2\sqrt{Dt})^2} \tag{A.6}$$

This is the gaussian distribution given in Eq. (4.51); it is useful for describing base diffusion profiles in a transistor (see Fig. 4.17).

A.1.2 Extended Initial Conditions

It is also possible to use Eq. (A.5) to determine the solution for other initial conditions. Consider, for example, diffusion from an *infinite* source into an infinite region ($-\infty < x < \infty$). Assume that this source has a concentration N_0, extending over $-\infty \leq x \leq 0$. Any finite element of this source, of thickness $d\xi$, contains a quantity of impurity $N_0\, d\xi$, at a distance ξ from any given plane P.

The concentration of P due to this sheet source is

$$\frac{N_0\, d\xi}{2\sqrt{\pi D t}} e^{-\xi^2/4Dt} \tag{A.7}$$

For an infinite source it follows that

$$N(x, t) = \frac{N_0}{2\sqrt{\pi D t}} \int_x^\infty e^{-\xi^2/4Dt}\, d\xi \tag{A.8}$$

The error function erf y is defined by

$$\operatorname{erf} y = \frac{2}{\sqrt{\pi}} \int_0^y e^{-z^2}\, dz \tag{A.9}$$

Values of erf y for different values of y are given in Table A.1. Combining with Eq. (A.8), gives

$$N(x, t) = \frac{N_0}{2}\left(1 - \operatorname{erf} \frac{x}{2\sqrt{Dt}}\right) \tag{A.10}$$

The function ($1 - \operatorname{erf} y$) is often written as the complementary error function, erfc y.

Again, the concept of reflection from a boundary can be used to determine the solution for diffusion into a semi-infinite region. It follows that, for this case ($\infty < x \leq 0$),

$$N(x, t) = N_0 \operatorname{erfc} \frac{x}{2\sqrt{Dt}} \tag{A.11}$$

This is the complementary error function distribution given in Eq. (4.49); it is useful for describing emitter diffusion profiles in a transistor. A plot of this function is also shown in Fig. 4.17.

Both Eqs. (A.6) and (A.11) show that a single dimensionless parameter $x/2\sqrt{Dt}$ is involved in describing these impurity distributions. It can be readily shown that this is true for any diffusion situation involving a semi-infinite medium with zero background concentration, whose surface is fixed. It follows that the penetration depth x is proportional to the square root of time in these situations. In practice, this relationship is found to hold for the vast majority of diffusion situations in silicon as well as in gallium arsenide. This is true for cases where D is concentration dependent, as well as for situations where D is constant.

A.1.3 Diffusion through a Narrow Slot

This serves as an example of direct solution for a three-dimensional situation. It is of practical importance in high-speed, shallow-diffused devices. The diffusion profile for this case is derived by first considering a limited point source. Then the equation describing diffusion from this source into an infinite volume is

$$\frac{\partial N}{\partial t} = D\left(\frac{\partial^2 N}{\partial x^2} + \frac{\partial^2 N}{\partial y^2} + \frac{\partial^2 N}{\partial z^2}\right) \tag{A.12}$$

The solution of this equation takes the form

$$N = \frac{A}{\sqrt{t}} e^{-(x^2+y^2+z^2)/4Dt} \tag{A.13}$$

If M is the amount of impurity at this point, then

$$M = \int_{-\infty}^{\infty}\int_{-\infty}^{\infty}\int_{-\infty}^{\infty} N\,dx\,dy\,dz = 8\pi DA \tag{A.14}$$

The concentration at a distance r from a point source in an infinite volume is obtained by combining Eqs. (A.13) and (A.14), to give

$$N(r, t) = \frac{M}{8\sqrt{\pi Dt}} e^{-r^2/4Dt} \tag{A.15}$$

Diffusion from a line source into an infinite volume can be determined by integrating this solution with respect to a space variable. Thus if diffusion is considered into an infinite volume from a line source along the y axis, then

$$N = \int_{-\infty}^{\infty} \frac{\delta}{4\pi Dt} e^{-(x^2+y^2+z^2)/4Dt}\, dy \tag{A.16}$$

where δ is the amount of diffusing impurity deposited initially, per unit length of the line source. Solving, the concentration at a distance r is obtained as

$$N(r, t) = \frac{\delta}{4\pi Dt} e^{-r^2/4Dt} \tag{A.17}$$

Solution for diffusion into a semi-infinite plane follows from the reflecting boundary method. For this situation,

$$N(r, t) = \frac{\delta}{2\pi Dt} e^{-r^2/4Dt} \tag{A.18}$$

Diffusion from a narrow slot is approximated by this distribution. In this case, $\delta \simeq QW$, where W is the width of the slot and Q is the surface density of dopant in atoms/cm^2.

A.1.4 Miscellaneous Useful Solutions

A number of solutions of the diffusion equation are available in the published literature. Many of these are of practical importance in integrated circuit fabrication technology. Some of these are now given, without their derivation.

1. *Diffusion from a limited source of finite width.* Consider an impurity of concentration N_0, deposited in a finite width of x_0. Then the resulting impurity profile in the semiconductor after time t is given by

$$N(x, t) = \frac{N_0}{2}\left[\operatorname{erf}\left(\frac{x - x_0}{2\sqrt{Dt}}\right) - \operatorname{erf}\left(\frac{x + x_0}{2\sqrt{Dt}}\right)\right] \tag{A.19}$$

   The surface concentration resulting from this diffusion is

$$N(0, t) = N_0\left[1 - \operatorname{erfc}\left(\frac{x_0}{2\sqrt{Dt}}\right)\right] \tag{A.20}$$

2. *Out-diffusion from a uniformly doped buried layer.* Consider a uniformly doped buried layer, of doping concentration N^+, penetrating

from $x = 0$ to $x = x^+$ into a substrate. An undoped epitaxial layer, from $x = 0$ to $x = -x_{epi}$, is grown on this substrate. This is subsequently processed by a series of high-temperature steps, resulting in an effective value of Dt. Then the out-diffusion of the buried layer into the epitaxial region ($-x_{epi} \leq x \leq 0$) is given by

$$N(x, t) = \frac{N^+}{2}\left[\operatorname{erfc}\left(\frac{-x}{2\sqrt{Dt}}\right) + \operatorname{erfc}\left(\frac{2x_{epi} + x}{2\sqrt{Dt}}\right) - \operatorname{erfc}\left(\frac{x^+ - x}{2\sqrt{Dt}}\right) - \operatorname{erfc}\left(\frac{2x_{epi} - x^+ + x}{2\sqrt{Dt}}\right)\right] \tag{A.21}$$

3. *Diffusion from a doped oxide source.* Consider an oxide, of thickness x_{ox}, having a dopant concentration of N_{ox}. Let D_{ox} and D be the diffusion coefficients of the impurity in the oxide and the semiconductor, respectively, and let m be the impurity segregation coefficient at the oxide–semiconductor interface (see Section 4.8.1). Then, for short diffusion times, such that $x_{ox} > 4\sqrt{Dt}$, the impurity concentration in the semiconductor is given by

$$N(x, t) \simeq \frac{N_{ox}(D_{ox}/D)^{1/2}}{1 + k} \operatorname{erfc}\left(\frac{x}{2\sqrt{Dt}}\right) \tag{A.22}$$

where

$$k = \frac{1}{m}\left(\frac{D_{ox}}{D}\right)^{1/2} \tag{A.23}$$

4. *Bilateral diffusion into a finite body.* Consider a slice of thickness x_0, into which diffusions are made from both sides. For all t let

$$N(0, t) = N(x_0, t) = N_0 \tag{A.24}$$

Then

$$N(x, t) = N_0\left[1 - \frac{4}{\pi} e^{-\pi^2 Dt/x_0^2} \sin\left(\frac{\pi x}{x_0}\right)\right] \tag{A.25}$$

Out-diffusion from two sides of a uniformly doped slice of thickness x_0 is given by

$$N(x, t) = \frac{4N_0}{\pi}\left[e^{-\pi^2 Dt/x_0^2} \sin\left(\frac{\pi x}{x_0}\right)\right] \tag{A.26}$$

where N_0 is the initial doping concentration.

A.1.5 Some Useful Error Function Relations

The following relations are useful in the solution of many diffusion problems:

1. $$\int_0^z e^{-y^2}\,dy = \frac{\sqrt{\pi}}{2}\,\mathrm{erf}\,z \tag{A.27}$$
2. $$\int_0^z e^{-y^2}\,dy = y - \frac{y^3}{3\times 1!} + \frac{y^5}{5\times 2!} - \cdots \tag{A.28}$$
3. $$\int_0^\infty e^{-y^2}\,dy = \frac{\sqrt{\pi}}{2} \tag{A.29}$$
4. $$\frac{d}{dz}(\mathrm{erf}\,z) = \frac{2}{\sqrt{\pi}}\,e^{-z^2} \tag{A.30}$$
5. $$\frac{d^2}{dz^2}(\mathrm{erf}\,z) = -\frac{4}{\sqrt{\pi}}\,z\,e^{-z^2} \tag{A.31}$$
6. $$\mathrm{erfc}\,z = 1 - \mathrm{erf}\,z \tag{A.32}$$
7. $$i\,\mathrm{erfc}\,z = \int_z^\infty \mathrm{erfc}\,z\,dz \tag{A.33}$$
8. $$i^2\,\mathrm{erfc}\,z = \int_z^\infty \int_z^\infty \mathrm{erf}\,z\,dz\,dz \tag{A.34}$$
9. $$\mathrm{erf}\,(-z) = -\mathrm{erf}\,z \tag{A.35}$$
10. $$\mathrm{erf}\,0 = 0 \tag{A.36}$$
11. $$\mathrm{erf}\,\infty = 1 \tag{A.37}$$

A short list of the function erfc z is given in Table A.1.

A.1.6 General Solution for Diffusion into a Semi-Infinite Body

We begin this solution with Fick's second law,

$$\frac{\partial N}{\partial t} = D\frac{\partial^2 N}{\partial x^2} \tag{A.38}$$

The initial distribution of the impurity is given by

$$N(x, 0) = f(x) \tag{A.39}$$

One solution of this equation may be written as the product of two functions, one of time and one of space. Thus let

$$N(x, t) = X(x)T(t) \tag{A.40}$$

Table A.1 Error Function erf z^a

z	erf(z)	z	erf(z)	z	erf(z)	z	erf(z)
0.00	0.000 000	0.50	0.520 500	1.00	0.842 701	1.50	0.966 105
0.01	0.011 283	0.51	0.529 244	1.01	0.846 810	1.51	0.967 277
0.02	0.022 565	0.52	0.537 899	1.02	0.850 838	1.52	0.968 413
0.03	0.033 841	0.53	0.546 464	1.03	0.854 784	1.53	0.969 516
0.04	0.045 111	0.54	0.554 939	1.04	0.858 650	1.54	0.970 586
0.05	0.056 372	0.55	0.563 323	1.05	0.862 436	1.55	0.971 623
0.06	0.067 622	0.56	0.571 616	1.06	0.866 144	1.56	0.972 628
0.07	0.078 858	0.57	0.579 816	1.07	0.869 773	1.57	0.973 603
0.08	0.090 078	0.58	0.587 923	1.08	0.873 326	1.58	0.974 547
0.09	0.101 281	0.59	0.595 936	1.09	0.876 803	1.59	0.975 462
0.10	0.112 463	0.60	0.603 856	1.10	0.880 205	1.60	0.976 348
0.11	0.123 623	0.61	0.611 681	1.11	0.883 533	1.61	0.977 207
0.12	0.134 758	0.62	0.619 411	1.12	0.886 788	1.62	0.978 038
0.13	0.145 867	0.63	0.627 046	1.13	0.889 971	1.63	0.978 843
0.14	0.156 947	0.64	0.634 586	1.14	0.893 082	1.64	0.979 622
0.15	0.167 996	0.65	0.642 029	1.15	0.896 124	1.65	0.980 376
0.16	0.179 012	0.66	0.649 377	1.16	0.899 096	1.66	0.981 105
0.17	0.189 992	0.67	0.656 628	1.17	0.902 000	1.67	0.981 810
0.18	0.200 936	0.68	0.663 782	1.18	0.904 837	1.68	0.982 493
0.19	0.211 840	0.69	0.670 840	1.19	0.907 608	1.69	0.983 153
0.20	0.222 703	0.70	0.677 801	1.20	0.910 314	1.70	0.983 790
0.21	0.233 522	0.71	0.684 666	1.21	0.912 956	1.71	0.984 407
0.22	0.244 296	0.72	0.691 433	1.22	0.915 534	1.72	0.985 003
0.23	0.255 023	0.73	0.698 104	1.23	0.918 050	1.73	0.985 578
0.24	0.265 700	0.74	0.704 678	1.24	0.920 505	1.74	0.986 135
0.25	0.276 326	0.75	0.711 156	1.25	0.922 900	1.75	0.986 672
0.26	0.286 900	0.76	0.717 537	1.26	0.925 236	1.76	0.987 190
0.27	0.297 418	0.77	0.723 822	1.27	0.927 514	1.77	0.987 691
0.28	0.307 880	0.78	0.730 010	1.28	0.929 734	1.78	0.988 174
0.29	0.318 283	0.79	0.736 103	1.29	0.931 899	1.79	0.988 641
0.30	0.328 627	0.80	0.742 101	1.30	0.934 008	1.80	0.989 091
0.31	0.338 908	0.81	0.748 003	1.31	0.936 063	1.81	0.989 525
0.32	0.349 126	0.82	0.753 811	1.32	0.938 065	1.82	0.989 943
0.33	0.359 279	0.83	0.759 524	1.33	0.940 015	1.83	0.990 347
0.34	0.369 365	0.84	0.765 143	1.34	0.941 914	1.84	0.990 736
0.35	0.379 382	0.85	0.770 668	1.35	0.943 762	1.85	0.991 111
0.36	0.389 330	0.86	0.776 100	1.36	0.945 561	1.86	0.991 472
0.37	0.399 206	0.87	0.781 440	1.37	0.947 312	1.87	0.991 821
0.38	0.409 009	0.88	0.786 687	1.38	0.949 016	1.88	0.992 156
0.39	0.418 739	0.89	0.791 843	1.39	0.950 673	1.89	0.992 479
0.40	0.428 392	0.90	0.796 908	1.40	0.952 285	1.90	0.992 790
0.41	0.437 969	0.91	0.801 883	1.41	0.953 852	1.91	0.993 090
0.42	0.447 468	0.92	0.806 768	1.42	0.955 376	1.92	0.993 378
0.43	0.456 887	0.93	0.811 564	1.43	0.956 857	1.93	0.993 656
0.44	0.466 225	0.94	0.816 271	1.44	0.958 297	1.94	0.993 923
0.45	0.475 482	0.95	0.820 891	1.45	0.959 695	1.95	0.994 179
0.46	0.484 655	0.96	0.825 424	1.46	0.961 054	1.96	0.994 426
0.47	0.493 745	0.97	0.829 870	1.47	0.962 373	1.97	0.994 664
0.48	0.502 750	0.98	0.834 232	1.48	0.963 654	1.98	0.994 892
0.49	0.511 668	0.99	0.838 508	1.49	0.964 898	1.99	0.995 111

Table A.1 (*Continued*)

2.00	0.995 322	2.50	0.999 593	3.00	0.999 977 91	3.50	0.999 999 257
2.01	0.995 525	2.51	0.999 614	3.01	0.999 979 26	3.51	0.999 999 309
2.02	0.995 719	2.52	9.999 634	3.02	0.999 980 53	3.52	0.999 999 358
2.03	0.995 906	2.53	0.999 654	3.03	0.999 981 73	3.53	0.999 999 403
2.04	0.996 086	2.54	0.999 672	3.04	0.999 982 86	3.54	0.999 999 445
2.05	0.996 258	2.55	0.999 689	3.05	0.999 983 92	3.55	0.999 999 485
2.06	0.996 423	2.56	0.999 706	3.06	0.999 984 92	3.56	0.999 999 521
2.07	0.996 582	2.57	0.999 722	3.07	0.999 985 86	3.57	0.999 999 555
2.08	0.996 734	2.58	0.999 736	3.08	0.999 986 74	3.58	0.999 999 587
2.09	0.996 880	2.59	0.999 751	3.09	0.999 987 57	3.59	0.999 999 617
2.10	0.997 021	2.60	0.999 764	3.10	0.999 988 35	3.60	0.999 999 644
2.11	0.997 155	2.61	0.999 777	3.11	0.999 989 08	3.61	0.999 999 670
2.12	0.997 284	2.62	0.999 789	3.12	0.999 989 77	3.62	0.999 999 694
2.13	0.997 407	2.63	0.999 800	3.13	0.999 990 42	3.63	0.999 999 716
2.14	0.997 525	2.64	0.999 811	3.14	0.999 991 03	3.64	0.999 999 736
2.15	0.997 639	2.65	0.999 822	3.15	0.999 991 60	3.65	0.999 999 756
2.16	0.997 747	2.66	0.999 831	3.16	0.999 992 14	3.66	0.999 999 773
2.17	0.997 851	2.67	0.999 841	3.17	0.999 992 64	3.67	0.999 999 790
2.18	0.997 951	2.68	0.999 849	3.18	0.999 993 11	3.68	0.999 999 805
2.19	0.998 046	2.69	0.999 858	3.19	0.999 993 56	3.69	0.999 999 820
2.20	0.998 137	2.70	0.999 866	3.20	0.999 993 97	3.70	0.999 999 833
2.21	0.998 224	2.71	0.999 873	3.21	0.999 994 36	3.71	0.999 999 845
2.22	0.998 308	2.72	0.999 880	3.22	0.999 994 73	3.72	0.999 999 857
2.23	0.998 388	2.73	0.999 887	3.23	0.999 995 07	3.73	0.999 999 867
2.24	0.998 464	2.74	0.999 893	3.24	0.999 995 40	3.74	0.999 999 877
2.25	0.998 537	2.75	0.999 899	3.25	0.999 995 70	3.75	0.999 999 886
2.26	0.998 607	2.76	0.999 905	3.26	0.999 995 98	3.76	0.999 999 895
2.27	0.998 674	2.77	0.999 910	3.27	0.999 996 24	3.77	0.999 999 903
2.28	0.998 738	2.78	0.999 916	3.28	0.999 996 49	3.78	0.999 999 910
2.29	0.998 799	2.79	0.999 920	3.29	0.999 996 72	3.79	0.999 999 917
2.30	0.998 857	2.80	0.999 925	3.30	0.999 996 94	3.80	0.999 999 923
2.31	0.998 912	2.81	0.999 929	3.31	0.999 997 15	3.81	0.999 999 929
2.32	0.998 966	2.82	0.999 933	3.32	0.999 997 34	3.82	0.999 999 934
2.33	0.999 016	2.83	0.999 937	3.33	0.999 997 51	3.83	0.999 999 939
2.34	0.999 065	2.84	0.999 941	3.34	0.999 997 68	3.84	0.999 999 944
2.35	0.999 111	2.85	0.999 944	3.35	0.999 997 838	3.85	0.999 999 948
2.36	0.999 155	2.86	0.999 948	3.36	0.999 997 983	3.86	0.999 999 952
2.37	0.999 197	2.87	0.999 951	3.37	0.999 998 120	3.87	0.999 999 956
2.38	0.999 237	2.88	0.999 954	3.38	0.999 998 247	3.88	0.999 999 959
2.39	0.999 275	2.89	0.999 956	3.39	0.999 998 367	3.89	0.999 999 962
2.40	0.999 311	2.90	0.999 959	3.40	0.999 998 478	3.90	0.999 999 965
2.41	0.999 346	2.91	0.999 961	3.41	0.999 998 582	3.91	0.999 999 968
2.42	0.999 379	2.92	0.999 964	3.42	0.999 998 679	3.92	0.999 999 970
2.43	0.999 411	2.93	0.999 966	3.43	0.999 998 770	3.93	0.999 999 973
2.44	0.999 441	2.94	0.999 968	3.44	0.999 998 855	3.94	0.999 999 975
2.45	0.999 469	2.95	0.999 970	3.45	0.999 998 934	3.95	0.999 999 977
2.46	0.999 497	2.96	0.999 972	3.46	0.999 999 008	3.96	0.999 999 979
2.47	0.999 523	2.97	0.999 973	3.47	0.999 999 077	3.97	0.999 999 980
2.48	0.999 547	2.98	0.999 975	3.48	0.999 999 141	3.98	0.999 999 982
2.49	0.999 571	2.99	0.999 976	3.49	0.999 999 201	3.99	0.999 999 983

[a] For a more complete table, see L. J. Comrie, *Chambers Six Figure Mathematical Tables*, Vol. 2, W. & R. Chambers, Edinburgh, 1949.

Substituting in Eq. (A.38) and rearranging terms,

$$\frac{1}{DT}\frac{dT}{dt} = \frac{1}{X}\frac{d^2X}{dx^2} \tag{A.41}$$

Since the left-hand side of this equation is only a function of t and the right-hand side only a function of x, each side must be equal to some constant that is independent of both t or x. Writing this constant as $-\lambda^2$, Eq. (A.40) can be broken into

$$\frac{dT}{T} = -\lambda^2 D\,dt \tag{A.42}$$

and

$$\frac{d^2X}{dx^2} = -\lambda^2 X \tag{A.43}$$

Solving,

$$T(t) = \gamma e^{-\lambda^2 Dt} \tag{A.44}$$

$$X(x) = \alpha \cos \lambda x + \beta \sin \lambda x \tag{A.45}$$

and

$$N(x, t) = \lambda e^{-\lambda^2 Dt}(A \cos \lambda x + B \sin \lambda x) \tag{A.46}$$

where γ, α, β are constants of integration, $A = \alpha\gamma$, and $B = \beta\gamma$.

The general solution of Eq. (A.38) can be written as a sum of partial solutions of this type because it is a linear equation. In addition, since the body has infinite dimensions, the choice of λ is quite arbitary, and the summation over discrete values of λ can be replaced by an integral. Hence this solution may be written as

$$N(x, t) = \int_{-\infty}^{+\infty} e^{-\lambda^2 Dt}(A \cos \lambda x + B \sin \lambda x)\,d\lambda \tag{A.47}$$

It is now necessary to solve for the various constants of integration. Fourier's integral theorem states that

$$f(x) = \frac{1}{2\pi}\int_{-\infty}^{+\infty}\int_{-\infty}^{+\infty} f(\xi)\cos[\lambda(\xi - x)]\,d\xi\,d\lambda \tag{A.48}$$

$$= \frac{1}{2\pi}\int_{-\infty}^{+\infty}\left\{\left[\int_{-\infty}^{+\infty} f(\xi)\cos\lambda\xi\,d\xi\right]\cos\lambda x\,dx + \left[\int_{-\infty}^{+\infty} f(\xi)\sin\lambda\xi\,d\xi\right]\sin\lambda x\,dx\right\}d\lambda \tag{A.49}$$

At time $t = 0$ the initial concentration is given by $f(x)$.

Substituting in Eq. (A.47),

$$f(x) = N(x, 0) = \int_{-\infty}^{\infty} (A \cos \lambda x + B \sin \lambda x)\, d\lambda \tag{A.50}$$

Comparing Eqs. (A.49) and (A.50),

$$A = \frac{1}{2\pi} \int_{-\infty}^{\infty} f(\xi) \cos \lambda\xi \, d\xi \tag{A.51}$$

$$B = \frac{1}{2\pi} \int_{-\infty}^{\infty} f(\xi) \sin \lambda\xi \, d\xi \tag{A.52}$$

Substituting these values of A and B in Eq. (A.47), gives

$$N(x, t) = \frac{1}{2\pi} \int_{-\infty}^{\infty} f(\xi) \left[\int_{-\infty}^{\infty} e^{-\lambda^2 Dt} \cos \lambda(\xi - x)\, d\lambda \right] d\xi \tag{A.53}$$

But

$$\int_{-\infty}^{\infty} e^{-\lambda^2 Dt} \cos \lambda(\xi - x)\, dx = \sqrt{\pi/Dt}\, e^{-(\xi - x)^2/4Dt} \tag{A.54}$$

Hence

$$N(x, t) = \frac{1}{2\sqrt{\pi Dt}} \int_{-\infty}^{\infty} f(\xi)\, e^{-(\xi - x)^2/4Dt}\, d\xi \tag{A.55}$$

This is the general solution of the diffusion equation for an infinite body. Here

$$f(\xi) = N(\xi, 0) \tag{A.56}$$

A.2 SOLUTION FOR A TIME-DEPENDENT DIFFUSION COEFFICIENT

Consider the situation where the diffusivity if given by

$$D = D(t) \tag{A.57}$$

Introducing a new time variable T, such that

$$\partial T = D(t)\, \partial t \tag{A.58}$$

the diffusion equation now becomes

$$\frac{\partial N}{\partial T} = \frac{\partial^2 N}{\partial x^2} \tag{A.59}$$

so that it can be solved as before. Note that this new time variable can be written as

$$T = (Dt)_{\text{eff}} = \int_0^{t_0} D(t)\, dt \tag{A.60}$$

where $(Dt)_{\text{eff}}$ is the effective Dt product, and t_0 is the time interval over which the diffusivity varies.

A.2.1 Ramping of a Diffusion Furnace

An important situation which occurs in practice is the ramping of a diffusion furnace to prevent warpage of slices. The effect of ramping can be readily estimated for a linear ramp rate C. Assuming a ramp-down situation, the furnace temperature is given by

$$T = T_0 - Ct \tag{A.61}$$

where T_0 is the initial temperature. Assume that ramp-down occurs for a time t_0. Then the effective Dt product during this time is given by

$$(Dt)_{\text{eff}} = \int_0^{t_0} D(t)\, dt \tag{A.62}$$

During a typical diffusion, ramping is carried out until the diffusivity is negligibly small. Thus this integration can be simplified with little error if the upper limit is taken as infinity. Since

$$\frac{1}{T} = \frac{1}{T_0 - Ct} \simeq \frac{1}{T_0}\left(1 + \frac{Ct}{T_0} + \cdots\right) \tag{A.63}$$

and

$$D = D_0\, e^{-E_0/kT} \tag{A.64}$$

we can write

$$\begin{aligned} D(t) &\simeq D_0 \exp\left[-\frac{E_0}{kT_0}\left(1 + \frac{Ct}{T_0} + \cdots\right)\right] \\ &= D(T_0)\, e^{-(CE_0/kT_0^2)t} \end{aligned} \tag{A.65}$$

where $D(T_0)$ is the diffusivity at T_0, and is given by

$$D(T_0) = D_0\, e^{-E_0/kT} \tag{A.66}$$

Substituting into Eq. (A.62) and using an upper integration limit of infinity, gives

$$(Dt)_{\text{eff}} \simeq D(T_0)\left(\frac{kT_0^2}{CE_0}\right) \tag{A.67}$$

Thus the ramp-down process results in an effective additional time equal to kT_0^2/CE_0 at the initial diffusion temperature T_0. A similar relationship can be derived for the ramp-up situation.

A.3 SOLUTION FOR CONCENTRATION-DEPENDENT DIFFUSION COEFFICIENTS

This situation is encountered often in practice, as shown in Chapter 4. The equation for diffusion is then

$$\frac{\partial N}{\partial t} = \frac{\partial}{\partial x}\left(D\frac{\partial N}{\partial x}\right) \tag{A.68}$$

where D is a function of N. This equation can be converted into an ordinary differential equation by means of Boltzmann's transformation. Let

$$\eta = x/2\sqrt{t} \tag{A.69}$$

Then

$$\frac{\partial N}{\partial x} = \frac{1}{2t^{1/2}}\frac{dN}{d\eta} \tag{A.70a}$$

$$\frac{\partial N}{\partial t} = -\frac{x}{4t^{3/2}}\frac{dN}{d\eta} \tag{A.70b}$$

so that Fick's second law reduces to

$$-2\frac{dN}{d\eta} = \frac{d}{d\eta}\left(D\frac{dN}{d\eta}\right) \tag{A.71}$$

This equation can be solved for those situations where the boundary conditions can be expressed in terms of η alone. An important case is that of diffusion in a semi-infinite medium with a constant surface concentration, such that

$$N = N_{\text{sur}}, \qquad x = 0,\ t > 0 \tag{A.72a}$$

$$N = N_1, \qquad x > 0,\ t = 0 \tag{A.72b}$$

Table A.2 Numerical Solutions to the Concentration Dependent Diffusion Equation[a,b]

y	N Case A	N Case B	N Case C
0.001	0.999	0.999	0.999
0.005	0.996	0.996	0.997
0.01	0.991	0.992	0.993
0.02	0.982	0.985	0.986
0.05	0.955	0.961	0.964
0.10	0.907	0.918	0.925
0.15	0.858	0.872	0.881
0.20	0.807	0.822	0.830
0.25	0.753	0.766	0.769
0.30	0.697	0.703	0.697
0.32	0.674	0.677	0.663
0.34	0.650	0.648	0.624
0.36	0.626	0.617	0.577
0.38	0.602	0.584	0.521
0.40	0.578	0.550	0.449
0.42	0.553	0.512	0.340
0.43			0.237
0.435			0.058
0.436			0.000
0.44	0.528	0.471	
0.46	0.503	0.425	
0.48	0.477	0.373	
0.50	0.451	0.312	
0.52	0.425	0.234	
0.54	0.398	0.107	
0.545		0.022	
0.546		0.000	
0.56	0.370		
0.58	0.343		
0.60	0.315		
0.62	0.287		
0.64	0.258		
0.66	0.229		
0.68	0.199		
0.70	0.169		
0.72	0.139		
0.74	0.108		
0.76	0.077		
0.78	0.045		
0.80	0.013		
0.808	0.000		

[a] Here $N = N(x, t)/N_{sur}$ and $y = x/2\sqrt{D_{sur}t}$.

[b] See reference 4.

For this situation the boundary conditions become

$$N = N_{sur}, \qquad \eta = 0 \tag{A.73a}$$

$$N = N_1, \qquad \eta = \infty \tag{A.73b}$$

and Eq. (A.71) can be used for this case.

The diffusion of many impurities in silicon and gallium arsenide was described in Section 4.5.2 by a diffusion coefficient D, such that

$$\text{Case } A: \qquad D = D_{sur}\left(\frac{N}{N_{sur}}\right) \tag{A.74a}$$

$$\text{Case } B: \qquad D = D_{sur}\left(\frac{N}{N_{sur}}\right)^2 \tag{A.74b}$$

$$\text{Case } C: \qquad D = D_{sur}\left(\frac{N}{N_{sur}}\right)^3 \tag{A.74c}$$

Numerical solutions for these cases have been made [4]. Tabulated values for N/N_{sur} versus $y\ (= x/2\sqrt{D_{sur}\,t})$ are given in Table A.2 for these solutions.

A.4 DETERMINATION OF THE DIFFUSION CONSTANT

The diffusion of impurities in semiconductors is characterized at any given temperature by a diffusion constant D, which may be a function of impurity concentration. Many diffusions in silicon, particularly those with low surface concentrations (base diffusions, for example), can be characterized by a concentration independent value of D. In general, this is not true for emitter diffusions, particularly for shallow emitters. Nevertheless, the assumption of concentration independence is often made for simplicity. Diffusions in gallium arsenide are usually described by a concentration dependent value of D.

A.4.1 The *p–n* Junction Method

This technique assumes that D is concentration independent. It is extremely popular because the experimental technique involves the formation of junctions. As a result, the answers obtained in this way are especially useful in device fabrication. The starting point for this technique is to make one or more diffusions into semiconductors of known background concentration and opposite impurity type. Assuming that diffusion takes place from a constant concentration impurity source into a background concentration

N_B, we obtain

$$N_B = N_0 \operatorname{erfc}\left(\frac{x_j}{2\sqrt{Dt}}\right) \tag{A.75}$$

where x_j is the junction depth and N_0 is the surface concentration. Both x_j and N_0 can be obtained by the methods outlined in Section 4.10. N_B is given from measurement of the specific resistivity of the original material.

An alternate method, which avoids determination of N_0, is to conduct simultaneous diffusions into two slices of background concentration N_{B1} and N_{B2} under identical conditions. If the junction depths are x_1 and x_2, respectively, then

$$\frac{N_{B1}}{N_{B2}} = \frac{\operatorname{erfc}(x_1/2\sqrt{Dt})}{\operatorname{erfc}(x_2/2\sqrt{Dt})} \tag{A.76}$$

This equation can be solved approximately with

$$D \simeq \frac{1}{4t}\left[\frac{x_1^2 - x_2^2}{\ln(N_{B2}x_1/N_{B1}x_2)}\right] \tag{A.77}$$

A more general method, for an arbitrary impurity profile, now follows.

A.4.2 The Boltzmann–Matano Method [5]

The starting point for this method is a knowledge of the impurity concentration as a function of diffusion depth. This can be obtained by conducting a diffusion with the impurity, followed by direct measurement by secondary ion mass spectrometry. An alternative technique is diffusion with a radioactive isotope of the impurity, and measurement of the radiation intensity as successive layers are removed. A third technique consists of measuring the sheet resistance upon successive removals of layers. Details of these techniques are beyond the scope of this book.

Once the concentration profile is known, it is necessary to establish that the diffusion process can be described by means of the Boltzmann transformation. This requires first that a linear relation exist between the impurity penetration depth and the square root of the diffusion time. This is true for the majority of diffusion situations in silicon as well as in gallium arsenide.

Consider the situation of a diffusing impurity into an undoped background. Here

$$N = N_0, \qquad x < 0, t = 0 \tag{A.78a}$$

$$N = 0, \qquad x > 0, t = 0 \tag{A.78b}$$

Using Boltzmann's transformation, Fick's second law is given by

$$-2\eta \frac{dN}{d\eta} = \frac{d}{d\eta}\left(D \frac{dN}{d\eta}\right) \tag{A.79}$$

so that

$$-2\int_0^{N_1} \eta \, dN = \left[D \frac{dN}{d\eta}\right]_{N=0}^{N=N_1} \tag{A.80}$$

But $D(dN/d\eta) = 0$ when $N = 0$. Making this substitution and setting $\eta = x/2\sqrt{t}$, gives

$$D_{N=N_1} = -\frac{1}{2t}\frac{dx}{dN}\int_0^{N_1} x \, dN \tag{A.81}$$

Eq. (A.81) shows that the diffusion constant can be determined at any given concentration N_1 from knowledge of the slope of the N versus x profile at $N = N_1$, and the total number of diffused impurities between $N = 0$ and $N = N_1$.

REFERENCES

1. B. I. Boltaks, *Diffusion in Semiconductors*, Academic, New York, 1963.
2. J. Crank, *The Mathematics of Diffusion*, Clarendon, Oxford, England, 1975.
3. W. Jost, *Diffusion in Solids, Liquids, and Gases*, Academic, New York, 1962.
4. L. R. Weisberg and J. Blanc, Diffusion with Interstitial-Substitutional Equilibrium. Zinc in GaAs, *Phys. Rev.* **131,** 1548 (15 Aug. 1963).
5. P. G. Shewmon, *Diffusion in Solids*, McGraw Hill, New York, 1963.

Index

Acetic acid, 480
Accelerator, 340
Activation energy of diffusion, 116, 132, 449
  GaAs, *140*
  Si, *129, 135*
Active parasitics, 593, 612
Adatoms, 216
Alloy spiking, 365, 455, 461
Al-Si system, *57*
Aluminum:
  diffusivity of Si, *457*
  etch, 447, 498, 509, *524*
  films, 443, 446, 453
  solubility of Si, *456*
Aluminum arsenide, 281
Aluminum gate transistor, 590
Amorphous, 323
Amorphous silicon, 430
Aphoteric dopants, 26, *29*, 275
Anisotropic etching, 487, 509
Annealing, 325, 332
Anode reaction, 479
Anodic oxidation, 401
Anodic oxies, properties, 409
Anodization:
  GaAs, 404, 408
  parameters, *407*
  Si, 403, 408
Antistructure defects, 16
Apparatus, Czochralski, 86
  float zone, 102
Aqueous reagents, *522*
Arsenic, *130*, 152, 170, *205*
Arsenic, partial pressure, 21
Arsenic trichloride, *219*, 248
Arsine, 170, 248, 253
As-Si system, *60*
Atom movement, 115, 151
Atomic number, *312*
Atomic weight, *312*
Au-Ga system, *65*
Au-Ge system, *59*
Au-Si system, *58*
Autocatalytic reaction, 480
Autodoping, 235
Autoepitaxy, 215
Axes, crystal, 8

Barrel reactor, 229
Barrier heights:
  GaAs, *466, 467*
  Si, *465*
Beam leads, 459
Beam scanning, 342
Beryllium, 31, 334
BHF, 580
Binary phase diagrams, 51
Bipolar transistor, 145, 357, 605
Bird beak, 580
Bird crest, 580
BJT devices, 145, 357, 605
Body centered crystal, 3
Boltzmann-Matano method, 654
Boltzmann's transformation, 651
Bonding:
  die, 461
  thermocompression, 444
  ultrasonic, 446
Borosilicate glass, 64, 440, *495, 524,* 546
Boundary layer, 95, 219, 271

All pages in *italics* refer to pertinent data concerning the item.

B_2O_3-SiO_2 system, *66*
Breakdown voltage, 25, 102, 610
Bridging oxygen, 373
Bridgman method, 85
Bromine-methanol, 483
Bubbler, 161, 217
Buffered hydrofluoric acid, 580
Buried layer, 606

Capacitor:
  diffused, 631
  MOS, 633
Capture rate constants:
  AU, *136*
Pt, *136*
Carbon, 26, 31, 256, 587
Caro's etch, *484*
Carrier concentration, intrinsic, *122*
Cathode reaction, 480
Cathode sputter etching, 512
CCl_4, 509
CF_4 plasma, 508
CF_4-H_2 plasma, 508
Channeling, 304, 553
Channel stopper:
  diffused, 578, 610
  implanted, 360
Charge states-silicon oxide, 394
Chlorine oxidation, 389, 588
Chlorobenzene, 550
Chromium, 31, 259, 275
Chromium etch, 498, 524
Chromium masks, 545
Citric acid etch, *486*
Cleavage planes, 11
Climb, 39
C-MOS, 593
Coefficient, distribution, 90, *91, 93,* 273
Cold wall reactor, 227, 249
Collector diffusion isolation, 619
Compensation ratio, 32
Complementary transistors, 593, 614
Complexes, 17
Complexing reaction, 480
Congruent transformations, 58
Constitutional supercooling, 98, 273
Contacts:
  Al, 453, 547
  Al-poly Si, 461
  Au-Ge, 454, 600
  implanted layers, 352
  multilayer, 458
  ohmic, 451
  silicide, 458
  single layer, 453
Coordination number, 4
COP, 560
Copper, 31, 163, 517
Cross-linking resists, 557
Crystal axes, 8
Crystal growth:
  requirements, 88
  types, 3
Crystal orientation, 250
Crystallographic etches, 490
  GaAs, *523*
  Si, *523*
Cubic crystals, 3
Czochralski, apparatus, 86
  liquid encapsulated, 87

Damage:
  annealing, 321
  diffusion, 158, 587
  doping, 165
  gettering, 58, 587
  implant, 321
Defects:
  antistructure, 16
  chemical, 23, 27
  Frenkel, 15
  line, 35
  mask, 561
  point, 14
  Schottky, 15
Degrees of freedom, 53, 71, 74
Density, *45, 312*
Density, packing, 5
Depletion mode transistors, 599
Diamond lattice, 3
Diborane, 167, 234, 432, 440
Dichlorosilane, *218*, 231
Dicing, wafer, 10
Die bonds, 461
Dielectric strength, oxides, *407,* 414
Diethylzinc, *219*
Diffused capacitor, 631
Diffused junctions, 142
Diffused resitors, 626
Diffusion, 111
  activation energy, *135*
  As in GaAs, *139*
  As in Si, *131,* 152, 170, 176, *205*
  Au in Si, *135, 138,* 171, 182
  B in Si, *130*, 153, 165, 173, *205*
  charge defect interactions, 123
  concentration-dependent, 149, 651
  concentration gradient, 118

constant source, 141
contamination control, 163, 587
dissociative, 117, 126
during annealing, 345
during epitaxy, 239
elemental source, 158
evaluation, 196
field-aided, 121
field enhancement factor, 123
flux density, 120
GA in GAas, *139*
grain boundary, 114
interstitial, 113
interstitialcy, 114
interstitial-substitutional, *136*, 151
ion implanted source, 162
lateral, 155
limited source, 143
liquid source, 161
mathematics, 640
narrow slot, 642
open tube, 158, 193
oxide source, 160
P in Si, 168, 178, *206*
Pt in Si, *136*, 171, 182
ramped, 159, 650
S in GaAs, 190
Sb in Si, 171
Sn in GaAs, 192
sealed tube, 158, 185
solid source, 160
substitutional, 114
successive, 148
systems, 157, 164
ternary considerations, 184, 191
two-step, 146
vacancy, *139*
Zn in GaAs, 151, 185, *208*
Diffusion behavior, Si, *129*
Diffusion coefficient, 120
extrinsic, 124
intrinsic, 124
time dependent, 649
Diffusion constant, 120
GaAs, *140*
vacancy, 132
Diffusion equation, 140, 640, *652*
Diffusion of reactants, 219, 269
Diffusivities in silica, *375*, 393
Diffusivity, Al, *457*
effective, 126
interstitial, 127
intrinsic, *132*
Diodes, junction, 142, 622
multiple, 625
parasitics, 624
Schottky, 463, 626
Direct compounding, 88
Dislocation:
movement, 38
multiplication, 40
Dislocations, 14, 35, 158, 587
edge, 37
electronic properties, 44
screw, 35
Distribution coefficient, 90, *91*, *93*
Doped oxide source, diffusion, 644
Doping:
diffusion, 160
LPE, 274
melt, 90, 266
polysilicon, 433
vapor phase, 234, 257
VPE, 234, 257
Doping profile:
bipolar device, 608
diffusion, *143*, *147*, *150*, 155, 175, 177, 183
implant, *315*
Double diffused epitaxial process, 575

E-beam printing, 550
Edge dislocation, 37
Electrolysis, 405
Electromigration, 448
Electron beam printing, 550
Electron beam writing, 538
Electronic stopping, 302, 311
Electron mobility, *27*
Electron wind effect, 448
Ellipsometer, 412
Emitter, washout, 576
Emitter-ballasted transistor, 629
Emitter push effects, 180
Endothermic, 226
Energy levels:
GaAs, *29*
Si, *24*
Engraving, 541
Engraving defects, 562
Enhancement-mode transistor:
GaAs, 602
Si, 590
Epitaxial devices, GaAs, 604
Epitaxial systems, 233, 252
Epitaxy, 214
liquid phase, 264
selective, 243
vapor phase, 216

Equilibrium cooling, 272
Equilibrium diagram, 50
Erfc diffusion, *143, 201*
Error function, *646*
Error function relations, *645*
Etch, Al, 498, 509, *524*
Etches:
  anisotropic, 488
  crystallographic, 490, *523*
  isotropic, 478, *522*
  non-crystalline films, 492, *524*
Etch profiles, 487
Eutectic diagrams, 55
Evaluation:
  diffused layers, 196
  epitaxial layers, 285
  oxide layers, 411
Exothermic, 226, 249

Face-centered crystal, 3
Faraday's law, 405
Fick's law, 140
Field effect transistor, 358, 363, 399, 411, 430, 463, 616, 619
  diffused, 616
  epitaxial, 601
  implanted, 363
  Schottky gate, 364
Field enhancement factor, 123
Field implant, 360
Field oxide, 580
Flaw, 123
Flicker noise, 12
Float-zone, apparatus, 102
Flow tube, 217
Fluctuations, thermal, 17
Folded resistor, 628
Four-point probe, 287
Free radicals, 502
Frenkel defects, 15, 19
Fully recessed oxide, 580

GaAs, *312*
Gain bandwidth product, 610
Gallium, 249, 253, 266
Gallium aluminum arsenide, 281
Gallium arsenide:
  anisotropic etches, 487
  annealing, 331
  anodization, *407*
  anodization systems, 408
  cleaning, 519
  cleaving, 14
  contacts, 454
  crystallographic etches, *522*
  equilibrium coefficients, *91*
  etches, 483
  mobility, *34*
  P-T diagram, *83*
  P-x diagram, *84*
  synthesis, 82
  thermal oxidation, 400
  vacancy concentration, 22
Gaussian diffusion, *143, 203*
Gettering, 162, 587
Glassy layers, 588
Gold, 26
  films, 443, 450
  etch, 498, *524*
  in silicon, 136
Gold-germanium contacts, 454
Growth, from melt, 84
Growth imperfections, 280
Growth pyramids, 264
Guard ring, 356, 594, 610

HMDS, 544, 584
Halide process, 248, 251, 260
Hall effect, 289
Halogenic oxidation, 388, 588
Halogenic sources, 163, 588
Heavy doping, 24, 200
Henry's law, 221, 379
Heteroepitaxy, 215, 281
Heterogeneous nucleation, 222
Hexamethyldisilizane, 544, 584
High-low junctions, 452
Hillocks, 263
Hole mobility, *27*
Homogeneous nucleation, 222
Horizontal reactor, 228
Hot electron effects, 399
Hot wall reactor, 227
Hydride process, 253
Hydrodynamic considerations, 276
Hydrogen, 227, 249
Hydrogen chloride, 235, 244, 253, 391
Hydroxyl groups, 376

Implanted devices, GaAs, 603
Impurities:
  GaAs, *29, 30*
  Si, 23, *24, 25*
Impurity profile, 291, 305
Impurity redistribution, 235
In-situ etching, 234, 273
Interconnections:
  multi-metal, 449
  single metal, 446
Interstital, 14, 115, 132

Intrinsic carrier concentration, *122*, *133*
heavy doping, *133*
Intrinsic gettering, 588
Ion beam printing, 552
Ion dose, 300, 339, 344
Ion implantation, 300
Ion implanted resistor, 630
Ion milling, *527*
Ion sources, 339
Iron, 31, 163, 517
Iron oxide masks, 536, 545
Isoetch curves, GaAs, 485
Si, *481*
Isolation, 568
Isomorphous diagram, 53
Isothermal sections, 71

J-FET diffused, 619
Jump frequency, 116
Junction:
delination, 197, *522*
diffused, 142
high-low, 452
implanted, 354
Junction depth
Junction formation, 142, 144
Junction isolation, 572
Junction lag, 239

Kinks, 216, 255
Kirkendall effects, 455
KTFR, 350, 557

Laminar flow, 220
Lateral autodoping, 242
Lateral diffusion, 155, 195
Lateral p-n-p transistor, 613, 621
Lateral transistor, oxide isolated, 621
Lattice constant, *45*
LEC, 87
Lever rule, 51
Lift-off process, 548, 600
Line defects, 39
Lineage, 42
Linear oxidation rate constant, *381*, *382*
Liquid encapsulation, 87
Liquid phase epitaxy, 264
Loading effects, 503
Locally oxidized transistor, 620
Local oxidation, 576
LOCOS, 576
Low angle grain boundary, 42
Low reverse grain transistors, 616

LPCVD, 230
oxides, 423
LPE systems, 268, 276

Magnetic analyzer, 341
Mask:
borosilicate glass, 536
chromium, 536, 545
defects, 561
emulsion, 536
iron oxide, 536, 545
master, 535
reticle, 536
working, 535
Mask alignment, 541, 544
Masked diffusions, 195, *393*
Masking, during implementation, 348, *350*
silicon oxides, *393*
Mass action law, 21, 125, 151, 179, 187
Mass flow controller, 217
Mass separation, 340
Mass transfer coefficient, 221, 226
Master mask, 535
Mean time to failure, 449
Melt grown crystals, properties, 98
Melt growth, 84
Memory devices, 595
Mesa isolation, 570
MESFET circuits, 598
Metallization:
lift-off, 548, 587
multi-level, 586
single level, 586
Metalorganic, 160, 167, 193. *219*, 254, 261, 422, 441
Metals, refractory, *437*
Miller planes, 8
MNOS transistors, 597
Mobility, *27*, *34*
measurement, 289
Molecular beam epitaxy, 215
Molybdenum:
etch, 498, 509, *524*
growth, 438
Momentum transfer, 509, 512
MOS capacitor, 633
MOS devices, diffused, 589
implanted, 358
transistor, 616
Movement of dislocations, 38
Multi-emitter transistor, 616
Multiple diodes, 625
Multiple implants, 346
Multiplication of dislocations, 40

Narrow slot, diffusion, 157, 642
n-Channel transistor, 591
Network formers, 375
Network modifiers, 376
Neutron doping, 105
Noise, flicker, 12
Non-bridging oxygen atom, 373
Non-crystalline films, etch, *524*
Non-volatile memory, 595
n^+-p-n^+ transistor, 609
Nuclear stopping, 302
Nucleation, 222, 268
Number, coordination, 4

Ohmic contacts, 451, 600
  GaAs, 454, 600
  Si, 451, 458
Optical printing, 542
Orange peel, 248
Organometallic, 160, 167, 193, *219*, 254, 261, 422, 441
Orientation effects, 260
Oxidation:
  GaAs, 400
  halogenic, 388, 588
  Si, 146, 373
  silicides, 436
  thermal, 373
Oxidation-induced stacking faults, 391
Oxidation rate constant, *381*, 382
Oxidation systems, 385
Oxide:
  color chart, *413*
  implanatation through, 353, 593
  masking properties, *350*, *393*, *394*
Oxide growth, 377
  doping dependence, 383
  plasma enhanced, 424
Oxide growth rate, *386-389*
Oxide isolation, 570, 576
Oxide layers, evaluation, 411
Oxygen, 25, 31, 249, 587
Oxygen content, 100, 108
Oxygen ion vacancy, 374, 397

Packing density, 5
Pad oxide, 580
Palladium, etch, 499, *524*
Palladium silicide, 458
Parabolic oxidation rate constant, *381*, 382
Parasitics, 601
  active, 612
Partially recessed oxide, 580
Partial pressure, As, *21*, 251, 258
Partial stirring, 95
Pattern delineation, 504
Pattern generation, 535
  electron beam, 540
Pattern shift, 244
p-Channel transistors, 589
Penetration depth, 301, 353, 455
Peritectic reaction, 62
Phase diagram, 50
  binary, 51
  eutectic, 55
  isomorphous, 53
  oxide systems, 68
  ternary, 71
  unitary, 50
Phase rule, 52
Phosphoric acid etch, *485*
Phosphosilicate glass, 398, 424, *426*, 441, 584
  etch, *426*, 494, *524*
Photoelectron printing, 554
Photolithography, 534
Photoresist, 350, 542, 556
  coating, 542
  masking properties, *350*
  negative, 350, 544, 557
  positive, 544, 557
  removal, 502, 547
Pinch resistor, 630
Pinch voltage, 599
Planarization, 582
Planar reactor, 501
Planes:
  cleavage, 11
  crystal, 8
  Miller, 8
Plasma anodization, 410
Plasma ashing, 502, 547
Plasma-assisted etching, 510
Plasma etching, 499
Platinum, 27, *136*
Platinum etch, 499, *524*
Platinum silicide, 458
PMMA, 559
p-n Junction method, 196, 291, 653
p-n-p transistor, 609, 613
  locally oxidized, 620
Point defects, 14
  electronic properties, 44
  GaAs, 43
Polishing etches:
  GaAs, 483
  Si, *522*
Polymethyl methacrylate, 559
Polymide, 584, 587
Polysilicon, 432, 460, 592
  doping, 433

etch, 496, 509, 524
gate, 592
growth, 432
Printing, 541
contact, 547
electron beam, 550
ion beam, 552
optical, 542
photoelectron, 554
projection, 547
X-ray, 555
Printing defects, 562
Projected range, 302, *316*
Projection printing, 545
Properties, melt-growth crystals, 98
Si and GaAs, *45*
Proton bombardment, 336, 346, 362
Proximity printing, 544
PSG, 398, 424, *426*, 441, *524*, 584
etch, 494, *524*
P-T diagram, GaAs, *83*
Purple plague, 50, 64
P-x diagram, GaAs, *84*

Radial resistivity, 100, 107
Radius tetrahedral, 5, *6*, *7*, 165, 168
Ramping, 159, 650
Range, 302, 313, *316*
Rapid stirring, 92
Reaction:
autocatalytic, 480
endothermic, 226
exothermic, 226, 249, 260
heterogeneous, 224, 231, 267
homogeneous, 222, 232
kinetic controlled, 249
mass transfer limited, 226
peritectic, 62
pyrolytic, 232, 260
reversible, 231, 236
surface catalyzed, 231
Reaction rate constant, 222
Reactive ion etching, 515
Reactive ion milling, 517
Reactor:
barrel, 229
low pressure, 230
planar, 501
tunnel, 500
Recessed oxide, 580
Redistribution effects, 172, 235
Refining, zone, 102
Reflow techniques, 584, 597
Refractory metals, *437*
Registration:
E-beam, 539
lack of, 562
optical, 546
Relations, error function, *645*
Resist, 556
inorganic, 560
plasma-developed, 560
Resistivity, Si, *28*
Resistors, 355, 626, 630
diffused, 626
implanted, 355, 630
pinch, 630
Reticle mask, 535
Reversible reactions, 231, 236
Reynolds number, 220

Sapphire, 281
Schottky contacts, 352, 454, 459
Schottky defects, 15, 18, 116
Schottky diodes:
clamp, 464, 468
cleaning, 520
films, 463
GaAs, 463, 600
hyperabrupt, 358
Schottky gates, 463, 520
Scission, 558
Screw dislocations, 35
Scribe lines, 11
Sealed junction technology, 459
Selective epitaxy, 243, 262
Selective etching, 10
Selenium, 30, 335
Self-alignment, 430, 575
Self-diffusion, coefficient, 127
Self-isolated transistor, GaAs, 604
Si, 589, 618
Self-isolation, 573
Semi-insulating, 31, 336, 599
Shallow impurities, 23, *24*, *25*
Sheet resistance, 198, 285
base, 199
emitter, 199
Silane, 232, 422, 427, 581
Silica etch, 493, *524*
Silica glass, 373
Silica polyhedra, 374
Silicide:
etch, 491, 510
growth, 435
oxidation, 436
properties, *436*
Silicon:
anisotropic etches, 488

annealing, 327
anodization, *407*
cleaning, 517
cleaving, 11
crystallographic etches, *522*
equilibrium coefficients, *91*
etch rates, 479, 481
in GaAs, 30
polishing etches, *522*
on sapphire, 281, 577, 596

Silicon dioxide:
charge states, 395
color chart, *413*
dielectric strength, *414*
diffusivities in, *393*
etch, 493, 506, *524*
growth, 377, 422
hot electron effects, 399
impurities through, 353
masking properties, *350, 393*
oxidation induced stacking faults, 391, 398
oxygen ion vacancies, 397
properties, 392, *407*

Silicon nitride, *312*, 332, 427, 577
etch, 495, 506, *524*
growth, 427
masking properties, *351*, 577

Silicon oxynitride, 332, 429
Silicon tetrachloride, *218*, 231
Silver, etch, 498, *524*
Si_3N_4, *312*, 427, 577
masking properties, *351*, 577

SiO_2, *312*, 353, 359, 374, 422, 577
masking properties, *350, 393*

Sipos, 434
Slip planes, 39
Smith integral, *148*
Sodium, 163, *375*, 389, 397, 518
Solid solubility, *70, 134*, 165, 168, 170
Solubility:
Al in Si, *456*
Au in Si, *137*

Solution growth, 100, 264
SOS transistors, 596
Source-drain contacts, 600
Sputter cleaning, 586
Sputtering, 510
Sputtering yields, 526
Stacking faults, 246, 587
oxidation induced, 391

Stagnant layer, 219, 477
Step-and-repeat, 536
Step cooling, 271
Step coverage, 564

Stirring:
partial, 95
rapid, 92

Straggle, 302, *317*
transverse, *317*, 319

Substitutional diffusers, *129, 140*
Substitutional impurities, 17, *24*, 29
in SiO_2, 375

Sulfur, 29, 190, 258, 335
Sulfur hexafluoride, 235
Superbeta transistors, 615
Supercooling, constitutional, 98, 273
Supersaturation, 222, 269
Surface inversion, 394, 610
Susceptor, 228, 234
System:
Al-Au, *62*
Al-Si, *57*
As-Si, *60*
Au-Ga, *65*
Au-Ge, *59*
Au-Si, *58*
B_2O_3-SiO_2, *66*
Ga-As, *60*
Ga-As-Ag, *76*
Ga-As-Au, *76*
GaAs-Gap, *55*
Ga-As-Ge, *77*
Ga-As-S, *77*
Ga-As-Sn, *78*
Ga-As-Te, *79*
Ga-As-Zn, *79*
Ge-As, 64
Ge-Ga, *60*
Ge-Si, *54*
P_2O_5-SiO_2, *66*
Pb-Sn, *56*
Pt-Si, 63
SnO_2-SiO_2, *68*
ZnO-SiO_2, ion implantation, 337

Taper control, 582
Tellurium, 30, 258, *318*
Terminations, resistor, 628
Ternary diagrams, 71, 184
Tetraethylorthosilane, 422
Tetrahedral radius, 5, *6, 7*, 165, 168, 170
Tetramethyltin, *219*, 254
Thermal effects during implementation, 343
Thermal expansion coefficients, *45*
Thermal fluctuations, 17
Thermocompression bonding, 444
Thin film resistor, 631
Thomas-Fermi function, 310

Tin, 30, 266
Tin oxide, 194, 442
Transistor:
  bipolar, 145, 180, 357, 605
  emitter ballasted, 629
  field effect, 589, 598, 616
  locally oxidized, 620
  low reverse gain, 616
  MOS, 616
  multi-emitter, 617
  self-isolated, 618
  superbeta, 615
Transistor type, choice, 609
Transmutation doping, 105
Transport of reactants, 217, 268
Transverse straggle, *319*
Trichloroethylene, 390, 517
Trichlorosilane, *218*, 231
Triethylgallium, 219, 249
Trimethylaluminum, *219*, 249
Trimethylgallium, 219, 254
Triply diffused process, 574
Tripyramidal defects, 247
Tungsten growth, 438
Tunneling contacts, 451
Tunnel reactor, 500
Turbulent flow, 220
Twin planes, 41
Two-step diffusions, *148*

Ultrasonic bonding, 446

VPE:
  GaAs, 248
  Si, 231
Vacancy, 14, 114, 118, *126*
  interaction, 123
  Si, *126*
Vapor pressure:
  arsenic trichloride, *219*
  dichlorosilane, *218*
  silicon tetrachloride, *218*
  tetramethyltin, *219*
  trichlorosilane, *218*
  trimethylgallium, *219*
Vapor pressure effects, 20
van der Pauw technique, 289
Vertical p-n-p transistor, 613
Vertical reactor, 229
Viscosity, 220
Voltage, breakdown, 25, 102, 610

Wafer, dicing, 10
Washout, 244
Washout emitter, 576
  locally oxidized, 579
Water, 376, 395, 519
Wien filter, 342
Working mask, 535

X-ray printing, 555

Yield, 561

Zinc, 29, 151, 185
Zincblende, 3
Zincblende lattice, 12, 16
Zinc oxide, 193, 441
Zone leveling, 104
Zone processed material, properties, 107
Zone processes, 101
Zone refining, 102

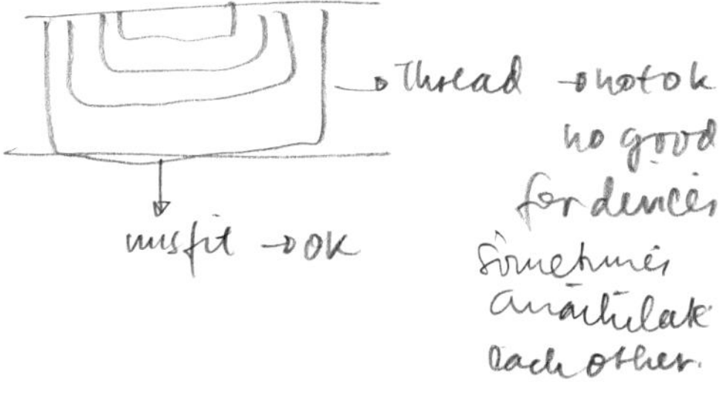
→ thread → not ok
no good
for devices
sometimes
annihilate
each other.
misfit → OK

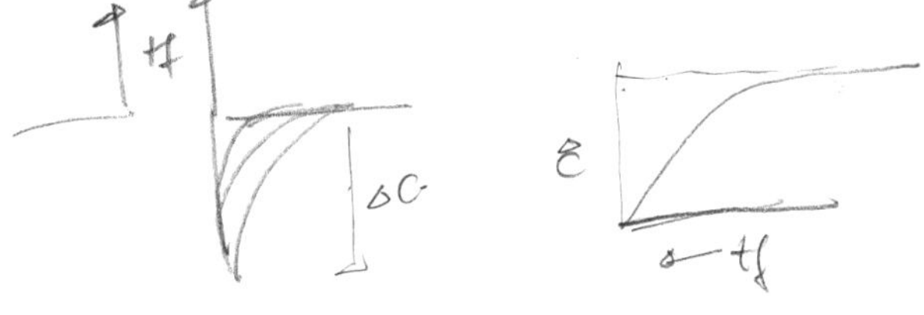
tf
ΔC
ε
→ tf

Useful Physical Constants

Boltzmann's constant, k	8.62×10^{-5} eV/°K
Electron rest mass, m_0	9.11×10^{-28} g
Energy associated with 1 eV	1.6×10^{-12} erg
Magnitude of the electronic charge, q	1.6×10^{-19} C
Permittivity of free space, ε_0	8.86×10^{-14} F/cm
Planck's constant, h	4.14×10^{-15} eV-s
Room temperature value of kT	0.0259 eV
Velocity of light, c	3×10^{10} cm/s
Angstrom unit, Å	10^{-8} cm
Micron., μm	10^{-4} cm
Thousandth of an inch, mil	25.4 μm
Avogadro's number	6.023×10^{23} molecules/g mole
Kilocalorie/mole	0.0434 eV/molecule
Calorie	4.184 J
Gram-mole	6.023×10^{23} molecules
Torr	1 mm Hg
Atmosphere	760 mm Hg